Spon's
Civil Engineering and
Highway Works
Price Book

2012

Spon's
Civil Engineering and Highway Works
Price Book

Edited by

Davis Langdon, An AECOM Company

2012

Twenty-sixth edition

Spon Press
an imprint of Taylor & Francis
LONDON AND NEW YORK

First edition 1984
Twenty-sixth edition published 2012
by Spon Press
2 Park Square, Milton Park, Abingdon, Oxon, OX14 4RN

Simultaneously published in the USA and Canada
by Spon Press
711 Third Avenue, New York, NY 10017

Spon Press is an imprint of the Taylor & Francis Group, an informa business

© 2012 Spon Press

British Library Cataloguing in Publication Data
A catalogue record for this book is available from the British Library

ISBN 13: 978-0-415-68064-6 (hardback)
ISBN 13: 978-0-203-15716-9 (ebook)

ISSN: 0957-171X

Typeset in Arial by Taylor & Francis Books

MIX
Paper from
responsible sources
FSC
www.fsc.org FSC® C004839

Printed and bound in Great Britain by
TJ International Ltd, Padstow, Cornwall

Contents

PART 5: RESOURCES

PART 6: UNIT COSTS – CIVIL ENGINEERING WORKS

PART 7: UNIT COSTS – HIGHWAY WORKS

Preface

At the time of writing the construction industry continues to experience a marked downturn especially in the Private Sector. Whilst raw materials, supplier and manufacturer prices are currently rising Tender Prices continue at the best to flat-line. This plus the effect of the continuing current tight credit market price levels for the coming year continue to be extremely difficult to forecast. Whilst infrastructure works associated with the London 2012 Olympics are providing a base load for civil engineering contractors this is beginning to tail and have yet to be replaced in volume by other significant projects such as London Crossrail. Out of London infrastructure work levels continue to be low.

The drop in demand for steel products in 2009/10 has flattened and there is evidence of hardening of prices. Oil prices have rising although there appears to be a softening in the second quarter of 2011. Petroleum and diesel prices continue to be pressed higher by continued demand in the Near and Far East. These prices continue to impact on manufacturing, energy and transportation costs and the effect this continues to make it extremely difficult to predict price levels for 2011 / 2012. Economic turmoil in the Euro Zone of countries is an added uncertainty to future construction activity in the region and elsewhere.

Thus under current market conditions the prices in the various Parts of this publication can only be taken as a guide to actual costs and any sustained upturn is liable to see marked increases in costs as manufacturers and specialists seek to recover margins lost in the downturn.

For the 2012 edition, we have carried out a general review of all prices up to May/June 2011 in consultation with leading manufacturers, suppliers and specialist contractors and included revisions as necessary. Our efforts have been directed at reviewing, revising and expanding the scope, range and detail of information to help enable the reader to compare or adjust any unit costs with reference to allocated resources or outputs.

The rates, prices and outputs included in the Resources and Unit Cost calculations, including allowances for wastage, normal productivity and efficiency, are based on medium sized Civil Engineering schemes of about £8 - £10 million in value, with no acute access or ground condition problems. However, they are equally applicable, with little or no adjustment, to a wide range of construction projects from £2 million to £50 million. Where suitable, tables of multipliers have been given to enable easy adjustment to outputs or costs for varying work conditions.

As with all attempts to provide price guidance on a general basis, this must be loaded with caveats. In applying the rates to any specific project, the user must take into account the general nature of the project, i.e. matters such as scale, site difficulties, locale, tender climate etc. This book aims at providing as much information as possible about the nature of the rate so as to assist the user to adapt it if necessary.

This edition continues to provide the reader with cost guidance at a number of levels, varying from the more general functional costs shown in Part 4, through the detailed unit costs in Parts 6 and 7 which relate respectively to the CESMM3 and the Highways Method of Measurement bills of quantities formats, down to the detailed resource costings given in Part 5 supplemented by the further advice on output factors in Part 10.

The Unit Costs sections (Parts 4, 6 and 7) cover a wide range of items and, where appropriate, notes are included detailing assumptions on composition of labour gang, plant resources and materials waste factors. Part 5 (Resources) includes detailed analysis of labour and plant costs, allowing the user to adjust unit costs to individual requirements by the substitution of alternative labour, materials or plant costs. Unit Costs are obviously dependent upon the outputs or man-hours used to calculate them. The outputs used in this work have been compiled in detail from the editors' wide ranging experience and are based almost exclusively on time and motion studies and records derived from a large number of recent Civil Engineering schemes. This information is constantly being re-appraised to ensure consistency with current practice. A number of prices and outputs are based upon detailed specialist advice and acknowledgements to the main contributors are included on page xv.

The market could change very rapidly and to monitor this and maintain accuracy levels readers should use the free price book update service, which advises of any significant changes to the published information over the period covered by this edition. The Update is posted free every three months on the publishers' website, until the publication of the next annual Price Book, to those readers who have registered with the publishers; in order to do so readers should follow the advice given on the coloured card bound within this volume.

Whilst all efforts are made to ensure the accuracy of the data and information used in the book, neither the editors nor the publishers can in any way accept liability for loss of any kind resulting from the use made by any person of such information.

Davis Langdon, An AECOM Company
MidCity Place
71 High Holborn
London
WC1V 6QS

Abbreviations

BS	British Standard	m³	cubic metre
BSS	British Standard Specification	μu	micron (10^{-3} millimetre)
DERV	diesel engine road vehicle	mm	millimetre
DfT	Department for Transport	mm²	square millimetre
Ha	hectare	N	newton
TSO	The Stationery Office	ne	not exceeding
Hp	horsepower	nr	number
Hr	hour	pa	per annum
In	inch	PC	Prime Cost
kg	kilogramme	sq	square
km	kilometre	t	tonne
kVA	kilovolt ampere	wk	week
kw	kilowatt	yd	yard
M	metre	yd³	cubic yard
m²	square metre		

Spon's Asia-Pacific Construction Costs Handbook

4th Edition

Edited by **Davis Langdon & Seah**

Spon's Asia Pacific Construction Costs Handbook includes construction cost data for 20 countries. This new edition has been extended to include Pakistan and Cambodia. Australia, UK and America are also included, to facilitate comparison with construction costs elsewhere. Information is presented for each country in the same way, as follows:

• key data on the main economic and construction indicators.

• an outline of the national construction industry, covering structure, tendering and contract procedures, materials cost data, regulations and standards

• labour and materials cost data

• Measured rates for a range of standard construction work items

• Approximate estimating costs per unit area for a range of building types

• price index data and exchange rate movements against £ sterling, $US and Japanese Yen

The book also includes a Comparative Data section to facilitate country-to-country comparisons. Figures from the national sections are grouped in tables according to national indicators, construction output, input costs and costs per square metre for factories, offices, warehouses, hospitals, schools, theatres, sports halls, hotels and housing.

This unique handbook will be an essential reference for all construction professionals involved in work outside their own country and for all developers or multinational companies assessing comparative development costs.

April 2010: 234x156: 480pp
Hb: 978-0-415-46565-6: **£120.00**

Acknowledgements

The Editors wish to record their appreciation of the assistance given by many individuals and organisations in the compilation of this edition.

Materials suppliers and sub-contractors who have contributed this year include

Abacus Lighting Ltd
Oddicroft Lane
Sutton-in-Ashfield
Notts NG17 5FT
Lighting and street furniture
Tel: 01623 511111
Fax: 01623 552133
Website: www.abacuslighting.com
e-mail: sales@abacuslighting.com

ACO Technologies Plc
Hitchin Road
Shefford
Beds SG17 5TE
Drainage systems
Tel: 01462 816666
Fax: 01462 815895
Website: www.aco.co.uk

Anderson Tree Care Ltd
Garden Cottage
Park Street
Barlborough
Landscape Management
Chesterfield, S43 4TJ
Tel: 01246 570044

Aggregate Industries
Hulland Ward
Ashbourne
Derby DE6 3ET
Kerbs, edgings and pavings
Tel: 01335 372222
Fax: 01335 370074
Website: www.aggregate.com

AMEC Specialist Businesses
Cold Meece
Swynnerton
Stone
Staffordshire ST15 0QX
Tunnelling
Tel: 01785 760022
Fax: 01785 760762
Website: www.amec.com

Arcelor Commercial RPS UK Ltd
Arcelor House
4 Princes Way
Solihull
West Midlands B91 3AL
Steel sheet piling
Tel: 0870 770 8057
Fax: 0870 770 8058
Website: www.sheet-piling.arcelor.com
e-mail: sheet-piling@arcelor.com

Arthur Fischer (UK) Ltd
Hithercroft Road
Wallingford
Oxon OX10 9AT
Fixings
Tel: 01491 833000
Fax: 01491 827953
Website: www.fischer.co.uk
e-mail: sales@fischer.co.uk

Asset International Ltd
Stephenson Street
Newport
South Wales NP19 0XH
Steel culverts
Tel: 01633 271906
Fax: 01633 290519
Website: www.assetint.co.uk

Avon Manufacturing (Warwick) Ltd
Avon House,
Kineton Road,
Southam
Warwickshire
CV47 2DG
Fixings
Tel: 01926 817292
Fax: 01926 814156
Website: www.avonmanufacturing.co.uk/

Bachy Soletanche Ltd
Foundation Court
Godalming Business Centre
Catteshall Lane
Godalming
Surrey GU7 1XW
Diaphragm walling
Tel: 01483 427311
Fax: 01483 417021
Website: www.bacsol.co.uk

BRC Special Products
Astonfield Industrial Estate
Stafford ST16 3BP
Masonry reinforcement and accessories
Tel: 01785 222288
Fax: 01785 240029
Website: www.brc-special-products.co.uk

Capital Demolition
Capital House
Woodham Park Road
Addlestone
Surrey KT15 3TG
Demolition and clearance
Tel: 01932 346222
Fax: 01932 340244
Website: www.capitaldemolition.co.uk

Caradon Parker Ltd
Units 1-2 Longwall Avenue
Queen's Drive Industrial Estate
Nottingham NG2 1NA
Geotextiles
Tel: 0115 986 2121

Castle Cement
Park Square
3160 Solihull Parkway
Birmingham Business Park
Birmingham B37 7YN
Cement products
Tel: 0121 606 4000
Fax: 0121 606 1412
Website: www.castlecement.co.uk
email: information@castlecement.co.uk

CCL Stressing Systems
Unit 4
Park 2000
Millennium Way
Westland Road
Leeds LS11 5AL
Bridge bearings
Tel: 0113 270 1221
Fax: 0113 271 4184
Website: www.ccclstressing.com
e-mail: sales@cclstressing.com

Celsa Steel (UK) Ltd
Building 58
East Moors Road
Cardiff CF24 5NN
Bar reinforcement
Tel: 0292 035 1800
Fax: 0292 035 1801
Website: www.celsauk.com

Cementation Foundations Skanska Ltd
Maple Cross House
Denham Way
Rickmansworth
Herts WD3 9AS
Piling
Tel: 01923 423 100
Fax: 01923 777 834
Website: www.cementationfoundations.
skanska.co.uk
e-mail: cementation.foundations@
skanska.co.uk

Civils and Lintels
Stock Lane
Chadderton,
Oldham, OL9 9EY
Drainage Materials
Tel: 01616 270888
Fax: 01616271888
Website: www.civilsandlintels.co.uk
Email: oldham@civilsandlintels.co.uk

Cogne UK LTD
Uniformity Steel Works
19 Don Road
Newhall
Sheffield
S9 2UD
Tel: 01142212020
Fax: 01142213030

Corus Construction & Industrial
PO Box 1
Brigg Road
Scunthorpe
North Lincoln DN16 1BP
Structural steel sections
Tel: 01724 404040
Fax: 0870 902 3133
Website: www.corusgroup.com
e-mail: corusconstruction@corusgroup.com

Corus Construction & Industrial
PO Box 1
Brigg Road
Scunthorpe
North Lincoln DN16 1BP
Structural steel sections
Tel: 01724 404040
Fax: 0870 902 3133
Website: www.corusgroup.com
e-mail: corusconstruction@corusgroup.com

Craigton Industries
Craigton Works
Milngavie
Glasgow
G62 7HF
Precast Concrete Manholes etc
Tel: 0141 956 6585
Fax: 0141 956 3757

Don & Low Ltd
Newford Park House
Glamis Road
Forfar
Angus DD8 1FR
Geotextiles
Tel: 01307 452200
Fax: 01307 452422
Website: www.donlow.co.uk

Dorsey Construction Materials Ltd
Unit 11 Nimrod Industrial Estate
Elgar Road
Reading
Berks RG2 0EB
Water Stops
Tel: 0118 975 3377
Fax: 0118 975 3393

Euromix Concrete Ltd
Chelmsford
Unit 1 Boreham Industrial Estate
Waltham Road
Chelmsford, CM3 3AW
Ready Mix Concrete
Tel: 01245 464545
Fax: 01245 451507

Fosroc Ltd
Coleshill Road
Tamworth
Staffordshire B78 3TL
Waterproofing and expansion joints
Tel: 01827 262222
Fax: 01827 262444
Website: www.fosrocuk.com
e-mail: sales@fosrocuk.com

Freyssinet Ltd
Hollinswood Court
Stafford Park
Telford
Shropshire TF3 3DE
Reinforced earth systems
Tel: 01952 201901
Fax: 01952 201753
Website: www.freyssinet.co.uk

Grace Construction Products Ltd
Ajax Avenue
Slough
Berks SL1 4BH
Waterproofing/expansion joints
Tel: 01753 692929
Fax: 01753 691623
Website: www.graceconstruction.com

Greenham Trading Ltd
Greenham House
671 London Road
Isleworth
Middx TW7 4EX
Contractors site equipment
Tel: 020 8560 1244
Fax: 020 8568 8423
Website: www.greenham.com
e-mail: isleworth.sales@greenham.com

Hanson Building Products
Stewartby
Bedford
Beds MK43 9LZ
Bricks
Tel: 08705 258258
Fax: 01234 762040
e-mail: info@hansonbrick.com

Hanson Concrete Products
PO Box 14
Appleford Road
Sutton Courtney
Abingdon
Oxfordshire OX14 4UB
Blockwork and precast concrete products
Tel: 01235 848808
Fax: 01235 846613
Website: www.hanson-europe.com

Hepworth Building Products Ltd
Hazlehead
Crow Edge
Sheffield S36 4HG
Clayware pipes
Tel: 01226 763561
Fax: 01226 764827
Website: www.hepworthdrainage.co.uk
e-mail: info@hepworthdrainage.co.uk

Hughes Concrete Ltd
Barnfields
Leek
Staffs ST13 5QG
Precast concrete manhole units
Tel: 01538 380500
Fax: 01538 380510
Website: www.hughesconcrete.co.uk

Inform UK Ltd
Industrial Park
Ely Road
Waterbeach
Cambridge CB5 9PG
Formwork and accessories
Tel: 01223 862230
Fax: 01223 440246
Website: www.inform.co.uk
e-mail: general@inform.co.uk

Jacksons Fencing Ltd
Stowting Common
Ashford
Kent
TN25 6BN
Tel: 01233 750393
Fax: 01233 750403
Website: www.jackson-fencing.co.uk
Email: sales@jacksons-fencing.co.uk

James Latham Plc
Unit 3
Finway Road
Hemel Hempstead HP2 7QU
Timber merchants
Tel: 01442 849100
Fax: 01442 267241
Website: www.lathamtimber.co.uk
e-mail: plc@lathams.co.uk

K B Rebar Ltd
Unit 5, Dobson Park Industrial Estate
Dobson
Park Way, Ince, Wigan
WN2 2DY
Steel Reinforcement
Tel: 01617908635
www.kbrebar.co.uk

Maccaferri Ltd
The Quorum
Oxford Business Park Nr
Garsington Road
Oxford OX4 2JZ
River and sea gabions
Tel: 01865 770555
Fax: 01865 774550
Website: www.maccaferri.co.uk
e-mail: oxford@maccaferri.co.uk

Marley Extrusions LTD
Dickley Lane
Lenham
Maidstone
Kent, ME17 2ED
Drainage products
Tel: 01622 858888
Fax: 01622 858725
Website: www.marley.co.uk

Marley Plumbing & Drainage Ltd
Dickley Lane
Lenham
Maidstone
Kent ME17 2DE
UPVC drainage and rainwater systems
Tel: 01622 858888
Fax: 01622 858725
Website: www.marley.co.uk

Marshalls Mono Ltd
Southowram
Halifax
West Yorkshire HX3 9SY
Kerbs, edgings and pavings
Tel: 01422 312000
Fax: 01422 330185
Website: www.marshalls.co.uk

Milton Pipes Limited
Milton Regis
Sittingbourne
Kent
ME10 2QF
Pre-cast Concrete Solutions
Tel: 01795 425191
Fax: 01795 478232
Website: www.miltonpipes.co.uk

Morelock Signs
Morelock House
Strawberry Lane
Willenhall
Wolverhampton, WV13 3RS
Traffic Signs
Tel: 01902 637575
Fax: 01902 637576
Website: morelock.co.uk

Naylor Industries Ltd
Clough Green
Cawthorne
Barnsley
South Yorkshire S75 4AD
Clayware pipes and fittings
Tel: 01226 790591
Fax: 01226 790531
Website: www.naylor.co.uk
e-mail: sales@naylor.co.uk

PHI Group Ltd
Harcourt House, Royal Crescent
Cheltenham
Glos GL50 3DA
Retaining walls, gabions
Tel: 0870 3334120
Fax: 0870 3334121
Website: www.phigroup.co.uk
e-mail: info@phigroup.co.uk

Platipus Anchors Ltd
Kingsfield Business Centre
Philanthropic Road
Redhill
Surrey RH1 4DP
Earth anchors
Tel: 01737 762300
Fax: 01737 773395
Website: www.platipus-anchors.com
e-mail: info@platipus- anchors.com

Roger- Bullivant Ltd
Innovation House
7 Kingdom Avenue
North acre Industrial Park
Westbury
Wiltshire, BA13 4FG
Pilling and Foundations
Tel: 01373 865012
Fax: 0 1373 865559
Website: www.roger-bullivant.co.uk

ROM Ltd
710 Rightside lane
Sheffield
South Yorkshire S9 2BR
Formwork and accessories
Tel: 0114 231 7900
Fax: 0114 231 7905
Website: www.rom.co.uk
e-mail: sales@rom.co.uk

ROM (NE) Ltd
Unit 10
Blaydon Industrial Estate
Chainbridge Road
Blaydon-on-Tyne
Northumberland NE21 5AB
Reinforcement
Tel: 0191 414 9600
Fax: 0191 414 9650
Website: www.rom.co.uk

Saint-Gobain Pipelines PLC
Lows Lane
Stanton-By-Dale
Derbyshire DE7 4QU
**Cast iron pipes and fittings,
polymer concrete channels,
manhole covers and street furniture**
Tel: 0115 9305000
Fax: 0115 9329513
Website. www.saint-gobain-pipelines.co.uk

Stent
Pavilion C2
Ashwood Park
Ashwood Way
Basingstoke RG23 8BG
Piling and foundations
Tel: 01256 763161
Fax: 01256 768614
Website: www.stent.co.uk

Stocksigns Ltd
Ormside Way
Redhill
Surrey RH1 2LG
Road signs and posts
Tel: 01737 764764
Fax: 01737 763763
Website: www.stocksigns.co.uk

Structural Soil Ltd
The Old School
Fell Bank
Bristol,
BS3 4EB.
Ground Investigation
Tel: 0117-947-1000
Fax: 0117-947-1004,
Email: admin@soils.co.uk
Website: www.soils.co.uk

Tarmac Precast Concrete Ltd
Tallington
Barhom Road
Stamford
Lincs PE9 4RL
Prestressed concrete beams
Tel: 01778 381000
Fax: 01778 348041
Website: www.tarmacprecast.com
e-mail: tall@tarmac.co.uk

Tarmac Topblock Ltd
Airfield Industrial Estate
Ford
Arundel
West Sussex BN18 0HY
Blockwork
Tel: 01903 723333
Fax: 01903 711043
Website: www.topblock.co.uk

Tensar International
Cunningham Court
Shadsworth Business Park
Blackburn
BB1 2QX
Soil Reinforcement
Tel: +44(0)1254 262431
Fax: +44(0)1254 26686
www.tensar.co.uk

Terry's Timber Ltd
35 Regent Road
Liverpool L5 9SR
Timber
Tel: 0151 207 4444
Fax: 0151 298 1443
Website: www.terrystimber.co.uk

Travis Perkins Trading Co. Ltd
Baltic Wharf
Boyn Valley Road
Maidenhead
Berkshire SL6 4EE
Builders merchants
Tel: 01628 770577
Fax: 01628 625919
Website: www.travisperkins.co.uk

Varley and Gulliver Ltd
Alfred Street
Sparkbrook
Birmingham B12 8JR
Bridge parapets
Tel: 0121 773 2441
Fax: 0121 766 6875
Website: www.v-and-g.co.uk
e-mail: sales@v-and-g.co.uk

Vibro Projects Ltd
Unit D3
Red Scar Business Park
Longridge Road
Preston PR2 5NQ
Ground consolidation - vibroflotation
Tel: 01772 703113
Fax: 01772 703114
Website: www.vibro.co.uk
e-mail: sales@vibro.co.uk

Volvo Construction Equipment Ltd
Duxford
Cambridge CB2 4QX
Plant manufacturers
Tel: 01223 836636
Fax: 01223 832357
Website: www.volvo.com
e-mail: sales@vcebg@volvo.com

Wavin Plastics Ltd
Parsonage Way
Chippenham
Wilts SN15 5PN
UPVC drain pipes and fittings
Tel: 01249 654121
Fax: 01249 443286
Website: www.wavin.co.uk

Wells Spiral Tubes Ltd
Prospect Works
Airedale Road
Keighley
West Yorkshire BD21 4LW
Steel culverts
Tel: 01535 664231
Fax: 01535 664235
Website: www.wells-spiral.co.uk
e-mail: sales@wells-spiral.co.uk

Winn & Coales (Denso) Ltd
Denso House
33-35 Chapel Road
West Norwood
London SE27 0TR
Anti-corrosion and sealing products
Tel: 0208 670 7511
Fax: 0208 761 2456
Website: www.denso.net
e-mail: mail@denso.net

Davis Langdon
An AECOM Company

A leading light in
our industry

Davis Langdon,
An AECOM Company

Best Tall Buildings 2011

A. Wood

Best Tall Buildings 2011 contains overviews of the nominees and winning projects from the Council on Tall Buildings and Urban Habitat's annual awards. The book provides vital statistics, images, plans and in depth accounts of the architectural design and sustainability features of the year's best tall buildings. Additionally included is the official list of the World's Tallest Buildings with the judging criteria from the foremost authority on tall buildings. Full colour illustrations and photographs throughout.

December 2011: 200pp
Hb: 978-0-415-68326-5: **£29.99**

To Order: Tel: +44 (0) 1235 400524 **Fax:** +44 (0) 1235 400525
or Post: Taylor and Francis Customer Services,
Bookpoint Ltd, Unit T1, 200 Milton Park, Abingdon, Oxon, OX14 4TA UK
Email: book.orders@tandf.co.uk

For a complete listing of all our titles visit:
www.tandf.co.uk

Art of Structures

A. Muttoni

This book describes the complete panorama of supporting structures and their function by describing how loads are sustained and transmitted to the ground. With a minimum of mathematics, the reader is guided through the analysis of some of the world's most famous designs and structures from a civil-engineering perspective.

An intuitive approach is taken – the basics of equilibrium analysis are explained by visualizing the internal forces of specific structures with the aid of simple graphical tools.

Ideal for anyone who needs an intuitive and practical approach to the design and appropriate sizing of load-bearing structures.

February 2011: 280pp
Pb: 978-0-415-61029-2: **£34.99**

To Order: Tel: +44 (0) 1235 400524 **Fax:** +44 (0) 1235 400525
or Post: Taylor and Francis Customer Services,
Bookpoint Ltd, Unit T1, 200 Milton Park, Abingdon, Oxon, OX14 4TA UK
Email: book.orders@tandf.co.uk

For a complete listing of all our titles visit:
www.tandf.co.uk

General

Manual for Detailing Reinforced Concrete Structures to EC2

J. Calavera

Detailing 210 structural elements, each with a double page spread, this book is an indispensible guide to designing for reinforced concrete. Drawing on Jose Calavera's many years of experience, each entry provides commentary and recommendations for the element concerned, as 0077ell as summarising the appropriate Eurocode legislation and referring to further standards and literature.

Includes a CD ROM with AutoCad files of all of the models, which you can adapt to suit your own purposes.

Section 1. General Rules for Bending, Placing and Lapping of Reinforcement
Section 2. Constructive Details

1. Foundations 2. Walls 3. Columns and Joints 4. Walls Subjected to Axial Loads. Shear Walls and Cores 5. Beams and Lintels 6. Slabs 7. Flat Slabs 8. Stairs 9. Supports 10. Brackets 11. Pavements and Galleries 12. Cylindrical Chimneys, Towers and Hollow Piles 13. Rectangular Silos, Caissons and Hollow Piles 14. Reservoirs, Tanks and Swimming Pools 15. Special Details for Seismic Areas

October 2011: 512pp
Hb: 978-0-415-66348-9: **£80.00**

To Order: Tel: +44 (0) 1235 400524 **Fax:** +44 (0) 1235 400525
or Post: Taylor and Francis Customer Services,
Bookpoint Ltd, Unit T1, 200 Milton Park, Abingdon, Oxon, OX14 4TA UK
Email: book.orders@tandf.co.uk

For a complete listing of all our titles visit:
www.tandf.co.uk

PURPOSE AND CONTENT OF THE BOOK

For many years the Editors have compiled a price book for use in the building industry with, more recently, companion volumes for use in connection with engineering services contracts and landscaping work. All of these price books take their reliability from the established practice within these sectors of the construction industry of pricing work by the application of unit rates to quantities measured from the designer's drawings. This practice is valid because most building work can be carried out under similar circumstances regardless of site location; a comparatively low proportion of contract value is subject to the risks that attend upon work below ground level; and once the building envelope is complete most trades can proceed without serious disruption from the weather.

This is not, however, the general method of pricing Civil Engineering work: the volume of work below ground, increased exposure to weather and the tremendous variety of projects, in terms of type, complexity and scale, makes the straightforward use of unit rates less reliable. So, whilst even in building work similar or identical measured items attract a fairly broad range of prices, the range is much greater in Civil Engineering Bills. This uncertainty is compounded by the lower number of bill items generated under Civil Engineering Methods of Measurement, so that the precise nature of the work is less apparent from the bill descriptions and the statistical effect of 'swings and roundabouts' has less scope to average out extremes of pricing.

To prepare a price for a Civil Engineering project, then, it is necessary to have regard to the method to be adopted in executing the work, draw up a detailed programme and then cost out the resources necessary to prosecute the chosen method. Because the first part of this process is the province of the Contractor's planner, there has been a tendency to postpone detailed estimating until the tendering stage itself, with the employer relying up to that point upon an estimate prepared on a 'broad brush' basis.

The result has been a growing pressure on the part of project sponsors for an improvement in budgetary advice, so that a decision to commit expenditure to a particular project is taken on firmer grounds. The absence of a detailed pricing method during the pre-contract phase also inhibits the accurate costing of alternative designs and regular cost checking to ensure that the design is being developed within the employer's budget.

This book therefore seeks to draw together the information appropriate to two methods of pricing: the cost of resources for use where an operational plan has been outlined, and unit rates for use where quantities can be taken from available drawings.

To take some note of the range of unit rates that might apply to an item, the rates themselves are in some cases related to working method – for example by identifying the different types of plant that would suit varying circumstances. Nonetheless, it would be folly to propose that all types of Civil Engineering work could be covered by a price book such as this. The Editors have therefore had in mind the type and scale of work commissioned by a local authority, a public corporation or a large private company.

This does embrace the great majority of work undertaken by the industry each year. Although almost all projects will have individual features that require careful attention in pricing, there will be some projects that are so specialist that they will not conform to standard pricing information at all.

But for most projects, within the range of work covered, this book should provide a firm foundation of cost information upon which a job-specific estimate can be built.

The contents of the book are therefore set out in a form that permits the user to follow the estimating process through in a structured way, as follows:

Part 1: General

The balance of this section describes in narrative form the work stages normally followed in a Contractor's office from receipt of the tender documents through to the submission of the tender.

The Land Remediation section reviews the general background of ground contamination and discusses the impact of the introduction of the Landfill Directive in July 2004 and recent changes thereto.

Part 2: On Costs and Profit

Having produced an estimate for the predicted cost of the work, being the sum of the preliminaries and the measured work, the estimate must be converted to a tender by the application of any adjustment made by management (which follows the Management Appraisal described later in this part of the book) and by additions for financing charges, head office overheads and profit. These additions are discussed in this section and also included is a worked example of a tender summary.

Part 3: Costs and Tender Prices Indices

The cost and tender price indices included in this part of the book provide a basis for updating historical cost or price information, by presenting changes in the indices since 1988. Caution must be taken when applying these indices as individual price fluctuations outside the general trend may have significant effect on contract cost.

Part 4: Approximate Estimating

The prices in this section have been assembled from a number of sources, including the relevant items in the unit costs section and recovered data from recent projects. They are intended to give broad price guides or to assist in comparison exercises.

Unit Costs (Ancillary Building Work)

This section is to be utilized in conjunction with Parts 4 and 5 to enable the user to incorporate within the estimate items more normally associated with Building Work rather than Civil Engineering and which do not fall readily under recognized methods of measurement for Civil Engineering Work. Due to the diversity of items that fall under such a definition, because of specification differences, the format for this section is structured to incorporate a range of items to allow the production of the estimate for such items prior to detailed design information being available. Additionally this section includes, in the same format, items covering simple Building Works which occur in connection with a Civil Engineering Contract but which do not form a significant proportion of the overall value and therefore do not need to be estimated in great detail using Parts 4 and 5 unit rates.

Preliminaries and General Items

Containing a checklist of items to be priced with Preliminaries and General Items (or Method Related Charges) and a worked example containing specific cost information.

Part 5: Resources

This deals with the basic cost of resources, so that a resource-based system of estimating can be adopted where it is possible to develop an outline programme and method statement. Reference to this section will also assist the user to make adjustments to unit rates where different labour or material costs are thought to apply and to calculate analogous rates for work based on the hypothetical examples given. It is stressed that all of the costs given in this section are exclusive of the items costed with the preliminaries and of financing charges, head office overheads and profit. The materials and plant costs as shown are gross, with no deduction of discount.

Parts 6 and 7: Unit Costs

These sections are structured around methods of measurement for Civil Engineering Work and Highway Works and gives 'trade by trade' unit rates for those circumstances where the application of unit rates to measured quantities is possible and practical. Again, it is stressed that the rates are exclusive of the items costed with pre-liminaries and of financing charges, head office overheads and profit. Both materials and plant costs are adjusted to allow a normal level of discount, with allowances for materials wastage and plant usage factors.

Part 8: Daywork

Including details of the CECA dayworks schedule and advice on costing excluded items.

Part 9: Professional Fees

These contain reference to standard documentation relating to professional fees for Consulting Engineers and Quantity Surveyors.

Part 10: Outputs

Scheduled here are various types of operations and the outputs expected of them. Also listed are man hours for various trades found in Civil Engineering.

Part 11: Tables and Memoranda

These include conversion tables, formulae, and a series of reference tables structured around trade headings. It also includes a review of current Capital Allowances, Value Added Tax and The Aggregates Levy.

OUTLINE OF THE TENDERING AND ESTIMATING PROCESS

This section of the book outlines the nature and purpose of Civil Engineering estimating and provides background information for users. It comprises an outline of the estimating and tendering process with supporting notes and commentaries on particular aspects. Some worked examples on tender preparation referred to in this part are included at the end of Part 2.

It must be emphasized that the main purpose of this book is to aid the estimating process. Thus it is concerned more with the predicted cost of Civil Engineering work than with the prices in a bill of quantities. To ensure the correct interpretation of the information provided it is important to distinguish clearly between estimating and ten-dering; the following definitions are followed throughout.

The estimate is the prediction of the cost of a project to the Contractor. The Tender is the price submitted by the Contractor to the Employer.

The tender is based on the estimate with adjustments being made after review by management; these include allowances for risk, overheads, profit and finance charges. As discussed later in this section, prices inserted against individual items in a bill of quantities may not necessarily reflect the true cost of the work so described due to the view taken by the Contractor on the risks and financial aspects involved in executing the work.

Whilst projects are now constructed using many different forms of contract the core estimating process falls into two main divisions namely 'Design & Construct' and 'Construct only'. The following list summarizes the activities involved in the preparation of a tender for a typical construct only Civil Engineering project where the client issues full drawings, specifications and Bills describing the extent of the works to be priced. The diagram that follows illustrates the relationships between the activities. After the diagram, notes on factors affecting each stage of the process are given.

1.   An overall appraisal of the project is made including any variations to the standard contract form, insurance provisions and any other unusual or onerous requirement.
2.   Material requirements are abstracted from the tender documents and prices are obtained from suppliers. Details of work normally subcontracted are abstracted, together with relevant extracts from the tender documents and prices are obtained from subcontractors.
3.   The site of the works and the surrounding area is visited and studied. Local information is obtained on factors affecting the execution of the contract.
4.   A programme and detailed method of working for the execution of the contract is prepared, to include details of plant requirements, temporary works, unmeasured work, appropriate supervisory staff, etc.
5.   Designs are made for temporary works and other features left to be designed by the Contractor, and quantities are taken off for costing.
6.   Major quantities given in the tender documents are checked.
7.   The cost estimate for the project is prepared by costing out all of the resources identified by stages 2 to 6. A more detailed report is made on the conditions of contract, financial requirements, etc. and an assessment of risk/opportunity is prepared.
     The tender documents are priced.
8.   Management reviews the estimate, evaluates the risks and makes allowances for overheads, profit and finance.
9.   The tender is completed and submitted.

1. INITIAL APPRAISAL

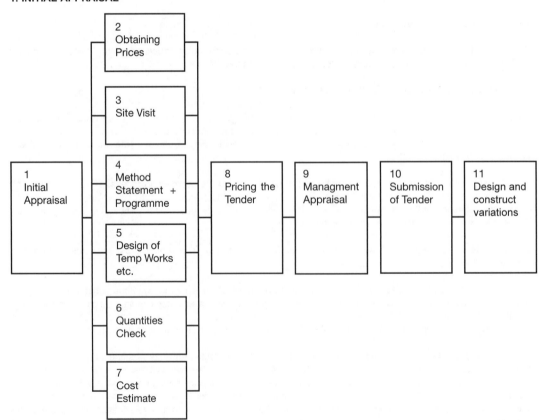

The purpose of the initial overall appraisal is to highlight any high value areas or any particular problems which may require specialist attention; it can also identify possible alternative methods of construction or temporary works.

Other points to be considered:

- the location and type of project and its suitability to the tenderer's particular expertise
- the size of project, its financing requirement, the proportion of annual turnover it would represent and the availability of resources
- the size of the tender list and the nature of the competition
- the identity of the employer and his professional consultants
- the adequacy of the tender documents
- an initial appraisal of risk and opportunity (see Section 8)

Corporate governance requires that the directors of companies are aware of all the liabilities inherent in the contract being sought. Few contracts are offered on totally un-amended standard forms and specifications and it is the estimator's duty to ensure that a report is prepared to advise on the precise terms and conditions being offered and their potential implications.

It is essential for the estimator to study the contract documents issued with the enquiry or made available for inspection and to note those parts which will affect pricing or involve the Contractor in liabilities which must be evaluated and drawn to the attention of management. The following comments are indicative only:

Conditions of Contract

For Civil Engineering work these are normally, but not exclusively, based on either the I.C.E. or ECC (NEC) standard forms. However these forms are rarely offered without addition and amendment and it is imperative that the full implications are understood and directors informed. Any required bonds, guarantees and warranties must be identified, reported and included in subcontract enquiries where appropriate. Insurance requirements and excesses must be checked against company policies.

Bill of Quantities

Where a Bill is provided it serves three purposes: first and foremost it must be prepared with the objective of providing the estimator with as accurate a picture of the project as possible, so as to provide a proper basis for pricing. Second, it should enable the employer to compare tenders on an equal basis and third it will be used to evaluate the work executed for payment purposes. Individual items in the Bill do not necessarily describe in detail the work to be carried out under a particular item; reference must be made to the specification and the drawings to ascertain the full scope of the work involved.

The method of preparing the Bill may be based on the 'Civil Engineering Standard Method of Measurement' issued by the Institution of Civil Engineers or the 'Method of Measurement for Highway Works' issued by the Highways Agency, but some employing authorities have evolved their own methods and it is important for the estimator to study the Bill and its preambles to ensure that his rates and prices are comprehensive.

In all cases the quantities given in the bill are not a guarantee and the drawings usually have precedence. The estimator must understand whether he is pricing a re-measurable or fixed price contract and make due allowance.

Specification

This gives a detailed description of the workmanship, finish and materials to be used in the construction of the work. It may also give completion periods for sections and/or the whole of the work together with details of the requirements for the employer and/or the Consulting Engineer in connection with their supervision on site.

Water Authority and Highway Works in particular are based around a standard specification. However even standard specifications will have contract specific appendices and tables. The estimator must take due note of these requirements and ensure that this information is issued and taken into account by all potential subcontractors and material suppliers.

Drawings

These give details of the work to be carried out and must be read in conjunction with the specification. It is important for the estimator to study the notes and descriptions given on the drawings as these amplify the specification.

Should the estimator discover any conflict between the various documents, it is important to have such discrepancies clarified by the Employer or Engineer prior to submission of any offer.

2. OBTAINING PRICES

(a) Materials

When pricing materials, the following points must be noted:

- checks must be made to ensure that the quality of materials to be supplied meets with the requirements of the specification. If necessary, samples should be obtained and tested for compliance
- checks must be made to ensure that the rates of delivery and fabrication periods can meet the demands of the programme. It is sometimes necessary to use more than one supplier, with differing prices, to ensure a sufficient flow of materials
- tests should be carried out to ascertain allowances to be made for operations such as compaction of soils and aggregates. Records of past contracts using similar materials can give this information, providing such records are accurate and reliable

(b) Subcontractors

It is common practice among Civil Engineering Contractors to subcontract a significant proportion of their work.

Subcontracted work can represent the bulk of the value of measured work.

When utilizing subcontractors' prices it is extremely important to ensure that the rates given cover the full extent of the work described in the main contract, and that the subcontractors quotation allows for meeting the main Tender programme and methods of working.

Unless an exclusive relationship has been entered into prior to the tendering process it is likely that the same subcontractor will submit prices to a number of competing tenderers. It is important for the estimator to ensure that the price offered represents the method intended and is not a generic sum which is subject to variation if successful.

3. SITE VISIT

Factors to check during the site visit include:

- access
- limitations of working space
- existing overhead and underground services
- nearby public transport routes
- availability of services – water, gas, electricity, telephone, etc.
- availability of labour and subcontractors
- availability of materials – particularly aggregates and fill materials and location of nearest tipping facility
- nature of the ground, including inspection of trial bores / pits if dug ground water level
- presence of other Contractors working on or adjacent to the site.

4. METHOD STATEMENT AND PROGRAMME

As previously stated whilst an estimate is being prepared it is necessary that a detailed method of working and a programme for the execution of the works is drawn up; the latter can take the form of a bar chart or, for large and more complex projects, may be prepared on more sophisticated computerised platforms. Compliance with the employer's target completion dates is, of course, essential. The method of working will depend on this programme in so far as the type and size of plant and the gang sizes to be used must be capable of achieving the output necessary to meet the programmed times. Allowance must be made for delays due to adverse weather, other hazards particular to the site and the requirements of the specification, particularly with regard to periods to be allowed for service diversions and other employer's requirements.

A method statement is prepared in conjunction with the programme, setting out the resources required, outputs to be achieved and the requirements in respect of temporary works, etc.

At the same time, separate bar charts may be produced giving:

- plant requirements
- staff and site supervision requirements
- labour requirements

These programmes and method statements will form the basis of the actual contract programme should the tender be accepted. They will also enable the Contractor to assure himself that he has available or can gain access to the necessary resources in plant, labour, materials and supervision to carry out the work should he be awarded the contract.

5. DESIGN OF TEMPORARY WORKS, ETC

Normally the period of time allowed for tendering is relatively short and therefore it is important that those aspects requiring design work are recognized as early as possible in the tender period. Design can be carried out either by the Contractor using his own engineers or by utilizing the services of a Consulting Engineer. There are three aspects of design to be considered:

A. Temporary works to the contractor's requirements to enable the works to be constructed

Design of temporary works covers the design of those parts of the work for which the contractor accepts full responsibility in respect of stability and safety during construction. Such parts include support structures, coffer-dams, temporary bridges, jetties and river diversions, special shuttering, scaffolding, haul roads and hardstandings, compounds, traffic management etc. Design must be in sufficient detail to enable materials, quantities and work content to be assessed and priced by the estimator. In designing such work, it is important that adequate attention is given to working platforms and access for labour and plant and also to ease of dismantling and re-erection for further uses without damage.

It should be noted that many specialist subcontractors will provide a design service when submitting quotations. For example, scaffolding contractors will design suitable support work for soffit shuttering, etc.

B. Specific items of the permanent works to meet a performance specification set out by the client or the Consulting Engineer

It is common practice for certain parts of the work to be specified by means of a performance specification. For example, concrete is specified by strength only, piles by load carrying capacity, etc. It is then left to the contractor to use those materials, workmanship and design which he feels are most suited to the particular site and conditions.

In many cases such design will be carried out by specialist suppliers or subcontractors.

C. Alternative designs for sections of the permanent works where the contractor's experience leads him to consider that a more economical design could be used

It is possible for a Contractor to use his expertise and experience to design and submit alternative proposals and prices for complete sections of the permanent work without, of course, altering the basic requirements of the original design. Examples would be foundations in difficult ground conditions, bridge superstructures, use of pre-cast in place of in situ concrete, etc. Such designs may be carried out by the contractor's own staff or in conjunction with Consulting Engineers or specialist contractors

Obviously this is only done when the Contractor can offer considerable savings in cost and/or a reduced construction period. It is necessary to include a price for the original design in the tender, but the decision to submit a keen tender may be underpinned by the hope of sharing such savings with the Employer.

In all cases, the Contractor's designs and calculations must be checked and approved by the Employer's Consulting Engineer, but such checks and approvals do not relieve the Contractor of his responsibilities for the safety and stability of the work.

Where any design work is undertaken it is important to ensure that adequate design insurance is maintained.

6. QUANTITIES CHECK

Working within obvious time constraints, the estimating team will endeavour to complete a quantities check on at least the major and high price quantities of the Bill, as this could affect pricing strategy, for example, in pricing provisional items. Any major discrepancies noted should be referred to the management appraisal agenda.

7. COST ESTIMATE

At this stage the estimator draws together the information and prepares a cost estimate made up of:

- preliminaries and general items (see Part 2)
- temporary works
- labour, costed by reference to the method statement, with appropriate allowances for labour oncosts (non-productive overtime, travelling time / fares, subsistence, guaranteed time, bonus, Employer's Liability Insurance, training, statutory payments, etc.)
- material costs taken from price lists or suppliers' quotations, with appropriate allowances for waste
- plant, whether owned or hired, with appropriate allowances for transport to / from site, erection/dismantling, maintenance, fuel and operation. Heavy static plant (batching plant, tower cranes, etc.) will normally be priced with general items; the remainder will normally be allocated to the unit rates
- sublet work, as quoted by specialist subcontractors, with appropriate allowances for all attendances required to be provided by the main contractor

At the same time, a preliminary assessment of risk/opportunity will be made for consideration with the Management Appraisal (see Section 8 below). On major tenders a formal quantified risk assessment (QRA) will be undertaken – the estimator will be expected to evaluate the effects of the risks for inclusion in the calculations. This will include a look at:

- weather conditions – costs not recoverable by claims under the Conditions of Contract, ground becoming unsuitable for working due to the effect of weather, etc.
- flooding – liability of site to flooding and the consequent costs
- suitability of materials – particular risk can arise if prices are based on the use of borrow pits or quarries and inadequate investigation has been carried out or development is refused by the local authorities
- reliability of subcontractors – failure of a subcontractor to perform can result in higher costs through delays to other operations and employing an alternative subcontractor at higher cost
- non-recoverable costs – such as excesses on insurance claims
- estimator's ability – e.g. outputs allowed. This can only be gauged from experience
- cost increase allowances for fixed price contract
- terms and conditions contained in the contract documents
- ability to meet specification requirements for the prices allowed
- availability of adequate and suitable labour

Risk is, of course, balanced by opportunity and consideration needs to be given to areas for which particular expertise possessed by the Contractor will lead to a price advantage against other tenderers.

8. PRICING THE TENDER

Once the cost estimate is complete, the estimator prices the items in the Bill. The rate to be entered against the items at this stage should be the correct rate for doing the job, whatever the quantity shown.

Where the overall operation covers a number of differing Bill items, the estimator will allocate the cost to the various items in reasonable proportions; the majority of the work is priced in this manner. The remaining items are normally priced by unit rate calculation.

Resources schedules, based on the programme and giving details of plant, labour and staff, perform an important role in enabling the estimator to check that he has included the total cost of resources for the period they are required on site. It is not unusual for an item of plant to be used intermittently for more than one operation. A reconciliation of the total time for which the cost has been included in the estimate against the total time the item is needed on site as shown on the programme, gives a period of non-productive time. The cost of this is normally included in site oncosts and preliminaries. A similar situation can arise in the cases of skilled labour, craftsmen and plant drivers.

Having priced the Bill at cost, there will remain a sum to be spread over the Bill items. The way in which this is done depends on the view taken by the Contractor of the project; for example:

- sums can be put against the listed general items in the Preliminaries Bill
- a fixed 'Adjustment Item' can be included in the Bill for the convenience of the estimator. This can be used for the adjustment made following the Management Appraisal, and for taking advantage of any late but favourable quotations received from subcontractors or suppliers
- the balance can be spread over the rates by equal percentage on all items, or by unequal percentages to assist in financing the contract or to take advantage of possible contract variations, or expected quantity changes (see notes against 7)

The Contractor will normally assess the financial advantages to be gained from submitting his bid in this manner and possibly enabling him to submit a more competitive offer.

After completing the pricing of all aspects of the tender, the total costs are summarized and profit, risk, etc., added to arrive at the total value of the Tender. A suggested form of this summary is set out in Part 8: Oncosts and Profit.

Finally, reasonable forecasts of cash flow and finance requirements are essential for the successful result of the project. Preliminary assessments may have been made for the information of management, but contract cash flow and the amount of investment required to finance the work can now be estimated by plotting income against expenditure using the programme of work and the priced Bill of Quantities. Payment in arrears and retentions, both from the Employer and to the suppliers and subcontractors, must be taken into account.

It is unlikely that sufficient time will be available during the tender period to produce such information accurately, but an approximate figure, for use as a guide for finance requirements, can be assessed.

A worked example is set out in Part 8: Oncosts and Profit.

9. MANAGEMENT APPRAISAL

Clearly, as far as the detail of the tender build-up is concerned, management must rely upon its established tendering procedures and upon the experience and skill of its estimators. However the comprehensive review of tenders prior to submission is an onerous duty and the estimator should look upon the process as an opportunity to demonstrate his skill. The Management Appraisal will include a review of:

- the major quantities
- the programme and method statement
- plant usage
- major suppliers and/or subcontractors, and discounts
- the nature of the competition
- risk and opportunity
- contract conditions, including in particular the level of damages for late completion, the minimum amount of certificates, retention and bonding requirements
- cash flow and finance
- margin for head office overheads and profit
- the weighting and spreading of the cost estimate over the measured items in the Bill

10. SUBMISSION OF TENDER

On completion of the tender, the documents are read over, comp checked and then despatched to the employer in accordance with the conditions set down in the invitation letter.

A complete copy of the tender as submitted should be retained by the contractor. Drawings on which the offer has been based should be clearly marked 'Tender Copy', their numbers recorded and the drawings filed for future reference. These documents will then form the basis for price variations should the design be amended during the currency of the contract.

The contractor may wish to qualify his offer to clarify the basis of his price. Normally such qualifications are included in a letter accompanying the tender. Legally the Form of Tender constitutes the offer and it is important that reference to such a letter is made on the Form of Tender to ensure that it forms part of the offer. Wording such as 'and our letter dated..., reference...' should be added prior to quoting the Tender Sum.

Before any qualifications are quoted, careful note must be taken of the 'instructions to tenderers', as qualifications, or at least qualifications submitted without a conforming tender, may be forbidden.

11. DESIGN AND CONSTRUCT VARIATIONS

Where once the design and construct tender was the province of only very major projects, this form of procurement is now much more widespread throughout the entire spectrum of Civil Engineering works. The key difference to the construct only tender is self evident: the bidder is required to produce and price designs that will satisfy the employer's stated requirements and often need to result in a scheme in line with indicative plans.

In practice, this process runs best when managed by an overall bid manager, with the estimator organizing the bill preparation and pricing from the submitted designs. The bid timescales are such that an orderly progression from design through bill preparation to pricing is rarely achieved. This process places far greater demands on the estimator's flexibility and management skills, but can prove ultimately more rewarding.

The Aggregates Levy

The Aggregates Levy came into operation on 1 April 2002 in the UK, except for Northern Ireland where it has been phased in over five years from 2003.

It was introduced to ensure that the external costs associated with the exploitation of aggregates are reflected in the price of aggregate, and to encourage the use of recycled aggregate. There continues to be strong evidence that the levy is achieving its environmental objectives, with sales of primary aggregate down and production of recycled aggregate up. The Government expects that the rates of the levy will at least keep pace with inflation over time, although it accepts that the levy is still bedding in.

The rate of the levy increased to £2.00 per tonne from 1 April 2009 and is levied on anyone considered to be responsible for commercially exploiting virgin aggregates in the UK and should naturally be passed by price increase to the ultimate user. The proposed rate increase to £2.10 per tonne was deferred within the 2011 Budget and will now take effect from 1 April 2012.

All materials falling within the definition of Aggregates are subject to the levy unless specifically exempted.

It does not apply to clay, soil, vegetable or other organic matter.

The intention is that it will:

- Encourage the use of alternative materials that would otherwise be disposed of to landfill sites.
- Promote development of new recycling processes, such as using waste tyres and glass
- Promote greater efficiency in the use of virgin aggregates
- Reduce noise and vibration, dust and other emissions to air, visual intrusion, loss of amenity and damage to wildlife habitats

Definitions

'Aggregates' means any rock, gravel or sand which is extracted or dredged in the UK for aggregates use. It includes whatever substances are for the time being incorporated in it or naturally occur mixed with it.

'Exploitation' is defined as involving any one or a combination of any of the following:

- Being removed from its original site
- Becoming subject to a contract or other agreement to supply to any person
- Being used for construction purposes
- Being mixed with any material or substance other than water, except in permitted circumstances

Incidence

It is a tax on primary aggregates production – i.e. virgin aggregates won from a source and used in a location within the UK territorial boundaries (land or sea). The tax is not levied on aggregates which are exported or on aggregates imported from outside the UK territorial boundaries.

It is levied at the point of sale.

Exemption from tax

An aggregate is exempt from the levy if it is:

- Material which has previously been used for construction purposes
- Aggregate that has already been subject to a charge to the Aggregates Levy
- Aggregate which was previously removed from its originating site before the start date of the levy
- Aggregate which is being returned to the land from which it was won
- Aggregate won from a farm land or forest where used on that farm or forest
- Rock which has not been subjected to an industrial crushing process
- Aggregate won by being removed from the ground on the site of any building or proposed building in the course of excavations carried out in connection with the modification or erection of the building and exclusively for the purpose of laying foundations or of laying any pipe or cable
- Aggregate won by being removed from the bed of any river, canal or watercourse or channel in or approach to any port or harbour (natural or artificial), in the course of carrying out any dredging exclusively for the purpose of creating, restoring, improving or maintaining that body of water
- Aggregate won by being removed from the ground along the line of any highway or proposed highway in the course of excavations for improving, maintaining or constructing the highway otherwise than purely to extract the aggregate
- Drill cuttings from petroleum operations on land and on the seabed
- Aggregate resulting from works carried out in exercise of powers under the New Road and Street Works Act 1991, the Roads (Northern Ireland) Order 1993 or the Street Works (Northern Ireland) Order 1995
- Aggregate removed for the purpose of cutting of rock to produce dimension stone, or the production of lime or cement from limestone.
- Aggregate arising as a waste material during the processing of the following industrial minerals:
    - ball clay
    - barytes
    - calcite
    - china clay
    - coal, lignite, slate or shale
    - feldspar
    - flint
    - fluorspar
    - fuller's earth
    - gems and semi-precious stones
    - gypsum
    - any metal or the ore of any metal
    - muscovite
    - perlite
    - potash
    - pumice
    - rock phosphates
    - sodium chloride
    - talc
    - vermiculite

    However, the levy is still chargeable on any aggregates arising as the spoil or waste from or the by-products of the above exempt processes. This includes quarry overburden.

    Anything that consists 'wholly or mainly' of the following is exempt from the levy (note that 'wholly' is defined as 100% but 'mainly' as more than 50%, thus exempting any contained aggregates amounting to less than 50% of the original volumes:

- clay, soil, vegetable or other organic matter
- coal, slate or shale
- china clay waste and ball clay waste

Relief from the levy either in the form of credit or repayment is obtainable where:

- it is subsequently exported from the UK in the form of aggregate
- it is used in an exempt process
- where it is used in a prescribed industrial or agricultural process
- it is waste aggregate disposed of by dumping or otherwise, e.g. sent to landfill or returned to the originating site

The Aggregates Levy Credit Scheme (ALCS) for Northern Ireland was suspended with effect from 1 December 2010 following a ruling by the European General Court.

A new exemption for aggregate obtained as a by-product of railway, tramway and monorail improvement, maintenance and construction was introduced in 2007.

Discounts
From 1 July 2005 the standard added water percentage discounts listed below can be used. Alternatively a more exact percentage can be agreed and this must be done for dust dampening of aggregates.

- washed sand 7%
- washed gravel 3.5%
- washed rock/aggregate 4%

Impact
The British Aggregates Association suggests that the additional cost imposed by quarries is more likely to be in the order of £2.765 per tonne on mainstream products, applying an above average rate on these in order that by-products and low grade waste products can be held at competitive rates, as well as making some allowance for administration and increased finance charges.

With many gravel aggregates costing in the region of £16.00 to £18.00 per tonne, there is a significant impact on construction costs.

Avoidance
An alternative to using new aggregates in filling operations is to crush and screen rubble which may become available during the process of demolition and site clearance as well as removal of obstacles during the excavation processes.

Example: Assuming that the material would be suitable for fill material under buildings or roads, a simple cost comparison would be as follows (note that for the purpose of the exercise, the material is taken to be 1.80 tonne per m³ and the total quantity involved less than 1,000 m³):

Specimen Unit Rates	£/m³	£/tonne
Importing fill material:		
Cost of 'new' aggregates delivered to site	31.23	17.35
Addition for Aggregates Tax	3.51	1.95
Total cost of importing fill materials	34.74	19.30
Disposing of site material:		
Cost of removing materials from site materials	21.52	11.95
Crushing site materials:		
Transportation of material from excavations or demolition to stockpiles	3.00	1.67
Transportation of material from temporary stockpiles to the crushing plant	4.00	2.22
Establishing plant and equipment on site; removing on completion	2.00	1.11
Maintain and operate plant	9.00	5.00
Crushing hard materials on site	13.00	7.22
Screening material on site	2.00	1.11
Total cost of crushing site materials	**33.00**	**18.33**

From the above it can be seen that potentially there is a great benefit in crushing site materials for filling rather than importing fill materials.

Setting the cost of crushing against the import price would produce a saving of £1.74 per m³. If the site materials were otherwise intended to be removed from the site, then the cost benefit increases by the saved disposal cost to £23.26 per m³.

Even if there is no call for any or all of the crushed material on site, it ought to be regarded as a useful asset and either sold on in crushed form or else sold with the prospects of crushing elsewhere.

Specimen Unit rates	unit³	£
Establishing plant and equipment on site; removing on completion		
Crushing plant	trip	1,200.00
Screening plant	trip	600.00
Maintain and operate plant		
Crushing plant	week	7,200.00
Screening plant	week	1,800.00
Transportation of material from excavations or demolition places to temporary stockpiles	m³	3.00
Transportation of material from temporary stockpiles to the crushing plant	m³	2.40
Breaking up material on site using impact breakers		
mass concrete	m³	14.00
reinforced concrete	m³	16.00
brickwork	m³	6.00
Crushing material on site		
mass concrete not exceeding 1000m³	m³	13.00
mass concrete 1000 – 5000m³	m³	12.00
mass concrete over 5000m³	m³	11.00
reinforced concrete not exceeding 1000m³	m³	15.00
reinforced concrete 1000 – 5000m³	m³	14.00
reinforced concrete over 5000m³	m³	13.00
brickwork not exceeding 1000m³	m³	12.00
brickwork 1000 – 5000m³	m³	11.00
brickwork over 5000m³	m³	10.00
Screening material on site	m³	2.00

More detailed information can be found on the HMRC website (www.hmrc.gov.uk) in Notice AGL 1 Aggregates Levy published in May 2009.

Capital Allowances

Introduction

Capital Allowances provide tax relief by prescribing a statutory rate of depreciation for tax purposes in place of that used for accounting purposes. They are utilized by government to provide an incentive to invest in capital equipment, including assets within commercial property, by allowing the majority of taxpayers a deduction from taxable profits for certain types of capital expenditure, thereby reducing or deferring tax liabilities.

The capital allowances most commonly applicable to real estate are those given for capital expenditure on existing commercial buildings in disadvantaged areas, and plant and machinery in all buildings other than residential dwellings. Relief for certain expenditure on industrial buildings and hotels was withdrawn from April 2011, although the ability to claim plant and machinery remains.

Enterprise Zone Allowances are also available for capital expenditure within designated areas only where there is a focus on high value manufacturing. Enhanced rates of allowances are available on certain types of energy and water saving plant and machinery assets, whilst reduced rates apply to 'integral features' and items with an expected economic life of more than 25 years.

The Act

The primary legislation is contained in the Capital Allowances Act 2001. Major changes to the system were announced by the Government in 2007 and there have been further changes in subsequent Finance Acts.

The Act is arranged in 12 Parts (plus two addenda) and was published with an accompanying set of Explanatory Notes.

Plant and Machinery

The Finance Act 1994 introduced major changes to the availability of Capital Allowances on real estate. A definition was introduced which precludes expenditure on the provision of a building from qualifying for plant and machinery, with prescribed exceptions.

List A in Section 21 of the 2001 Act sets out those assets treated as parts of buildings: -

- Walls, floors, ceilings, doors, gates, shutters, windows and stairs.
- Mains services, and systems, for water, electricity and gas.
- Waste disposal systems.
- Sewerage and drainage systems.
- Shafts or other structures in which lifts, hoists, escalators and moving walkways are installed.
- Fire safety systems.

Similarly, List B in Section 22 identifies excluded structures and other assets.

Both sections are, however, subject to Section 23. This section sets out expenditure, which although being part of a building, may still be expenditure on the provision of Plant and Machinery.

List C in Section 23 is reproduced below:

Sections 21 and 22 do not affect the question whether expenditure on any item in List C is expenditure on the provision of Plant or Machinery

1.   Machinery (including devices for providing motive power) not within any other item in this list.
2.   Gas and sewerage systems provided mainly –
     a.   to meet the particular requirements of the qualifying activity, or
     b.   to serve particular plant or machinery used for the purposes of the qualifying activity.
3.   Omitted
4.   Manufacturing or processing equipment; storage equipment (including cold rooms); display equipment; and counters, checkouts and similar equipment.
5.   Cookers, washing machines, dishwashers, refrigerators and similar equipment; washbasins, sinks, baths, showers, sanitary ware and similar equipment; and furniture and furnishings.
6.   Hoists.
7.   Sound insulation provided mainly to meet the particular requirements of the qualifying activity.
8.   Computer, telecommunication and surveillance systems (including their wiring or other links).
9.   Refrigeration or cooling equipment.
10. Fire alarm systems; sprinkler and other equipment for extinguishing or containing fires.
11. Burglar alarm systems.
12. Strong rooms in bank or building society premises; safes.
13. Partition walls, where moveable and intended to be moved in the course of the qualifying activity.
14. Decorative assets provided for the enjoyment of the public in hotel, restaurant or similar trades.
15. Advertising hoardings; signs, displays and similar assets.
16. Swimming pools (including diving boards, slides & structures on which such boards or slides are mounted).
17. Any glasshouse constructed so that the required environment (namely, air, heat, light, irrigation and temperature) for the growing of plants is provided automatically by means of devices forming an integral part of its structure.
18. Cold stores.
19. Caravans provided mainly for holiday lettings.
20. Buildings provided for testing aircraft engines run within the buildings.
21. Moveable buildings intended to be moved in the course of the qualifying activity.
22. The alteration of land for the purpose only of installing Plant or Machinery.
23. The provision of dry docks.
24. The provision of any jetty or similar structure provided mainly to carry Plant or Machinery.
25. The provision of pipelines or underground ducts or tunnels with a primary purpose of carrying utility conduits.
26. The provision of towers to support floodlights.
27. The provision of –
     a.   any reservoir incorporated into a water treatment works, or
     b.   any service reservoir of treated water for supply within any housing estate or other particular locality.
28. The provision of –
     a.   silos provided for temporary storage, or
     b.   storage tanks.
29. The provision of slurry pits or silage clamps.
30. The provision of fish tanks or fish ponds.
31. The provision of rails, sleepers and ballast for a railway or tramway.
32. The provision of structures and other assets for providing the setting for any ride at an amusement park or exhibition.
33. The provision of fixed zoo cages.

Capital Allowances on plant and machinery are given in the form of writing down allowances at the rate of 20% per annum on a reducing balance basis. For every £100 of qualifying expenditure £20 is claimable in year 1, £16 in year 2 and so on until either the all the allowances have been claimed or the asset is sold. This rate reduces to 18% from April 2012.

Integral Features

The category of qualifying expenditure on 'integral features' was introduced with effect from April 2008. The following items are integral features:

- An electrical system (including a lighting system)
- A cold water system
- A space or water heating system, a powered system of ventilation, air cooling or air purification, and any floor or ceiling comprised in such a system
- A lift, an escalator or a moving walkway
- External solar shading

The draft legislation also included active facades but these were subsequently omitted, the explanation being that allowances are already given on the additional inner skin because it is part of the air conditioning system.

A reduced writing down allowance of 10% per annum is available on integral features. This rate reduces to 8% from April 2012.

The legislation also includes a rule that prevents a revenue deduction being obtained where expenditure is incurred that is more than 50% of the cost of replacing the integral feature.

Thermal Insulation

For many years the addition of thermal insulation to an existing industrial building has been treated as qualifying for plant and machinery allowances. From April 2008 this has been extended to include all commercial buildings but not residential buildings.

A reduced writing down allowance of 10% per annum is available on thermal insulation. This rate reduces to 8% from April 2012.

Long Life Assets

A reduced writing down allowance of 10% per annum is available on long-life assets (reducing to 8% from April 2012). Allowances were given at the rate of 6% before April 2008.

A long-life asset is defined as plant and machinery that can reasonably be expected to have a useful economic life of at least 25 years. The useful economic life is taken as the period from first use until it is likely to cease to be used as a fixed asset of any business. It is important to note that this likely to be a shorter period than an item's physical life.

Plant and machinery provided for use in a building used wholly or mainly as dwelling house, showroom, hotel, office or retail shop or similar premises, or for purposes ancillary to such use, cannot be long-life assets.

In contrast plant and machinery assets in buildings such as factories, cinemas, hospitals and so on are all potentially long-life assets.

Case Law

The fact that an item appears in List C does not automatically mean that it will qualify for capital allowances. It only means that it may potentially qualify.

Guidance about the meaning of plant has to be found in case law. The cases go back a long way, beginning in 1887. The current state of the law on the meaning of plant derives from the decision in the case of *Wimpy International Ltd and Associated Restaurants Ltd v Warland* in the late 1980s.

The Judge in that case said that there were three tests to be applied when considering whether or not an item is plant.

1.   Is the item stock in trade? If the answer yes, then the item is not plant.
2.   Is the item used for carrying on the business? In order to pass the business use test the item must be employed in carrying on the business; it is not enough for the asset to be simply used in the business. For example, product display lighting in a retail store may be plant but general lighting in a warehouse would fail the test.
3.   Is the item the business premises or part of the business premises? An item cannot be plant if it fails the premises test, i.e. if the business use is as the premises (or part of the premises) or place on which the business is conducted. The meaning of part of the premises in this context should not be confused with the law of real property. The Inland Revenue's internal manuals suggest there are four general factors to be considered, each of which is a question of fact and degree:
•   Does the item appear visually to retain a separate identity
•   With what degree of permanence has it been attached to the building
•   To what extent is the structure complete without it
•   To what extent is it intended to be permanent or alternatively is it likely to be replaced within a short period

There is obviously a core list of items that will usually qualify in the majority of cases. However, many other still need to be looked at on a case-by-case basis. For example, decorative assets in a hotel restaurant may be plant but similar assets in an office reception area would almost certainly not be.

One of the benefits of the integral features rules, apart from simplification, is that items that did not qualify by applying these rules, such as general lighting in an office building, will now qualify albeit at a reduced rate.

Refurbishment Schemes

Building refurbishment projects will typically be a mixture of capital costs and revenue expenses, unless the works are so extensive that they are more appropriately classified a redevelopment. A straightforward repair or a 'like for like' replacement of part of an asset would be a revenue expense, meaning that the entire amount can be deducted from taxable profits in the same year.

Where capital expenditure is incurred that is incidental to the installation of plant or machinery then Section 25 of the 2001 Act allows it to be treated as part of the expenditure on the qualifying item. Incidental expenditure will often include parts of the building that would be otherwise disallowed, as shown in the Lists reproduced above. For example, the cost of forming a lift shaft inside an existing building would be deemed to be part of the expenditure on the provision of the lift.

The extent of the application of section 25 was reviewed for the first time by the Special Commissioners in December 2007 and by the First Tier Tribunal (Tax Chamber) in December 2009, in the case of JD Wetherspoon. The key areas of expenditure considered were overheads and preliminaries where it was held that such costs could be allocated on a pro-rata basis; decorative timber panelling which was found to be part of the premises and so ineligible for allowances; toilet lighting which was considered to provide an attractive ambience and qualified for allowances; and incidental building alterations of which enclosing walls to toilets and kitchens and floor finishes did not qualify but tiled splash backs, toilet cubicles and drainage did qualify along with the related sanitary fittings and kitchen equipment.

Annual Investment Allowance

The annual investment allowance is available to all businesses of any size and allows a deduction for the whole of the first £100,000 (£50,000 before April 2010) of qualifying expenditure on plant and machinery, including integral features and long life assets. The allowance reduces to £25,000 from April 2011.

The Enhanced Capital Allowances Scheme

The scheme is one of a series of measures introduced to ensure that the UK meets its target for reducing green-house gases under the Kyoto Protocol. 100% first year allowances are available on products included on the Energy Technology List published on the website at www.eca.gov.uk and other technologies supported by the scheme. All businesses will be able to claim the enhanced allowances, but only investments in new and unused machinery and plant can qualify.

There are currently 14 technologies and 53 sub-technologies currently covered by the scheme.

- Air-to-air energy recovery.
- Automatic monitoring and targeting (AMT)
- Boiler equipment
- Combined heat and power (CHP)
- Compressed air equipment
- Heat pumps for space heating
- Heating ventilation and air conditioning equipment
- Lighting
- Motors and drives
- Pipework insulation
- Radiant and warm air heaters
- Refrigeration equipment
- Solar thermal systems
- Uninterruptible power supplies (UPS)

Compact heat exchangers were removed in 2010.

The Finance Act 2003 introduced a new category of environmentally beneficial plant and machinery qualifying for 100% first-year allowances. The Water Technology List includes 14 technologies,

- Cleaning in place equipment.
- Efficient showers
- Efficient taps.
- Efficient toilets.
- Efficient washing machines.
- Flow controllers.
- Leakage detection equipment.
- Meters and monitoring equipment.
- Rainwater harvesting equipment.
- Small scale slurry and sludge dewatering equipment.
- Vehicle wash water reclaim units.
- Water efficient industrial cleaning equipment.
- Water management equipment for mechanical seals.
- Water reuse systems.

Buildings and structures and long life assets as defined above cannot qualify under the scheme. However, following the introduction of the integral features rules lighting in any non residential building may potentially qualify for enhanced capital allowances if it meets the relevant criteria.

A limited payable ECA tax credit equal to 19% of the loss surrendered was also introduced for UK companies in April 2008.

Industrial Buildings Allowances

Industrial Buildings Allowances were withdrawn from April 2011.

For expenditure incurred prior to this date, an industrial building (or structure) is defined in Sections 271 and 274 of the 2001 Act and includes buildings used for qualifying purposes, including manufacturing, processing and storage. Additional uses include agricultural contracting, working foreign plantations, fishing and mineral extraction.

The following undertakings are also qualifying trades:

- Electricity
- Water
- Hydraulic power
- Sewerage
- Transport
- Highway undertakings
- Tunnels
- Bridges
- Inland navigation
- Docks

The definition extended to include buildings provided for the welfare of workers in a qualifying trade and sports pavilions provided and used for the welfare of workers in any trade. Vehicle repair workshops and roads on industrial estates may also form part of the qualifying expenditure.

Retail shops, showrooms, offices, dwelling houses and buildings used ancillary to a retail purpose are specifically excluded.

Writing-Down Allowances

Allowances are given on qualifying expenditure on a straight-line basis over 25 years. The allowance is given if the building is being used for a qualifying purpose on the last day of the accounting period. Where the building is used for a non-qualifying purpose the year's allowance is lost.

As part of the phased abolition, rates were reduced to 3% for 2008-09, 2% for 2009-10, 1% for 2010-11 and 0% for 2011 onwards.

From 21 March 2007 a balancing adjustment is no longer made on the sale of an industrial building. A purchaser of the used industrial building will be entitled to allowances based on the vendor's tax written down value, rather than the original construction cost adjusted for any periods of non-qualifying use.

The allowances will still be spread equally over the remaining period to the date twenty-five years after first use. However, even if the building was acquired prior to 21 March 2007, whatever the annual allowance given in 2007-08, it will be reduced to ¾ of that amount in 2008-09, ½ in 2009-10, ¼ in 2010-11 and zero from 2011 onwards.

Hotel Allowances

Industrial Buildings Allowances are also available on capital expenditure incurred on constructing a 'qualifying hotel'. The building must not only be a 'hotel' in the normal sense of the word, but must also be a 'qualifying hotel' as defined in Section 279 of the 2001 Act, which means satisfying the following conditions:

- The accommodation is in buildings of a permanent nature
- It is open for at least 4 months in the season (April to October)
- It has 10 or more letting bedrooms

- The sleeping accommodation consists wholly or mainly of letting bedrooms
- The services that it provides include breakfast and an evening meal (i.e. there must be a restaurant), the making of beds and cleaning of rooms.

A hotel may be in more than one building and swimming pools, car parks and similar amenities are included in the definition.

Enterprise Zones

The creation of 21 new Enterprise Zones was announced in the 2011 Budget. Originally introduced in the early 1980s as a stimulus to commercial development and investment, they had virtually faded from the real estate psyche.

The original zones benefited from a 100% first year allowance on capital expenditure incurred on the construction (or the purchase within two years of first use) of any commercial building within a designated enterprise zone, within ten years of the site being so designated. Like other allowances given under the industrial buildings code the building has a life of twenty-five years for tax purposes.

The majority of these enterprise zones had reached the end of their ten-year life by 1993. However, in certain very limited circumstances it may still be possible to claim these allowances up to twenty years after the site was first designated.

The precise nature of the new zones is not available at the time of press; however, enhanced allowances will only be available in assisted areas, where there is a strong focus on high value manufacturing.

Flat Conversion Allowances

Tax relief is available on capital expenditure incurred on or after 11 May 2001 on the renovation or conversion of vacant or underused space above shops and other commercial premises to provide flats for rent.

In order to qualify the property must have been built before 1980 and the expenditure incurred on, or in connection with:

- Converting part of a qualifying building into a qualifying flat.
- Renovating an existing flat in a qualifying building if the flat is, or will be a qualifying flat.
- Repairs incidental to conversion or renovation of a qualifying flat, and
- The cost of providing access to the flat(s).

The property must not have more than four storeys above the ground floor and it must appear that, when the property was constructed, the floors above the ground floor were primarily for residential use. The ground floor must be authorized for business use at the time of the conversion work and for the period during which the flat is held for letting. Each new flat must be a self-contained dwelling, with external access separate from the ground-floor premises. It must have no more than 4 rooms, excluding kitchen and bathroom. None of the flats can be 'high value' flats, as defined in the legislation. The new flats must be available for letting as a dwelling for a period of not more than 5 years.

An initial allowance of 100 per cent is available or, alternatively, a lower amount may be claimed, in which case the balance may be claimed at a rate of 25 per cent per annum in subsequent a years. The allowances may be recovered if the flat is sold or ceases to be let within 7 years.

Business Premises Renovation Allowance

The Business Premises Renovation Allowance (BPRA) was first announced in December 2003. The idea behind the scheme is to bring long-term vacant properties back into productive use by providing 100 per cent capital allowances for the cost of renovating and converting unused premises in disadvantaged areas. The legislation was included in Finance Act 2005 and was finally implemented on 11 April 2007 following EU state aid approval.

The legislation is identical in many respects to that for flat conversion allowances. The scheme will apply to properties within one of the areas specified in the Assisted Areas Order 2007 and Northern Ireland.

BPRA will be available to both individuals and companies who own or lease business property that has been unused for 12 months or more. Allowances will be available to a person who incurs qualifying capital expenditure on the renovation of business premises.

An announcement to extend the scheme by a further five years to 2017 was made within the 2011 Budget.

Agricultural Building Allowances

Allowances are available on capital expenditure incurred on the construction of buildings and works for the purposes of husbandry on land in the UK. Agricultural building means a building such as a farmhouse or farm building, a fence or other works. A maximum of only one-third of the expenditure on a farmhouse may qualify.

Husbandry includes any method of intensive rearing of livestock or fish on a commercial basis for the production of food for human consumption, and the cultivation of short rotation coppice. Over the years the Courts have held that sheep grazing and poultry farming are husbandry, and that a dairy business and the rearing of pheasants for sport are not. Where the use is partly for other purposes the expenditure can be apportioned.

The rate of allowances available and the way in which the system operates is very similar to that described above for industrial buildings. However, no allowance is ever given if the first use of the building is not for husbandry. A different treatment is also applied following acquisition of a used building unless the parties to the transaction elect otherwise.

Other Capital Allowances

Other types of allowances include those available for capital expenditure on Mineral Extraction, Research and Development, Know-How, Patents, Dredging and Assured Tenancy.

Value Added Tax

Introduction

Value Added Tax (VAT) is a tax on the consumption of goods and services. The UK adopted VAT when it joined the European Community in 1973. The principal source of European law in relation to VAT is Council Directive 2006/112/EC, a recast of Directive 77/388/EEC which is currently restated and consolidated in the UK through the VAT Act 1994 and various Statutory Instruments, as amended by subsequent Finance Acts.

VAT Notice 708: Buildings and construction (June 2007) gives an interpretation of the law in connection with construction works from the point of view of HM Revenue & Customs. VAT tribunals and court decisions since the date of this publication will affect the application of the law in certain instances. The Notice is available on HM Revenue & Customs website at www.hmrc.gov.uk.

The scope of VAT

VAT is payable on:

- Supplies of goods and services made in the UK
- By a taxable person
- In the course or furtherance of business; and
- Which are not specifically exempted or zero-rated.

Rates of VAT

There are three rates of VAT:

- A standard rate, currently 20% since January 2011
- A reduced rate, currently 5%; and
- A zero rate.

Additionally some supplies are exempt from VAT and others are outside the scope of VAT.

Recovery of VAT

When a taxpayer makes taxable supplies he must account for VAT at the appropriate rate of either 20% or 5%. This VAT then has to be passed to HM Revenue & Customs and will normally be charged to the taxpayer's customers.

As a VAT registered person, the taxpayer can reclaim from HM Revenue & Customs as much of the VAT incurred on their purchases as relates to the standard-rated, reduced-rated and zero-rated onward supplies they make. A person cannot however reclaim VAT that relates to any non-business activities (but see below) or to any exempt supplies they make.

At predetermined intervals the taxpayer will pay to HM Revenue & Customs the excess of VAT collected over the VAT they can reclaim. However if the VAT reclaimed is more than the VAT collected, the taxpayer can reclaim the difference from HM Revenue & Customs.

Example

X Ltd constructs a block of flats. It sells long leases to buyers for a premium. X Ltd has constructed a new building designed as a dwelling and will have granted a long lease. This sale of a long lease is VAT zero-rated. This means any VAT incurred in connection with the development which X Ltd will have paid (e.g. payments for consultants and certain preliminary services) will be reclaimable. For reasons detailed below the contractor employed by X Ltd will not have charged VAT on his construction services.

Use for Business and Non Business Activities

Where VAT relates partly to business use and partly to non-business use then the basic rule is that it must be apportioned so that only the business element is potentially recoverable. In some cases VAT on land, buildings and certain construction services purchased for both business and non-business use could be recovered in full by applying what is known as 'Lennartz' accounting to reclaim VAT relating to the non-business use and account for VAT on the non business use over a maximum period of 10 years. HM Revenue & Customs revised their policy in January 2010 following an ECJ case restricting the scope and its application to immovable property will be removed completely from January 2011 when the UK law is amended to comply with EU Council Directive 2009/162/EU.

Taxable Persons

A taxable person is an individual, firm, company etc who is required to be registered for VAT. A person who makes taxable supplies above certain value limits is required to be registered. The current registration limit is £73,000 for 2011-12. The threshold is exceeded if at the end of any month the value of taxable supplies in the period of one year then ending is over the limit, or at any time, if there are reasonable grounds for believing that the value of the taxable supplies in the period of 30 days than beginning will exceed £73,000.

A person who makes taxable supplies below these limits is entitled to be registered on a voluntary basis if they wish, for example in order to recover VAT incurred in relation to those taxable supplies.

In addition, a person who is not registered for VAT in the UK but acquires goods from another EC member state, or make distance sales in the UK, above certain value limits may be required to register for VAT in the UK.

VAT Exempt Supplies

If a supply is exempt from VAT this means that no tax is payable – but equally the person making the exempt supply cannot normally recover any of the VAT on their own costs relating to that supply.

Generally property transactions such as leasing of land and buildings are exempt unless a landlord chooses to standard-rate its supplies by a process known as opting to tax. This means that VAT is added to rental income and also that VAT incurred, on say, an expensive refurbishment, is recoverable.

Supplies outside the scope of VAT

Supplies are outside the scope of VAT if they are:

- Made by someone who is not a taxable person
- Made outside the UK; or
- Not made in the course or furtherance of business

In course or furtherance of business

VAT must be accounted for on all taxable supplies made in the course or furtherance of business with the corresponding recovery of VAT on expenditure incurred.

If a taxpayer also carries out non-business activities then VAT incurred in relation to such supplies is generally not recoverable.

In VAT terms, business means any activity continuously performed which is mainly concerned with making supplies for a consideration. This includes:

- Any one carrying on a trade, vocation or profession;
- The provision of membership benefits by clubs, associations and similar bodies in return for a subscription or other consideration; and
- Admission to premises for a charge.

It may also include the activities of other bodies including charities and non-profit making organizations.

Examples of non-business activities are:

- Providing free services or information;
- Maintaining some museums or particular historic sites;
- Publishing religious or political views.

Construction Services

In general the provision of construction services by a contractor will be VAT standard rated at 20%, however, there are a number of exceptions for construction services provided in relation to certain residential and charitable use buildings.

The supply of building materials is VAT standard rated at 20%, however, where these materials are supplied as part of the construction services the VAT liability of those materials follows that of the construction services supplied.

Zero-rated construction services

The following construction services are VAT zero-rated including the supply of related building materials.

The construction of new dwellings

The supply of services in the course of the construction of a building designed for use as a dwelling or number of dwellings is zero-rated other than the services of an architect, surveyor or any other person acting as a consultant or in a supervisory capacity.

The following conditions must be satisfied in order for the works to qualify for zero-rating:

- The work must not amount to the conversion, reconstruction or alteration of an existing building;
- The work must not be an enlargement of, or extension to, an existing building except to the extent that the enlargement or extension creates an additional dwelling or dwellings;
- The building must be designed as a dwelling or number of dwellings. Each dwelling must consist of self-contained living accommodation with no provision for direct internal access from the dwelling to any other dwelling or part of a dwelling;
- Statutory planning consent must have been granted for the construction of the dwelling, and construction carried out in accordance with that consent;
- Separate use or disposal of the dwelling must not be prohibited by the terms of any covenant, statutory planning consent or similar provision.

The construction of a garage at the same time as the dwelling can also be zero-rated as can the demolition of any existing building on the site of the new dwelling

A building only ceases to be an existing building (see points 1. and 2. above) when it is:

- Demolished completely to ground level; or when
- The part remaining above ground level consists of no more than a single facade (or a double facade on a corner site) the retention of which is a condition or requirement of statutory planning consent or similar permission.

The construction of a new building for 'relevant residential or charitable' use

The supply of services in the course of the construction of a building designed for use as a relevant residential or charitable building is zero-rated other than the services of an architect, surveyor or any other person acting as a consultant or in a supervisory capacity.

- A 'relevant residential' use building means:
- A home or other institution providing residential accommodation for children;
- A home or other institution providing residential accommodation with personal care for persons in need of personal care by reason of old age, disablement, past or present dependence on alcohol or drugs or past or present mental disorder;
- A hospice;
- Residential accommodation for students or school pupils
- Residential accommodation for members of any of the armed forces;
- A monastery, nunnery, or similar establishment; or
- An institution which is the sole or main residence of at least 90% of its residents.

The construction of a new building for 'relevant residential or charitable' use – continued

A 'relevant residential' purpose building does not include use as a hospital, a prison or similar institution or as a hotel, inn or similar establishment.

A 'relevant charitable' use means use by a charity:

- Otherwise than in the course or furtherance of a business; or
- As a village hall or similarly in providing social or recreational facilities for a local community.

Non qualifying use which is not expected to exceed 10% of the time the building is normally available for use can be ignored. The calculation of business use can be based on time, floor area or head count subject to approval being acquired from HM Revenue & Customs.

The construction services can only be zero-rated if a certificate is given by the end user to the contractor carrying out the works confirming that the building is to be used for a qualifying purpose i.e. for a 'relevant residential or charitable' purpose. It follows that such services can only be zero-rated when supplied to the end user and, unlike supplies relating to dwellings, supplies by subcontractors cannot be zero-rated.

The construction of an annex used for a 'relevant charitable' purpose

Construction services provided in the course of construction of an annexe for use entirely or partly for a 'relevant charitable' purpose can be zero-rated.

In order to qualify the annexe must:

- Be capable of functioning independently from the existing building;
- Have its own main entrance; and
- Be covered by a qualifying use certificate.

The conversion of a non-residential building into dwellings or the conversion of a building from non-residential use to 'relevant residential' use where the supply is to a 'relevant' housing association

The supply to a 'relevant' housing association in the course of conversion of a non-residential building or non-residential part of a building into:

- A building or part of a building designed as a dwelling or number of dwellings; or
- A building or part of a building for use solely for a relevant residential purpose, of any services related to the conversion other than the services of an architect, surveyor or any person acting as a consultant or in a supervisory capacity are zero-rated.

A 'relevant' housing association is defined as:

- A private registered provider of social housing
- A registered social landlord within the meaning of Part I of the Housing Act 1996 (Welsh registered social landlords)
- A registered social landlord within the meaning of the Housing (Scotland) Act 2001 (Scottish registered social landlords), or
- A registered housing association within the meaning of Part II of the Housing (Northern Ireland) Order 1992 (Northern Irish registered housing associations).

If the building is to be used for a 'relevant residential' purpose the housing association should issue a qualifying use certificate to the contractor completing the works.

The construction of a permanent park for residential caravans

The supply in the course of the construction of any civil engineering work 'necessary for' the development of a permanent park for residential caravans of any services related to the construction can be VAT zero-rated. This includes access roads, paths, drainage, sewerage and the installation of mains water, power and gas supplies.

Certain building alterations for 'disabled' persons

Certain goods and services supplied to a disabled person, or a charity making these items and services available to 'disabled' persons can be zero-rated. The recipient of these goods or services needs to give the supplier an appropriate written declaration that they are entitled to benefit from zero rating.

The following services (amongst others) are zero-rated:

- the installation of specialist lifts and hoists and their repair and maintenance
- the construction of ramps, widening doorways or passageways including any preparatory work and making good work
- the provision, extension and adaptation of a bathroom, washroom or lavatory; and
- emergency alarm call systems

Approved alterations to protected buildings

A supply in the course of an 'approved alteration' to a 'protected building' of any services other than the services of an architect, surveyor or any person acting as consultant or in a supervisory capacity can be zero-rated.

A 'protected building' is defined as a building that is:

- designed to remain as or become a dwelling or number of dwellings after the alterations; or
- is intended for use for a 'relevant residential or charitable purpose' after the alterations; and which is;
- a listed building or scheduled ancient monument.

A listed building does not include buildings that are in conservation areas, but not on the statutory list, or buildings included in non-statutory local lists.

An 'approved alteration' is an alteration to a 'protected building' that requires and has obtained listed building consent or scheduled monument consent. This consent is necessary for any works that affect the character of a building of special architectural or historic interest.

It is important to note that 'approved alterations' do not include any works of repair or maintenance or any incidental alteration to the fabric of a building that results from the carrying out of repairs or maintenance work.

A 'protected building' that is intended for use for a 'relevant residential or charitable purpose' will require the production of a qualifying use certificate by the end user to the contractor providing the alteration services.

Listed Churches are 'relevant charitable' use buildings and where 'approved alterations' are being carried out zero-rate VAT can be applied. Additionally since 1 April 2004, listed places of worship can apply for a grant for repair and maintenance works equal to the full amount of VAT paid on eligible works carried out on or after 1 April 2004. Information relating to the scheme can be obtained from the website at www.lpwscheme.org.uk.

DIY Builders and Converters

Private individuals who decide to construct their own home are able to reclaim VAT they pay on goods they use to construct their home by use of a special refund mechanism made by way of an application to HM Revenue & Customs. This also applies to services provided in the conversion of an existing non-residential building to form a new dwelling.

The scheme is meant to ensure that private individuals do not suffer the burden of VAT if they decide to construct their own home.

Charities may also qualify for a refund on the purchase of materials incorporated into a building used for non-business purposes where they provide their own free labour for the construction of a 'relevant charitable' use building.

Reduced-rated construction services

The following construction services are subject to the reduced rate of VAT of 5%, including the supply of related building materials.

A changed number of dwellings conversion

In order to qualify for the 5% rate there must be a different number of 'single household dwellings' within a building than there were before commencement of the conversion works. A 'single household dwelling' is defined as a dwelling that is designed for occupation by a single household.

These conversions can be from 'relevant residential' purpose buildings, non-residential buildings and houses in multiple occupation.

A house in multiple occupation conversion

This relates to construction services provided in the course of converting a 'single household dwelling', a number of 'single household dwellings', a non-residential building or a 'relevant residential' purpose building into a house for multiple occupation such as a bed sit accommodation.

A special residential conversion

A special residential conversion involves the conversion of a single household dwelling, a house in multiple occupation or a non-residential building into a 'relevant residential' purpose building such as student accommodation or a care home.

Renovation of derelict dwellings

The provision of renovation services in connection with a dwelling or 'relevant residential' purpose building that has been empty for two or more years prior to the date of commencement of construction works can be carried out at a reduced rate of VAT of 5%.

Installation of energy saving materials

The supply and installation of certain energy saving materials including insulation, draught stripping, central heating and hot water controls and solar panels in a residential building or a building used for a relevant charitable purpose.

Grant-funded of heating equipment or connection of a gas supply

The grant funded supply and installation of heating appliances, connection of a mains gas supply, supply, installation, maintenance and repair of central heating systems, and supply and installation of renewable source heating systems, to qualifying persons. A qualifying person is someone aged 60 or over or is in receipt of various specified benefits.

Installation of security goods

The grant funded supply and installation of security goods to a qualifying person.

Housing alterations for the elderly

Certain home adaptations that support the needs of elderly people were reduced rated with effect from 1 July 2007.

Building Contracts

Design and build contracts

If a contractor provides a design and build service relating to works to which the reduced or zero rate of VAT is applicable then any design costs incurred by the contractor will follow the VAT liability of the principal supply of construction services.

Management contracts

A management contractor acts as a main contractor for VAT purposes and the VAT liability of his services will follow that of the construction services provided. If the management contractor only provides advice without engaging trade contractors his services will be VAT standard rated.

Construction Management and Project Management

The project manager or construction manager is appointed by the client to plan, manage and co-ordinate a construction project. This will involve establishing competitive bids for all the elements of the work and the appointment of trade contractors. The trade contractors are engaged directly by the client for their services.

The VAT liability of the trade contractors will be determined by the nature of the construction services they provide and the building being constructed.

The fees of the construction manager or project manager will be VAT standard rated. If the construction manager also provides some construction services these works may be zero or reduced rated if the works qualify.

Liquidated and Ascertained Damages

Liquidated damages are outside of the scope of VAT as compensation. The employer should not reduce the VAT amount due on a payment under a building contract on account of a deduction of damages. In contrast an agreed reduction in the contract price will reduce the VAT amount.

Similarly, in certain circumstances HM Revenue & Customs may agree that a claim by a contractor under a JCT or other form of contract is also compensation payment and outside the scope of VAT.

Land Remediation

GUIDANCE NOTES

Generally

The introduction of the Landfill Directive in July 2004 has had a considerable impact on the cost of remediation works in general and particularly the practice of dig-and-dump. The number of landfill sites licensed to accept hazardous waste has drastically reduced and inevitably this has led to increased costs.

Market forces will determine future increases in cost and the cost guidance given will require review

It must be emphasized that the cost advice given is an average and that costs can and do vary considerable from contract to contract depending on individual contractors, site conditions, type and extent of contamination, methods of working and various other factors as diverse as difficulty of site access and distance from approved tip.

Adjustments should be made to the rates shown for time, location, local conditions, site constraints and any other factors likely to affect the costs of a specific scheme.

Method of Measurement

Although this part of the book is primarily based on CESMM3, the specific rules have been varied from in cases where it has been felt that an alternative presentation would be of value to the book's main purpose of providing guidance on prices. This is especially so with a number of specialist contractors.

DEFINITION

Generally

Contaminated land refers generally to land which contains contaminants in sufficient quantities to harm people, animals, the environment or structures. There is now a statutory definition of 'contaminated land' contained within Part IIA of the Environmental Protection Act.

Contaminants comprise hazardous substances (solids, liquids or gases) that are not naturally occurring in the site. They arise from previous site usage, although sites can be affected by pollutants arising from adjoining sites through the movement of water and air. A contaminated site can similarly pose a risk to surrounding land by offsite migration of contaminants.

Contaminated sites can be sold on, although the new owner would take on the responsibility for the contamination and would obviously take this into account in the offered price.

The extent of remediation works required to address contamination varies dependent on the intended future use of the site – with industrial uses calling for a lower level of work than if the site were intended to be used for residential or agricultural purposes. In a commercial world, expenditure on decontaminating the land would usually need to be balanced against the release of the latent site value – unless of course the contamination contravened statutory limits.

Hazardous contaminants fall into three broad categories:

- Chemical contamination of land or water
- Biological contamination of land or water (e.g. samples containing pathological bacteria potentially harmful to health).
- Contamination of a physical nature (e.g. radioactive material, unsuitable fill materials, flammable gas or combustible material e.g. wood dust)

These can also be listed in the following sub-groups:

Gases	Toxic, flammable and explosive gases
	e.g. hydrogen cyanide
	hydrogen sulphide
	Flammable and explosive gases
	e.g. methane
Liquids	Flammable liquids and solids
	Fuels, oils and other hydrocarbons
	Solvents
Combustible materials	Timber
	Ash
	Coal residue
Heavy metals	Arsenic, lead, mercury, cadmium, chromium
Corrosive substances	Acids
	Alkalis
Toxic substances	Hydrocarbons
	Inorganic salts
Asbestos	
Substituted aromatic compounds	PCBs
	Dioxins
	Furans
Biological agents	Anthrax
	Tetanus
	BSE
	Genetically modified organisms

The following is a brief list of some of the main industrial sectors and their potential contaminants.

Sector	Contaminant type	Example
Gasworks	Coal tar	polyaromatic hydrocarbons (PAH's)
		phenol
	Cyanide	free / complex
	Sulphur	sulphide / sulphate
	Metals	lead, cadmium, mercury
	Aromatic hydrocarbons	benzene
Iron + Steel works	Metals	copper, nickel, lead
	Acids	sulphuric, hydrochloric
	Mineral oils	-
	Coking works residues	(as for gasworks)
Metal finishing	Metals	cadmium, chromium, copper, nickel, zinc
	Acids	sulphuric, hydrochloric
	Plating salts	cyanide
	Aromatic hydrocarbons	benzene
	Chlorinated hydrocarbons	trichloroethane
Non ferrous metal processing	Metals	Copper, cadmium, lead, zinc
	Impurity metals	antimony, arsenic

Sector	Contaminant type	Example
	Other wastes	battery cases, acids
Oil refineries	Hydrocarbons	various fractions
	Acids, alkalis	sulphuric, caustic soda
	Lagging, insulation	asbestos
	Spent catalysts	lead, nickel, chromium
Paints	Metals	lead, cadmium, barium
	Alcohols	toluol, xylol
	Chlorinated hydrocarbons	methylene chloride
	Fillers, extenders	silica, titanium dioxide, talc
Petrochemical plants	Acids, alkalis	sulphuric, caustic soda
	Metals	copper, cadmium, mercury
	Reactive monomers	styrene, acrylate, VCM
	Cyanide	toluene di-isocyanate
	Amines	analine
	Aromatic hydrocarbons	benzene, toluene
Petrol stations	Metals	copper, cadmium, lead, nickel, zinc
	Aromatic hydrocarbons	benzene
	Octane boosters	lead, MTBE
	Mineral oil	-
	Paint, plastic residues	barium, cadmium, lead
Rubber processing	Metals	zinc, lead
	Sulphur compounds	sulphur, thiocarbonate
	Reactive monomers	isoprene, isobutylene
	Acids	sulphuric. hydrochloric
	Aromatic hydrocarbons	xylene, Toluene
Semi-conductors	Metals	copper, nickel, cadmium
	Metalloids	arsenic, antimony, zinc
	Acids	nitric, hydrofluoric
	Chlorinated hydrocarbons	trichloroethylene
	Alcohols	methanol
	Aromatic hydrocarbons	xylene, toluene
Tanneries	Acids	hydrochloric
	Metals	trivalent chromium
	Salts	chlorides, sulphides
	Solvents	kerosene, white spirit
	Cyanide	methyl isocyanate
	Degreasers	trichloroethylene
	Dyestuff residues	cadmium, benzidine
Textiles	Metals	aluminium, tin, titanium, zinc
	Acids, alkalines	sulphuric, caustic soda
	Salts	sodium hypochlorite
	Chlorinated hydrocarbons	perchloroethylene
	Aromatic hydrocarbons	phenol
	Pesticides	dieldrin, aldrin, endrine
	Dyestuff residues	cadmium, Benzedrine
Wood processing	Coal tar based preservatives	creosote
	Chlorinated hydrocarbons	pentachlorophenol
	Metalloids / metals	arsenic, copper, chromium
Hat making	Mercury	
Tin smelting	Radioactivity	
Vehicle parking areas	Metals	copper, cadmium, lead, nickel, zinc
	Aromatic hydrocarbons	benzene
	Octane boosters	lead, MTBE
	Mineral oil	-

BACKGROUND

Statutory framework

April 2000 saw the introduction of new contaminated land provisions, contained in Part IIA of the Environmental Protection Act 1990. A primary objective of the legislation is to identify and remediate contaminated land.

Under the Act action to remediate land is required only where there are unacceptable actual or potential risks to health or the environment. Sites that have been polluted from previous land use may not need remediating until the land use is changed to a more sensitive end-use. In addition, it may be necessary to take action only where there are appropriate, cost-effective remediation processes that take the use of the site into account.

The Environmental Act 1995 amended the Environment Protection Act 1990 by introducing a new regime for dealing with 'contaminated land' as defined. The regime became operational on 1 April 2000. Local authorities are the main regulators of the new regime although the Environment Agency regulates seriously contaminated sites which are known as 'special sites'.

The contaminated land regime incorporates statutory guidance on the inspection, definition, remediation, apportionment of liabilities and recovery of costs of remediation. The measures are to be applied in accordance with the following criteria:

• the planning system
• the standard of remediation should relate to the present use
• the costs of remediation should be reasonable in relation to the seriousness of the potential harm
• the proposals should be practical in relation to the availability of remediation technology, impact of site constraints and the effectiveness of the proposed clean-up method.

The contaminated land provisions of the Environmental Protection Act 1990 are only one element of a series of statutory measures dealing with pollution and land remediation that have been introduced. Others include:

• groundwater regulations, including pollution prevention measures
• an integrated prevention and control regime for pollution
• sections of the Water Resources Act 1991, which deals with works notices for site controls, restoration and clean up.

The risks involved in the purchase of potentially contaminated sites are high, particularly considering that a transaction can result in the transfer of liability for historic contamination from the vendor to the purchaser.

The ability to forecast the extent and cost of remedial measures is essential for both parties, so that they can be accurately reflected in the price of the land.

The EU Landfill Directive

The Landfill (England and Wales) Regulations 2002 came into force on 15 June 2002 followed by Amendments in 2004 and 2005. These new regulations implement the Landfill Directive (Council Directive 1999/31/EC), which aims to prevent, or to reduce as far as possible, the negative environmental effects of landfill. These regulations have had a major impact on waste regulation and the waste management industry in the UK.

The Scottish Executive and the Northern Ireland Assembly will be bringing forward separate legislation to implement the Directive within their regions.

In summary, the Directive requires that:

- Sites are to be classified into one of three categories: hazardous, non-hazardous or inert, according to the type of waste they will receive.
- Higher engineering and operating standards will be followed.
- Biodegradable waste will be progressively diverted away from landfills.
- Certain hazardous and other wastes, including liquids, explosive waste and tyres will be prohibited from landfills.
- Pre-treatment of wastes prior to landfilling will become a requirement.

On 15 July 2004 the co-disposal of hazardous and non-hazardous waste in the same landfill site ended and in July 2005 new waste acceptance criteria (WAC) were introduced which also prevents the disposal of materials contaminated by coal tar.

The effect of this Directive has been to dramatically reduce the hazardous disposal capacity post July 2004, resulting in a **SIGNIFICANT** increase in remediating costs. There are now approximately 20 commercial landfills licensed to accept hazardous waste as a direct result of the implementation of the Directive. There is one site in Scotland, none in Wales and only limited capacity in the South of England. This has significantly increased travelling distance and cost for disposal to landfill. The increase in operating expenses incurred by the landfill operators has also resulted in higher tipping costs.

All hazardous materials designated for disposal off-site are subject to WAC tests. Samples of these materials are taken from site to laboratories in order to classify the nature of the contaminants. These tests, which cost approximately £200 each, have resulted in increased costs for site investigations and as the results may take up to 3 weeks this can have a detrimental effect on programme.

As from 1 July 2008 the WAC derogations which have allowed oil contaminated wastes to be disposed in landfills with other inert substances were withdrawn. As a result the cost of disposing oil contaminated solids has increased.

There has been a marked slowdown in brownfield development in the UK with higher remediationcosts, longer clean-up programmes and a lack of viable treatment options for some wastes.

The UK Government established the Hazardous Waste Forum in December 2002 to bring together key stakeholders to advise on the way forward on the management of hazardous waste.

Effect on Disposal Costs

Although most landfills are reluctant to commit to tipping prices tipping costs have generally stabilized. However, there are significant geographical variances, with landfill tip costs in the North of England typically being less than their counterparts in the Southern regions.

For most projects to remain viable there will be an increasing need to treat soil in-situ by bioremediation, soil washing or other alternative long-term remediation measures. Waste untreatable on-site such as coal tar remains a problem. Development costs and programmes will need to reflect this change in methodology.

Types of hazardous waste

- Sludges, acids and contaminated wastes from the oil and gas industry
- Acids and toxic chemicals from chemical and electronics industries
- Pesticides from the agrochemical industry
- Solvents, dyes and sludges from leather and textile industries
- Hazardous compounds from metal industries
- Oil, oil filters and brake fluids from vehicles and machines
- Mercury-contaminated waste from crematoria
- Explosives from old ammunition, fireworks and airbags

- Lead, nickel, cadmium and mercury from batteries
- Asbestos from the building industry
- Amalgam from dentists
- Veterinary medicines

[Source: Sepa]

Foam insulation materials containing ODP (Ozone Depletant Potential) are also considered as hazardous waste under the EC Regulation 2037/2000.

INITIAL STUDY

Approach

Part IIA of the Environmental Protection Act (1990) [EPA], which was introduced by Section 57 of the Environment Act 1995, requires an overall risk-based approach to dealing with contaminated sites, which is consistent with the general good practice approach to managing land subject to contamination.

The regulatory regime set out in Part IIA is based on the following activities:

- identify the problem
- assess the risks
- determine the appropriate remediation requirements
- consider the costs
- establish who should pay
- implementation and remediation

These are examined more fully below:

Identify the problem

Site investigations comprising desk study research and intrusive investigations are necessary to provide information on the soil conditions and possible contaminants located on the site in order that an assessment of the risks may be carried out. Based on these, it should then be possible to ensure that the most appropriate remedial measures are used. From a purely cost angle, the more complete the study is at this stage, the greater the reliability of the cost estimate.

Initially:

- research previous use(s) of the site by reference to historical maps, local records, interviewing local inhabitants, previous employees etc.
- study geological maps and local records to determine ground strata, water table, underground aquifers, direction of movement of ground water, presence of water extraction wells locally which may be at risk from contamination.
- examine local records to try and determine underground and above ground service routes and whether still live

Ideally such studies should also take in surrounding land as the site may be at risk of contamination from an adjoining site problem.

The aim would be to establish the previous pattern of development of the site. Based on the sites various uses, areas of previous development and likely contaminants may be identifiable. It may well be possible to categorize the site into areas each with possibly differing problems and occurring at differing depths. In all probability, a proportion of even a contaminated site could prove to be trouble free.

Even though a site may have no history of potentially contaminative use, previous owners may have inadvertently created a problem by importing fill materials from a contaminated source.

A site 'conceptual model' should be developed from the desk study research, showing the 'receptors' potentially at risk and the 'pathway' between the contamination source and each receptor.

The next step should be soil sampling of the various areas and laboratory analysis of the materials to determine if the soil contains a contaminant (or contaminants) as well as obtaining an idea of the distribution of the contaminants over the site (in areas as well as depth) and the associated concentrations.

The number of sampling points taken on any given site will largely depend on engineering judgement, economics and time. The use of an accurate and regularly spaced sampling grid allows the site to be categorized into areas of high and low risk (i.e. areas exceeding or falling below the trigger values). It may then be possible to use different remedial measures for the differing areas of concentration.

Consideration should be given to the proposed development. There is little point in carrying out expensive remedial works to a site merely for it to be then subjected to the same use. On the other hand, work is very much called for if the proposed use is for agricultural or recreational purpose or else the contaminants threaten the water regime. If the contaminants are static and not putting development at risk, there may be a strong economic argument for leaving them alone.

Bearing in mind the history of the site, the nature and concentration of the contaminant(s) and the most likely path, it should then be possible to choose an appropriate remedial technique.

Assess the risks

With an understanding of the site geology and of the nature, distribution and magnitude of contaminants within the site, an assessment can then be made of the risk that they pose, which will depend largely on the sensitivity of the site and the future land use.

A qualitative risk assessment used to involve a comparison of the observed contaminant concentrations with screening values, such as the ICRCL trigger concentrations. A quantitative risk assessment looks at the particular site, the potential of the hazards to migrate and then a calculation of the likely dose of contaminant at the receptor to compare with EEC and other Regulator limits.

It can be difficult to arrive at an adequate definition of the acceptable levels of risk for the particular site, in view of the complications caused by differences in geology on a regional scale and local site variations. Individual site variations can affect chemical and physical properties and substances themselves can interact to increase or even reduce the risk.

Threats to aquifers would force a vigorous approach to treatment.

Determine the appropriate remediation requirements

Consideration should be given to the most appropriate treatment suited to the location and type of contaminant, the end use of the site and the risks.

Consider the costs

The estimate should review the chosen treatment and take into account:

- a careful measurement/estimate of the quantities involved, e.g. the volume(s) of contaminants which may need removal, length and depth of containment walls, size distribution and depth of boreholes for the extraction of contaminated liquids
- site location – a study of the surrounding roads, built-up areas etc. which may restrict access to and from the site for construction vehicles; this could well influence the size of the vehicles used
- appropriate landfill sites / haulage distance – a review of suitable tips which can take the contaminants, checking their rate of accepting such material, which may well be limited by their licence; bear in mind that tipping charges can vary significantly over a period of time but preferential rates can be obtained for a programme of tipping. Large quantities of material may force a proportion being disposed of not just at the closest, but also at the more distant tips as well – if time is not at a premium, it may well prove cheaper to extend the length of the programme.
- the location / haulage distance of sites of suitable fill materials
- a review of any set time scales and a calculation of a practical contract period, perhaps taking into account any requirements for phased hand over of parts of the site.
- an assessment of the implications of the contaminants on the site establishment – protective clothing and footwear and shower and messing facilities for personnel, wheel wash facilities for vehicles and the careful disposal of contaminated wash water etc.

It is essential that the estimate is a practical all-embracing exercise to help ensure that a realistic budget is set for the project. Once the work commences on site, it cannot be halted even if costs overrun the anticipated budget.

If it does not prove possible to survey the site fully, part or all of the estimate could be at risk and the client should be made aware of the potential risk.

Cost minimization

Apart from careful measurement and rate evaluation, there are a number of methods of helping to minimize costs, such as:

- On-site testing to reduce off site volumes
- To help ensure that low risk materials are not removed unnecessarily
- Correct classification of waste
- To help ensure that the contaminant incurs the lowest tipping charge
- Use of clean site rubble as far as possible for fill
- Use of existing contaminated solids for the stabilization of contaminated liquids/sludges.
- The removal of such a mixture by lorry may be much cheaper than removing liquid by tanker.
- Ensure the fill and rate of compaction are suitable for the end use
- To avoid future cost.
- Encapsulation of non-mobile contaminants
- Cheaper than removing from site or treating.
- Back-hauling of fill materials

Where possible, organizing the truck taking the material to the tip to return with the fill materials, avoiding the cost of travelling empty.

Establish who should pay

Liability for the costs of remediation rests with either the party that 'caused or knowingly permitted' contamination, or with the current owners or occupiers of the land.

Apportionment of liability, where shared, is determined by the local authority. Although owners or occupiers become liable only if the polluter cannot be identified, the liability for contamination is commonly passed on when land is sold.

If neither the polluter nor owner can be found, the clean-up is funded from public resources.

Implementation and remediation

Tenders should be sought on clear documentation with the ability for the subsequent work to be remeasured and revalued.

There should be a stringent site monitoring system implemented. This should include monitoring the effectiveness of operations (especially in the case of remedial operations other than dig and dump) as well as checking the chemical profile of all imported fill materials to avoid the obvious.

Approvals involve both the Environment Agency and the local authority Environmental Health Officer / Contaminated Land Officer.

A Health and Safety file handed to the site owner at completion should cover all the checking procedures in detail. Arrange for suitable warranties, which would be required by any future purchaser of the site.

LAND REMEDIATION TECHNIQUES

There are two principal approaches to remediation – dealing with the contamination in situ or off site. The selection of the approach will be influenced by factors such as: initial and long term cost, timeframe for remediation, types of contamination present, depth and distribution of contamination, the existing and planned topography, adjacent land uses, patterns of surface drainage, the location of existing on-site services, depth of excavation necessary for foundations and below-ground services, environmental impact and safety, prospects for future changes in land use and long-term monitoring and maintenance of in situ treatment.

On most sites, contamination can be restricted to the top couple of metres, although gasholder foundations for example can go down 8 or 10 metres. Underground structures can interfere with the normal water regime and trap water pockets.

There could be a problem if contaminants get into fissures in bedrock.

In situ techniques

A range of in situ techniques is available for dealing with contaminants, including:

- Clean cover – a layer of clean soil is used to segregate contamination from receptor. This technique is best suited to sites with widely dispersed contamination. Costs will vary according to the need for barrier layers to prevent migration of the contaminant.
- On-site encapsulation – the physical containment of contaminants using barriers such as slurry trench cut-off walls. The cost of on-site encapsulation varies in relation to the type and extent of barriers required, the costs of which range from £50/m² to more than £175/m².

There are also in situ techniques for treating more specific contaminants, including:

- Bio-remediation – for removal of oily, organic contaminants through natural digestion by micro-organisms. Most bio-remediation is ex-situ, i.e. it is dug out and then treated on site in bio-piles. The process can be slow, perhaps taking as much as one to three years depending upon the scale of the problem, but is particularly effective for the long-term improvement of a site, prior to a change of use.
- Soil washing – involving the separation of a contaminated soil fraction or oily residue through a washing process. This also involves the excavation of the material for washing ex-situ. The de-watered contaminant still requires disposal to landfill. In order to be cost effective, 70 – 90% of soil mass needs to be recovered. It

will involve constructing a hard area for the washing, intercepting the now-contaminated water and taking it away in tankers.

- Vacuum extraction – involving the extraction of volatile organic compounds (e.g. benzene) from soil and groundwater by vacuum.
- Thermal treatment – the incineration of contaminated soils on site. The uncontaminated soil residue can be recycled. By-products of incineration can create air pollution and exhaust air treatment may be necessary.
- Stabilization – cement or lime, is used to physically or chemically bound oily or metal contaminants to prevent leaching or migration. Stabilization can be used in both in situ and off-site locations.
- Aeration – if the ground contamination is highly volatile, e.g. fuel oils, then the ground can be ploughed and rotovated to allow the substance to vaporize.
- Air sparging – the injection of contaminant-free air into the sub-surface enabling a phase transfer of hydrocarbons from a dissolved state to a vapour phase.
- Chemical oxidization – the injection of reactive chemical oxidants directly into the soil for the rapid destruction of contaminants.
- Pumping – to remove liquid contaminants from boreholes or excavations. Contaminated water can be pumped into holding tanks and allowed to settle; testing may well prove it to be suitable for discharging into the foul sewer subject to payment of a discharge fee to the local authority of 35p to 65p per m³. It may be necessary to process the water through an approved water treatment system to render it suitable for discharge.

Off-site techniques

Removal for landfill disposal has, historically, been the most common and cost-effective approach to remediation in the UK, providing a broad spectrum solution by dealing with all contaminants. As discussed above, the implementation of the Landfill Directive will result in other techniques becoming more competitive and enjoy a wider usage. Removal to Landfill is suited to sites where sources of contamination can be easily identified and it is local to an approved Landfill site.

If used in combination with material-handling techniques such as soil washing, the volume of material disposed at landfill sites can be significantly reduced. The disadvantages of the technique include the fact that the contamination is not destroyed, there are risks of pollution during excavation and transfer; road haulage may also cause a local nuisance.

Soil treatment centres are now beginning to be established. These use a combination of treatment technologies to maximize the potential recovery of soils and aggregates and render them suitable for disposal to the landfill. The technologies include:

- Physico-chemical treatment – a method which uses the difference in grain size and density of the materials to separate the different fractions by means of screens, hydrocyclones and upstream classification
- Bioremediation – the aerobic biodegradation of contaminants by naturally occurring micro-organisms
- Stabilization/solidification – a cement stabilization unit capable of immobilizing persistent leachable components.

COST CONSIDERATIONS

Cost drivers

Cost drivers relate to the selected remediation technique, site conditions and the size and location of a project. The wide variation of indicative costs of land remediation techniques shown below is largely because of differing site conditions.

Indicative costs of land remediation techniques for 2011 (excluding General Items, testing, landfill tax and backfilling)

Remediation technique	Unit	Rate (£/unit)
Removal – non-hazardous	disposed material (m³)	30 – 60
Removal – hazardous	disposed material (m³)	75 – 200
Note: excluding any pre-treatment of material		
Clean cover	surface area of site (m²)	20 – 45
On-site encapsulation	encapsulated material (m³)	25 – 95
Bio-remediation (in-situ)	treated material (m³)	15 – 45
Bio-remediation (ex-situ)	treated material (m³)	20 – 50
Chemical oxidation	treated material (m³)	30 – 85
Stabilization/solidification	treated material (m³)	25 – 65
Vacuum extraction	treated material (m³)	25 – 70
Soil washing	treated material (m³)	25 – 80
Thermal treatment	treated material (m³)	100 – 375

Many other on-site techniques deal with the removal of the contaminant from the soil particles and not the wholesale treatment of bulk volumes. Costs for these alternative techniques are very much Engineer designed and site specific.

Factors that need to be considered include:

- waste classification of the material
- underground obstructions, pockets of contamination and live services
- ground water flows and the requirement for barriers to prevent the migration of contaminants
- health and safety requirements and environmental protection measures
- location, ownership and land use of adjoining sites
- distance from landfill tips, capacity of the tip to accept contaminated materials, and transport restrictions
- the cost of diesel fuel, currently approximately £1.40 per litre (at August 2011 prices)

Other project related variables include size, access to disposal sites and tipping charges; the interaction of these factors can have a substantial impact on overall unit rates.

The tables below set out the costs of remediation using dig-and-dump methods for different sizes of project, differentiated by the disposal of non-hazardous and hazardous material. Variation in site establishment and disposal cost accounts for 60 – 70% of the range in cost.

Variation in the costs of land remediation by removal: Non-hazardous Waste

Item	Disposal Volume (less than 3000 m³) (£/m³)	Disposal Volume (3000 – 10 000 m³) (£/m³)	Disposal Volume (more than 10 000 m³) (£/m³)
General items and site organizationcosts	55 – 90	25 – 40	7 – 20
Site investigation and testing	5 – 12	2 – 7	2 – 6
Excavation and backfill	18 – 35	12 – 25	10 – 20
Disposal costs (including tipping charges but not landfill tax)	20 – 35	20 – 35	20 – 35
Haulage	15 – 35	15 – 35	15 – 35
Total (£/m³)	**113 – 207**	**74 – 142**	**54 – 116**
Allowance for site abnormals	*0 – 10 +*	*0 – 15 +*	*0 – 10 +*

Variation in the costs of land remediation by removal: Hazardous Waste

Item	Disposal Volume (less than 3000 m³) (£/m³)	Disposal Volume (3000 – 10 000 m³) (£/m³)	Disposal Volume (more than 10 000 m³) (£/m³)
General items and site organizationcosts	55 – 90	25 – 40	7 – 20
Site investigation and testing	10 – 18	5 – 12	5 – 12
Excavation and backfill	18 – 35	12 – 25	10 – 20
Disposal costs (including tipping charges but not landfill tax)	80 – 170	80 – 170	80 – 170
Haulage	25 – 120	25 – 120	25 – 120
Total (£/m³)	**188 – 433**	**147 – 367**	**127 – 342**
Allowance for site abnormals	*0 – 10 +*	*0 – 15 +*	*0 – 10 +*

The strict Health and Safety requirements of remediation works can produce quite high site organizationcosts as a % of the overall project cost (see the table above). A high proportion of these costs are fixed and, as a result, the unit costs of site organization increase disproportionately on smaller projects.

Haulage costs are largely determined by the distances to a licensed tip. Current average haulage rates, based on a return journey range from £1.75 to £3.25 per mile. Short journeys to tips, which involve proportionally longer standing times, typically incur higher mileage rates, up to £9.00 per mile.

A further source of cost variation relates to tipping charges. The table below summarizes indicative tipping charges for 2011, exclusive of landfill tax:

Typical 2011 tipping charges (excluding landfill tax)

Waste classification	Charges (£/tonne)
Non-hazardous wastes	10 – 25
Hazardous wastes	25 – 85
Contaminated liquid	40 – 75
Contaminated sludge	55 – 200

Tipping charges fluctuate in relation to the grades of material a tip can accept at any point in time. This fluctuation is a further source of cost risk. Furthermore, tipping charges in the North of England are generally less than in the rest of the country.

Prices at licensed tips can vary by as much as 50%. In addition, landfill tips generally charge a tip administration fee of approximately £25 per load, equivalent to £1.25 per tonne. This charge does not apply to non-hazardous wastes.

Cost Studies

Site study 1

A recently completed project involving site remedial work to a former gas works site by the dig and dump approach (1,000m^3 sent to hazardous landfill) analyses as follows (the Class references are from CESMM3):

Class A	General Items	23%
Class B	Ground Investigation	8%
Class C	Geotechnical Services	0%
Class D	Demolition and Site Clearance	1%
Class E	Earthworks	
	Excavation	4%
	Haulage	21%
	Disposal	31%
	Backfilling	9%
Class F-X	(A number of minor work classes)	3%
	Provisional Sums	0%
	Abnormal Costs	0%
		100%

Site study 2

A recently completed project involving site remedial work to a former gas works site by the dig and dump and soil washing approach (1,500 m3 sent to hazardous landfill) analyses as follows (the Class references are from CESMM3):

Class A	General Items	19%
Class B	Ground Investigation	6%
Class C	Geotechnical Services	40%
Class D	Demolition and Site Clearance	4%
Class E	Earthworks	
	Excavation	5%
	Haulage	5%
	Disposal	12%
	Backfilling	4%
Class F-X	(A number of minor work classes)	0%
	Provisional Sums	5%
	Abnormal Costs	1%
		100%

Site study 3

A recently completed project involving site remedial work to a former gas works site by the dig and dump and soil washing/bioremediation approach (5,000m³ sent to hazardous landfill) analyses as follows (the Class references are from CESMM3):

Class A	General Items	14%
Class B	Ground Investigation	5%
Class C	Geotechnical Services	34%
Class D	Demolition and Site Clearance	5%
Class E	Earthworks	
	Excavation	6%
	Haulage	7%
	Disposal	15%
	Backfilling	5%
Class F-X	(A number of minor work classes)	2%
	Provisional Sums	6%
	Abnormal Costs	1%
		100%

Class A General Items

Remedial works contracts generally show a high level of preliminaries, perhaps around the 30-40% mark for relatively small projects, mainly due to the high costs of Health and Safety and temporary works when dealing with contamination. The tables below demonstrate the spread of costs on three site studies included as examples. Over and above the normal site establishment costs included within Part 4 Class A; a number of the following items may need to be included.

- Protective clothing and footwear / site safety inductions
- Hygiene / Decontamination Unit
- Occupational health checks (office staff as well as site labour)
- Bath wheel wash facility
- Weighbridge facility with auto ticketing
- Scaffold gantry for safe covering of wagons
- Steel storage tanks for contaminated liquids
- Temporary fencing between clean and dirty areas and around excavations
- Administration connected with special waste taxes, licenses, etc.
- Portable on-site laboratory
- Dust suppression
- Odour suppression
- Vibration monitoring

Class B Ground Investigation

A further 3 – 9% can be spent on carrying out the site testing and validation sampling. There is a direct connection between a client spending more money on adequate site investigations and reducing the risk of there being something unforeseen and untoward on the site. This section covers items such as trial pits and trenches, light cable percussion boreholes, laboratory testing which are carried out as part of the remediation contract to prove the ground has been cleaned to an acceptable standard. Indicative costs for testing can be found in Part 4 Class B. Long term ground water monitoring may also be required.

Class C Geotechnical Services

As part of the remediation strategy it may be deemed necessary by the site engineer to use diaphragm walls, ground anchorages, ground consolidation techniques, grout holes, etc. Indicative costs are included within Part 4 Class C.

Class D Demolition and Site Clearance

The costs associated with this section are site dependent and can found in Part 4 Class D.

Class E Excavation

Excavation and backfilling costs can be found within Part 4 Class E. As part of the remediation strategy contaminated material may be deemed acceptable as backfill by mixing with inert material. Costs for the rotovation of material within stockpiles are approximately £1.00/m³. Use of clean material to stabilize contaminated sludges/liquids to allow transportation off site will vary depending on the ratio. As a guide, mixing on a 1:1 ratio will cost approximately £4/m³. Costs for crushing and screening excavated material are usually expected to be around £6/m³ and £2/m³ respectively.

Haulage costs are largely determined by the distances to a licensed tip. A frequently used haulage vehicle will be a 19 tonne payload articulated vehicle, costing say £660 per day including driver and fuel. The cost per load naturally reflects the difficulty of the route. An average 75 miles round trip, with 5 trips being carried out per day would cost £132 per load, or £1.75 per mile. For short distances, the cost per mile could rise to £9.00, reflecting the greater number of trips and hence a greater amount of time in loading and dumping. The haulage cost per m³ of the disposed material naturally depends on its density. Disposal costs need to be established with the nearest licensed landfill site as they vary depending on the locality and waste classification. On top of this must be included a Tip Administration fee of approximately £25 per load for hazardous wastes and consideration as to whether landfill tax is applicable.

THE LANDFILL TAX

The Tax

The Landfill Tax came into operation on 1 October 1996. It is levied on operators of licensed landfill sites at the following rates with effect from 1 April 2011:

Inactive or inert wastes.	£2.50 per tonne	Included are soil, stones, brick, plain and reinforced concrete, plaster and glass – lower rate
All other taxable wastes.	£48.00 per tonne	Included are timber, paint and other organic wastes generally found in demolition work, builders skips – standard rate

The standard (higher) rate for 'all other taxable wastes' will be increased by £8 per tonne each year at least until 2014 when the rate will be £80 per tonne. The lower rate for 'inactive or inert wastes' will be frozen at £2.50 per tonee to 31 March 2012.

The Landfill Tax (Qualifying Material) Order 2011 came into force on 1 April 2011. This has amended the qualifying criteria that are eligible for the lower rate of landfill tax. The revisions introduced arose primarily from the need to reflect the changes in wider environmental policy and legislation since 1996, such as the implementation of the European Landfill Directive.

A waste will be lower rated for landfill tax from 1 April 2011 only if it is listed as a qualifying material in the Landfill Tax (Qualifying Material) Order 2011.

The principle changes to qualifying material are:

- Rocks and sub-soils that are currently lower rated will remain so.
- Topsoil and peat will be removed from the lower rate, as these are natural resources that can always be recycled/reused.
- Used foundry sand, which has in practice been lower rated by extra-statutory concession since the tax's introduction in 1996, will now be included in the lower rate Order.
- Definitions of qualifying ash arising from the burning of coal and petroleum coke (including when burnt with biomass) will be clarified.
- The residue from titanium dioxide manufacture will qualify, rather than titanium dioxide itself, reflecting industry views.
- Minor changes will be made to the wording of the calcium sulphate group of wastes to reflect the implementation of the Landfill Directive since 2001.
- Water will be removed from the lower rate – water is now banned from landfill so its inclusion in the lower rated wastes is unncessary; where water is used as a waste carrier the water is not waste and therefore not taxable.

Exemptions

The following disposals are exempt from Landfill Tax subject to meeting certain conditions:

- dredgings which arise from the maintenance of inland waterways and harbours
- naturally occurring materials arising from mining or quarrying operations
- pet cemetaries
- reclamation of contaminated land
- inert waste used to restore landfill sites and to fill working and old quarries where a planning condition or obligation is in existence

The exemption for waste from contaminated land will be phased out completely by 1 April 2012 and no new applications for landfill tax exemption are now accepted.

For further information contact the National Advisory Service, Telephone: 0845 010 9000.

Volume to weight conversion factors (for estimating purposes)

To convert inactive or inert waste (i.e. largely water insoluble and non or very slowly biodegradable: e.g. sand, subsoil, concrete, bricks, mineral fibres, fibreglass etc.), multiply the measured volume in cubic metres by 1.9 to calculate the weight in tonnes.

Calculating the weight of waste

There are two options:

- If licensed sites have a weighbridge, tax will be levied on the actual weight of waste.
- If licensed sites do not have a weighbridge, tax will be levied on the permitted weight of the lorry based on an alternative method of calculation based on volume to weight factors for various categories of waste.

Effect on prices

The tax is paid by Landfill site operators only. Tipping charges reflect this additional cost.

Apart from the possible incidence of Landfill Tax, the cost of disposal will generally comprise the haulage cost plus a tipping charge which will vary according to the toxicity of the material.

Tax Relief for Remediation of Contaminated Land

The Finance Act 2001 included provisions that allow companies (but not individuals or partnerships) to claim tax relief on capital and revenue expenditure on the 'remediation of contaminated land' in the United Kingdom. The relief is available for expenditure incurred on or after 11 May 2001.

From 1 April 2009 there was an increase in the scope of costs that qualify for Land Remediation Relief where they are incurred on long-term derelict land. The list includes costs that the Treasury believe to be primarily responsible for causing dereliction, such as additional costs for removing building foundations and machine bases. However, while there is provision for the list to be extended, the additional condition for the site to have remained derelict since 1998 is likely to render this relief redundant in all but a handful of cases. The other positive change is the fact that Japanese Knotweed removal and treatment (on-site only) will now qualify for the relief under the existing legislation, thereby allowing companies to make retrospective claims for any costs incurred since May 2001 – provided all other entitlement conditions are met.

A company is able to claim an additional 50% deduction for 'qualifying land remediation expenditure' allowed as a deduction in computing taxable profits, and may elect for the same treatment to be applied to qualifying capital expenditure.

With Landfill Tax exemption (LTE) being phased out, Land Remediation Relief (LRR) for contaminated and derelict land is now the Government's primary tool to create incentives for brownfield development. LRR is available to companies engaged in land remediation that are not responsible for the original contamination.

Over 7 million tonnes of waste each year were being exempted from Landfill Tax in England alone, so this change could have a major impact on the remediation industry. The modified LRR scheme, which provides Corporation Tax relief on any costs incurred on qualifying land remediation expenditure, is in the long run designed to yield benefits roughly equal to those lost through the withdrawal of LTE.

However, with much remediation undertaken by polluters or public authorities, who cannot benefit from tax relief benefits, the change could result in a net withdrawal of Treasury support to a vital sector. Lobbying and consultation continues to ensure the Treasury maintains its support for remediation.

While there are no financial penalties for not carrying out remediation, a steep escalator now affects the rate of Landfill Tax for waste material other than inert or inactive wastes, which will rise at the rate of £8/tonne per year until 2014. By then, the rate will be £80/tonne. This means that for schemes where there is no alternative to dig and dump and no pre-existing LTE, the cost of remediation could rise to prohibitive levels.

Existing Landfill Tax exemptions are only valid until 31 March 2012, and it is also foreseeable that there will be a rise in the volume of exempted waste material being sent to landfill, which in turn could increase disposal prices ahead of April 2012.

Looking forward, tax-relief benefits under LRR could provide a significant cash contribution to remediation. Careful planning is the key to ensure that maximum benefits are realised, with actions taken at the points of purchase, formation of JV arrangements, procurement of the works and formulation of the Final Account (including apportionment of risk premium) all influencing the final value of the claim agreed with HM Revenue & Customs.

Following the withdrawal of LTE, the land remediation relief regime is also being expanded to create incentives for companies to remove features such as underground structures, or redundant services that might cause a site to become derelict. Conditions to qualify for derelict land are fairly onerous, requiring sites to have been derelict since 1998 and for it to be shown that the site would not be capable of reuse without the removal of buildings or other structures.

The Relief

Qualifying expenditure may be deducted at 150% of the actual amount expended in computing profits for the year in which it is incurred.

For example, a property trading company may buy contaminated land for redevelopment and incurs £250,000 on qualifying land remediation expenditure that is an allowable for tax purposes. It can claim an additional deduction of £125,000, making a total deduction of £375,000. Similarly, a company incurring qualifying capital expenditure on a fixed asset of the business is able to claim the same deduction provided it makes the relevant election within 2 years.

What is Remediation?

Land remediation is defined as the doing of works including preparatory activities such as condition surveys, to the land in question, any controlled waters affected by the land, or adjoining or adjacent land for the purpose of:

- Preventing or minimizing, or remedying or mitigating the effects of, any relevant harm, or any pollution of controlled waters, by reason of which the land is in a contaminated state.

Definitions

Contaminated land is defined as land that, because of substances on or under it, is in such a condition that relevant harm is or has the significant possibility of relevant harm being caused to: –

- The health of living organisms
- Ecological systems
- Quality of controlled waters
- Property

Relevant harm is defined as meaning: -

- death of living organisms or significant injury or damage to living organisms,
- significant pollution of controlled waters,
- a significant adverse impact on the ecosystem, or
- structural or other significant damage to buildings or other structures or interference with buildings or other structures that significantly compromises their use.

Land includes buildings on the land, and expenditure on asbestos removal is expected to qualify for this tax relief. It should be noted that the definition is not the same as that used in the Environmental Protection Act Part 11A.

Sites with a nuclear license are specifically excluded.

Conditions

To be entitled to claim LRR, the general conditions for all sites, which must all be met are: -

- Must be a company
- Must be land in the United Kingdom
- Must acquire an interest in the land
- Must not be the polluter or have a relevant
- connection to the polluter
- Must not be in receipt of a subsidy.
- Must not also qualify for Capital Allowances (particular to capital expenditure only)

Additional conditions introduced since1 April 2009: -

- The interest in land must be major – freehold or leasehold longer than 7 years
- Must not be obligated to carry out remediation under a statutory notice.

Additional condition particular to derelict land:

- Must not be in or have been in productive use at any time since at least 1 April 1998.
- Must not be able to be in productive use without the removal of buildings or other structures

In order for expenditure to become qualifying, it must relate to substances present at the point of acquisition. Furthermore, it must be demonstrated that the expenditure would not have been incurred had those substances not been present.

Construction Contracts Questions and Answers

2nd Edition

By **David Chappell**

What they said about the first edition: "A fascinating concept, full of knowledgeable gems put in the most frank of styles... A book to sample when the time is right and to come back to when another time is right, maybe again and again." – *David A Simmonds,* Building Engineer *magazine*

- Is there a difference between inspecting and supervising?
- What does 'time-barred' mean?
- Is the contractor entitled to take possession of a section of the work even though it is the contractor's fault that possession is not practicable?

Construction law can be a minefield. Professionals need answers which are pithy and straightforward, as well as legally rigorous. The two hundred questions in the book are real questions, picked from the thousands of telephone enquiries David Chappell has received as a Specialist Adviser to the Royal Institute of British Architects. Although the enquiries were originally from architects, the answers to most of them are of interest to project managers, contractors, QSs, employers and others involved in construction.

The material is considerably updated from the first edition – weeded, extended and almost doubled in coverage. The questions range in content from extensions of time, liquidated damages and loss and/or expense to issues of warranties, bonds, novation, practical completion, defects, valuation, certificates and payment, architects' instructions, adjudication and fees. Brief footnotes and a table of cases will be retained for those who may wish to investigate further.

August 2010: 216x138: 352pp
Pb: 978-0-415-56650-6: **£34.99**

To Order: Tel: +44 (0) 1235 400524 **Fax:** +44 (0) 1235 400525
or Post: Taylor and Francis Customer Services,
Bookpoint Ltd, Unit T1, 200 Milton Park, Abingdon, Oxon, OX14 4TA UK
Email: book.orders@tandf.co.uk

For a complete listing of all our titles visit:
www.tandf.co.uk

PART 2

On Costs and Profit

In Part 1 of this book, it is stressed that the cost information given in Parts 3 to 7 leads to a cost estimate that requires further adjustment before it is submitted as a tender. This part deals with those adjustments and includes a worked example of a Tender Summary.

Handbook of Road Technology

4th Edition

M. Lay

"Widely considered to be the single most comprehensive and informative road technology reference text." – *Highway Engineering in Australia*

"A fully revised edition..." – *Transport and Road Update*

"This handbook has its origins in Max Lay's work over many decades in Australia, but reflects a breadth of knowledge far beyond that continent. In that sense it is perhaps the most international in its coverage of practice of such works."
– *Mike Chrimes, Librarian, Institution of Civil Engineers, MyICE newsletter*

This fully revised fourth edition of Max Lay's well-established reference work covers all aspects of the technology of roads and road transport, and urban and rural road technology. It forms a comprehensive but accessible reference for all professionals and students interested in roads, road transport and the wide range of disciplines involved with roads.

International in scope, it begins with the preliminary construction procedures; from road planning policies and design considerations to the selection of materials and the building of roads and bridges. It then explores road operating environments that include driver behaviour, traffic flow, lighting and maintenance, and assesses the cost, economics, transport implications and environmental impact of road use.

June 2009: 944pp
Hb: 978-0-415-47265-4: **£150.00**

To Order: Tel: +44 (0) 1235 400524 **Fax:** +44 (0) 1235 400525
or Post: Taylor and Francis Customer Services,
Bookpoint Ltd, Unit T1, 200 Milton Park, Abingdon, Oxon, OX14 4TA UK
Email: book.orders@tandf.co.uk

For a complete listing of all our titles visit:
www.tandf.co.uk

RISK / OPPORTUNITY

The factors to be taken into account when gauging the possibility of the Estimator's prediction of cost being inadequate or excessive are given in Part 2. Clearly it is considered in parallel with profit and it is not possible to give any indicative guidance on the level of adjustment that might result. For the purpose of a preliminary estimate, it is suggested that no adjustment is made to the costs generated by the other parts of this book.

At the same time as making a general appraisal of risk/opportunity, management will look at major quantities and may suggest amendments to the unit rates proposed.

HEAD OFFICE OVERHEADS

An addition needs to be made to the net estimate to cover all costs incurred in operating the central services provided by head office. Apart from general management and accountancy, this will normally include the departments dealing with:

- estimating
- planning and design
- purchasing
- surveying
- insurance
- wages and bonus
- site safety.

The appropriate addition varies with the extent of services provided centrally, rather than on site, and with size of organization, but a range of 4% to 8% on turnover would cover most circumstances.

Some organizations would include finance costs with head office overheads, as a general charge to the company, but for the purposes of this book finance costs are treated separately (see below).

PROFIT

Obviously, the level of profit is governed by the degree of competition applicable to the job – which is in turn a function of the industry's current workload. Again, the appropriate addition is highly variable, but for the purposes of a preliminary estimate an addition of 2% to 5% onto nett turnover is suggested.

FINANCE COSTS – ASSESSMENT OF CONTRACT INVESTMENT

The following procedure may be followed to give an indication of the average amount of capital investment required to finance the contract. It must be emphasized that this method will not give an accurate investment as this can only be done by preparing a detailed cash flow related to the programme for the contract. The example is based on the same theoretical contract used for the worked example in Part 2 and should be read in conjunction with the Tender Summary that follows.

The average monthly income must first be assessed. This is done by deducting from the Tender total the contingency items and those items for which immediate payment is necessary.

	£	£
Tender total (excluding finance charges)		10,396,313
Deduct		
Subcontractors	2,000,000	
Prime cost sums	100,000	
Employer's contingencies	245,500	2,345,500
Amount to be financed	£	8,050,813

The average monthly income is this sum (£8,050,813) divided by the contract period (12 months), that is £670,901.

The average contract investment may now be calculated as follows:

	£	£
Plant and equipment to be purchased		90,000
Non time related		
Contractor	240,000	
Employer	8,000	
Other services, charges and fees	NIL	
Subtotal £	248,000	
Take 50% as an average [1]		124,000
Stores and unfixed materials on site		20,000
Work done but not paid for		
2½ months at £670,901 (see table above) [2]	1,677,253	
Less retention at 5% [3]	83,863	1,593,390
Retention (5% with limit of 3%)		
Average retention [4] 3% of £ 8,050,813 (see table above)		241,524
Subtotal £		2,068,914
Deduct		
Advance payment by client	NIL	
Bill loading [5]	180,000	
Creditors (suppliers)	500,000	
Average contract investment	£	1,388,914

The interest charges that must be added to the Tender price (or absorbed from profit if capital needs to be borrowed) are therefore:

    £ 1,388,914 × say 3.5% [6] × 1 year = say – £ 48,612

Notes

1.   These non time related oncosts and services are incurred as lump sums during the contract and, therefore, only 50% of such costs are taken for investment purposes.
2.   This period depends on the terms of payment set out in the contract.
3.   Retention is deducted as full retention is taken into account later.
4.   Average retention will depend on the retention condition set out in the contract, taking into account any partial completion dates.
5.   The contractor assesses here any financial advantage he may obtain by varying his items.
6.   Assumed Rate – current market varies greatly

VALUE ADDED TAX

All of the figures quoted in this book exclude value added tax, which the conditions of contract normally make the subject of a separate invoicing procedure between the contractor and the employer.

Value Added Tax will be chargeable at the standard rate, currently 15%, on supplies of services in the course of:

7. The construction of a non-domestic building
8. The construction or demolition of a civil engineering work
9. The demolition of any building, and
10. The approved alteration of a non-domestic protected building

TENDER SUMMARY

This summary sets out a suggested method of collecting together the various costs and other items and sums which, together, make up the total Tender sum for the example contract.

	£	£
Preliminaries and General Items (from Part 2)		
Contractor's site oncosts – time related	909,935	
Contractor's site oncosts – non time related	238,502	
Employer's requirements – time related	145,300	
Employer's requirements – non time related	7,700	
Other services, charges and fees	NIL	
Temporary works not included in unit costs	123,920	
Plant not included in unit costs	210,693	1,636,050
Estimated net cost of measured work, priced at unit costs		5,280,000
	£	6,916,050
Allowance for fixed price contract		
6% on labour (assumed to be £1,400,000)	84,000	
4% on materials (assumed to be £2,500 000)	100,000	
4% on plant (assumed to be £400,000)		
5% on staff, overheads etc. (assumed to be £600,000)	30,000	230,000
Subcontractors (net)		2,000,000
Prime cost sums		100,000
Adjustments made at Management Appraisal		
price adjustments, add say	75,000	
risk evaluation, add say	50,000	
	£	9,246,050
Head office overheads and profit at 6 %		554,763
Finance costs (from previous page)		48,612
Provisional Sums		175,000
Dayworks Bill		175,000
	£	10,199,425
Employer's contingencies		245,500
TENDER TOTAL	**£**	**10,444,925**

Reynolds's Reinforced Concrete Designer Handbook

11th Edition

A. Threlfall et al.

This classic and essential work has been thoroughly revised and updated in line with the requirements of new codes and standards which have been introduced in recent years, including the new Eurocode as well as up-to-date British Standards.

It provides a general introduction along with details of analysis and design of a wide range of structures and examination of design according to British and then European Codes.

Highly illustrated with numerous line diagrams, tables and worked examples, *Reynolds's Reinforced Concrete Designer's Handbook* is a unique resource providing comprehensive guidance that enables the engineer to analyze and design reinforced concrete buildings, bridges, retaining walls, and containment structures.

Written for structural engineers, contractors, consulting engineers, local and health authorities, and utilities, this is also excellent for civil and architecture departments in universities and FE colleges.

July 2007: 416pp
Hb: 978-0-419-25820-9: **£99.99**
Pb: 978-0-419-25830-8: **£41.99**

To Order: Tel: +44 (0) 1235 400524 **Fax:** +44 (0) 1235 400525
or Post: Taylor and Francis Customer Services,
Bookpoint Ltd, Unit T1, 200 Milton Park, Abingdon, Oxon, OX14 4TA UK
Email: book.orders@tandf.co.uk

For a complete listing of all our titles visit:
www.tandf.co.uk

Costs and Tender Prices Indices

Decoding Eurocode 7

A. Bond et al.

'Well presented, clear and unambiguous ... I have no hesitation in recommending Decoding Eurocode 7 to both students and practitioners as an authoritative, practical text, representing excellent value for money.' – *Tony Bracegirdle, Engineering Structures*

'A beautifully written and presented book.' – *Yul Tammo, Cornwall County Council*

Decoding Eurocode 7 provides a detailed examination of Eurocode 7 Parts 1 and 2 and an overview of the associated European and International standards. The detail of the code is set out in summary tables and diagrams, with extensive. Fully annotated worked examples demonstrate how to apply it to real designs. Flow diagrams explain how reliability is introduced into design and mind maps gather related information into a coherent framework.

Written by authors who specialise in lecturing on the subject, *Decoding Eurocode 7* explains the key principles and application rules of Eurocode 7 in a logical and simple manner. Invaluable for practitioners, as well as for high-level students and researchers working in geotechnical fields.

August 2008: 616pp
hbk: 978-0-415-40948-3: **£80.00**

To Order: Tel: +44 (0) 1235 400524 **Fax:** +44 (0) 1235 400525
or Post: Taylor and Francis Customer Services,
Bookpoint Ltd, Unit T1, 200 Milton Park, Abingdon, Oxon, OX14 4TA UK
Email: book.orders@tandf.co.uk

For a complete listing of all our titles visit:
www.tandf.co.uk

The purpose of this part is to present historic changes in Civil Engineering costs and tender prices. It gives published and constructed indices and diagrammatic comparisons between building and Civil Engineering costs and tender prices and the retail price index, and will provide a basis for updating historical cost or tender price information.

INTRODUCTION

It is important to distinguish between costs and tender prices; Civil Engineering costs are the costs incurred by a Contractor in the course of his business; Civil Engineering tender prices are the prices for which a Contractor undertakes work. Tender prices will be based on Contractor's costs but will also take into account market considerations such as the availability of labour and materials and the prevailing workload for Civil Engineering Contractors. This can mean that in a period when work is scarce tender prices may fall as costs are rising while when there is plenty of work prices will tend to increase at a faster rate than costs. This section comprises published Civil Engineering cost and tender indices, a constructed Civil Engineering cost index and comparisons of these with building cost and tender indices and with the retail price index.

PRICE ADJUSTMENT FORMULA INDICES

The indices were originally introduced in the 1970s to reduce the time and effort in allowing for fluctuations in labour and materials in contracts with a Variation of Price clause.

Used in conjunction with the formulae price adjustment method of adjusting building and civil and specialist engineering contracts to allow for changes in the cost of labour, plant and materials. They are familiarly known as NEDO Indices, Baxter Indices or Osborne Indices and are widely used primarily on variation of price contracts.

BCIS calls them the Price Adjustment Formulae Indices (PAFI) but will recognize them by any of their previous aliases.

Index 1970 Series Title		Index 1990 Series Title	
1	Labour and supervision in Civil Engineering	1	Labour and supervision
2	Plant and road vehicles: provision and maintenance	2	Plant and road vehicles
3	Aggregates	3	Aggregates
4	Bricks and clay products	4	Bricks and clay products
5	Cements	5	Cements
6	Cast iron products	6	Ready mixed concrete
7	Coated roadstone for road pavement & bituminous products	7	Cast and spun iron products
8	DERV fuel	8	Plastic products
9	Gas oil fuel	9	Coated macadam and bituminous products
10	Timber	10	DERV fuel
11A	Steel for reinforcement	11	Gas oil fuel
11B	Metal sections	12	Timber
12	Fabricated structural steel	13	Steel for reinforcement
13	Labour and supervision in fabricating and erecting steelwork	14	Metal sections
		15	Sheet steel piling

Price Adjustment Formulae for Construction Contracts
Monthly Bulletin of Indices are published by BCIS

Costs and Tender Prices Indices

Quarterly values of price adjustment formula 1970 Series. Base: – 1970 = 100 (except for index 11A which has a base date of July 1976).

Year	Q	1	2	3	4	5	6	7	9	10	11a
2005	1	1507	1057	1657	1873	1171	1826	3022	3291	977	508
	2	1523	1063	1705	1919	1238	1815	3079	3573	986	459
	3	1639	1105	1661	1889	1234	1820	3081	4179	988	446
	4	1639	1106	1619	1909	1243	1839	3140	3868	1020	461
2006	1	1639	1112	1659	1962	1329	1854	3292	4044	1009	474
	2	1648	1114	1689	1970	1328	1853	3346	4465	1014	521
	3	1696	1128	1656	2000	1326	1916	3527	4332	1056	592
	4	1696	1130	1614	2013	1350	1947	3500	3437	1076	594
2007	1	1696	1131	1709	2034	1395	1967	3623	3982	1100	605
	2	1710	1137	1667	2028	1404	1890	3610	4359	1156	635
	3	1768	1158	1651	2084	1416	1919	3546	4638	1194	629
	4	1768	1159	1656	2013	1408	1944	3646	5440	1190	593
2008	1	1768	1174	1735	2099	1552	1938	4051	6305	1168	647
	2	1770	1184	1890	2066	1549	2021	4280	7799	1133	830
	3	1868	1218	1832	1987	1591	2165	4318	6830	1131	879
	4	1868	1230	1780	2032	1589	2265	4369	4576	1187	703
2009	1	1869	1249	1883	2195	1733	2290	4401	4288	1158	617
	2	1867	1249	1902	2157	1686	2379	4537	4983	1127	588
	3	1867	1251	1951	2166	1657	2386	4539	5048	1253	581
	4	1867	1253	1863	2159	1638	2389	4941	5390	1318	572
2010	1	1869	1268	1863	2200	1651	2384	4774	6060	1374	632
	2	1869	1275	1887	2223	1650	2485	4966	6128	1462	727
	3	1869	1274	1977	2231	1644	2485	4857	6027	1541	709
	*4	1869	1273	1962	2208	1640	2485	4939	6657	1537	695
2011	*1	1869	1282	2021	2204	1649	2634	5202	7921	1499	759
	*2	1874	1287	1955	2223	1662	2634	5230	7669	1500	829

* Provisional

Note: The indices shown are for the third month of each quarter

A CONSTRUCTED COST INDEX BASED ON THE PRICE ADJUSTMENT FORMULA INDICES

Although the above indices are prepared and published in order to provide a common basis for calculating reimbursement of increased costs during the course of a contract, they also present time series of cost indices for the main components of Civil Engineering work. They can therefore be used as the basis of an index for Civil Engineering work. The method used here is to construct a composite index by allocating weightings to each of the 10 indices, the weightings being established from an analysis of actual projects. The composite index is calculated by applying these weightings to the appropriate price adjustment formula indices and totalling the results; this index is again presented with a base date of 1970.

Constructed Civil Engineering Cost Index base: 1970 = 100

Year	First quarter	Second quarter	Third quarter	Fourth quarter	Annual average
2005	1611	1652	1722	1718	1676
2006	1757	1785	1842	1813	1799
2007	1841	1862	1895	1939	1884
2008	2039	2168	2215	2128	2138
2009	2099	2129	2167	2225	2155
2010	2226	2284	2285	2316*	2278
2011	2400*				

Note: * Provisional

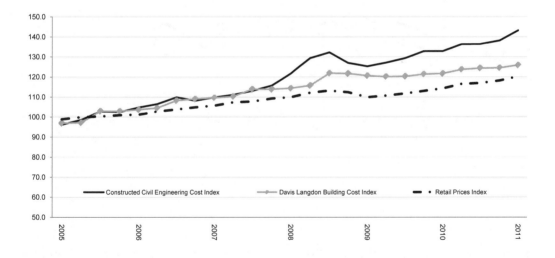

Source: Davis Langdon based on BIS figures.

The chart illustrates the movement of the Constructed Civil Engineering Index, the Davis Langdon Building Cost Index and the Retail Price Index.

THE ROAD CONSTRUCTION TENDER PRICE INDEX

Civil Engineering work generally does not lend itself easily to the preparation of tender price indices in the same way as building work. There is, however, a published tender index for road construction and this is reproduced below with the permission of BIS. The index is intended to indicate the movement in tender prices for road construction contracts. It is based on priced rates contained in accepted tenders for Road Construction, Motorway Widening and Major Maintenance Schemes. It is published with a base of 1995=100.

Tender Price Index of Road Construction – Base: 1995 = 100

Year	First quarter	Second quarter	Third quarter	Fourth quarter	Annual average
2005	129	133	137	142	135
2006	144	150	153	152	150
2007	159	153	151	149	153
2008	145	144	146	147	145
2009	137	145	151	152	
2010	157	163	167*	169*	
2011	170*				

Note: * Provisional

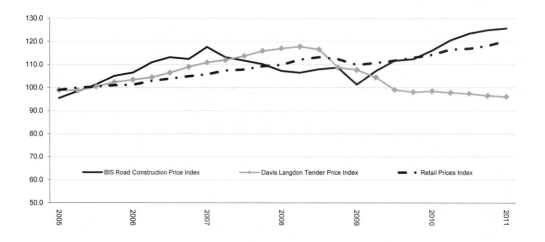

Source: BIS

The chart illustrates the movement of the BIS Road Construction Price Index; the Davis Langdon Tender Price Index and the Retail Price Index.

PART 4

Approximate Estimating

Estimating by means of priced approximate quantities is always more accurate than by using overall prices per square metre. Prices given in this section, which is arranged in elemental order, also include for all the incidental items and labours which are normally measured separately in Bills of Quantities. They include overheads and profit but do not include for preliminaries.

Whilst every effort is made to ensure the accuracy of these figures, they have been prepared for approximate estimating purposes only and on no account should they be used for the preparation of tenders.

Unless otherwise described units denoted as m² refer to appropriate unit areas (rather than gross floor areas).

As elsewhere in this edition prices do not include Value Added Tax, which should be applied at the current rate.

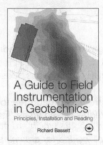

Building Prices per Square Metre

Prices given under this heading are average prices on a 'fluctuating basis' for typical buildings based on the provisional third quarter 2011 price level. Unless otherwise stated, prices INCLUDE overheads and profit but EXCLUDE preliminaries, loose or special equipment and fees for professional services.

Prices are based upon the total floor area of all storeys, measured between external walls and without deduction for internal walls, columns, stairwells, liftwells and the like.

As in previous editions it is emphasized that the prices must be treated with reserve in that they represent the average of prices from our records and cannot provide more than a rough guide to the probable cost of a building or structure.

In many instances normal commercial pressures together with a limited range of available specifications ensure that a single rate is sufficient to indicate the prevailing average price. However, where such restrictions do not apply a range has been given; this is not to suggest that figures outside this range will not be encountered, but simply that the calibre of such a type of building can itself vary significantly.

As elsewhere in this edition, prices do not include Value Added Tax, which should be applied at the current rate where applicable.

	Unit	Cost
Surface car parking	m²	52 to 80
Multi-storey car parks		
split level	m²	230 to 310
split level with brick facades	m²	320 to 440
flat slab	m²	320 to 460
warped	m²	350 to 410
Underground car parks		
partially underground under buildings	m²	420 to 500
completely underground under buildings	m²	630 to 860
completely underground with landscaped roof	m²	900 to 1070
Railway stations	m²	1670 to 2810
Bus and coach stations	m²	670 to 1440
Bus garages	m²	780 to 860
Garage showrooms	m²	800 to 1150
Garages, domestic	m²	430 to 870
Airport facilities (excluding aprons)		
airport terminals	m²	1470 to 3230
airport piers/satellites	m²	1800 to 4030
apron/runway – varying infrastructure content	m²	102 to 190
Airport campus facilities		
cargo handling bases	m²	530 to 880
distribution centres	m²	270 to 530
hangars (type C and D aircraft)	m²	1140 to 1340
hangars (type E aircraft)	m²	1580 to 3970
TV, radio and video studios	m²	1020 to 2020
Telephone exchanges	m²	820 to 1250
Telephone engineering centres	m²	680 to 770

Approximate Estimating

BUILDING PRICES PER SQUARE METRE

	Unit	Cost
Branch Post Offices	m²	860 to 1160
Postal delivery offices/Sorting offices	m²	870 to 1190
Mortuaries	m²	1560 to 2160
Sub – stations	m²	1160 to 1750
INDUSTRIAL FACILITIES		
Agricultural storage buildings	m²	410 to 700
Factories		
for letting (incoming services only)	m²	290 to 410
for letting (including lighting, power and heating)	m²	390 to 520
nursery units (including lighting, power and heating)	m²	470 to 700
workshops	m²	450 to 870
maintenance/motor transport workshops	m²	530 to 920
owner occupation – for light industrial use	m²	550 to 700
owner occupation – for heavy industrial use	m²	910 to 1060
Factory/office buildings – high technology production		
for letting (shell and core only)	m²	510 to 700
for letting (ground floor shell, first floor offices)	m²	1010 to 1310
for owner occupation (controlled environment,		
fully finished)	m²	1100 to 1470
B1 Light industrial/office building		
economical shell and core with heating only	m²	470 to 820
medium shell with heating and ventilation	m²	720 to 1080
high quality shell and core with air conditioning	m²	960 to 1750
Warehouse & Distribution centres		
high bay (10-15 m high) for owner occupation (no heating) up to 10,000m²	m²	260 to 330
high bay (10-15 m high) for owner occupation (no heating)10,000 m² to 20,000m²	m²	190 to 260
high bay (16-24 m high) for owner occupation (no heating)10,000 m² to 20,000m²	m²	280 to 370
high bay (16-24 m high) for owner occupation (no heating) over 20,000m²	m²	210 to 330
Fit out cold stores, refrigerated stores inside warehouse	m²	400 to 760
WATER AND TREATMENT FACILITIES		
Reinforced Concrete tanks		
including all excavation; fill; structural work; valves; penstocks; pipeworks per		
m³ of concrete in structure. treatment tanks	m³	540 to 610
fire ponds and lagoons	m³	470 to 550
reservoirs	m³	480 to 560
Reinforced Concrete river weirs, quay and wave walls including temporary dams; caissons overpumping, anchorages,		
all structural works	m²	370 to 570
Reinforced Concrete dams (up to 12 m high)		
arch dam, including excavation anchorages and structural work only. m³ of structure	m³	1230 to 1800
flat slab dam including excavation anchorages and structural work only. m³ of structure	m³	930 to 1460
Earth dams		
rock fill with concrete core m³ of completed structure	m³	240 to 850
hydraulic fill embankment dam m³ of completed structure	m³	130 to 380

Approximate Estimating Rates

FOUNDATIONS FOR STRUCTURES	Unit	£
In-situ concrete foundations complete with associated excavation, formwork, reinforcement and disposal		
trench fill strip foundation; 450 × 1000 mm deep	m	110
trench fill strip foundation; 600 × 1200 mm deep	m	260
strip foundation; 900 mm deep	m	220
strip foundation; 1200 mm deep	m	330
column/stanchion base; unreinforced; 900 × 900 × 450 mm	nr	360
column/stanchion base; reinforced 50 kg/m³; 1750 × 1750 × 500 mm	nr	770
column/stanchion base; reinforced 50 kg/m³; 2200 × 2200 × 600 mm	nr	1190
column/stanchion base; reinforced 75 kg/m³; 2400 × 2400 × 600 mm	nr	1430
pile cap; reinforced 75 kg/m³; 900 × 900 × 1400 mm; for one pile	nr	800
pile cap; reinforced 75 kg/m³; 2700 × 900 × 1400 mm; for two piles	nr	1670
pile cap; reinforced 75 kg/m³; 2700 × 2700 × 1400 mm; for four piles	nr	3560
pile cap; reinforced 75 kg/m³; 3700 × 2700 × 1400 mm; for six piles	nr	4610
In-situ concrete slabs comprising concrete C30 slab with tamped finish; reinforced with two layers of A393 fabric; laid in bays size 20 m × 10m on subbase of 250 mm granular material; including all associated excavation (250 mm depth), formwork and formed expansion joints with filler		
200 mm thick	m²	94
250 mm thick	m²	102
275 mm thick	m²	106
300 mm thick	m²	110
extra over cost of slab		
for hand trowel finish	m²	1.00
for power float finish	m²	2.40
for wire brush finish	m²	1.80
for additional subbase depth (including excavation and disposal)		
50 mm	m²	3.40
100 mm	m²	6.80
150 mm	m²	10.20
The prices following have been assembled from the relevant items in the unit costs section. They are intended to give a broad overall price or help in comparisons between a number of basic construction procedures		
These approximate estimates are for construction only. They do not include for preliminaries, design/supervisioncosts, land purchase or OH&P etc		
Prices in this section are based on the same information and outputs as used in the unit costs section.		
Costs per m² or m³ of completed structure		

APPROXIMATE ESTIMATING RATES

EARTH RETENTION AND STABILISATION	Unit	£
Reinforced in-situ concrete retaining wall		
(including excavation; reinforcement; formwork; expansion joints; granular backfill and		
100 mm land drain; profiled formwork finish to one side typical retaining wall, allowing for		
profiling finishes)		
1.0 m high	m²	498
3.0 m high	m²	436
6.0 m high	m²	523
9.0 m high	m²	585
Precast concrete block, earth retaining wall		
(including granular fill; earth anchors and proprietary units or concrete panels, strips, fixings		
and accessories (Reinforced Earth))		
1.5 m high	m²	370
3.0 m high	m²	340
6.0 m high	m²	370
Precast, reinforced concrete unit retaining wall (in-situ foundation):		
(including excavation and fill; reinforced concrete foundation; pre-cast concrete units, joints)		
1.00 high	m²	450
2.00 high	m²	460
3.00 high	m²	485
Precast concrete crib wall *(including excavation; stabilization and foundation work)*		
up to 1.0 m high	m²	295
up to 1.5 m high	m²	290
up to 2.5 m high	m²	285
up to 4.0 m high	m²	270
Timber crib walling		
up to 1.5 m high	m²	145
up to 3.7 m high	m²	180
up to 5.9 m high	m²	225
up to 7.4 m high	m²	260
Rock gabions		
including preparation; excluding anchoring		
1 m thick	m²	80

BRIDGEWORKS	Unit	Span	Range £
The following prices are based on recovered data and information from approximately 50 separate structures completed as part of actual projects Prices include for the works described to the bridge decks and abutments, but exclude any approach works			
Road bridges			
Reinforced in-situ concrete viaduct			
(including excavation; reinforcement; formwork; concrete; bearings;			
expansion joints; deck waterproofing; deck finishings; P1 parapet)			
per m² of deck maximum span between piers or abutments	m²	15 m	1860 to 3830
	m²	20 m	1770 to 3640
	m²	25 m +	1720 to 3540
Reinforced concrete bridge with precast beams			
(including excavation; reinforcement; formwork; concrete; bearings;			
expansion joints; deck waterproofing; deck finishings; P1 parapet)			
per m² of deck maximum span between piers or abutments	m²	12 m	2230 to 3320
	m²	17 m	2090 to 3110
	m²	22 m	1990 to 2970
	m²	30 m +	1900 to 2840

APPROXIMATE ESTIMATING RATES

BRIDGEWORKS	Unit	Span	Range £
Reinforced concrete bridge with pre-fabricated steel beams			
(including excavation; reinforcement; formwork; concrete; bearings			
expansion joints; deck waterproofing; deck finishings; P1 parapet)			
per m² of deck maximum span between piers or abutments	m²	20 m	1980 to 4410
	m²	30 m	1870 to 4160
	m²	40 m	1820 to 4060
Footbridges			
Reinforced in situ concrete with precast beams			
(including excavation; reinforcement; formwork; concrete; bearings			
expansion joints; deck waterproofing; deck finishings; P6 parapet)			
widths up to 6 m wide; per m² of deck maximum span between			
piers or abutments	m²	5 m	2130 to 4750
	m²	10 m	2070 to 4610
	m²	20 m +	2020 to 4510
Structural steel bridge with concrete foundations width up to 4 m wide per m² of deck			
maximum span: -	m²	10 m	1980 to 2640
	m²	12 m	1970 to 2620
	m²	16 m	1960 to 2610
	m²	20 m	1980 to 2640
Timber footbridge (stress graded with concrete piers) per m² of deck			
maximum span: -	m²	12 m	1050 to 1160
	m²	18 m	1030 to 1190
HIGHWAY WORKS			
The following prices are the approximate costs per metre run of roadway, and are based on information from a number of sources including engineers estimates, tenders, final account values etc on a large number of highways contracts.			
Motorway and All Purpose Road prices include for earthworks, structures, drainage, pavements, line markings, reflective studs, footways signs, lighting, motorway communications, fencing and barrier works as well as allowance for accommodation works, statutory undertakings and landscaping as appropriate to the type and location of the carriageway. The earthworks elements can be adjusted by reference to factors detailed at the end of this sub-section.			
Motorway and All Purpose Road prices do NOT include for the cost of associated features such as side roads, interchanges, underbridges, overbridges, culverts, sub-ways, gantries and retaining walls. These are shown separately beside the cost range for each road type, based on statistical frequency norms.			

APPROXIMATE ESTIMATING RATES

MOTORWAYS	Unit	Feature £	Range £
The following costs are based on a 850 mm construction comprising 40 mm wearing course, 60 mm base course, 250 mm road base, 200 mm subbase and 350 mm capping layer; central reserve incorporating two 0.7 m wide hardstrips; no footpaths or cycle paths included			
Rural motorways			
grassed central reserve and outer berms; no kerbs or edgings; assumption that 30% of excavated material is unsuitable for filling; costs allow for forming embankments for 50% of highway length average 4.70 m high and 50% of length in cuttings average 3.90 m deep; accommodation fencing each side; allowance of 25% of length having crash barriers and 20% of length having lighting			
dual two lane (D2M_R); 25.20 m overall width; each carriageway 7.30 m with 3.30 m hard shoulder; 4.00 m central reserve	m	1760	3010 to 3820
dual three lane (D3M_R); 32.60 m overall width; each carriageway 11.00 m with 3.30 m hard shoulder; 4.00 m central reserve	m	1920	3780 to 4820
dual four lane (D4M_R); 39.80 m overall width; each carriageway 14.60 m with 3.30 m hard shoulder; 4.00 m central reserve	m	2050	4520 to 5780
Urban motorways			
hard paved central reserve and outer berms; precast concrete kerbs; assumption that 30% of excavated material is unsuitable for filling; costs allow for forming embankments for 50% of highway length average 3.20 m high and 50% of length in cuttings average 1.60 m accommodation fencing each side; allowance of 25% of length having crash barriers and 20% of length having lighting dual two lane (D2M_U); 23.10 m overall width; each carriageway			
7.30 m with 2.75m hard shoulder; 3.00 m central reserve	m	2970	2850 to 3690
dual three lane (D3M_U); 30.50 m overall width; each carriageway 11.00 m with 2.75 m hard shoulder; 3.00 m central reserve	m	3130	3570 to 4630
dual four lane (D4M_U); 37.70 m overall width; each carriageway 14.60 m with 2.75 m hard shoulder; 3.00 m central reserve	m	3280	4270 to 5550
ALL-PURPOSE ROADS			
The following costs are based on a 800 mm construction comprising 40 mm wearing course, 60 mm base course, 200 mm road base, 150 mm subbase and 350 mm capping layer; no footpaths or cycle paths included			
Rural all-purpose roads			
grassed central reserve; no kerbs or edgings; assumption that 30% of excavated material is unsuitable for filling; costs allow for forming embankments for 50% of highway length average 4.00m high and 50% of length in cuttings average 3.75m deep; allowance of 25% of length having crash barriers and 20% of length having lighting			
dual two lane (D2AP_R); 18.60m overall width; each carriageway 7.30 m; 4.00m central reserve	m	1260	2190 to 2630
dual three lane (D3AP_R); 26.00m overall width; each carriageway 11.00m ; 4.00m central reserve	m	1300	2530 to 344
Urban all-purpose roads			
hard paved central reserve; precast concrete kerbs; assumption that 30% of excavated material is unsuitable for filling; costs allow for forming embankments for 50% of highway length in average 2.2 m high and 50% of length in cuttings average 1.62 m deep; allowance of 25% of length having crash barriers and 20% of length having lighting			
dual two lane (D2AP_U); 23.10m overall width; each carriageway 7.30 m with 2.75m hard shoulder; 3.00m central reserve	m	2150	2810 to 3190
dual three lane (D3AP_U); 30.50m overall width; each carriageway 11.00 m C456with 2.75m hard shoulder; 3.00m central reserve	m	2330	3570 to 4200

APPROXIMATE ESTIMATING RATES

MOTORWAYS	Unit	Feature £	Range £
Other roads			
Rural All-Purpose Roads			
Single carriageway all-purpose road (carriageway is 7.3 m wide)	m	700	1010 to 1170
Wide single carriageway all-purpose road (carriageway is 10.0 m wide)	m	760	1260 to 1470
Rural Link Roads			
Two lane link road (carriageway is 7.3 m wide)	m	760	910 to 1110
Single lane link road (carriageway is 3.7 m wide)	m	490	490 to 730
Rural Motorway or Dual Carriageway Slip Roads			
Single carriageway slip road (carriageway is 5.0 m wide)	m	620	600 to 870
Urban Motorway or Dual Carriageway Sliproads			
Single carriageway slip road (carriageway is 6.0 m wide)	m	990	760 to 1070
Wide single carriageway all-purpose road with footway each side			
(carriageway is 10.0 m wide each footway is 3.0 m wide)	m	1120	1200 to 1620
Nominal 3.0 m cycle track to all-purpose roads (one side only)	m		160 to 200
Urban Link Roads			
Two lane link road (carriageway is 7.3 m wide)	m	1130	1000 to 1,220
Single lane link road (carriageway is 3.7 m wide)	m	750	540 to 790

The following are approximate costs for the installation of roads and drains to serve as part of the development of infrastructure for Housing, Retail or Industrial development. The density (i.e. percentage of area used for roads, etc., is also given to enable adjustments). NB: excludes car parking.

Type A Construction =	medium duty carriageway consisting of 100 mm surfacing, roadbase up to 115 mm subbase of 150 mm
Type B Construction =	heavy duty carriageway consisting of reinforced concrete slab 225 mm thick subbase 130 mm thick capping layer 280 mm thick

Density of facility		Cost/unit £		Cost/hectare £		Density of road
per hectare	per acre	Type A	Type B	Type A	Type B	per hectare
5	(2)	4300	6250	21500	31260	2.00%
15	(6)	2510	3650	37625	54705	3.50%
20	(8)	2040	2970	40850	59394	3.80%
25	(10)	1760	2560	44075	64083	4.10%
30	(12)	1610	2340	48375	70335	4.50%
37	(15)	1510	2200	55900	81276	5.20%
50	(20)	1400	2030	69875	101595	6.50%

	Unit	£
Turning or passing bay = 35 m² overall area; suitable for cars, vans		
Type A construction	nr	4630
Type B construction	nr	6650
Turning or passing bay = 100 m² overall area; suitable for semi trailer		
Type A construction	nr	11630
Type B construction	nr	16900
Bus lay-by = 40 m² overall area		
Type B construction	nr	6600
Parking lay-by = 200 m² overall area		
Type A construction	nr	22400
Type B construction	nr	32600
Vehicle crossing verge/footway/central reserve	m²	153
Footway construction (bit-mac plus edgings)	m²	51

APPROXIMATE ESTIMATING RATES

CIVIL ENGINEERING WORKS SITE UTILITIES AND INFRASTRUCTURE

The following prices have been compiled for various services and provide average costs for site utilities and infrastructure works

UNDERPASSES	Unit	To estate road £	To major road £
Provision of underpasses to new roads, constructed as part of a road building programme			
pedestrian underpass 3 m wide × 2.5 m high	m	3900	4700
vehicle underpass 7 m wide × 5 m high	m	16000	21000
vehicle underpass 14 m wide × 5 m high	m	38500	48500
ROAD CROSSINGS			
Costs include road markings, beacons, lights, signs, advance danger signs etc but exclude associated service trenches, ducts etc.			
4 way traffic signal installation	nr	40000	68000
zebra crossing	nr	4750	6100
pelican crossing	nr	18000	22750
pedestrian guard railing	m	120	175

STREET FURNITURE	Unit	£
There is an almost infinite variety of items with which local authorities and private developers can enhance the roadside and public areas, including statutory requirements such as lighting. It is therefore impossible to price all the different street furniture available. The following however gives a selection of the more common items: -		
reflectorised traffic signs 0.25-0.75 m² area on steel post	nr	120 to 200
internally illuminated traffic signs (dependent on area)	nr	225 to 300
externally illuminated traffic signs (up to 4 m²)	nr	550 to 1500
lighting to pedestrian areas and estates roads on 4-6 m columns with up to 70 W lamps	nr	280 to 350
lighting to main roads on 10-12 m columns with 250 W lamps	nr	1030 to 1290
lighting to main roads on 12-15 m columns with 400 W high pressure sodium lighting	nr	1380 to 1730
benches – hardwood and precast concrete	nr	220 to 950
litter bins		
precast concrete	nr	220 to 390
hardwood slatted	nr	140 to 155
cast iron	nr	255 to 525
large all-steel	nr	790
concrete bollards	nr	120 to 275
steel bollards	nr	160 to 175
bus stops	nr	385
bus stops inc basic shelter	nr	950
pillar box on post	nr	315
Hi mast radio/beacon – 60 m	nr	32500
automatic barrier equipment	nr	20000

FOOTPATHS AND PAVINGS	Unit	£
Costs include excavation, base course as necessary and precast concrete edgings on foundations to one or both sides		
bitumen macadam footpath	m²	45 to 50
precast concrete paving flags	m²	30 to 70
precast concrete block pavings	m²	45 to 70
clay brick paving	m²	55 to 165
granite setts	m²	90 to 110
cobbled paving	m²	80 to 095

APPROXIMATE ESTIMATING RATES

SURFACE CAR PARKS		
Surface car parking is the cheapest way of providing car parking. The cost of rooftop or basement car parking can be 10 to 15 times the cost of the surface car parking. The vogue in car parking for all except industrial schemes is for concrete block pavers, with different colours marking out parking bays and zones.		
Guideline figures for parking requirements are shown below: -		
one car space per:		
offices	22 to 25	m² gross floor area
industrial – factories	45 to 55	m² gross floor area
- warehouses	200	m² gross floor area
shops	20 to 25	m² gross floor area
superstores	10	m² gross floor area
cinemas, theatres	3 to 5	seats
hospital	3	hospital beds
residential	1 to 2	dwellings
For surface level car parking the area required per car will not generally show much variation. Site shape, position of the building on the site, and parking configuration will be the main determinants of the area to be allowed per car. A fairly low range of 20-23 m² / car will usually suffice.		
Typical costs for surface car parking including lighting and drainage are illustrated below: -		
	£/m²	£/car
Tarmacadam surfaced, marked out with thermoplastic road paint	100	2200 to 2500
Interlocking or herringbone concrete block paving, marked out with coloured blocks	143	3150 to 3580
Grassblock recycled polythene units filled with top soil and grass seeded	76	1670 to 1900
Note: Costs include forecourts, aprons and access areas but not approach roads.		
SERVICES	Unit	£
Costs of services to a site are built up of connection charges and service runs. This can vary significantly depending upon the availability or otherwise of a suitably sized main in the neighbourhood of the site. However, typical service charges for an estate of 200 houses might be as follows: -		
Charge per house		
Water		525 to 1,100
Electric		275 if all electric
		525 gas/electric
(plus cost of substation £ 12,500 – 21,000 total)		
Gas		525 to 675
(plus cost of governing station £ 19,000 – 22,500 total)		
Telephone		175
Sewerage		400 to 525
Mains laid in trenches including excavation and filling		
Water main		
75 mm PVC-U main in 225 mm ductile iron pipe as duct	m	95
Electric main		
600/1000 volt cables. Two core 25 mm cable including 100 mm clayware duct	m	40
Gas main	m	75
150 mm ductile or cast iron gas pipe		

APPROXIMATE ESTIMATING RATES

SERVICES	Unit	£
Telephone	m	22
British Telecom installation in 110 mm PVC-U duct		
Drainage		
100/150 mm clay or PVC-U pipes on granular bed and surround up to 3 m deep	m	45
100/150 mm cast iron pipes on concrete beds up to 3 m deep	m	90
300 mm clay drain pipe on granular bed and surround up to 3 m deep	m	165
450 mm concrete pipe on concrete beds up to 3 m deep	m	135
900 mm diameter concrete sewage pipe on granular bed up to 3.5 m deep	m	270
Brick manholes in commons, rendered internally, clay channel with three branches, concrete cover slab, cast iron cover and frame 900 – 1500 mm deep	nr	1440
Precast concrete manholes with precast concrete rings up to 1200 mm diameter, channel with three branches, concrete cover slab and frame 900-1500 mm deep	nr	1490
Vitrified clay or precast concrete road gully 450 × 900 mm deep with concrete surround, brick seating and cast iron grating	nr	465
LANDSCAPING		
Landscaping as a subject matter is sufficiently large to fill a book (see Spon's External Works and Landscape Price Book). However, there are certain items that arise on a majority of projects and other discrete items for which indicative costs can be produced. The following rates include for normal surface excavation but exclude bulk excavation, levelling or earth shifting and land drain provision.		
Soft landscaping		
cultivate ground, remove rubbish and plant with grass seed	m²	5.90
ditto and turf	m²	11.70
shrubbed areas	m²	66.70
shrubbed areas including allowance for small trees	m²	71.20
standard tree in tree pit including stake	nr	58.70
ditto but with tree guard and precast tree grid slabs	nr	223.00
Sports pitches		
The provision of sports facilities will involve different techniques of earth shifting and cultivation and usually will be carried out by specialist landscaping contractors. Costs include for cultivating ground, bringing to appropriate levels for the specified game, applying fertilizer, weedkiller, seeding and rolling and white line marking with nets, posts etc as required.		
football pitch (114 × 72 m)	nr	19100
artificial football pitch including subbase, bitumen macadam open textured base and heavy duty astro-turf carpet	nr	450000
cricket outfield (160 × 142 m)	nr	52400
cricket square (20 × 20 m) including imported marl or clay loam bringing to accurate levels, seeding with cricket square type grass	nr	14800
bowling green (38 × 38 m) rink, including French drain and gravel path on four sides	nr	53600
grass tennis courts 1 court (35 × 17 m) including bringing to accurate levels, chain link perimeter fence and gate, tennis posts and net	nr	22600
two grass tennis courts (35 × 32 m) ditto	pair	38400
artificial surface tennis courts (35 × 17 m) including chain link fencing, gate, tennis posts and net	nr	42560
two courts (45 × 32 m) ditto	pair	103000
golf putting green	hole	1900
pitch and putt course	hole	6000 to 9100
full length golf course, full specifications inc watering system	hole	21200 to 40100
championship course	hole	up to 149000

APPROXIMATE ESTIMATING RATES

SERVICES	Unit	£
Parklands		
As with all sports pitches, parklands will involve different techniques of earth shifting and cultivation.		
The following rates include for normal surface excavation.		
parklands, including cultivating ground applying fertilizer etc. and seeding with parks type grass	ha	17300
general sportsfield	ha	20500
lakes including excavation up to 10 m deep, laying 1.5 mm thick butyl rubber sheet and spreading top soil evenly on top to depth of 300 mm		
under 1 ha in area	ha	376800
between 1 and 5 ha in area	ha	352000
extra for planting aquatic plants in lake top soil	m²	60
Playground equipment		
Modern swings with flat rubber safety seats four seats, two bays	nr	3160
Stainless steel slide, 3.40 m long	nr	2730
Climbing frame – Lappset Playhouse	nr	3350
Seesaw comprising timber plank on sealed ball bearings 3960 × 230 × 70 mm thick – no-bump	nr	2230
Wicksteed safety tiles surfacing around play equipment	m²	56
Playbark particles type safety surfacing 150 mm thick on hardcore bed	m²	52
Land drainage		
The above rates exclude provision of any land drainage. If land drainage is required on a project, the propensity of the land to flood will decide the spacing of the land drains. However, some indicative figures can be given for land drainage.		
Costs include for excavation and backfilling of trenches and laying agricultural clay drain pipes with 75 mm diameter lateral runs average 600 mm deep and 100 mm diameter main runs average 750 mm deep.		
land drainage to parkland with laterals at 30 m centres and main runs at 100 m centres	ha	3460
land drainage to sportsfields with laterals at 10 m centres and main runs at 33 mm centres	ha	16570
TEMPORARY WORKS		
Bailey bridges		
Installation of temporary Bailey bridges		
(including temporary concrete abutments; erection maintenance; dismantling)		
hire costs (for 52 weeks) @	12890	
delivery/collect @	1110	
erect/dismantle @	4120	
concrete abutments @	5060	
demolish after dismantling @	840	
allowance for maintenance, etc @	*2480*	
span up to 10 m	nr	26500
span 15 m	nr	30000
span 20 m	nr	39000
span 25 m	nr	42000

APPROXIMATE ESTIMATING RATES

Cofferdams	Unit	£ range
Installation of cofferdams (based on driven steel sections with recovery value) *(including all plant, for installation and dismantling; loss of materials; pumping and maintenance excluding excavation and disposal of material – backfilling on completion)*		
Cost range based on 12 weeks installation		
Depth of drive up to 5 m, diameter or side length		
up to 2 m	nr	3900 to 4100
up to 10 m	nr	18800 to 20900
up to 20 m	nr	33,100 to 41,700
Depth of drive 5 – 10 m, diameter or side length		
up to 2 m	nr	5500 to 7900
up to 10 m	nr	24500 to 39000
up to 20 m	nr	44900 to 77700
Depth of drive 10 – 15 m, diameter or side length		
up to 2 m	nr	7900 to 10800
up to 10 m	nr	37800 to 54100
up to 20 m	nr	70700 to 108300
Depth of drive 15 – 20 m, diameter or side length		
up to 2 m	nr	13300 to 16700
up to 10 m	nr	37700 to 66700
up to 20 m	nr	94500 to 133000
Above based on soft-medium ground conditions		
add for medium-hard		+ 20%
hard but not rock		+ 33%
Access scaffolding	**Unit**	**£**
The following are guideline costs for the hire and erection of proprietary scaffold systems (tube and coupling). Costs are very general and are based on a minimum area of 360 m² at 1.80m deep		
Approximate hire (supply and fix) of patent scaffold per 4 week (cost dependent upon length of hire, quality of system, number of toe boards, handrails etc.)	m²	5.00 to 6.60
Based on this a typical cost of a 60 × 10m (600 m²) area for 8 weeks would be about £7,200		
Approximate hire (supply and fix of mobile access towers per 4 week hire (refer also to plant hire in Section 3)	m²	16.90 to 25.30
Additional costs of Pole ladder access, per 4.0 m high, per 4 week	nr	6.40 to 7.30
Additional cost of stair towers extra over the cost of scaffold system per 2 m rise, per 4 week	m²	71.00
Additional cost of hoarding around base perimeter (using multi-use ply sheeting) per 4 week period	m²	1.00 to 1.80
Additional cost of polythene debris netting (no reuse)	m²	0.20
Additional cost of Monaflex 'T' plus weather-proof sheeting including anchors and straps (based on 3 uses)	m²	1.20
Erection of scaffolding system is based on 3 men erecting a 16 m² bay in about 1 hour and dismantling the same in 20 minutes (i.e. experienced scaffolders)		
Note: Although scaffolding is essentially plant hire, allowance must be made for inevitable loss and damage to fittings, for consumables used and for maintenance during hire periods		

APPROXIMATE ESTIMATING RATES

Access scaffolding	Unit	£
Earthwork support		
The following are comparative costs for earthwork and trench support based on the hire of modular systems (trench box hydraulic) with an allowance for consumable materials (maximum 5 day hire allowance).		
Earthwork support not exceeding 1m deep, distance between opposing faces not exceeding 2m	m²	5.60
Earthwork support not exceeding 1m deep distance between opposing faces 2-4m	m²	9.60
Earthwork support not exceeding 2m deep distance between opposing faces not exceeding 2m	m²	10.80
Earthwork support not exceeding 2m deep distance between opposing faces 2-4m	m²	19.10
Earthwork support not exceeding 4m deep distance between opposing faces not exceeding 2m	m²	27.20
Earthwork support not exceeding 4m deep distance between opposing face 2-4m	m²	39.30
For larger excavations requiring Earthwork support refer to cofferdam estimates; or sheet piling within unit cost sections		
The following are approximate weekly hire costs for a range of basic support equipment used on site (see also Plant Costs Section – Part 5)	Unit	£/week
Steel sheet piling		
AU or AZ series section	tonne	31.70 to 39.60
Trench sheeting		
Standard overlapping sheets	m²	1.50 to 1.90
Interlocking type	m²	2.90 to 4.20
Heavy duty overlapping sheets		
6 mm thick	m²	4.80 to 5.50
8 mm thick	m²	4.90
driving cap	nr	13.20
extraction clamp	nr	26.40
Trench box with hydraulic wallings – as above		
Trench struts	nr	0.70
No. 0 – 0.3 – 0.40 m	nr	0.70
No. 1 – 0.5 – 0.75 m	nr	0.70
No. 2 – 0.7 – 1.14 m	nr	0.90
No. 3 – 1.0 – 1.75 m		

ESSENTIAL READING FROM TAYLOR AND FRANCIS

Failures in Concrete Structures: Case Studies in Reinforced and Prestressed Concrete

R. Whittle

Some lessons are only learned from mistakes. But, it's much cheaper to learn from someone else's mistakes than to have to do so from your own. Drawing on over 50 years of working with concrete structures, Robin Whittle examines the problems which he has seen occur and shows how they could have been avoided.

The first and largest part of the book tells the stories of a number of cases where things have gone wrong with concrete structures. Each case is analysed to identify its cause and how it might have been prevented. It then looks at how failures in structural modelling can lead to big problems if they are not identified before construction is undertaken. Beyond this it examines how contract arrangements can encourage or prevent problems in the designing and building processes. It concludes with an examination of the role research and development in preventing failures.

By identifying the differences between shoddy economisations and genuine efficiency savings, this book offers savings in the short term which won't be at the expense of a structure's long-term performance. Invaluable reading if you're designing or building concrete structures and want to avoid problems which could be expensive or embarrassing further down the line.

December 2011: 208pp
Hb: 978-0-415-56701-5: **£60.00**

To Order: Tel: +44 (0) 1235 400524 **Fax:** +44 (0) 1235 400525
or Post: Taylor and Francis Customer Services,
Bookpoint Ltd, Unit T1, 200 Milton Park, Abingdon, Oxon, OX14 4TA UK
Email: book.orders@tandf.co.uk

For a complete listing of all our titles visit:
www.tandf.co.uk

Unit Costs – Ancillary Building Works

This part enables the user to include within the estimate for Ancillary Building Works which may be associated with a Civil Engineering project but, because of the diversity which can occur on the specification for these works cannot be priced as accurately as unit cost items.

Additionally, as such works form only a minor percentage of an overall Civil Engineering budget, and then the need for such accuracy is not as critical. Therefore the rates given within this part are based upon an average range for each item to allow the user discretionary use based upon more detailed knowledge of the specific project. Should however more detailed pricing information be required then reference should be made to the latest editions of:

> ***SPON'S ARCHITECTS' AND BUILDERS' PRICE BOOK***
> ***SPON'S EXTERNAL WORKS AND LANDSCAPE PRICE BOOK***
> ***SPON'S MECHANICAL AND ELECTRICAL SERVICES PRICE BOOK***

Also included in this Part are items which allow the user to prepare order of cost estimates for various areas of Civil Engineering Works more accurately than by using Approximate Estimates but without the necessity to complete a full cost estimate.

Adjustments should be made to the rates shown to allow for time, location, local conditions, site constraints and any other factors likely to affect the cost of the specific project.

ANCILLARY BUILDING WORKS

Item	Unit	Range £		
1 SUBSTRUCTURE				
Trench fill foundations				
Machine excavation, disposal, plain in situ concrete 21 N/mm² – 20 mm aggregate (1:2:4) trench fill, 300 mm high brickwork in cement mortar (1:3), pitch polymer damp-proof course with common bricks:				
width and depth of concrete 600mm × 1200mm				
half brick wall	m	88.00	to	110.00
one brick wall	m	100.00	to	130.00
cavity wall	m	100.00	to	130.00
width and depth of concrete 750mm × 1200mm				
half brick wall	m	110.00	to	140.00
one brick wall	m	120.00	to	150.00
cavity wall	m	120.00	to	150.00
Strip foundations				
Excavate trench, partial backfill, partial disposal, earthwork support (risk item), compact base of trench, plain in-situ concrete 21 N/mm² – 20 mm aggregate (1:2:4) 250 mm thick, formwork, common brickwork in cement mortar (1:3), pitch polymer damp-proof course; one brick thick, 600 mm foundation, and english bond				
hand excavation, depth of wall:				
600 mm	m	88.00	to	110.00
900 mm	m	110.00	to	140.00
1200 mm	m	140.00	to	180.00
1500 mm	m	165.00	to	210.00
machine excavation, depth of wall:				
600 mm	m	59.00	to	75.00
900 mm	m	77.00	to	99.00
1200 mm	m	88.00	to	110.00
1500 mm	m	145.00	to	185.00
cavity, 750 mm foundation and stretcher bond				
Hand excavation, depth of wall:				
600 mm	m	87.00	to	110.00
900 mm	m	130.00	to	165.00
1200 mm	m	165.00	to	210.00
1500 mm	m	200.00	to	250.00
machine excavation, depth of wall:				
600 mm	m	88.00	to	110.00
900 mm	m	110.00	to	145.00
1200 mm	m	145.00	to	185.00
1500 mm	m	170.00	to	220.00
extra over for three courses of facing bricks	m	6.20	to	7.90
Column bases				
Excavate pit in firm ground, partial backfill, partial disposal, earth work support, compact base of pit, plain in-situ concrete 21 N/mm² – 20 mm aggregate (1:2:4), formwork				

ANCILLARY BUILDING WORKS

Item	Unit	Range £		
1 SUBSTRUCTURE – cont				
Column bases – cont				
depth of pit – 1200 mm				
hand excavation, base size:				
600 × 600 × 300 mm	nr	67.00	to	86.00
900 × 900 × 450 mm	nr	140.00	to	175.00
1200 × 1200 × 450 mm	nr	210.00	to	270.00
1500 × 1500 × 600 mm	nr	380.00	to	485.00
machine excavation, base size:				
600 × 600 × 300 mm	nr	53.00	to	68.00
900 × 900 × 450 mm	nr	115.00	to	150.00
1200 × 1200 × 450 mm	nr	170.00	to	215.00
1500 × 1500 × 600 mm	nr	315.00	to	405.00
depth of pit – 1800 mm				
hand excavation, base size:				
600 × 600 × 300 mm	nr	85.00	to	110.00
900 × 900 × 450 mm	nr	160.00	to	200.00
1200 × 1200 × 450 mm	nr	265.00	to	340.00
1500 × 1500 × 600 mm	nr	440.00	to	570.00
machine excavation, base size:				
600 × 600 × 300 mm	nr	66.00	to	84.00
900 × 900 × 450 mm	nr	125.00	to	160.00
1200 × 1200 × 450 mm	nr	190.00	to	240.00
1500 × 1500 × 600 mm	nr	345.00	to	440.00
Excavate pit in firm ground by machine, partial backfill, partial disposal, earth work support, compact base of pit, plain in-situ concrete 21 N/mm² – 20 mm aggregate (1:2:4), formwork				
depth of pit – 1200 mm				
reinforcement at 50 kg/m³ concrete, base size:				
1750 × 1750 × 500 mm	nr	370.00	to	470.00
2000 × 2000 × 500 mm	nr	455.00	to	580.00
2200 × 2200 × 600 mm	nr	670.00	to	860.00
2400 × 2400 × 600 mm	nr	780.00	to	1000.00
reinforcement at 75 kg/m³ concrete, base size				
1750 × 1750 × 500 mm	nr	390.00	to	500.00
2000 × 2000 × 500 mm	nr	490.00	to	620.00
2200 × 2200 × 600 mm	nr	700.00	to	900.00
2400 × 2400 × 600 mm	nr	820.00	to	1050.00
depth of pit – 1800 mm				
reinforcement at 50 kg/m³ concrete, base size:				
1750 × 1750 × 500 mm	nr	410.00	to	530.00
2000 × 2000 × 500 mm	nr	510.00	to	650.00
2200 × 2200 × 600 mm	nr	740.00	to	940.00
2400 × 2400 × 600 mm	nr	850.00	to	1075.00
reinforcement at 75 kg/m³ concrete, base size				
1750 × 1750 × 500 mm	nr	430.00	to	550.00
2000 × 2000 × 500 mm	nr	540.00	to	690.00
2200 × 2200 × 600 mm	nr	760.00	to	970.00
2400 × 2400 × 600 mm	nr	890.00	to	1125.00

ANCILLARY BUILDING WORKS

Item	Unit	Range £		
Pile caps				
Excavate pit in firm ground by machine, partial backfill, partial disposal, earthwork support, compact base of pit, cut off top of pile and prepare reinforcement, reinforced in situ concrete 26 N/mm² – 20 mm aggregate (1:2:4), formwork				
depth of pit – 1500mm				
reinforcement at 50 kg/m³ concrete, cap size				
900 × 900 × 1400 mm, one pile	nr	345.00	to	440.00
2700 × 900 × 1400 mm, two piles	nr	870.00	to	1100.00
*2700 × 2475 × 1400 mm, three piles	nr	1575.00	to	2000.00
2700 × 2700 × 1400 mm, four piles	nr	2000.00	to	2550.00
3700 × 2700 × 1400 mm, six piles	nr	2700.00	to	3450.00
reinforcement at 75 kg/m³ concrete, cap size				
900 × 900 × 1400 mm, one pile	nr	345.00	to	440.00
2700 × 900 × 1400 mm, two piles	nr	870.00	to	1100.00
*2700 × 2475 × 1400 mm, three piles	nr	1575.00	to	2025.00
2700 × 2700 × 1400 mm, four piles	nr	2175.00	to	2750.00
3700 × 2700 × 1400 mm, six piles	nr	2850.00	to	3650.00
depth of pit – 2100mm				
reinforcement at 50 kg/m³ concrete, cap size				
900 × 900 × 1400 mm, one pile	nr	370.00	to	470.00
2700 × 900 × 1400 mm, two piles	nr	890.00	to	1125.00
*2700 × 2475 × 1400 mm, three piles	nr	1625.00	to	2075.00
2700 × 2700 × 1400 mm, four piles	nr	2075.00	to	2650.00
3700 × 2700 × 1400 mm, six piles	nr	2850.00	to	3650.00
reinforcement at 75 kg/m³ concrete, cap size				
900 × 900 × 1400 mm, one pile	nr	365.00	to	470.00
2700 × 900 × 1400 mm, two piles	nr	920.00	to	1175.00
*2700 × 2475 × 1400 mm, three piles	nr	1725.00	to	2200.00
2700 × 2700 × 1400 mm, four piles	nr	2175.00	to	2800.00
3700 × 2700 × 1400 mm, six piles	nr	3000.00	to	3900.00
* = triangular on plan, overall size given				
additional cost of alternative strength concrete				
30N/mm²	m³	1.50	to	1.90
40N/mm²	m³	3.70	to	4.70
Strip or base foundations				
Foundations in good ground; reinforced concrete bed; for one storey development				
shallow foundations per m² ground floor plan area	m²	76.00	to	96.00
deep foundations per m² ground floor plan area	m²	125.00	to	160.00
Foundations in good ground; reinforced concrete bed; for two storey development				
shallow foundations per m² ground floor plan area	m²	84.00	to	130.00
deep foundations per m² ground floor plan area	m²	130.00	to	205.00
Extra for each additional storey	m²	29.00	to	33.50

ANCILLARY BUILDING WORKS

Item	Unit	Range £		
1 SUBSTRUCTURE – cont				
Raft foundations				
Raft on poor ground for development				
one storey per m² ground floor plan area	m²	87.00	to	110.00
two storey per m² ground floor plan area	m²	140.00	to	180.00
Extra for each additional storey	m²	140.00	to	180.00
Piled foundations				
Foundations in poor ground; reinforced concrete slab; for one storey commercial development per m² ground floor plan area				
short bore piles to columns only	m²	135.00	to	175.00
short bore piles	m²	160.00	to	210.00
fully piled	m²	220.00	to	285.00
Alternative concrete mixes in lieu of 21.00N/m² – 20 mm aggregate (1:2:4); per m² ground floor plan area				
25 N/mm²	m²	0.70	to	0.85
30 N/mm²	m²	0.90	to	1.20
40 N/mm²	m²	1.65	to	2.10
Ground slabs				
(per m² ground floor area)				
Mechanical excavation to reduce levels, disposal, level and compact, hardcore bed blinded with sand, 1200 gauge polythene damp proff membrane, concrete 21.00N/mm² – 20 mm aggregate (1:2:4) ground slab, tamped bed				
200 mm thick concrete slab; thickness of hardcore bed				
150 mm	m²	44.00	to	56.00
175 mm	m²	46.00	to	59.00
200 mm	m²	46.00	to	59.00
300 mm thick concrete slab; thickness of hardcore bed				
150 mm	m²	52.00	to	66.00
175 mm	m²	54.00	to	69.00
200 mm	m²	54.00	to	69.00
450 mm thick concrete slab; thickness of hardcore bed				
150 mm	m²	68.00	to	87.00
175 mm	m²	70.00	to	89.00
200 mm	m²	70.00	to	89.00
Add to the foregoing prices for fabric reinforcement BS 4483, lapped; per m² ground floor plan area				
A142 (2.22 kg/m²); 1 layer	m²	4.40	to	5.60
A142 (2.22 kg/m²); 2 layers	m²	6.40	to	8.20
A193 (3.02 kg/m²); 1 layer	m²	4.40	to	5.60
A193 (3.02 kg/m²); 2 layers	m²	7.50	to	9.60
A252 (3.95 kg/m²); 1 layer	m²	5.45	to	7.00
A252 (3.95 kg/m²); 2 layers	m²	8.75	to	11.20
A393 (6.16 kg/m²); 1 layer	m²	6.40	to	8.20
A393 (6.16 kg/m²); 2 layers	m²	12.90	to	16.40

ANCILLARY BUILDING WORKS

Item	Unit	Range £		
High yield bent bar reinforcement BS 4449 ; per m² ground floor plan area at a rate of				
25 kg/m²	m²	5.45	to	7.00
50 kg/m²	m²	9.85	to	12.60
75 kg/m²	m²	15.20	to	19.40
100 kg/m²	m²	19.60	to	25.00
Other foundations/alternative slabs/extras				
Reinforced concrete bed including excavation and hardcore; per m² ground floor plan area				
150 mm thick	m²	42.50	to	55.00
200 mm thick	m²	57.00	to	73.00
300 mm thick	m²	71.00	to	100.00
Extra per m² ground floor plan area for				
sound reducing quil in screed	m²	4.50	to	7.95
50 mm inslutaion under slab and at edge	m²	7.35	to	10.80
75 mm inslutaion under slab and at edge	m²	9.15	to	13.60
suspended precast concrete slabs in lieu of in-situ slab	m²	19.30	to	24.50
2A FRAME AND 2B UPPER FLOORS				
FRAME AND UPPER FLOORS				
Reinforced in-situ concrete columns at 30 N/mm², bar reinforcement at 200 kg/m³; basic formwork (assumed four uses) ; column size				
225 × 225 mm	m	50.00	to	64.00
300 × 300 mm	m	70.00	to	90.00
300 × 450 mm	m	97.00	to	125.00
300 × 600 mm	m	120.00	to	150.00
450 × 450 mm	m	125.00	to	160.00
In-situ concrete casing to steel column at 30 N/mm², basic framework (assumed four uses), column size				
225 × 225 mm	m	47.00	to	60.00
300 × 300 mm	m	66.00	to	84.00
300 × 450 mm	m	88.00	to	110.00
300 × 600 mm	m	105.00	to	135.00
450 × 450 mm	m	105.00	to	135.00
450 × 600 mm	m	130.00	to	170.00
450 × 900 mm	m	180.00	to	230.00
Reinforced in-situ concrete isloated beams at 30 N/mm²; bar reinforcement at 200 kg/m³ basic formwork (assumed four uses); beam size				
225 × 450 mm	m	88.00	to	110.00
225 × 600 mm	m	110.00	to	140.00
300 × 600 mm	m	125.00	to	160.00
300 × 900 mm	m	180.00	to	230.00
300 × 1200 mm	m	225.00	to	290.00
450 × 600 mm	m	160.00	to	205.00
450 × 900 mm	m	220.00	to	285.00
450 × 1200 mm	m	285.00	to	360.00
600 × 600 mm	m	200.00	to	255.00
600 × 900 mm	m	270.00	to	340.00
600 × 1200 mm	m	340.00	to	440.00

ANCILLARY BUILDING WORKS

Item	Unit	Range £		
2A FRAME AND 2B UPPER FLOORS – cont				
FRAME AND UPPER FLOORS – cont				
In-situ concrete casing to steel attached beams at 30 N/mm³; basic formwork (assumed four uses) ; beam size				
225 × 450 mm	m	80.00	to	100.00
225 × 600 mm	m	100.00	to	130.00
300 × 600 mm	m	115.00	to	145.00
300 × 900 mm	m	160.00	to	200.00
300 × 1200 mm	m	200.00	to	255.00
450 × 600 mm	m	140.00	to	180.00
450 × 900 mm	m	190.00	to	245.00
450 × 1200 mm	m	240.00	to	310.00
600 × 600 mm	m	180.00	to	230.00
600 × 900 mm	m	230.00	to	290.00
600 × 1200 mm	m	285.00	to	360.00
Extra smooth finish formwork; all categories	m	7.10	to	15.10
Softwood joisted floor; no frame				
Joisted floor; no frame; 22 mm chipboard t & g flooring; herring bone strutting; no coverings or finishes ; per m² of upper floor area				
150 × 50 mm joists	m²	29.50	to	38.00
175 × 50 mm joists	m²	34.00	to	43.00
200 × 50 mm joists	m²	35.00	to	45.00
225 × 50 mm joists	m²	38.00	to	48.00
250 × 50 mm joists	m²	40.50	to	52.00
275 × 50 mm joists	m²	44.50	to	57.00
Softwood joisted floor; average depth; platerboard; skim; emulsion; vinyl flooring and painted softwood skirtings ; per m² of upper floor area	m²	70.00	to	90.00
Joisted floor; no frame; 22 mm chipboard t & g flooring; herring bone strutting; no coverings or finishes ; per m² of upper floor area				
150 × 50 mm joists	m²	29.50	to	38.00
175 × 50 mm joists	m²	34.00	to	43.00
200 × 50 mm joists	m²	35.00	to	45.00
225 × 50 mm joists	m²	38.00	to	48.00
250 × 50 mm joists	m²	41.00	to	52.00
275 × 50 mm joists	m²	44.50	to	57.00
Softwood joisted floor; average depth; plasterboard; skim; emulsion; vinyle flooring and painted softwood skirtings ; per m² of upper floor area				
Reinforced concrete floors; no frame				
Suspended slab; no coverings or finishes; per m² of upper floor area				
2.75 m span; 8.00 KN/m² loading	m²	58.00	to	74.00
3.35 m span; 8.00 KN/m² loading	m²	66.00	to	84.00
4.25 m span; 8.00 KN/m² loading	m²	82.00	to	105.00
Suspended slab; no coverings or finishes; per m² of upper floor area				
150 mm thick	m²	78.00	to	100.00
225 mm Thick	m²	120.00	to	155.00

ANCILLARY BUILDING WORKS

Item	Unit	Range £		
Reinforced concrete floors and frame				
Suspended slab; average depth; no coverings or finishes ; per m² of upper floor area				
up to six storeys	m²	150.00	to	190.00
Wide span suspended slab with frame; per m² of upper floor area				
up to six storeys	m²	170.00	to	215.00
Reinforced concrete floors; steel frame				
Suspended slab; average depth; 'Holorib' permanent steel shuttering; protected steel frame; no coverings or finishes; per m² of upper floor area				
up to six storeys	m²	200.00	to	255.00
Extra for spans 7.5 to 15 m	m²	23.00	to	29.00
Suspended slab; average depth; prtected steel frame; no coverings or finishes; per m² of upper floor area				
up to six storeys	m²	190.00	to	245.00
Suspended slab; 75 mm screed; no coverings or finishes ; per m² of upper floor area				
3 m span; 8.50 KN/m² loading	m²	62.00	to	79.00
6 m span; 8.50 KN/m² loading	m²	66.00	to	84.00
7.5 m span; 8.50 KN/m² loading	m²	69.00	to	88.00
3 m span; 12.50 KN/m² loading	m²	74.00	to	95.00
6 m span; 12.50 KN/m² loading	m²	61.00	to	78.00
Precast concrete floors; reinforced concrete frame				
Suspended slab; average depth; no coverings or finishes ; per m² of upper floor area	m²	98.00	to	125.00
Precast concrete floors; steele frame				
Suspended slabs; average depth; protected steel frame; no coverings or finishes; per m² of upper floor area				
up to six storeys	m²	180.00	to	230.00
Extra per m² of upper floor area for				
wrought formwork	m²	4.20	to	5.35
sound reducing quilt in screed	m²	4.45	to	5.70
insulation to avoid cold bridging	m²	7.95	to	10.20
2C ROOF				
Softwood flat roofs				
Roof joists; average depth; 25 mm softwood boarding; PVC rainwater goods; plasterboard; skim and emulsion ; per m² of roof plan area				
three layer felt and chippings	m²	110.00	to	130.00
two coat asphalt and chippings	m²	110.00	to	140.00
Softwood trussed pitched roofs				
Structure only comprising 100 × 38 mm Fink trusses @ 600 mm centres (measured on plan) ; per m² of roof plan area				
30° pitch	m²	25.00	to	30.00
35° pitch	m²	25.00	to	31.00
40° pitch	m²	28.00	to	34.00

ANCILLARY BUILDING WORKS

Item	Unit	Range £		
2C ROOF – cont				
Softwood trussed pitched roofs – cont				
Fink roof trusses; narrow span; 100 mm insulation; PVC rainwater goods; plasterboard; skim and emulsion per m² or roof plan area				
concrete interlocking tile coverings	m²	96.00	to	120.00
clay pantile coverings	m²	105.00	to	130.00
composition slate coverings	m²	110.00	to	130.00
plain clay tile coverings	m²	130.00	to	160.00
natural slate coverings	m²	140.00	to	170.00
reconstructed stone coverings	m²	110.00	to	140.00
Monopitch roof trusses; 100 mm insulation; PVC rainwater goods; platerboard; skim and emulsion ; per m² of roof plan area				
concrete interlocking tile coverings	m²	115.00	to	140.00
clay pantile coverings	m²	110.00	to	140.00
composition slate coverings	m²	120.00	to	145.00
plain clay tile coverings	m²	140.00	to	170.00
natural slate coverings	m²	140.00	to	175.00
reconstructed stone coverings	m²	120.00	to	140.00
Steel trussed pitched roofs				
Steel roof trusses and beams; thermal and accoustic insulations; per m² of roof plan area				
aluminium profiled composite cladding	m²	250.00	to	300.00
Steel roof and glulam beams; thermal and accoustic insulation; per m² of roof plan area				
aluminium profiled composite cladding	m²	250.00	to	300.00
Concrete flat roofs				
Structure only comprising reinforced concrete suspended slab; no coverings or finishes ; per m² of roof plan area				
3.65 m span; 8.00 KN/m² loading	m²	61.00	to	75.00
4.25 m span; 8.00 KN/m² loading	m²	73.00	to	90.00
Precast concrete suspended slab; average depth; 100 mm insulation; PVC rainwater goods; per m² of roof plan area				
two coat asphalt coverings and chippings	m²	110.00	to	140.00
Reinforced concrete or waffle suspended slabs; average depth; 100 mm insulation; PVC rainwater goods; per m² of roof plan area				
two coat asphalt coverings and chippings	m²	120.00	to	150.00
Reinforced concrete suspended slabs; on 'Holorib' permanent steel shuttering; average depth; 100 mm insulation; PVC rainwater goods; per m² of roof area				
two coat asphalt coverings and chippings	m²	110.00	to	140.00
Flat roof decking and finishes				
Woodwool roof decking; per m² of roof plan area				
50 mm thick; two coat asphalt coverings to BS 6925 and chippings	m²	51.00	to	62.00
Galvanized steel roof decking; 100 mm insulation; three layer felt roofing and chippings ; per m² of roof plan area				
0.7 mm thick; 3.74 m span	m²	56.00	to	69.00
0.7 mm thick; 5.13 m span	m²	57.00	to	70.00

ANCILLARY BUILDING WORKS

Item	Unit	Range £		
Aluminium roof decking; 100 mm insulation; three layer felt roofing and chippings ; per m² of roof plan area				
0.9 mm thick; 2.34 m span	m²	65.00	to	80.00
Metal decking; 100 mm insulation; on wood/steel open lattice beams; per m² of roof plan area				
three layer felt roofing and chippings	m²	97.00	to	120.00
two layer higher performance felt roofing and chippings	m²	100.00	to	120.00
Roof claddings				
Non-asbestos profiled cladding; per roof plan area				
'profile 3'; natural	m²	16.80	to	20.50
'profile 6'; natural	m²	18.70	to	23.00
'profile 6'; natural; insulated; inner lining panel	m²	21.00	to	25.50
Extra for colours	m²	1.90	to	2.30
Non-asbestos profiled cladding on steel purlins; per m² of roof plan area				
Insulated	m²	33.00	to	40.00
Insulated; with 10% translucent sheets	m²	37.00	to	45.00
Insulated; platerboard inner lining on metal tees	m²	54.00	to	66.00
PVF2 coated galvanized steel profiled cladding on steel purlins; per m² of roof plan area				
cladding only; 0.72 mm thick	m²	23.00	to	28.00
Insulated	m²	40.00	to	49.00
Insulated; plasterboard inner lining on metal tees	m²	60.00	to	73.00
Insulated; plasterboard inner lining on metal tees; with 1% fire vents	m²	70.00	to	86.00
Insulated; plasterboard inner lining on metal tees; with 2.5% fire vents	m²	87.00	to	105.00
Insulated; coloured inner lining panel	m²	62.00	to	76.00
Insulated; coloured inner lining panel; with 1% fire vents	m²	70.00	to	86.00
Insulated; coloured inner lining panel; with 2.5% fire vents	m²	87.00	to	105.00
Insulated; sandwich panel	m²	130.00	to	160.00
Rooflights/patent glazing and glazed roofs				
Rooflights				
standard pvc	m²	145.00	to	180.00
feature/ventilating	m²	260.00	to	320.00
Patent glazing; including flashings				
Standard aluminium georgian wired; single glazed	m²	185.00	to	230.00
Standard aluminium georgian wired; double glazed	m²	22.00	to	27.00
Comparitive over/underlays				
Roofing felt; unreinforced				
sloping (measured on face)	m²	1.65	to	2.00
Roofing felt; reinforced				
sloping (measured on face)	m²	2.05	to	2.50
sloping (measured on plan) 20° pitch	m²	2.40	to	2.90
sloping (measured on plan) 30° pitch	m²	2.60	to	3.20
sloping (measured on plan) 40° pitch	m²	3.35	to	4.10
Building paper; per m² of roof plan area	m²	1.45	to	1.75
Vapour barrier; per m² of roof plan area	m²	2.20	to	2.65

ANCILLARY BUILDING WORKS

Item	Unit	Range £	
2C ROOF – cont			
Comparitive over/underlays – cont			
Insulation quilt; laid over ceiling joists ; per m² of roof plan area			
100 mm thick	m²	4.85 to	5.95
150 mm thick	m²	6.70 to	8.20
200 mm thick	m²	8.90 to	10.90
Wood fibre insulation boards; impregnated; density 220 – 350 kg/m³; per m² of roof plan area			
12.7 mm thick	m²	6.00 to	7.30
Limestone ballast; per m² of roof plan area	m²	6.55 to	8.00
Polyurethane insulation boards; density 32 kg/m³; per m² of roof plan area			
30 mm thick	m²	9.35 to	11.40
50 mm thick	m²	11.20 to	13.70
Glass fibre insulation boards; density 120 – 130 kg/m²; per m² of roof plan area			
60 mm thick	m²	17.20 to	21.00
Extruded polystyrene foam boards; per m² of roof plan area			
50 mm thick	m²	16.30 to	19.90
50 mm thick; with cement topping	m²	26.50 to	32.00
75 mm thick	m²	21.00 to	26.00
Screeds to receive roof coverings; per m² of roof plan area			
50 mm cement and sand screed	m²	10.80 to	13.20
60 mm (av.) 'isocrete K' screed; density 500 kg/m³	m²	11.60 to	14.20
75 mm lightweight bituminous screed and vapour barrier	m²	18.20 to	22.00
100 mm lightweight bituminous screed and vapour barrier	m²	22.00 to	27.00
50 mm woodwool slabs; unreinforced			
Sloping (measured on face)	m²	9.35 to	11.40
Sloping (measured on plan); 20° pitch	m²	10.30 to	12.60
Sloping (measured on plan); 30° pitch	m²	11.30 to	13.90
Sloping (measured on plan); 40° pitch	m²	14.00 to	17.10
50 mm woodwool slabs; unreinforced; on and including steel purlins at 600 mm centres	m²	11.60 to	14.20
25 mm 'Tanalized' softwood boarding			
Sloping (measured on face)	m²	15.90 to	19.40
Sloping (measured on plan); 20° pitch	m²	17.20 to	21.00
Sloping (measured on plan); 30° pitch	m²	21.00 to	25.50
Sloping (measured on plan); 40° pitch	m²	25.00 to	30.00
18 mm external quality plywood boarding			
Sloping (measured on face)	m²	20.00 to	24.50
Sloping (measured on plan); 20° pitch	m²	22.00 to	27.00
Sloping (measured on plan); 30° pitch	m²	27.00 to	33.00
Sloping (measured on plan); 40° pitch	m²	32.00 to	39.00
Comparative tiling and slating finishes/perimeter treatments (including underfelt, battening, eaves courses and ridges)			
Concrete troughed interlocking tiles; 413 × 300 mm; 75 mm lap			
Sloping (measured on face)	m²	19.20 to	23.50
Sloping (measured on plan); 30° pitch	m²	25.00 to	30.00
Sloping (measured on plan); 40° pitch	m²	30.00 to	36.50

ANCILLARY BUILDING WORKS

Item	Unit	Range £		
Concrete troughed interlocking slates; 430 × 330 mm; 75 mm lap				
Sloping (measured on face)	m²	20.00	to	24.50
Sloping (measured on plan); 30° pitch	m²	22.00	to	27.00
Sloping (measured on plan); 40° pitch	m²	30.00	to	36.50
Concrete bold roll interlocking tiles; 418 × 332 mm; 75 mm lap				
Sloping (measured on face)	m²	19.20	to	23.50
Sloping (measured on plan); 30° pitch	m²	25.00	to	31.00
Sloping (measured on plan); 40° pitch	m²	28.50	to	35.00
Natural red pantiles; 337 × 241 mm; 76 mm head and 38 mm side laps				
Sloping (measured on face)	m²	32.00	to	39.00
Sloping (measured on plan); 30° pitch	m²	41.50	to	51.00
Sloping (measured on plan); 40° pitch	m²	47.00	to	58.00
Blue composition (non asbestos) slates; 600 × 300 mm; 75 mm lap				
Sloping (measured on face)	m²	33.00	to	40.00
Sloping to mansard (measured on face)	m²	47.00	to	57.00
Sloping (measured on plan); 30° pitch	m²	43.50	to	53.00
Sloping (measured on plan); 40° pitch	m²	49.00	to	60.00
Concrete plain tiles; 267 × 165 mm; 64 mm lap				
Sloping (measured on face)	m²	38.50	to	47.00
Sloping (measured on plan); 30° pitch	m²	47.50	to	58.00
Sloping (measured on plan); 40° pitch	m²	58.00	to	70.00
Machine-made clay plain tiles; 267 × 165 mm; 64 mm lap				
Sloping (measured on face)	m²	50.00	to	62.00
Sloping (measured on plan); 30° pitch	m²	62.00	to	76.00
Sloping (measured on plan); 40° pitch	m²	76.00	to	92.00
Welsh natural slates; 510 × 255 mm; 76 mm lap				
Sloping (measured on face)	m²	56.00	to	68.00
Sloping (measured on plan); 30° pitch	m²	73.00	to	89.00
Sloping (measured on plan); 40° pitch	m²	85.00	to	105.00
Reconstructed stone slates; random lengths; 80 mm lap				
Sloping (measured on face)	m²	35.00	to	43.00
Sloping (measured on plan); 30° pitch	m²	43.50	to	53.00
Sloping (measured on plan); 40° pitch	m²	52.00	to	63.00
Verges to sloping roofs; 200 × 25 mm painted softwood bargeboard; measured perimeter length				
6 mm 'Masterboard' soffit lining 150 mm wide	m	18.20	to	22.00
19 × 150 mm painted softwood soffit	m	22.00	to	27.00
Eaves to sloping roofs; 200 × 25 mm painted softwood fascia; 6 mm 'Masterboard' soffit lining 225 mm wide; measured perimeter length				
100 mm PVC gutter	m	25.00	to	30.00
150 mm PVC gutter	m	31.50	to	38.50
100 mm cast iron gutter; decorated	m	39.50	to	48.50
150 mm cast iron gutter; decorated	m	48.00	to	59.00
Eaves to sloping roofs; 200 × 25 mm painted softwood fascia; 19 × 225 mm painted softwood soffit; measured perimeter length				
100 mm PVC gutter	m	30.00	to	36.50
150 mm PVC gutter	m	37.00	to	45.00
100 mm cast iron gutter; decorated	m	45.00	to	55.00
150 mm cast iron gutter; decorated	m	52.00	to	64.00

ANCILLARY BUILDING WORKS

Item	Unit	Range £		

2C ROOF – cont

Comparative tiling and slating finishes/perimeter treatments (including underfelt, battening, eaves courses and ridges) – cont

Rainwater pipes; fixed to backgrounds; including offsets and shoe measured length of pipes

Item	Unit		Range £	
68 mm PVC	m	8.70	to	10.60
110 mm PVC	m	13.00	to	15.80
75 mm cast iron; decorated	m	31.50	to	38.50
100 mm cast iron; decorated	m	37.00	to	45.00
Ridges measured length of ridge				
concrete helf round tiles	m	17.60	to	21.50
machine-made clay half round tiles	m	22.00	to	27.00
Hips; including mitring roof tiles measured length of hip				
concrete half round tiles	m	23.00	to	28.00
machine-made clay half round tiles	m	32.50	to	39.50
Comparative cladding finishes (including underfelt, labours etc.)				
0.91 mm Aluminium roofing; commercial grade				
flat	m²	51.00	to	62.00
0.91 mm Aluminium roofing; commercial grade ; fixed to boarding (included)				
Sloping (measured on face)	m²	53.00	to	65.00
Sloping (measured on plan); 20° pitch	m²	60.00	to	73.00
Sloping (measured on plan); 30° pitch	m²	72.00	to	88.00
Sloping (measured on plan); 40° pitch	m²	85.00	to	105.00
Comparative waterproof finishes/perimeter treatments				
Liquid applied coatings				
solar reflected paint	m²	1.90	to	2.30
solar applied bitumen	m²	7.45	to	9.10
solar applied co-polymer	m²	8.80	to	10.80
solar applied polyurethane	m²	14.50	to	17.80
20 mm Two coat asphalt roofing; laid flat; on felt underlay				
to BS 6925	m²	15.50	to	18.90
to BS 6577	m²	22.00	to	27.00
Extra for				
Solar reflective paint	m²	2.50	to	3.10
limestone chipping finish	m²	3.20	to	3.90
grp tiles in hot bitumen	m²	35.00	to	43.00
20 mm Two coat reinforced asphaltic compound; laid flat; on felt underlay to BS 6577				
three layer glass fibre roofing	m²	21.00	to	25.50
three layer asbestos based roofing	m²	26.00	to	31.50
Extra for granite chipping finish	m²	3.20	to	3.90
Built-up self-finished asbestos based bitumen felt roofing; laid sloping				
two layer roofing (measured on face)	m²	28.00	to	34.00
two layer roofing (measured on plan); 40 degree pitch	m²	43.00	to	52.00
three layer roofing (measured on face)	m²	37.00	to	45.00
three layer roofing (measured on plan); 20 degree pitch	m²	56.00	to	68.00
three layer roofing (measured on plan); 30 degree pitch	m²	58.00	to	71.00

ANCILLARY BUILDING WORKS

Item	Unit	Range £		
Elastomeric single ply roofing; laid flat				
EDPM membrane; laid loose	m²	24.50	to	30.00
Butyl rubber membrane; laid loose	m²	24.50	to	30.00
Extra for ballast	m²	7.45	to	9.10
Thermoplastic single ply roofing; laid flat				
laid loose	m²	23.00	to	28.50
mechanically fixed	m²	30.00	to	36.50
fully adhered	m²	33.00	to	40.00
CPE membrane; laid loose	m²	27.00	to	33.00
CSPG membrane; fully adhered	m²	27.00	to	33.00
PIB membrane; laid loose	m²	31.00	to	38.00
Extra for ballast	m²	7.45	to	9.10
High performance built-up felt roofing; laid flat				
three layer 'Ruberglas 120 GP' felt roofing; granite chipping finish	m²	33.00	to	40.00
'Andersons' three layer self-finish polyester-based bitumen felt roofing	m²	33.00	to	40.00
High performance built-up felt roofing; laid flat				
three layer polyester-based modified bitumen felt roofing	m²	34.00	to	41.50
three layer 'Ruberfort HP 350' felt roofing; granite chipping finish	m²	41.50	to	51.00
three layer 'Hyload 150E' elastomeric roofing; granite chipping finish	m²	41.50	to	51.00
three layer 'Polybit 350' elastomeric roofing; granite chipping finish	m²	43.50	to	53.00
Torch on roofing; laid flat				
three layer polyester-based modified bitumen roofing	m²	30.00	to	36.50
two layer polymeric isotropic roofing	m²	30.00	to	36.50
Extra granite chipping finish	m²	2.80	to	3.40
Edges to flat felt roofs; softwood splayed fillet; 280 × 25 mm painted softwood fascia; no gutter				
aluminium edge trim	m	35.00	to	43.00
Edges to flat roofs; code 4 lead drip dresser into gutter; 230 × 25 mm painted softwood fascia				
100 mm PVC gutter	m	33.00	to	40.00
150 mm PVC gutter	m	41.50	to	51.00
100 mm cast iron gutter; decorated	m	50.00	to	61.00
150 mm cast iron gutter; decorated	m	62.00	to	76.00

2D STAIRS

Timber construction

Item	Unit	Range £		
Softwood staircase; softwood balustrades and hardwood handrail; plasterboard; skim and emulsion to soffit				
2.6 m rise; standard; straight flight	nr	740.00	to	1025.00
2.6 m rise; standard; top three treads winding	nr	910.00	to	1275.00
2.6 m rise; standard; dogleg	nr	1050.00	to	1450.00

Reinforced concrete construction

Item	Unit	Range £		
Escape staircase; granolithic finish; mild steel balustrades and handrails				
3 m rise; dogleg	nr	3900.00	to	5400.00
Plus or minus for each 300 mm variation in storey height	nr	380.00	to	520.00

ANCILLARY BUILDING WORKS

Item	Unit	Range £		

2D STAIRS – cont

Reinforced concrete construction – cont
Staircase; terrazzo finish; mild steel balustrades and handrails; plastered
soffit; balustrades and staircase soffit decorated

Item	Unit	Range £		
3 m rise; dogleg	nr	7800.00	to	11000.00
Plus or minus for each 300 mm variation in storey height	nr	780.00	to	1075.00

Metal construction
Steel access/fire ladder

3 m high	nr	590.00	to	820.00
4 m high; expoxide finished	nr	820.00	to	1150.00

Light duty metal staircase; glavanised finish; perforated treads; no risers;
balustrades and handrails; decorated

3 m rise; straight; 760 mm wide	nr	2900.00	to	4000.00
Plus or minus for each 300 mm variation in storey height	nr	290.00	to	405.00

Heavy duty cast iron staircase; perforated treads; no risers; balustrades and
handrails; decorated

3 m rise; straight; 760 mm wide	nr	4550.00	to	6300.00
Plus or minus for each 300 mm variation in storey height	nr	455.00	to	630.00

Galvanized steel catwalk; nylon coated balustrading

450 mm wide	nr	320.00	to	445.00

Finishes to treads and risers

PVC floor tiles including screeds	storey	780.00	to	1075.00
granolithic	storey	1175.00	to	1650.00
heavy duty carpet	storey	1650.00	to	2275.00
terrazzo	storey	3300.00	to	4600.00

Comparative finishes/balustrading
Wall handrails

PVC mild steel rail on brackets	storey	310.00	to	430.00
hardwood handrail on brackets	storey	850.00	to	1175.00
stainless stell handrail on brackets	storey	2700.00	to	3800.00

Balustrading and handrails

mild steel balustrading and PVC covered handrails	storey	970.00	to	1350.00
mild steel balustrading and hardwood handrails	storey	1750.00	to	2450.00
stainless steel balustrading and handrails	storey	7500.00	to	10000.00
stainless steel and glass balustrading	storey	6500.00	to	9100.00

2E EXTERNAL WALLS

Brick/block walling
Dense aggregate block walls

100 mm thick	m²	23.00	to	28.00
140 mm thick	m²	33.00	to	40.00

Common brick solid walls

half brick thick	m²	35.50	to	43.00
one brick thick	m²	64.00	to	77.00
one and a half brick thick	m²	93.00	to	110.00

ANCILLARY BUILDING WORKS

Item	Unit	Range £		
Add or deduct for each variation of £10.00/1000 in pc value				
half brick thick	m²	1.00	to	1.25
one brick thick	m²	2.10	to	2.50
one and a half brick thick	m²	3.10	to	3.70
Extra for fair face one side	m²	2.10	to	2.50
Engineering brick walls; class B				
half brick thick	m²	43.50	to	52.00
one brick thick	m²	83.00	to	99.00
Facing brick walls; machine-made facings				
half brick thick; pointed one side	m²	64.00	to	77.00
half brick thick; built against concrete	m²	69.00	to	83.00
one brick thick; pointed both sides	m²	125.00	to	150.00
Add or deduct for each variation of £10.00/1000 in pc value				
half brick thick	m²	1.00	to	1.25
one brick thick	m²	2.10	to	2.50
Composite solid walls; facing brick on outside; and common brick on outside				
one brick thick; pointed one side	m²	100.00	to	120.00
Extra for weather pointing as a seperate operation	m²	5.50	to	6.60
one and a half brick thick; pointed one side	m²	125.00	to	150.00
Composite cavity wall; block outer skin; 50 mm insulation; lightwieght block inner skin				
outer block rendered	m²	71.00	to	85.00
Extra for				
heavyweight block inner skin	m²	2.80	to	3.40
fair face one side	m²	13.30	to	16.00
75 mm cavity insulation	m²	71.00	to	85.00
100 mm cavity insulation	m²	77.00	to	93.00
plaster and emulsion	m²	88.00	to	105.00
outer block rendered; no insulation; inner skin insulating	m²	1.50	to	1.80
outer block roughcast	m²	0.75	to	0.90
coloured masonry outer block	m²	1.50	to	1.80
Composite cavity wall; facing brick outer skin; 50 mm insulation; common brick inner skin; fair face on inside				
machine-made facings	m²	110.00	to	130.00
Composite cavity wall; facing brick outer skin; 50 mm insulation; common brick inner skin; plaster and emulsion				
machine-made facings	m²	100.00	to	125.00
Composite cavity wall; coloured masonry block; outer and inner skins; fair faced both sides	m²	1.00	to	1.25
Reinforced concrete walling				
In-situ reinforced concrete 25.5 N/m³; 13 kg/m² reinforcement; formwork both sides				
150 mm thick	m²	110.00	to	135.00
225 mm thick	m²	140.00	to	165.00

ANCILLARY BUILDING WORKS

Item	Unit	Range £		
2E EXTERNAL WALLS – cont				
Wall claddings				
Non-asbestos profiled cladding				
'profile 3'; natural	m²	24.00	to	29.00
'profile 3'; coloured	m²	26.50	to	32.00
'profile 6'; natural	m²	26.50	to	32.00
'profile 6'; coloured	m²	27.50	to	33.00
insulated; inner lining of plasterboard	m²	47.50	to	57.00
'profile 6'; natural; insulated; inner lining panel	m²	47.50	to	57.00
insulated; with 2.8 m high inner skin; emulsion	m²	41.00	to	49.50
insulated; with 2.8 m high inner skin; plasterboard lining on metal tees;				
emulsion	m²	56.00	to	68.00
PVF2 coated galvanized steel profiled cladding				
0.60 mm thick; 'profile 20B'; corrugations verticle	m²	33.00	to	40.00
0.60 mm thick; 'profile 30'; corrugations verticle	m²	33.00	to	40.00
0.60 mm thick; 'profile TOP 40'; corrugations verticle	m²	31.00	to	37.00
0.60 mm thick; 'profile 60B'; corrugations verticle	m²	39.00	to	46.50
0.60 mm thick; 'profile 30'; corrugations horizontal	m²	34.50	to	41.00
0.60 mm thick; 'profile 60B'; corrugations horizontal	m²	39.00	to	46.50
Extra for				
80 mm insulation and 0.4 mm thick coated inner lining sheet	m²	18.70	to	22.50
PVF2 coated galvanized steel profiled cladding on steel rails				
2.8 m high insulating block inner skin; plasterboard lining on metal tees;				
emulsion	m²	68.00	to	82.00
2.8 m high insulated block inner skin; plasterboard lining on metal tees;				
emulsion	m²	77.00	to	93.00
insulated; coloured inner lining panel	m²	77.00	to	93.00
insulated; full-height insulating block inner skin; plaster and emulsion	m²	96.00	to	115.00
insulated; metal sandwich panel system	m²	195.00	to	235.00
Other cladding systems				
vitreous enamelled insulated steel sandwich panel system; with				
non-asbestos fibre insulating board on inner face	m²	185.00	to	225.00
Formalux sandwich panel system; with coloured lining tray; on steel				
cladding rails	m²	220.00	to	260.00
aluminium over-cladding on system rain screen	m²	275.00	to	330.00
natural stone cladding on full-height insulating block inner skin; plaster				
and emulsion	m²	530.00	to	630.00
Curtain/glazed walling				
Single galzed polyester powder coated aluminium curtain walling				
economical; including part-height block back-up wall; plaster and emulsion	m²	415.00	to	500.00
Extra over single 6 mm float glass for				
double glazing unit with two 6 mm float glass skins	m²	44.50	to	53.00
double glazing unit with one 6 mm 'Antisun' skin and one 6 mm float				
glass skin	m²	86.00	to	100.00
look-a-like' non-vision spandrel panels	m²	56.00	to	68.00
10% opening lights	m²	13.30	to	16.00
economical; including infill panels	m²	400.00	to	475.00
Extra for				
50 mm insulation	m²	24.00	to	28.50
anodized finish in lieu of polyester powder coating	m²	28.00	to	34.00
bronze anodizing in lieu of polyester powder coating	m²	56.00	to	68.00

ANCILLARY BUILDING WORKS

Item	Unit	Range £		
good quality	m²	650.00	to	780.00
good quality; 35% opening lights	m²	820.00	to	980.00
Extra over single 6 mm float glass for double glazing unit with low 'E' and tinted glass	m²	85.00	to	100.00
High quality structural glazing to entrance elevation	m²	890.00	to	1075.00
Patent glazing systems; excluding opening lights and flashings				
7 mm georgian wired cast glass, aluminium alloy bars spanning up to 3 mat 600–625 mm spacing	m²	130.00	to	160.00
6.4 mm laminated safety glass polyester powder coated aluminium capped glazing bars spanning up to 3 m at 600–625 mm spacing	m²	415.00	to	500.00
Comparative external finishes				
Comparative concrete wall finishes				
wrought formwork one side including rubbing down	m²	3.85	to	4.60
shotblasting to expose aggregate	m²	4.90	to	5.85
bush hammering to expose aggregate	m²	15.70	to	18.90
two coats 'Sandex Matt' cement paint	m²	9.20	to	11.00
cement and sand plain face rendering	m²	16.20	to	19.40
three-coat Tyrolean rendering; including backing	m²	39.00	to	47.00
Mineralite' decorative rendering; including backing	m²	78.00	to	93.00
2F WINDOWS AND EXTERNAL DOORS				
Softwood windows and external doors				
Standard windows; painted				
single glazed	m²	260.00	to	320.00
double glazed	m²	340.00	to	410.00
Standard external softwood doors and hardwood frames; doors painted; including ironmongery				
solid flush door	nr	760.00	to	930.00
heavy duty solid flush door; single leaf	nr	870.00	to	1075.00
heavy duty solid flush door; double leaf	nr	1475.00	to	1800.00
Extra for				
emergency fire exit door	nr	330.00	to	410.00
Steel windows and external doors				
Standard windows				
single glazed; galvanized; painted	m²	270.00	to	330.00
single glazed; powder-coated	m²	275.00	to	335.00
double glazed; galvanized; painted	m²	340.00	to	420.00
double glazed; powder coated	m²	340.00	to	420.00
Steel roller shutters				
Shutters; galvanized				
manual	m²	270.00	to	330.00
electric	m²	350.00	to	430.00
manual; insulated	m²	430.00	to	530.00
electric; insulated	m²	530.00	to	640.00
electric; insulated; fire-resistant	m²	1225.00	to	1475.00

ANCILLARY BUILDING WORKS

Item	Unit	Range £		
2F WINDOWS AND EXTERNAL DOORS – cont				
Hardwood windows and external doors				
Standard windows; stained or UPVC coated				
single glazed	m²	425.00	to	520.00
double glazed	m²	460.00	to	560.00
Pvc-U windows and external doors				
Purpose-made windows				
double glazed	m²	730.00	to	890.00
Aluminium windows and external doors				
Standard windows; anodized finish				
single glazed; horizontal sliding sash	m²	360.00	to	440.00
single glazed; vertical sliding sash	m²	510.00	to	630.00
single glazed; casement; in hardwood sub-frame	m²	410.00	to	500.00
double glazed; vertical sliding sash	m²	560.00	to	680.00
double glazed; casement; in hardwood sub-frame	m²	500.00	to	610.00
Purpose-made entrance screens and doors				
double glazed	m²	1050.00	to	1275.00
Stainless steel entrance screens and doors				
Purpose-made screen; double glazed				
with manual doors	m²	1675.00	to	2025.00
with automatic doors	m²	2025.00	to	2475.00
Shop fronts, shutters and grilles				
Shutters and grilles per meter of plan length				
Grilles or shutters	m	1000.00	to	1225.00
Fire shutters; power-operated	m	1425.00	to	1750.00
2G INTERNAL WALLS, PARTITIONS AND DOORS				
Timber or metal stud partitions and doors				
Softwood stud and plasterboard partitions				
100 mm partition; skim and emulsioned both sides	m²	56.00	to	69.00
150 mm partition as party wall; skim and emulsioned both sides	m²	69.00	to	84.00
Metal stud and plasterboard partitions				
170 mm partition; one hour; taped joints; emulsioned both sides	m²	59.00	to	72.00
200 mm partition; two hour; taped joints; emulsioned both sides	m²	79.00	to	96.00
Metal stud and plasterboard partitions; emulsioned both sides; softwood doors and frames; painted				
170 mm partition	m²	83.00	to	100.00
200 mm partition; insulated	m²	105.00	to	130.00
Stud or plasterboard partitions; softwood doors and frames; painted				
partition; plastered and emulsioned both sides	m²	91.00	to	110.00
Stud or plasterboard partitions; hardwood doors and frames; painted				
partition; plastered and emulsioned both sides	m²	130.00	to	160.00

ANCILLARY BUILDING WORKS

Item	Unit	Range £		
Brick/block partitions and doors				
Autoclaved aerated/lightweight block partitions				
75 mm thick	m²	18.20	to	22.00
100 mm thick	m²	25.00	to	30.50
130 mm thick; insulating	m²	28.50	to	35.00
150 mm thick	m²	31.00	to	38.00
190 mm thick	m²	40.00	to	49.00
Extra for				
fair face both sides	m²	4.50	to	5.50
curved work	m²	3.65	to	4.45
average thickness; fair face both sides	m²	33.50	to	41.00
average thickness; fair face and emulsioned both sides	m²	39.50	to	48.00
average thickness; plastered and emulsioned both sides	m²	59.00	to	72.00
Dense aggregate block partitions				
average thickness; fair face both sides	m²	39.00	to	48.00
average thickness; fair face and emulsioned both sides	m²	44.50	to	55.00
average thickness; plastered and emulsioned both sides	m²	64.00	to	78.00
Common brick partitions;				
half brick thick	m²	33.50	to	41.00
half brick thick; fair face both sides	m²	39.00	to	47.50
half brick thick; fair face and emulsioned both sides	m²	44.50	to	55.00
half brick thick; plastered and emulsioned both sides	m²	67.00	to	81.00
one brick thick	m²	64.00	to	78.00
one brick thick; fair face both sides	m²	69.00	to	85.00
one brick thick; fair face and emulsioned both sides	m²	76.00	to	92.00
one brick thick; plastered and emulsioned both sides	m²	97.00	to	120.00
Block partitions; softwood doors and frames; painted				
partition	m²	59.00	to	72.00
partition; fair face both sides	m²	61.00	to	75.00
partition; fair face and emulsioned both sides	m²	68.00	to	84.00
partition; plastered and emulsioned both sides	m²	89.00	to	110.00
Block partitions; hardwood doors and frames				
partition	m²	85.00	to	100.00
partition; platered and emulsioned both sides	m²	115.00	to	140.00
Reinforced concrete walls				
Walls				
150 mm thick	m²	115.00	to	140.00
150 mm thick; plastered and emulsioned both sides	m²	150.00	to	190.00
Special partitioning and doors				
Demountable partitioning; hardwood doors				
medium quality; vinyl-faced	m²	140.00	to	170.00
high quality; vinyl-faced	m²	185.00	to	225.00
Aluminium internal patent glazing				
single glazed	m²	135.00	to	160.00
double glazed	m²	215.00	to	260.00
Demountable steel partitioning and doors				
medium quality	m²	230.00	to	285.00
high quality	m²	290.00	to	360.00

ANCILLARY BUILDING WORKS

Item	Unit	Range £		
2G INTERNAL WALLS, PARTITIONS AND DOORS – cont				
Special partitioning and doors – cont				
Demountable fire partitions				
enamelled steel; half hour	m²	890.00	to	1075.00
stainless steel; half hour	m²	1200.00	to	1475.00
Soundproof partitions; hardwood doors				
luxury veneered	m²	420.00	to	510.00
WC/Changing cubicles				
W/C cubicles cost per cubicle	nr	540.00	to	650.00
INTERNAL DOORS				
Comparative doors/door linings/frames				
Standard softwood doors and frames; including ironmongery and painting				
flush; hollow core	nr	280.00	to	340.00
flush; hollow core; hardwood faced	nr	280.00	to	340.00
flush; solid core				
single leaf	nr	330.00	to	405.00
double leaf	nr	495.00	to	610.00
flush; solid core; hardwood faced	nr	370.00	to	450.00
four panel door	nr	470.00	to	580.00
Purpose-made softwood doors and hardwood frames; including ironmongery; painting and polishing				
flush; solid core; heavy duty				
single leaf	nr	780.00	to	950.00
duoble leaf	nr	1100.00	to	1350.00
flush; solid core; heavy duty; plastic laminate faced				
single leaf	nr	920.00	to	1125.00
double leaf	nr	1250.00	to	1550.00
Purpose-made softwood fire doors and hardwood frames; including ironmongery; painting and polishing				
flush; one hour fire-resisting				
single leaf	nr	1000.00	to	1225.00
double leaf	nr	1250.00	to	1550.00
flush; one hour fire-resisting; plastic laminated faced				
single leaf	nr	1225.00	to	1475.00
double leaf	nr	1575.00	to	1900.00
purpose-made softwood doors and pressed steel frames;				
flush; half hour fire check; plastic laminate faced	nr	1175.00	to	1475.00
Perimeter treatments				
Precast concrete lintels; in block walls				
75 mm wide	m	15.80	to	19.30
100 mm wide	m	19.70	to	24.00
Precast concrete lintels; in brick walls				
half brick thick	m	19.70	to	24.00
one brick thick	m	31.00	to	37.50
Purpose-made softwood architraves; painted; including grounds				
25 × 50 mm; to both sides of openings				
726 × 2040 mm opening	nr	150.00	to	180.00
826 × 2040 mm opening	nr	100.00	to	125.00

ANCILLARY BUILDING WORKS

Item	Unit	Range £		
3A WALL FINISHES				
WALL FINISHES				
Dry plasterboard lining, taped joints; for direct decoration				
9.5 mm Gyproc Wallboard	m²	11.10	to	13.60
Extra for insulating grade	m²	0.75	to	0.90
Extra for insulating grade;plastic faced	m²	2.30	to	2.80
12.5 mm Gyproc Wallboard (half-hour fire-resisting)	m²	12.80	to	15.60
Extra for insulating grade	m²	0.75	to	0.95
Extra for insulating grade; plastic faced	m²	2.00	to	2.45
two layers of 12.5 mm Gyproc Wallboard (one houre fire-resisting)	m²	21.00	to	26.00
9 mm Supalux (half-hour fire-resisting)	m²	21.00	to	26.00
Dry plasterboard lining, taped joints; for direct decoration; fixed to wall on dabs				
9.5 mm Gyproc Wallboard	m²	12.80	to	15.60
Dry plasterboard lining, taped joints; for direct decoration; including metal tees				
9.5 mm Gyproc Wallboard	m²	24.50	to	30.00
12.5 mm Gyproc Wallboard	m²	26.00	to	31.50
Dry lining/sheet panelling; including battens; plugged to wall				
6.4 mm hardboard	m²	12.40	to	15.20
9.5 mm Gyproc Wallboard	m²	19.60	to	24.00
6 mm birch faced plywood	m²	15.10	to	18.50
6 mm WAM plywood	m²	24.00	to	30.00
15 mm chipboard	m²	180.00	to	220.00
15 mm melamine faced chipboard	m²	32.00	to	39.00
13.2 mm 'Formica' faced chipboard	m²	48.50	to	59.00
In situ wall finishes				
Extra over common brickwork for fair face and pointing both sides	m²	4.40	to	5.40
Comparative finishes				
one mist and two coats emuslion	m²	2.80	to	3.40
mutli-coloured gloss paint	m²	5.95	to	7.25
two coats of lightweight plaster	m²	11.40	to	14.00
9.5 mm Gyproc Wallboard and skim coat	m²	14.60	to	17.90
12.5 mm Gyproc Wallboard and skim coat	m²	16.50	to	20.00
two coates of 'Thistle' plaster	m²	15.30	to	18.70
plaster and emulsion	m²	16.30	to	20.00
Extra for gloss paint in lieu of emulsion	m²	2.55	to	3.15
two coat render and emulsion	m²	26.00	to	32.00
Ceramic wall tiles; including backing				
economical	m²	35.00	to	42.50
medium quality	m²	63.00	to	78.00
high quality; to toilet blocks; kitchens and first aid rooms	m²	85.00	to	100.00
Decorations				
Emulsion				
two coats	m²	2.40	to	2.95
one mist and two coats	m²	3.15	to	3.85
Artex plastic compund				
one coat; textured	m²	4.10	to	5.00
Wall paper	m²	6.50	to	7.90
Gloss				
primer and two coats	m²	4.80	to	5.80
primer and three coats	m²	4.10	to	5.00

ANCILLARY BUILDING WORKS

Item	Unit	Range £		
3A WALL FINISHES – cont				
Decorations – cont				
Comparative steel/metalwork finishes				
primer only	m²	1.20	to	1.50
grit blast and one coat zinc chromate primer	m²	2.20	to	2.70
touch up primer and one coat of two pack epoxy zinc phosphate	m²	2.70	to	3.35
primer gloss three coats	m²	6.10	to	7.40
sprayed mineral fibre; one hour	m²	13.20	to	16.20
sprayed mineral fibre; two hour	m²	14.50	to	17.80
sprayed vermiculite cement; one hour	m²	14.60	to	17.90
sprayed vermiculite cement; two hour	m²	17.40	to	21.00
intumescent coating with decorative top seal; half hour	m²	19.80	to	24.00
intumescent coating with decorative top seal; one hour	m²	33.00	to	40.50
Comparative woodword finishes				
primer only	m²	1.50	to	1.90
two coats gloss; touch up primer	m²	3.05	to	3.75
three coats gloss; touch up primer	m²	4.60	to	5.60
primer and two coat gloss	m²	5.10	to	6.20
primer and three coat gloss	m²	6.10	to	7.50
polyurethene lacquer two coats	m²	6.40	to	7.80
polyurethene lacquer three coats	m²	5.10	to	6.20
flame-retardent pain three coats	m²	7.30	to	8.90
3B FLOOR FINISHES				
FLOOR FINISHES				
Chipboard flooring; t & g joints				
18 mm thick	m²	10.30	to	12.70
22 mm thick	m²	12.50	to	15.30
Wrought softwood flooring				
25 mm thick; butt joints; cleaned off and polished	m²	25.00	to	31.00
25 mm thick; t & g joints; cleaned off and polished	m²	29.00	to	35.00
Extra over concrete floor for				
power floating	m²	7.00	to	8.60
power floating; surface hardener	m²	11.30	to	13.80
Latex cement screeds				
3 mm thick; one coat	m²	5.60	to	6.80
5 mm thick; two coat	m²	7.40	to	9.05
Rubber latex non-slip solution and epox sealant	m²	14.80	to	18.00
Cement and sand (1:3) screeds				
25 mm thick	m²	10.90	to	13.30
50 mm thick	m²	13.50	to	16.50
75 mm thick	m²	18.70	to	23.00
Cement and sand (1:3) paving				
32 mm thick	m²	10.70	to	13.00
32 mm thick; surface hardener	m²	14.00	to	17.20
Screed only (for subsequent finish)	m²	17.10	to	21.00
Screed only (for subsequent finish); allowance for skirtings	m²	20.00	to	25.00
Mastic asphalt paving				
20 mm thick; BS 6925; black	m²	23.00	to	28.50
20 mm thick; BS 6925; red	m²	27.50	to	33.50

ANCILLARY BUILDING WORKS

Item	Unit	Range £		
Granolithic				
20 mm thick	m²	15.30	to	18.70
25 mm thick	m²	18.20	to	22.00
25 mm thick; including screed	m²	27.50	to	34.00
38 mm thick; including screed	m²	40.00	to	49.00
Synthanite; on and including building paper				
25 mm thick	m²	26.00	to	32.00
50 mm thick	m²	37.00	to	45.00
75 mm thick	m²	4.50	to	5.50
Acrylic polymer floor finish				
10 mm thick	m²	27.00	to	33.00
Epoxy floor finish				
1.5 – 2 mm thick	m²	29.00	to	35.00
5 – 6 mm thick	m²	53.00	to	65.00
Polyester resin floor finish				
5 – 9 mm thick	m²	59.00	to	73.00
Quarry tile flooring				
150 × 150 × 12.5 mm thick; red	m²	30.50	to	37.50
150 × 150 × 12.5 mm thick; brown	m²	39.50	to	48.00
200 × 200 × 19 mm thick; brown	m²	48.00	to	58.00
average tiling	m²	40.00	to	49.00
tiling; including screed	m²	54.00	to	66.00
tiling; including screed and allowance for skirtings	m²	69.00	to	84.00
Glazed ceramic tile flooring				
150 × 150 × 12 mm thick; black	m²	40.50	to	49.50
150 × 150 × 12 mm thick; antislip	m²	45.00	to	55.00
fully vitrified	m²	56.00	to	69.00
fully vitrified; including screed	m²	70.00	to	86.00
fully vitrified; including screed and allowance for skirtings	m²	82.00	to	100.00
Composition block flooring				
174 × 57 mm blocks	m²	70.00	to	85.00
Vinyl tile flooring				
2 mm thick; semi-flexible tiles	m²	11.30	to	13.80
2 mm thick; fully flexible tiles	m²	10.60	to	13.00
2.5 mm thick; semi-flexible tiles	m²	13.50	to	16.50
tiling; including screed	m²	28.00	to	34.00
tiling; including screed and allowance for skirtings	m²	49.50	to	61.00
tiling; antistatic	m²	47.00	to	57.00
tiling; antistatic; including screed	m²	59.00	to	72.00
Vinyl sheet flooring; heavy duty				
2 mm thick	m²	17.60	to	21.50
2.5 mm thick	m²	20.00	to	25.00
3 mm thick; needle felt backed	m²	13.50	to	16.50
3 mm thick; foam backed	m²	18.90	to	23.00
Sheeting; including screed and allowances for skirtings				
'Altro' safety flooring	m²	34.00	to	42.00
2 mm thick; Marine T20	m²	27.00	to	33.00
2.5 mm thick; Classic D25	m²	32.50	to	39.50
3.5 mm thick; stonghold	m²	40.50	to	49.50
flooring	m²	34.00	to	42.00
flooring; including screed	m²	48.00	to	58.00

ANCILLARY BUILDING WORKS

Item	Unit	Range £		
3B FLOOR FINISHES – cont				
FLOOR FINISHES – cont				
Rubber tile flooring; smooth; ribbed or studded tiles				
2.5 mm thick	m²	33.00	to	40.00
5 mm thick	m²	38.00	to	46.00
5 mm thick; including screed	m²	54.00	to	66.00
Carpet tiles/Underlay				
Underlay	m²	6.75	to	8.25
nylon needlepunch (stick down)	m²	16.20	to	19.80
80% animal hair; 20% wool cord	m²	25.00	to	31.00
100% wool	m²	37.00	to	45.00
80% wool; 20% nylon antistatic	m²	42.00	to	52.00
economical; including screed and allowance for skirtings	m²	39.00	to	47.00
good quality	m²	41.00	to	50.00
good quality; including screed	m²	53.00	to	65.00
good quality; including screed and allowance for skirtings	m²	58.00	to	70.00
Access floors; excluding finish				
600 × 600 mm chipboard panels; faced both sides with galvanized steel sheet; on adjustable steel/aluminium pedestals; cavity height 100–300 mm high				
light grade duty	m²	54.00	to	66.00
medium grade duty	m²	60.00	to	74.00
heavy grade duty	m²	79.00	to	97.00
extra heavy grade duty	m²	85.00	to	100.00
600 × 600 mm chipboard panels; faced both sides with galvanized steel sheet; on adjustable steel/aluminium pedestals; cavity height 300–600 mm high				
medium grade duty	m²	68.00	to	84.00
heavy grade duty	m²	81.00	to	99.00
extra heavy grade duty	m²	85.00	to	100.00
Common floor coverings bonded to access floor panels				
heavy-duty fully flexible vinyl to BS 3261, type A	m²	16.80	to	20.50
anti-static grade sheet PVC to BS 3261	m²	21.00	to	25.00
Comparative skirtings				
25 × 75 mm softwood skirting; painted; including grounds	m	11.60	to	14.20
12.5 × 150 mm Quarry tile skirting; including backing	m	16.90	to	20.50
13 × 75 mm granolithic skirting; including backing	m	24.00	to	29.00
Entrance matting in aluminium-framed				
matwell	m²	435.00	to	530.00

ANCILLARY BUILDING WORKS

Item	Unit	Range £		
3C CEILING FINISHES				
CEILING FINISHES				
Plastered ceilings				
Plaster to soffits				
lightweight plaster	m²	12.20	to	15.00
plaster and emulsion	m²	17.10	to	21.00
Extra for gloss paint in lieu of emulsion	m²	2.50	to	3.00
Plasterboard to soffits				
9.5 mm Gyproc lath and skim coat	m²	15.80	to	19.40
9.5 mm Gyproc insulating lath and skim coat	m²	16.60	to	20.00
plasterboard, skim and emulsion	m²	18.90	to	23.00
Extra for gloss paint in lieu of emulsion	m²	2.50	to	3.00
plasterboard and artex	m²	12.60	to	15.40
plasterboard, artex and emulsion	m²	16.20	to	19.80
plaster and emulsion; including metal lathing	m²	27.00	to	33.00
Other board finishes; with fire-resisting properties; excluding decoration				
12.5 Gyproc Fireline; half hour	m²	13.50	to	16.50
6 mm Supalux; half hour	m²	16.60	to	20.00
two layers of 12.5 mm Gyproc Wallboard; half hour	m²	18.40	to	22.50
two layers of 12.5 mm Gyproc Fireline; one hour	m²	22.00	to	27.00
9 mm Supalux; one hour; on fillets	m²	25.00	to	30.00
Specialist plasters; to soffits				
sprayed accoustic plaster; self finished	m²	36.00	to	44.00
rendering; 'Tyrolean' finish	m²	37.00	to	45.50
Other ceiling finishes				
50 mm wood wool slabs as permanent lining	m²	15.80	to	19.40
12 mm Pine tongued and grooved boardings	m²	21.00	to	25.00
16 mm Softwood tongued and grooved boardings	m²	25.00	to	30.50
Suspended and intergrated ceilings				
Suspended ceilings				
economical; exposed grid	m²	28.00	to	34.00
jointless; plasterboard	m²	34.00	to	41.00
semi- concealed grid	m²	36.50	to	44.50
medium quality; 'Minatone'; concealed grid	m²	41.50	to	51.00
high quality; 'Travertone'; concealed grid	m²	46.00	to	57.00
metal linear strip; 'Dampa'/'Luxalon'	m²	49.50	to	61.00
metal tray	m²	53.00	to	65.00
egg-crate	m²	70.00	to	110.00
open grid; 'Formalux'/'Dimension'	m²	100.00	to	125.00
Intergrated ceilings				
coffered; with steel services	m²	120.00	to	190.00

ANCILLARY BUILDING WORKS

Item	Unit	Range £		
4A FITTINGS AND FURNISHINGS				
FITTINGS AND FURNISHINGS				
Comparative wrought softwood shelving				
25 × 225 mm; including black japanned brackets	m	13.90	to	17.10
25 mm thick slatted shelving; including bearers	m²	55.00	to	67.00
25 mm thick cross-tongued shelving; including bearers	m²	72.00	to	88.00
Allowances per gross floor area for:				
reception desk, shelves and cupboards for general areas				
economical	m²	6.00	to	10.00
medium quality	m²	11.30	to	13.80
high quality	m²	18.00	to	22.00
Extra for				
high quality finishes to reception areas	m²	9.00	to	11.00
full kitchen equipment	m²	12.20	to	14.80
furniture and fittings to general office areas				
economical	m²	10.10	to	12.40
medium quality	m²	15.30	to	18.70
high quality	m²	26.00	to	32.00
General fittings and equipment				
internal planting	m²	27.00	to	33.00
signs, notice-boards, shelving, fixed seating, curtains and blinds	m²	13.90	to	17.10
5A SANITARY AND DISPOSAL INSTALLATIONS				
SANITARY AND DISPOSAL INSTILLATIONS				
Note: Material prices vary considerably, the following composite rates are based on average prices for mid priced fittings.				
Lavatory basins; vitreous china; chromium plated taps, waste, chain and plug, cantilever brackets	nr	215.00	to	290.00
Low level WC's; vitreous china pan and cistern; black plastic seat; low pressure ball valve; plastic flush pipe; fixing brackets				
on ground floor	nr	190.00	to	260.00
one of a range; on upper floors	nr	375.00	to	510.00
Bowl type wall urinal; white glazed vitreous china flushing cistern; chromium plated flush pipes and spreaders; fixing brackets				
white	nr	185.00	to	250.00
Shower tray; glazed fireclay; chromium plated waste, chain and plug, riser pipe, rose and mixing valve				
white	nr	520.00	to	700.00
Sink; glazed fireclay; chromium plated waste, chain and plug; fixing brackets				
white	nr	425.00	to	580.00
Sink; stainless steel; chromium plated waste, chain and self coloured plug				
single drainer	nr	230.00	to	310.00
double drainer	nr	260.00	to	355.00
Soil waste stacks; 3.15 m storey height; branch and connection to drain				
110 mm PVC	nr	330.00	to	440.00
Extra for additional floors	nr	170.00	to	230.00
100 mm cast iron; decorated	nr	650.00	to	870.00
Extra for additional floors	nr	330.00	to	450.00

ANCILLARY BUILDING WORKS

Item	Unit	Range £		
Industrial buildings Allowance per m² of floor area				
minimum provision	m²	12.80	to	17.30
high provision	m²	17.90	to	24.00
Hot and cold water installations – Allowance per m² of floor area				
Complete installations	m²	21.00	to	29.00
5F HEATING, AIR-CON AND VENTILATION				
SERVICE INSTALLATIONS				
[HEATING, AIR-CONDITIONING AND VENTILATING INSTALLATIONS]				
Gas or oil-fired radiator heating				
gas-fired hot water service and central heating for				
three radiators	nr	2800.00	to	3800.00
extra for additional radiator	nr	320.00	to	440.00
LPHW radiator system – Allowance per m² of floor area	m²	73.00	to	99.00
Ventilation systems				
Local ventilation to				
WC's	nr	280.00	to	380.00
toilet areas – Allowance per m² of floor area	m²	7.45	to	10.10
Air extract system – Allowance per m² of floor area	m²	46.00	to	62.00
Air supply and extract system – Allowance per m² of floor area	m²	71.00	to	97.00
Heating and ventilation – Allowance per m² of floor area	m²	71.00	to	96.00
Warm air heating and ventilation – Allowance per m² of floor area	m²	140.00	to	190.00
Comfort cooling systems				
Stand-alone air conditioning unit systems				
air supply and extract – Allowance per m² of floor area	m²	265.00	to	360.00
Air conditioning systems				
fully air conditioning with dust and humidity control – Allowance per m² of floor area	m²	230.00	to	315.00
5H ELECTRICAL INSTALLATIONS				
ELECTRICAL INSTALLATIONS				
Based upon gross internal area serviced				
Mains and sub-mains switchgear and distribution				
mains intake only	m²	2.70	to	3.60
mains switchgear only	m²	6.15	to	8.35
Mains and sub-mains distribution				
to floors only	m²	8.05	to	10.90
to floors; including small power and supplies to equipment	m²	17.00	to	23.00
to floors; including lighting and power and supplies to equipment	m²	16.60	to	22.50
Lighting installation				
general lighting; including luminaries	m²	38.00	to	52.00
emergency lighting	m²	12.80	to	17.30
standby generators only	m²	7.20	to	9.80
Lighting and power installations to buildings				
plant area	m²	64.00	to	86.00
plant are; high provision	m²	82.00	to	110.00
office area	m²	115.00	to	155.00
office area; high provision	m²	155.00	to	210.00

ANCILLARY BUILDING WORKS

Item	Unit	Range £		
5H ELECTRICAL INSTALLATIONS – cont				
Comparative fittings/rates per each point				
Fittings; excluding lamps or light fittings				
lighting point; PVC cables	nr	58.00	to	78.00
lighting point; PVC cables in screwed conduits	nr	150.00	to	210.00
lighting point; MICC cables	nr	135.00	to	180.00
Switch socket outlet; PVC cables				
single	nr	54.00	to	72.00
double	nr	76.00	to	100.00
Switch socket outlet; PVC cables in screwed conduit				
single	nr	94.00	to	125.00
double	nr	100.00	to	135.00
Based upon gross internal area serviced				
Fittings; excluding lamps or light fittings				
Switch socket outlet; MICC cables				
single	nr	95.00	to	130.00
double	nr	105.00	to	140.00
Immersion heater point (excluding heater)	nr	92.00	to	125.00
Cooker point; including control unit	nr	160.00	to	215.00
5I GAS INSTALLATIONS				
GAS INSTALLATION				
Connection charge	nr	720.00	to	970.00
5J LIFT AND CONVEYOR INSTALLATIONS				
LIFT AND CONVEYOR INSTALLATIONS				
Goods lifts				
hoist	nr	11000.00	to	34000.00
Electric heavy duty goods lifts				
500 kg 2 – 3 levels	nr	38000.00	to	52000.00
1000 kg; 2 – 3 levels	nr	48500.00	to	66000.00
1500 kg; 3 levels	nr	110000.00	to	140000.00
2000 kg; 2 levels	nr	110000.00	to	140000.00
2000 kg; 3 levels	nr	130000.00	to	170000.00
3000 kg; 2 levels	nr	115000.00	to	160000.00
3000 kg; 3 levels	nr	140000.00	to	190000.00
Oil hydraulic heavy duty goods lifts				
500 kg; 3 levels	nr	110000.00	to	145000.00
1000 kg; 3 levels	nr	110000.00	to	150000.00
2000 kg; 3 levels	nr	130000.00	to	170000.00
Dock levellers				
dock levellers	nr	13000.00	to	30000.00
dock levellers and canopy	nr	17000.00	to	36000.00

ANCILLARY BUILDING WORKS

Item	Unit	Range £		

5K PROTECTIVE, COMMUNICATION

PROTECTIVE, COMMUNICATION AND SPECIAL INSTALLATIONS
Based upon gross internal area serviced
Fire fighting/protective installations

smoke detectors, alarms and controls	m^2	5.80	to	9.05
hosereels, dry risers and extinguishers	m^2	8.80	to	13.80

Sprinkler installations

single level sprinkler systems, alarms and smoke detectors; low hazard	m^2	16.10	to	22.00
single level sprinkler systems, alarms and smoke detectors; ordinary hazard	m^2	23.00	to	28.50
double level sprinkler systems, alarms and smoke detectors; high hazard	m^2	34.00	to	42.00
Lightning protection	m^2	1.00	to	3.00

Security/communication installations

security alarm system	m^2	2.50	to	3.10
telephone system	m^2	1.70	to	2.50
public address, television ariel and clocks	m^2	3.75	to	5.60
closed-circuit television and public address system	m^2	5.10	to	6.20

5N GENERAL BWIC WITH SERVICES

BUILDERS' WORK IN CONNECTION WITH SERVICES
General builders work to

mains supplies; lighting and power to landlords areas	m^2	2.25	to	6.75
central heating and electrical installation	m^2	5.50	to	16.50
central heating; electrical and lift installations	m^2	7.90	to	18.50
air conditioning	m^2	20.00	to	27.00
air conditioning and electrical installation	m^2	21.00	to	28.00
air conditioning; electrical and lift installations	m^2	23.00	to	31.00

General builders work; including allowance for plant rooms; to

central heating and electrical installations	m^2	31.50	to	38.50
central heating, electrical and lift installations	m^2	39.50	to	48.50
air conditioning	m^2	54.00	to	66.00
air conditioning and electrical installation	m^2	67.00	to	81.00
air conditioning; electrical and lift installations	m^2	82.00	to	100.00

6A SITE WORK

SITE WORK

LANDSCAPING AND EXTERNAL WORKS

Seeded and planted areas
Plant supply, planting, maintenance and 12 months guarantee

seeded areas	m^2	5.20	to	7.10
turfed areas	m^2	6.80	to	9.20

Planted areas (per m^2 of suface area)

herbaceous plants	m^2	4.60	to	6.20
climbing plants	m^2	7.20	to	9.80
general planting	m^2	13.60	to	18.40
woodland	m^2	36.00	to	48.00
shrubbed planting	m^2	44.00	to	60.00
dense planting	m^2	44.00	to	60.00
shrubbed area including allowances for small trees	m^2	60.00	to	81.00

ANCILLARY BUILDING WORKS

Item	Unit	Range £		
6A SITE WORK – cont				
Seeded and planted areas – cont				
Trees				
advanced nursery stock trees (12–20 cm girth)	tree	160.00	to	220.00
semi-mature trees; 5 – 8 m high – conferous	tree	820.00	to	1100.00
semi-mature trees; 5 – 8 m high – deciduous	tree	1700.00	to	2300.00
Paved areas				
Precast concrete paving slabs				
50 mm thick	m²	15.70	to	21.00
50 mm thick 'Texitone' slabs	m²	17.50	to	23.50
slabs on subbase; including excavation	m²	28.50	to	38.50
Precast concrete block paviors				
65 mm 'Keyblock' grey paving	m²	22.00	to	29.50
65 mm 'Mount Sorrel' grey paving	m²	21.50	to	29.00
65 mm 'Intersett' paving	m²	24.00	to	32.00
60 mm 'Pedesta' paving	m²	20.00	to	27.00
paviors on subbase; including excavations	m²	34.00	to	46.00
Brick paviors; 229 × 114 × 38 mm paving bricks				
laid flat	m²	36.50	to	49.50
laid to herring-bone pattern	m²	56.00	to	75.00
paviors on subbase; including excavation	m²	66.00	to	89.00
Granite setts				
200 × 100 × 100 mm setts	m²	85.00	to	115.00
setts on subbase; including excavation	m²	100.00	to	140.00
York stone slab paving				
paving on subbase; including excavation	m²	120.00	to	160.00
Cobblestone paving				
50 – 75 mm	m²	68.00	to	92.00
cobblestones on subbase; including excavation	m²	82.00	to	110.00
Car parking alternatives				
Surface level parking; including lighting and drainage				
tarmacadam on subbase	car	1350.00	to	1800.00
concrete interlocking blocks	car	1400.00	to	1875.00
Grasscrete precast concrete units filled with top soil and grass seed	car	810.00	to	1075.00
At ground level with deck or building over	car	6200.00	to	8300.00
Garages etc				
single car park	nr	960.00	to	1275.00
single; traditional construction; in a block	nr	2850.00	to	3850.00
single; traditional construction; pitched roof	nr	4600.00	to	6300.00
double; traditional construction; pitched roof	nr	8700.00	to	12000.00
External furniture				
Guard rails and parking bollards etc.				
open metal post and rail fencing 1 m high	m	140.00	to	190.00
galvanized steel post and rail fencing 2 m high	m	170.00	to	230.00
steel guard rails and vehicle barriers	m	59.00	to	79.00
Parking bollards				
precast concrete	nr	115.00	to	155.00
steel	nr	200.00	to	270.00
cast iron	nr	240.00	to	320.00

ANCILLARY BUILDING WORKS

Item	Unit	Range £		
Vehicle control barrier; manual pole	nr	960.00	to	1275.00
Glavanized steel cycle stand	nr	44.50	to	60.00
Galvanized steel flag staff	nr	1175.00	to	1575.00
Benches – hardwood and precast concrete	nr	230.00	to	310.00
Litter bins				
precast concrete	nr	210.00	to	290.00
hardwood slatted	nr	94.00	to	125.00
cast iron	nr	360.00	to	480.00
large aluminium	nr	630.00	to	850.00
Bus stops	nr	380.00	to	520.00
Bus stops incl basic shelter	nr	880.00	to	1175.00
Pillar box	nr	320.00	to	430.00
Telephone box	nr	3450.00	to	4700.00
Fencing and screen walls				
Chain link fencing; plastic coated				
1.2 m high	m	17.90	to	24.00
1.8 m high	m	25.00	to	33.50
Timber fencing				
1.2 m high chestnut pale fencing	m	19.80	to	27.00
1.8 m high close-boarded fencing	m	53.00	to	71.00
Screen walls; one brick thick; including foundations etc				
1.8 m high facing brick screen wall	m	260.00	to	355.00
1.8 m high coloured masonry block boundry wall	m	300.00	to	400.00
6B DRAINAGE				
DRAINAGE				
Overall £ /m^2 allowances based on gross areas				
sire drainage (per m^2 of paved areas)	m^2	7.30	to	22.00
building drainage (per m^2 of gross floor area)	m^2	8.10	to	18.90
drainage work beyond the boundary of the site and final connection	nr	1800.00	to	12500.00
6C EXTERNAL SERVICES				
EXTERNAL SERVICES				
Service runs laid in trenches including excavation				
Water main				
75 mm uPVC main in 225 mm ductile iron pipe as duct	m	53.00	to	71.00
Electric main				
600/1000 volt cables. Two core 25 mm cable including				
100 mm clayware duct	m	34.50	to	46.50
Gas main				
150 mm ductile or cast iron gas pipe	m	53.00	to	71.00
Telephone				
British Telecom installation in 100 mm uPVC duct	m	21.00	to	29.00
External lighting (per m^2 of lighted area)	m^2	3.30	to	4.50

Best Tall Buildings 2011

A. Wood

Best Tall Buildings 2011 contains overviews of the nominees and winning projects from the Council on Tall Buildings and Urban Habitat's annual awards. The book provides vital statistics, images, plans and in depth accounts of the architectural design and sustainability features of the year's best tall buildings. Additionally included is the official list of the World's Tallest Buildings with the judging criteria from the foremost authority on tall buildings. Full colour illustrations and photographs throughout.

December 2011: 200pp
Hb: 978-0-415-68326-5: **£29.99**

To Order: Tel: +44 (0) 1235 400524 **Fax:** +44 (0) 1235 400525
or Post: Taylor and Francis Customer Services,
Bookpoint Ltd, Unit T1, 200 Milton Park, Abingdon, Oxon, OX14 4TA UK
Email: book.orders@tandf.co.uk

For a complete listing of all our titles visit:
www.tandf.co.uk

Preliminaries and General Items

This part deals with that portion of Civil Engineering costs not, or only indirectly, related to the actual quantity of work being carried out. It comprises a definition of Method Related Charges, a checklist of items to be accounted for on a typical Civil Engineering contract and a worked example illustrating how the various items on the checklist can be dealt with.

GENERAL REQUIREMENTS AND METHOD RELATED CHARGES

Although the more familiar terminology of Preliminaries and General Items is used in this book the principle of Method Related Charges – separating non quantity related charges from quantity related charges – is adopted. Generally the former are dealt with in this part, while the latter are dealt with in Part 4: Unit Costs. In this part non quantity related charges are further subdivided into those that are time related and those that are non time related.

The concept of METHOD RELATED CHARGES can be summarized as follows:

In commissioning Civil Engineering work the Employer buys the materials left behind, but only hires from the Contractor the men and machines which manipulate them, and the management skills to manipulate them effectively. It is logical to assess their values in the same terms as the origin of their costs. It is illogical not to do so if the Employer is to retain the right at any time to vary what is left behind and if the financial uncertainties affecting Employer and Contractors are to be minimized.

Tenderers have the option to define a group of bill items and insert charges against them to cover those expected costs which are not proportional to the quantities of Permanent Works. To distinguish these items they are called Method Related Charges. They are themselves divided into charges for recurrent or time related cost elements, such as maintaining site facilities or operating major plant, and charges for elements which are neither recurrent nor directly related to quantities, such as setting up, bringing plant to site and Temporary Works.

Another hope expressed with the introduction of Method Related Charges was that they should accurately reflect the work described in the item and that they should not, as had become the practice with some of the vague general items frequently included in Civil Engineering Bills, be used as a home for lump sum tender adjustments quite unrelated to the item. Where cost information is given in the worked example presented at the end of this part of the book, therefore, it must be stressed that only direct and relevant costs are quoted.

Where no detailed information is available, it is suggested that when preparing a preliminary estimate an addition of between 12.5% and 25% of net contract value is made to cover Contractor's Site Oncosts, both time and non time related. The higher percentages' should apply to specialist works such as railways, marine and tunnelling.

CHECKLIST OF ITEMS

The following checklist is representative but not exhaustive. It lists and describes the major preliminary and general items which are included, implicitly or explicitly, in a typical Civil Engineering contract and, where appropriate, gives an indication of how they might be costed. Generally contract documents give detailed requirements for the facilities and equipment to be provided for the Employer and for the Engineer's representative and Bills of Quantities produced in conformity with CESMM3 Class A provide items against which these may be priced; no such items are provided for Contractor's site oncosts or, usually, for temporary works and general purpose plant. For completeness a checklist of both types of item is given here under the following main headings:

Approximate Estimating Rates

PRELIMINARIES AND GENERAL ITEMS

Contractor's site oncosts	time related
	non time related
Employer's and consultants' site requirements	time related
	non time related
Site requirements	time related
	non time related
Other services, charges and fees	
Temporary works (other than those included in unit costs)	
General purpose plant (other than that included in unit costs)	

CONTRACTOR'S SITE ONCOSTS – TIME RELATED

Site staff salaries

All non-productive supervisory staff on site including: agent, sub-agent, engineers, general foremen, non-productive section foremen, clerks, typists, timekeepers, checkers, quantity surveyors, cost engineers, security guards, etc. Cost includes salaries, subsistence allowance, National Health Insurance and Pension Scheme contributions, etc. Average cost approximately 3% to 5% of contract value.

Site staff expenses

Travelling, hotel and other incidental expenses incurred by staff. Average cost approximately 1% of staff salaries.

Attendant labour

Chainmen, storemen, drivers for staff vehicles, watchmen, cleaners, etc.

General yard labour

Labour employed on loading and offloading stores, general site cleaning, removal of rubbish, etc.

Plant maintenance

Fitters, electricians, and assistants engaged on general plant maintenance on site. This excludes drivers and banksmen who are provided for specifically in the Unit Costs Sections.

Site transport for staff and general use

Vehicles provided for use of staff and others including running costs, licence and insurance and maintenance if not carried out by site fitters.

Transport for labour to and from site

Buses or coaches provided for transporting employees to and from site including cost of drivers and running costs, etc., or charges by coach hire company for providing this service.

Contractor's office rental

This includes:

- rental charges for provision of offices for Contractor's staff
- main office
- section offices
- timekeepers, checkers and security
- laboratory, etc.
- an allowance of approximately 8 m² per staff member should be made

PRELIMINARIES AND GENERAL ITEMS

Contractor's site huts

Rental charges for stores and other general-use site huts

Canteen and welfare huts

Rental charges for canteen and huts for other welfare facilities required under Rule XVI of the Working Rule Agreement

Rates

Chargeable by local authorities on any site, temporary buildings or quarry

General office expenditure

Provision of postage, stationery and other consumables for general office use

Telecommunications

Rental charges and charges for calls

Furniture and equipment

Rental charges for office furniture and equipment including photocopiers, calculators, personal computers and laser printers, etc.

Surveying equipment

Rental

Canteen and welfare equipment

Rental charges for canteen and other welfare equipment

Radio communication equipment

Rental

Testing and laboratory equipment

Rental

Lighting and heating for offices and huts

Electricity, gas or other charges in connection with lighting and heating site offices and hutting

Site lighting electrical consumption

Electricity charges in connection with general external site lighting

Water consumption

Water rates and charges

Canteen operation

Labour, consumables and subsidy costs in operating site canteens

Carpenter's shop equipment

Rental of building and mechanical equipment

PRELIMINARIES AND GENERAL ITEMS

Fitter's shop equipment

Rental of building and equipment

Small tools

Provision of small tools and equipment for general use on site. Average cost 5% of total labour cost

Personal protective equipment

Provision of protective clothing for labour including boots, safety helmets, etc. Average cost 2% of total labour cost

Traffic control

Hire and operation of traffic lights

Road lighting

Hire and operation of road lighting and traffic warning lights

Cleaning vehicles

Equipment and labour cleaning vehicles before entering public roads

Cleaning roads and footpaths

Equipment and labour cleaning public roads and footpaths

Progress photographs for Contractor's records

Cost of taking and processing photographs to demonstrate progress

CONTRACTOR'S SITE ONCOSTS – NON TIME RELATED

Rent of additional land

For Contractor's use for erection of huts, storage of soil and other materials, etc.

Staff removal expenses

Costs of staff moving house to new location. Generally only applies on longer-term contracts.

Erection of offices including drainage, paths, etc.

Construction of foundations, drainage, footpaths and parking areas, erection of huts, installation of electric wiring, in situ fittings and decorating, etc.

Dismantle offices and restore site on completion

Dismantling and taking away huts and furniture, disconnecting and removing services, removing temporary foundations etc. and re-instating ground surface to condition prevailing before construction.

Erection of general site huts

Construction of foundations, drainage, footpaths and parking areas, erection of huts, installation of electric wiring, in situ fittings and decorating, etc.

Dismantle general site huts

Dismantling and taking away huts and furniture, disconnecting and removing services, removing temporary foundations etc. and re-instating ground surface to condition prevailing before construction

PRELIMINARIES AND GENERAL ITEMS

Erection of canteen and welfare huts

Construction of foundations, drainage, footpaths and parking areas, erection of huts, installation of electric wiring, in situ fittings and decorating, etc.

Dismantle canteen and welfare huts

Dismantling and taking away huts and furniture, disconnecting and removing services, removing temporary foundations etc. and re-instating ground surface to condition prevailing before construction

Caravan site construction and clearance

Construction of site for employees' caravans including provision for water, electricity and drainage, and subsequently clear away and restore site on completion, allow credit for any charges to be levied

Telecommunications

Charges for initial installation and removal

Furniture and equipment

Purchase costs of furniture and equipment, allow for residual sale value

Survey equipment

Purchase costs of survey equipment including pegs, profiles, paint, etc., for setting out

Canteen and welfare equipment

Purchase costs of equipment

Testing and laboratory equipment

Purchase cost of equipment

Radio communication

Installation costs

Electrical connection and installation

Initial charges for connections to mains supply

Electrical connection site plant

Connection to site mains supply and final disconnection and removal

Electrical connection site lighting

Connection to site mains supply and final disconnection and removal

Water supply

Installation on site and connection charges

Haulage plant

Cost of transport of plant and equipment to and from site

PRELIMINARIES AND GENERAL ITEMS

Progress photographs

Depot loading and unloading charges

Carpenter's shop

Erection of building, installation of equipment including electrical installation, etc. Dismantle and clear away on completion

Fitter's shop

Erection of building, installation of equipment including electrical installation, etc. Dismantle and clear away on completion

Stores compound

Erect and dismantle stores compound

Notice boards and signs

Supply, erect and remove Contractor's signboards, traffic control signs, etc.

Insurances

Payment of premiums for all Contractor's insurance obligations (see separate section on insurances and bond below)

Bond

Charges for provision of bond (see separate section on insurances and bond below)

Plant erection

Cost of erection of Contractor's plant on site including foundations, hardstandings, drainage, etc.

Plant dismantling

Cost of removal of Contractor's plant on site including foundations, hardstandings, drainage, etc.

Clear site on completion including removal of rubbish and reinstatement

EMPLOYER'S AND CONSULTANTS' SITE REQUIREMENTS – TIME RELATED

Office and other huts

Rental of office accommodation, sub offices, laboratory, etc.

Office and site attendant labour

Office cleaning, chainmen, laboratory assistants, etc.

Site transport

Rental for vehicles for use of client and engineer

Telecommunications

Rental and cost of calls (if to be borne by Contractor)

PRELIMINARIES AND GENERAL ITEMS

Furniture and equipment

Rental of office furniture and equipment

Survey equipment

Rental of surveying equipment

Testing and laboratory equipment

Rental of testing and laboratory equipment

Radio communication equipment

Rental and maintenance

Office lighting and heating

Cost of heating and lighting all offices and huts

Office consumables

Cost of office consumables to be provided by the Contractor

EMPLOYER'S AND CONSULTANTS' SITE REQUIREMENTS – NON TIME RELATED

Erection of huts and offices

Client's and engineer's offices and other huts including foundations, pathways, parking area, electrical installation and drainage, etc.

Dismantling huts and offices

Restoration of site on completion

Telecommunications installation charges

Furniture and equipment purchase

Purchase cost for furniture and equipment

Survey equipment purchase

Testing and laboratory equipment purchase

Radio communication equipment installation

Progress photographs

Cost of professional photographer and supplying prints as required

OTHER SERVICES, CHARGES AND FEES

Design fees for alternative designs for permanent works

Design and drawing office costs and charges for preparing alternative designs and specifications and bill of quantities for alternative designs for permanent works

PRELIMINARIES AND GENERAL ITEMS

Design and design office charges for temporary works

Design and drawing office costs and charges for preparing designs and drawings for temporary works

Preparation of bending schedules

Drawing office charges for preparation of bending schedules

Fees to local authorities

Legal advice and fees

Fees and charges from legal adviser

Fencing

Traffic diversions

Lighting

Traffic signs

Traffic control

Footpath diversions

Stream or river diversion

Cofferdam installation

Cofferdam removal

Support works

Jetties

Bridges

De-watering

General pumping

Including construction of collecting sumps, etc.

Site access roads and maintenance

Scaffolding

GENERAL PURPOSE PLANT – OTHER THAN THAT INCLUDED IN UNIT COSTS

Lorries and dumpers for general transport around site

Tractors and trailers

Craneage for general use

Compressed air plant

PRELIMINARIES AND GENERAL ITEMS

Pumps

Bowsers for fuelling plant

Bowsers for water supply

Non productive time for plant on site

Obtained by comparing plant requirements as on programme with the plant time included in the build up of bill rates

Note: For all items of plant listed above the cost of drivers and other attendants must be allowed for together with consumables and other operating costs

INSURANCES AND BOND

Contractors are legally required to insure against liability which may be incurred when employees are injured at work and when individuals are injured by owners' vehicles. There is also a statutory requirement for certain types of machinery to be inspected at regular intervals. In addition to legal requirements, companies insure against possible loss due to fire, explosion, fraud, liability incurred as a result of damage to the property of others and through serious injury to individuals.

Certain risks are excluded from insurers' policies; these include war, revolution, etc, contamination by radioactivity and risks which arise from bad management.

Generally insurance companies take into account the claims record of a Contractor when assessing premiums payable on a particular contract or policy. Premiums are related to the risks involved and, on large Civil Engineering contracts, the insurers will require full details of the work, the methods of construction, plant used and risks involved due to flood, ground conditions, etc. Insurance companies or brokers should be consulted before submitting a tender for major Civil Engineering work.

The following gives an outline of the items to be allowed for in a tender.

Employer's liability insurance

This provides indemnity to the Employer against legal liability for death of or injury to employees sustained in the course of their employment. The cost is normally allowed for in the build-up of the 'all-in' labour rate as a percentage addition to the gross cost. This will vary, depending on the Contractor's record. An allowance of 2% has been made in this book.

Vehicle insurance

This can cover individual vehicles or fleets. The cost is normally covered in the rate charged to the contract for the use of the vehicles.

All risks insurance

This provides for loss of or damage to permanent and/or temporary works being executed on the contract. It also covers plant, materials, etc. The cost is allowed for in the tender as a percentage of the total contract value. This will vary depending on the Contractor's record and type of contract undertaken and also on the value of excesses included in the policy.

Public liability insurance

This provides indemnity against legal liability which arises out of business activities resulting in bodily injury to any person (other than employees), loss of or damage to property (not owned or under the control of the company), obstruction, trespass or the like.

Such insurance can be extended to include labour-only subcontractors and self-employed persons if required. The cost is generally included with head office overheads.

PRELIMINARIES AND GENERAL ITEMS

Professional indemnity insurance

This provides against liability arising out of claims made against the conduct and execution of the business. This covers such items as design liability, etc. The cost may be with head office overheads where such insurance is considered desirable.

Loss of money insurance

This covers loss of money and other negotiable items and loss or damage to safes and strong rooms as a result of theft. It is necessary to cover cash in transit for wages, etc. The cost is included with head office overheads.

Fidelity guarantee insurance

This covers loss by reason of any act of fraud or dishonesty committed by employees. The cost is included with head office overheads.

Other insurances

Other insurances which may be carried by a Contractor include fire insurance on his permanent premises and contents, consequential loss insurance in relation to his permanent premises, personal accident insurance on a 24 hour per day basis for employees.

Contract bond

Where the contract calls for a bond to be provided, this is normally given by either banks or insurance companies. The total value of these guarantees available to any company is limited, depending on the goodwill and assets of the company. It will also affect the borrowing facilities available to the company and therefore, to some extent, can restrict his trading. An allowance should be made at a rate of 1½% per annum on the amount of the bond for the construction period plus a rate of ½% per annum for the maintenance period.

PRELIMINARIES AND GENERAL ITEMS

Worked Example – Programme of Activities, Staffing and Plant

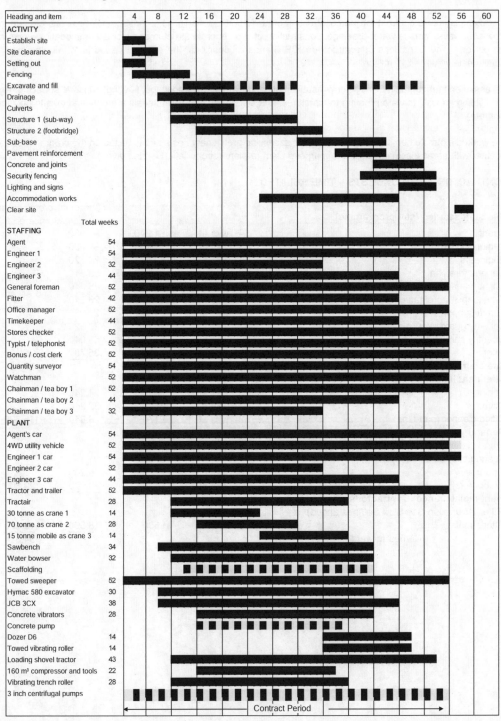

Heading and item	Total weeks
ACTIVITY	
Establish site	
Site clearance	
Setting out	
Fencing	
Excavate and fill	
Drainage	
Culverts	
Structure 1 (sub-way)	
Structure 2 (footbridge)	
Sub-base	
Pavement reinforcement	
Concrete and joints	
Security fencing	
Lighting and signs	
Accommodation works	
Clear site	
STAFFING	
Agent	54
Engineer 1	54
Engineer 2	32
Engineer 3	44
General foreman	52
Fitter	42
Office manager	52
Timekeeper	44
Stores checker	52
Typist / telephonist	52
Bonus / cost clerk	52
Quantity surveyor	54
Watchman	52
Chainman / tea boy 1	52
Chainman / tea boy 2	44
Chainman / tea boy 3	32
PLANT	
Agent's car	54
4WD utility vehicle	52
Engineer 1 car	54
Engineer 2 car	32
Engineer 3 car	44
Tractor and trailer	52
Tractair	28
30 tonne as crane 1	14
70 tonne as crane 2	28
15 tonne mobile as crane 3	14
Sawbench	34
Water bowser	32
Scaffolding	
Towed sweeper	52
Hymac 580 excavator	30
JCB 3CX	38
Concrete vibrators	28
Concrete pump	
Dozer D6	14
Towed vibrating roller	14
Loading shovel tractor	43
160 m³ compressor and tools	22
Vibrating trench roller	28
3 inch centrifugal pumps	

Contract Period

Approximate Estimating Rates

PRELIMINARIES AND GENERAL ITEMS

WORKED EXAMPLE

The example is of a contract for the construction of an Airport extension. The contract includes concrete surfaced aprons/runways, surface water drainage, construction of two concrete structures, minor accommodation works and two culverts. The Conditions of contract are ICE Standard conditions; the contract period is 12 months; and the approximate value is £11 million.

It is assumed that the main Contractor will sublet bulk earthworks and landscape, fencing, concrete surfacing, signs and lighting as well as waterproofing to structures. It is also taken that all materials are obtained off site (including concrete).

The worked example demonstrates a method of assessing preliminary costs and is based on the programme below together with an assessment of general purpose plant (a plant reconciliation is also given).

CONTRACTOR'S SITE ONCOSTS – TIME RELATED

Site staff salaries (see programme)				£	£
Agent	54	wks at	£1,840	99,360	
Senior engineer	54	wks at	£1,630	88,020	
Engineers	76	wks at	£970	73,720	
General foreman	52	wks at	£1,020	53,040	
Office manager/cost clerk	52	wks at	£870	45,240	
Timekeeper/storeman/checker	44	wks at	£610	26,840	
Admin support	52	wks at	£4600	23,920	
Security guard	52	wks at	£510	26,520	
Quantity surveyor	52	wks at	£1,020	53,040	
Fitter	42	wks at	£850	35,700	525,400
Site staff expenses (1% staff salaries)					5,254
Attendant labour					
Chainman	128	wks at	£580	74,240	
Driver	52	wks at	£620	32,240	
Office cleaner (part-time)	52	wks at	£170	8,840	115,320
General yard labour					
(Part-time involvement in loading and offloading, clearing site rubbish etc.)					
1 ganger	10	wks at	£620	6,200	
4 labourers	40	wks at	£590	23,600	29,800
Plant maintenance (Contractor's own plant)					
(Fitter included in Site Staff Salaries above)					
Fitter's mate	32	wks at	£580	18,560	18,560
Carried forward				£	694,334

PRELIMINARIES AND GENERAL ITEMS

Site staff salaries (see programme)				£	£
Brought forward				£	694,334
Site transport for staff and general use					
QS/Agent's cars	54	wks at	£160	8,640	
Engineers' cars (contribution)	130	wks at	£80	10,400	
Land Rover or similar SWB	54	wks at	£390	21,060	40,100
Site transport – Labour	48	wks at	£610	29,280	29,280
Transport for labour to and from site					NIL
Contractor's office rental					
Mobile offices (10 staff × 8 m²) = 80 m²	52	wks at	£205	10,660	
Section offices (2 nr at 10 m²) = 20 m²	52	wks at	£50	2,600	13,260
Contractor's site huts					
Stores hut, 22 m²	52	wks at	£56	2,912	2,912
Canteen and welfare huts					
Canteen 70 m² (assume 70 men)	52	wks at	£180	9,360	
Washroom 30 m²	52	wks at	£70	3,640	
Staff toilets	52	wks at	£180	9,360	
Site toilets	52	wks at	£240	12,480	34,840
Rates					NIL
General office expenditure etc.					
Postage, stationery and other consumables	52	wks at	£130	6,760	
Telephone / fax calls / e-mail and rental	52	wks at	£140	7,280	
Furniture and equipment rental	52	wks at	£70	3,640	
Personal computers, laser printers, scanners rental	52	wks at	£140	7,280	
Surveying equipment rental	52	wks at	£60	3,120	
Canteen and welfare equipment rental	52	wks at	£80	4,160	
Photocopier rental	52	wks at	£120	6,240	
Testing equipment rental	52	wks at	£50		
Lighting and heating offices and huts (200m²)	52	wks at	£125	6,500	47,580
Water consumption					
2,250,000 litres at £2.75 per 5,000 litres	450	units at	£2.75	1,238	1,238
Small tools					
1% on labour costs of say £1,400,000					
Protective clothing					
½% on labour costs of say £1,400,000					
Cleaning vehicles					
Cleaning roads					
Towed sweeper (tractors elsewhere)	52	wks at	£150	7,800	
Brushes			say,	1,000	
Labour (skill rate 4)	52	wks at	£638	33,176	
Progress photographs			say,		1,000
Total Contractor's site oncosts – Time related				£	**917,608**

PRELIMINARIES AND GENERAL ITEMS

CONTRACTOR'S SITE ONCOSTS – NON TIME RELATED

				£	£
Erect and dismantle offices					
Mobile	108	m² at	£9.00	972	
Site works			say,	1000	
Toilets			say,	300	
Wiring, water, etc.			say,	750	3,022
Erect and dismantle other buildings					
Stores and welfare	130	m² at	£16.00	2,080	
Site works			say,	1000	
Toilets			say,	700	3,780
Telephone installation					500
Survey equipment and setting out					
Purchase cost including pegs, profiles, paint ranging rods, etc					1,200
Canteen and welfare equipment					
Purchase cost less residual value					1,500
Electrical installation					2,500
Water supply					
Connection charges				1,000	
Site installation				1,000	
Transport of plant and equipment					3,500
Stores compound and huts					1,000
Sign boards and traffic signs					1,000
Insurances (dependent on Contractor's policy and record)					
Contractor's all risks 2.5% on £7,800,000				195,000	
Allow for excesses				20,000	215,000
General site clearance					3,500
Total Contractor's site oncosts – Non time related				**£**	**238,502**

PRELIMINARIES AND GENERAL ITEMS

EMPLOYER'S AND CONSULTANTS' SITE REQUIREMENTS – TIME RELATED

(details of requirements will be defined in the contract documents)

Offices (50 m²)	52	wks at	£110	5,720
Site attendant labour (man weeks)	150	wks at	£580	87,000
Site transport (2 Land Rovers or similar)	52	wks at	£780	40,560
Telephones and calls	52	wks at	£120	6,240
Furniture and equipment	52	wks at	£50	2,600
Survey equipment	52	wks at	£60	3,120
Office heating and lighting (50 m²)	52	wks at	£45	2,340
Office consumables (provided by Contractor)	52	wks at	£50	2,600
Total employer's and consultants' requirements – time related			£	150,180

EMPLOYER'S AND CONSULTANTS' SITE REQUIREMENTS – NON TIME RELATED

(details of requirements will be defined in the contract documents)

Erection and dismantling of huts and offices (50 m²)	50	m² at	£9.00	450
Site works, toilets, etc.				1,000
Telephone installation				500
Electrical installation				1,000
Furniture and equipment Purchase cost less residual value				750
Progress photographs	100	sets at	£40	4,000
Total employer's and consultants' requirements – non time related			£	7,700

OTHER SERVICES, CHARGES AND FEES

Not applicable to this cost model.

TEMPORARY WORKS – OTHER THAN THOSE INCLUDED IN UNIT COSTS

Temporary Fencing					
1200 mm chestnut fencing					
Materials	5,000	m at	£6.94	34,700	
Labour	500	hrs at	£12.70	6,350	41,050
Traffic diversions					
Structure No. 1				8,000	
Structure No. 2				6,500	
Footpath diversion					5,000
Stream diversion					15,500
Site access roads	400	m at	£120.00		
Total temporary works				£	124,050

PRELIMINARIES AND GENERAL ITEMS

GENERAL PURPOSE PLANT – OTHER THAN THAT INCLUDED IN UNIT COSTS

Description				Labour	Plant	Fuel etc.	
Wheeled tractor							
Hire charge	52	wks at	£580		30,160		30,160
driver (skill rate 4)	52	wks at	£580	30,160			30,160
consumables	52	wks at	£80			4,160	4,160
Trailer							
Hire charge	52	wks at	£25		1,300		1,300
10 t Crawler Crane							
Hire charge	40	wks at	£1,000				
driver (skill rate 3)	40	wks at	£713	28,520			28,520
consumables	40	wks at	£45			1,800	1,800
14.5 t hydraulic backacter							
Hire charge	6	wks at	£1,010		6,060		6,060
Driver (skill rate 3)	6	wks at	£713	4,278			4,278
Banksman (skill rate 4)	6	wks at	£639	3,834			3,834
consumables	6	wks at	£90			540	540
Concrete vibrators (two)							
Hire charge in total	24	wks at	£100		2,400		2,400
D6 Dozer or similar							
Hire charge	4	wks at	£1,550		6,200		6,200
driver (skill rate 2)	4	wks at	£796	3,184			3,184
consumables	4	wks at	£195			780	780
Towed roller BW6 or similar							
Hire charge	6	wks at	£350		2,100		2,100
consumables	6	wks at	£60			360	360
Loading shovel Cat 939 or similar							
Hire charge	16	wks at	£1,100		17,600		17,600
driver (skill rate 2)	16	wks at	£796	12,736			12,736
consumables	16	wks at	£95			1,520	1,520
Compressor 22.1m³/min (silenced)							
Hire charge	12	wks at	£492		5,904		5,904
consumables	12	wks at	£410			4,920	4,920
Plate compactor (180 kg)							
Hire charge	12	wks at	£50		600		600
consumables	12	wks at	£15			180	180
75 mm 750 l/min pumps							
Hire charge	25	wks at	£79		1,975		1,975
consumables	25	wks at	£15			375	375
Total costs			£	**82,712**	**114,299**	**14,635**	**211,646**

SUMMARY OF PRELIMINARIES AND GENERAL ITEMS

Contractor's site oncosts – Time related	917,608
Contractor's site oncosts – Non time related	238,502
Employer's and consultants' requirements on site – Time related	150180
Employer's and consultants' requirements on site – Non Time related	7,700
Other services, charges and fees	
Temporary works not included in unit costs	124,050
General purpose plant and plant not included in unit costs	211,646
Total of Preliminaries and General Items	**£649,686**

PART 5

Resources

This part comprises sections on labour, materials and plant for civil engineering work. These resources form the basis of the unit costs in Parts 6 and 7 and are given here so that users of the book may:

> Calculate rates for work similar to, but differing in detail from, the unit costs given in
> Compare the costs given here with those used in their own organization
> Calculate the effects of changes in wage rates, material prices, etc.

Adjustments should be made to the rates shown to allow for time, location, local conditions, site constraints and any other factors likely to affect the cost of the specific scheme.

Art of Structures

A. Muttoni

This book describes the complete panorama of supporting structures and their function by describing how loads are sustained and transmitted to the ground. With a minimum of mathematics, the reader is guided through the analysis of some of the world's most famous designs and structures from a civil-engineering perspective.

An intuitive approach is taken – the basics of equilibrium analysis are explained by visualizing the internal forces of specific structures with the aid of simple graphical tools.

Ideal for anyone who needs an intuitive and practical approach to the design and appropriate sizing of load-bearing structures.

February 2011: 280pp
Pb: 978-0-415-61029-2: **£34.99**

To Order: Tel: +44 (0) 1235 400524 **Fax:** +44 (0) 1235 400525
or Post: Taylor and Francis Customer Services,
Bookpoint Ltd, Unit T1, 200 Milton Park, Abingdon, Oxon, OX14 4TA UK
Email: book.orders@tandf.co.uk

For a complete listing of all our titles visit:
www.tandf.co.uk

BASIS OF THIS SECTION

The following are brief details of the Construction Industry Joint Council agreement on pay and conditions for the Building and Civil Engineering Industry, effective from July 2010.
Copies of the Working Rule Agreement may be obtained from:

> Construction Industry Joint Council,
> 55 Tufton Street,
> London,
> SW1P 3QL

Rates of Pay (Rule WR.1)

The basic and additional hourly rates of pay are:

Labourer / General Operative	£7.91
Skill Rate 4	£8.52
Skill Rate 3	£9.05
Skill Rate 2	£9.65
Skill Rate 1	£10.02
Craft Operative	£10.51

Additional Payment for Skilled Work (Rule WR.1.2.2)

Skilled Operative Additional Rate:

> These rates were discontinued with effect from 26 June 2006.

Additional Payment (Rule WR.1.4)

The WRA recognizes an entitlement to additional payments to operatives employing intermittent skill, responsibility or working in adverse conditions, for each hour so engaged as defined in Schedule 2 as follows:

	Extra £/hr		
A	£0.17	Tunnels	Operatives (other than Tunnel Machine Operators, Tunnel Miners and Tunnel Miners' Mates) wholly or mainly engaged in work of actual construction including the removal and dumping of mined materials but excluding operatives whose employment in the tunnel is occasional and temporary.
B	£0.28	Sewer Work	Operatives working within a totally enclosed active surface water sewer of any nature or condition.
C	£0.37	Sewer Work	Operatives working outside existing sewers excavating or removing foul materials emanating from existing sewers.
D	£0.38	Working at Height	Operatives (including drivers of tower cranes, but excluding the drivers of power driven derricks on high stages and linesmen-erectors and their mates and scaffolders) employed on detached work.
E	£0.43	Sewer Work	Operatives working within a totally enclosed active foul sewer of any nature or condition.
F	£0.65	Stone Cleaning	Operatives other than craft operatives dry-cleaning stonework by mechanical process for the removal of protective material and/or discolouration.

Bonus (Rule WR.2)

The Working Rule leaves it open to employers and employees to agree a bonus scheme based on measured output and productivity for any operation or operations on a job.

Working Hours (Rule WR.3)

Normal working hours are unchanged at 39 hour week – Monday to Thursday at 8 hours per day and Friday at 7 hours. The working hours for operatives working shifts are 8 hours per weekday, 40 hours per week.

Rest/Meal Breaks (Rule WR.3.1)

Times fixed by the employer, not to exceed 1 hour per day in aggregate, including a meal break of not less than half an hour.

Overtime Rates (Rule WR.4)

During the period of Monday to Friday, the first 4 hours after normal working day is paid at time and a half, after 4 hours double time shall be paid. Saturday is paid at time and a half until completion of the first four hours. Remainder of Saturday and all Sunday paid at double time.

Daily Fare and Travelling Allowances (Rule WR.5)

This applies only to distances from home to the site of between 15 and 75 km, giving rates of 99p to £7.53 for Travelling Allowance (taxed) and £3.60 to £12.74 for Fare Allowance (not taxed). For the one way distance of 15km used in the following example we have assumed a rate of £3.60 with effect from 30th June 2008.

Rotary Shift Working (Rule WR.6)

This relates to the situation where more than one shift of minimum 8 hours is worked on a job in a 24 hour period, and the operative rotates between the shifts either in the same or different pay weeks.
The basic rate shall be the operative's normal hourly rate plus 14%. Overtime beyond the 8 hour shift shall be at time and a half for the first 4 hours at normal rate plus 14% thereafter double normal rate.

Night Work (Rule WR.7)

Providing night work is carried out by a separate gang from those working during daytime, an addition of 25% shall be paid on top of the normal hourly rate.
Overtime payments: during the period of Monday to Friday, the first 4 hours after normal working day is paid at time and a half plus the 25% addition of normal hourly rate, after 4 hours double time shall be paid. All hours worked on Saturday and all Sunday paid at double time.

Tide Work (Rule WR.9)

Where the operative is also employed on other work during the day, additional time beyond the normal working day shall be paid in accordance with the rules for overtime payments.
Where the operative is solely involved in work governed by tidal conditions, he shall be paid a minimum of 6 hours at normal rates for each tide. Payment for hours worked in excess of 8 over two tides shall be calculated proportionately.

$$\frac{(\text{Total Hours Worked} - 8 \text{ hours}) \times \text{Total Hours Worked}}{8 \text{ hours}}$$

Work done after 4pm Saturday and during Sunday shall be at double time. Operatives are guaranteed 8 hours at ordinary rate for any time worked between 4pm and midnight on Saturdays and 16 hours for two tides worked on a Sunday.

Tunnel Work (Rule WR.10)

The first part of a shift equivalent to the length of the normal working day shall be paid at the appropriate normal rate, the first four hours thereafter at time and a half and thereafter at double time. In the case of shifts on a Saturday, the first 4 hours are at time and a half, thereafter at double time. All shifts on a Sunday at double time.

Saturday shifts extending into Sunday are at double time. Sunday shifts extending into Monday, time after midnight: first 4 hours at time and a half, thereafter at double time.

Subsistence Allowance (Rule WR.15)

The 2010 allowance will increase to £29.66 per night.

Annual Holidays Allowance (Rules WR.18 and 21)

Annual and Public Holiday Pay is included in accordance with the Building & Civil Engineering Benefit Schemes (B&CE) Template Scheme. Allowances are calculated on total weekly earnings inclusive of overtime and bonus payments and the labour cost calculation assumes 21 days (4.2 weeks) annual and 8 days (1.6 weeks) public holidays.

Easy Build Pension Contributions (Rule WR 21.3)

The minimum employer contribution is £3.00 per week. Where the operative contributes between £3.01 and £10.00 per week the employer will increase the minimum contribution to match that of the operative up to a maximum of £10.00 per week. The calculations below assume 1 in 10 employees make contributions of £10.00 per week.

LABOUR COSTS

CALCULATION OF LABOUR COSTS

This section sets out a method used in calculating all-in labour rates used within this book. The calculations are based on the wage rates, plus rates and other conditions of the Working Rule Agreement (WRA); important points are discussed below. The calculations can be used as a model to enable the user to make adjustments to suit specific job conditions in respect of plus rate, bonus, subsistence and overhead allowances together with working hours etc. to produce alternative all-in rates which can be substituted for those printed.
All-in labour costs are calculated on the pages following for six categories of labour reflecting the different classifications set out in the WRA.

AVERAGE WORKING WEEK has been calculated to an equal balance between winter and summer (46.20 working weeks per year).

SUBSISTENCE PAYMENTS are included for key men only: for a stated percentage of the workforce.

TRAVELLING ALLOWANCES are based on rate payable for a journey of 15 Km at £3.60 per day. These allowances are adjusted to cater for operatives receiving Subsistence Payments.

BONUS PAYMENTS reflect current average payments in the industry.

WORKING HOURS & NON-PRODUCTIVE OVERTIME are calculated thus:

SUMMER (hours worked 8.00 am – 6.00pm, half hour for lunch)

	Mon	Tue	Wed	Thu	Fri	Sat	Total Paid Hours	Deductions Lost & Wet time	Paid Breaks	Effective Total Hours
Total hrs	9.50	9.50	9.50	9.50	9.50	4.00	51.50	0.50	1.00	**50.00**
Overtime	0.75	0.75	0.75	0.75	1.25	2.00	6.25			
TOTAL						PAID:	**57.75**	EFFECTIVE		**50.00**

WINTER (hours worked 7.30 am – 4.30pm, half hour for lunch)

	Mon	Tue	Wed	Thu	Fri	Sat	Total Paid Hours	Deductions: Lost & Wet time	Paid Breaks	Effective Total Hours
Total hrs	8.50	8.50	8.50	8.50	7.50	4.00	45.50	1.50	–	**44.00**
Overtime	0.25	0.25	0.25	0.25	0.25	2.00	3.25	–	–	–
TOTAL						PAID	**48.75**	EFFECTIVE		**44.00**

AVERAGE TOTAL HOURS PAID	(57.75 + 48.75 / 2)	=	53.25	hours per week
AVERAGE EFFECTIVE HOURS WORKED	(50.00 + 44.00 / 2)	=	47.00	hours per week

Payment details		Categories of Labour					
		general	skill rate 4	skill rate 3	skill rate 2	skill rate 1	craft rate
		£	£	£	£	£	£
Basic rate per hour		7.91	8.52	9.05	9.65	10.02	10.51
53.25 Hours Paid @ Total Rate		421.21	453.69	481.91	513.86	533.57	559.66
Weekly Bonus Allowance	*	31.00	31.00	62.00	52.10	93.30	124.29
TOTAL WEEKLY EARNINGS	£	452.21	484.69	543.91	565.96	626.87	683.94
Travelling Allowances (say, 15km per day, i.e. £3.60 × 6 days)							
general, skill rate 4, skill rate 3	=100%	21.60	21.60	21.60	–	–	–
skill rate 2, skill rate 1	=80%	–	–	–	17.28	17.28	–
craft rate	=90%	–	–	–	–	–	19.44
Subsistence allowance (Average 6.66 nights × £29.16, + 7.00% to cover periodic travel)							
general, skill rate 4, skill rate 3	=0%	–	–	–	–	–	–
skill rate 2, skill rate 1	=20%	–	–	–	41.56	41.56	–
craft rate	=10%	–	–	–	–	–	20.78
TOTAL WAGES	£	473.81	506.29	565.51	624.80	685.71	724.16
National Insurance contributions – (12.8% above Earnings Threshold except Fares)		44.79	48.96	56.41	62.94	72.58	77.17
Annual Holiday Allowance – 4.2 weeks @ Total Weekly Earnings over 46.2 weeks		41.09	44.04	49.34	52.38	56.97	62.16
Public Holidays with Pay – 1.6 weeks @ Total Weekly Earnings over 46.2 weeks		15.65	16.78	18.80	19.95	21.71	23.68
Easy Build pension contribution		4.00	4.00	4.00	4.00	4.00	4.00
CITB Levy (0.50% Wage Bill)		2.37	2.53	2.83	3.18	3.44	3.62
	£	581.71	622.60	696.89	767.25	844.41	894.79
Allowance for Employer's liability & third party insurances, safety officer's time							
QA policy / inspection and all other costs & overheads:		14.55	15.58	17.40	19.47	21.13	22.39
TOTAL WEEKLY COST (47 HOURS)	£	596.26	638.18	714.29	786.72	856.54	917.98
COST PER HOUR	£	12.69	13.57	15.20	16.74	18.42	21.07
Plant Operators rates: Addition to Cost per Hour for Rule WR.11 – plant servicing time (6 hrs)	£	1.17	1.26	1.34	1.43	1.48	1.56
COST PER HOUR	£	13.86	14.84	16.54	18.17	19.90	22.63
Additional Payments under WR 1.4 (see above)		A	B	C	D	E	F
Extra rate per hour	£	0.17	0.29	0.37	0.38	0.43	1.65

* The bonus levels have been assessed to reflect the general position as at May/June 2010 regarding bonus payments limited to key personnel, as well as being site induced.

LABOUR CATEGORIES

Schedule 1 to the WRA lists specified work establishing entitlement to the Skilled Operative Pay Rate 4, 3, 2, 1 or the Craft Rate as follows:

General Operative

Unskilled general labour

Skilled Operative Rate 4

supervision	Gangers + trade chargehands
plant	Contractors plant mechanics mate; Greaser
transport	Dumper <7t; Agricultural Tractor (towing use); Road going motor vehicle <10t; Loco driver
scaffolding	Trainee scaffolder
drilling	Attendee on drilling
explosives	Attendee on shot firer
piling	General skilled piling operative
tunnels	Tunnel Miner's mates; Operatives driving headings over 2 m in length from the entrance (inwith drain, cable & mains laying)
excavation	Banksmen for crane / hoist / derrick; Attendee at loading / tipping; Drag shovel; Trenching machine (multi bucket) <30hp; Power roller <4t; Timberman's attendee
coal	Opencast coal washeries and screening plants
tools	Compressor/generator operator; Power driven tools (breakers, tamping machines etc.); Power-driven pumps; air compressors 10 KW+; Power driven winches
concrete	Concrete leveller/vibrator operator/screeder + surface finisher, concrete placer; Mixer < 21/14 or 400 litres; Pumps / booms operator
linesmen	Linesmen-erector's mate
timber	Carpentry 1st year trainee
pipes	Pipe layers preparing beds and laying pipes <300 mm diameter; Pipe jointers, stoneware or concrete pipes; pipe jointers flexible or lead joints <300 mm diameter
paving	Rolled asphalt, tar and/or bitumen surfacing: Mixing Platform Chargehand, Chipper or Tamperman; Paviors; Rammerman, kerb & paving jointer
dry lining	Trainee dry liners
cranes	Forklift truck <3t; Crane < 5t; Tower crane <2t; Mobile crane / hoists / fork-lifts <5t; Power driven hoist / crane

Skilled Operative Rate 3

transport	Road going motor vehicle 10t+
drilling	Drilling operator
explosives	Explosives/shot firer
piling	Piling ganger / chargehand; Pile frame winch driver
tunnels	Tunnel Miner (working at face/operating drifter type machine)
excavation	Tractors (wheeled / tracked with/without equipment) <100hp; Excavator <0.6 m³ bucket; Trenching machine (multi bucket) 30–70hp; Dumper 7–16t; Power roller 4t+; Timberman
concrete	Mixer 400–1500 litres; Mobile concrete pump with / without concrete placing boom; Hydraulic jacks & other tensioning devices in post-tensioning and / or prestressing
formwork	Formwork carpenter 2nd year trainee
masonry	Face pitching or dry walling
linesmen	Linesmen-erectors 2nd grade
steelwork	Steelwork fixing simple work; Plate layer
pipes	Pipe jointers flexible or lead joints (300–535 mm diameter)
paving	Rolled asphalt, tar and/or bitumen surfacing: Raker, Power roller 4t+, mechanical spreader operator/ leveller
dry lining	Certified dry liners
cranes	Cranes with grabs fitted; Crane 3–10t; Mobile crane / hoist / fork-lift 4–10t; Overhead / gantry crane <10t; Power-driven derrick <20t;Tower crane 2–10t; Forklift 3t+

Skilled Operative Rate 2

plant	Maintenance mechanic; Tyre fitter on heavy earthmover
transport	Dumper 16–60t
scaffolding	Scaffolder < 2 years scaffolding experience and < 1 year as Basic Scaffolder
piling	Rotary or specialist mobile piling rig driver
tunnels	Face tunnelling machine
excavation	Tractors (wheeled/tracked with / without equipment) 100–400hp; Excavator 0.6–3.85 m³ bucket; Trenching machine (multi bucket) 70hp+; Motorized scraper; Motor grader
concrete	Mixer 1500 litres+; Mixer, mobile self-loading and batching <2500 litres
linesmen	Linesmen-erectors 1st grade
welding	Gas or electric arc welder up to normal standards
pipes	Pipe jointers flexible or lead joints over 535 mm diameter
cranes	Mobile cranes / hoists / fork-lifts 5–10t; Power-driven derrick 20t+, with grab <20t; Tower crane 10–20t

Skilled Operative Rate 1

plant	Contractors plant mechanic
transport	LGV driver; Lorry driver Class C+E licence; Dumper 60–125t
excavation	Excavator 3.85–7.65m³; Tractors (wheeled / tracked with / without equipment) 400–650hp
concrete	
linesmen	
steelwork	Steelwork assembly, erection and fixing of steel framed construction
welding	Electric arc welder up to highest standards for structural fabrication & simple pressure vessels (air receivers including CO_2 processes)
drilling	Drilling rig operator
cranes	Power-driven derrick with grab 20t+; Tower crane 20t+

Craft Operative

transport	Dumper 125t+
scaffolding	Scaffolder a least 2 years scaffolding experience and at least 1 year as Basic and Advanced Scaffolder
excavation	Excavator 7.65t+ (see WR 1.2.2 above); Tractors (wheeled/tracked with / without equipment) 650hp+
concrete	Reinforcement bender & fixer
formwork	Formwork carpenter
welding	Electric arc welder capable of all welding processes on all weldable materials including working on own initiative from drawings
cranes	Crane 10t+; Mobile cranes / hoists / fork-lifts 10t+

BASIC MATERIAL PRICES

This section comprises a price list for materials which assumes that the materials would be in quantities as required for a medium sized civil engineering project of, say, £10–12 million and the location of the works would be neither city centre nor excessively remote.

The material prices have been obtained from manufacturers and suppliers and are generally those prevailing at the time of preparing this edition (April/May 2010). In view of the current high demand and raw materials prices, it is important to note that steel prices are based at May 2010. Prices are given for the units in which the materials are sold, which may not necessarily be the units used in the Unit Costs sections.

In effect these prices (except where noted) reflect the normal 'list price'; there are NO adjustments for the following:

* waste or loss or any offloading or distribution charges, unless specifically noted
* trade discounts pertinent to the market, type of work involved and the relationship between contractor and supplier

In addition, all prices quoted throughout this book are exclusive of Value Added Tax.

BAR AND FABRIC REINFORCEMENT

		Grade 500C deformed reinforcing bars		Stainless steel reinforcing bars	
		High Tensile Steel to BS 4449		Stainless steel to EN 1.4301	
		Straight	Bent	Straight	Bent
dia. (mm)	Unit	£	£	£	£
8	t	600	640	2920	2970
10	t	590	635	3140	3190
12	t	580	625	2880	2930
16	t	575	620	2790	2820
20	t	575	620	2700	2750
25	t	575	620	2700	2750
32	t	580	625	2700	2750
40	t	585	630		

Welded fabric to BS 4483 in sheets 4.80 × 2.40 metres	Unit	£
A98 (1.54 kg per m²)	m²	0.87
A142 (2.22 kg per m²)	m²	1.24
A193 (3.02 kg per m²)	m²	1.69
A252 (3.95 kg per m²)	m²	2.21
A393 (6.16 kg per m²)	m²	3.46
B196 (3.05 kg per m²)	m²	1.71
B283 (3.73 kg per m²)	m²	2.09
B385 (4.53 kg per m²)	m²	2.54
B503 (5.93 kg per m²)	m²	3.33
B785 (8.14 kg per m²)	m²	4.57
B1131 (10.90 kg per m²)	m²	6.12
C283 (2.61 kg per m²)	m²	1.55
C385 (3.41 kg per m²)	m²	2.01
C503 (4.34 kg per m²)	m²	2.53
C636 (5.55 kg per m²)	m²	3.11
C785 (6.72 kg per m²)	m²	3.77
D49 (0.77 kg per m²)	m²	0.46
D98 (0.77 kg per m²)	m²	0.87

The prices shown above include delivery but a charge of £30.00 is applicable for each delivery of under 8 tonnes

Black annealed tying wire		
16 swg tying wire (coil)	25 kg	40.80
16 swg tying wire (coil)	2 kg	7.65
Stainless steel tying wire		
18 swg tying wire (coil)	25 kg	178.00

BRICKWORK AND BLOCKWORK	Unit	£
Clay bricks to BS EN 771-1		
Commons		
flettons	1000	234–275
non-flettons	1000	175–240
Facings		
flettons, pressed, sand-faced	1000	250–395
machine moulded	1000	285–460
extruded wire-cut	1000	225–520
pressed, repressed	1000	380–510
handmade	1000	460–900
Engineering, Class A		
perforated	1000	290–350
solid	1000	400–475
solid facing	1000	460–575
Engineering, Class B		
perforated	1000	200–275
solid	1000	350–400
solid facing	1000	450–530
Concrete bricks to BS EN 771-3		
Commons	1000	225–290
Facings	1000	340–400
Engineering	1000	160–190
Calcium silicate bricks to BS EN 771-2		
Commons	1000	230–290
Facings	1000	290–330
Glazed bricks to BS EN 771-1		
Stretcher face only	1000	2600–4300
Stretcher and header faces	1000	3300–5200
Refractory bricks	1000	3600–6200
Note: Prices are ex-works for full loads		
Concrete blocks to BS EN 771-3		
Aerated concrete blocks		
(Strength 2.8 N/m²)		
100 mm	m²	12.25–14.50
140 mm	m²	17.50–19.00
190 mm	m²	22.50–24.00
(Strength 4.0 N/m²)		
100 mm	m²	10.00–12.50
140 mm	m²	13.50–16.50
190 mm	m²	20.00–22.50

(Strength 7.0 N/m²)		
100 mm	m²	14.50–16.00
140 mm	m²	18.50–21.00
190 mm	m²	26.50–28.00
Lightweight aggregate medium density blocks		
(Strength 3.5 N/m²)		
100 mm	m²	10.00–12.00
140 mm	m²	14.00–16.00
190 mm	m²	20.00–22.00
Dense aggregate blocks		
(Strength 7.01 N/m²)		
100 mm	m²	9.00–11.00
140 mm	m²	13.00–15.00
190 mm	m²	18.00–20.00
Coloured dense concrete masonry blocks		
Hollow		
100 mm	m²	23.50–29.00
140 mm	m²	28.50–36.00
190 mm	m²	32.00–46.00
Solid		
100 mm	m²	27.00–35.50
140 mm	m²	35.50–50.00
190 mm	m²	54.50–68.50

Note: Prices are for full loads delivered to site

CAST IRON PIPES AND FITTINGS	Unit	£	£	£	£	£	£
Diameter in mm		50	70	100	125	150	200
Lengths in m		3	3	3	3	3	3
Cast and ductile iron above ground pipes and fittings to BS EN 877, St Gobian							
Plain ended cast iron pipes	m	–	17.91	27.56	–	54.90	136.43
Cast iron couplings complete with stainless steel nuts and bolts and synthetic rubber gaskets	nr	–	7.66	9.97	–	19.98	64.00
Bends, 45 degree	nr	–	13.55	16.04	–	28.61	167.39
Bends, short radius, 87.5°	nr	–	13.55	16.04	–	28.61	167.39
Bends, medium radius, 87.5°	nr	–	–	–	–	104.03	–
Bends, long radius, 87.5°	nr	–	–	74.71	–	–	–
Single Junction × 45°	nr	–	20.39	30.22	–	72.39	436.22
Double Junction 100mm × 100mm × 45°	nr	–	–	38.78	–	–	–
Taper reducer pipes	nr	–	18.52	21.78	–	41.80	162.81

CLAYWARE PIPES AND FITTINGS	Unit	£	£	£	£	£	£
Vitrified clay socketed pipes and fittings to BS EN 295							
Diameter in mm		100	150	225	300	400	450
Standard pipe length in m		1.60	1.75	1.75	2.00	2.50	2.50
Straight pipes	m	4.02	7.01	18.97	24.03	80.89	113.14
Bends	nr	13.38	42.77	91.14	173.99	–	–
Bends, 22.5°	nr	–	–	–	–	327.37	431.03
Bends, 45°	nr	–	–	–	–	467.57	615.71
Bends, 90°	nr	–	–	–	–	654.62	862.03
Rest bends	nr	23.13	51.98	99.67	263.44	–	–
Saddles							
oblique/square	nr	18.40	27.38	101.86	177.26	–	–
Junctions							
Single Junction	nr	28.21	55.90	161.90	346.50	–	–
Double Junction	nr	–	85.87	200.58	413.45	–	–
Tapers							
150mm – 100mm	nr	–	22.14	–	–	–	–
225mm – 150mm	nr	–	–	69.04	–	–	–
300mm – 225mm	nr	–	–	–	190.56	–	–
Vitrified clay plain ended pipes to BS EN 295							
Diameter in mm		100	150	225	300		
Standard pipe length in m		1.60	1.75	1.75	2.00		
Straight pipes	m	5.86	11.65	18.97	24.03		
Polypropylene coupling	nr	4.74	14.26	26.48	50.58		
Bends	nr	8.68	17.87	77.89	116.91		
Rest bends	nr	15.36	22.96	99.67	283.44		
Single Junctions	nr	18.74	26.25	96.68	203.98		
Saddles							
Tapers							
150mm – 100mm	nr	–	22.14	–	–		
225mm – 150mm	nr	–	–	69.04	–		
Unglazed dense-vitrified clay plain ended cable conduit to BS EN 295							
Straight conduits	m	8.31	17.77				
Bends/Bellmouths	nr	9.51	19.54				
Diameter in mm		75	100	150	225		
Standard pipe length in m		3	3	3	3		
Land drain to BS 1196 (Hepworth)							
Straight pipes	nr	1.56	3.08	6.30	16.25		
Diameter in mm			100	150	225	300	
Standard pipe length in m			1.60	1.75	1.85	2.00	
Straight pipes	m	–	16.78	19.59	26.60	40.37	
Bends , 45°	nr	–	8.68	17.87	91.14	173.09	
Single Junctions	nr	–	18.74	26.25	127.23	326.58	
Vitrified clay road gullies, etc							
Road Gulley, 450 mm dia. × 900 mm Deep	nr	213.18					
Cast Iron Gulley Grating 434mm × 434 mm to above	nr	140.67					
Yard Gulley, 225 mm dia. including grating	nr	214.46					
Grease Interceptor 600mm × 400mm	nr	1,090.27					

CONCRETE AND CEMENT

Ready mixed concrete supplied in full loads delivered to site within 5 miles (8 km) radius of concrete mixing plant. The figures below include for the incidence of the Aggregate Tax.

Designed mixes	Unit	Aggregate size 10 mm £	20 mm £	40 mm £	Standard mixes	Unit	£
Grade C7.5	m³	87.10	85.30	84.80	ST1	m³	82.90
Grade C10	m³	87.90	86.20	85.20	ST2	m³	83.40
Grade C15	m³	88.30	85.60	85.60	ST3	m³	83.80
Grade C20	m³	88.75	87.00	86.00	ST4	m³	87.10
Grade C25	m³	91.20	89.40	88.40	ST5	m³	88.55
Grade C30	m³	96.70	89.80	97.90			
Grade C40	m³	98.70	88.90	–	Lean Mix C20P 20mm	m³	77.20
Grade C50	m³	98.50	96.60	–			
Grade C60	m³	98.70	98.40	–			

Prescribed mixes: – Add to the above designated mix prices approximately £5.00/m³

Add to the above prices for	Unit	£
rapid-hardening cement to BS EN 197-1	m³	4.63
sulphate-resisting cement to BS 4027	m³	11.25
polypropylene fibre additive	m³	12.50
distance per mile in excess of 5 miles (8 km)	m³	1.65
air entrained concrete	m³	6.50
water repellent additive	m³	6.55
part loads per m³ below full load	m³	34.55
waiting time (in excess of 6 mins. / m³ unloading norm)	hr	75.00
returned concrete	m³	135.00

Cements	bulk £ / t	bagged £/25 kg
ordinary Portland to BS EN 197-1	120.00	4.40
high alumina	1126.28	23.77
sulphate-resisting	121.32	4.47
rapid hardening	120.00	4.60
white Portland	504.82	10.49
masonry	161.03	4.37

Cement admixtures	Unit	£
Febtone colourant, brown / black	kg	12.61
Febproof waterproofer	5 l	17.16
Febspeed frostproofer	5 l	9.60
Febbond PVA bonding agent	5 l	26.99
Febmix plus plasticizer	5 l	9.13

Resources – Materials

CONCRETE MANHOLES AND GULLIES	Unit	£
Shaft and chamber rings		
900 mm diameter – unreinforced	m	48.21
1050 mm diameter – unreinforced	m	50.90
1200 mm diameter – unreinforced	m	62.01
1350 mm diameter – reinforced	m	92.33
1500 mm diameter – reinforced	m	103.40
1800 mm diameter – reinforced	m	144.90
2100 mm diameter – reinforced	m	284.70
2400 mm diameter – reinforced	m	366.45
2700 mm diameter – reinforced	m	428.19
3000 mm diameter – reinforced	m	601.68
Short length surcharge on 250 mm depth – 100%		
Short length surcharge on 500 mm depth – 50%		
Double steps (built-in)	nr	5.75
Soakaway perforations (75 mm dia)	nr	4.80
Reducing slabs (900 mm diameter access)		
1350 mm diameter	nr	147.48
1500 mm diameter	nr	170.12
1800 mm diameter – 1200 mm diameter access, 250 mm deep	nr	226.43
2100 mm diameter – 1200 mm diameter access, 250 mm deep	nr	483.54
2400 mm diameter – 1200 mm diameter access, 250 mm deep	nr	635.36
2700 mm diameter – 1200 mm diameter access, 250 mm deep	nr	1037.00
3000 mm diameter – 1200 mm diameter access, 250 mm deep	nr	1322.29
Landing slabs (900 mm diameter access)		
1500 mm diameter	nr	170.12
1800 mm diameter	nr	226.03
2100 mm diameter	nr	483.16
2400 mm diameter	nr	635.36
2700 mm diameter	nr	1037.00
3000 mm diameter	nr	1322.29
Cover slabs (heavy duty)		
900 mm diameter, 600 mm square access	nr	57.75
1050 mm diameter, 600 mm square access	nr	62.10
1200 mm diameter, 600 or 750 mm square access	nr	75.21
1350 mm diameter, 600 or 750 mm square access	nr	114.03
1500 mm diameter, 600 or 750 mm square access	nr	130.78
1800 mm diameter, 600 or 750 mm square access	nr	191.52
Concrete Road Gulleys		
Road Gulley, 450 mm dia. × 750 mm Deep	nr	63.05
Road Gulley, 450 mm dia. × 900 mm Deep	nr	65.58
Road Gulley, 450 mm dia. × 1050 mm Deep	nr	67.08
Cast Iron Gulley Grating 434mm × 434 mm to above	nr	140.67

CONCRETE PIPES AND FITTINGS	Unit	£	£	£	£	£	£
Concrete pipes with flexible joints to BS5911							
Straight pipes (pipe length 2.5m)							
Diameter in mm		300	375	450	525	750	900
Class L	m	6.45	7.95	9.59	15.46	30.40	40.82
Diameter in mm		1200	1500	1800	2100		
Class L	m	70.16	123.52	171.72	220.90		
Diameter in mm		300	375	450	525	750	900
Bends (22.5 or 45°) Class L	nr	64.50	79.50	95.90	154.60	304.00	612.30
Diameter in mm		1200	1500	1800	2100		
Bends (22.5 or 45°) Class L	nr	701.60	1235.20	1717.20	2209.00		
Lubricant	kg	4.00					
Junctions (extra over)							
300mm	nr	19.35					
375mm	nr	23.85					
375 × 150mm	nr	23.85					
450mm	nr	28.77					
450 × 150mm	nr	28.77					
525mm	nr	46.38					
525 × 150mm	nr	46.38					
750mm	nr	91.20					
750 × 150mm	nr	91.20					
900mm	nr	122.46					

CONSUMABLE STORES	Unit	£
Cutting discs		
for disc cutters (230 × 6mm)	nr	3.04
for angle grinders (115 × 6mm)	nr	1.50
for floor saws (230mm) – diamond blade	nr	10.31
Diamond Tipped Cutting Discs; 50 mm Deep Cut	nr	3.56
Diamond Tipped Cutting Discs; 100 mm Deep Cut	nr	3.75
Diamond Tipped Cutting Discs; 150 mm Deep Cut	nr	3.82

CONTRACTORS' SITE EQUIPMENT (PURCHASED)	Unit	£
Traffic cones and cylinders		
thermoplastic cone 18 inches with reflective sleeve	nr	3.10
thermoplastic cone 30 inches with reflective sleeve	nr	4.15
medium density polyethylene cone 30 inches with reflective sleeve	nr	4.15
high visibility motorway cone 1 m	nr	7.30
verge post complete with metal fixing stack	nr	13.60
KINGPIN traffic cylinder 30 inches with twist and lock base	nr	6.25
'No waiting' cones (standard Mk 111)	nr	3.10
'No waiting' cones (police Mk 111)	nr	3.41
Pendant barrier markers on 26 m cord	nr	4.15
Crowd Barrier		
fencing 1m × 50m roll, non corrodible, reusable	nr	15.70
thermoplastic fencing foot, 720 × 230 × 160 mm	nr	12.55
galvanized crowd contro; barrier 2.5 × 1.1 m high	nr	36.70
Temporary road signs and frames		
triangular 600 mm	nr	14.24
rectangular 1050 × 750 mm	nr	27.46
circular 1050 mm	nr	45.69
Ecolite road lamp	nr	8.87
Battery operated 'Unilamp', flashing	nr	8.87
Battery operated 'Unilite', photocell flashing	nr	9.40
Heavy duty ballasted traffic barrier 400mm × 1.0m high, 2.0m long, red/white	nr	204.75
Water ballasted traffic separator 650 × 550 mm, 1.0 m long, red / white	nr	38.80
Hydrant stand pipe	nr	29.93
Hydrant key	nr	6.83
Drain stoppers		
100 mm steel with plastic cap	nr	3.94
150 mm steel with plastic cap	nr	6.14
225 mm steel with plastic cap	nr	15.38
300 mm steel with plastic cap	nr	24.36
500 mm aluminium with metal cap	nr	82.95
525 mm aluminium with metal cap	nr	85.68
675 mm aluminium with metal cap	nr	149.57
Air bag stopper		
100 mm	nr	10.19
150 mm	nr	11.76
225 mm	nr	13.44
300 mm	nr	19.27
Drain testing manometer kit (air gauge)	nr	33.09

CULVERTS	Unit	£	£	£	£
Asset International corrugated steel, galvanized bitumen coated culverts, supplied in 3, 4, and 6 m lengths					
Diameter in mm		1000	1600	2000	2200
Thickness		1.6 mm	2.0 mm	2.0 mm	2.0 mm
1.6 mm thick	m	118.06	188.23	406.80	459.23
coupling	nr	27.64	53.14	58.44	61.94
Precast concrete rectangular culverts					
1000 × 500 mm × 2.0 m long (MC. 10.05)	nr	577.50			
1500 × 1000 mm × 2.0 m long (MC. 15.10)	nr	945.00			
1500 × 1500 mm × 2.0 m long (MC. 15.15)	nr	1102.50			
2000 × 1000 mm × 2.0 m long (MC. 20.10)	nr	1207.50			
2000 × 1500 mm × 2.0 m long (MC. 20.15)	nr	1365.00			
2500 × 1000 mm × 2.0 m long (MC. 25.10)	nr	1837.50			
2500 × 1750 mm × 1.5 m long (MC. 25.17)	nr	1627.50			
3000 × 2000 mm × 1.0 m long (MC. 30.20)	nr	1575.00			
3000 × 2750 mm × 1.0 m long (MC. 30.27)	nr	1785.00			
4000 × 2500 mm × 1.0 m long (MC. 40.25)	nr	2100.00			

EARTH RETENTION – GABIONS			
Woven wire mesh gabions in mesh size 75 × 75 mm, wire diameter 2.7 mm galvanized to BS EN 10244			
PVC coated galvanized wire; overall diameter 3.2 mm		Galvanized only	
Units size	£/each	Unit Size	£/each
3 × 1 × 1 m	62.99	3 × 1 × 1 m	47.99
2 × 1 × 0.5 m	31.42	2 × 1 × 0.5 m	24.30

EARTH RETENTION – CRIB WALLING	Unit	£
Pre-cast concrete crib units		
Andacrib super maxi system	m²	104.60
Andacrib maxi system	m²	83.69
Timber crib walling system		
Permacrib 1050 for retaining walls upto 3.5 m high	m²	90.67
Permacrib 1650 for retaining walls upto 5.9 m high	m²	103.69
Precast concrete wall units		
1000 mm high × 1000 mm long	nr	88.15
1250 mm high × 1000 mm long	nr	123.10
1500 mm high × 1000 mm long	nr	137.09
1750 mm high × 1000 mm long	nr	147.68
2400 mm high × 1000 mm long	nr	216.20
2690 mm high × 1000 mm long	nr	271.23
3000 mm high × 1000 mm long	nr	285.95
3750 mm high × 1000 mm long	nr	427.29

	Unit	Toe	Heel
		£	£
Filler units			
for 1000 mm high wall unit	nr	44.92	36.29
for 1500 mm high wall unit	nr	44.92	44.61
for 1750 mm high wall unit	nr	44.92	58.02
for 2400 mm high wall unit	nr	44.92	87.86
for 3000 mm high wall unit	nr	44.92	122.64
for 3600 mm high wall unit	nr	44.92	163.51
Mild steel straps and bolts for filler units			
1000 to 3000 mm high units	nr	131.08	147.09
3600 mm high units	nr	131.08	220.57
Extra over for sulphate-resisting cement = 10 %			

FENCING (SITE BOUNDARY FENCING)	Unit	£
Chainlink fencing (25 m rolls), including line wire		
galvanized mild steel, wire 3.55 mm thick, 50 mm mesh, 1800 mm high	m	5.85
plastic coated bright steel, wire 2.44/3.15 mm thick, 50 mm mesh, 1800 mm high	m	12.89
plastic coated galvanized mild steel, wire, 50 mm mesh, 1800 mm high	m	14.18
plastic coated galvanized mild steel, 50 mm mesh, 2400 mm high	m	18.91
Metal posts		
1800 mm		
standard	nr	21.26
strainer	nr	51.05
end	nr	68.08
corner	nr	68.08
2400 mm		
standard	nr	26.78
strainer	nr	62.96
end	nr	77.87
corner	nr	77.87
Wire		
barbed wire (200 m roll)	m	0.17
razor wire concertina (10m concertina coils)	m	2.62
Plain wire (75 m roll)	m	0.25
Chestnut pale fencing		
900 mm wide	m	3.32
1200 mm wide	m	4.59
1500 mm wide	m	6.35
Chestnut posts		
posts 1650 mm long	nr	1.71
posts 1830 mm long	nr	1.99
Treated softwood fencing		
100 × 100 mm posts	m	3.95
75 × 75 mm posts	m	2.47
75 × 25 mm rails	m	0.83
150 × 25 mm rails	m	1.43
81 × 38 mm rails	m	1.04
150 mm feather edge rails	m	0.76

GEOTECHNICAL INVESTIGATION	Unit	£
Rotary Drilled Boreholes		
Establishment Of Plant And Equipment, Inc Removal	sum	1000.00
Setting Up At Each Borehole Position	nr	80.00
Boring Without Core Recovery; 0 m – 10 m Deep	m	32.00
Boring Without Core Recovery; Ne 10 m – 30 m Deep	m	33.00
Boring With Core Recovery; 75 mm dia.; 0 m -10 m Deep	m	84.00
Boring With Core Recovery; 75 mm dia.; Ne 10 m – 20 m Deep	m	92.00
Boring With Core Recovery; 75 mm dia.; Ne 20 m -30 m Deep	m	104.00
Extra For Use Of Semi-Rigid Plastic Core Barrel Liner	m	8.00
Samples		
Boreholes; Open Tube (U100)	nr	18.00
Boreholes; Disturbed	nr	2.50
Boreholes; Stationary Piston	nr	35.00
Boreholes; Swedish Foil / Delft / Bishop Sand	nr	25.00
Boreholes; Groundwater	nr	4.00
Instrumental Observations		
Pressure Head; Standpipe 75 mm Hdpe	m	29.00
Pressure Head; Piezometers	m	41.77
Pressure Head; Install Protective Cover	nr	80.00
Pressure Head; Readings	nr	20.00
Laboratory Tests		Cost Range
Class: Moisture Content	nr	5.00 to 6.00
Class: Specific Gravity	nr	8.00 to 9.00
Class: Particle Size Analysis By Sieve	nr	34.00 to 37.00
Class: Particle Size Analysis By Pipette/Hydrometer	nr	45.00 to 50.00
Chem: Organic Matter	nr	35.00 to 40.00
Chem: Sulphate	nr	25.00 to 27.00
Chem: Ph Value	nr	8.00 to 9.00
Chem: Contaminants – Icrcl Maxi Comprehensive	nr	290.00 to 310.00
Chem: Contaminants – Icrcl Midi Abbreviated	nr	210.00 to 225.00
Chem: Contaminants – Icrcl Mini Screening	nr	160.00 to 180.00
Chem: Contaminants – Nitrogen Herbicides	nr	150.00 to 160.00
Chem: Contaminants – Organophosphorus/chlorine Pesticides	nr	125.00 to 130.00
Compact: Standard or Heavy	nr	150.00 to 160.00
Compact: Vibratory	nr	160.00 to 180.00
Permeability: Falling Head	nr	65.00 to 75.00
Soil: Quick Undrained Triaxial	nr	30.00 to 40.00
Soil: Shear Box; Peak Only	nr	40.00 to 45.00
Soil: Cbr	nr	50.00 to 55.00
Rock: Point Load	nr	80.00 to 90.00

GEOTEXTILES	Unit	£
Enkamat		
Enkamat Erosion Control Matting (Including Fixing Pegs)	m²	7.35
Netlon		
Ce131 Geotextile (30 m × 2 m roll)	roll	248.43
Tensar		
SS20 (75 × 4 m roll) for weak soils	roll	425.25
TX160 geogrid; (75 × 4 m roll)	roll	561.00
TX170 geogrid (50 × 4 m roll)	roll	478.00
Mat 200 (30 × 3 m roll) for erosion protection	roll	259.20
Mat 400 (30 × 3 m roll) for erosion protection	roll	324.90
Terram		
2000 Permeable Membrane (100 m × 4.5 m roll)	roll	539.37
1000 Permeable Membrane (100 m × 4.5 m roll)	roll	208.00
Terram 1B1 (25 m × 2 m roll) laminated filter drainage membrane	roll	145.00
Terram 1Bz (25 m × 2 m roll) laminated filter drainage membrane	roll	266.00
Typar		
3207; Geotextile; (200 m × 4.5 m roll)	roll	237.00
3407; Geotextile (100 m × 4.5 m roll)	roll	130.00
3607-3; Geotextile (100 m × 4.5 m roll)	roll	199.00
For waterproofing polythene sheeting see Waterproofing Section Later		

GULLEY GRATINGS AND FRAMES TO BS EN 124	Unit	£
Group 4; ductile iron, heavy duty pattern		
370 × 430 × 100 mm depth; hinged, non-rocking; D400	nr	120.75
450 × 450 × 100 mm depth; non-rocking; D400	nr	173.78
600 × 600 × 100 mm depth; D400	nr	423.15
600 diameter × 100 mm depth; D400	nr	274.05
Group 3; ductile iron, non-rocking		
400 × 432 × 75 mm depth; C250	nr	180.60
400 × 380 × 75 mm depth; C250	nr	111.83
Group 3; ductile iron, hinged		
325 × 312 × 75 mm depth; C250	nr	103.95
325 × 437 × 75 mm depth; C250	nr	117.08
400 × 432 × 75 mm depth; C250	nr	130.73
370 × 305 × 100 mm depth; C250	nr	129.15
510 × 360 × 100 mm depth; C250	nr	129.15
370 × 430 × 100 mm depth; C250	nr	129.15
Group 3; ductile iron, lockable, steel sping clips		
325 × 312 × 75 mm depth; C250	nr	152.78
325 × 437 × 75 mm depth; C250	nr	166.43
400 × 432 × 75 mm depth; C250	nr	180.60
370 × 305 × 100 mm depth; C250	nr	178.50
510 × 360 × 100 mm depth; C250	nr	178.50
370 × 430 × 100 mm depth; C250	nr	178.50
Group 3; ductile iron, pedestrian grating		
325 × 312 × 75 mm depth; C250	nr	171.15
Group 3; ductile iron, pavement gratings		
250 × 250 × 39 mm depth; C250	nr	71.41
300 × 300 × 39 mm depth; C250	nr	91.78
400 × 400 × 39 mm depth; C250	nr	106.53
700 × 700 × 39 mm depth; C250	nr	324.08

GULLEY GRATINGS AND FRAMES TO BS EN 124	Unit	£
Group 3; ductile iron, pavement gratings, concave		
300 × 300 × 58 mm depth; C250	nr	98.59
400 × 400 × 63 mm depth; C250	nr	129.17
500 × 500 × 68 mm depth; C250	nr	196.02
600 × 600 × 73 mm depth; C250	nr	301.42
700 × 700 × 78 mm depth; C250	nr	344.05
Group 3; ductile iron, kerb type pattern – to suit standard kerb profile		
385 × 502 mm; 37 deg kerb profile, C250.	nr	223.13

INTERLOCKING STEEL SHEET PILING AND UNIVERSAL BEARING PILES
The following prices are ex works in lots of 25 tonnes at one time, for delivery from one works to one destination.
Carriage from Mill to destination is included on CIP terms.

Section Arcelor Mittal All Sections – Hot Rolled	Unit	EN10248-1995 Grade S270GP £	EN10248-1995 Grade S355GP £
Standard Lengths to 18 m	tonne	810.00	815.00
Long Lengths 18 m to 24 m	tonne	820.00	825.00
Long Lengths 24 m to 29 m	tonne	830.00	835.00
Extra For Small Loads 5-10 t	tonne	35.00	35.00
One Coat Tar Vinyl (Pc1) Protection	m^2	3.68	3.68
One Coat Epoxy Pitch (Pc2) Protection	m^2	5.78	5.78

UNIVERSAL BEARING PILES (UKBP)			EN10025-2: 2004	EN10025-2: 2004
Size mm	Weight Kg/m	Unit	Grade S275JR £	Grade S355JO £
203 × 203	45, 54	tonne	840.00	880.00
254 × 254	63, 71, 85	tonne	840.00	880.00
305 × 305	79 through 223	tonne	840.00	880.00
356 × 368	109, 133	tonne	885.00	925.00

Steel piling sections supplied in lengths of 5 to 15 metres Material ex basing point - Middlesborough Railway Station - Scunthorpe Railway Station			
Add to the above prices for:		Unit	£
Quantity			
under 5 tonnes		tonne	35.00
under 10 tonnes to 5 tonnes		tonne	15.00
over 10 tonnes		tonne	basis
Size			
length 3 m to under 9 m		tonne	15.00
length over 18.5 m to 24 m		tonne	5.00
length over 24 m		tonne	POA
Holes			
lifting holes		nr	10.00

JOINT FILLERS AND WATERSTOPS	Unit	£
Flexible epoxy resin joint sealant (Expoflex 800)		
2 litre tin	litre	18.85
Non absorbent joint filler (Hydrocell XL)		
thickness 10 mm (1000 mm × 2000 mm sheet)	m²	16.75
thickness 15 mm (1000 mm × 2000 mm sheet)	m²	19.55
thickness 20 mm (1000 mm × 2000 mm sheet)	m²	22.35
thickness 25 mm (1000 mm × 2000 mm sheet)	m²	27.95
Polysulphide sealant (Thioflex 600)		
gun grade – all colours	litre	18.54
grey pouring grade	litre	19.33
Hot applied sealant		
hand applied, Plastijoint (5 litre)	litre	6.94
poured, low extension grade, Pliastic N2 (15 kg sack)	kg	2.45
poured, hard grade, Pliastic 77 (15 kg sack)	kg	2.50
Epoxy resin mortars (Expocrete)		
UA (13.5 litre pack)	litre	18.43
High duty elastomeric pavement sealant (Colpor 200PF)	litre	4.81
Non extruding expansion joint filler (Fibreboard)		
thickness 12.5 mm (2440 mm × 1200 mm sheet)	m²	4.86
thickness 20.0 mm (2440 mm × 1200 mm sheet)	m²	8.32
thickness 25.0 mm (2440 mm × 1200 mm sheet)	m²	10.08
Flexible expansion joint strip membrane (Expoband H45)		
100 mm (25 m roll)	m	7.58
200 mm (25 m roll)	m	11.51
Waterstop		
PVC 150 mm Centre Bulb	m	7.53
PVC 230 mm Centre Bulb	m	7.79
PVC 305 mm Centre Bulb	m	8.11
PVC 170 mm Flat Dumbell	m	7.39
PVC 210 mm Flat Dumbell	m	7.75
PVC 250 mm Flat Dumbell	m	8.10
PVC 150 mm Centre Bulb; Junction Piece; Vert T	nr	65.23
PVC 230 mm Centre Bulb; Junction Piece; Vert T	nr	128.10
PVC 260 mm Centre Bulb; Junction Piece; Vert T	nr	129.48
PVC 170 mm Flat Dumbell; Junction Piece	nr	114.24
PVC 210 mm Flat Dumbell; Junction Piece	nr	128.53
PVC 250 mm Flat Dumbell; Junction Piece	nr	181.09
Hydrophilic waterstops		
Supercast SW10 (15 m roll)	m	6.39
Supercast SW20 (5 m roll)	m	6.49
ADCOR 500S 25 mm × 20 mm strip (5 m coil)	m	48.37
Servistrip AH 205 20 mm × 5 mm strip (10 m coil)	m	8.19

PRECAST CONCRETE KERBS, EDGINGS AND PAVING SLABS	Unit	£
Precast Concrete Kerbs		
914 mm Lengths; 125 mm × 150 mm	m	3.55
914 mm Lengths; 125 mm × 255 mm	m	3.55
914 mm Lengths; 150 mm × 305 mm	m	4.19
Quadrants; 305 mm × 305 mm × 150 mm	nr	8.48
Quadrants; 455 mm × 455 mm × 255 mm	nr	13.78
Drop Kerbs; 125 mm × 255 mm	nr	5.19
Drop Kerbs; 150 mm × 305 mm	nr	8.37
Machine Laid In situ Kerbs & Channels		
Kerb		
75 mm straight or curved to radius over 12 m	m	8.05
100 mm straight or curved to radius over 12 m	m	11.99
125 mm straight or curved to radius over 12 m	m	25.26
75 mm curved to radius ne 12 m	m	15.18
100 mm curved to radius ne 12 m	m	11.99
125 mm curved to radius ne 12 m	m	14.57
Channel		
250 mm straight or curved to radius over 12 m	m	17.93
300 mm straight or curved to radius over 12 m	m	17.93
250 mm curved to radius ne 12 m	m	17.86
300 mm curved to radius ne 12 m	m	22.91
Safety Kerbs		
Ellis Trief; Kerb 415 mm × 380 mm, >12 m radius	nr	48.75
Ellis Trief; Kerb 415 mm × 380 mm, <12 m radius	nr	48.75
Beany block type kerbs		
Standard Unit Comprising A+C+D	nr	54.49
Deep Unit Comprising A+C+E	nr	68.11
Shallow Unit Comprising A+C+F	nr	51.65
Standard Curved Unit Comprising G+H+I	nr	76.29
Deep Curved Unit Comprising G+H+J	nr	95.36
Shallow Curved Unit Comprising G+H+K	nr	72.31
Base Block Depth Tapers	nr	56.76
Precast Edgings		
914 mm Lengths; 50 × 150 mm	m	2.47
914 mm Lengths; '50 × 205 mm	m	3.67
914 mm Lengths; 50 × 255 mm	m	3.88

PRECAST CONCRETE KERBS, EDGINGS AND PAVING SLABS	Unit	Natural £	Coloured £
Paving flags BS 7263-1			
PCC Flags; Grey; 600 × 450 × 50mm;	m²	9.42	13.18
PCC Flags; Grey; 600 × 600 × 50mm	m²	7.51	11.27
PCC Flags; Grey; 900 × 600 × 50mm	m²	5.99	9.75
PCC Flags; Grey; 600 × 600 × 63mm	m²	12.22	
PCC Flags; Grey; 900 × 600 × 63mm	m²	9.81	
PCC Recangular Paving; Grey; 200 × 100 × 80 mm	m²	20.70	23.00
Block paving BS 6717: 2001			
rectangular block paving (Charcon Europa)			
200 × 100 × 60 mm	m²	12.48	13.32
200 × 100 × 80 mm	m²	15.06	15.62
Deterrent paving (Charcon Elite)			
format 1–200 × 144 × 80 mm	m²	54.04	–
format 12–298 × 132 × 80 mm	m²	68.28	–
format 13–298 × 132 × 80 mm	m²	57.18	–
Block starter units			
herringbone 65 mm thick	pack of 80	95.29	116.08
herringbone 80 mm thick	pack of 80	120.10	124.77
interlocking 65 mm thick	pack of 98	63.70	74.26
interlocking 80 mm thick	pack of 98	70.04	–
Landscape paving			
textured (exposed aggregate finish)			
400 × 400 × 65 mm	m²	27.51	31.68
450 × 450 × 70 mm	m²	22.40	25.36
600 × 600 × 50 mm	m²	18.16	21.12
ground (smooth finish)			
400 × 400 × 65 mm	m²	28.35	30.85
600 × 600 × 50 mm	m²	25.57	26.68
non slip (pimpled finish)			
400 × 400 × 50 mm	m²	–	22.93
Grassgrid			
366 × 274 × 100 mm	m²	22.01	
Stone paving			
granite setts new 100 × 100 mm (approx)	t	188.00	
granite setts reclaimed random sized	t	108.00	
granite setts reclaimed 100 × 100 mm sorted	t	127.00	
random size cobbles reclaimed	t	35.00	

LANDSCAPING / SOILING

For a more detailed appraisal of landscaping prices, please refer to Spon's External Works and Landscape Price Book.

CULTIVATED TURF	Unit	Over 1200 m²
Rolawn		
RB Medallion	m²	3.79
Inturf		
Inturf SS2 Standard Turf	m²	2.92

IMPORTED TOPSOIL

The price for topsoil is so variable depending on site location, quantity available, season, etc. The following prices are a guide for use in the unit cost build ups and reflect reasonable quality soils from a single source delivered to site in 20 tonne tipper trucks, within a 20 km radius of source.

	Unit	Under 1000 m² £	1000 m² to 5000 m² £	Over 5000 m² £
Subsoil	m³	7.15	5.72	5.72
Reasonable quality topsoil for seeding and general, sward establishment (1.6 t/m³)	m³	17.89	16.10	15.30
Better quality topsoil for species planting and for use in raised beds, planters, etc. (1.6 t/m³)	m³	36.08	28.87	25.98

GROWING MEDIUMS/SOIL IMPROVERS	Unit	£
Mulches (delivered in 70 m³ loads)		
Graded bark flakes	m³	39.58
Bark nuggets	m³	36.08
Ornamental bark mulch	m³	40.85
Amenity bark mulch	m³	21.53
Forest biomulch	m³	19.31
Decorative biomulch	m³	22.49
Rustic biomulch	m³	27.25
Mulchip	m³	24.27
Pulverized bark	m³	22.40
Woodland mulch	m³	16.22
Woodfibre mulch	m³	20.53

GRASS SEEDS/FERTILIZERS ETC	Unit	£
Soil ameliorants:		
Landscape amenity	m³	20.56
Spent Mushroom Compost	m³	9.22
Super humus	m³	16.31
Topgrow	m³	18.98
Grass seed mixtures, British Seed Houses		
(Reclamation) BSH ref. A15, sowing rate 15–25 g/m²	kg	4.59
(Country Park) BSH ref. A16, sowing rate 8–20 g/m²	kg	5.00
(Road Verges) BSH ref. A18, sowing rate 6 g/m²	kg	4.94
(Landscape) BSH ref. A3, sowing rate 25–50 g/m²	kg	4.76
(Parkland) BSH ref. A4, sowing rate 17–35 g/m²	kg	4.87
Fertilizers		
Fisons PS5 pre-seeding fertilizer	kg	0.45
Fisons Ficote fertilizer	kg	3.57

GRASS SEEDS/FERTILIZERS ETC	Unit	£
BSH amenity granular and slow release fertilizers		
BSH 1 pre-seeding granular (reasonable) 6-9-6	kg	0.86
BSH 2 pre-seeding granular (impoverished) 10-15-10	kg	1.01
BSH 4 pre-seeding mini granular (spring/summer) 11-5-5	kg	0.89
BSH 5 pre-seeding granular (spring/summer) 9-7-7	kg	0.88
BSH 7 pre-seeding mini granular (autumn/winter) 3-10-5.2	kg	0.94
BSH 8 outfield granular (autumn/winter fertilizer) 3-12-12	kg	0.92
Growmore granular fertilizer (per 25 kg)	kg	0.62
Compost		
tree planting and mulching compost	m³	23.06
seeding	m³	18.09
Selective weedkiller	litre	8.18
Hydroseeding – specialist process; refer to unit cost landscaping section for typical specification and situation		

MANHOLE COVERS AND FRAMES TO BS EN 124	Unit	£
Manhole covers and frames (all dimensions are clear opening sizes)		
Group 5, ductile iron		
600 × 600 × 150 mm depth; Super Heavy Duty; E600	nr	342.00
1210 × 685 × 150 mm depth; Super Heavy Duty; E600	nr	570.00
1825 × 685 × 150 mm depth; Super Heavy Duty; E600	nr	1220.00
Group 4, ductile iron, double triangular		
600 × 600 × 100 mm depth; D400	nr	181.00
600 × 600 × 100 mm depth; ventilated; D400	nr	215.00
675 × 675 × 100 mm depth; D400	nr	250.00
Group 4, ductile iron, double triangular, hinged, lockable		
600 × 600 × 100 mm depth; D400	nr	168.00
Group 4, ductile iron, double triangular		
600 mm diameter × 100 mm depth; D400	nr	138.00
600 × 600 × 100 mm depth; D400	nr	160.00
675 mm diameter × 100 mm depth; D400	nr	217.00
675 × 675 × 100 mm depth; D400	nr	225.00
900 × 600 × 100 mm depth; D400	nr	344.00
900 × 900 × 125 mm depth; D400	nr	564.00
1250 × 675 × 100 mm depth; D400	nr	365.00
Group 4, ductile iron, single triangular, hinged		
600 mm × 120 mm depth; D400	nr	262.00
Group 4, ductile iron, circular cover, hinged		
600 mm diameter × 100 mm depth; square frame; D400	nr	133.00
Group 2, ductile iron, single seal		
600 × 450 × 75 mm depth; B125	nr	0.00
600 × 600 × 75 mm depth; B125	nr	122.00
Group 2, ductile iron, single seal, ventilated		
600 × 450 × 75 mm depth: B125	nr	137.00
600 × 600 × 75 mm depth: B125	nr	191.00
Group 2, ductile iron, double seal, single piece cover		
600 × 600 mm; B125	nr	257.00
Group 2, ductile iron, double seal, single piece cover, recessed		
600 × 450 mm; B125	nr	197.00
600 × 600 mm; B125	nr	237.00
MANHOLE STEP IRONS		
Step irons, galvanized malleable iron	nr	3.60

MISCELLANEOUS AND MINOR ITEMS	delivery unit	Unit	£
Concrete spacer blocks for reinforcement			
bars; 30 mm cover	500	nr	0.06
bars; 50 mm cover	200	nr	0.09
bars; 75 mm cover	50	nr	0.19
mesh; 50 mm cover	100	nr	0.21
Wire spacers			
60 mm height		m	1.41
90 mm height		m	1.44
120 mm height		m	1.50
150 mm height		m	1.62
180 mm height		m	1.92
Tying wire; 16 gauge black annealed	20 kg coil	kg	1.40

	Coupler A £ / each	Coupler P £ / each	Thread bar (at factory) £ / end
ROM Lenton reinforcement couplers (Coupler A for use where one bar can be rotated, Coupler P where neither can be)			
12 mm diameter bar	3.30	20.00	2.90
16 mm diameter bar	4.20	23.00	3.10
20 mm diameter bar	7.50	25.00	3.60
25 mm diameter bar	11.70	29.00	4.10
32 mm diameter bar	16.00	39.00	5.60
40 mm diameter bar	23.00	59.00	7.20

	Unit	£
Dowel bars round		
16 × 800 mm	nr	1.50
20 × 800 mm	nr	2.30
25 × 800 mm	nr	3.60
32 × 800 mm	nr	5.90
Expansion dowel caps		
12 × 100 mm	nr	0.29
16 × 100 mm	nr	0.32
20 × 100 mm	nr	0.37
25 × 100 mm	nr	0.37
32 × 100 mm	nr	0.49
Debonding sleeves for dowel bars		
12 × 450 mm	nr	0.34
16 × 200 mm	nr	0.36
20 × 300 mm	nr	0.37
25 × 300 mm	nr	0.49
32 × 375 mm	nr	0.64
Debonding compound	litre	4.23
Crack inducers (5 m lengths)		
10 mm wide × 50 mm deep top type (two piece)	m	2.07
10 mm wide × 75 mm deep top type (two piece)	m	3.46
10 mm wide × 40 mm deep bottom Y type	m	1.56
10 mm wide × 75 mm deep bottom Y type	m	2.93
Silver sand (bagged)	kg	0.18

	Unit	£
Formwork release agents		
General purpose mould oil	litre	1.19
Chemical release agent	litre	1.16
ROMLEASE chemical release agent (35–50 m²/l)	litre	1.31
Retarders		
ROMTARD CF retarder, subsequently brush to expose aggregate/form key (5m²/l)	litre	3.32
ROMTARD MA retarder, apply to face of formwork (15–16 m²/l)	litre	6.61
Surface hardeners and sealers		
ROMTUF concrete surface hardener and dustproofer (2 coats, 6m²/l/coat)	litre	2.90
ROMCURE Sealer RS resin based concrete floor sealer (2 coats, 5.5 m²/l/coat)	litre	4.19
ROMCURE Standard, resin based curing membrane (5.5 m²/l)	litre	2.26
Mortars and grouts		
Epoxy grout high density	kg	4.80
Epoxy mortar general purpose	kg	5.33
Non-shrink grout	kg	4.76
ROMGROUT grout admixture for infilling bolt boxes (1 packet / 50 kg OPC)	per 50kg	54.00
Air entraining agents		
Extra for Air Entrained concrete	kg	5.52
Rockbond Admix RB A2001, flowing/air entraining	kg	5.18
Resin bonded anchors:		
Resin capsules		
R-HAC 8 mm (10 per box)	nr	1.42
R-HAC 10 mm (10 per box)	nr	1.50
R-HAC 12 mm (10 per box)	nr	1.81
R-HAC 16 mm (10 per box)	nr	2.25
R-HAC 20 mm (6 per box)	nr	2.60
R-HAC 24 mm (6 per box)	nr	3.23
R-HAC 30 mm (6 per box)	nr	6.17
Zinc plated steel threaded rod with nut and washer		
8 × 110 mm (10 per box)	nr	0.66
10 × 130 mm (10 per box)	nr	1.03
12 × 165 mm (10 per box)	nr	1.61
16 × 190 mm (10 per box)	nr	2.96
20 × 250 mm (6 per box)	nr	5.74
24 × 300 mm (6 per box)	nr	10.70
30 × 380 mm (6 per box)	nr	29.90
Stainless steel threaded rod with nut and washer		
8 × 110 mm (10 per box)	nr	1.70
10 × 130 mm (10 per box)	nr	2.60
12 × 165 mm (10 per box)	nr	4.10
16 × 190 mm (10 per box)	nr	8.60
20 × 250 mm (6 per box)	nr	15.70
24 × 300 mm (6 per box)	nr	29.50

	Unit	£
Rawl SafetyPlus Anchors		
Loose Bolt Anchors		
M8 15L (50 per box)	nr	1.40
M8 40L (50 per box)	nr	1.70
M10 40L (50 per box)	nr	2.20
M12 25L (25 per box)	nr	2.80
M16 25L (10 per box)	nr	5.50
M16 50L (10 per box)	nr	6.40
M20 30L (10 per box)	nr	10.70
Bolt Projecting type Anchors		
M8 15P (50 per box)	nr	1.60
M10 20P (50 per box)	nr	2.20
M12 25P (25 per box)	nr	3.10
M12 50P (25 per box)	nr	3.50
M16 25Pm (10 per box)	nr	6.30
M16 50P (10 per box)	nr	7.10
M20 30P (10 per box)	nr	11.70
Hyrib permanent formwork system		
ref. 2411	m²	23.40
ref. 2611	m²	18.30
ref. 2811	m²	17.10
Exmet galvanized brickwork reinforcement (20 m rolls)		
65 mm wide	m	0.73
115 mm wide	m	1.26
175 mm wide	m	1.96
225 mm wide	m	2.52
305 mm wide	m	3.18
Expamet stainless steel bed joint brickwork reinforcement (3050 mm rolls)		
65 mm wide	m	1.27
115 mm wide	m	2.12
175 mm wide	m	3.43
225 mm wide	m	4.67
Polystyrene sheets		
sheet 25 × 2440 × 1220mm	nr	4.70
sheet 50 × 2440 × 1220mm	nr	9.30
sheet 75 × 2440 × 1220mm	nr	13.90
sheet 100 × 2440 × 1220mm	nr	18.50
sheet 150 × 2440 × 1220mm	nr	27.70

PAINT / STAINS / PROTECTIVE COATINGS	Unit	£
Trade gloss		
magnolia / white / brilliant white	litre	11.20
colours	litre	14.70
Trade undercoat		
white	litre	10.00
colours	litre	13.10
Trade matt emulsion		
magnolia / white / brilliant white	litre	10.90
colours	litre	14.20
Trade masonry paint, colours	litre	16.50
Trade Timberguard	litre	14.30
Creosote; dark / light	litre	1.50
Wood preservers		
wood preservers	litre	8.70
Exterior Wood; garden shed + fence preservers	litre	11.00
Woodstains		
Select Woodstain; standard finishes	litre	10.90
Trade varnishes		
Gloss varnish	litre	10.90

PROTECTIVE COATINGS	Unit	£	£ per m²
For fabricated steel, lighting columns, parapets and guard railings, etc.			
Prefabrication primers	litre	14.40 to 36.10	1.30 to 1.70
Alkyd primers	litre	14.90 to 19.20	2.30 to 2.60
Alkyd intermediates & top coats	litre	16.20 to 25.10	1.10 to 2.30
Protective finishes – alkyd	litre	25.60 to 31.40	3.60 to 5.80
Protective finishes – epoxy	litre	36.40 to 42.70	4.60 to 8.90
Fire protection	litre	61.00	Varies
Concrete protection			
RIW liquid asphatic composition (two coats) (25 litre drum)	litre	9.60	8.40
RIW Heviseal (two coats) (20 litre drum)	litre	10.90	13.90
RIW hydrocoat (25 litre drum)	litre	4.40	1.30
RIW bitumen protection board, 1.625 × 1.285 m × 3 mm thick sheet	nr	27.20	–
RIW sheet seal, grade 9000/300, 20 × 0.3 m roll	nr	172.10	–

POLYMER CHANNELS AND FITTINGS	Unit	£
ACO S100 interlocking,stepped channel, locked ductile iron grating; Class F900		
length 1000 mm	nr	104.00
length 500 mm	nr	74.00
ACO N100 KS interlocking, pre-sloped channel, SS frame; Class C250		
length 1000 mm	nr	61.00
length 500 mm	nr	43.00
ACO ParkDrain one piece polymer concrete channel system with integral grating; Class C250		
length 500 mm	nr	42.00
ACO KerbDrain 305 one piece polymer concrete combined kerb drainage system; Class D400		
length 500 mm	nr	34.00
access unit	nr	110.00
drop kerb unit	nr	110.00
mitre unit	nr	38.00
endcaps; closing	nr	16.20

	Unit	S100 & S100K (heavy duty Class F) £	N100K (normal duty Class A-C) £
ACO end caps			
end cap	nr	11.20	20.50
outlet end cap	nr	19.70	27.50
inlet end cap	nr	19.70	27.50

	Unit	£
ACO universal gullies complete with grating and galvanized bucket		
gulley	nr	326.00
ACO sump units with galvanized steel bucket		
for S100 channel with ductile iron grating	nr	182.00
for N100 KS channel with stainless steel frame	nr	132.00
for KerbDrain 305 deep trapped gully base	nr	110.00
ACO Quicklock composite grating; Class C250		
galvanized steel perforated, length 500mm	nr	43.00
ACO KerbDrain 480; Class D400		
length 500mm	nr	31.00
access unit	nr	103.00
drop kerb unit; length 915mm	nr	59.00
mitre unit	nr	34.00
endcaps	nr	16.20
ACO RoadDrain interlocking channel system		
RoadDrain 100; length 500mm	nr	69.00
RoadDrain 200; length 500mm	nr	88.00
End caps		
RoadDrain 100	nr	12.90
RoadDrain 200	nr	28.00
Step connectors	nr	11.80
Sump unit c/w sediment bucket	nr	139.00

	Unit	£
PVC accessories		
110 mm drain union	nr	2.80
160 mm drain union	nr	12.70
200 mm drain union	nr	17.00
160 mm oval to round union	nr	14.20
110 mm foul air trap	nr	21.90
160 mm foul air trap	nr	41.00
Channel gratings		
Class C ductile iron, slotted, 500 mm long	nr	18.50
Class E ductile iron, slotted, 500 mm long	nr	20.50
Class A galvanized steel, slotted, 500 mm long	nr	10.20
Class A galvanized steel, slotted, 1000 mm long	nr	19.10
Class C galvanized steel, mesh, 500 mm long	nr	19.80
Class C galvanized steel, mesh, 1000 mm long	nr	27.00
Class A stainless steel, slotted, 500 mm long	nr	34.00
Class A stainless steel, slotted, 1000 mm long	nr	52.00
Class C perforated stainless steel, 500 mm long	nr	57.00
Class C perforated stainless steel, 1000 mm long	nr	111.00
Ductile iron or steel locking bar and mild steel bolt	nr	3.20
Steel locking bar to suit mesh gratings	nr	3.20
ACO Mini ParkDrain		
length 1000 mm	nr	81.00
access, length 1000 mm	nr	100.00
endcap	nr	11.10
ACO LightWeight Kerb		
length 915 mm	nr	11.40
drop unit, length 915 mm	nr	12.70
centre stone unit, length 915 mm	nr	12.70

QUARRY PRODUCTS, AGGREGATES, ETC.

Cost per tonne delivered to site within a 20 mile radius of quarry or production centres, cost based on most economically available materials (average prices – see footnote).

	Scotland, North, & Wales	Midlands & South West	South General
Graded granular material			
natural gravels (HA class 1 A, B, or C) general fill	13.20	14.50	16.50
crushed gravel or rock ditto	15.30	16.80	19.20
sand	17.90	19.60	22.40
topsoil	15.40	16.90	19.30
Reclaimed material (HA class 2E)			
quarry waste	7.50	8.20	9.40
imported broken rick/concrete rubble	7.50	8.20	9.40
hardcore	11.90	13.10	14.90
Well graded granular material			
natural gravel or sand (HA class 6 A) fill below water	14.70	16.10	18.40
crushed gravel or rock; ditto (free draining)	16.00	17.50	20.00
Selected granular material			
natural gravel (HA class 6F) capping layers	14.30	15.70	17.90
crushed rock; ditto	16.00	17.50	20.00
Rock fill			
crushed rock, core fill, rock punching	14.03	15.40	17.60
Armour stone single size (minimum costs) sea defence shore protection			
0.5 tonne	18.70	20.50	23.40
1 tonne	27.40	30.10	34.30
3 tonne	43.40	47.60	54.40
Granular material single sized aggregate			
natural gravel (HA Clause NG503) pipe bedding/haunching	10.30	11.30	12.90
crushed rock; ditto	14.03	15.40	17.60
Filter material			
natural or crushed other than unburned slag or chalk (HA Clause NG505) backfill to drain trenches and filter drains			
type A	13.20	14.50	16.50
type B	12.00	13.20	15.00
type C	13.60	14.90	17.00
Graded granular material			
type 1/2	13.30	14.60	16.70
Concrete aggregates			
all in aggregate			
10 mm for concrete production	15.10	16.60	18.90
20 mm for concrete production	14.70	16.10	18.40
40 mm for concrete production	15.10	16.60	18.90
fine screen sharp sand for concrete production	18.00	19.70	22.60

Footnote:

The above prices are indicative, and are based on typical supplier's quotations. An average of these prices is used in Parts 4 and 5 (Unit Cost compilation), and provides a reasonable guide as to comparative costs of common or HA Specification materials.

It must be remembered that transport costs form a high proportion of delivered stone prices: if a 20 tonne tipper costs £50.00 per hour to operate then it follows that for each hour journey time £2.50 will be added to each tonne delivered.

The above costs are based on approximately £3.00 per tonne transport cost. Minimum 16 tonne loads.

Allow 19-25p / tonne per mile adjustment to 20 mile inclusion if required.

Market conditions and commercial considerations could also affect these prices to a significant degree – up to double for the more selective materials. We have included such variations only where we feel that there is a sound reason for them and so have attempted to produce a reasonable guide to market conditions.

If specific job requirements are known, then local quarries are usually helpful in supplying information and guide prices.

SCAFFOLDING AND PROPS	Unit	£
Scaffold tube		
black tube	m	3.36
galvanized tube	m	3.36
alloy tube	m	6.72
Scaffold fittings		
double	nr	1.69
single	nr	1.13
swivel	nr	1.90
sleeve	nr	1.64
joint pin	nr	1.54
baseplate	nr	0.41
Scaffold boards		
grade A selected (3.97 m long)	nr	8.20
grade A selected (3.00 m long)	nr	7.20
Scaffold towers, alloy,coming with ladder access		
2.5 × 1.35 × 6.20 m platform height	nr	1,227.00
2.5 × 1.35 × 8.2 m platform height	nr	1,631.00
Adjustable struts		
size 0	nr	18.00
size 1	nr	23.00
size 2	nr	26.00
size 3	nr	28.00
Adjustable props		
size 0	nr	28.00
size 1	nr	33.00
size 2	nr	32.00
size 3	nr	34.00
size 4	nr	41.00
Lightweight stagings		
3.60 m	nr	153.00
4.20 m	nr	173.00
4.80 m	nr	229.00
5.40 m	nr	255.00
6.10 m	nr	288.00
6.70 m	nr	317.00
7.30 m	nr	345.00
Reinforced timber pole ladders		
4 m	nr	85.00
5 m	nr	103.00
6 m	nr	123.00
7 m	nr	144.00
8 m	nr	164.00
9 m	nr	185.00
10 m	nr	205.00
Health & Safety Signs	nr	2.50–25.00
Trench sheeting		
standard overlapping	m	11.00
L8 interlocking	m	18.00
Fence panels		
anti-climb, 2m high × 3.5m with block and coupler	nr	72.00

SEPTIC TANKS, CESSPOOLS AND INTERCEPTORS	Unit	£
Klargester 'Alpha' GRP septic tanks excluding cover and frame		
2800 litre capacity; 1000 mm invert	nr	630.00
2800 litre capacity; 1500 mm invert	nr	680.00
3800 litre capacity; 1000 mm invert	nr	860.00
3800 litre capacity; 1500 mm invert	nr	910.00
4600 litre capacity; 1000 mm invert	nr	970.00
4600 litre capacity; 1500 mm invert	nr	1,020.00
pedestrian cover and frame	nr	67.00
Klargester GRP cesspools including covers and frames		
18200 litre capacity; 1000 mm invert	nr	2,300.00
Klargester 3 stage GRP petrol interceptors		
2000 litre capacity	nr	1,000.00
2500 litre capacity	nr	1,260.00
4000 litre capacity	nr	1,450.00
Klargester Biodisc; self contained sewage treatment plant		
population equivalent of 6, 450 mm invert	nr	3,010.00
population equivalent of 6, 750 mm invert	nr	3,100.00
population equivalent of 6, 1250 mm invert	nr	3,160.00
population equivalent of 12, 450 mmm invert	nr	3,920.00
population equivalent of 12, 750 mm invert	nr	4,010.00
population equivalent of 12, 1250 mm invert	nr	4,080.00
population equivalent of 18, 600 mm invert	nr	5,010.00
population equivalent of 18, 1100 mm invert	nr	5,450.00
Klargester Single Effluent Pump Stations, GRP sump and cover		
1000 mm diameter, 1.0 m invert, 2340 mm total depth	nr	740.00
Klargester Bypass Interceptors, suitable for areas of low risk such as carparks		
NSB3 160 mm pvc bypass separator; 1,670 m² drainage area	nr	740.00
NSB4 160 mm pvc bypass separator; 2,500 m² drainage area	nr	830.00
NSB6 300 mm grp bypass separator; 3,335 m² drainage area	nr	890.00
NSB8 300 mm grp bypass separator; 4,445 m² drainage area	nr	1,090.00

SHUTTERING, TIMBER AND NAILS	Unit	£
Plywood exterior WBP (Hardwood ply) size 1220 × 2440 mm		
4 mm thick	m²	6.00
6 mm thick	m²	5.71
12 mm thick	m²	9.44
18 mm thick	m²	14.15
24 mm thick	m²	18.86
Plywood shuttering (Softwood ply) size 1220 × 2440 mm,		
12.0 mm thick	m²	10.00
18.0 mm thick	m²	13.00

Sawn softwood random lengths; good quality timber for formwork, falseworks, temporary works and trench supports, delivered to site

Thickness	£/m 25 mm	£/m 47 mm	£/m 75 mm
50 mm wide	0.40	0.67	–
75 mm wide	0.62	0.89	1.42
100 mm wide	0.82	1.18	1.90
150 mm wide	1.22	1.78	2.84
200 mm wide	1.63	2.37	3.79

Thickness	£/m 25 mm	£/m 38 mm	£/m 50 mm	£/m 75 mm	£/m 100 mm
Wrought (planed) timber; European soft-wood					
25 mm wide	0.38	–	–	–	–
38 mm wide	0.54	–	–	–	–
50 mm wide	0.58	0.93	1.04	–	–
75 mm wide	0.90	1.41	1.53	2.41	–
100 mm wide	1.12	1.81	2.04	3.22	5.31
125 mm wide	1.43	2.34	2.66	4.02	6.19
150 mm wide	1.79	2.90	3.28	4.83	7.44
175 mm wide	2.08	3.37	3.81	5.64	8.67
200 mm wide	2.46	3.97	4.46	6.44	9.81
225 mm wide	3.01	–	–	7.19	–

Structural softwood graded SC3 / 4				Unit	£
100 × 75 mm				m³	275.00
150 × 75 mm				m³	275.00
200 × 100 mm				m³	329.00
200 × 200 mm				m³	389.00
300 × 200 mm				m³	413.00
300 × 300 mm				m³	424.00

	Unit	Greenheart £	Opepe £	Ekki £
Construction hardwood (for piers, jetties, groynes, dolphins, etc.) Note: Medium price shown, actual can vary considerably dependent on quantity and specification; larger sizes may prove difficult if not impossible to obtain				
50 mm thick	m²	11.80	15.40	13.20
75 mm thick	m²	18.70	22.00	19.80
100 mm thick	m²	25.60	29.20	27.00

			Unit	£
Used timber beams, baulks, pit props (used for kerbing timbers, temporary barriers, plant supports, etc)				135–165
Nails; steel				
annular ring shank				
100 mm long			kg	2.23
50 mm long			kg	2.29
round plain head				
150 mm long			kg	1.82
100 mm long			kg	1.88
oval lost head				
50 mm long			kg	2.23
75 mm long			kg	1.99
oval brad head				
150 mm long			kg	2.68
100 mm long			kg	2.44
clout, plain head nails				
75 mm long			kg	3.92
panel pins				
20 mm long			kg	6.05
40 mm long			kg	6.05

	Unit	£
Coach screws; stainless steel; hexagon head		
M6 × 70 mm	nr	0.05
M8 × 70 mm	nr	0.06
M8 × 100 mm	nr	0.11
M10 × 100 mm	nr	0.14
M10 × 160 mm	nr	0.23
Metric high tensile steel bolts and nuts		
M6 × 50 mm	nr	0.05
M6 × 100 mm	nr	0.09
M8 × 50 mm	nr	0.08
M8 × 100 mm	nr	0.14
M10 × 50 mm	nr	0.13
M10 × 100 mm	nr	0.22
M12 × 50 mm	nr	0.18
M12 × 100 mm	nr	0.28
M12 × 200 mm	nr	0.75
M12 × 300 mm	nr	1.33
M16 × 50 mm	nr	0.32
M16 × 100 mm	nr	0.47
M16 × 200 mm	nr	1.10
M16 × 300 mm	nr	1.85
M20 × 50 mm	nr	0.56
Metric High Tensile steel bolts and nuts – continued		
M20 × 100 mm	nr	0.78
M20 × 200 mm	nr	1.63
M20 × 300 mm	nr	2.62
M24 × 50 mm	nr	1.23
M24 × 100 mm	nr	1.47
M24 × 200 mm	nr	2.39
M24 × 300 mm	nr	3.78
M30 × 100 mm	nr	3.96
M30 × 150 mm	nr	4.55
M30 × 200 mm	nr	5.13
Washers for metric mild steel bolts and nuts		
M8 Washer	nr	0.01
M10 Washer	nr	0.02
M12 Washer	nr	0.03
M16 Washer	nr	0.05
M20 Washer	nr	0.07
M24 Washer	nr	0.12
M30 Washer	nr	0.12
Metric mild steel setscrews		
M6 × 25 mm	nr	0.04
M6 × 60 mm	nr	0.05
M8 × 25 mm	nr	0.05
M8 × 40 mm	nr	0.06
M10 × 20 mm	nr	0.07
M10 × 100 mm	nr	0.16
M12 × 50 mm	nr	0.08
M12 × 75 mm	nr	0.20
M16 × 50 mm	nr	0.28
M16 × 100 mm	nr	0.47

	Unit	£
M20 × 50 mm	nr	0.50
M20 × 60 mm	nr	0.61
M24 × 50 mm	nr	0.73
M24 × 80 mm	nr	1.17
Stainless Steel Straps		
30 × 2.50 mm vertical; stainless steel; restraint strap		
800 mm girth	nr	6.90
1000 mm girth	nr	8.53
1200 mm girth	nr	10.08
1600 mm girth	nr	13.18
30 × 2.5 mm lateral; stainless steel; restraint strap		
800 mm girth	nr	10.66
1000 mm girth	nr	13.33
1200 mm girth	nr	16.00
1600 mm girth	nr	21.39
Joist fittings (galvanized)		
Splice plate		
400 × 57 × 18 mm	nr	1.22
550 × 82 × 18 mm	nr	1.34
560 × 98 × 18 mm	nr	2.00
Junction clip		
dry wall construction	nr	0.34
Toothplate single sided timber connector (galvanized)		
50 mm diameter	nr	0.34
63 mm diameter	nr	0.34
75 mm diameter	nr	0.34
Toothplate double sided timber connector (galvanized)		
50 mm diameter	nr	0.56
63 mm diameter	nr	0.81
75 mm diameter	nr	1.13
Split ring connector (galvanized)		
3 mm diameter × 5 mm thick	nr	2.36
101 mm diameter × 5 mm thick	nr	6.30
Shear plate connector (galvanized)		
67 mm diameter × 4 mm thick	nr	3.78
101 mm diameter × 6 mm thick	nr	16.37
Bonded polystyrene void formers circular in section		
225 mm diameter	m	1.02
300 mm diameter	m	1.81
450 mm diameter	m	4.08
600 mm diameter	m	7.25
750 mm diameter	m	11.33
Expanded polystyrene in blocks	m³	20.52

STEEL PIPES AND FITTINGS	Unit	£	£	£	£	£	£
Diameter in mm		100	125	150	200	250	300
Steel pipe to BS EN 10216 or equivalent	m	21.25	28.48	32.73	34.43	43.35	49.73
Fittings to BS 1640, standard strength							
90° bend, long radius	nr	18.70	41.65	45.48	90.53	164.48	242.68
45° bend, long radius	nr	14.03	29.75	34.00	71.40	129.20	174.25
single junction, equal	nr	47.18	83.30	85.43	167.88	320.45	499.80
single junction, 100 mm branch	nr	–	–	98.60	200.60	–	–
single junction, 150 mm branch	nr	–	–	–	200.60	–	–
single junction, 200 mm branch	nr	–	–	–	–	–	–
concentric reducers to 100 mm	nr	–	–	25.50	59.93	–	–
concentric reducers to 150 mm	nr	–	–	–	39.53	72.68	–
concentric reducers to 200 mm	nr	–	–	–	–	65.03	113.90
concentric reducers to 250 mm	nr	–	–	–	–	–	114.75
	Unit	£	£	£	£	£	£
Diameter in mm		75	110	160	200	250	
Stainless steel pipe, grade AISI 316	m	23.80	34.00	52.70	78.20		
Fittings							
90 degree bend	m	30.18	38.25	96.90	218.45	257.00	
45 degree bend	m	22.10	29.33	82.45	155.98	183.50	
90 degree branch	m	32.73	41.65	99.88	147.05	173.00	
	Unit		£	£	£	£	£
Diameter in mm			100	150	200	250	300
Fusion bonded epoxy coated steel pipe	m		169.50	99.00	169.50	191.00	242.50
Fittings							
90° bend; flanged	nr		219.00	266.00	401.00	932.50	821.00
45° bend; flanged	nr		135.00	240.00	921.00	542.00	684.00
Single junctions; flanged							
branch size 100 mm	nr		366.00	427.00	563.00	690.00	
branch size 150 mm	nr		90.00	510.00	658.00	722.00	
branch size 200 mm	nr		–	564.00	709.50	842.50	
branch size 250 mm	nr		–	–	791.00	926.00	
branch size 300 mm	nr		–	–	–	997.00	

STRUCTURAL STEELWORK

Note: The following basic prices are for basic quantities of BS EN10025-2 grade S275JR steel. In view of firming prices in the steel market, we would note that these prices are based at June 2010 and may be subject to surcharge. Transport charges are additional.

Based on delivery Middlesborough/Scunthorpe Railway Stations – refer to supplier for section availability at each location.

Quality	£/tonne		£/tonne
Basis specification and quality extras			
EN10025-2			
S275JR	Basis	S355JR	30.00
S275J0	25.00	S355J0	40.00
S275J2	30.00	S355J2	85.00
EN10025-4			
All Grades	P.O.A.		

Offshore			£/tonne				£/tonne
EN10225	Min Yield	Charpy V			Min Yield	Charpy V	
S355G1	355 MPa	-20°C	85.00	S355G11	355 MPa	-20°C	125.00
S355G4	355 MPa	-20°C	90.00	S355G11+M	355 MPa	-20°C	125.00
S355G4+M	355 MPa	-20°C	90.00	S355G12	355 MPa	-20°C	135.00
Shipbuilding							
Lloyds Register Grade A			Basis				
Other qualities			P.O.A.				

Universal Beams (UKB) (kg/m)			
1016 × 305 mm (222,249,272,314,349,393,438,487)	1350.00	406 × 178 mm (54,60,67,74,85)	840.00
914 × 419 mm (343,388)	880.00	406 × 140 mm (39,46,54)	840.00
914 × 305 mm (201,224,253,289)	880.00	356 × 171 mm (45,51,57,67)	840.00
838 × 292 mm (176,194,226)	870.00	356 × 127 mm (33,39)	840.00
762 × 267 mm (134,147,173,197)	870.00	305 × 165 mm (40,46,54)	830.00
686 × 254 mm (125,140,152,170)	870.00	305 × 127 mm (37,42,48)	830.00
610 × 305 mm (149,179,238)	860.00	305 × 102 mm (25,28,33)	830.00
610 × 229 mm (101,113,125,140)	840.00	254 × 146 mm (31,37,43)	820.00
610 × 178 mm (82,92,100)	840.00	254 × 102 mm (22,25,28)	830.00
533 × 312 mm (150,182,219,272)	840.00	203 × 133 mm (25,30)	820.00
533 × 210 mm (82,92,101,109,122,138)	840.00	203 × 102 mm (23)	820.00
533 × 165 mm (66,74,85)	840.00	178 × 102 mm (19)	820.00
457 × 191 mm (67,74,82,89,98,106,133,161)	830.00	152 × 89 mm (16)	830.00
457 × 152 mm (52,60,67,74,82)	840.00	127 × 76 mm (13)	830.00

Universal Columns,UKC (kg/m)	
356 × 406 mm (551,634)	895.00
356 × 406 mm (235,287,340,393,467)	885.00
356 × 368 mm (129,153,177,202)	885.00
305 × 305 mm (97,118,137,158,198,240,283)	840.00
254 × 254 mm (73,89,107,132,167)	830.00
203 × 203 mm (46,52,60,71,86,100,113,127)	830.00
152 × 152 mm (23,30,37,44,51)	820.00
Parallel Flange Channels, UKPFC (kg/m)	
Category A sizes	
125 × 65 mm (15), 150 × 75 mm (18), 180 × 75 mm (20), 200 × 75 mm (23)	805.00
Category B sizes	
430 × 100 mm (64)	1070.00
380 × 100 mm (54)	940.00
230 × 75 mm (26), 230 × 90 mm (32), 260 × 75 mm (28), 260 × 90 mm (35), 300 × 90 mm (41),	
300 × 100 mm (46)	910.00
150 × 90 mm (24), 180 x90 mm (26), 200 × 90 mm (30)	840.00
Equal & Unequal Angles (UKA) (thickness mm)	
Category A sizes	
100 × 100 mm (8,10,12,15), 125 × 75 mm (8,10,12), 120 × 120 mm (8,10,12,15)	790.00
Category B sizes	
150 × 75 mm (10,12,15), 150 × 150 mm (10,12,15,18), 150 × 90 mm (10,12,15)	800.00
200 × 200 mm (16,18,20,24), 200 × 100 mm (10,12,15)	880.00
200 × 150 mm (12,15,18)	1280.00
Asymmetric Slimflor Beams, ASB (S355JR)	
280 ASB (74,100,105,124,136)	1050.00
300 ASB (153,155,185,196,237,249)	1050.00
Quantity	
Each order or specification of one quality, one serial size of section and one thickness for one delivery to one destination	
10 tonnes and over	Basis
Under 10 tonnes to 5 tonnes	10.00
Under 5 tonnes to 2 tonnes	25.00
Under 2 tonnes (not normally available – refer to supplier)	50.00
Length	
Universal Beams, Columns and Bearing Piles (UKB, UKC and UKBP)	
6,000 mm to under 9,000 mm in 100 mm increments	15.00
9,000 mm to 18,500 mm in 100 mm increments	Basis
Over 18,500 mm to 24,000 mm in 100 mm increments	P.O.A.
Lengths over 24,000 mm – refer to supplier	–
Channels, Asymmetric Slimfor Beams and Angles (UKPFC, ASB and UKA)	
6,000 mm to 18,500 mm in 100 mm increments	Basis
Over 18,500 mm to 24,000 mm in 100 mm increments	P.O.A.
Lengths over 24,000 mm – refer to supplier	–

		E/O £/tonne
Tees cut from universal beams and columns and joists		
weight per metre of rolled section before splitting		
Less than 25 kg per metre		150.00
25–40 kg per metre		120.00
40–73 kg per metre		100.00
73–125 kg per metre		90.00
Over 125 kg per metre		85.00
Impact Testing – refer to supplier		P.O.A.
	Unit	£
Shotblasting and priming		
Epoxy Zinc Phosphate primer to universal beams and columns	m²	2.75
Epoxy Zinc Phosphate primer to channels and angles	m²	2.80
Zinc rich epoxy primer to universal beams and columns	m²	3.80
Zinc rich epoxy primer to channels and angles	m²	3.70

Note: The following prices include end user discounts as advised at May 2010 for quantities of 10 tonnes and over in one size, thickness, length, steel grade and surface finish and include delivery to mainland of Great Britain to one destination. Additional costs for variations to these factors vary between sections and should be ascertained from the supplier.

The following lists are not fully comprehensive and for other sections manufacturer's price lists should be consulted.

Hot formed structural hollow section	kg/m	Approx metres per tonne (m)	S355J2H £/tonne	Grade 50D £/100m
Square				
100 × 100 × 10.0 mm	27.40	36.5	930	2536
120 × 120 × 5.0 mm	17.80	56.2	900	1597
120 × 120 × 6.3 mm	22.20	45.0	900	1991
120 × 120 × 8.0 mm	27.60	36.2	900	2475
120 × 120 × 10.0 mm	33.70	29.7	930	3119
120 × 120 × 12.5 mm	40.90	24.4	1020	4155
140 × 140 × 5.0 mm	21.00	47.6	900	1884
140 × 140 × 6.3 mm	26.10	38.3	900	2341
140 × 140 × 8.0 mm	32.60	30.7	900	2924
140 × 140 × 10.0 mm	40.00	25.0	930	3703
140 × 140 × 12.5 mm	48.70	20.5	1020	4947
150 × 150 × 5.0 mm	22.60	44.2	900	2027
150 × 150 × 6.3 mm	28.10	35.6	900	2520
150 × 150 × 8.0 mm	35.10	28.5	900	3148
150 × 150 × 10.0 mm	43.10	23.2	930	3990
150 × 150 × 12.5 mm	52.70	19.0	1020	5353
160 × 160 × 5.0 mm	24.10	41.5	950	2287
160 × 160 × 6.3 mm	30.10	33.2	950	2856
160 × 160 × 8.0 mm	37.60	26.6	950	3567
160 × 160 × 10.0 mm	46.30	21.6	950	4393
160 × 160 × 12.5 mm	56.60	17.7	990	5588
180 × 180 × 6.3 mm	34.00	29.4	950	3226
180 × 180 × 8.0 mm	42.70	23.4	950	4051
180 × 180 × 10.0 mm	52.50	19.0	950	4981
180 × 180 × 12.5 mm	64.40	15.5	990	6358
180 × 180 × 16.0 mm	80.20	12.5	1010	8128
200 × 200 × 5.0 mm	30.40	32.9	950	2884
200 × 200 × 6.3 mm	38.00	26.3	950	3605
200 × 200 × 8.0 mm	47.70	21.0	950	4526

Hot formed structural hollow section	kg/m	Approx metres per tonne (m)	S355J2H £/tonne	Grade 50D £/100m
200 × 200 × 10.0 mm	58.80	17.0	950	5579
200 × 200 × 12.5 mm	72.30	13.8	990	7138
200 × 200 × 16.0 mm	90.30	11.1	1010	9152
250 × 250 × 6.3 mm	47.90	20.9	950	4545
250 × 250 × 8.0 mm	60.30	16.6	950	5721
250 × 250 × 10.0 mm	74.50	13.4	950	7066
250 × 250 × 12.5 mm	91.90	10.9	990	9073
250 × 250 × 16.0 mm	115.00	8.7	1010	11655
300 × 300 × 6.3 mm	57.80	17.3	950	5484
300 × 300 × 8.0 mm	72.80	13.7	9480	69007
300 × 300 × 10.0 mm	90.20	11.1	950	8558
300 × 300 × 12.5 mm	112.00	8.9	990	11058
300 × 300 × 16.0 mm	141.00	7.1	1010	14291
350 × 350 × 8.0 mm	85.40	11.7	950	8103
350 × 350 × 10.0 mm	106.00	9.4	950	10057
350 × 350 × 12.5 mm	131.00	7.6	990	12934
350 × 350 × 16.0 mm	166.00	6.0	1010	16824
400 × 400 × 10.0 mm	122.00	8.2	950	11575
400 × 400 × 12.5 mm	151.00	6.6	990	14908
400 × 400 × 16.0 mm	191.00	5.2	1010	19358
Rectangular				
120 × 80 × 10.0 mm	27.40	36.5	930	2536
150 × 100 × 5.0 mm	18.60	53.8	900	1668
150 × 100 × 6.3 mm	23.10	43.3	900	2072
150 × 100 × 8.0 mm	28.90	34.6	900	2592
150 × 100 × 10.0 mm	35.30	28.3	930	3268
150 × 100 × 12.5 mm	42.80	23.4	1020	4348
160 × 80 × 4.0 mm	14.40	69.4	900	1292
160 × 80 × 5.0 mm	17.80	56.2	900	1597
160 × 80 × 6.3 mm	22.20	45.0	900	1991
160 × 80 × 8.0 mm	27.60	36.2	900	2475
160 × 80 × 10.0 mm	33.70	29.7	930	3119
200 × 100 × 5.0 mm	22.60	44.2	900	2027
200 × 100 × 6.3 mm	28.10	35.6	900	2520
200 × 100 × 8.0 mm	35.10	28.5	900	3148
200 × 100 × 10.0 mm	43.10	23.2	930	3990
200 × 100 × 12.5 mm	52.70	19.0	1020	5353
200 × 120 × 5.0 mm	24.10	41.5	950	2287
200 × 120 × 6.3 mm	30.10	33.2	950	2856
200 × 120 × 8.0 mm	37.60	26.6	950	3567
200 × 120 × 10.0 mm	46.30	21.6	950	4393
250 × 100 × 10.0 mm	41.40	24.2	950	3928
250 × 100 × 12.5 mm	58.80	17.0	820	4839
250 × 150 × 8.0 mm	47.70	21.0	820	3928
250 × 150 × 10.0 mm	58.80	17.0	820	4839
300 × 200 × 6.3 mm	47.90	20.9	950	4545
300 × 200 × 8.0 mm	60.30	16.6	950	5721
300 × 200 × 10.0 mm	74.50	13.4	950	7068
300 × 200 × 12.5 mm	91.90	10.9	990	9073
300 × 200 × 16.0 mm	115.00	8.7	1010	11655
400 × 200 × 8.0 mm	72.80	13.7	950	6907
400 × 200 × 10.0 mm	90.20	11.1	950	8558
400 × 200 × 12.5 mm	112.00	8.9	990	11058

Hot formed structural hollow section	kg/m	Approx metres per tonne (m)	S355J2H £/tonne	Grade 50D £/100m
400 × 200 × 16.0 mm	141.00	7.1	1010	14291
450 × 250 × 8.0 mm	85.40	11.7	950	8103
450 × 250 × 10.0 mm	106.00	9.4	950	10057
450 × 250 x12.5 mm	131.00	7.6	990	12934
450 × 250 x16.0 mm	166.00	6.0	1010	16824
500 × 300 × 8.0 mm	98.00	10.2	950	9298
500 × 300 × 10.0 mm	122.00	8.2	950	11575
500 × 300 × 12.5 mm	151.00	6.6	990	14908
500 × 300 × 16.0 mm	191.00	5.2	1010	19358
Circular				
219.1 × 5.0 mm	26.40	37.9	1020	2689
219.1 × 6.3 mm	33.10	30.2	1020	3371
219.1 × 8.0 mm	41.60	24.0	1020	4237
219.1 × 10.0 mm	51.60	19.4	1020	5255
219.1 × 12.5 mm	63.70	15.7	1060	6776
244.5 × 8.0 mm	46.70	21.4	1020	4756
244.5 × 10.0 mm	57.80	17.3	1020	5886
244.5 × 12.5 mm	71.50	14.0	1060	7605
244.5 × 16.0 mm	90.20	11.1	1100	9880
273.0 × 6.3 mm	41.40	24.2	1020	4216
273.0 × 8.0 mm	52.30	19.1	1020	5326
273.0 × 10.0 mm	64.90	15.4	1020	6609
273.0 × 12.5 mm	80.30	12.5	1060	8541
273.0 × 16.0 mm	101.00	9.9	1100	11063
323.9 × 6.3 mm	49.30	20.3	1020	5021
323.9 × 8.0 mm	62.30	16.1	1020	6345
323.9 × 10.0 mm	77.70	12.9	1020	7913
323.9 × 12.5 mm	96.00	10.4	1060	10211
323.9 × 16.0 mm	121.00	8.3	1100	13254
355.6 × 16.0 mm	134.00	7.5	1100	14678
406.4 × 10.0 mm	97.80	10.2	1020	9960
406.4 × 12.5 mm	121.00	8.3	1060	12870
406.4 × 16.0 mm	154.00	6.5	1100	16868
457.0 × 10.0 mm	110.00	9.1	1020	11202
457.0 × 12.5 mm	137.00	7.3	1060	14572
457.0 × 16.0 mm	174.00	5.7	1100	19059
508.0 × 10.0 mm	123.00	8.1	1020	12526
508.0 × 12.5 mm	153.00	6.5	1060	16274
508.0 × 16.0 mm	194.00	5.2	1100	21250
Ovals				
150 × 75 × 4.0 mm	10.70	93.5	1020	1090
150 × 75 × 5.0 mm	13.30	75.2	1020	1354
150 × 75 × 6.3 mm	16.50	60.6	1020	1680
200 × 100 × 5.0 mm	17.90	55.9	1020	1823
200 × 100 × 6.3 mm	22.30	44.8	1020	2271
200 × 100 × 8.0 mm	28.50	35.1	1020	2902
200 × 100 × 10.0 mm	34.50	29.0	1020	3513
200 × 100 × 12.5 mm	42.40	23.6	1060	4510
250 × 125 × 6.3 mm	28.20	35.5	1020	2872
250 × 125 × 8.0 mm	34.40	29.1	1020	3503
250 × 125 × 10.0 mm	43.80	22.8	1020	4461
250 × 125 × 12.5 mm	53.90	18.6	1060	5733
300 × 150 × 8.0 mm	42.80	23.4	1020	4359

Hot formed structural hollow section	kg/m	Approx metres per tonne (m)	S355J2H £/tonne	Grade 50D £/100m
300 × 150 × 10.0 mm	53.00	18.9	1020	5398
300 × 150 × 12.5 mm	65.50	15.3	1060	6967
300 × 150 × 16.0 mm	82.50	12.1	1100	9037
400 × 200 × 8.0 mm	57.60	17.4	1020	5866
400 × 200 × 10.0 mm	71.50	14.0	1020	7282
400 × 200 × 12.5 mm	88.60	11.3	1060	9424
400 × 200 × 16.0 mm	112.00	8.9	1100	12268
500 × 250 × 10.0 mm	90.00	11.1	1020	9166
500 × 250 × 12.5 mm	112.00	8.9	1060	11913
500 × 250 × 16 mm	142.00	7.0	1100	15554

EARTH RETENTION – CRIB WALLING

Add to the aforementioned prices for:	Percentage extra (%)
Quantity	
- Work despatches	
Orders for the following hollow sections of one size, thickness, length, steel grade and surface finish.	
(a) circular hollow sections over 200 mm diameter	
(b) square hollow sections over 150 × 150 mm	
(c) rectangular hollow sections over 600 mm girth	
4 tonnes to under 10 tonnes	15%
2 tonnes to under 4 tonnes	20%
1 tonne to under 2 tonnes	25%
orders under 1 tonne are not supplied	
- Warehouse despatches	
Orders for the following hollow sections in one steel grade, for delivery to one destination in one assignment.	
(a) circular hollow sections less than 200 mm diameter	
(b) square hollow sections up to and including 150 × 150 mm	
(c) rectangular hollow sections up to and including 600 mm girth	
10 tonnes and over	7.5%
4 tonnes to under 10 tonnes	10%
2 tonnes to under 4 tonnes	12.5%
1 tonne to under 2 tonnes	17.5%
500 kg to under 1 tonne	23%
250 kg to under 500 kg	35%
100 kg to under 250 kg	50%
under 100 kg	100%
(a) circular hollow sections over 200 mm diameter	
(b) square hollow sections over 150 × 150 mm	
(c) rectangular hollow sections over 600 mm girth	
10 tonnes and over	12.5%
4 tonnes to under 10 tonnes	15%
2 tonnes to under 4 tonnes	20%
1 tonne to under 2 tonnes	25%
500 kg to under 1 tonne	45%
250 kg to under 500 kg	75%
Finish	
Self colour is supplied unless otherwise specified.	
Transit primer painted (All sections except for circular hollow sections over 200 mm dia)	5%
Test Certificates	
Test certificates will be charged at a rate of 'price on application'	

PVC-U DRAIN PIPES AND FITTINGS	Unit	£	£	£
Diameter in mm		82	110	160
Plain ended pipes (Osmadrain)				
3 m lengths	m	15.46	9.45	21.73
Couplers				
for jointing pipes	nr	12.23	12.63	18.35
for new branch entry connections	nr	19.32	17.03	33.95
Reducers, single socket	nr	–	23.41	30.12
Junctions, equal, single socket				
87.5°	nr	32.22	35.18	114.84
45°	nr	36.72	104.95	95.41
Short radius bends, single socket				
up to 30°	nr	–	21.25	63.58
45 to 90°	nr	26.22	24.92	63.69
Suspended bracketing system adjustable pipe bracket assembly (pack containing threaded bracket, bracket plate and pipe/socket bracket)	nr	–	30.44	43.40
Adjustable socket bracket and brace assembly (pack containing threaded rod, threaded bracket, bracket plate, two adjustable braces and pipe/socket bracket)	nr	–	62.65	73.57
Marley underground drainage system				
Straight pipes				
ring seal socket 6 m	m	–	10.04	26.76
ring seal socket 3 m	m	15.62	10.86	30.17
double spigot 6 m	m	–	9.26	23.33
double spigot 3 m	m	–	9.26	–
Couplings				
double ring seal straight, polypropylene	nr	10.29	8.55	28.44
loose pipe socket	nr	10.29	8.55	22.84
triple socket	nr	–	27.60	–
Bends, socket/spigot				
short radius (87.5°)	nr	20.78	21.75	54.02
short radius (45°)	nr	20.78	23.52	48.85
adjustable (11 to 87.5°)	nr	35.07	–	–
adjustable (21 to 90°)	nr	–	41.30	–
adjustable (15 to 90°)	nr	–	–	78.90
Diameter in mm		82	110	160
Bends, socket/socket				
short radius (87.5°)	nr	–	21.87	62.16
short radius (15, 30, 45°)	nr	–	18.88	56.11
long radius (87.5°)	nr	–	37.89	–
Branches				
socket/spigot 45° equal	nr	32.88	32.41	63.22
socket/spigot 87.5° equal	nr	–	32.41	63.15
socket/spigot 45°, 160 × 110 mm	nr	–	–	54.84
socket/socket 87.5° equal	nr	–	35.08	88.54
socket/socket 45° equal	nr	–	35.08	88.54

PVC-U DRAIN PIPES AND FITTINGS	Unit	£	£	£
Access components				
socket/spigot pipe	nr	–	81.80	–
socket/spigot bend 87.5° rear access	nr	–	43.94	–
socket/spigot branches				
45° branch	nr	–	120.28	–
45° double branch	nr	–	114.23	–
rodding point 45° socketed	nr	–	71.01	184.54
cap and pressure plug	nr	–	22.40	57.12
cap	nr	18.66	–	–
pressure plug	nr	–	17.14	29.62
socket plug, 1 boss upstand	nr	–	11.28	–
Open channels				
straight double spigot (1500 mm long)	nr	–	68.05	–
long radius channel bend 87.5°	nr	–	97.60	–
slipper bend	nr	–	31.39	–
Bottle Gully				
bottle gully	nr	–	81.83	–
sealed access lid	nr	–	13.25	–
Reducers				
level invert				
110 mm spigot to 82 mm socket	nr	–	20.78	–
160 mm spigot to 110 mm socket	nr	–	-	23.31
eccentric				
82 mm socket to boss upstand	nr	17.46	-	–
82 mm socket to 68 mm socket	nr	7.48	-	–
110 mm socket to boss upstand	nr	–	12.66	–
110 mm socket to 68 mm socket	nr	–	10.46	–
concentric				
110 mm socket to boss upstand	nr	–	11.28	–
Diameter in mm		82	110	160
Adapters				
PVC-U spigot to salt glazed/pitch fibre socket	nr	–	14.73	–
PVC-U spigot to salt glazed/cast iron to pvc-u socket	nr	–	28.62	–
PVC-U spigot to thin wall clay spigot	nr	–	33.47	–
PVC-U spigot to thick wall clay spigot	nr	–	33.47	–
Access point covers				
450 mm lid and frame (A15 loading)	nr	–	98.92	–
Inspection chambers				
450 mm chamber base 230 mm high	nr	–	180.87	225.60
450 mm chamber riser 400 mm high	nr	–	68.43	–
450 mm cast iron cover and frame	nr	–	115.75	–
450 mm ductile iron lid and cast iron frame	nr	–	165.82	–
250 mm equal double branch chamber base	nr	–	122.64	–
250 mm double branch chamber base	nr	–	99.35	–
250 mm bottom outlet chamber body 4 × 110 mm upstands	nr	–	124.91	–
250 mm pressure plug	nr	–	61.02	–
250 mm chamber riser 375 mm long	nr	–	37.66	–
lifting handle	nr	–	20.18	–
square lid and frame	nr	–	99.29	–
PVC-U lid and frame	nr	–	48.86	–

PVC-U DRAIN PIPES AND FITTINGS	Unit	£	£	£
Gully components				
compact gully, 45° outlet	nr	–	94.14	–
gully trap base, 45° outlet	nr	–	24.24	–
P' trap gully 81.5° outlet	nr	–	60.82	–
hoppers (rectangular)	nr	–	24.24	–
hoppers (square)	nr	–	18.64	–
inlet raising pieces				
2 × 82 mm upstands	nr	–	12.02	–
4 boss upstands	nr	–	20.75	–
grating assembly	nr	–	13.00	–
Diameter in mm		150	225	300
Marley Quantum underground drainage system				
Straight pipes				
plain pipe 6 m	m	7.82	22.32	32.82
pipe 6 m with coupling and seals	m	8.90	23.70	36.05
Couplings				
straight or slip	nr	12.37	30.20	58.66
Bends				
double socket 87.5°	nr	24.89	119.64	214.51
double socket, 15, 30 and 45°	nr	21.62	85.84	159.64
Branches				
all socket, equal	nr	43.25	201.61	466.28
all socket, 150 × 110 mm	nr	37.99	-	–
all socket, 225 × 110 mm	nr	–	123.13	–
all socket, 225 × 150 mm	nr	–	102.47	–
all socket, 300 × 110 mm	nr	–	–	194.49
all socket, 300 × 150 mm	nr	–	–	166.67
all socket, 300 × 225 mm	nr	–	–	466.28
Reducers				
level invert 225 mm to 150 mm	nr	–	62.59	–
level invert 300 mm to 225 mm	nr	–	–	135.88
Plugs				
socket plug	nr	16.78	–	–
end cap	8.16	8.16	48.06	114.67
Adapters				
pipe to clayware	nr	34.53	–	–
pipe spigot to pvc socket	nr	21.46	–	–
flexible adaptors	nr	69.98	86.12	214.26

PVC-U SUBSOIL DRAIN PIPES AND FITTINGS

		Unit	£	
Diameter in mm		80	100	
WavinCoil subsoil drainage system				
S/S coils, 25 m coils	m	1.73	2.83	
end caps	nr	3.00	3.50	
couplers	nr	2.609	2.90	
reducers 100 to 80 mm	nr	–	3.20	
equal junctions, 67.5 degree	nr	7.40	8.30	
unequal junctions, 67.5 degree	nr	–	7.90	

	Unit	£	£	£
Diameter in mm		150	225	300
Ultra-Rib system				
S/S pipe (3 m length)	m	11.17	27.68	41.53
D/S pipe coupler	nr	22.00	49.40	98.50
D/S short radius bend, 45 degree	nr	64.20	149.00	235.00
D/S equal junction, 45 degree	nr	66.60	160.00	486.00
Ultra-Rib Inspection Chambers, 450 mm diameter				
D/S base, 150 × 150 mm	nr	193.00	–	–
shaft, 230 mm effective length	nr	53.40	–	–
DI cover and frame, single seal, Class BS EN 124-B125				
medium duty	nr	168.00	–	–
steel cover and polypropylene frame, to 25kN loading	nr	101.00	–	–
Ultra-Rib manhole bases, 750 mm diameter				
P/E unequal, 150 × 110 mm	nr	341.59	–	–
P/E unequal, 225 × 150 mm	nr	–	453.19	–
P/E equal, 150 × 150 mm	nr	369.36	–	–
P/E equal, 225 × 225 mm	nr	–	365.69	–
P/E channel access pipe	nr	75.64	173.40	291.26
Ultra-Rib sealed rodding eye				
oval cover, up to 35 kN loading	nr	178.23	–	–

WATERPROOFING

	Unit	£
Visqueen		
Visqueen Polythene Dpm, 6/83 A, 250 Micron (100 m² Roll)	roll	75.94
Visqueen Polythene Dpm, 6/83 A, 500 Micron (50 m² Roll)	roll	91.55
Aquaseal		
Aquaseal; 2 Coats (5 Litre)	litre	12.00
Ventrot		
Ventrot Primer, 1 Coat (25 Litre)	litre	3.17
Ventrot , 1 Coat (100 Kg Drum)	kg	1.80
Mastic Asphalt		
Mastic Asphalt Bs 6925, Type T 1097	t	219.02
Mastic Asphalt Bs 6925, Type R 988	t	290.37
Mastic Asphalt Ancilliaries		
Limestone Chippings; 10 mm	t	36.95
Bitumen	litre	6.72
Solar Reflective Paint	litre	34.16
300 × 300 × 8 GRP Tiles	nr	2.73
Cold applied waterproofing membrane (Mulseal DP)		
200 litre drum	litre	1.43
Hot bonded reinforced bitumen sheet waterproofing membrane (Famguard GS100)		
1 m × 8 m roll	m²	7.60
Bituthene rolls		
2000 grade (20 m² – over 75 Rolls)	m²	3.80
3000 grade (20 m² – over 75 Rolls)	m²	5.36
4000 grade (20 m² – over 75 Rolls)	m²	7.06
8000 grade (20 m² – over 75 Rolls)	m²	7.59
B2 Primer (5 litre can) over 40 cans	litre	3.86
Primer No 3 (5 litre)	litre	5.89
MR dpc (30 m × 600 mm wide)	m	14.20
Biodegration resistant sealant (Nitoseal 12)		
sealant (2 litre tin) gun grade	tin	47.57
accelerator (50 ml tin) gun grade	tin	13.29
Cold applied bituminous sheet waterproofing membrane (Proofex)		
3000, non reinforced (20 m² roll)	m²	7.00
3000MR, reinforced (20 m² roll)	m²	10.46
primer (5 litre can)	litre	24.77

Spon's First Stage Estimating Handbook

3rd Edition

By **Bryan Spain**

Have you ever had to provide accurate costs for a new supermarket or a pub "just an idea…a ballpark figure…" ?

The earlier a pricing decision has to be made, the more difficult it is to estimate the cost and the more likely the design and the specs are to change. And yet a rough-and-ready estimate is more likely to get set in stone.

Spon's First Stage Estimating Handbook is the only comprehensive and reliable source of first stage estimating costs. Covering the whole spectrum of building costs and a wide range of related M&E work and landscaping work, vital cost data is presented as:

- Costs per square metre
- Principal rates
- Elemental cost analyses
- Composite rates

Compact and clear, *Spon's First Stage Estimating Handbook* is ideal for those key early meetings with clients. And with additional sections on whole life costing and general information, this is an essential reference for all construction professionals and clients making early judgements on the viability of new projects.

January 2010: 216x138: 244pp
Pb: 978-0-415-54715-4: **£45.00**

INTRODUCTION

The information, rates and prices included in this section are calculated examples of actual owning and operating costs of a range of construction plant and equipment. To an extent they will serve as a guide to prevailing commercial plant hire rates, but be aware that many factors will influence actual hire rates. For example, rates could be lower because of long term hire, use of older or second hand machines, low market demands/loss leaders, easy tasks or sites near to the depot. Rates could be higher due to short term hire, high production demands, low utilization factors, specialist operations, restricted working and restricted accessibility problems, high profit or overhead demands and finance fluctuations. The use factors will mean a hired machine may not be used productively for each hour on site, it is therefore up to the estimator to allow for reasonable outputs in his unit rate calculations.

These and other considerations must be borne in mind if using these costs as a comparison to hire rates, especially operated plant.

All rates quoted EXCLUDE the following:

• Cost of drivers / operators
• VAT

CONTRACTOR OWNED PLANT

Plant owned by a Contractor generally falls under two headings:

• Small plant and tools which are the subject of a direct charge to contracts and for estimating purposes are normally allowed for as a percentage of the labour cost in site oncosts (see Preliminaries and General Items), although many items are shown in this section for information.
• Power driven plant and major non-mechanical plant such as steel trestling, scaffolding, gantries etc. Such plant is normally charged to the contract on a rental basis except in the case of purpose-made or special plant bought specifically for a particular operation; this latter is normally charged in full to contracts and allowance made for disposal on completion (often at scrap value).

A very wide range of plant is readily available from plant hire companies; it is not usually economical for Contractors to own plant unless they can ensure at least 75 to 80% utilization factor based on the Contractor's normal working hours. Where a Contractor does own plant, however, it is essential that he maintains reasonably accurate records of the working hours and detailed costs of maintenance and repairs in order that he may estimate the charges to be made for each item of plant. For this reason, where a Contractor owns a large quantity of plant, it is normal for a separate plant hire department or company to be formed.

This department or company should be financially self supporting and may hire plant to other Contractors when it is not needed. Maintenance of contractor owned plant can be carried out on site; it is not necessary to maintain a centrally based repair workshop. It is, however, desirable to have some storage facilities available for plant when not in use. The cost of owning plant and hence the rental charges to be made to contracts must take into account:

• Capital cost
• Depreciation charges
• Maintenance and repairs
• Cost of finance
• Insurances and licences
• Administration, head office, depot and other overhead charges

Calculated examples of hourly owning costs of a range of plant and equipment.

PLANT ITEMS	A UTILISATION FACTOR %	B P.A. hours	C PLANT years	D LIFE hours	E PURCHASE PRICE £	F RESALE %	G RESALE PRICE £	H LOSS IN VALUE £	I DEPREC £	J MAINTEN £	K FINANCE @ 5% £	L INSUR. £	M ADMIN £	N TOTAL COST/HOUR £
ACCESS PLATFORMS														
JLG 450AJ, diesel	65	1,365	10	13,650	35,000	20.00	7,000	28,000	2.05	2.56	0.51	0.51	1.13	6.76
JLG 600AJ, diesel	65	1,365	10	13,650	55,000	20.00	11,000	44,000	3.22	4.03	0.81	0.81	1.77	10.64
ASPHALT PAVERS														
BK 151	65	1,365	7	9,555	75,000	27.50	20,625	54,375	5.69	5.49	1.00	1.10	2.66	15.94
BK 181	65	1,365	7	9,555	115,000	27.50	31,625	83,375	8.73	8.42	1.53	1.68	4.07	24.43
BK 191	65	1,365	7	9,555	120,000	27.50	33,000	87,000	9.11	8.79	1.59	1.76	4.25	25.50
COMPRESSORS (PORTABLE)														
17 m³/min, electric motor	85	1,785	9	16,065	50,000	20.00	10,000	40,000	2.49	2.80	0.56	0.56	1.28	7.69
7.3 m³/min, electric motor	85	1,785	9	16,065	25,000	20.00	5,000	20,000	1.24	1.40	0.28	0.28	0.64	3.84
CRANES														
Kato 16 t mobile	60	1,260	12	15,120	180,000	25.00	45,000	135,000	8.93	14.29	2.68	2.86	5.75	34.51
Grove 35 t mobile	60	1,260	15	18,900	200,000	27.50	55,000	145,000	7.67	15.87	2.88	3.17	5.92	35.51
GENERATORS														
10 kVA, water cooled	85	1,785	5	8,925	6,000	15.00	900	5,100	0.57	0.34	0.07	0.07	0.21	1.26
10 kVA, air cooled	85	1,785	5	8,925	5,000	15.00	750	4,250	0.48	0.28	0.06	0.06	0.18	1.06
EXCAVATORS														
Hitachi ZX130 B/A track	80	1,680	9	15,120	80,000	27.50	22,000	58,000	3.84	4.76	0.86	0.95	2.08	12.49
Hitachi ZX800 B/A track	75	1,575	9	14,175	375,000	27.50	103,125	271,875	19.18	23.81	4.32	4.76	10.41	62.48
JCB 3CX	85	1,785	6	10,710	48,000	27.50	13,200	34,800	3.25	2.69	0.49	0.54	1.39	8.36
ROLLERS														
Bomag BW65H	85	1,785	6	10,710	15,000	27.50	4,125	10,875	1.02	0.84	0.15	0.17	0.44	2.62
Bomag BW80ADH	85	1,785	6	10,710	28,000	27.50	7,700	20,300	1.90	1.57	0.28	0.31	0.81	4.87
(narrow, heavy duty)														
Bomag BW75H	85	1,785	6	10,710	28,000	27.50	7,700	20,300	1.90	1.57	0.28	0.31	0.81	4.87
CRAWLER LOADERS														
CAT 963 C	80	1,680	9	15,120	140,000	27.50	38,500	101,500	6.71	8.33	1.51	1.67	3.64	21.86
TIPPERS														
Scania P94 CB 4	90	1,890	6	11,340	45,000	22.00	9,900	35,100	3.10	2.38	0.46	0.48	1.28	7.70
Scania P114 CB 8	90	1,890	6	11,340	70,000	22.00	15,400	54,600	4.81	3.70	0.72	0.74	1.99	11.96
DUMP TRUCKS														
Volvo A35 D	75	1,575	10	15,750	290,000	30.00	87,000	203,000	12.89	18.41	3.22	3.68	7.64	45.84

NOTES: The price of foreign manufactured plant will vary according to the strength of the £ Sterling internationally.
No allowance is included for Road Fund Licence on road-going vehicles.

Notes – example plant ownership hourly costs

- The above costs would be updated annually or bi-annually, the rental rate being revised to ensure complete recovery of the costs associated with the item of plant and return of capital to enable the machine to be replaced at the end of its life with the Contractor. The purchase price must also be adjusted to ensure recovery of the replacement cost and not the original cost of purchase.
- Driver's wages and costs should be charged direct to site wages.
- Column A: The utilization factors used in conjunction with the period of ownership is the percentage of time (per annum) that an item of plant can be expected to be used productively on a site or job and therefore is a very important influence of hourly costs

(Note: Utilization factors are not the same as site utilization rates – see notes regarding Fuel Consumption)

Economical owning periods/use factors:	Life (years)	Usage (%)
Hydraulic Excavators, large	9	75
Hydraulic Excavators, medium	7	76
Hydraulic Excavators, mini	6	77
Dozers/Scrapers	10	78
Loaders/Shovels	9	79
Mobile Cranes	12	80
Crawler Cranes	15	81
Dump Trucks	8	82
Dumpers	6	83
Rollers/Compaction	6	84
Compressors/Generators	7 – 9	85
Diesel engine road vehicles	6 – 7	86
Petrol engine road vehicles	5 – 6	87

- Column B: The result of applying the utilization factor to 2,100 hours per annum.
- Columns C and D: The period over which the Contractor owns the plant and during which time the items of plant can be maintained at reasonable efficiency will vary according to the Contractor's experience, the type of plant under consideration and the work on which is it employed. The owning period of an item of plant can vary considerably on this basis. The data we have used in our calculations varies considerably (see the tables shown above and on the next page).
- Column E: The list price being quoted by dealers, against which should be allowed a level of discount suited to the availability of the plant and the bargaining power between the purchaser and the agent.
- Column J: Maintenance costs cover major overhauls and replacement costs for wear items only, excluding insurable damage and will vary with each type of plant. This does not include day to day general servicing on site.
- Column K: Finance interest charges are taken on the average base rate over the previous 12 months plus 3%.
- Column L: Insurance premiums have been taken as 2% on the replacement value of the plant (purchase price)
- Column M: Administration, head office, depot and other overhead charges have been taken at +20%.
- Consumables should be charged direct to site costs. The costs included with this section are based upon manufacturers' data and are used in good faith. (For outline estimates fuel consumption can be taken as approximately 0.15 – 0.20 litres per kW per working hour; lube oils, filters, grease etc., can be taken as 2% to 10% of the fuel cost, depending upon the working conditions of the machine and its attachments). See below for details.
- Please note that the costs in this section have been reviewed for recent editions, based on the latest available advice and data from manufacturers and Plant Hire firms, the basis of calculation, including retention period of the plant, has been changed in a number of cases, providing a realistic overall owning and operating cost of the machine.

Resources – Plant

- These examples are for machine costs only and do not include for operator costs or profit element or for transport costs to and from site.
- Fuel consumption is based on the following site utilization rates. (Site utilization being the percentage of time that the machine is operating at its average fuel consumption during a working day).

Typical utilization factors:	Utilization factor	Typical utilization factors:	Utilization factor
Excavators	0.75	Compressors, mixers, generators	
Dozers/Scrapers	0.80	Tractors	0.75
Loaders/Shovels/Graders	0.75	Hoists, etc.	0.65
Mobile Cranes	0.50	Drills and Saws	0.90
Crawler Cranes	0.25	Rollers/Compaction Plant	0.75
Dump Trucks (not Tippers)	0.75	Piling/Asphalt Equipment	0.65
Dumpers	0.80	All other items	1.00

Approximate fuel consumption in litres per hour and percentage addition for consumables (examples only) based on the above site utilization rates, assuming that reasonably new and well maintained plant would be used.

		Fuel consumables				Fuel consumables	
		l/hr	% of fuel			l/hr	% of fuel
Tractor	116 hp	9.0	2.0%	Rollers	BW 90AD	2.0	7.5%
Paver	Bitelli BB650	16.0	5.0%		BW 120AD	4.4	7.5%
Compactor plate	188kg	0.9	3.0%		BW6 (towed)	7.6	7.5%
Compressor	3.5 m³/min	5.9	3.0%	Dozers	CAT D6N	21.0	6.0%
	10 m³/min	18.0	3.0%		CAT D8T	38.0	3.0%
	Tractor mounted	9.0	5.0%	Loaders	CAT 953C	18.0	4.0%
Mixer	5/3.5	1.2	2.0%	Scrapers	CAT 621G	41.5	3.0%
Mobile crane	15t	5.1	10.0%		CAT 657E	125.0	6.0%
Mobile crane	40t	6.3	10.0%	Skidsteer	Bobcat 553	3.4	5.0%
Generator	4 kVA	1.6	3.0%	Graders	CAT 140H	19.0	5.0%
	25 kVA	4.9	5.0%	Dumpers	2 tonne	3.0	3.0%
Backacter	11.5 t	9.5	7.5%		Volvo A25D	17.0	3.0%
	19 t	14.0	7.5%		36 tonne	25.0	5.0%
	30 t	24.0	7.5%	Pumps	75 mm 65 m³/hr	1.1	2.0%
Backhoes	JCB 3CX	7.5	6.0%		150 mm 360 m³/hr	4.2	3.0%

Note that the above consumption figures represent medium plant operation, i.e. the plant would not be operating at full throttle and working conditions such as the soil and grades are average. A study of manufacturers' figures has indicated that 'high' or 'low' usage can affect consumption on average by ± 25%; high usage would involve full loads, continuous working at full throttle with difficult ground and adverse grades, whereas low usage would involve more intermittent working with perhaps considerable idling periods, more easily worked ground and easier grades

PLANT COSTS

The costs included in this section are based upon the methodology of the worked examples, these have then been compared to average costs, allowances and hire rates in force at the time of writing (May 2010).

These costs then form the basis for plant costs included in parts 4 and 5 (except for specialist advice). Weekly costs are based upon a 40 hour week, daily costs are based on an 8 hour day, all costs are exclusive of labour for operation. Also given here is a reference (DSR) to the July 2009 CECA Daywork Schedule (Plant Section).
Rates given for consumables are in line with the notes on the preceding page without loss or wastage and are based on the following (All plant except those noted are priced on Gas Oil usage):

Petrol Unleaded	126.00	Pence per litre
Petrol Unleaded Super	128.00	Pence per litre
Fuel Oil (Taxed for use in licensed vehicles DERV	132.00	Pence per litre
Fuel Oil (Lower tax for use on site) Gas Oil	65.00	Pence per litre
Lubrication Oil (15/40)	66.80	Pence per litre
Mains Electric Power	12.50	Pence per kW hr
Oxygen	£20.92	per 10m^3
	21.00	Pence per litre
Acetylene	£59.60	per 8.4 m^3
	71.00	Pence per litre
Propane	120.50	Pence per kg

Also included here are allowances for Transmission Oils, Hydraulic Oil, Filters, Grease, etc.
N/A indicates that information not available.

Description	Hire Rate £	Hire Period	Cost per working hour					
			Plant £	Usage %	Fuel l/hr	Oil etc. % fuel	Total £	DSR
ACCESS PLATFORMS								
Access platform, scissor type,								
towed; height of platform 7 m	4.54	hour	4.54	–	–	–	4.54	1(8)
Access platform, scissor type,								
electric; height of platform 5 m	4.63	hour	4.63	–	–	–	4.63	1(1)
Access platform, scissor type,								
rough terrain, petrol driven; height of platform 9 m	9.62	hour	9.62	50	1.5	2	10.53	1(6)
Access platform, telescopic,								
towed; height of platform 10 m	4.63	hour	4.63	–	–	–	4.63	1(13)
Access platform telescopic,								
petrol; height of platform 15 m	18.08	hour	18.08	50	2	3	19.3	1(11)
Access platform, telescopic,								
vehicle mounted (DERV)								
height of platform 10 m	18.83	hour	18.83	50	2	3	20.08	1(16)
height of platform 20 m	28.93	hour	28.93	50	3	3	30.8	1(*)
Access platform under bridge type								
(Simon UB40)	35.08	hour	35.08	–	–	–	35.08	1(13)

Description	Hire Rate £	Hire Period	Cost per working hour					
			Plant £	Usage %	Fuel l/hr	Oil etc. % fuel	Total £	DSR
AGRICULTURAL TYPE TRACTORS								
Tractor 4.82 t, 4WD (75 kW)	18.38	hour	18.38	75	9	2	22.17	32(63)
Tractor (50 kW) c/w 1T Hiab lift	20.32	hour	20.32	75	9	2	24.11	32(62)
Tractor (50 kW)								
with front end loading bucket (0.33 m³)	8.36	hour	8.36	75	9	2	12.15	37(51)
with Hydro seeding equipment	8.36	hour	8.36	75	9	2	12.15	37(*)
Tractor (50 kW) includes 2 tool								
incl 2 tool compressor &| front end loading bucket	17.53	hour	17.53	75	10.5	3.5	22.01	–
Tractor (68HP) with Fencing Auger	8.36	hour	8.36	75	9	3.5	12.2	–
ASPHALT/ROAD CONSTRUCTION								
Asphalt pavers								
maximum paving width 3.60 m, 37 kW engine	23.48	hour	23.48	80	6.2	4	26.32	2(1)
extending up to 4 m, 56 kW engine	31.11	hour	31.11	80	8.2	5	34.9	2(1)
maximum paving width 9 m, 82 kW engine	41.51	hour	41.51	80	17	3	49.21	2(3)
Associated equipment								
Tar boiler & sprayer (1000 litre)	5.27	hour	5.27	80	9	3	9.35	29(6)
self propelled chip spreader	21.25	hour	21.25	80	9	3	25.33	2(9)
heating iron	1.22	hour	1.22	–	–	–	1.22	–
insulated (24.5 tonne) tipper (DERV)	28.03	hour	28.03	80	14	7	42.53	14(18)
3 point roller 8.5 t	18.4	hour	18.4	80	6	7.5	21.24	23(4)
Surface planers, cold plane								
planing width 0.5 m	31.93	hour	31.93	80	6	5	34.7	2(12)
planing width up to 2.25 m	150.51	hour	150.51	80	12	5	156.05	2(14)
Surface planers, heat plane								
planing width up to 4.5 m remixer	165.56	hour	165.56	80	13	5	171.57	2(-)
Concrete pavers								
depth 300 mm, 82 kW	46.53	hour	46.53	80	17	3	54.23	–
depth 500 mm 150 kW	63.14	hour	63.14	80	72	3	95.77	–
Concrete Slipform paver/trimmer								
maximum paving width 5 m	86.84	hour	86.84	80	24	3	97.72	–
Concrete Slipform joint and bar inserter	10.33	hour	10.33	80	4	3	12.14	–
Concrete Slipform finisher	15.6	hour	15.6	80	9	3	19.68	–
COMPACTION								
Plate compactors								
Vibrating compaction plate 80kg unit weight 360 mm wide (petrol)	80.67	week	2.02	85	0.8	3	2.85	22(6)
Vibrating compaction plate 150kg unit weight 400 mm wide (petrol)	78.67	week	1.97	85	1.2	3	3.22	22(8)
Vibrating compaction plate 300kg unit weight 600 mm wide	97.67	week	2.44	85	0.9	3	2.87	22(10)
Vibrating compaction plate 500 kg unit weight 600 mm wide	142.33	week	3.56	85	1.2	3	4.14	22(12)
Vibrating compaction trench plate 300 mm wide	80.67	week	2.02	85	0.8	3	2.41	22(*)
Tamper 60 kg (petrol)	75.33	week	1.88	85	0.8	3	2.71	22(1)

Description	Hire Rate £	Hire Period	Cost per working hour					
			Plant £	Usage %	Fuel l/hr	Oil etc. % fuel	Total £	DSR
CLEANERS/SWEEPERS								
Front Tractor or Truck mounted 2 m wide (excluding vehicle)	117.67	week	2.94	–	–	–	2.94	14(36)
Towed sweeper diesel engined	110.13	week	2.75	75	4	4	6.53	–
Glutton type mobile gully sucker	24.48	hour	24.48	75	12	3	35.7	–
MOBILE COMPRESSORS (SILENCED OR SUPER SILENCED)								
Compressors, single tool								
2.5 m³/min	274.33	week	6.86	85	3.2	4	8.42	4(1)
Compressors, two tool								
3.0 m³/min, two tool	312	week	7.8	85	4.5	3	9.97	4(2)
4.0 m³/min, two tool	331.67	week	8.29	85	5.9	3	11.13	4(3)
5.0 m³/min, two tool	376	week	9.4	85	6.8	3	12.67	4(4)
8.0 m³/min, two tool	480	week	12	85	10.2	3	16.91	4(6)
Compressors, four tool								
11.00 m³/min, four tool	644.67	week	16.12	85	19	3	25.27	4(7)
14.00 m³/min, four tool	875.33	week	21.88	85	32	3	37.29	4(8)
17.00 m³/min, four tool	1,140.67	week	28.52	85	32	3	43.93	4(9)
20.00 m³/min, four tool	1,502.33	week	37.56	85	47	3	60.19	4(10)
26.00 m³/min, four tool	2,076.67	week	51.92	85	43	3	72.63	4(11)
Petrol driven hydraulic power unit								
c/w breaker (63 kg)	0	week	0	85	1.5	3	0.72	–
c/w breaker (63 kg)	75.36	week	1.88	85	3	3	3.32	–
Tractair, wheeled loader mounted compressor, c/w hoses, 4 steels and 2 tarmac cutters	34.56	hour	34.56	85	12	3	40.34	4(12)
COMPRESSOR TOOLS								
(c/w 50 ft/15 m HOSE)								
chipping hammer, 0.7 m³/min	55.33	week	1.38	–	–	–	1.38	30(4)
brick hammer/demolition pick,1.4 m³/ min	13.45	week	0.34	–	–	–	0.34	–
clay spade,1.6 m³/min	65	week	1.63	–	–	–	1.63	30(3)
road breaker, 2.4 m³/min	69	week	1.73	–	–	–	1.73	30(1)
hand hammer drill, 0.9 m³/min	140.81	week	3.52	–	–	–	3.52	30(*)
light rock drill 16-20 kg	58.67	week	1.47	–	–	–	1.47	30(5)
medium rock drill up to 24 kg	73	week	1.83	–	–	–	1.83	30(6)
heavy rock drill up to 32 kg	51.33	week	1.28	–	–	–	1.28	30(7)
scabbler 1 head hand, 0.3 m³/min	45	week	1.13	–	–	–	1.13	–
scabbler 3 head hand, 0.7 m³/min	81	week	2.03	–	–	–	2.03	–
scabbler 5 head trolley, 3.1 m³/min	202.5	week	5.06	–	–	–	5.06	–
scabbler 7 head trolley, 4.5 m³/min	252	week	6.3	–	–	–	6.3	–
needle gun/descaler, 0.3 m³/min	45	week	1.13	–	–	–	1.13	30(15)
concrete crack cutter, 0.7 m³/min	108.72	week	2.72	–	–	–	2.72	–
rammer, 1.5 m³/min	75.33	week	1.88	–	–	–	1.88	22(1)
air lance, variable valve, 0.3 m³/min	17.26	week	0.43	–	–	–	0.43	–
impact wrench 12 mm drive, 0.4m³/min	40.36	week	1.01	–	–	–	1.01	–
impact wrench 25 mm drive, 0.4m³/min	68.29	week	1.71	–	–	–	1.71	–
chipping hammer, 0.7 m³/min	55.33	week	1.38	–	–	–	1.38	30(4)
brick hammer/demolition pick,1.4 m³/ min	13.45	week	0.34	–	–	–	0.34	–
clay spade,1.6 m³/min	65	week	1.63	–	–	–	1.63	30(3)
road breaker, 2.4 m³/min	69	week	1.73	–	–	–	1.73	30(1)
hand hammer drill, 0.9 m³/min	140.81	week	3.52	–	–	–	3.52	30(*)

Description	Hire Rate £	Hire Period	Cost per working hour					
			Plant £	Usage %	Fuel l/hr	Oil etc. % fuel	Total £	DSR
light rock drill 16-20 kg	58.67	week	1.47	–	–	–	1.47	30(5)
medium rock drill up to 24 kg	73	week	1.83	–	–	–	1.83	30(6)
heavy rock drill up to 32 kg	51.33	week	1.28	–	–	–	1.28	30(7)
scabbler 1 head hand, 0.3 m³/min	45	week	1.13	–	–	–	1.13	–
scabbler 3 head hand, 0.7 m³/min	81	week	2.03	–	–	–	2.03	–
scabbler 5 head trolley, 3.1 m³/min	202.5	week	5.06	–	–	–	5.06	–
scabbler 7 head trolley, 4.5 m³/min	252	week	6.3	–	–	–	6.3	–
needle gun/descaler, 0.3 m³/min	45	week	1.13	–	–	–	1.13	30(15)
concrete crack cutter, 0.7 m³/min	108.72	week	2.72	–	–	–	2.72	–
rammer, 1.5 m³/min	75.33	week	1.88	–	–	–	1.88	22(1)
air lance, variable valve, 0.3 m³/min	17.26	week	0.43	–	–	–	0.43	–
impact wrench 12 mm drive, 0.4m³/min	40.36	week	1.01	–	–	–	1.01	–
impact wrench 25 mm drive, 0.4m³/min	68.29	week	1.71	–	–	–	1.71	–
trench sheet driver, 2.4 m³/min	88.02	week	2.2	–	–	–	2.2	–
steel post driver (crash barrier), 2.4 m³/min	88.18	week	2.2	–	–	–	2.2	–
angle grinder 9", 0.8 m³/min	24	week	0.6	–	–	–	0.6	–
disc cutter 12", 1.4 m³/min	57.71	week	1.44	–	–	–	1.44	–
poker vibrator P35 (35 mm), 2.4 m³/min	41.33	week	1.03	–	–	–	1.03	6(24)
poker vibrator P54 (54 mm), 2.4 m³/min	49.6	week	1.24	–	–	–	1.24	6(24)
poker vibrator P70 (70 mm), 2.4 m³/min	57.87	week	1.45	–	–	–	1.45	6(24)
rotary drill, 0.3 m³/min	20.67	week	0.52	–	–	–	0.52	–
extra air hose, 25mm × 15 m hose	2	week	0.05	–	–	–	0.05	–
extra air hose, 38mm × 15 m hose	3	week	0.08	–	–	–	0.08	–
CONCRETE EQUIPMENT								
Poker vibrators								
air pokers (see compressor tables)								
50 mm, petrol	106	week	2.65	75	2	2	4.58	6(21)
50 mm, electric	60.33	week	1.51	–	–	–	1.51	6(23)
75 mm, diesel	104.67	week	2.62	75	1.5	2	3.25	6(22)
75 mm, electric	66.37	week	1.66	–	–	–	1.66	6(23)
Tampers and screeders								
vibrating tamper and handles (petrol)	70.67	week	1.77	75	2	2	3.7	6(32)
razor-back screeder (per metre)	44.24	week	1.11	75	1.5	2	1.74	–
double beam screeder (7.6 m wide)	85.33	week	2.13	75	1.5	2	2.76	6(34)
Rotary power float (petrol),								
687 mm diameter	89.67	week	2.24	75	2	2	4.17	6(36)
865 mm diameter	98.63	week	2.47	75	2	2	4.4	6(36)
Reinforcement power bar cropper (electric)	213.33	week	5.33	–	–	–	5.33	3(6)
Reinforcement power bar bender (electric)	220	week	5.5	–	–	–	5.5	3(3)
Compaction stud roller	158.67	week	3.97	75	1.5	2	4.6	–
MIXERS								
Concrete mixer								
tilting drum mixer 90 litres	70	week	1.75	–	–	–	1.75	5(1)
tilting drum mixer 150 itres	2.46	week	0.06	75	1.5	2	1.5	5(3)
tilting drum mixer 200 litres	107.67	week	2.69	75	1.2	2	3.19	5(5)
non tilting drum mixer 400 litres	151	week	3.78	75	1.3	2	4.33	5(7)
non tilting drum mixer 500 litres	170.33	week	4.26	75	1.2	2	4.76	5(8)
Truck mixer 6 m³	34.16	hour	34.16	75	14	3	40.11	5(35)

Description	Hire Rate £	Hire Period	Cost per working hour					
			Plant £	Usage %	Fuel l/hr	Oil etc. % fuel	Total £	DSR
CONCRETE PUMPS								
Concrete pump skid/trailer mounted 40m³/hr, excl. piping	18.85	hour	18.85	74	4	2	20.51	6(2)
Concrete pump Truck mounted 80 m³/hr	144.88	hour	144.88	75	8	3	148.28	6(7)
Schwing slurry pump	12.37	hour	12.37	75	1.5	3	13.01	6(1)
CONCRETE SKIPS								
Rollover type								
0.5 m³	46.67	week	1.17	–	–	–	1.17	8(29)
Dual flow type								
0.5 m³	45.67	week	1.14	–	–	–	1.14	8(34)
0.75 m³	58.33	week	1.46	–	–	–	1.46	8(35)
1.5 m³	80	week	2	–	–	–	2	8(37)
CRANES								
up to 18t mobile crane (site use)	39.55	hour	39.55	60	3.4	10	40.78	7(2)
Truck mounted mobile crane (wheeled)								
12 t SWL	65.48	hour	65.48	60	5.1	10	67.33	7(14)
20 t SWL	76.27	hour	76.27	60	5.5	10	78.27	7(15)
26 t SWL	85.93	hour	85.93	60	5.8	10	88.04	7(16)
45 t SWL	114.92	hour	114.92	60	6.3	10	117.21	7(18)
60 t SWL	136.28	hour	136.28	60	8.5	10	139.37	7(19)
90 t SWL	174.78	hour	174.78	60	10	10	178.41	7(20)
135 t SWL	236.81	hour	236.81	60	12	10	241.17	7(22)
Crawler cranes								
28 t SWL	42.82	hour	42.82	60	5	10	44.64	7(7)
56 t SWL	72.27	hour	72.27	60	8	10	75.17	7(9)
85 t SWL	101.51	hour	101.51	60	10	10	105.14	7(11)
150 t SWL	155.11	hour	155.11	61	12	10	159.54	7(13)
DIESEL GENERATORS								
500 w Petrol 110 or 240 V	61.33	week	1.53	85	0.8	3	2.41	11(1)
2.0 kVA Petrol 110 or 240 V	122.67	week	3.07	85	1.6	3	4.83	11(1)
4 kVA diesel dual voltage	152	week	3.8	85	1.7	3	4.62	11(2)
10 kVA diesel dual voltage	147.67	week	3.69	85	1.8	3	4.56	11(3)
16 kVA diesel 3 phase 440 V	174	week	4.35	85	1.9	3	5.26	11(4)
30 kVA diesel 3 phase 440 V	230.33	week	5.76	85	2	3	6.72	11(5)
50 kVA diesel 3 phase 440 V	303	week	7.58	85	2.5	4	8.8	11(6)
100 kVA diesel 3 phase 440 V	508.67	week	12.72	85	3	4	14.18	11(9)
125 kVA diesel 3 phase 440 V	691	week	17.28	85	4.9	5	19.69	11(11)
DIESEL GENERATORS								
150 kVA diesel 3 phase 440 V	691	week	17.28	85	10.9	5	22.63	11(12)
175 kVA diesel 3 phase 440 V	768	week	19.2	85	17	5	27.54	11(12)
200 kVA diesel 3 phase 440 V	845	week	21.13	85	21.8	5	31.83	11(12)
250 kVA diesel 3 phase 440 V	1,019.00	week	25.48	85	40	5	45.12	11(13)
330 kVA diesel 3 phase 440 V	1,195.00	week	29.88	85	48	5	53.44	11(14)

Description			Hire Rate £	Hire Period	Cost per working hour					
					Plant £	Usage %	Fuel l/hr	Oil etc. % fuel	Total £	DSR
EXCAVATORS										
Bucket capacity refers to SAE heaped										
Mini Excavators (tracked)										
Up to 2000 kg (0.04 m³)			418.67	week	10.47	85	4	5	12.43	10(15)
Up to 2000 kg (0.045 m³)			418.67	week	10.47	85	4	5	12.43	10(15)
Up to 3000 kg (0.075 m³)			519.33	week	12.98	85	5	5	15.43	10(16)
Up to 4000 kg (0.12m³)			591.67	week	14.79	85	5	5	17.24	10(17)
Mini Excavators (wheeled)										
Up to 3000 kg (0.075 m³)			519.33	week	12.98	85	5	5	15.43	10(15)
Hydraulic crawler mounted backacter approximate data:										
weight up to	SAE bucket	kW								
4.0 tonne	0.10 m³	25	14.79	hour	14.79	80	4	7.5	16.68	10(17)
6.0 tonne	0.25 m³	40	17.45	hour	17.45	80	5	7.5	19.82	10(18)
11.0 tonne	0.30 m³	50	21.99	hour	21.99	80	8	7.5	25.77	10(19)
14.0 tonne	0.40 m³	63	26.21	hour	26.21	80	9.5	7.5	30.7	10(20)
17.0 tonne	0.45 m³	70	29.33	hour	29.33	80	12	7.5	35.01	10(21)
21.0 tonne	1.00 m³	100	33.25	hour	33.25	80	16	7.5	40.82	10(22)
25.0 tonne	1.00 m³	115	37.25	hour	37.25	80	19	7.5	46.24	10(23)
30.0 tonne	1.20 m³	130	47.03	hour	47.03	75	24	7.5	57.67	10(24)
38.0 tonne	1.20 m³	160	60.99	hour	60.99	75	31	7.5	74.74	10(25)
55.0 tonne	2.00 m³	205	81.32	hour	81.32	75	45	7.5	101.27	10(26)
75.0 tonne	2.50 m³	235	108.2	hour	108.2	75	52	7.5	131.26	10(27)
Hydraulic wheeled backacter										
weight up to	SAE bucket									
4.0 tonne	0.10 m³		14.79	hour	14.79	80	4	7.5	16.68	10(17)
6.0 tonne	0.27 m³		17.45	hour	17.45	80	5	7.5	19.82	10(18)
11 tonne	0.34 m³		21.99	hour	21.99	80	8	7.5	25.77	10(19)
14 tonne	0.43 m³		26.21	hour	26.21	80	9.5	7.5	30.7	10(20)
17 tonne	0.47 m³		29.33	hour	29.33	80	10.5	7.5	34.3	10(21)
21 tonne	1.00 m³		33.25	hour	33.25	80	14	7.5	39.87	10(22)
Backhoe loader, wheeled										
JCB 3CX, 1.1 m³			20.98	hour	20.98	75	7.5	6	24.26	10(36)
Volvo BL61, 1.0 m³			20.98	hour	20.98	75	7.5	6	24.26	10(36)
JCB 2CX, 0.6 m³			14.89	hour	14.89	75	7.5	6	18.17	10(34)
Hymac 180C, 0.95 m³			15.16	hour	15.16	75	7.5	6	18.44	10(36)
Percussion breaker attachments										
Montabert 140 (380 kg) with steels			322.33	week	8.06	–	–	–	8.06	10(30)
Montabert 300 (606 kg) with steels			322.33	week	8.06	–	–	–	8.06	10(30)
Montabert 900 (1133 kg) with steels			470	week	11.75	–	–	–	11.75	10(31)
Montabert V 1200 (1570 kg) with steels			617.67	week	15.44	–	–	–	15.44	10(32)
Other hydraulic attachments										
Rock backacter bucket (ripper teeth)			123.66	week	3.09	–	–	–	3.09	–
Ditch bucket (c/w side cutters)			83.03	week	2.08	–	–	–	2.08	–
Trapezoidal bucket			91.85	week	2.3	–	–	–	2.3	–
Clamshell Grab (1.0 m³)			166.04	week	4.15	–	–	–	4.15	–
Rock handling Grab (up to 3 t)			155.45	week	3.89	–	–	–	3.89	–
Scrap handling Grab (up to 1 t)			155.45	week	3.89	–	–	–	3.89	–
Scrap shears			206.68	week	5.17	–	–	–	5.17	–
Grab bucket			91.85	week	2.3	–	–	–	2.3	–
Demolition ball			30.04	week	0.75	–	–	–	0.75	–

Description	Hire Rate £	Hire Period	Cost per working hour					
			Plant £	Usage %	Fuel l/hr	Oil etc. % fuel	Total £	DSR
Large scoop bucket 1.40 m³	74.19	week	1.85	–	–	–	1.85	–
Single shank ripper	38.87	week	0.97	–	–	–	0.97	–
Flat track shoes (highway use)	21.2	week	0.53	–	–	–	0.53	–
HOISTS, LIFTING AND HANDLING								
Telescopic handler 4WD lift up to 2.50 t	14.55	hour	14.55	25	75	4	25.28	15(12)
Electric forklift 2.5 tonnes	8.74	hour	8.74	–	–	–	8.74	
Rough terrain forklift 5.0 tonnes	16.98	hour	16.98	75	7.5	4	20.2	15(20)
Lorry (8T) with 1T Hiab	15.73	hour	15.73	75	7.5	4	18.95	14(10)
Mast section per 1.5 m (max × 91m)	10.06	week	0.25	–	–	–	0.25	–
Tirfor winch TU 32 (3T SWL)	59	week	1.48	–	–	–	1.48	36(14)
Tirfor winch TU 16	50	week	1.25	–	–	–	1.25	36(12)
Chain hoist up to 2000 kg capacity	47	week	1.18	–	–	–	1.18	–
Chain hoist up to 3000 kg capacity	561	week	14.03	–	–	–	14.03	–
Electric cable hoist 200 kg with gantry	146	week	3.65	–	–	–	3.65	36(6)
PILING PLANT								
(weight = piston weight)								
Piling hammer, double acting (air)								
weight up to 3,000 kg	364.33	week	9.11	–	–	–	9.11	17(5)
weight up to 4,000 kg	401.33	week	10.03	–	–	–	10.03	17(6)
weight up to 6,000 kg	440.67	week	11.02	–	–	–	11.02	17(7)
Piling hammer, double acting hydraulic								
weight up to 5,000 kg	2,059.33	week	51.48	–	–	–	51.48	–
weight up to 10,000 kg	2,265.27	week	56.63	–	–	–	56.63	–
weight up to 50,000 kg	2,491.79	week	62.29	–	–	–	62.29	–
weight up to 100,000 kg	2,740.97	week	68.52	–	–	–	68.52	–
Piling hammer, single acting diesel								
weight up to 1,500 kg	667	week	16.68	75	6	5	19.28	17(30)
weight up to 2,500 kg	977	week	24.43	75	11	5	29.19	17(32)
weight up to 3,500 kg	1,267.00	week	31.68	75	14	5	37.74	17(34)
weight up to 5,000 to 6,200 kg	2,482.33	week	62.06	75	16	5	68.99	–
weight up to 8,000 kg	2,856.67	week	71.42	75	30	5	84.41	17(37)
weight up to 10,000 kg	3,177.33	week	79.43	75	35	5	94.59	17(38)
Hanging leaders for diesel hammers								
length 8.5 m	183.33	week	4.58	–	–	–	4.58	17(*)
length 15.5 m	183.33	week	4.58	–	–	–	4.58	17(53)
length 22.5 m	204.33	week	5.11	–	–	–	5.11	17(54)
Flexible hose for compressed air (per 30 ft length)								
38 mm diameter	62.41	week	1.56	–	–	–	1.56	17(24)
50 mm diameter	63.48	week	1.59	–	–	–	1.59	17(25)
Piling extractors (including compressor)								
BSP HD10 unit weight 3,000 kg	1,195.00	week	29.88	75	20	4	38.46	17(20)
BSP HD15 unit weight 4,580 kg	1,638.00	week	40.95	75	30	4	53.82	17(21)
Vibratory hammer/extractor, hydraulic								
cent. force 16 t, pulling force 12.5 t	1,102.00	week	27.55	75	30	7.5	40.85	17(39)
cent. force 59 t, pulling force 36 t	1,657.67	week	41.44	75	30	7.5	54.74	17(41)
cent. force 200 t, pulling force 80 t	4,281.00	week	107.03	75	30	7.5	120.33	17(43)

Description	Hire Rate £	Hire Period	Cost per working hour					
			Plant £	Usage %	Fuel l/hr	Oil etc. % fuel	Total £	DSR
ROLLERS								
Pedestrian rollers								
475 kg trench compactor	2.47	hour	2.47	80	0.9	7.5	2.89	23(11)
600 kg double drum vibratory	5.13	hour	5.13	80	1	7.5	5.6	23(11)
928 kg double drum vibratory	6.61	hour	6.61	80	1.6	7.5	7.36	23(12)
1300 kg double drum vibratory	6.61	hour	6.61	80	2.3	7.5	7.7	23(13)
159 kg single drum vibratory	3.97	hour	3.97	80	0.8	7.5	4.35	23(9)
466 kg single drum vibratory	4.4	hour	4.4	80	0.9	7.5	4.83	23(9)
Self propelled rollers								
2½ tonne double vibratory	9.03	hour	9.03	80	3.6	7.5	10.75	23(21)
5 tonne double vibratory	13.73	hour	13.73	80	3.8	7.5	15.55	23(22)
Self propelled rollers								
9 tonne single drum vibratory	22.2	hour	22.2	80	10	7.5	26.93	23(23)
Bomag BW55E single drum vibratory	1.22	hour	1.22	80	11	7.5	6.42	23(25)
Bomag BW71E single drum vibratory	2.47	hour	2.47	80	16.3	7.5	10.18	23(26)
Bomag BW217D single drum vibratory	15.03	hour	15.03	80	23.8	7.5	26.27	23(*)
Bomag BC672 RB Refuse compactor	12.7	hour	12.7	80	42	7.5	32.57	23(*)
3 wheel dead weight								
10 tonne	19.61	hour	19.61	80	16	7.5	27.18	23(3)
12 tonne	21.77	hour	21.77	80	17	7.5	29.81	23(4)
Rubber tyred rollers								
10 tonne	26.96	hour	26.96	80	17	7.5	35	23(7)
6 tonne	21.36	hour	21.36	80	12	7.5	27.04	23(6)
Towed vibratory rollers								
6 tonne	13.69	hour	13.69	80	7.5	7.5	17.24	23(17)
Sheepsfoot	16.02	hour	16.02	80	4	7.5	17.91	23(30)
Dual Purpose Rollers with breaker	2.39	hour	2.39	80	4	7.5	4.28	–
Bomag BW138 with breaker	13.73	hour	13.73	80	4	7.5	15.62	–
SCAFFOLDING AND ACCESSORIES								
Tube alloy per metre	0.2	4 wk	0.05	–	–	–	0.05	25(2)
Tube galvanized per metre	0.1	4 wk	0.03	–	–	–	0.03	25(1)
Boards – per 4 metre length	0.27	4 wk	0.07	–	–	–	0.07	25(19)
Base plates adjustable each	0.23	4 wk	0.06	–	–	–	0.06	25(8)
Base plates fixed each	0.07	4 wk	0.02	–	–	–	0.02	25(7)
Clips each (all types)	0.07	4 wk	0.02	–	–	–	0.02	25(7)
Couplers each (all types)	0.07	4 wk	0.02	–	–	–	0.02	25(7)
Spigots each ledgers andbraces/m	0.2	4 wk	0.05	–	–	–	0.05	25(7)
Putlog 1.5 m blade end each steel	0.23	4 wk	0.06	–	–	–	0.06	25(3)
Putlog 1.8 m blade end each steel	0.3	4 wk	0.08	–	–	–	0.08	25(5)
Reveal screws each	0.15	4 wk	0.04	–	–	–	0.04	25(7)
Castor wheel rubber tyred w/brake	5.47	4 wk	1.37	–	–	–	1.37	25(16)
ACCESS STAGING AND TOWERS								
Alloy towers base size 1.8 × 0.80height to platform 2.6 m	191.1	week	4.78	–	–	–	4.78	–
Extra sections per 2 m rise (max 7)	49.73	week	49.73	–	–	–	49.73	–
Alloy stairwell tower per 2 m rise	91.65	week	91.65	–	–	–	91.65	–
Steel towers base size 1.80 × 1.80 height to platform 1.80 m	95.55	week	95.55	–	–	–	95.55	–
Extra sections per 0.60 m rise	20.48	week	20.48	–	–	–	20.48	–

Description	Hire Rate £	Hire Period	Cost per working hour					
			Plant £	Usage %	Fuel l/hr	Oil etc. % fuel	Total £	DSR
Steel towers base size 3.80 × 3.80 height to platform 1.80 m	425.85	week	425.85	–	–	–	425.85	–
Extra sections per 1.20 m rise	182.51	week	182.51	–	–	–	182.51	–
Castor wheels	21.87	4 wk	5.47	–	–	–	5.47	25(14)
Toe boards per 4 metre length	0.27	4 wk	0.07	–	–	–	0.07	25(19)
Toe boards hinged per 4 metre length	0.27	4 wk	0.07	–	–	–	0.07	–
Handrail (included with towers)	0.25	4 wk	0.06	–	–	–	0.06	–
Stair tread complete	0.41	4 wk	0.1	–	–	–	0.1	–
Brick guard	3.56	4 wk	0.89	–	–	–	0.89	–
Bridging Unit per 1.00 m	4.63	week	4.63	–	–	–	4.63	–
TRACTORS								
Tractor dozers with single equipment								
50 kW	24.98	hour	24.98	80	11	7.5	30.18	32(1)
74 kW	32.85	hour	32.85	80	14.5	7.5	39.71	32(2)
104 kW	47.53	hour	47.53	80	21	6	57.32	32(4)
212 kW	82	hour	82	80	38	6	99.72	32(7)
276 kW	136.08	hour	136.08	80	53	7.5	161.15	32(9)
67 kW	24.98	hour	24.98	80	14.5	7.5	31.84	32(1)
123 kW	51.38	hour	51.38	80	25	7.5	63.21	32(5)
Dozer attachments								
single shank ripper	3.75	hour	3.75	–	–	–	3.75	–
triple shank ripper	6.83	hour	6.83	–	–	–	6.83	–
'U' dozer blade	2.05	hour	2.05	–	–	–	2.05	–
angle dozer blade	2.13	hour	2.13	–	–	–	2.13	–
skeleton blade	2.13	hour	2.13	–	–	–	2.13	–
Tractor loaders								
0.8 m³	22.89	hour	22.89	75	11	7.5	27.77	32(11)
1.5 m³	44.41	hour	44.41	75	19	7.5	52.84	32(15)
2.0 m³	48.66	hour	48.66	75	25	7.5	59.75	32(16)
3.0 m³	74.2	hour	74.2	75	34	7.5	89.28	32(18)
3.0 m³	74.2	hour	74.2	75	34	7.5	83.25	32(18)
Wheeled loaders								
0.55 m³ 33 kW	16.03	hour	16.03	75	8	7.5	19.58	32(28)
0.75 m³ 41 kW	16.03	hour	16.03	75	9.5	7.5	20.24	32(28)
1.60 m³ 83 kW	21.58	hour	21.58	75	11	7.5	26.46	32(32)
2.70 m³ 115 kW	31.89	hour	31.89	75	19	7.5	40.32	32(36)
3.10 m³ 135 kW	36.78	hour	36.78	75	23	7.5	46.98	32(37)
3.85 m³ 194 kW	47.56	hour	47.56	80	30	7.5	61.75	32(39)
6.00 m³ 310 kW	139.48	hour	139.48	80	53	7.5	164.55	32(41)
Wheeled loaders 10.50 m³ 588 kW	115.93	hour	115.93	80	90	7.5	158.5	32(*)
Long reach loader; 1.00 m³, 54 kW	31.62	hour	31.62	75	11	7.5	36.5	–
Wheel dozer, blade 7.5 m³ 338 kW	64.12	hour	64.12	80	68	7.5	96.28	32(9)
Loader Attachments								
4 in 1 bucket	2.75	hour	2.75	–	–	–	2.75	–
side dumping bucket	2.69	hour	2.69	–	–	–	2.69	–
rock bucket	2.33	hour	2.33	–	–	–	2.33	–
skeleton bucket	2.41	hour	2.41	–	–	–	2.41	–
single shank ripper (D6)	3.93	hour	3.93	–	–	–	3.93	–
multi shank ripper (D8)	7.15	hour	7.15	–	–	–	7.15	–
rear mounted backacter	10.2	hour	10.2	–	–	–	10.2	–

Description	Hire Rate £	Hire Period	Cost per working hour					
			Plant £	Usage %	Fuel l/hr	Oil etc. % fuel	Total £	DSR
Hydraulic face shovels								
32 tonne, 1.50 m³	60.99	hour	60.99	80	34	7.5	77.07	10(25)
42 tonne, 2.60 m³	81.32	hour	81.32	80	44	7.5	102.13	10(26)
62 tonne, 1.30 – 3.40 m³	108.2	hour	108.2	80	64	7.5	138.47	10(27)
Skid Steer Loaders								
Bobcat 463 or similar	4.63	hour	4.63	75	3.5	5	6.15	32(48)
Bobcat 553 or similar	4.88	hour	4.88	75	4.5	5	6.83	32(49)
Case 85 XT or similar	25.4	hour	25.4	75	8.3	5	28.99	32(49)
Motor Scraper – single engine								
15.30 m³	83.43	hour	83.43	80	44	3	103.37	32(69)
33.60 m³	181.46	hour	181.46	80	20	3	190.52	32(*)
Motor Scraper – twin engine								
16.00 m³	106.04	hour	106.04	80	70	3	137.76	32(71)
33.60 m³	238.93	hour	238.93	80	125.5	6	297.46	32(73)
Motor scraper – elevating								
8.40 m³	65.23	hour	65.23	80	25	3	76.56	32(74)
13.00 m³	99.75	hour	99.75	80	36	3	116.07	32(75)
16.80 m³	130.98	hour	130.98	80	46	3	151.83	32(76)
Tractor (tracked) pipe layer complete with counterweight and boom								
up to 145 kW lifting max 28 t	45.75	hour	45.75	80	17	4	53.53	–
up to 149 kW lifting max 41 t	53.77	hour	53.77	80	17	4	61.55	–
up to 224 kW lifting max 70 t	65.69	hour	65.69	80	26	4	77.59	–
Tractor (tracked) no equipment, for agro/towing use								
75 kW	14.32	hour	14.32	75	10	5	18.65	–
90 kW	18.23	hour	18.23	75	12	5	23.43	–
125 kW	24.16	hour	24.16	75	18	5	31.96	–
Motor Grader (6 wheel) (Cat or similar)								
up to 93kW blade 3.66m	40.7	hour	40.7	80	17	7.5	48.74	32(55)
up to 113kW blade 3.66m	45.99	hour	45.99	80	23	7.5	56.87	32(56)
up to 158kW blade 3.96m	72.87	hour	72.87	80	27	7.5	85.64	32(57)
up to 205kW blade 4.88m	119.16	hour	119.16	80	32	7.5	134.3	33(*)
Trench excavator (Trencher) (Vermeer or similar)								
up to 100kW width 0.6m & depth 2.4m	31.92	hour	31.92	75	22	3	41.27	33(5)
up to 300kW width 1.1m & depth 3.7m	85.81	hour	85.81	75	40	3	102.81	33(7)
Trench excavator agro tractor towed type max width 0.13m & depth 0.55m	9.74	hour	9.74	75	3	3	11.01	33(*)
TRANSPORT (TIPPERS AND DUMPERS)								
Tipping lorries 4 × 2 wheel tippers								
gross weight up to 11.0 tonnes								
payload up to 5.5 tonnes – side tipping	16.07	hour	16.07	80	30	4	29.8	14(15)
payload up to 8 tonnes	13.39	hour	13.39	80	25	4	24.83	14(15)
gross weight up to 14 tonnes								
payload up to 10 tonnes	20.37	hour	20.37	80	30	4	34.1	14(16)
gross weight up to 17 tonnes								
payload up to 12 tonnes	26.66	hour	26.66	80	33	4	41.76	14(16)
Tipping lorries 6 × 4 wheel tippers								
gross weight up to 25 tonnes								
payload up to 18 tonnes	26.66	hour	26.66	80	35	4	42.68	14(17)
gross weight up to 31 tonnes								
payload up to 20 tonnes	32.13	hour	32.13	80	40	4	50.43	14(18)

Description	Hire Rate £	Hire Period	Cost per working hour					
			Plant £	Usage %	Fuel l/hr	Oil etc. % fuel	Total £	DSR
Tipping lorries 8 × 4 wheel tippers								
gross weight up to 33 tonnes								
payload up to 20 tonnes	32.13	hour	32.13	80	40	4	50.43	14(18)
payload up to 22 tonnes	32.13	hour	32.13	80	42	4	51.35	14(18)
payload up to 24 tonnes	32.13	hour	32.13	80	44	4	52.26	14(*)
Articulated tipper trailers (Highway type excluding tractor)								
twin axle up to 20 t (12.2 m length)	29.59	hour	29.59	80	40	4	47.89	–
triple axle up to 25 t (15.3 m length)	33.75	hour	33.75	80	44	4	53.88	–
Dump truck (on site use only)								
10.5 m³ heaped capacity, 15.5 t payload	24.28	hour	24.28	80	15	5	31.21	9(12)
14.0 m³ heaped capacity, 18 t payload	34.8	hour	34.8	80	31	5	49.12	9(14)
19.0 m³ heaped capacity, 31 t payload	60.08	hour	60.08	80	37.5	5	77.41	9(17)
24.0 m³ heaped capacity, 36 t payload	74.57	hour	74.57	80	37.9	5	92.08	9(18)
57.0 m³ heaped capacity, 88 t payload	82.77	hour	82.77	80	112	5	134.51	9(*)
78.0 m³ heaped capacity, 136 t payload	97.2	hour	97.2	80	122	5	153.56	9(*)
Dump truck (articulated type ADT's)								
Volvo 6 × 6 or similar								
18.5 t payload, 11 m³ heaped capacity	40.24	hour	40.24	80	21.4	3	49.94	9(22)
22.5 t payload, 12 m³ heaped capacity	46.46	hour	46.46	80	29.2	3	59.69	9(23)
Dump truck (articulated type ADT's)								
23 tonne payload, 13 m³ heaped capacity	46.46	hour	46.46	80	26.9	3	58.65	9(23)
25 tonne payload, 13.6 m³ heaped capacity	54	hour	54	80	29.5	3	67.37	9(24)
25 tonne payload, 14.2 m³ heaped capacity	54	hour	54	80	25	3	65.33	9(24)
32 tonne payload, 19 m³ heaped capacity	62.09	hour	62.09	80	36.5	3	78.63	9(25)
Small dumpers								
750kg payload, heaped capacity 0.55 m³	5.28	hour	5.28	80	1.5	3	5.96	9(3)
1000kg payload, heaped capacity 0.75 m³	5.28	hour	5.28	80	2	3	6.19	9(3)
Thwaites 4 × 4								
2000kg payload, heaped capacity 1.55 m³	6.27	hour	6.27	80	3	3	7.63	9(5)
2500kg payload, heaped capacity 1.80 m³	6.89	hour	6.89	80	4	3	8.7	9(6)
3000kg payload, heaped capacity 2.00 m³	7.61	hour	7.61	80	5	3	9.88	9(7)
4000kg payload, heaped capacity 2.50 m³	8.37	hour	8.37	80	7	3	11.54	9(8)
Power barrow	2.86	hour	2.86	80	0.8	2	3.22	6(49)
Transport (trucks and vans)								
Truck, 5t chassis	11.92	hour	11.92	80	7	3	15.09	
Truck, 8t chassis	16.35	hour	16.35	80	8	3	19.98	
Truck, 10t chassis	16.35	hour	16.35	80	10	3	20.88	
Truck, 16t chassis	19.48	hour	19.48	80	15	3	26.28	
Skip Loader, 17t chassis	19.48	hour	19.48	80	15	3	26.28	
Small van (petrol)	6.68	hour	6.68	80	8	3	10.31	14(20)
Pick up (DERV) (1.10 tonnes)	11.76	hour	11.76	80	8	3	15.39	14(21)
Personnel carrier (12 seat transit)	16.91	hour	16.91	80	10	3	21.44	14(30)
SWB utility Land Rover type (DERV)	15.08	hour	15.08	80	7	3	18.25	14(26)
LWB utility Land Rover type (DERV)	16.91	hour	16.91	80	7	3	20.08	14(27)
Trailers								
Flat trailer (2 axles) up tp 2 t	11.58	day	1.45	–	–	–	1.45	15 (1)
Drop side trailer power tipping up to 4 t	15.4	day	1.93	–	–	–	1.93	15 (4)
Plant Trailer up to 10 t	19.62	day	2.45	–	–	–	2.45	15 (6)
Plant Trailer up to 20 t	23.35	day	2.92	–	–	–	2.92	15 (7)

Description	Hire Rate £	Hire Period	Cost per working hour					
			Plant £	Usage %	Fuel l/hr	Oil etc. % fuel	Total £	DSR
TRENCH SHEETS ETC. (MINIMUM 50)								
Trench struts (each)								
Nr 0; range 0.30 m – 0.40 m	1.21	week	1.21	–	–	–	1.21	–
Nr 1; range 0.46 m – 0.71 m	1.21	week	1.21	–	–	–	1.21	26(3)
Nr 2; range 0.70 m – 1.10 m	1.24	week	1.24	–	–	–	1.24	26(4)
Nr 3; range 1.04 m – 1.70 m	1.31	week	1.31	–	–	–	1.31	26(5)
Standard props (each)								
Nr 0; 1.07 m – 1.83 m	1.47	week	1.47	–	–	–	1.47	–
Nr 1; 1.75 m – 3.13 m	1.54	week	1.54	–	–	–	1.54	–
Nr 2; 1.98 m – 3.36 m	1.58	week	1.58	–	–	–	1.58	–
Nr 3; 2.59 m – 3.97 m	1.63	week	1.63	–	–	–	1.63	–
Nr 4; 3.20 m – 4.88 m	1.8	week	1.8	–	–	–	1.8	–
Nr 5; 4.88 m – 6.41 m	1.99	week	1.99	–	–	–	1.99	–
Beam heads (each)								
Nr 1	2.6	week	2.6	–	–	–	2.6	–
Nr 2	2.64	week	2.64	–	–	–	2.64	–
Nr 3	2.68	week	2.68	–	–	–	2.68	–
Nr 4	2.72	week	2.72	–	–	–	2.72	–
Split heads (each)								
Nr 1 1ft 9in – 2ft 9in	1.08	week	1.08	–	–	–	1.08	–
Nr 2 3ft 9in – 4ft 6in	1.08	week	1.08	–	–	–	1.08	–
Nr 3 4ft 6in – 6ft 3in	1.09	week	1.09	–	–	–	1.09	–
Trench Sheets								
1.83 m	1.26	week	1.33	–	–	–	1.33	–
2.44 m	1.44	week	1.51	–	–	–	1.51	–
3.05 m	1.78	week	1.87	–	–	–	1.87	–
3.66 m	2.23	week	2.34	–	–	–	2.34	–
4.27 m	2.5	week	2.63	–	–	–	2.63	–
4.88 m	2.87	week	3.01	–	–	–	3.01	–
Trench boxes (approximate cost per m²) – refer to Approximate Estimates								
PUMPING AND DEWATERING								
Self priming centrifugal diesel pumps (inc. 8 m suction and delivery pipes)								
50 mm 550 l/min	91.23	week	2.28	80	0.8	2	2.64	20(10)
75 mm Flygt submersible 1050 l/min (4 kW)	60.13	week	1.5	80	–	–	1.5	20(24)
75 mm 750 l/min	130.61	week	3.27	80	1	2	3.72	20(11)
100 mm 1500 l/min	159.66	week	3.99	80	2.8	2	5.25	20(12)
150 mm Sykes UV0 head pump / high capacity	199.09	week	4.98	80	6.8	2	8.03	20(13)
76 mm Sykes VCD3 jetting pump	241.37	week	6.03	80	2	2	6.93	–
100 mm Sykes WP100/60 wellpointing pump	222.58	week	5.56	80	4.5	2	7.58	–
150 mm Sykes WP150/60 wellpointing pump	274.89	week	6.87	80	6.8	2	9.92	–
Extra hose (per metre)								
50 mm Suction	0.41	week	0.01	–	–	–	0.01	21(2)
50 mm Delivery	0.64	week	0.02	–	–	–	0.02	21(8)
75 mm Suction	0.64	week	0.02	–	–	–	0.02	21(3)
75 mm Delivery	0.99	week	0.02	–	–	–	0.02	21(9)
100 mm Suction	0.82	week	0.02	–	–	–	0.02	21(4)
100 mm Delivery	1.28	week	0.03	–	–	–	0.03	21(10)
150 mm Suction	1.23	week	0.03	–	–	–	0.03	21(5)
150 mm Delivery	1.93	week	0.05	–	–	–	0.05	21(11)

Description	Hire Rate £	Hire Period	Plant £	Usage %	Fuel l/hr	Oil etc. % fuel	Total £	DSR
			Cost per working hour					
150 mm Duraline (lay flat fine hose)		week	(included with Jetting Pump)					
150 mm Jetting tube		week	(included with Jetting Pump)					
150 mm PVC header pipe and attachments at 1.50m centres including jet wells for wellpointing to 6.5 m deep (per well-point)	4.39	week	0.11	–	–	–	0.11	–
MISCELLANEOUS								
Road form 10m long (each)								
102 mm high	0.63	day	0.08	–	–	–	0.08	27(2)
152 mm high	0.69	day	0.09	–	–	–	0.09	27(3)
203 mm high	0.81	day	0.1	–	–	–	0.1	27(4)
203 mm high (heavy section)	0.81	day	0.1	–	–	–	0.1	27(5)
203 mm high (heavy section with rail)	1.11	day	0.14	–	–	–	0.14	27(6)
254 mm high (heavy section with rail)	1.36	day	0.17	–	–	–	0.17	27(7)
3.05 mm high (heavy section with rail)	1.49	day	0.19	–	–	–	0.19	27(8)
Flexible road form 3m long (each)								
150 mm high	2.04	day	0.26	–	–	–	0.26	–
200 mm high	2.13	day	0.27	–	–	–	0.27	–
225 mm high	2.21	day	0.28	–	–	–	0.28	–
Road signs on stands (each)								
600 mm diameter	0.2	day	0.03	–	–	–	0.03	37(43)
750 mm diameter	0.31	day	0.04	–	–	–	0.04	37(44)
900 mm diameter	0.45	day	0.06	–	–	–	0.06	37(45)
1200 mm diameter	0.73	day	0.09	–	–	–	0.09	37(46)
Road safety								
flashing hazard lamps (Batteries & tripod inc.)	3.32	day	0.42	–	–	–	0.42	37(22)
traffic lamps static	4.56	day	0.57	–	–	–	0.57	37(10)
standard cone	0.42	day	0.05	–	–	–	0.05	37(39)
road pins	0.77	week	0.02	–	–	–	0.02	–
PVC-U barrier	4.08	day	0.51	–	–	–	0.51	37(6)
cone converter	1.92	week	0.05	–	–	–	0.05	–
railway sleepers – baulks – each	2.9	week	0.07	–	–	–	0.07	–
traffic light systems								
two way radar, 110V	175	week	4.38	–	–	–	4.38	–
two way, mains, timer operated	125	week	3.13	–	–	–	3.13	37(9)
Transformers/cables								
up to 2.5 kVA	0.73	hour	0.73	–	–	–	0.73	11(16)
up to 5.0 kVA	0.81	hour	0.81	–	–	–	0.81	11(17)
up to 7.5 kVA	0.89	hour	0.89	–	–	–	0.89	11(18)
240V junction box	11.63	week	0.29	–	–	–	0.29	–
110V junction box	11.63	week	0.29	–	–	–	0.29	–
Extension cable 240V/110V								
15 m plug and socket	6.45	week	0.16	–	–	–	0.16	–
30 m plug and socket	8.07	week	0.2	–	–	–	0.2	–
15 m 4 mm	11.13	week	0.28	–	–	–	0.28	–
Welding & cutting sets								
ARC Mains 150 amps portable	1.88	hour	1.88	–	–	–	1.88	35(5)
ARC diesel 150 amps mobile	2.42	hour	2.42	80	1.8	–	4.23	35(2)
ARC Diesel 300 amps trailer mounted	3.75	hour	3.75	80	2.3	–	4.76	35(3)
MIG welder 120 amp	2.64	hour	2.64	–	–	–	2.64	–
Oxy acetylene set portable (cutting)	2.11	hour	2.11	75	650/325-		5.13	35(1)
Oxy acetylene set portable (welding)	2.11	hour	2.11	75	500/325-		5.13	35(1)

Description	Hire Rate £	Hire Period	Cost per working hour					
			Plant £	Usage %	Fuel l/hr	Oil etc. % fuel	Total £	DSR
Lighting								
mobile lighting 8m c/w 2 light tungsten incl 1.25 kVA generator	4.45	hour	4.45	80	0.8	3	4.76	11(29)
mobile lighting 16m c/w 4 light mercury vapour incl 6.25 kVA generator	6.33	hour	6.33	80	1.6	3	6.96	11(30)
mobile lighting 20m c/w 4 light tungsten halogen incl 7.5 kVA generator	6.69	hour	6.69	80	1.7	3	7.36	11(31)
festoon lighting set (34m)	33.44	week	0.84	–	–	–	0.84	–
tripod floodlight 1.8m stand 500 watt	32.93	week	0.82	–	–	–	0.82	–
Drain tools								
drain rods (10m set) incl. equipment	0.08	day	0.01	–	–	–	0.01	37(24)
drain plug (rubber diaphragm) 100	0.05	day	0.01	–	–	–	0.01	37(26)
drain plug (rubber diaphragm) 150	0.07	day	0.01	–	–	–	0.01	37(27)
drain plug (rubber diaphragm) 300	0.23	day	0.03	–	–	–	0.03	37(28)
mains powered drain cleaner	69.26	week	1.73	–	–	–	1.73	–
drain tester – smoke	9.24	week	0.23	–	–	–	0.23	–
drain tester – U gauge	10.39	week	0.26	–	–	–	0.26	–
drain tester – pressure pump	30.78	week	0.77	–	–	–	0.77	–
drain tester – mandrel	5.77	week	0.14	–	–	–	0.14	–
stilson wrench 0.600 m	8	week	0.2	–	–	–	0.2	–
stilson wrench 0.915 m	12.31	week	0.31	–	–	–	0.31	–
clay pipe cutter to 0.225 m	23.08	week	0.58	–	–	–	0.58	–
steel pipe cutter to 150 mm	17.5	week	0.44	–	–	–	0.44	–
drain bag 100 mm or 150 mm	7.7	week	0.19	–	–	–	0.19	–
Tar boilers								
hand operated 1000 litres	5.27	hour	5.27	–	–	–	5.27	29(1)
power operated 2000 litres	6.38	hour	6.38	–	–	–	6.38	29(2)
Shot blasting equipment								
67 kg	133.04	week	3.33	–	–	–	3.33	–
156 kg	183.9	week	4.6	–	–	–	4.6	–
Grit blaster ICE160, 72kg	148.34	week	3.71	–	–	–	3.71	–
Stone splitter 600 mm × 100 mm	58.29	week	1.46	–	–	–	1.46	–
Block splitter	30.78	week	0.77	–	–	–	0.77	–
Slab splitter	38.48	week	0.96	–	–	–	0.96	–
Slab lifter	8.09	week	0.20	–	–	–	0.20	–
Cartridge guns								
Hilti DX76	32.97	week	0.82	–	–	–	0.82	–
Hilti DX460	29.33	week	0.73	–	–	–	0.73	–
Heating/Drying					Propane			
up to 30.000 BTU/hhour	1.33	hour	1.33	80	0.3	–	2.03	37(11)
up to 63.000 BTU/hhour	1.58	hour	1.58	80	1.2	–	4.96	37(12)
up to 150.000 BTU/hhour	2.4	hour	2.4	80	3	–	10.85	37(14)
up to 325.000 BTU/hhour	6.45	hour	6.45	80	6.5	–	24.75	37(16)
dehumidifier (68 litres/day)	6.95	day	0.87	–	–	–	0.87	37(34)
fume extractor (150 mm diameter)	76.34	week	1.91	–	–	–	1.91	–
Drills/Saws/Tools								
diamond drill, 150 mm max. diameter	96	week	2.40	–	–	–	2.40	–
masonry drill & core bits, 10mm	0.32	hour	0.32	–	–	–	0.32	31(1)
masonry drill & core bits, 19mm	0.42	hour	0.42	–	–	–	0.42	31(3)
masonry drill & core bits, 32mm	0.55	hour	0.55	–	–	–	0.55	31(4)
Roto broach, milling to 34mm	66.27	week	1.66	–	–	–	1.66	–
rotary hammer drill, light weight	0.72	hour	0.72	–	–	–	0.72	31(15)

Description	Hire Rate £	Hire Period	Plant £	Usage %	Fuel l/hr	Oil etc. % fuel	Total £	DSR
rotary hammer drill, heavy weight	0.95	hour	0.95	–	–	–	0.95	31(16)
demolition hammer, heavy duty	0.95	hour	0.95	–	–	–	0.95	–
grinder, 225 mm, electric	0.56	hour	0.56	–	–	–	0.56	–
grinder, 114 mm, electric	0.45	hour	0.45	–	–	–	0.45	31(11)
saw bench, 410 mm diameter, petrol	1.88	hour	1.88	75	1	2	2.84	24(10)
saw bench, 610 mm diameter, diesel	2.48	hour	2.48	75	1	2	2.9	24(13)
chain saw, up to 410mm	1.94	hour	1.94	75	0.4	2	2.11	24(2)
225 mm electric grinder (angle)	0.37	hour	0.37	–	–	–	0.37	31(10)
450 mm conc saw 25hp diesel	3.38	hour	3.38	75	0.3	2	3.51	6(41)
600 mm conc saw 40hp diesel	6.86	hour	6.86	75	0.3	2	6.99	6(42)
Stihl saw disc 305 mm petrol	1.07	hour	1.07	75	0.2	2	1.26	37(18)
Stihl saw disc 305 mm air	1.07	hour	1.07	–	–	–	1.07	37(18)
230 mm floor sander (electric)	0.38	hour	0.38	–	–	–	0.38	31(13)
Water/Fuel Supply								
towed (trailer) water/diesel bowser,								
1500 litres – water	4.19	day	0.52	–	–	–	0.52	34(10)
3000 litres – water	5.1	day	0.64	–	–	–	0.64	34(11)
1500 litres – fuel	7.18	day	0.9	–	–	–	0.9	34(14)
lorry-mounted tanker								
4500 litres (DERV) 8.5t chassis	14.2	hour	14.2	75	9	2	22.53	34(22)
7000 litres (DERV) 12t chassis	18.37	hour	18.37	75	13	2	30.4	34(15)
storage tanks, up to 1500 litres water	1.85	day	0.23	–	–	–	0.23	34(1)
storage tanks, up to 6000 litres water	6.09	day	0.76	–	–	–	0.76	34(3)
up to 1500 litres fuel incl bund	9.78	day	1.22	–	–	–	1.22	34(5)
up to 6000 litres fuel incl bund	18.51	day	2.31	–	–	–	2.31	34(6)
Landscaping Items and Tools								
post hole auger 150 / 225 mm hand	11.17	week	0.28	–	–	–	0.28	–
powered auger petrol	73.3	week	1.83	75	2	2	3.76	–
post driver for 200 mm post	9.16	week	0.23	–	–	–	0.23	–
post driver for 150 mm posts	10.09	week	0.25	–	–	–	0.25	–
linemarker 50, 75 or 100 mm wide	20.52	week	0.51	–	–	–	0.51	
bolt croppers 600 mm or 900 mm	12.82	week	0.32	–	–	–	0.32	
tipping and loading skip 2.0m³	1.32	hour	1.32	–	–	–	1.32	8(28)
cultivator self propelled	161.23	week	4.03	75	1	2	4.45	–
Boilers and Sprays								
bitumen boiler 10 gallon (45 litres)	29.33	week	0.73	75	1	2	1.69	29(*)
bitumen boiler 15 gallon (68 litres)	52.19	week	1.3	75	1	2	2.26	29(*)
cold tar spray 700 litres	1.9	hour	1.9	–	–	–	1.9	29(6)
Asphalt road burner (portable)	10.33	hour	10.33	75	N/a	N/a	10.33	2(14)
Air tool bits (see also compressor tools)								
shove holer	10.54	week	0.26	–	–	–	0.26	–
points and chisels	1.74	week	0.04	–	–	–	0.04	–
clayspade	1.63	hour	1.63	–	–	–	1.63	30(3)
tarmac cutter	4.11	week	0.1	–	–	–	0.1	–
comb holder	1.83	week	0.05	–	–	–	0.05	–
self-drill anchor holder	2.38	week	0.06	–	–	–	0.06	–
rammer foot	3.5	week	0.09	–	–	–	0.09	–
Floor sander (belt)	0.38	hour	0.38	–	–	–	0.38	
Pinking roller	18.32	week	0.46	–	–	–	0.46	–
Floor grinder/scarifier, electric	69.63	week	1.74	–	–	–	1.74	–
Floor saw, 450 mm petrol	2.73	hour	2.73	75	1	2	3.69	6(38)
Floor plane, petrol	82.45	week	2.06	75	1	2	3.02	–

Description	Hire Rate £	Hire Period	Cost per working hour					
			Plant £	Usage %	Fuel l/hr	Oil etc. % fuel	Total £	DSR
Floor scaler, petrol	100.78	week	2.52	75	1	2	3.48	–
Rubbish chute (sections)	5.85	week	0.15	–	–	–	0.15	–
Diesel elevators								
up to 15 m	76.77	week	1.92	75	3	2	3.18	–
up to 30 m	105.59	week	2.64	75	5	2	4.74	–
Formwork Equipment								
Adjustable props								
Nr 0 1.07 – 1.82	2.22	week	2.22	–	–	–	2.22	27(12)
Nr 1 1.75 – 3.12	2.29	week	2.29	–	–	–	2.29	27(13)
Nr 2 1.98 – 3.35	2.33	week	2.33	–	–	–	2.33	27(14)
Nr 3 2.59 – 3.95	2.34	week	2.34	–	–	–	2.34	27(15)
Nr 4 3.20 – 4.88	2.38	week	2.38	–	–	–	2.38	27(16)
Formwork Miscellaneous								
Decking per SM (Kwikform) (excluding props)	9.47	week	9.47	–	–	–	9.47	
Column clamps (set)								
Nr 1; 254mm – 508mm	2.4	week	2.4	–	–	–	2.4	27(17)
Nr 2 ; 406mm – 813mm	2.53	week	2.53	–	–	–	2.53	27(18)
Nr 3; 610mm – 1220mm	2.77	week	2.77	–	–	–	2.77	27(19)
Beam Clamps								
300 mm arms	2.6	week	2.6	–	–	–	2.6	27(20)
450 mm arms	2.64	week	2.64	–	–	–	2.64	27(21)
600 mm arms	2.68	week	2.68	–	–	–	2.68	27(22)
Manhole shutters – 1800 mm high								
675 mm internal diameter	39.31	week	39.31	–	–	–	39.31	
900 mm internal diameter	47.18	week	47.18	–	–	–	47.18	
1050 mm internal diameter	50.84	week	50.84	–	–	–	50.84	
1200 mm internal diameter	55.03	week	55.03	–	–	–	55.03	
1500 mm internal diameter	62.9	week	62.9	–	–	–	62.9	
1800 mm internal diameter	66.57	week	66.57	–	–	–	66.57	
2100 mm internal diameter	78.62	week	78.62	–	–	–	78.62	
2400 mm internal diameter	86.48	week	86.48	–	–	–	86.48	
2700 mm internal diameter	90.16	week	90.16	–	–	–	90.16	
SITE ACCOMODATION								
(including electrical fittings but excluding additional security items, shutters, grilles etc. per unit delivered and set up on site)								
Offices, stores etc. on site								
Offices per 10m² floor area	42.17	week	42.17	–	–	–	42.17	16(1)
Mess room per 10m² floor area	37.88	week	37.88	–	–	–	37.88	16(2)
Stores per 10m² floor area	34.89	week	34.89	–	–	–	34.89	16(3)
Mobile office up to 5m long	79.58	week	79.58	–	–	–	79.58	16(8)
Mobile office up to 7m long	103.39	week	103.39	–	–	–	103.39	16(9)
Stores		week						
containers used as tool stores, fitting shop, etc up to 13m long	25.92	week	25.92	–	–	–	25.92	16(10)
Toilets		week						
Toilet unit, single chemical type	19.59	week	19.59	–	–	–	19.59	16(4)
Toilet unit mains flushing type with wash basin		week						
single	24.08	week	24.08	–	–	–	24.08	16(5)
double	36.49	week	36.49	–	–	–	36.49	16(6)
treble	58.3	week	58.3	–	–	–	58.3	16(7)
Sundries		week						
Pollution Decontamination unit 6.75m X 2.34m	256.25	week	256.25	–	–	–	256.25	–

PART 6

Unit Costs – Civil Engineering Works

INTRODUCTORY NOTES

The Unit Costs in this part represent the net cost to the Contractor of executing the work on site; they are not the prices which would be entered in a tender Bill of Quantities.

It must be emphasised that the unit rates are averages calculated on unit outputs for typical site conditions. Costs can vary considerably from contract to contract depending on individual Contractors, site conditions, methods of working and various other factors. Reference should be made to Part 1 for a general discussion on Civil Engineering Estimating.

Guidance prices are included for work normally executed by specialists, with a brief description where necessary of the assumptions upon which the costs have been based. Should the actual circumstances differ, it would be prudent to obtain check prices from the specialists concerned, on the basis of actual / likely quantity of the work, nature of site conditions, geographical location, time constraints, etc.

The method of measurement adopted in this section is the **CESMM3***, subject to variances where this has been felt to produce more helpful price guidance.*

We have structured this Unit Costs section to cover as many aspects of Civil works as possible.

The Gang hours column shows the output per measured unit in actual time, not the total labour hours; thus for an item involving a gang of 5 men each for 0.3 hours, the total labour hours would naturally be 1.5, whereas the Gang hours shown would be 0.3.

This section is structured to provide the User with adequate background information on how the rates have been calculated, so as to allow them to be readily adjusted to suit other conditions to the example presented:

- ☐ Alternative gang structures as well as the effect of varying bonus levels, travelling costs etc.
- ☐ Other types of plant or else different running costs from the medium usage presumed
- ☐ Other types of materials or else different discount / waste allowances from the levels presumed

Reference to Part 3 giving basic costs of labour, materials and plant together with Parts 13 and 14 will assist the reader in making adjustments to the unit costs.

GUIDANCE NOTES

Generally

Adjustments should be made to the rates shown for time, location, local conditions, site constraints and any other factors likely to affect the costs of a specific scheme.

Method of Measurement

Although this part of the book is primarily based on CESMM3, the specific rules have been varied from in cases where it has been felt that an alternative presentation would be of value to the book's main purpose of providing guidance on prices. This is especially so with a number of specialist contractors but also in the cases of work where a more detailed presentation will enable the user to allow for ancillary items.

Materials cost

Materials costs within the rates have been calculated using the 'list prices' contained in Part 3: Resources, with an index appearing on page 39), adjusted to allow for delivery charges (if any) and a 'reasonable' level of discount obtainable by the contractor. This will vary very much depending on the contractor's standing, the potential size of the order and the supplier's eagerness and will vary also between raw traded goods such as timber which will attract a low discount of perhaps 3%, if at all, and manufactured goods where the room for bargaining is much greater and can reach levels of 30% to 40%. High demand for a product at the time of pricing can dramatically reduce the potential discount, as can the world economy in the case of imported goods such as timber and copper. Allowance has also been made for wastage on site (generally 2½% to 5%), dependent upon the risk of damage, the actual level should take account of the nature of the material and its method of storage and distribution about the site.

Labour cost

The composition of the labour and type of plant is generally stated at the beginning of each section, more detailed information on the calculation of the labour rates is given in Part 3: Resources. In addition there is a summary of labour grades and responsibilities extracted from the Working Rule Agreement. Within Parts 4 and 5, each section is prefaced by a detailed build-up of the labour gang assumed for each type of work. This should allow the user to see the cost impact of a different gang as well as different levels of bonus payments, allowances for skills and conditions, travelling allowances etc. The user should be aware that the output constants are based on the gangs shown and would also need to be changed.

Plant cost

A rate build-up of suitable plant is generally stated at the beginning of each section, with more detailed information on alternative machines and their average fuel costs being given in Part 3: Resources. Within Parts 4 and 5, each section is prefaced by a detailed build-up of plant assumed for each type of work. This should allow the user to see the cost impact of using alternative plant as well as different levels of usage. The user should be aware that the output constants are based on the plant shown and would also need to be changed.

Outputs

The user is directed to Part 13: Outputs which contains a selection of output constants, in particular a chart of haulage times for various capacities of Tippers.

CLASS A: GENERAL ITEMS

Item	Gang hours	Labour £	Plant £	Material £	Unit	Total rate £
METHOD RELATED CHARGES						
NOTES						
Refer also to the example calculation of Preliminaries in Part 2 and also to Part 8 Oncosts and Profit.						
CONTRACTUAL REQUIREMENTS						
Performance bond						
The cost of the bond will relate to the nature and degree of difficulty and risk inherent in the type of works intended, the perceived ability and determination of the Contractor to complete them, his financial status and whether he has a proven track record in this field with the provider.						
Refer to the discussion of the matter in Part 2: Preliminaries and General Matters.						
Insurance of the Works						
Refer to the discussion of the various insurances in Part 2: Preliminaries and General Matters.						
Third party insurance						
Refer to the discussion of the various insurances in Part 2: Preliminaries and General Matters.						
SPECIFIED REQUIREMENTS						
General						
This section entails the listing of services and facilities over and above the 'Permanent Works' which the Contractor would be instructed to provide in the Contract Documents.						
Accommodation for the Engineer's Staff						
Refer to resources - Plant page 145 and 146 for a list of accommodation types.						
Services for the Engineer's staff						
Transport vehicles						
4 WD utility short wheelbase	–	–	565.33	–	week	**565.33**
4 WD long wheelbase	–	–	465.73	–	week	**465.73**
(for other vehicles, refer to Resources - Plant)						
Telephones						
(allow for connection charges and usage of telephones required for use by the Engineer's Staff)						
Equipment for use by the Engineer's staff						
Allow for equipment specifically required; entailing Office Equipment, Laboratory Equipment and Surveying Equipment.						

CLASS A: GENERAL ITEMS

Item	Gang hours	Labour £	Plant £	Material £	Unit	Total rate £
Attendance upon the Engineer's staff						
Driver	40.00	528.00	–	–	week	**528.00**
Chainmen	40.00	528.00	–	–	week	**528.00**
Laboratory assistants	40.00	300.00	–	–	week	**300.00**
Testing of materials						
Testing of the Works						
Temporary Works						
Temporary Works relate to other work which the contractor may need to carry out due to his construction method. These are highly specific to the particular project and envisaged construction method. Examples are:						
* access roads and hardstandings/ bases for plant and accommodation as well as to the assembly/ working area generally						
* constructing ramps for access to low excavations						
* steel sheet, diaphragm wall or secant pile cofferdam walling to large excavations subject to strong forces / water penetration						
* bridges						
* temporary support works and decking						
The need for these items should be carefully considered and reference made to the other sections of the book for guidance on what costs should be set against the design assumptions made. Extensive works could well call for the involvement of a contractor's temporary works engineer for realistic advice.						
METHOD RELATED CHARGES						
Accommodation and buildings						
Offices; establishment and removal; Fixed Charge						
80 m² mobile unit (10 staff × 8 m²)	–	–	1680.00	–	sum	**1680.00**
10 m² section units; two	–	–	175.49	–	sum	**175.49**
Offices; maintaining; Time Related Charge						
80 m² mobile unit (10 staff × 8 m²)	–	–	472.50	–	week	**472.50**
10 m² section units; two	–	–	262.50	–	week	**262.50**
Stores; establishment and removal; Fixed Charge						
22 m² section unit	–	–	630.00	–	sum	**630.00**
Stores; maintaining; Time Related Charge						
22 m² section unit	–	–	105.00	–	sum	**105.00**
Canteens and messrooms; establishment and removal; Fixed Charge						
70 m² mobile unit (70 men)	–	–	1680.00	–	sum	**1680.00**
Canteens and messrooms; maintaining; Time-Related Charge						
70 m² mobile unit (70 men)	–	–	472.50	–	week	**472.50**

Unit Costs – Civil Engineering Works

CLASS A: GENERAL ITEMS

Item	Gang hours	Labour £	Plant £	Material £	Unit	Total rate £
METHOD RELATED CHARGES – cont						
Plant						
General purpose plant not included in Unit Costs:						
Transport						
wheeled tractor	40.00	497.60	500.80	–	week	998.40
trailer	–	–	13.86	–	week	13.86
Tractair						
hire charge	–	–	549.60	–	week	549.60
driver (skill rate 4)	40.00	582.40	–	–	week	582.40
fuel and consumables	–	–	205.80	–	week	205.80
Cranes						
10t Crane	40.00	647.60	1048.65	–	week	1696.25
Miscellaneous						
sawbench	–	–	73.82	–	week	73.82
concrete vibrator	–	–	76.66	–	week	76.66
75 mm 750 l/min pump	–	–	94.82	–	week	94.82
compressor	–	–	3527.90	–	week	3527.90
plate compactor; 180 kg	–	–	64.97	–	week	64.97
towed roller; BW6	–	–	429.12	–	week	429.12
Excavators etc.						
hydraulic backacter; 14.5 tonne; driver +						
banksman	80.00	1180.40	1025.81	–	week	2206.21
bulldozer; D6; driver	40.00	722.00	1845.38	–	week	2567.38
loading shovel; CAT 939; driver	40.00	647.60	951.86	–	week	1599.46
Temporary Works						
Supervision and labour						
Supervision for the duration of construction; Time Related Charge						
Agent	40.00	1320.00	–	–	week	1320.00
Senior Engineer	40.00	960.00	–	–	week	960.00
Engineers	40.00	720.00	–	–	week	720.00
General Foreman	40.00	880.00	–	–	week	880.00
Administration for the duration of construction; Time Related Charge						
Office manager / Cost clerk	40.00	650.00	–	–	week	650.00
Timekeeper / Storeman / Checker	40.00	450.00	–	–	week	450.00
Typist / Telephonist	40.00	300.00	–	–	week	300.00
Security guard	40.00	450.00	–	–	week	450.00
Quantity Surveyor	40.00	1091.60	–	–	week	1091.60
Labour teams for the duration of construction; Time Related Charge						
General yard labour (part time); loading and offloading, clearing site rubbish etc.; ganger	40.00	660.00	–	–	week	660.00
General yard labour (part time); loading and offloading, clearing site rubbish etc.; four unskilled operatives	160.00	1777.60	–	–	week	1777.60
Maintenance of contractor's own plant; Fitter	40.00	700.00	–	–	week	700.00
Maintenance of contractor's own plant; Fitter's Mate	40.00	594.40	–	–	week	594.40

CLASS B: GROUND INVESTIGATION

Item	Gang hours	Labour £	Plant £	Material £	Unit	Total rate £
RESOURCES - LABOUR						
Trial hole gang						
1 ganger or chargehand (skill rate 4)		13.32				
1 skilled operative (skill rate 4)		13.32				
2 unskilled operatives (general)		24.88				
1 plant operator (skill rate 3)		16.19				
Total Gang Rate / Hour	£	**67.71**				
RESOURCES - PLANT						
Trial holes						
8 tonne wheeled backacter			20.11			
3 tonne dumper			8.91			
3.7 m³/min compressor, 2 tool			9.42			
two 2.4m³/min road breakers			3.92			
extra 50 ft / 15m hose			0.31			
plate compactors; vibrating compaction; plate 180kg/ 600mm			1.62			
Total Rate / Hour		£	**44.30**			
TRIAL PITS AND TRENCHES						
The following costs assume the use of mechanical plant and excavating and backfilling on the same day						
Trial holes measured by number						
Excavating trial hole; plan size 1.0×2.0m; supports, backfilling						
ne 1.0m deep	0.24	16.58	10.63	–	nr	**27.21**
1.0–2.0m deep	0.47	32.46	20.84	–	nr	**53.30**
over 2.0m deep	0.53	36.60	23.50	–	nr	**60.10**
Excavating trial hole in rock or similar; plan size 1.0×2.0m; supports, backfilling						
ne 1.0m deep	0.28	19.34	12.40	–	nr	**31.74**
1.0–2.0m deep	0.51	35.22	22.61	–	nr	**57.83**
over 2.0m deep	0.58	38.95	25.69	–	nr	**64.64**
Trial holes measured by depth						
Excavating trial hole; plan size 1.0×2.0m; supports, backfilling						
ne 1.0m deep	0.24	16.12	10.63	–	m	**26.75**
1.0–2.0m deep	0.53	35.59	23.50	–	m	**59.09**
2.0–3.0m deep	0.59	39.62	26.15	–	m	**65.77**
3.0–5.0m deep	0.65	43.65	28.81	–	m	**72.46**
Excavating trial hole in rock or similar; plan size 1.0×2.0m; supports, backfilling						
ne 1.0m deep	2.95	198.09	130.70	–	m	**328.79**
1.0–2.0m deep	3.65	245.10	161.70	–	m	**406.80**
2.0–3.0m deep	3.95	265.24	174.99	–	m	**440.23**
3.0–5.0m deep	4.50	302.18	199.34	–	m	**501.52**

CLASS B: GROUND INVESTIGATION

Item	Gang hours	Labour £	Plant £	Material £	Unit	Total rate £
TRIAL PITS AND TRENCHES – cont						
Sundries in trial holes						
Removal of obstructions from trial holes						
irrespective of depth	1.00	67.15	44.30	–	hr	111.45
Pumping; maximum depth 4.0 m						
minimum 750 litres per hour	0.06	1.28	5.65	–	hr	6.93
LIGHT CABLE PERCUSSION BOREHOLES						
The following costs are based on using						
Specialist Contractors and are for guidance only.						
Establishment of standard plant and equipment and						
removal on completion	–	–	–	–	sum	450.00
Number; 150 mm nominal diameter at base	–	–	–	–	nr	50.00
Depth; 150 mm nominal diameter of base						
in holes of maximum depth not exceeding 5 m	–	–	–	–	m	18.00
in holes of maximum depth 5–10 m	–	–	–	–	m	18.00
in holes of maximum depth 10–20 m	–	–	–	–	m	23.00
in holes of maximum depth 20–30 m	–	–	–	–	m	29.00
Depth backfilled; imported pulverized fuel ash	–	–	–	–	m	5.00
Depth backfilled; imported gravel	–	–	–	–	m	7.00
Depth backfilled; bentonite grout	–	–	–	–	m	10.00
Chiselling to prove rock or to penetrate obstructions	–	–	–	–	h	48.00
Standing time of rig and crew	–	–	–	–	h	48.00
ROTARY DRILLED BOREHOLES						
The following costs are based on using						
Specialist Contractors and are for guidance only.						
Establishment of standard plant and equipment and						
removal on completion	–	–	–	–	sum	1000.00
Setting up at each borehole position	–	–	–	–	nr	8.00
Depth without core recovery; nominal minimum core diameter 100 mm						
ne 5.0 m deep	–	–	–	–	m	32.00
5–10 m deep	–	–	–	–	m	32.00
10–20 m deep	–	–	–	–	m	33.00
20–30 m deep	–	–	–	–	m	33.00
Depth with core recovery; nominal minimum core diameter 75 mm						
ne 5.0 m deep	–	–	–	–	m	70.00
5–10 m deep	–	–	–	–	m	70.00
10–20 m deep	–	–	–	–	m	78.00
20–30 m deep	–	–	–	–	m	88.00
Depth cased; semi-rigid plastic core barrel liner	–	–	–	–	m	6.00

CLASS B: GROUND INVESTIGATION

Item	Gang hours	Labour £	Plant £	Material £	Unit	Total rate £
SAMPLES						
From the surface or from trial pits and trenches						
undisturbed soft material; minimum 200 mm cube	–	–	–	–	nr	16.00
disturbed soft material; minimum 5 kg	–	–	–	–	nr	2.50
rock; minimum 5 kg	–	–	–	–	nr	10.00
groundwater; miniumum 1 l	–	–	–	–	nr	4.00
From boreholes						
open tube; 100 mm diameter × 450 mm long	–	–	–	–	nr	16.00
disturbed; minimum 5 kg	–	–	–	–	nr	2.50
groundwater; minimum 1 l	–	–	–	–	nr	4.00
stationary piston	–	–	–	–	nr	4.00
Swedish foil	–	–	–	–	nr	25.00
Delft	–	–	–	–	nr	25.00
Bishop sand	–	–	–	–	nr	25.00
SITE TESTS AND OBSERVATIONS						
Groundwater level						
Standard penetration						
in light cable percussion boreholes	–	–	–	–	nr	–
Vane in borehole						
Plate bearing						
in pits and trenches; loading table	–	–	–	–	nr	–
in pits and trenches; hydraulic jack and kentledge	–	–	–	–	nr	–
California bearing ratio	–	–	–	–	nr	–
Mackintosh probe						
Hand auger borehole						
mm minimum diameter; 6 m maximum depth	–	–	–	–	nr	–
INSTRUMENTAL OBSERVATIONS						
General						
Pressure head						
standpipe; 75 mm diameter HDPE pipe	–	–	–	–	m	29.00
piezometer	–	–	–	–	m	41.77
install protective cover	–	–	–	–	nr	80.00
readings	–	–	–	–	nr	20.00
LABORATORY TESTS						
General						
Classification						
moisture content	–	–	–	–	nr	4.24
specific gravity	–	–	–	–	nr	7.07
particle size analysis by sieve	–	–	–	–	nr	28.29
particle size analysis by pipette or hydrometer	–	–	–	–	nr	39.62

CLASS B: GROUND INVESTIGATION

Item	Gang hours	Labour £	Plant £	Material £	Unit	Total rate £
LABORATORY TESTS – cont						
General – cont						
Chemical content						
organic matter	–	–	–	–	nr	31.12
sulphate	–	–	–	–	nr	21.22
pH value	–	–	–	–	nr	6.84
contaminants; Comprehensive	–	–	–	–	nr	247.48
contaminants; Abbreviated	–	–	–	–	nr	176.86
contaminants; Mini, Screening	–	–	–	–	nr	141.48
contaminants; nitrogen herbicides	–	–	–	–	nr	127.34
contaminants; organophosphorus pesticides	–	–	–	–	nr	106.12
contaminants; organochlorine pesticides	–	–	–	–	nr	106.12
Compaction						
standard	–	–	–	–	nr	127.33
heavy	–	–	–	–	nr	127.33
vibratory	–	–	–	–	nr	141.48
Permeability						
falling head	–	–	–	–	nr	56.59
Soil strength						
quick undrained triaxial; set of three 38 mm diameter specimens	–	–	–	–	nr	28.29
shear box; peak only; size of shearbox 100 × 100 mm	–	–	–	–	nr	35.38
California bearing ratio; typical	–	–	–	–	nr	42.45
Rock strength						
point load test; minimum 5 kg sample	–	–	–	–	nr	70.74
PROFESSIONAL SERVICES						
General						
Technician	–	–	–	–	hr	34.00
Technician engineer	–	–	–	–	hr	46.00
Engineer or geologist						
graduate	–	–	–	–	hr	46.00
chartered	–	–	–	–	hr	59.00
principal or consultant	–	–	–	–	hr	79.00
Visits to the Site						
technician	–	–	–	–	nr	34.00
technician engineer / graduate engineer or geologist	–	–	–	–	nr	46.00
chartered engineer	–	–	–	–	nr	59.00
principal or consultant	–	–	–	–	nr	79.00
Overnight stays in connection with visits to the site						
technician	–	–	–	–	nr	65.00
technician engineer / graduate engineer or geologist	–	–	–	–	nr	65.00
chartered engineer / senior geologist	–	–	–	–	nr	85.00
principal or consultant	–	–	–	–	nr	85.00

CLASS C: GEOTECHNICAL AND OTHER SPECIALIST PROCESSES

Item	Gang hours	Labour £	Plant £	Material £	Unit	Total rate £
DRILLING FOR GROUT HOLES						
NOTE						
The processes referred to in this Section are generally carried out by Specialist Contractors and the costs are therefore an indication of the probable costs based on average site conditions.						
The following unit costs are based on drilling 100 grout holes on a clear site with reasonable access						
Establishment of standard drilling plant and equipment and removal on completion	–	–	–	–	sum	8750.00
Standing time	–	–	–	–	hour	175.00
Drilling through material other than rock or artificial hard material						
vertically downwards						
depth ne 5 m	–	–	–	–	m	16.65
depth 5–10 m	–	–	–	–	m	19.32
depth 10–20 m	–	–	–	–	m	22.65
depth 20–30 m	–	–	–	–	m	27.97
downwards at an angle 0-45° to the vertical						
depth ne 5 m	–	–	–	–	m	16.65
depth 5–10 m	–	–	–	–	m	19.32
depth 10–20 m	–	–	–	–	m	22.65
depth 20–30 m	–	–	–	–	m	27.97
horizontally or downwards at an angle less than 45° to the horizontal						
depth ne 5 m	–	–	–	–	m	16.65
depth 5–10 m	–	–	–	–	m	19.32
depth 10–20 m	–	–	–	–	m	22.65
depth 20–30 m	–	–	–	–	m	27.97
upwards at an angle 0-45° to the horizontal						
depth ne 5 m	–	–	–	–	m	27.97
depth 5–10 m	–	–	–	–	m	31.98
depth 10–20 m	–	–	–	–	m	34.65
depth 20–30 m	–	–	–	–	m	37.98
upwards at an angle less than 45° to the vertical						
depth ne 5 m	–	–	–	–	m	27.97
depth 5–10 m	–	–	–	–	m	31.98
depth 10–20 m	–	–	–	–	m	34.65
depth 20–30 m	–	–	–	–	m	37.98
Drilling through rock or artificial hard material						
Vertically downwards						
depth ne 5 m	–	–	–	–	m	18.50
depth 5–10 m	–	–	–	–	m	22.70
depth 10–20 m	–	–	–	–	m	27.78
depth 20–30 m	–	–	–	–	m	34.47

CLASS C: GEOTECHNICAL AND OTHER SPECIALIST PROCESSES

Item	Gang hours	Labour £	Plant £	Material £	Unit	Total rate £
DRILLING FOR GROUT HOLES – cont						
Drilling through rock or artificial hard material – cont						
Downwards at an angle 0-45° to the vertical						
depth ne 5m	–	–	–	–	m	**18.50**
depth 5–10m	–	–	–	–	m	**22.70**
depth 10–20m	–	–	–	–	m	**27.78**
depth 20–30m	–	–	–	–	m	**34.47**
Horizontally or downwards at an angle less than 45° to the horizontal						
depth ne 5m	–	–	–	–	m	**18.50**
depth 5–10m	–	–	–	–	m	**22.70**
depth 10–20m	–	–	–	–	m	**27.78**
depth 20–30m	–	–	–	–	m	**34.47**
Upwards at an angle 0-45° to the horizontal						
depth ne 5m	–	–	–	–	m	**33.65**
depth 5–10m	–	–	–	–	m	**37.84**
depth 10–20m	–	–	–	–	m	**42.89**
depth 20–30m	–	–	–	–	m	**47.93**
Upwards at an angle less than 45° to the horizontal						
depth ne 5m	–	–	–	–	m	**33.65**
depth 5–10m	–	–	–	–	m	**37.84**
depth 10–20m	–	–	–	–	m	**42.89**
depth 20–30m	–	–	–	–	m	**47.93**
GROUT HOLES						
The following unit costs are based on drilling 100 grout holes on a clear site with reasonable access						
Grout holes						
number of holes	–	–	–	–	nr	**92.04**
multiple water pressure tests	–	–	–	–	nr	**6.59**
GROUT MATERIALS AND INJECTION						
The following unit costs are based on drilling 100 grout holes on a clear site with reasonable access						
Materials						
ordinary portland cement	–	–	–	–	t	**110.00**
sulphate resistant cement	–	–	–	–	t	**112.00**
cement grout	–	–	–	–	t	**136.14**
pulverized fuel ash	–	–	–	–	t	**15.95**
sand	–	–	–	–	t	**15.38**
pea gravel	–	–	–	–	t	**15.38**
bentonite (2:1)	–	–	–	–	t	**141.90**

CLASS C: GEOTECHNICAL AND OTHER SPECIALIST PROCESSES

Item	Gang hours	Labour £	Plant £	Material £	Unit	Total rate £
Injection						
Establishment of standard injection plant and removal on completion	–	–	–	–	sum	8750.00
Standing time	–	–	–	–	hr	120.45
number of injections	–	–	–	–	nr	56.53
neat cement grout	–	–	–	–	t	98.25
cement / P.F.A. grout	–	–	–	–	t	57.08

DIAPHRAGM WALLS

Notes
Diaphragm walls are the construction of vertical walls, cast in place in a trench excavation. They can be formed in reinforced concrete to provide structural elements for temporary or permanent retaining walls. Wall thicknesses of 500 mm to 1.50 m and up to 40 m deep may be constructed. Special equipment such as the Hydrofraise can construct walls up to 100 m deep. Restricted urban sites will significantly increase the costs.

The following costs are based on constructing a diaphragm wall with an excavated volume of 4000 m³ using standard equipment Typical progress would be up to 500 m per week.

Item	Gang hours	Labour £	Plant £	Material £	Unit	Total rate £
Establishment of standard plant and equipment including bentonite storage tanks and removal on completion	–	–	–	–	sum	108000.00
Standing time	–	–	–	–	hr	500.00
Excavation, disposal of soil and placing of concrete	–	–	–	–	m³	440.00
Provide and place reinforcement cages	–	–	–	–	t	750.00
Excavate/chisel in hard material/rock	–	–	–	–	hr	975.00
Waterproofed joints	–	–	–	–	m	6.50
Guide walls						
guide walls (twin)	–	–	–	–	m	115.00

CLASS C: GEOTECHNICAL AND OTHER SPECIALIST PROCESSES

Item	Gang hours	Labour £	Plant £	Material £	Unit	Total rate £
GROUND ANCHORAGES						
Notes						
Ground anchorages consist of the installation of a cable or solid bar tendon fixed in the ground by grouting and tensioned to exceed the working load to be carried. Ground anchors may be of a permanent or temporary nature and can be used in conjunction with diaphragm walls or sheet piling to eliminate the use of strutting etc..						
The following costs are based on the installation of 50 nr ground anchors.						
Establishment of standard plant and equipment and removal on completion	–	–	–	–	sum	10500.00
Standing time	–	–	–	–	hr	160.00
Ground anchorages; temporary or permanent						
15.0 m maximum depth; in rock, alluvial or clay; 0–50 t load	–	–	–	–	m	35.00
15.0 m maximum depth; in rock or alluvial; 50–90 t load	–	–	–	–	m	50.00
15.0 m maximum depth; in rock only; 90–150 t load	–	–	–	–	m	105.00
Temporary tendons						
in rock, alluvial or clay; 0–50 t load	–	–	–	–	m	26.65
in rock or alluvial; 50–90 t load	–	–	–	–	m	39.98
in rock only; 90–150 t load	–	–	–	–	m	54.33
Permanent tendons						
in rock, alluvial or clay; 0–50 t load	–	–	–	–	m	40.00
in rock or alluvial; 50–90 t load	–	–	–	–	m	54.00
in rock only; 90–150 t load	–	–	–	–	m	69.00

CLASS C: GEOTECHNICAL AND OTHER SPECIALIST PROCESSES

Item	Gang hours	Labour £	Plant £	Material £	Unit	Total rate £
SAND, BAND AND WICK DRAINS						
Notes						
Vertical drains are a technique by which the rate of consolidation of fine grained soils can be considerably increased by the installation of vertical drainage paths commonly in the form of columns formed by a high-quality plastic material encased in a filter sleeve. Columns of sand are rarely used in this country these days.						
Band drains are generally 100 mm wide and 3–5 mm thick. water is extracted through the drain from the soft soils when the surface is surcharged. The rate of consolidation is dependent on the drain spacing and the height of surcharge.						
Drains are usually quickly installed up to depths of 25 m by special lances either pulled or vibrated into the ground. typical drain spacing would be one per 1–2 m with the rate of installation varying between 1,500 to 6,000 m per day depending on ground conditions and depths.						
The following costs are based on the installation of 2,000 nr vertical band drains to a depth of 12 m						
Establishment of standard plant and equipment and removal on completion	–	–	–	–	sum	6000.00
Standing time	–	–	–	–	hr	175.00
Set up installation equipment at each drain position	–	–	–	–	nr	0.95
Install drains maximum depth 10–15 m	–	–	–	–	m	0.82
Additional costs in pre-drilling through hard upper strata at each drain position :						
establishment of standard drilling plant and equipment and removal on completion	–	–	–	–	sum	2175.00
set up at each drain position	–	–	–	–	nr	5.63
drilling for vertical band drains up to a maximum depth of 3 m	–	–	–	–	m	12.38

CLASS C: GEOTECHNICAL AND OTHER SPECIALIST PROCESSES

Item	Gang hours	Labour £	Plant £	Material £	Unit	Total rate £
GROUND CONSOLIDATION - VIBRO REPLACEMENT						
Notes						
Vibroreplacement is a method of considerably increasing the ground bearing pressure and consists of a specifically designed powerful poker vibrator penetrating vertically into the ground hyderaulically. Air and water jets may be used to assist penetration. In cohesive soils a hole is formed into which granular backfill is placed and compacted by the poker, forming a dense stone column. In natural sands and gravels the existing loose deposits may be compacted without the addition of extra material other than making up levels after settlement resulting from compaction.						
There are many considerations regarding the soil types to be treaten, whether cohesive or non-cohesive, made-up or natural ground, which influence the choice of wet or dry processes, pure densification or stone column techniques with added granular backfill. It is therefore possible only to give indicative costs; a Specialist Contractor should be consulted for more accurate costs for a particular site.						
Testing of conditions after consolidation can be static or dynamic penetration tests, plate bearing tests or zone bearing tests. A frequently adopted specification calls for plate bearing tests at 1 per 1000 stone columns. Allowable bearing pressures of up to $400\,kN/m^2$ by the installation of stone columns in made or natural ground.						
The following costs are typical rates for this sort of work						
Establishment of standard plant and equipment and removal on completion	–	–	–	–	sum	4071.00
Standing time	–	–	–	–	hr	218.50
Construct stone columns to a depth ne 4 m						
dry formed	–	–	–	–	m	19.24
water jet formed	–	–	–	–	m	28.08
Plate bearing test						
ne 11 t or 2 hour duration	–	–	–	–	nr	468.70
Zone loading test to specification	–	–	–	–	nr	8965.00

CLASS C: GEOTECHNICAL AND OTHER SPECIALIST PROCESSES

Item	Gang hours	Labour £	Plant £	Material £	Unit	Total rate £
GROUND CONSOLIDATION - DYNAMIC COMPACTION						
Notes						
Ground consolidation by dynamic compaction is a technique which involves the dropping of a steel or concrete pounder several times in each location on a grid pattern that covers the whole site. For ground compaction up to 10 m, a 15 t pounder from a free fall of 20 m would be typical. Several passes over the site are normally required to achieve full compaction. The process is recommended for naturally cohesive soils and is usually uneconomic for areas of less than 4,000 m² for sites with granular or mixed granular cohesive soils and 6,000 m for a site with weak cohesive soils. The main considerations to be taken into account when using this method of consolidation are:						
* sufficient area to be viable						
* proximity and condition of adjacent property and services						
* need for blanket layer of granular material for a working surface and as backfill to offset induced settlement						
* water table level						
The final bearing capacity and settlement criteria that can be achieved depends on the nature of the material being compacted. Allowable bearing capacity may be increased by up to twice the pre-treated value for the same settlement. Control testing can be by crater volume measurements, site levelling between passes, penetration tests or plate loading tests.						
The following range of costs are average based on treating an area of about 10,000 m² for a 5–6 m compaction depth. Typical progress would be 1,500–2,000 m² per week						
Establishment of standard plant and equipment and removal on completion.	–	–	–	–	sum	21750.00
Ground treatment	–	–	–	–	m²	1.50
Laying free-draining granular blanket layer as both working surface and backfill material (300 mm thickness required of filter material)	–	–	–	–	m²	8.25
Control testing including levelling, piezometers and penetrameter testing	–	–	–	–	m²	3.50
Kentledge load test	–	–	–	–	nr	8000.00

CLASS C: GEOTECHNICAL AND OTHER SPECIALIST PROCESSES

Item	Gang hours	Labour £	Plant £	Material £	Unit	Total rate £
CONSOLIDATION OF ABANDONED MINE WORKINGS						
The following costs are based on using						
Specialist Contractors and are for guidance only						
Transport plant, labour and all equipment to and from site (max. 100 miles)	–	–	–	–	sum	310.00
Drilling bore holes						
move to each seperate bore position; erect equipment; dismantle prior to next move	–	–	–	–	nr	32.26
drill 50 mm diameter bore holes	–	–	–	–	m	9.56
drill 100 mm diameter bore holes for pea gravel injection	–	–	–	–	m	15.86
extra for casing, when required	–	–	–	–	m	13.06
standing time for drilling rig and crew	–	–	–	–	hr	83.99
Grouting drilled bore holes						
connecting grout lines	–	–	–	–	m	16.68
injection of grout	–	–	–	–	t	65.32
add for pea gravel injection	–	–	–	–	t	78.39
standing time for grouting rig and crew	–	–	–	–	hr	80.48
Provide materials for grouting						
ordinary portland cement	–	–	–	–	t	110.00
sulphate resistant cement	–	–	–	–	t	112.00
pulverized fuel ash (PFA)	–	–	–	–	t	15.95
sand	–	–	–	–	t	15.38
pea gravel	–	–	–	–	t	15.39
bentonite (2:1)	–	–	–	–	t	141.90
cement grout	–	–	–	–	t	136.14
Capping to old shafts or similar; reinforced concrete grade C20P, 20 mm aggregate; thickness						
ne 150 mm	–	–	–	–	m^3	176.00
150-300 mm	–	–	–	–	m^3	170.50
300-500 mm	–	–	–	–	m^3	165.00
over 500 mm	–	–	–	–	m^3	132.00
Mild steel bars BS4449; supplied in bent and cut lengths						
6 mm nominal size	–	–	–	–	t	820.00
8 mm nominal size	–	–	–	–	t	697.00
10 mm nominal size	–	–	–	–	t	668.00
12 mm nominal size	–	–	–	–	t	654.00
16 mm nominal size	–	–	–	–	t	625.00
20 mm nominal size	–	–	–	–	t	564.00
High yield steel bars BS4449 or 4461; supplied in bent and cut lengths						
6 mm nominal size	–	–	–	–	t	900.00
8 mm nominal size	–	–	–	–	t	900.00
10 mm nominal size	–	–	–	–	t	900.00
12 mm nominal size	–	–	–	–	t	900.00
16 mm nominal size	–	–	–	–	t	900.00
20 mm nominal size	–	–	–	–	t	900.00

CLASS D: DEMOLITION AND SITE CLEARANCE

Item	Gang hours	Labour £	Plant £	Material £	Unit	Total rate £
GENERAL CLEARANCE						
The rates for site clearance include for all sundry items, small trees (i.e. under 500 mm diameter), hedges etc., but exclude items that are measured separately; examples of which are given in this section						
Clear site vegetation						
generally	–	–	–	–	ha	1027.40
wooded areas	–	–	–	–	ha	3162.48
areas below tidal level	–	–	–	–	ha	3850.00
TREES						
The following rates are based on removing a minimum of 100 trees, generally in a group. Cutting down a single tree on a site would be many times these costs						
Remove trees						
girth 500 mm–1 m	–	–	–	–	nr	41.54
girth 1–2 m	–	–	–	–	nr	67.78
girth 2–3 m	–	–	–	–	nr	245.96
girth 3–5 m	–	–	–	–	nr	1010.50
girth 7 m	–	–	–	–	nr	1295.78
STUMPS						
Clearance of stumps						
diameter 150–500 mm	–	–	–	–	nr	29.91
diameter 500 mm–1 m	–	–	–	–	nr	54.94
diameter 2 m	–	–	–	–	nr	120.91
Clearance of stumps; backfilling holes with topsoil from site						
diameter 150–500 mm	–	–	–	–	nr	34.83
diameter 500 mm–1 m	–	–	–	–	nr	88.01
diameter 2 m	–	–	–	–	nr	278.04
Clearance of stumps; backfilling holes with imported hardcore						
diameter 150–500 mm	–	–	–	–	nr	56.66
diameter 500 mm–1 m	–	–	–	–	nr	211.86
diameter 2 m	–	–	–	–	nr	749.00

CLASS D: DEMOLITION AND SITE CLEARANCE

Item	Gang hours	Labour £	Plant £	Material £	Unit	Total rate £
BUILDINGS						
The following rates are based on assuming a non urban location where structure does not take up a significant area of the site						
Demolish building to ground level and dispose off site						
brickwork with timber floor and roof	–	–	–	–	m³	6.56
brickwork with concrete floor and roof	–	–	–	–	m³	10.93
masonry with timber floor and roof	–	–	–	–	m³	8.57
reinforced concrete frame with brick infill	–	–	–	–	m³	11.39
steel frame with brick cladding	–	–	–	–	m³	6.20
steel frame with sheet cladding	–	–	–	–	m³	5.90
timber	–	–	–	–	m³	5.31
Demolish buildings with asbestos linings to ground level and dispose off site						
brick with concrete floor and roof	–	–	–	–	m³	26.57
reinforced concrete frame with brick infill	–	–	–	–	m³	27.69
steel frame with brick cladding	–	–	–	–	m³	15.14
steel frame with sheet cladding	–	–	–	–	m³	14.62
OTHER STRUCTURES						
The following rates are based on assuming a non urban location where structure does not take up a significant area of the site						
Demolish walls to ground level and dispose off site						
reinforced concrete wall	–	–	–	–	m³	150.76
brick or masonry wall	–	–	–	–	m³	67.85
brick or masonry retaining wall	–	–	–	–	m³	82.91
PIPELINES						
Removal of redundant services						
electric cable; LV	–	–	–	–	m	2.81
75mm diameter water main; low pressure	–	–	–	–	m	5.63
150mm diameter gas main; low pressure	–	–	–	–	m	6.75
earthenware ducts; one way	–	–	–	–	m	2.81
earthenware ducts; two way	–	–	–	–	m	5.63
100 or 150mm diameter sewer or drain	–	–	–	–	m	11.25
225mm diameter sewer or drain	–	–	–	–	m	16.88
300mm diameter sewer or drain	–	–	–	–	m	19.69
450mm diameter sewer or drain	–	–	–	–	m	22.50
750mm diameter sewer or drain	–	–	–	–	m	28.13
Extra for breaking up concrete surround	–	–	–	–	m	28.13
Grouting redundant drains or sewers						
100mm diameter	–	–	–	–	m	7.05
150mm diameter	–	–	–	–	m	15.86
225mm diameter	–	–	–	–	m	35.25
manhole chambers	–	–	–	–	m³	881.25

CLASS E: EXCAVATION

Item	Gang hours	Labour £	Plant £	Material £	Unit	Total rate £
NOTES						
Ground conditions						
The following unit costs for 'excavation in material other than topsoil, rock or artificial hard material' are based on excavation in firm sand and gravel soils. For alternative types of soil, multiply the following rates by:						
Scrapers						
Stiff clay 1.5						
Chalk 2.5						
Soft rock 3.5						
Broken rock 3.7						
Tractor dozers and loaders						
Stiff clay 2.0						
Chalk 3.0						
Soft rock 2.5						
Broken rock 2.5						
Backacter (minimum bucket size 0.5 m^3)						
Stiff clay 1.7						
Chalk 2.0						
Soft rock 2.0						
Broken rock 1.7						
Basis of disposal rates						
All pricing and estimating for disposal is based on the volume of solid material excavated and rates for disposal should be adjusted by the following factors for bulkage. Multiply the rates by :						
Sand bulkage 1.10						
Gravel bulkage 1.20						
Compacted soil bulkage 1.30						
Compacted sub-base, suitable fill etc. bulkage 1.30						
Stiff clay bulkage 1.20						
See also Part 14: Tables and Memoranda						
Basis of rates generally						
To provide an overall cost comparison, rates, prices and outputs have been based on a medium sized Civil Engineering project of £ 10–12 million, location neither in city centre nor excessively remote, with no abnormal ground conditions that would affect the stated output and consistency of work produced. The rates are optimum rates and assume continuous output with no delays caused by other operations or works.						

CLASS E: EXCAVATION

Item	Gang hours	Labour £	Plant £	Material £	Unit	Total rate £
RESOURCES - LABOUR						
Excavation for cuttings gang						
1 plant operator (skill rate 1) - 33% of time		19.54				
1 plant operator (skill rate 2) - 66% of time		11.91				
1 banksman (skill rate 4)		13.32				
Total Gang Rate / Hour	£	**30.58**				
Excavation for foundations gang						
1 plant operator (skill rate 3) - 33% of time		5.34				
1 plant operator (skill rate 2) - 66% of time		11.91				
1 banksman (skill rate 4)		13.32				
Total Gang Rate / Hour	£	**30.58**				
General excavation gang						
1 plant operator (skill rate 3)		16.19				
1 plant operator (skill rate 3) - 25% of time		4.05				
1 banksman (skill rate 4)		13.32				
Total Gang Rate / Hour	£	**33.56**				
Filling gang						
1 plant operator (skill rate 4)		14.56				
2 unskilled operatives (general)		24.88				
Total Gang Rate / Hour	£	**39.44**				
Treatment of filled surfaces gang						
1 plant operator (skill rate 2)		18.05				
Total Gang Rate / Hour	£	**18.05**				
Geotextiles (light sheets) gang						
1 ganger/chargehand (skill rate 4) - 20% of time		2.85				
2 unskilled operatives (general)		24.88				
Total Gang Rate / Hour	£	**27.73**			s	
Geotextiles (medium sheets) gang						
1 ganger/chargehand (skill rate 4) - 20% of time		2.85				
3 unskilled operatives (general)		37.32				
Total Gang Rate / Hour	£	**40.17**			s	
Geotextiles (heavy sheets) gang						
1 ganger/chargehand (skill rate 4) - 20% of time		2.85				
2 unskilled operatives (general)		24.88				
1 plant operator (skill rate 3)		–	14.36			
Total Gang Rate / Hour	£	**43.92**				
Horticultural works gang						
1 skilled operative (skill rate 4)		13.32				
1 unskilled operative (general)		12.44				
Total Gang Rate / Hour	£	**25.76**				

CLASS E: EXCAVATION

Item	Gang hours	Labour £	Plant £	Material £	Unit	Total rate £
RESOURCES - PLANT						
Excavation for foundations						
Hydraulic Backacter - 21 tonne (33% of time)			13.35			
Hydraulic Backacter - 14.5 tonne (33% of time)			8.55			
1000 kg hydraulic breaker (33% of time)			6.03			
Tractor loader - 0.80 m³ (33% of time)			7.93			
Total Rate / Hour		£	**35.86**			
Filling						
1.5 m³ tractor loader			38.52			
6t vibratory roller			10.73			
Pedestrian Roller, Bomag BW35			2.86			
Total Rate / Hour		£	**52.11**			
Treatment of filled surfaces						
1.5 m³ tractor loader			38.52			
Pedestrian Roller, Bomag BW35			2.86			
Total Rate / Hour		£	**41.38**			
Geotextiles (heavy sheets)						
1.5 m³ tractor loader			38.52			
Total Rate / Hour		£	**38.52**			

CLASS E: EXCAVATION

Item	Gang hours	Labour £	Plant £	Material £	Unit	Total rate £
EXCAVATION BY DREDGING						
Notes						
Dredging can be carried out by land based machines or by floating plant. The cost of the former can be assessed by reference to the excavation costs of the various types of plant given below, suitably adjusted to take account of the type of material to be excavated, depth of water and method of disposal. The cost of the latter is governed by many factors which affect the rates and leads to wide variations. For reliable estimates it is advisable to seek the advice of a Specialist Contractor. The prices included here are for some typical dredging situations and are shown for a cost comparison and EXCLUDE initial mobilization charges which can range widely between £3,000 and £10,000 depending on plant, travelling distance etc.. Some clients schedule operations for when the plant is passing so as to avoid the large travelling cost. Of the factors affecting the cost of floating plant, the matter of working hours is by far one of the most important. The customary practice in the dredging industry is to work 24 hours per day, 7 days per week. Local constraints, particularly noise restrictions, will have a significant impact. Other major factors affecting the cost of floating plant are :						
* type of material to be dredged						
* depth of water						
* depth of cut						
* tidal range						
* disposal location						
* size and type of plant required						
* current location of plant						
* method of disposal of dredged material						
In tidal locations, creating new channels on approaches to quays and similar locations or within dock systems						
Backhoe dredger loading material onto two hopper barges with bottom dumping facility maximum water depth 15 m, distance to disposal site less than 20 miles						
approximate daily cost of backhoe dredger and two hopper barges £10,500; average production 100 m³/hr, 1,500 m³/day; locate, load, deposit and relocate	–	–	–	–	m³	**7.51**

CLASS E: EXCAVATION

Item	Gang hours	Labour £	Plant £	Material £	Unit	Total rate £
For general bed lowering or in maintaining shipping channels in rivers, estuaries or deltas Trailer suction hopper dredger excavating non-cohesive sands, grits or silts; hopper capacity 2000m^3, capable of dredging to depths of 25m with ability either to dump at disposal site or pump ashore for reclamation. Costs totally dependent on nature of material excavated and method of disposal. Approximate cost per tonne £2.50, which should be adjusted by the relative density of the material for conversion to m^3						
Locate, load, deposit and relocate	–	–	–	–	tonne	**2.26**
Harbour bed control Maintenance dredging of this nature would in most cases be carried out by trailing suction hopper dredger as detailed above, at similar rates. The majority of the present generation of trailers have the ability to dump at sea or discharge ashore. Floating craft using diesel driven suction method with a 750mm diameter flexible pipe for a maximum distance of up to 5000m from point of suction to point of discharge using a booster (standing alone, the cutter suction craft should be able to pump up to 2000m). Maximum height of lift 10m.						
Average pumping capacity of silt/sand type materials containing maximum 30% volume of solids would be about 8,000m^3/day based on 24 hour working, Daily cost (hire basis) in the region of £20,000 including all floating equipmnet and discharge pipes, maintenance and all labour and plant to service but excluding mobilization/initial set-up and demobilization costs (minimum £10,000).	–	–	–	–	m^3	**2.91**
For use in lakes, canals, rivers, industrial lagoons and from silted locations in dock systems Floating craft using diesel driven suction method with a 200mm diameter flexible pipe maximum distance from point of suction to point of discharge 1500m. Maximum dredge depth 5m.						
Average pumping output 30m^3 per hour and approximate average costs (excluding mobilization and demobilization costs ranging between £3,000and £6,000 apiece)	–	–	–	–	m^3	**5.41**

CLASS E: EXCAVATION

Item	Gang hours	Labour £	Plant £	Material £	Unit	Total rate £
EXCAVATION FOR CUTTINGS						
The following unit costs are based on backacter and tractor loader machines						
Excavate topsoil						
maximum depth ne 0.25m	0.03	0.97	1.11	–	m³	2.08
Excavate material other than topsoil, rock or artificial hard material						
ne 0.25m maximum depth	0.03	0.97	1.11	–	m³	2.08
0.25–0.5m maximum depth	0.03	0.97	1.11	–	m³	2.08
0.5–1.0m maximum depth	0.04	1.29	1.41	–	m³	2.70
1.0–2.0m maximum depth	0.06	1.93	2.15	–	m³	4.08
2.0–5.0m maximum depth	0.10	3.22	3.56	–	m³	6.78
5.0–10.0m maximum depth	0.20	6.44	7.20	–	m³	13.64
10.0–15.0m maximum depth	0.29	9.34	10.40	–	m³	19.74
The following unit costs are based on backacter machines fitted with hydraulic breakers and tractor loader machines						
Excavate rock (medium hard)						
ne 0.25m maximum depth	0.31	9.98	11.13	–	m³	21.11
0.25–0.5m maximum depth	0.43	13.84	15.44	–	m³	29.28
0.5–1.0m maximum depth	0.58	18.67	20.79	–	m³	39.46
1.0–2.0m maximum depth	0.80	25.75	28.76	–	m³	54.51
Excavate unreinforced concrete exposed at the commencing surface						
ne 0.25m maximum depth	0.57	18.35	20.49	–	m³	38.84
0.25–0.5m maximum depth	0.60	19.31	21.53	–	m³	40.84
0.5 -1.0m maximum depth	0.67	21.57	24.05	–	m³	45.62
1.0–2.0m maximum depth	0.70	22.53	25.09	–	m³	47.62
Excavate reinforced concrete exposed at the commencing surface						
ne 0.25m maximum depth	0.90	28.97	32.30	–	m³	61.27
0.25–0.5m maximum depth	0.90	28.97	32.30	–	m³	61.27
0.5–1.0m maximum depth	0.95	30.58	34.08	–	m³	64.66
1.0–2.0m maximum depth	1.08	34.77	38.76	–	m³	73.53
Excavate unreinforced concrete not exposed at the commencing surface						
ne 0.25m maximum depth	0.65	20.92	23.32	–	m³	44.24
0.25–0.5m maximum depth	0.67	21.57	24.05	–	m³	45.62
0.5 -1.0m maximum depth	0.69	22.21	24.80	–	m³	47.01
1.0–2.0m maximum depth	0.72	23.18	25.84	–	m³	49.02

CLASS E: EXCAVATION

Item	Gang hours	Labour £	Plant £	Material £	Unit	Total rate £
EXCAVATION FOR FOUNDATIONS						
The following unit costs are based on the use of backacter machines						
Excavate topsoil						
maximum depth ne 0.25 m	0.03	0.94	1.11	–	m³	**2.05**
Excavate material other than topsoil, rock or artificial hard material						
0.25–0.5 m deep	0.05	1.56	1.81	–	m³	**3.37**
0.5–1.0 m deep	0.06	1.82	2.15	–	m³	**3.97**
1.0–2.0 m deep	0.07	2.18	2.49	–	m³	**4.67**
2.0–5.0 m deep	0.12	3.74	4.31	–	m³	**8.05**
5.0–10.0 m deep	0.23	7.17	8.27	–	m³	**15.44**
The following unit costs are based on backacter machines fitted with hydraulic breakers						
Excavate unreinforced concrete exposed at the commencing surface						
ne 0.25 m maximum depth	0.60	18.71	21.53	–	m³	**40.24**
0.25–0.5 m maximum depth	0.66	20.58	23.69	–	m³	**44.27**
0.5–1.0 m maximum depth	0.72	22.45	25.84	–	m³	**48.29**
1.0–2.0 m maximum depth	0.80	24.95	28.76	–	m³	**53.71**
Excavate reinforced concrete exposed at the commencing surface						
ne 0.25 m maximum depth	0.94	29.32	33.68	–	m³	**63.00**
0.25–0.5 m maximum depth	0.98	30.56	35.22	–	m³	**65.78**
0.5–1.0 m maximum depth	1.00	31.19	35.86	–	m³	**67.05**
1.0–2.0 m maximum depth	1.04	32.43	37.38	–	m³	**69.81**
Excavate tarmacadam exposed at the commencing surface						
ne 0.25 m maximum depth	0.30	9.36	10.77	–	m³	**20.13**
0.25–0.5 m maximum depth	0.34	10.60	12.15	–	m³	**22.75**
0.5–1.0 m maximum depth	0.37	11.54	13.26	–	m³	**24.80**
1.0–2.0 m maximum depth	0.40	12.47	14.30	–	m³	**26.77**
Excavate unreinforced concrete not exposed at the commencing surface						
ne 0.25 m maximum depth	0.70	21.83	25.07	–	m³	**46.90**
0.25–0.5 m maximum depth	0.85	26.51	30.49	–	m³	**57.00**
0.5 -1.0 m maximum depth	0.96	29.94	34.45	–	m³	**64.39**
1.0–2.0 m maximum depth	1.02	31.81	36.60	–	m³	**68.41**

CLASS E: EXCAVATION

Item	Gang hours	Labour £	Plant £	Material £	Unit	Total rate £
GENERAL EXCAVATION						
The following unit costs are based on backacter and tractor loader machines						
Excavate topsoil						
maximum depth ne 0.25 m	0.03	1.03	1.11	–	m³	**2.14**
Excavate material other than topsoil, rock or artificial hard material						
ne 0.25 m maximum depth	0.03	1.03	1.09	–	m³	**2.12**
0.25–0.5 m maximum depth	0.03	1.03	1.11	–	m³	**2.14**
0.5–1.0 m maximum depth	0.04	1.37	1.41	–	m³	**2.78**
1.0–2.0 m maximum depth	0.06	2.05	2.15	–	m³	**4.20**
2.0–5.0 m maximum depth	0.10	3.42	3.56	–	m³	**6.98**
5.0–10.0 m maximum depth	0.20	6.85	7.20	–	m³	**14.05**
10.0–15.0 m maximum depth	0.29	9.93	10.40	–	m³	**20.33**
The following unit costs are based on backacter machines fitted with hydraulic breakers and tractor loader machines						
Excavate rock (medium hard)						
ne 0.25 m maximum depth	0.31	10.61	11.13	–	m³	**21.74**
0.25–0.5 m maximum depth	0.43	14.72	15.44	–	m³	**30.16**
0.5–1.0 m maximum depth	0.58	19.85	20.79	–	m³	**40.64**
1.0–2.0 m maximum depth	0.80	27.38	28.76	–	m³	**56.14**
Excavate unreinforced concrete exposed at the commencing surface						
ne 0.25 m maximum depth	0.57	19.51	20.49	–	m³	**40.00**
0.25–0.5 m maximum depth	0.60	20.54	21.53	–	m³	**42.07**
0.5 -1.0 m maximum depth	0.67	22.93	24.05	–	m³	**46.98**
1.0–2.0 m maximum depth	0.70	23.96	25.09	–	m³	**49.05**
Excavate reinforced concrete exposed at the commencing surface						
ne 0.25 m maximum depth	0.90	30.81	32.30	–	m³	**63.11**
0.25–0.5 m maximum depth	0.90	30.81	32.30	–	m³	**63.11**
0.5–1.0 m maximum depth	0.95	32.52	34.08	–	m³	**66.60**
1.0–2.0 m maximum depth	1.08	36.97	38.76	–	m³	**75.73**
Excavate unreinforced concrete not exposed at the commencing surface						
ne 0.25 m maximum depth	0.65	22.25	23.32	–	m³	**45.57**
0.25–0.5 m maximum depth	0.67	22.93	24.05	–	m³	**46.98**
0.5 -1.0 m maximum depth	0.69	23.62	24.80	–	m³	**48.42**
1.0–2.0 m maximum depth	0.72	24.64	25.84	–	m³	**50.48**

CLASS E: EXCAVATION

Item	Gang hours	Labour £	Plant £	Material £	Unit	Total rate £
EXCAVATION ANCILLARIES						
The following unit costs are for various machines appropriate to the work						
Trimming topsoil; using D4H dozer						
horizontal	0.04	0.59	0.43	–	m²	1.02
10–45° to horizontal	0.05	0.74	0.54	–	m²	1.28
Trimming material other than topsoil, rock or artificial hard material; using D4H dozer, tractor loader or motor grader average rate						
horizontal	0.04	0.59	0.86	–	m²	1.45
10–45° to horizontal	0.04	0.59	0.86	–	m²	1.45
45–90° to horizontal	0.06	0.89	1.28	–	m²	2.17
Trimming rock; using D6E dozer						
horizontal	0.08	1.18	18.45	–	m²	19.63
10–45° to horizontal	0.80	11.80	18.45	–	m²	30.25
45–90° to horizontal	1.06	15.64	24.45	–	m²	40.09
vertical	1.06	15.64	24.45	–	m²	40.09
Preparation of topsoil; using D4H dozer, tractor loader or motor grader average rate						
horizontal	0.06	0.89	0.90	–	m²	1.79
10–45° to horizontal	0.06	0.89	0.90	–	m²	1.79
Preparation of material other than rock or artificial hard material; using D6E dozer, tractor loader or motor grader average rate						
horizontal	0.06	0.89	1.53	–	m²	2.42
10–45° to horizontal	0.06	0.89	1.53	–	m²	2.42
45–90° to horizontal	0.06	0.89	1.53	–	m²	2.42
Preparation of rock; using D6E dozer						
horizontal	0.80	11.80	18.45	–	m²	30.25
10–45° to horizontal	0.60	8.85	13.84	–	m²	22.69
45–90° to horizontal	0.60	8.85	13.84	–	m²	22.69
The following unit costs for disposal are based on using a 22.5 t ADT for site work and 20 t tipper for off-site work. The distances used in the calculation are quoted to assist estimating, although this goes beyond the specific requirements of CESMM3						
Disposal of excavated topsoil						
storage on site; 100 m maximum distance	0.05	0.90	2.38	–	m³	3.28
removal; 5 km distance	0.11	1.99	5.87	–	m³	7.86
removal; 15 km distance	0.21	3.79	11.21	–	m³	15.00
Disposal of excavated earth other than rock or artificial hard material						
storage on site; 100 m maximum distance; using 22.5 t ADT	0.05	0.90	2.38	–	m³	3.28
removal; 5 km distance; using 20 t tipper	0.12	2.17	6.41	–	m³	8.58
removal; 15 km distance; using 20 t tipper	0.22	3.97	11.74	–	m³	15.71

CLASS E: EXCAVATION

Item	Gang hours	Labour £	Plant £	Material £	Unit	Total rate £
EXCAVATION ANCILLARIES – cont						
The following unit costs for disposal are based on using a 22.5 t ADT for site work and 20 t tipper for off-site work. – cont						
Disposal of excavated rock or artificial hard material						
storage on site; 100 m maximum distance; using						
22.5 t ADT	0.06	1.08	2.85	–	m^3	3.93
removal; 5 km distance; using 20 t tipper	0.13	2.35	6.94	–	m^3	9.29
removal; 15 km distance; using 20 t tipper	0.23	4.15	12.28	–	m^3	16.43
Add to the above rates where tipping charges apply:						
non-hazardous waste	–	–	–	–	m^3	36.30
hazardous waste	–	–	–	–	m^3	74.25
special waste	–	–	–	–	m^3	90.75
contaminated liquid	–	–	–	–	m^3	99.00
contaminated sludge	–	–	–	–	m^3	132.00
Add to the above rates where Landfill Tax applies:						
exempted material	–	–	–	–	m^3	–
inactive or inert material	–	–	–	–	m^3	3.75
other material	–	–	–	–	m^3	84.00
The following unit costs for double handling are based on using a 1.5 m^3 tractor loader and 22.5 t ADT. A range of distances is listed to assist in estimating, although this goes beyond the specific requirements of CESMM3.						
Double handling of excavated topsoil; using 1.5 m^3 tractor loader and 22.5 t ADT						
300 m average distance moved	0.08	1.44	3.44	–	m^3	4.88
600 m average distance moved	0.10	1.80	4.30	–	m^3	6.10
1000 m average distance moved	0.12	2.17	5.16	–	m^3	7.33
Double handling of excavated earth other than rock or artificial hard material; using 1.5 m^3 tractor loader and 22.5 t ADT						
300 m average distance moved	0.08	1.44	3.44	–	m^3	4.88
600 m average distance moved	0.10	1.80	4.30	–	m^3	6.10
1000 m average distance moved	0.12	2.17	5.16	–	m^3	7.33
Double handling of rock or artificial hard material; using 1.5 m^3 tractor loader and 22.5 t ADT						
300 m average distance moved	0.16	2.89	6.88	–	m^3	9.77
600 m average distance moved	0.18	3.25	7.74	–	m^3	10.99
1000 m average distance moved	0.20	3.61	8.60	–	m^3	12.21
The following unit rates for excavation below Final Surface are based on using a 16 t backacter machine						
Excavation of material below the Final Surface and replacement of with:						
granular fill	1.20	17.18	17.93	30.55	m^3	65.66
concrete Grade C7.5P	0.60	8.59	8.97	90.99	m^3	108.55
Misc						
Timber supports left in	0.32	6.03	–	7.97	m^2	14.00
Metal supports left in	0.89	16.40	–	15.89	m^2	32.29

CLASS E: EXCAVATION

Item	Gang hours	Labour £	Plant £	Material £	Unit	Total rate £
FILLING						
Excavated topsoil; DfT specified type 5A						
Filling						
to structures	0.07	2.80	3.73	–	m³	**6.53**
embankments	0.03	0.98	1.30	–	m³	**2.28**
general	0.02	0.82	1.09	–	m³	**1.91**
150 mm thick	0.01	0.27	0.36	–	m²	**0.63**
250 mm thick	0.01	0.47	0.63	–	m²	**1.10**
400 mm thick	0.02	0.66	0.89	–	m²	**1.55**
600 mm thick	0.02	0.94	1.25	–	m²	**2.19**
Imported topsoil DfT specified type 5B;						
Filling						
embankments	0.03	0.98	1.30	24.59	m³	**26.87**
general	0.02	0.82	1.09	24.59	m³	**26.50**
150 mm thick	0.01	0.27	0.36	3.69	m²	**4.32**
250 mm thick	0.01	0.47	0.63	6.15	m²	**7.25**
400 mm thick	0.02	0.66	0.89	9.84	m²	**11.39**
600 mm thick	0.02	0.94	1.25	14.76	m²	**16.95**
Non-selected excavated material other than topsoil or rock						
Filling						
to structures	0.04	1.41	1.88	–	m³	**3.29**
embankments	0.02	0.59	0.78	–	m³	**1.37**
general	0.01	0.43	0.57	–	m³	**1.00**
150 mm thick	–	0.16	0.21	–	m²	**0.37**
250 mm thick	0.01	0.23	0.31	–	m²	**0.54**
400 mm thick	0.01	0.35	0.47	–	m²	**0.82**
600 mm thick	0.01	0.47	0.63	–	m²	**1.10**
Selected excavated material other than topsoil or rock						
Filling						
to structures	0.04	1.60	2.14	–	m³	**3.74**
embankments	0.01	0.55	0.73	–	m³	**1.28**
general	0.01	0.47	0.63	–	m³	**1.10**
150 mm thick	–	0.16	0.21	–	m²	**0.37**
250 mm thick	0.01	0.27	0.36	–	m²	**0.63**
400 mm thick	0.01	0.39	0.52	–	m²	**0.91**
600 mm thick	0.01	0.51	0.68	–	m²	**1.19**
Imported natural material other than topsoil or rock; subsoil						
Filling						
to structures	0.04	1.41	1.88	25.00	m³	**28.29**
embankments	0.02	0.59	0.78	25.00	m³	**26.37**
general	0.02	0.59	0.78	25.00	m³	**26.37**
150 mm thick	0.01	0.35	0.47	3.75	m²	**4.57**
250 mm thick	0.01	0.51	0.68	6.25	m²	**7.44**
400 mm thick	0.02	0.59	0.78	10.00	m²	**11.37**
600 mm thick	0.02	0.66	0.89	15.00	m²	**16.55**

CLASS E: EXCAVATION

Item	Gang hours	Labour £	Plant £	Material £	Unit	Total rate £
FILLING – cont						
Imported natural material other than topsoil or						
rock; granular graded material						
Filling						
to structures	0.04	1.41	1.88	30.99	m^3	**34.28**
embankments	0.02	0.66	0.89	30.99	m^3	**32.54**
general	0.02	0.66	0.89	30.99	m^3	**32.54**
150 mm thick	0.01	0.47	0.63	4.65	m^2	**5.75**
250 mm thick	0.02	0.66	0.89	7.75	m^2	**9.30**
400 mm thick	0.02	0.78	1.04	12.40	m^2	**14.22**
600 mm thick	0.02	0.86	1.15	18.60	m^2	**20.61**
Imported natural material other than topsoil or						
rock; granular selected material						
Filling						
to structures	0.04	1.41	1.88	26.63	m^3	**29.92**
embankments	0.02	0.66	0.89	26.63	m^3	**28.18**
general	0.02	0.66	0.89	26.63	m^3	**28.18**
150 mm thick	0.01	0.47	0.63	4.00	m^2	**5.10**
250 mm thick	0.02	0.66	0.89	6.66	m^2	**8.21**
400 mm thick	0.02	0.78	1.04	10.65	m^2	**12.47**
600 mm thick	0.02	0.86	1.15	15.98	m^2	**17.99**
Excavated rock						
Filling						
to structures	0.04	1.56	2.08	–	m^3	**3.64**
embankments	0.05	1.96	2.61	–	m^3	**4.57**
general	0.05	1.96	2.61	–	m^3	**4.57**
150 mm thick	0.02	0.66	0.89	–	m^2	**1.55**
250 mm thick	0.03	1.02	1.35	–	m^2	**2.37**
400 mm thick	0.04	1.56	2.08	–	m^2	**3.64**
600 mm thick	0.06	2.15	2.87	–	m^2	**5.02**
Imported rock						
Filling						
to structures	0.04	1.56	2.08	33.31	m^3	**36.95**
embankments	0.02	0.94	1.25	33.31	m^3	**35.50**
general	0.02	0.78	1.04	33.31	m^3	**35.13**
150 mm thick	0.01	0.47	0.63	5.00	m^2	**6.10**
250 mm thick	0.02	0.78	1.04	8.33	m^2	**10.15**
400 mm thick	0.03	1.17	1.56	13.32	m^2	**16.05**
600 mm thick	0.03	1.17	1.56	19.99	m^2	**22.72**

CLASS E: EXCAVATION

Item	Gang hours	Labour £	Plant £	Material £	Unit	Total rate £
FILLING ANCILLARIES						
Trimming of filled surfaces						
Topsoil						
horizontal	0.02	0.38	0.87	–	m²	1.25
inclined at an angle of 10–45° to horizontal	0.02	0.38	0.87	–	m²	1.25
inclined at an angle of 45–90° to horizontal	0.03	0.48	1.12	–	m²	1.60
Material other than topsoil, rock or artificial hard material						
horizontal	0.02	0.38	0.87	–	m²	1.25
inclined at an angle of 10–45° to horizontal	0.02	0.38	0.87	–	m²	1.25
inclined at an angle of 45–90° to horizontal	0.03	0.48	1.12	–	m²	1.60
Rock						
horizontal	0.51	9.13	21.10	–	m²	30.23
inclined at an angle of 10–45° to horizontal	0.52	9.31	21.52	–	m²	30.83
inclined at an angle of 45–90° to horizontal	0.70	12.53	28.97	–	m²	41.50
Preparation of filled surfaces						
Topsoil						
horizontal	0.03	0.54	1.24	–	m²	1.78
inclined at an angle of 10–45° to horizontal	0.03	0.54	1.24	–	m²	1.78
inclined at an angle of 45–90° to horizontal	0.04	0.67	1.53	–	m²	2.20
Material other than topsoil, rock or artificial hard material						
horizontal	0.03	0.54	1.24	–	m²	1.78
inclined at an angle of 10–45° to horizontal	0.03	0.54	1.24	–	m²	1.78
inclined at an angle of 45–90° to horizontal	0.04	0.67	1.53	–	m²	2.20
Rock						
horizontal	0.33	5.96	13.65	–	m²	19.61
inclined at an angle of 10–45° to horizontal	0.33	5.96	13.65	–	m²	19.61
inclined at an angle of 45–90° to horizontal	0.52	9.39	21.52	–	m²	30.91
GEOTEXTILES						
NOTES						
The geotextile products mentioned below are not specifically confined to the individual uses stated but are examples of one of many scenarios to which they may be applied. Conversely, the scenarios are not limited to the geotextile used as an example.						
Geotextiles; stabilization applications for reinforcement of granular sub-bases, capping layers and railway ballast placed over weak and variable soils						
For use over weak soils with moderate traffic intensities e.g. car parks, light access roads; Tensar SS20 Polypropylene Geogrid						
horizontal	0.04	1.33	–	2.11	m²	3.44
inclined at an angle of 10 to 45° to the horizontal	0.05	1.68	–	2.11	m²	3.79
For use over weak soils with high traffic intensities and/or high axle loadings; Tensar SS30 Polypropylene Geogrid						
horizontal	0.05	1.79	–	2.40	m²	4.19
inclined at an angle of 10 to 45° to the horizontal	0.06	2.23	–	2.40	m²	4.63

CLASS E: EXCAVATION

Item	Gang hours	Labour £	Plant £	Material £	Unit	Total rate £
GEOTEXTILES – cont						
Geotextiles – cont						
For use over very weak soils e.g. alluvium, marsh or peat or firmer soil subject to exceptionally high axle loadings;Tensar SS40 Polypropylene Geogrid						
horizontal	0.05	2.08	1.73	3.71	m²	**7.52**
inclined at an angle of 10 to 45° to the horizontal	0.06	2.59	2.16	3.71	m²	**8.46**
For trafficked areas where fill comprises aggregate exceeding 100mm; Tensar SSLA20 Polypropylene Geogrid						
horizontal	0.04	1.33	–	3.09	m²	**4.42**
inclined at an angle 10–45° to the horizontal	0.05	1.68	–	3.09	m²	**4.77**
Stabilization and separation of granular fill from soft sub grade to prevent intermixing: Terram 1000						
horizontal	0.05	2.03	–	0.40	m²	**2.43**
inclined at an angle of 10 to 45° to the horizontal	0.06	2.55	–	0.40	m²	**2.95**
Stabilization and separation of granular fill from soft sub grade to prevent intermixing: Terram 2000						
horizontal	0.04	1.94	1.62	1.22	m²	**4.78**
inclined at an angle of 10 to 45° to the horizontal	0.05	2.45	2.04	1.22	m²	**5.71**
Geotextiles; reinforcement applications for asphalt pavements						
For roads, hardstandings and airfield pavements; Tensar AR-G composite comprising Tensar AR-1 grid bonded to a geotextile, laid within asphalt						
horizontal	0.05	1.79	–	2.78	m²	**4.57**
inclined at an angle of 10 to 45° to the horizontal	0.06	2.23	–	2.78	m²	**5.01**
Geotextiles; slope reinforcement and embankment support; for use where soils can only withstand limited shear stresses, therefore steep slopes require external support						
Paragrid 30/155; 330g/m²						
horizontal	0.04	1.33	–	2.46	m²	**3.79**
inclined at an angle of 10-45° to the horizontal	0.05	1.68	–	2.46	m²	**4.14**
Paragrid 100/255; 330g/m²						
horizontal	0.04	1.33	–	3.20	m²	**4.53**
inclined at an angle of 10-45° to the						
horizontalhorizontal	0.05	1.68	–	3.20	m²	**4.88**
Paralink 200s; 1120g/m²						
horizontal	0.05	2.49	2.08	4.87	m²	**9.44**
inclined at an angle of 10-45° to the horizontal	0.07	3.14	2.62	4.87	m²	**10.63**
Paralink 600s; 2040g/m²						
horizontal	0.06	2.91	2.43	11.13	m²	**16.47**
inclined at an angle of 10-45° to the horizontal	0.08	3.65	3.04	11.13	m²	**17.82**
Terram grid 30/30						
horizontal	0.06	2.63	2.20	0.01	m²	**4.84**
inclined at an angle of 10-45° to the horizontal	0.07	3.28	2.73	0.01	m²	**6.02**

CLASS E: EXCAVATION

Item	Gang hours	Labour £	Plant £	Material £	Unit	Total rate £
Geotextiles; scour and erosion protection						
For use where erosion protection is required to the surface of a slope once its geotechnical stability has been achieved, and to allow grass establishment; Tensar 'Mat' Polyethelene mesh; fixed with Tensar pegs						
horizontal	0.04	1.63	–	4.70	m²	**6.33**
inclined at an angle of 10-45° to the horizontal	0.05	2.03	–	4.70	m²	**6.73**
For use where hydraulic action exists, such as coastline protection from pressures exerted by waves, currents and tides; Typar SF56						
horizontal	0.05	2.36	1.96	0.54	m²	**4.86**
inclined at an angle of 10-45° to the horizontal	0.06	2.96	2.46	0.54	m²	**5.96**
For protection against puncturing to reservoir liner; Typar SF563						
horizontal	0.05	2.36	1.96	0.54	m²	**4.86**
inclined at an angle of 10-45° to the horizontal	0.06	2.96	2.46	0.54	m²	**5.96**
Geotextiles; temporary parking areas						
For reinforcement of grassed areas subject to wear from excessive pedestrian and light motor vehicle traffic; Netlon CE131 high density polyethelyene geogrid						
horizontal	0.04	1.40	–	3.79	m²	**5.19**
Geotextiles; landscaping applications						
For prevention of weed growth in planted areas by incorporating a geotextile over top soil and below mulch or gravel; Typar SF20						
horizontal	0.08	2.99	–	0.32	m²	**3.31**
inclined at an angle of 10-45° to the horizontal	0.09	3.75	–	0.32	m²	**4.07**
For root growth control-Prevention of lateral spread of roots and mixing of road base and humus; Typar SF20						
horizontal	0.08	2.99	–	0.32	m²	**3.31**
inclined at an angle of 10-45° to the horizontal	0.09	3.75	–	0.32	m²	**4.07**
Geotextiles; drainage applications						
For clean installation of pipe support material and to prevent silting of the drainage pipe and minimising differential settlement; Typar SF10						
horizontal	0.04	1.63	–	0.35	m²	**1.98**
inclined at an angle of 10-45° to the horizontal	0.05	2.03	–	0.35	m²	**2.38**
For wrapping to prevent clogging of drainage pipes surrounded by fine soil; Typar SF10						
sheeting	0.08	3.19	–	0.69	m²	**3.88**
For wrapping to prevent clogging of drainage pipes surrounded by fine soil; Terram 1000						
sheeting	0.08	3.19	–	0.43	m²	**3.62**
For vertical structure drainage to sub-surface walls, roofs and foundations; Filtram 1B1						
sheeting	0.08	3.19	–	3.52	m²	**6.71**

CLASS E: EXCAVATION

Item	Gang hours	Labour £	Plant £	Material £	Unit	Total rate £
GEOTEXTILES – cont						
Geotextiles – cont						
For waterproofing to tunnels, buried structures, etc. where the membrane is buried, forming part of the drainage system; Filtram 1BZ						
sheeting	0.08	3.19	–	6.46	m²	**9.65**
Geotextiles; roofing insulation and protection						
Protection of waterproofing membrane from physical damage and puncturing; Typar SF56						
sheeting	0.05	2.36	1.96	0.54	m²	**4.86**
Internal reinforcement of in situ spread waterproof bitumen emulsion; Typar SF10						
sheeting	0.07	2.63	–	0.35	m²	**2.98**
LANDSCAPING						
Preparatory operations prior to landscaping						
Supply and apply granular cultivation treatments by hand						
35 grammes/m²	0.50	12.77	–	5.13	100 m²	**17.90**
50 grammes/m²	0.65	16.60	–	6.66	100 m²	**23.26**
75 grammes/m²	0.85	21.71	–	9.33	100 m²	**31.04**
100 grammes/m²	1.00	25.54	–	10.76	100 m²	**36.30**
Supply and apply granular cultivation treatments by machine in suitable economically large areas						
100 grammes/m²	0.25	3.64	0.99	10.76	100 m²	**15.39**
Supply and incorperate cultivation additives into top 150 mm of topsoil by hand						
1 m³ / 10 m²	20.00	510.80	–	48.38	100 m²	**559.18**
1 m³ / 13 m²	20.00	510.80	–	37.20	100 m²	**548.00**
1 m³ / 20 m²	19.00	485.26	–	24.19	100 m²	**509.45**
1 m³ / 40 m²	17.00	434.18	–	12.09	100 m²	**446.27**
Supply and incorporate cultivation additives into top 150 mm of topsoil by machine in suitable economically large areas						
1 m³ / 10 m²	–	–	147.17	48.38	100 m²	**195.55**
1 m³ / 13 m²	–	–	135.84	37.20	100 m²	**173.04**
1 m³ / 20 m²	–	–	120.76	24.19	100 m²	**144.95**
1 m³ / 40 m²	–	–	111.32	12.09	100 m²	**123.41**
Turfing						
Turfing						
horizontal	0.12	3.06	–	6.21	m²	**9.27**
10–45 degress to horizontal	0.17	4.34	–	6.21	m²	**10.55**
45–90° to horizontal; pegging down	0.19	4.85	–	6.21	m²	**11.06**
Hydraulic mulch grass seeding						
Grass seeding						
horizontal	0.01	0.13	–	1.64	m²	**1.77**
10–45° to horizontal	0.01	0.18	–	1.64	m²	**1.82**
45–90° to horizontal	0.01	0.23	–	1.64	m²	**1.87**

CLASS E: EXCAVATION

Item	Gang hours	Labour £	Plant £	Material £	Unit	Total rate £
Selected grass seeding						
Grass seeding; sowing at the rate of 0.050 kg/m² in two operations						
horizontal	0.01	0.26	–	0.22	m²	0.48
10–45° to horizontal	0.02	0.38	–	0.22	m²	0.60
45–90° to horizontal	0.02	0.51	–	0.22	m²	0.73
Plants						
Form planting hole in previously cultivated area, supply and plant specified herbaceous plants and backfill with excavated material						
5 plants/m²	0.01	0.26	–	5.08	m²	5.34
10 plants/m²	0.02	0.56	–	18.60	m²	19.16
25 plants/m²	0.05	1.28	–	43.32	m²	44.60
35 plants/m²	0.07	1.79	–	60.64	m²	62.43
50 plants/m²	0.10	2.55	–	86.64	m²	89.19
Supply and fix plant support netting on 50 mm diameter stakes 750 mm long driven into the ground at 1.5 m centres						
1.15m high green extruded plastic mesh, 125mm square mesh	0.06	1.53	–	4.82	m²	6.35
Form planting hole in previously cultivated area; supply and plant bulbs and backfill with excavated material						
small	0.01	0.26	–	0.17	each	0.43
medium	0.01	0.26	–	0.27	each	0.53
large	0.01	0.26	–	0.32	each	0.58
Supply and plant bulbs in grassed area using bulb planter and backfill with screened topsoil or peat and cut turf plug						
small	0.01	0.26	–	0.17	each	0.43
medium	0.01	0.26	–	0.27	each	0.53
large	0.01	0.26	–	0.32	each	0.58
Shrubs						
Form planting hole in previously cultivated area, supply and plant specified shrub and backfill with excavated material						
shrub 300 mm high	0.01	0.26	–	2.86	each	3.12
shrub 600 mm high	0.01	0.26	–	4.24	each	4.50
shrub 900 mm high	0.01	0.26	–	4.98	each	5.24
shrub 1 m high and over	0.01	0.26	–	6.42	each	6.68
Supply and fix shrub stake including two ties						
one stake; 1.5 m long, 75 mm diameter	0.12	3.06	–	5.95	each	9.01

CLASS E: EXCAVATION

Item	Gang hours	Labour £	Plant £	Material £	Unit	Total rate £
LANDSCAPING – cont						
Trees						
The cost of planting semi-mature trees will depend on the size and species, and on the access to the site for tree handling machines. Prices should be obtained for individual trees and planting.						
Break up subsoil to a depth of 200 mm						
in tree pit	0.05	1.28	–	–	each	**1.28**
Supply and plant tree in prepared pit; backfill with excavated topsoil minimum 600 mm deep						
light standard tree	0.25	6.38	–	79.71	each	**86.09**
standard tree	0.45	11.49	–	107.63	each	**119.12**
selected standard tree	0.75	19.16	–	121.53	each	**140.69**
heavy standard tree	0.85	21.71	–	166.04	each	**187.75**
extra heavy standard tree	1.50	38.31	–	206.86	each	**245.17**
Extra for filling with topsoil from spoil heap ne						
100 m distant	0.15	3.83	–	–	m³	**3.83**
Extra for filling with imported topsoil	0.08	2.04	–	26.91	m³	**28.95**
Supply tree stake and drive 500 mm into firm ground and trim to approved height, including two tree ties to approved pattern						
one stake; 2.4 m long, 100 mm diameter	0.16	4.09	–	6.88	each	**10.97**
one stake; 3.0 m long, 100 mm diameter	0.20	5.11	–	6.98	each	**12.09**
two stakes; 2.4 m long, 100 mm diameter	0.24	6.13	–	11.09	each	**17.22**
two stakes; 3.0 m long, 100 mm diameter	0.30	7.66	–	13.97	each	**21.63**
Supply and fit tree support comprising three collars and wire guys; including pickets						
galvanized steel 50 × 600 mm	1.50	38.31	–	23.23	each	**61.54**
hardwood 75 × 600 mm	1.50	38.31	–	23.76	each	**62.07**
Supply and fix standard steel tree guard	0.30	7.66	–	24.08	each	**31.74**
Hedges						
Excavate trench by hand for hedge and deposit soil alongside trench						
300 wide x 300 mm deep	0.10	2.55	–	–	m	**2.55**
450 wide x 300 mm deep	0.13	3.32	–	–	m	**3.32**
Excavate trench by machine for hedge and deposit soil alongside trench						
300 wide x 300 mm deep	0.02	0.51	–	–	m	**0.51**
450 wide x 300 mm deep	0.02	0.51	–	–	m	**0.51**
Set out, nick out and excavate trench and break up subsoil to minimum depth of 300 mm						
400 mm minimum deep	0.28	7.13	–	–	m	**7.13**
Supply and plant hedging plants, backfill with excavated topsoil						
single row; plants at 200 mm centres	0.25	6.38	–	10.74	m	**17.12**
single row; plants at 300 mm centres	0.17	4.34	–	7.15	m	**11.49**
single row; plants at 400 mm centres	0.13	3.19	–	5.37	m	**8.56**
single row; plants at 500 mm centres	0.10	2.55	–	4.29	m	**6.84**
single row; plants at 600 mm centres	0.08	2.04	–	3.56	m	**5.60**
double row; plants at 200 mm centres	0.50	12.77	–	21.47	m	**34.24**
double row; plants at 300 mm centres	0.34	8.68	–	14.30	m	**22.98**

CLASS E: EXCAVATION

Item	Gang hours	Labour £	Plant £	Material £	Unit	Total rate £
double row; plants at 400 mm centres	0.25	6.38	–	10.74	m	**17.12**
double row; plants at 500 mm centres	0.20	5.11	–	8.59	m	**13.70**
Extra for incorporating manure at the rate of 1m³						
per 30 m of trench	0.12	3.06	–	0.14	m	**3.20**

COMPARATIVE COSTS - EARTH MOVING

Notes
The cost of earth moving and other associated works is dependent on matching the overall quantities and production rate called for by the programme of works with the most appropriate plant and assessing the most suitable version of that plant which will:
 * deal with the site conditions (e.g. type of ground, type of excavation, length of haul, prevailing weather, etc.)
 * comply with the specification requirements (e.g. compaction, separation of materials, surface tolerances, etc.)
 * complete the work economically (e.g. provide surface tolerances which will avoid undue expense of imported materials)
Labour costs are based on a plant operative skill rate 3 unless otherwise stated

Comparative costs of excavation equipment
The following are comparative costs using various types of excavation equipment and include loading into transport. All costs assume 50 minutes productive work in every hour and adequate disposal transport being available to obviate any delay.

Item	Gang hours	Labour £	Plant £	Material £	Unit	Total rate £
Dragline (for excavations requiring a long reach and long discharge, mainly in clearing streams and rivers)						
bucket capacity ne 1.15 m³	0.06	1.08	3.76	–	m³	**4.84**
bucket capacity ne 2.00 m³	0.03	0.54	3.01	–	m³	**3.55**
bucket capacity ne 3.00 m³	0.02	0.36	2.78	–	m³	**3.14**
Hydraulic backacter (for all types of excavation and loading, including trenches, breaking hard ground, etc.)						
bucket capacity ne 0.40 m³	0.05	0.81	1.15	–	m³	**1.96**
bucket capacity ne 1.00 m³	0.02	0.36	0.80	–	m³	**1.16**
bucket capacity ne 1.60 m³	0.01	0.18	1.09	–	m³	**1.27**
Hydraulic face shovel (predominantly for excavating cuttings and embankments over 2 m high requiring high output)						
bucket capacity ne 1.50 m³	0.04	0.72	3.68	–	m³	**4.40**
bucket capacity ne 2.60 m³	0.02	0.36	2.44	–	m³	**2.80**
bucket capacity ne 3.40 m³	0.01	0.18	1.65	–	m³	**1.83**

Unit Costs – Civil Engineering Works

CLASS E: EXCAVATION

Item	Gang hours	Labour £	Plant £	Material £	Unit	Total rate £
COMPARATIVE COSTS - EARTH MOVING – cont						
Comparative costs of excavation equipment – cont						
Tractor loader (for loading, carrying, placing materials, spreading and levelling, some site clearance operations and reducing levels)						
bucket capacity ne 0.80 m³	0.02	0.36	0.48	–	m³	**0.84**
bucket capacity ne 1.50 m³	0.01	0.18	0.39	–	m³	**0.57**
bucket capacity ne 3.00 m³	0.01	0.13	0.76	–	m³	**0.89**
Multipurpose wheeled loader / backhoe (versatile machine for small to medium excavations, trenches, loading, carrying and back filling)						
bucket capacity ne 0.76 m³	0.08	1.44	1.00	–	m³	**2.44**
bucket capacity ne 1.00 m³	0.06	1.08	1.65	–	m³	**2.73**
Comparative costs of transportation equipment						
The following are comparative costs for using various types of transportation equipment to transport excavated loose material. The capacity of the transport must be suitable for the output of the loading machine. The cost will vary depending on the number of transport units required to meet the output of the loading unit and the distance to be travelled.						
Loading loose material into transport by wheeled loader						
ne 2.1 m³ capacity	0.02	0.36	0.97	–	m³	**1.33**
ne 5.4 m³ capacity	0.01	0.18	1.98	–	m³	**2.16**
ne 10.5 m³ capacity	0.01	0.13	1.03	–	m³	**1.16**
Transport material within site by dump truck (rear dump)						
ne 24 m³ heaped capacity, distance travelled ne 0.5 Km	0.03	0.54	3.11	–	m³	**3.65**
Add per 0.5 Km additional distance	0.01	0.18	1.04	–	m³	**1.22**
ne 57 m³ heaped capacity, distance travelled ne 0.5 Km	0.03	0.59	2.53	–	m³	**3.12**
Add per 0.5 Km additional distance	0.03	0.59	1.26	–	m³	**1.85**
Transport material within site by dump truck (articulated)						
ne 32 t payload, distance travelled ne 0.5 Km	0.04	0.78	3.52	–	m³	**4.30**
Add per 0.5 Km additional distance	0.02	0.39	1.76	–	m³	**2.15**
Transport material within or off site by tipping lorry						
ne 10 t payload, distance travelled ne 1 Km	0.05	0.81	1.36	–	m³	**2.17**
Add per 1 Km additional distance	0.03	0.40	0.68	–	m³	**1.08**
10–15 t payload, distance travelled ne 1 Km	0.04	0.65	1.77	–	m³	**2.42**
Add per 1 Km additional distance	0.02	0.32	0.88	–	m³	**1.20**
15–25 t payload, distance travelled ne 1 Km	0.03	0.54	1.47	–	m³	**2.01**
Add per 1 Km additional distance	0.02	0.27	0.73	–	m³	**1.00**

CLASS E: EXCAVATION

Item	Gang hours	Labour £	Plant £	Material £	Unit	Total rate £
Comparative costs of earth moving equipment The following are comparative costs using various types of earth moving equipment to include excavation, transport, spreading and levelling. Bulldozer up to 74 KW (CAT D4H sized machine used for site strip, reducing levels or grading and spreading materials over smaller sites)						
average push one way 10 m	0.03	0.49	0.64	–	m^3	1.13
average push one way 30 m	0.08	1.30	1.71	–	m^3	3.01
average push one way 50 m	0.14	2.27	3.00	–	m^3	5.27
average push one way 100 m	0.29	4.70	6.21	–	m^3	10.91
Bulldozer up to 104 KW (CAT D6E sized machine for reducing levels, excavating to greater depths at steeper inclines, grading surfaces, small cut and fill operations; maximum push 100 m)						
average push one way 10 m	0.03	0.54	1.38	–	m^3	1.92
average push one way 30 m	0.07	1.26	3.23	–	m^3	4.49
average push one way 50 m	0.09	1.62	4.15	–	m^3	5.77
average push one way 100 m	0.19	3.43	8.77	–	m^3	12.20
Bull or Angle Dozer up to 212 KW (CAT D8N sized machine for high output, ripping and excavating by reducing levels at steeper inclines or in harder material than with D6E, larger cut and fill operations used in conjunction with towed or S.P. scrapers. Spreading and grading over large areas; maximum push 100 m)						
average push one way 10 m	0.03	0.54	3.10	–	m^3	3.64
average push one way 30 m	0.06	1.08	6.20	–	m^3	7.28
average push one way 50 m	0.07	1.26	7.24	–	m^3	8.50
average push one way 100 m	0.11	1.99	11.37	–	m^3	13.36
Motorized scraper, 15 m^3 capacity (for excavating larger volumes over large haul lengths, excavating to reduce levels and also levelling ground, grading large sites, moving and tipping material including hard material - used in open cast sites)						
average haul one way 500 m	–	0.07	0.50	–	m^3	0.57
average haul one way 1,000 m	0.01	0.11	0.75	–	m^3	0.86
average haul one way 2,000 m	0.01	0.14	1.00	–	m^3	1.14
average haul one way 3,000 m	0.01	0.18	1.25	–	m^3	1.43
Twin engined motorized scraper, 16 m^3 capacity						
average haul one way 500 m	–	0.05	0.50	–	m^3	0.55
average haul one way 1,000 m	–	0.07	0.66	–	m^3	0.73
average haul one way 2,000 m	0.01	0.09	0.83	–	m^3	0.92
average haul one way 3,000 m	0.01	0.13	1.16	–	m^3	1.29
Twin engined motorized scraper, 34 m^3 capacity						
average haul one way 500 m	–	0.04	0.71	–	m^3	0.75
average haul one way 1,000 m	–	0.05	1.07	–	m^3	1.12
average haul one way 2,000 m	–	0.05	1.07	–	m^3	1.12
average haul one way 3,000 m	–	0.07	1.43	–	m^3	1.50

CLASS E: EXCAVATION

Item	Gang hours	Labour £	Plant £	Material £	Unit	Total rate £
COMPARATIVE COSTS - EARTH MOVING – cont						
Excavation by hand						
Desirable for work around live services or in areas of highly restricted access.						
Excavate and load into skip or dumper bucket						
loose material	2.02	25.13	–	–	m³	**25.13**
compacted soil or clay	3.15	39.19	–	–	m³	**39.19**
mass concrete or sandstone	3.00	37.32	–	–	m³	**37.32**
broken rock	2.95	36.70	–	–	m³	**36.70**
existing sub-base or pipe surrounds	3.12	38.81	–	–	m³	**38.81**
Excavate by hand using pneumatic equipment						
Excavate below ground using 1.80 m³/min single tool compressor and pneumatic breaker and load material into skip or dumper bucket						
rock (medium drill)	2.12	26.37	16.34	–	m³	**42.71**
brickwork or mass concrete	2.76	34.33	22.25	–	m³	**56.58**
reinforced concrete	3.87	48.14	29.36	–	m³	**77.50**
asphalt in carriageways	1.15	14.31	9.44	–	m³	**23.75**
Comparative prices for ancillary equipment						
Excavate using 6 tonne to break out (JCB 3CX and Montalbert 125 breaker)						
medium hard rock	0.43	6.96	9.81	–	m³	**16.77**
brickwork or mass concrete	0.54	8.74	11.95	–	m³	**20.69**
reinforced concrete	0.64	10.36	14.17	–	m³	**24.53**
Load material into skip or dumper						
Load material into skip or dumper bucket using a 11.5 tonne crawler backacter with 0.80 m³ rock bucket						
medium hard rock	0.07	1.13	1.82	–	m³	**2.95**
brickwork or mass concrete	0.08	1.30	2.07	–	m³	**3.37**
reinforced concrete	0.09	1.46	2.34	–	m³	**3.80**
DRILLING AND BLASTING IN ROCK						
The cost of blasting is controlled by the number of holes and the length of drilling required to achieve the tolerances and degree of shatter required, e.g. line drilling to trenches, depth of drilling to control horizontal overbreak.						
Drilling with rotary percussion drills						
105–110 mm diameter; hard rock	–	–	–	–	m	**14.21**
105–110 mm diameter; sandstone	–	–	–	–	m	**7.18**
125 mm diameter; hard rock	–	–	–	–	m	**11.28**
125 mm diameter; sandstone	–	–	–	–	m	**10.14**
165 mm diameter; hard rock	–	–	–	–	m	**12.93**
165 mm diameter; sandstone	–	–	–	–	m	**12.93**

CLASS E: EXCAVATION

Item	Gang hours	Labour £	Plant £	Material £	Unit	Total rate £
Drilling and blasting in open cut for bulk excavation excluding cost of excavatio or trimming						
hard rock	–	–	–	–	m	3.27
sandstone	–	–	–	–	m	4.42
Drilling and blasting for quarry operations with face height exceeding 10 m						
hard rock	–	–	–	–	m	3.13
sandstone	–	–	–	–	m	3.13
Drilling and blasting in trenches excluding cost of excavation or trimming						
trench width ne 1.0 m	–	–	–	–	m	24.53
trench width 1.0–1.5 m	–	–	–	–	m	21.27
trench width over 1.5 m	–	–	–	–	m	18.01
Secondary blasting to boulders						
pop shooting	–	–	–	–	m	7.68
plaster shooting	–	–	–	–	m	4.10
DEWATERING						
The following unit costs are for dewatering pervious ground only and are for sets of equipment comprising :						
1 nr diesel driven pump (WP 150/60 or similar) complete with allowance of £50 for fuel	–	–	–	–	day	91.44
50 m of 150 mm diameter header pipe	–	–	–	–	day	16.60
35 nr of disposable well points	–	–	–	–	buy	278.93
18 m of delivery pipe	–	–	–	–	day	7.33
1 nr diesel driven standby pump	–	–	–	–	day	57.71
1 nr jetting pump with hoses (for installation of wellpoints only)	–	–	–	–	-	96.01
attendant labour and plant (2 hrs per day) inclusive of small dumper and bowser)	–	–	–	–	-	27.47
Costs are based on 10 hr shifts with attendant labour and plant (specialist advice)						
Guide price for single set of equipment comprising pump, 150 mm diameter header pipe, 35 nr well points, delivery pipes and attendant labour and plant						
Bring to site equipment and remove upon completion	–	–	–	–	sum	2562.50
Installation costs						
hire of jetting pump with hoses; 1 day	–	–	–	–	sum	96.01
purchase of well points; 35 Nr	–	–	–	–	sum	278.93
labour and plant; 10 hours	–	–	–	–	sum	274.68
Operating costs						
hire of pump, header pipe, delivery pipe and standby pump complete with fuel etc. and 2 hours attendant labour and plant	–	–	–	–	day	228.01

CLASS F: IN SITU CONCRETE

Item	Gang hours	Labour £	Plant £	Material £	Unit	Total rate £
NOTES						
The unit rates in this section are based on nett measurements and appropriate adjustments should be made for unmeasured excess (e.g. additional blinding thickness as a result of ground conditions). The unit rates for the provision of concrete are based on ready mixed concrete in full loads delivered to site within 5 miles (8km) of the concrete mixing plant and include an allowance for wastage prior to placing.						
This section assumes optimum outputs of an efficiently controlled pour with no delays caused by out of sequence working and no abnormal conditions that would affect continuity of work.						
RESOURCES - LABOUR						
Concrete gang						
1 ganger or chargehand (skill rate 4)		14.23				
2 skilled operatives (skill rate 4)		26.64				
4 unskilled operatives (general)		49.76				
1 plant operator (skill rate 3) - 25% of time		4.05				
Total Gang Rate / Hour	£	**94.68**				
RESOURCES - MATERIALS						
Misc						
The following costs do not reflect in the rates and should be considered separately						
Delivery to site for each additional mile from the concrete plant further than 5 miles (8km)	–	–	–	1.57	m³	**1.57**
Mix design, per trial mix	–	–	–	234.00	mix	**234.00**
Part loads, per m³ below full load	–	–	–	31.80	m³	**31.80**
Waiting time (in excess of 6 mins/m³ 'norm' discharge time)	–	–	–	69.00	hr	**69.00**
Making and testing concrete cube	0.69	8.56	–	–	nr	**8.56**
Pumping from ready mix truck to point of placing at the rate of 25 m³ / hour	0.11	1.31	4.09	–	m³	**5.40**
Pumping from ready mix truck to point of placing at the rate of 45 m³ / hour	0.05	0.63	4.31	–	m³	**4.94**
RESOURCES - PLANT						
Misc						
Concrete						
10t Crane (50% of time)			13.11			
1.00 m³ concrete skip (50% of time)			0.95			
11.30 m³/min compressor, 2 tool			40.14			
four 54 mm poker vibrators			5.13			
Total Rate / Hour		£	**59.33**			

CLASS F: IN SITU CONCRETE

Item	Gang hours	Labour £	Plant £	Material £	Unit	Total rate £
PROVISION OF CONCRETE						
Standard mix; cement to BS EN 197-1						
Type ST1	–	–	–	115.76	m^3	**115.76**
Type ST2	–	–	–	128.90	m^3	**128.90**
Type ST3	–	–	–	129.53	m^3	**129.53**
Type ST4	–	–	–	134.59	m^3	**134.59**
Type ST5	–	–	–	136.85	m^3	**136.85**
Standard mix; sulphate resisting cement to BS 4027						
Type ST1	–	–	–	127.49	m^3	**127.49**
Type ST2	–	–	–	140.62	m^3	**140.62**
Type ST3	–	–	–	141.25	m^3	**141.25**
Type ST4	–	–	–	146.31	m^3	**146.31**
Type ST5	–	–	–	148.57	m^3	**148.57**
Designed mix; cement to BS EN 197-1						
Grade C7.5						
10 mm aggregate	–	–	–	135.90	m^3	**135.90**
20 mm aggregate	–	–	–	144.42	m^3	**144.42**
40 mm aggregate	–	–	–	160.17	m^3	**160.17**
Grade C10						
10 mm aggregate	–	–	–	138.15	m^3	**138.15**
20 mm aggregate	–	–	–	145.80	m^3	**145.80**
40 mm aggregate	–	–	–	161.55	m^3	**161.55**
Grade C15						
10 mm aggregate	–	–	–	140.82	m^3	**140.82**
20 mm aggregate	–	–	–	146.53	m^3	**146.53**
40 mm aggregate	–	–	–	162.28	m^3	**162.28**
Grade C20						
10 mm aggregate	–	–	–	142.56	m^3	**142.56**
20 mm aggregate	–	–	–	147.26	m^3	**147.26**
40 mm aggregate	–	–	–	163.01	m^3	**163.01**
Grade C25						
10 mm aggregate	–	–	–	144.14	m^3	**144.14**
20 mm aggregate	–	–	–	148.65	m^3	**148.65**
40 mm aggregate	–	–	–	164.40	m^3	**164.40**
Grade C30						
10 mm aggregate	–	–	–	145.60	m^3	**145.60**
20 mm aggregate	–	–	–	149.71	m^3	**149.71**
40 mm aggregate	–	–	–	165.46	m^3	**165.46**
Grade C40						
10 mm aggregate	–	–	–	148.12	m^3	**148.12**
20 mm aggregate	–	–	–	151.45	m^3	**151.45**
Grade C50						
10 mm aggregate	–	–	–	149.71	m^3	**149.71**
20 mm aggregate	–	–	–	152.87	m^3	**152.87**
Grade C60						
10 mm aggregate	–	–	–	152.23	m^3	**152.23**
20 mm aggregate	–	–	–	154.75	m^3	**154.75**

CLASS F: IN SITU CONCRETE

Item	Gang hours	Labour £	Plant £	Material £	Unit	Total rate £
PROVISION OF CONCRETE – cont						
Designed mix; sulphate resisting cement to BS 4027						
Grade C7.5						
10 mm aggregate	–	–	–	147.35	m^3	**147.35**
20 mm aggregate	–	–	–	155.87	m^3	**155.87**
40 mm aggregate	–	–	–	171.62	m^3	**171.62**
Grade C10						
10 mm aggregate	–	–	–	149.79	m^3	**149.79**
20 mm aggregate	–	–	–	157.44	m^3	**157.44**
40 mm aggregate	–	–	–	173.19	m^3	**173.19**
Grade C15						
10 mm aggregate	–	–	–	152.64	m^3	**152.64**
20 mm aggregate	–	–	–	158.36	m^3	**158.36**
40 mm aggregate	–	–	–	174.11	m^3	**174.11**
Grade C20						
10 mm aggregate	–	–	–	154.50	m^3	**154.50**
20 mm aggregate	–	–	–	159.21	m^3	**159.21**
40 mm aggregate	–	–	–	174.96	m^3	**174.96**
Grade C25						
10 mm aggregate	–	–	–	156.48	m^3	**156.48**
20 mm aggregate	–	–	–	160.99	m^3	**160.99**
40 mm aggregate	–	–	–	176.74	m^3	**176.74**
Grade C30						
10 mm aggregate	–	–	–	158.14	m^3	**158.14**
20 mm aggregate	–	–	–	162.24	m^3	**162.24**
40 mm aggregate	–	–	–	177.99	m^3	**177.99**
Prescribed mix; cement to BS EN 197-1						
Grade C7.5						
10 mm aggregate	–	–	–	135.90	m^3	**135.90**
20 mm aggregate	–	–	–	144.42	m^3	**144.42**
40 mm aggregate	–	–	–	160.17	m^3	**160.17**
Grade C10						
10 mm aggregate	–	–	–	138.15	m^3	**138.15**
20 mm aggregate	–	–	–	145.80	m^3	**145.80**
40 mm aggregate	–	–	–	161.55	m^3	**161.55**
Grade C15						
10 mm aggregate	–	–	–	140.82	m^3	**140.82**
20 mm aggregate	–	–	–	146.53	m^3	**146.53**
40 mm aggregate	–	–	–	162.28	m^3	**162.28**
Grade C20						
10 mm aggregate	–	–	–	142.56	m^3	**142.56**
20 mm aggregate	–	–	–	147.26	m^3	**147.26**
40 mm aggregate	–	–	–	163.01	m^3	**163.01**
Grade C25						
10 mm aggregate	–	–	–	144.14	m^3	**144.14**
20 mm aggregate	–	–	–	148.65	m^3	**148.65**
40 mm aggregate	–	–	–	164.40	m^3	**164.40**

CLASS F: IN SITU CONCRETE

Item	Gang hours	Labour £	Plant £	Material £	Unit	Total rate £
Grade C30						
10 mm aggregate	–	–	–	145.60	m³	**145.60**
20 mm aggregate	–	–	–	149.71	m³	**149.71**
40 mm aggregate	–	–	–	165.46	m³	**165.46**
Grade C40						
10 mm aggregate	–	–	–	148.12	m³	**148.12**
20 mm aggregate	–	–	–	151.45	m³	**151.45**
Grade C50						
10 mm aggregate	–	–	–	149.71	m³	**149.71**
20 mm aggregate	–	–	–	152.87	m³	**152.87**
Grade C60						
10 mm aggregate	–	–	–	152.23	m³	**152.23**
20 mm aggregate	–	–	–	154.75	m³	**154.75**
Prescribed mix; sulphate resisting cement to BS 4027						
Grade C7.5						
10 mm aggregate	–	–	–	147.35	m³	**147.35**
20 mm aggregate	–	–	–	155.87	m³	**155.87**
40 mm aggregate	–	–	–	171.62	m³	**171.62**
Grade C10						
10 mm aggregate	–	–	–	149.79	m³	**149.79**
20 mm aggregate	–	–	–	157.44	m³	**157.44**
40 mm aggregate	–	–	–	173.19	m³	**173.19**
Grade C15						
10 mm aggregate	–	–	–	152.64	m³	**152.64**
20 mm aggregate	–	–	–	158.36	m³	**158.36**
40 mm aggregate	–	–	–	174.11	m³	**174.11**
Grade C20						
10 mm aggregate	–	–	–	154.50	m³	**154.50**
20 mm aggregate	–	–	–	159.21	m³	**159.21**
40 mm aggregate	–	–	–	174.96	m³	**174.96**
Grade C25						
10 mm aggregate	–	–	–	156.48	m³	**156.48**
20 mm aggregate	–	–	–	160.99	m³	**160.99**
40 mm aggregate	–	–	–	176.74	m³	**176.74**
Grade C30						
10 mm aggregate	–	–	–	158.14	m³	**158.14**
20 mm aggregate	–	–	–	162.24	m³	**162.24**
40 mm aggregate	–	–	–	177.99	m³	**177.99**

CLASS F: IN SITU CONCRETE

Item	Gang hours	Labour £	Plant £	Material £	Unit	Total rate £
PLACING OF CONCRETE; MASS						
Blinding; thickness						
ne 150 mm	0.18	17.38	0.65	–	m³	18.03
150–300 mm	0.16	15.45	1.84	–	m³	17.29
300–500 mm	0.14	13.52	3.02	–	m³	16.54
exceeding 500 mm	0.12	11.59	4.21	–	m³	15.80
Bases, footings, pile caps and ground slabs; thickness						
ne 150 mm	0.20	19.31	11.87	–	m³	31.18
150–300 mm	0.17	16.42	10.14	–	m³	26.56
300–500 mm	0.15	14.49	8.96	–	m³	23.45
exceeding 500 mm	0.14	13.52	8.31	–	m³	21.83
ADD to the above for placing against an excavated surface	0.03	2.41	1.45	–	m³	3.86
Walls; thickness						
ne 150 mm	0.21	20.28	12.51	–	m³	32.79
150–300 mm	0.15	14.49	8.96	–	m³	23.45
300–500 mm	0.13	12.55	7.77	–	m³	20.32
exceeding 500 mm	0.12	11.59	7.12	–	m³	18.71
ADD to the above for placing against an excavated surface	0.03	2.90	1.84	–	m³	4.74
Other concrete forms						
plinth 1000 × 1000 x 600 mm	0.33	31.87	19.63	–	m³	51.50
plinth 1500 × 1500 x 750 mm	0.25	24.14	14.89	–	m³	39.03
plinth 2000 × 2000 x 600 mm	0.20	19.31	11.87	–	m³	31.18
surround to precast concrete manhole chambers 200 mm thick	0.29	28.01	17.26	–	m³	45.27
PLACING OF CONCRETE; REINFORCED						
Bases, footings, pile caps and ground slabs; thickness						
ne 150 mm	0.21	20.28	12.51	–	m³	32.79
150–300 mm	0.18	17.38	10.68	–	m³	28.06
300–500 mm	0.16	15.45	9.49	–	m³	24.94
exceeding 500 mm	0.15	14.49	8.96	–	m³	23.45
Suspended slabs; thickness						
ne 150 mm	0.27	26.07	16.08	–	m³	42.15
150–300 mm	0.21	20.28	12.51	–	m³	32.79
300–500 mm	0.19	18.35	11.33	–	m³	29.68
exceeding 500 mm	0.19	18.35	11.33	–	m³	29.68
Walls; thickness						
ne 150 mm	0.29	28.01	17.26	–	m³	45.27
150–300 mm	0.22	21.25	13.06	–	m³	34.31
300–500 mm	0.20	19.31	11.87	–	m³	31.18
exceeding 500 mm	0.20	19.31	11.87	–	m³	31.18
Columns and piers; cross-sectional area						
ne 0.03 m²	0.50	48.29	29.67	–	m³	77.96
0.03–0.10 m²	0.40	38.63	23.74	–	m³	62.37
0.10–0.25 m²	0.35	33.80	20.82	–	m³	54.62
0.25–1.00 m²	0.35	33.80	20.82	–	m³	54.62
exceeding 1 m²	0.28	27.04	16.61	–	m³	43.65

CLASS F: IN SITU CONCRETE

Item	Gang hours	Labour £	Plant £	Material £	Unit	Total rate £
Beams; cross-sectional area						
ne 0.03 m²	0.50	48.29	29.67	–	m³	**77.96**
0.03–0.10 m²	0.40	38.63	23.74	–	m³	**62.37**
0.10–0.25 m²	0.35	33.80	20.82	–	m³	**54.62**
0.25–1.00 m²	0.35	33.80	20.82	–	m³	**54.62**
exceeding 1 m²	0.28	27.04	16.61	–	m³	**43.65**
Casings to metal sections; cross-sectional area						
ne 0.03 m²	0.47	45.39	27.94	–	m³	**73.33**
0.03–0.10 m²	0.47	45.39	27.94	–	m³	**73.33**
0.10–0.25 m²	0.40	38.63	23.73	–	m³	**62.36**
0.25–1.00 m²	0.40	38.63	23.73	–	m³	**62.36**
exceeding 1 m²	0.35	33.80	20.82	–	m³	**54.62**
PLACING OF CONCRETE; PRESTRESSED						
Suspended slabs; thickness						
ne 150 mm	0.28	27.04	16.61	–	m³	**43.65**
150–300 mm	0.22	21.25	13.06	–	m³	**34.31**
300–500 mm	0.20	19.31	11.87	–	m³	**31.18**
exceeding 500 mm	0.19	18.35	11.33	–	m³	**29.68**
Beams; cross-sectional area						
ne 0.03 m²	0.50	48.29	29.67	–	m³	**77.96**
0.03–0.10 m²	0.40	38.63	23.74	–	m³	**62.37**
0.10–0.25 m²	0.35	33.80	20.82	–	m³	**54.62**
0.25–1.00 m²	0.35	33.80	20.82	–	m³	**54.62**
exceeding 1 m²	0.28	27.04	16.61	–	m³	**43.65**

CLASS G: CONCRETE ANCILLARIES

Item	Gang hours	Labour £	Plant £	Material £	Unit	Total rate £
FORMWORK; ROUGH FINISH						
Plane horizontal, width						
ne 0.1 m	0.10	6.26	1.84	2.80	m	10.90
0.1– 0.20 m	0.18	11.26	3.31	2.99	m	17.56
0.2–0.40 m	0.40	30.28	4.49	7.19	m²	41.96
0.4–1.22 m	0.40	30.28	4.49	7.19	m²	41.96
exceeding 1.22 m	0.38	28.76	4.27	7.19	m²	40.22
Plane sloping, width						
ne 0.1 m	0.11	6.88	2.02	2.80	m	11.70
0.1–0.20 m	0.20	12.51	3.67	2.99	m	19.17
0.2–0.40 m	0.43	32.55	4.78	10.81	m²	48.14
0.4–1.22 m	0.43	32.55	4.78	10.81	m²	48.14
exceeding 1.22 m	0.38	28.76	4.27	10.81	m²	43.84
Plane battered, width						
ne 0.1 m	0.12	7.51	2.20	5.37	m	15.08
0.1–0.20 m	0.20	12.51	3.67	5.56	m	21.74
0.2–0.40 m	0.43	32.55	4.90	14.28	m²	51.73
0.4–1.22 m	0.43	32.55	4.90	14.28	m²	51.73
exceeding 1.22 m	0.40	30.28	4.49	14.28	m²	49.05
Plane vertical, width						
ne 0.1 m	0.12	7.51	2.20	4.58	m	14.29
0.1– 0.20 m	0.19	11.89	3.49	4.78	m	20.16
0.2–0.40 m	0.70	52.98	7.86	11.69	m²	72.53
0.4–1.22 m	0.51	38.60	5.66	11.69	m²	55.95
exceeding 1.22 m	0.47	35.57	5.35	11.69	m²	52.61
Curved to one radius in one plane, 0.5 m radius, width						
ne 0.1 m	0.19	11.89	3.49	6.39	m	21.77
0.1–0.20 m	0.25	15.64	4.59	6.58	m	26.81
0.2–0.40 m	0.90	68.12	10.11	13.49	m²	91.72
0.4–1.22 m	0.72	54.50	8.08	13.49	m²	76.07
exceeding 1.22 m	0.65	49.20	7.23	13.49	m²	69.92
Curved to one radius in one plane, 2 m radius, width						
ne 0.1 m	0.18	11.26	3.31	6.39	m	20.96
0.1– 0.20 m	0.22	13.76	4.04	6.58	m	24.38
0.2–0.40 m	0.84	63.58	9.43	13.49	m²	86.50
0.4–1.22 m	0.66	49.96	7.41	13.49	m²	70.86
exceeding 1.22 m	0.52	39.36	5.84	13.49	m²	58.69
For voids						
small void; depth ne 0.5 m	0.07	4.38	–	3.31	nr	7.69
small void; depth 0.5–1.0 m	0.12	7.51	–	6.31	nr	13.82
small void; depth ne 1.0–2.0 m	0.16	10.01	–	11.61	nr	21.62
large void; depth ne 0.5 m	0.14	8.76	–	9.92	nr	18.68
large void; depth 0.5–1.0 m	0.33	20.64	–	21.14	nr	41.78
large void; depth 1.0–2.0 m	0.58	36.28	–	43.53	nr	79.81
For concrete components of constant cross-section						
beams; 200 × 200 mm	0.48	30.03	8.81	5.55	m	44.39
beams; 500 × 500 mm	0.55	34.41	10.10	10.57	m	55.08
beams; 500 × 800 mm	0.67	41.92	12.31	12.30	m	66.53
columns; 200 × 200 mm	0.55	34.41	10.10	7.70	m	52.21
columns; 300 × 300 mm	0.55	34.41	10.10	10.47	m	54.98
columns; 300 × 500 mm	0.62	38.79	11.39	11.91	m	62.09
to walls; 1.0 m high thickness 250 mm	1.10	83.26	12.35	13.56	m	109.17

CLASS G: CONCRETE ANCILLARIES

Item	Gang hours	Labour £	Plant £	Material £	Unit	Total rate £
to walls; 1.5 m high thickness 300 mm	1.50	113.53	16.84	19.04	m	**149.41**
box culvert; 2 × 2 m internally and wall thickness						
300 mm	4.50	340.61	50.53	143.95	m	**535.09**
projections (100 mm deep)	0.10	6.26	1.84	2.86	m	**10.96**
intrusions (100 mm deep)	0.10	6.26	1.84	2.86	m	**10.96**
Allowance for additional craneage and rub up where required						
ADD to items measures linear	0.04	2.50	0.73	0.03	m	**3.26**
ADD to items measures m²	0.12	7.51	2.20	0.10	m²	**9.81**
FORMWORK; FAIR FINISH						
Plane horizontal, width						
ne 0.1 m	0.10	6.26	1.84	4.32	m	**12.42**
0.1–0.20 m	0.18	11.26	3.31	4.73	m	**19.30**
0.2–0.40 m	0.40	30.28	4.49	14.53	m²	**49.30**
0.4–1.22 m	0.40	30.28	4.49	14.53	m²	**49.30**
exceeding 1.22 m	0.38	28.76	4.27	14.53	m²	**47.56**
Plane sloping, width						
ne 0.1 m	0.11	6.88	2.02	4.32	m	**13.22**
0.1–0.20 m	0.20	12.51	2.25	4.73	m	**19.49**
0.2–0.40 m	0.43	32.55	13.01	23.45	m²	**69.01**
0.4–1.22 m	0.43	32.55	13.01	23.45	m²	**69.01**
exceeding 1.22 m	0.38	28.76	4.27	23.45	m²	**56.48**
Plane battered, width						
ne 0.1 m	0.12	7.51	2.20	8.22	m	**17.93**
0.1–0.20 m	0.20	12.51	3.67	8.62	m	**24.80**
0.2–0.40 m	0.43	32.55	4.83	29.57	m²	**66.95**
0.4–1.22 m	0.43	32.55	4.83	29.57	m²	**66.95**
exceeding 1.22 m	0.40	30.28	4.49	29.57	m²	**64.34**
Plane vertical, width						
ne 0.1 m	0.12	7.51	2.20	7.44	m	**17.15**
0.1–0.20 m	0.20	12.51	3.67	7.84	m	**24.02**
0.2–0.40 m	0.72	54.50	8.08	24.33	m²	**86.91**
0.4–1.22 m	0.53	40.12	5.95	24.33	m²	**70.40**
exceeding 1.22 m	0.48	36.33	5.39	24.33	m²	**66.05**
Curved to one radius in one plane, 0.5 m radius, width						
ne 0.1 m	0.20	12.51	3.67	11.89	m	**28.07**
0.1–0.20 m	0.25	15.64	4.59	12.30	m	**32.53**
0.2–0.40 m	0.90	68.12	10.11	28.79	m²	**107.02**
0.4–1.22 m	0.72	54.50	8.08	28.79	m²	**91.37**
exceeding 1.22 m	0.65	49.20	7.30	28.79	m²	**85.29**
Curved to one radius in one plane, 2.0 m radius, width						
ne 0.1 m	0.18	11.26	3.31	11.89	m	**26.46**
0.1–0.20 m	0.22	13.76	4.04	12.30	m	**30.10**
0.2–0.40 m	0.84	63.58	9.43	28.79	m²	**101.80**
0.4–1.22 m	0.66	49.96	7.41	28.79	m²	**86.16**
exceeding 1.22 m	0.52	39.36	5.84	28.79	m²	**73.99**

CLASS G: CONCRETE ANCILLARIES

Item	Gang hours	Labour £	Plant £	Material £	Unit	Total rate £
FORMWORK; FAIR FINISH – cont						
For voids, using void former						
small void; depth ne 0.5 m	0.07	4.38	–	5.84	nr	**10.22**
small void; depth 0.5–1.0 m	0.12	7.51	–	11.17	nr	**18.68**
small void; depth ne 1.0–2.0 m	0.16	10.01	–	21.30	nr	**31.31**
large void; depth ne 0.5 m	0.14	8.76	–	12.32	nr	**21.08**
large void; depth 0.5–1.0 m	0.33	20.64	–	25.39	nr	**46.03**
large void; depth 1.0–2.0 m	0.58	36.28	–	45.45	nr	**81.73**
For concrete components of constant cross-section						
beams; 200 × 200 mm	0.49	30.65	9.00	12.07	m	**51.72**
beams; 500 × 500 mm	0.57	35.66	10.47	20.94	m	**67.07**
beams; 500 × 800 mm	0.69	43.17	12.67	26.57	m	**82.41**
columns; 200 × 200 mm	0.57	35.66	10.47	14.63	m	**60.76**
columns; 300 × 300 mm	0.57	35.66	10.47	20.65	m	**66.78**
columns; 300 × 500 mm	0.64	40.04	11.75	24.44	m	**76.23**
to walls; 1.0 m high thickness 250 mm	1.20	90.83	13.47	28.23	m	**132.53**
to walls; 1.5 m high thickness 300 mm	1.60	121.10	17.97	41.05	m	**180.12**
box culvert; 2 x 2 m internally and wall thickness 300 mm	4.60	348.17	51.65	298.15	m	**697.97**
projections (100 mm deep)	0.10	6.26	1.84	6.06	m	**14.16**
intrusions (100 mm deep)	0.10	6.26	1.84	6.06	m	**14.16**
Allowance for additional craneage and rub up where required						
ADD to items measures linear	0.14	8.76	2.57	0.04	m	**11.37**
ADD to items measures m²	0.12	7.51	2.20	0.13	m²	**9.84**
FORMWORK; EXTRA SMOOTH FINISH						
Plane horizontal, width						
ne 0.1 m	0.10	6.26	1.84	4.55	m	**12.65**
0.1–0.20 m	0.18	11.26	3.31	5.14	m	**19.71**
0.2–0.40 m	0.40	30.28	4.49	16.64	m²	**51.41**
0.4–1.22 m	0.40	30.28	4.49	16.64	m²	**51.41**
exceeding 1.22 m	0.38	28.76	4.26	16.64	m²	**49.66**
Plane sloping, width						
ne 0.1 m	0.11	6.88	2.02	4.55	m	**13.45**
0.1–0.20 m	0.20	12.51	3.67	5.14	m	**21.32**
0.2–0.40 m	0.43	32.55	4.85	25.56	m²	**62.96**
0.4–1.22 m	0.43	32.55	4.85	25.56	m²	**62.96**
exceeding 1.22 m	0.38	28.76	4.27	25.56	m²	**58.59**
Plane battered, width						
ne 0.1 m	0.12	7.51	2.20	8.44	m	**18.15**
0.1–0.20 m	0.20	12.51	3.67	9.04	m	**25.22**
0.2–0.40 m	0.43	32.55	4.85	31.68	m²	**69.08**
0.4–1.22 m	0.43	32.55	4.85	31.68	m²	**69.08**
exceeding 1.22 m	0.40	30.28	4.49	31.68	m²	**66.45**
Plane vertical, width						
ne 0.1 m	0.12	9.08	2.20	7.66	m	**18.94**
0.1–0.20 m	0.20	12.51	3.67	8.25	m	**24.43**
0.2–0.40 m	0.74	56.01	8.31	26.44	m²	**90.76**
0.4–1.22 m	0.53	40.12	5.97	26.44	m²	**72.53**
exceeding 1.22 m	0.48	36.33	5.39	26.44	m²	**68.16**

CLASS G: CONCRETE ANCILLARIES

Item	Gang hours	Labour £	Plant £	Material £	Unit	Total rate £
Curved to one radius in one plane, 0.5 m radius, width						
ne 0.1 m	0.20	12.51	3.67	12.12	m	28.30
0.1–0.20 m	0.25	15.64	4.59	8.25	m	28.48
0.2–0.40 m	0.90	68.12	10.11	30.90	m²	109.13
0.4–1.22 m	0.72	54.50	8.08	30.90	m²	93.48
exceeding 1.22 m	0.65	49.20	7.32	30.90	m²	87.42
Curved to one radius in one plane, 1.0 m radius, width						
ne 0.1 m	0.18	11.26	3.31	12.12	m	26.69
0.1–0.20 m	0.22	13.76	4.04	8.25	m	26.05
0.2–0.40 m	0.84	63.58	15.14	30.90	m²	109.62
0.4–1.22 m	0.66	49.96	7.41	30.90	m²	88.27
exceeding 1.22 m	0.52	39.36	5.84	30.90	m²	76.10
For voids, using void former						
small void; depth ne 0.5 m	0.07	4.38	–	9.92	nr	14.30
small void; depth 0.5–1.0 m	0.12	7.51	–	19.22	nr	26.73
small void; depth ne 1.0–2.0 m	0.16	10.01	–	36.42	nr	46.43
large void; depth ne 0.5 m	0.14	8.76	–	21.75	nr	30.51
large void; depth 0.5–1.0 m	0.33	20.64	–	42.58	nr	63.22
large void; depth 1.0–2.0 m	0.58	36.28	–	75.16	nr	111.44
For concrete components of constant cross-section						
beams; 200 × 200 mm	0.50	31.28	9.18	13.34	m	53.80
beams; 500 × 500 mm	0.57	35.66	6.39	24.10	m	66.15
beams; 500 × 800 mm	0.69	43.17	7.76	30.37	m	81.30
columns; 200 × 200 mm	0.57	35.66	6.39	16.32	m	58.37
columns; 300 × 300 mm	0.57	35.66	6.39	23.18	m	65.23
columns; 300 × 500 mm	0.64	40.04	7.19	27.82	m	75.05
to walls; 1.0 m high thickness 250 mm	1.20	90.83	13.47	32.45	m	136.75
to walls; 1.5 m high thickness 300 mm	1.60	121.10	17.97	47.38	m	186.45
box culvert; 2 × 2 m internally and wall thickness						
300 mm	4.60	348.17	51.65	320.55	m	720.37
projections (100 mm deep)	0.10	6.26	1.84	6.47	m	14.57
intrusions (100 mm deep)	0.10	6.26	1.84	6.47	m	14.57
Allowance for additional craneage and rub up where required						
ADD to items measures linear	0.14	8.76	2.57	0.04	m	11.37
ADD to items measures m²	0.12	7.51	2.20	0.17	m²	9.88
REINFORCEMENT						
Plain round mild steel bars to BS 4449						
Bars; supplied in straight lengths						
6 mm nominal size	8.00	909.76	115.34	432.68	tonne	1457.78
8 mm nominal size	6.74	766.47	97.18	498.62	tonne	1362.27
10 mm nominal size	6.74	766.47	97.18	496.56	tonne	1360.21
12 mm nominal size	6.74	766.47	97.18	494.50	tonne	1358.15
16 mm nominal size	6.15	699.38	88.68	479.04	tonne	1267.10
20 mm nominal size	4.44	504.92	64.01	561.46	tonne	1130.39
25 mm nominal size	4.44	504.92	64.01	569.70	tonne	1138.63
32 mm nominal size	4.44	504.92	64.01	566.61	tonne	1135.54
40 mm nominal size	4.44	504.92	64.01	561.46	tonne	1130.39

CLASS G: CONCRETE ANCILLARIES

Item	Gang hours	Labour £	Plant £	Material £	Unit	Total rate £
REINFORCEMENT – cont						
Plain round mild steel bars to BS 4449 – cont						
Bars; supplied in bent and cut lengths						
6 mm nominal size	8.00	909.76	115.34	643.88	tonne	**1668.98**
8 mm nominal size	6.74	766.47	97.18	638.72	tonne	**1502.37**
10 mm nominal size	6.74	766.47	97.18	571.76	tonne	**1435.41**
12 mm nominal size	6.74	766.47	97.18	540.86	tonne	**1404.51**
16 mm nominal size	6.15	699.38	88.68	412.08	tonne	**1200.14**
20 mm nominal size	4.44	504.92	64.01	379.11	tonne	**948.04**
25 mm nominal size	4.44	504.92	64.01	360.57	tonne	**929.50**
32 mm nominal size	4.44	504.92	64.01	355.42	tonne	**924.35**
40 mm nominal size	4.44	504.92	64.01	345.12	tonne	**914.05**
Deformed high yield steel bars to BS 4449						
Bars; supplied in straight lengths						
6 mm nominal size	8.00	909.76	115.34	614.68	tonne	**1639.78**
8 mm nominal size	6.74	766.47	97.18	609.76	tonne	**1473.41**
10 mm nominal size	6.74	766.47	97.18	545.84	tonne	**1409.49**
12 mm nominal size	6.74	766.47	97.18	536.01	tonne	**1399.66**
16 mm nominal size	6.15	699.38	88.68	516.34	tonne	**1304.40**
20 mm nominal size	4.44	504.92	64.01	585.18	tonne	**1154.11**
25 mm nominal size	4.44	504.92	64.01	590.09	tonne	**1159.02**
32 mm nominal size	4.44	504.92	64.01	595.01	tonne	**1163.94**
40 mm nominal size	4.44	504.92	64.01	604.85	tonne	**1173.78**
Bars; supplied in bent and cut lengths						
6 mm nominal size	8.00	909.76	115.34	647.03	tonne	**1672.13**
8 mm nominal size	6.74	766.47	97.18	641.86	tonne	**1505.51**
10 mm nominal size	6.74	766.47	97.18	574.56	tonne	**1438.21**
12 mm nominal size	6.74	766.47	97.18	564.21	tonne	**1427.86**
16 mm nominal size	6.15	699.38	88.68	543.51	tonne	**1331.57**
20 mm nominal size	4.44	504.92	64.01	615.97	tonne	**1184.90**
25 mm nominal size	4.44	504.92	64.01	621.15	tonne	**1190.08**
32 mm nominal size	4.44	504.92	64.01	626.33	tonne	**1195.26**
40 mm nominal size	4.44	504.92	64.01	636.68	tonne	**1205.61**
Stainless steel bars; Alloy 1.4301						
Bars; supplied in straight lengths						
8 mm nominal size	6.74	766.47	97.18	3300.00	tonne	**4163.65**
10 mm nominal size	6.74	766.47	97.18	3300.00	tonne	**4163.65**
12 mm nominal size	6.74	766.47	97.18	3300.00	tonne	**4163.65**
16 mm nominal size	6.15	699.38	88.68	3300.00	tonne	**4088.06**
20 mm nominal size	4.44	504.92	64.01	3300.00	tonne	**3868.93**
25 mm nominal size	4.44	504.92	64.01	3300.00	tonne	**3868.93**
32 mm nominal size	4.44	504.92	64.01	3300.00	tonne	**3868.93**
Bars; supplied in bent and cut lengths						
8 mm nominal size	6.74	766.47	97.18	3450.00	tonne	**4313.65**
10 mm nominal size	6.74	766.47	97.18	3450.00	tonne	**4313.65**
12 mm nominal size	6.74	766.47	97.18	3450.00	tonne	**4313.65**
16 mm nominal size	6.15	699.38	88.68	3450.00	tonne	**4238.06**
20 mm nominal size	4.44	504.92	64.01	3450.00	tonne	**4018.93**
25 mm nominal size	4.44	504.92	64.01	3450.00	tonne	**4018.93**
32 mm nominal size	4.44	504.92	64.01	3450.00	tonne	**4018.93**

CLASS G: CONCRETE ANCILLARIES

Item	Gang hours	Labour £	Plant £	Material £	Unit	Total rate £
Additional allowances to bar reinforcement						
Add to the above bars						
12-15 m long; mild steel to BS 4449	–	–	–	31.63	tonne	**31.63**
12-15 m long; high yield steel to BS 4449	–	–	–	19.80	tonne	**19.80**
Over 15 m long, per 500 mm increment; mild steel to BS 4449	–	–	–	12.65	tonne	**12.65**
Over 15 m long, per 500 mm increment; high yield steel to BS 4449	–	–	–	4.40	tonne	**4.40**
Add for cutting, bending, tagging and baling reinforcement on site						
6 mm nominal size	4.87	177.66	70.22	1.90	tonne	**249.78**
8 mm nominal size	4.58	167.08	66.03	1.90	tonne	**235.01**
10 mm nominal size	3.42	124.76	49.31	1.90	tonne	**175.97**
12 mm nominal size	2.55	93.02	36.77	1.90	tonne	**131.69**
16 mm nominal size	2.03	74.05	29.28	1.90	tonne	**105.23**
20 mm nominal size	1.68	61.29	24.22	1.90	tonne	**87.41**
25 mm nominal size	1.68	61.29	24.22	1.90	tonne	**87.41**
32 mm nominal size	1.39	50.71	20.05	1.90	tonne	**72.66**
40 mm nominal size	1.39	50.71	20.05	1.90	tonne	**72.66**
Special joints						
Lenton type A couplers; threaded ends on reinforcing bars						
12 mm	0.09	10.23	–	6.72	nr	**16.95**
16 mm	0.09	10.23	–	8.11	nr	**18.34**
20 mm	0.09	10.23	–	11.84	nr	**22.07**
25 mm	0.09	10.23	–	16.37	nr	**26.60**
32 mm	0.09	10.23	–	22.47	nr	**32.70**
40 mm	0.09	10.23	–	30.87	nr	**41.10**
Lenton type B couplers; threaded ends on reinforcing bars						
12 mm	0.09	10.23	–	21.59	nr	**31.82**
16 mm	0.09	10.23	–	25.02	nr	**35.25**
20 mm	0.09	10.23	–	27.75	nr	**37.98**
25 mm	0.09	10.23	–	32.04	nr	**42.27**
32 mm	0.09	10.23	–	43.30	nr	**53.53**
40 mm	0.09	10.23	–	63.48	nr	**73.71**
Steel fabric to BS 4483						
Fabric						
nominal mass 0.77 kg/m^2; ref D49	0.02	2.27	0.29	0.57	m^2	**3.13**
nominal mass 1.54 kg/m^2; ref D98	0.02	2.27	0.29	1.08	m^2	**3.64**
nominal mass 1.54 kg/m^2; ref A98	0.03	3.41	0.44	2.21	m^2	**6.06**
nominal mass 2.22 kg/m^2; ref A142	0.03	3.41	0.44	2.10	m^2	**5.95**
nominal mass 2.61 kg/m^2; ref C283	0.03	3.41	0.44	1.92	m^2	**5.77**
nominal mass 3.02 kg/m^2; ref A193	0.04	4.55	0.58	2.87	m^2	**8.00**
nominal mass 3.05 kg/m^2; ref B196	0.04	4.55	0.58	5.12	m^2	**10.25**
nominal mass 3.41 kg/m^2; ref C385	0.04	4.55	0.58	2.49	m^2	**7.62**
nominal mass 3.73 kg/m^2; ref B283	0.04	4.55	0.58	3.38	m^2	**8.51**
nominal mass 3.95 kg/m^2; ref A252	0.04	4.55	0.58	3.75	m^2	**8.88**
nominal mass 4.34 kg/m^2; ref C503	0.05	5.69	0.73	3.14	m^2	**9.56**

CLASS G: CONCRETE ANCILLARIES

Item	Gang hours	Labour £	Plant £	Material £	Unit	Total rate £
REINFORCEMENT – cont						
Steel fabric to BS 4483 – cont						
Fabric – cont						
nominal mass 4.35 kg/m^2; ref B385	0.05	5.69	0.73	4.12	m^2	**10.54**
nominal mass 5.55 kg/m^2; ref C636	0.05	5.69	0.73	3.86	m^2	**10.28**
nominal mass 5.93 kg/m^2; ref B503	0.05	5.69	0.73	3.75	m^2	**10.17**
nominal mass 6.16 kg/m^2; ref A393	0.07	7.96	1.02	5.27	m^2	**14.25**
nominal mass 6.72 kg/m^2; ref C785	0.07	7.96	1.02	4.68	m^2	**13.66**
nominal mass 8.14 kg/m^2; ref B785	0.08	9.10	1.15	5.15	m^2	**15.40**
nominal mass 10.90 kg/m^2; ref B1131	0.09	10.23	1.31	9.78	m^2	**21.32**
RESOURCES - LABOUR						
Formwork gang - small areas						
1 foreman joiner (craftsman)		22.32				
1 joiner (craftsman)		19.15				
1 unskilled operative (general)		12.44				
1 plant operator (craftsman) - 50% of time		10.34				
Total Gang Rate / Hour	£	**64.25**				
Formwork gang - large areas						
1 foreman joiner (craftsman)		22.32				
2 joiners (craftsman)		38.30				
1 unskilled operative (general)		12.44				
1 plant operator (craftsman) - 25% of time		5.17				
Total Gang Rate / Hour	£	**78.23**				
Reinforcement gang						
1 foreman steel fixer (craftsman)		22.32				
4 steel fixers (craftsman)		76.60				
1 unskilled operative (general)		12.44				
1 plant operator (craftsman) - 25% of time		5.17				
Total Gang Rate / Hour	£	**116.53**				
Reinforcement - on-site bending/baling gang						
1 steel fixer (craftsman)		19.15				
1 unskilled operative (general)		12.44				
1 plant operator (craftsman) - 25% of time		5.17				
Total Gang Rate / Hour	£	**36.76**				
Joints gang						
1 ganger/chargehand (skill rate 4)		13.32				
1 skilled operative (skill rate 4)		13.32				
1 unskilled operative (general)		12.44				
Total Gang Rate / Hour	£	**39.08**				
Concrete accessories gang						
1 ganger/chargehand (skill rate 4)		13.32				
1 skilled operative (skill rate 4)		13.32				
1 unskilled operative (general)		12.44				
Total Gang Rate / Hour	£	**39.08**				

CLASS G: CONCRETE ANCILLARIES

Item	Gang hours	Labour £	Plant £	Material £	Unit	Total rate £
RESOURCES - PLANT						
Formwork - small areas						
20t crawler crane - 50% of time			14.27			
22' diameter saw bench			1.85			
allowance for small power tools			2.25			
Total Rate / Hour		£	**18.36**			
Formwork - large areas						
20t crawler crane - 25% of time			7.14			
22' diameter saw bench			1.85			
allowance for small power tools			2.25			
Total Rate / Hour		£	**11.23**			
Reinforcement						
30 t crawler crane - 25% of time			7.79			
bar cropper			4.09			
small power tools			1.65			
support acrows, tirfors, kentledge, etc.			0.89			
Total Rate / Hour		£	**14.42**			

RESOURCES - MATERIALS

Formwork

Formwork materials include for shutter, bracing, ties, support, kentledge and all consumables.
The following unit costs do not include for formwork outside the payline and are based on an optimum of a minimum 8 uses with 10% per use towards the cost of repairs / replacement of components damaged during disassembly
ADD to formwork material costs generally depending on the number of uses :

Nr of uses	% Addition	% Waste
1	+ 90 to 170	+7
2	+ 50 to 80	+7
3	+ 15 to 30	+6
6	+ 5 to 10	+6
8	No change	+5
10	- 5 to 7	+5

Reinforcement

Reinforcement materials include for bars, tying wire, spacers, couplers and steel supports for bottom layer reinforcement (stools, chairs and risers).

CLASS G: CONCRETE ANCILLARIES

Item	Gang hours	Labour £	Plant £	Material £	Unit	Total rate £
JOINTS						
Misc						
Open surface plain; average width						
ne 0.5m; scabbling concrete for subsequent pour	0.04	1.15	1.73	–	m²	2.88
0.5–1m; scabbling concrete for subsequent pour	0.03	0.86	1.34	–	m²	2.20
Open surface with filler; average width						
ne 0.5m; 12mm Flexcell joint filler	0.04	1.15	1.73	3.60	m²	6.48
0.5–1m; 12mm Flexcell joint filler	0.04	1.15	1.73	3.60	m²	6.48
ne 0.5m; 19mm Flexcell joint filler	0.05	1.44	2.14	6.21	m²	9.79
0.5–1m; 19mm Flexcell joint filler	0.05	1.44	2.14	6.21	m²	9.79
Formed surface plain; average width (including formwork)						
ne 0.5m	0.24	6.90	10.41	11.22	m²	28.53
0.5–1.0m	0.24	6.90	10.41	11.22	m²	28.53
Formed surface with filler; average width						
ne 0.5m; 10mm Flexcell joint filler	0.40	11.49	13.07	14.83	m²	39.39
0.5–1m; 10mm Flexcell joint filler	0.41	11.78	17.76	14.83	m²	44.37
ne 0.5m; 19mm Flexcell joint filler	0.42	12.07	18.28	17.43	m²	47.78
0.5–1m; 19mm Flexcell joint filler	0.43	12.35	18.69	17.43	m²	48.47
ne 0.5m; 25mm Flexcell joint filler	0.45	12.93	19.49	19.10	m²	51.52
0.5–1m; 25mm Flexcell joint filler	0.47	13.50	20.42	19.10	m²	53.02
PVC						
Plastics or rubber waterstops						
160 mm centre bulb	0.04	1.15	–	8.10	m	9.25
junction piece	0.04	1.15	–	54.01	nr	55.16
210 mm centre bulb	0.05	1.44	–	8.39	m	9.83
junction piece	0.04	1.15	–	106.05	nr	107.20
260 mm centre bulb	0.05	1.44	–	8.72	m	10.16
junction piece	0.05	1.44	–	10.72	nr	12.16
170 mm flat dumbell	0.04	1.15	–	7.96	m	9.11
junction piece	0.05	1.44	–	94.58	nr	96.02
210 mm flat dumbell	0.04	1.15	–	8.34	m	9.49
junction piece	0.07	2.01	–	106.41	nr	108.42
250 mm flat dumbell	0.05	1.44	–	8.72	m	10.16
junction piece	0.09	2.59	–	149.92	nr	152.51
Polysulphide sealant; gun grade						
Sealed rebates or grooves						
10 × 20 mm	0.05	1.44	–	–	m	1.44
20 × 20 mm	0.07	2.01	–	0.01	m	2.02
25 × 20 mm	0.08	2.30	–	0.01	m	2.31
Mild steel						
Dowels, plain or greased						
12 mm diameter x 500 mm long	0.04	1.15	–	0.55	nr	1.70
16 mm diameter x 750 mm long	0.05	1.29	–	1.53	nr	2.82
20 mm diameter x 750 mm long	0.05	1.29	–	2.40	nr	3.69
25 mm diameter x 750 mm long	0.05	1.29	–	3.73	nr	5.02
32 mm diameter x 750 mm long	0.05	1.29	–	6.13	nr	7.42

CLASS G: CONCRETE ANCILLARIES

Item	Gang hours	Labour £	Plant £	Material £	Unit	Total rate £
Dowels, sleeved or capped						
12 mm diameter × 500 mm long, debonding agent for 250 mm and capped with pvc dowel cap	0.05	1.29	–	0.61	nr	**1.90**
16 mm diameter × 750 mm long, debonding agent for 375 mm and capped with pvc dowel cap	0.05	1.52	–	1.59	nr	**3.11**
20 mm diameter × 750 mm long, debonding agent for 375 mm and capped with pvc dowel cap	0.05	1.52	–	2.46	nr	**3.98**
25 mm diameter × 750 mm long, debonding agent for 375 mm and capped with pvc dowel cap	0.06	1.72	–	3.80	nr	**5.52**
32 mm diameter × 750 mm long, debonding agent for 375 mm and capped with pvc dowel cap	0.07	1.87	–	6.20	nr	**8.07**
POST-TENSIONED PRESTRESSING						
The design of prestressing is based on standard patented systems, each of which has produced its own method of anchoring, joining and stressing the cables or wires. The companies marketing the systems will either supply all the materails and fittings required together with the sale or hire of suitable jacks and equipment for prestressing and grouting or they will undertake to complete the work on a sub-contract basis. The rates given below are therefore indicative only of the probable labour and plant costs and do not include for any permanent materials. The advice of specialist contractors should be sought for more accurate rates based on the design for a particular contract. Pretensioned prestressing is normally used only in the manufacture of precast units utilising special beds set up in the supplier's factory.						
Labour and plant cost in post-tensioning; material cost excluded						
form ducts to profile including supports and fixings; 50 mm internal diameter	1.00	1.44	3.29	–	m	**4.73**
Extra for grout vents	1.00	7.16	–	–	nr	**7.16**
form ducts to profile including supports and fixings; 80 mm internal diameter	1.00	2.25	3.86	–	m	**6.11**
Extra for grout vents	1.00	7.16	–	–	nr	**7.16**
form ducts to profile including supports and fixings; 100 mm internal diameter	1.00	3.07	4.06	–	m	**7.13**
Extra for grout vents	1.00	7.16	–	–	nr	**7.16**
grout ducts including provision of equipment; 50 mm internal diameter	1.00	2.14	0.57	–	m	**2.71**
grout ducts including provision of equipment; 80 mm internal diameter	1.00	2.14	0.72	–	m	**2.86**
grout ducts including provision of equipment; 100 mm internal diameter	1.00	2.91	0.85	–	m	**3.76**
form tendons including spacers etc. and pull through ducts; 7 Nr strands	0.25	8.26	3.90	–	m	**12.16**
form tendons including spacers etc. and pull through ducts; 12 Nr strands	0.45	14.87	7.02	–	m	**21.89**

CLASS G: CONCRETE ANCILLARIES

Item	Gang hours	Labour £	Plant £	Material £	Unit	Total rate £
POST-TENSIONED PRESTRESSING – cont						
form tendons including spacers etc. and pull						
through ducts; 19 Nr strands	0.65	21.49	10.14	–	m	**31.63**
dead end anchorage; 7 Nr strands	0.75	24.79	11.70	–	nr	**36.49**
dead end anchorage; 12 Nr strands	0.95	31.40	14.82	–	nr	**46.22**
dead end anchorage; 19 Nr strands	1.15	38.01	17.94	–	nr	**55.95**
looped buried dead end anchorage; 7 Nr strands	0.50	16.53	7.80	–	nr	**24.33**
looped buried dead end anchorage; 12 Nr strands	0.66	21.82	10.29	–	nr	**32.11**
looped buried dead end anchorage; 19 Nr strands	0.89	29.42	13.88	–	nr	**43.30**
end anchorage including reinforcement; 7 Nr strands	1.67	55.20	26.05	–	nr	**81.25**
add to last for anchorage coupling	0.46	15.20	7.17	–	nr	**22.37**
end anchorage including reinforcement; 12 Nr						
strands	2.04	67.43	31.82	–	nr	**99.25**
add to last for anchorage coupling	0.79	26.11	12.32	–	nr	**38.43**
end anchorage including reinforcement; 19 Nr						
strands	2.56	84.62	39.93	–	nr	**124.55**
add to last for anchorage coupling	1.20	39.66	18.72	–	nr	**58.38**
stress and lock off including multimatic jack; 7 Nr						
strands	3.21	106.10	50.06	–	nr	**156.16**
stress and lock off including multimatic jack; 12 Nr						
strands	4.19	138.50	65.35	–	nr	**203.85**
stress and lock off including multimatic jack; 19 Nr						
strands	5.58	184.44	87.03	–	nr	**271.47**
cut off and seal ends of tendons; 7 Nr strands	0.20	6.61	3.12	–	nr	**9.73**
cut off and seal ends of tendons; 12 Nr strands	0.35	11.57	5.46	–	nr	**17.03**
cut off and seal ends of tendons; 19 Nr strands	0.55	18.18	8.58	–	nr	**26.76**
CONCRETE ACCESSORIES						
Finishing of top surfaces						
wood float; level	0.02	0.77	–	–	m²	**0.77**
wood float; falls or cross-falls	0.03	1.16	–	–	m²	**1.16**
steel trowel; level	0.03	1.16	–	–	m²	**1.16**
wood float; falls or cross-falls	0.03	1.16	–	–	m²	**1.16**
steel trowel; falls or cross-falls	0.05	1.94	–	–	m²	**1.94**
granolithic finish 20 mm thick laid monolithically	0.07	2.71	0.59	8.20	m²	**11.50**
Finishing of formed surfaces						
aggregate exposure using retarder	0.05	1.94	–	1.17	m²	**3.11**
bush hammering; kango hammer	0.28	10.84	3.98	–	m²	**14.82**
rubbing down concrete surfaces after striking						
shutters	0.02	0.77	–	1.34	m²	**2.11**
Inserts totally within the concrete volume						
HDPE conduit 20 mm diameter	0.10	3.87	–	0.82	m	**4.69**
black enamelled steel conduit 20 mm diameter	0.10	3.87	–	1.66	m	**5.53**
galvanized steel conduit 20 mm diameter	0.10	3.87	–	2.23	m	**6.10**
Unistrut channel type P3270	0.20	7.75	–	7.95	m	**15.70**
Unistrut channel type P3370	0.20	7.75	–	6.83	m	**14.58**

CLASS G: CONCRETE ANCILLARIES

Item	Gang hours	Labour £	Plant £	Material £	Unit	Total rate £
Inserts projecting from surface(s) of the concrete						
expanding bolt; 10 mm diameter x 25 mm deep	0.05	1.94	–	1.93	nr	3.87
holding down bolt; 16 mm diameter x 250 mm deep	0.25	9.68	–	2.50	nr	12.18
holding down bolt; 16 mm diameter x 350 mm deep	0.25	9.68	–	3.32	nr	13.00
holding down bolt; 20 mm diameter x 250 mm deep	0.25	9.68	–	3.11	nr	12.79
holding down bolt; 20 mm diameter x 450 mm deep	0.25	9.68	–	3.38	nr	13.06
vitrified clay pipe to BS 65; 100 mm diameter x						
1000 mm long	0.25	9.68	–	4.32	nr	14.00
cast iron pipe to BS 437; 100 mm diameter x						
1000 mm long	0.25	9.68	–	28.93	nr	38.61
Grouting under plates; cement and sand (1:3)						
area ne 0.1 m^2	0.10	3.87	–	0.23	nr	4.10
area 0.1–0.5 m^2	0.45	17.43	–	1.11	nr	18.54
area 0.5–1.0 m^2	0.78	30.21	–	2.21	nr	32.42
Grouting under plates; non-shrink cementitious grout						
area ne 0.1 m^2	0.10	3.87	–	4.04	nr	7.91
area 0.1–0.5 m^2	0.45	17.43	–	38.84	nr	56.27
area 0.5–1.0 m^2	0.78	30.21	–	77.69	nr	107.90

CLASS H: PRECAST CONCRETE

Item	Gang hours	Labour £	Plant £	Material £	Unit	Total rate £
NOTES						
The cost of precast concrete items is very much dependant on the complexity of the moulds, the number of units to be cast from each mould and the size and weight of the unit to be handled. The unit rates below are for standard precast items that are often to be found on a civil engineering project. It would be misleading to quote for indicative costs for tailor-made precast concrete units and it is advisable to contact Specialist Manufacturers for guide prices.						
BEAMS						
Concrete mix C20						
Beams						
100 × 150 × 1050 mm long	1.00	5.84	–	22.05	nr	**27.89**
225 × 150 × 1200 mm long	1.00	8.15	2.00	30.45	nr	**40.60**
225 × 225 × 1800 mm long	1.00	10.40	4.16	68.25	nr	**82.81**
PRESTRESSED PRE-TENSIONED BEAMS						
Concrete mix C20						
Beams						
100 × 65 × 1050 mm long	1.00	3.63	–	5.81	nr	**9.44**
265 × 65 × 1800 mm long	1.00	4.56	3.77	22.38	nr	**30.71**
Bridge beams						
Inverted 'T' Beams, flange width 495 mm						
section T1; 8 m long, 380 mm deep; mass 1.88t	–	–	–	–	nr	**888.75**
section T2; 9 m long, 420 mm deep; mass 2.29t	–	–	–	–	nr	**1068.75**
section T3; 11 m long, 535 mm deep; mass 3.02t	–	–	–	–	nr	**1254.38**
section T4; 12 m long, 575 mm deep; mass 3.54t	–	–	–	–	nr	**1395.00**
section T5; 13 m long, 615 mm deep; mass 4.08t	–	–	–	–	nr	**1440.00**
section T6; 13 m long, 655 mm deep; mass 4.33t	–	–	–	–	nr	**1440.00**
section T7; 14 m long, 695 mm deep; mass 4.95t	–	–	–	–	nr	**1670.63**
section T8; 15 m long, 735 mm deep; mass 5.60t	–	–	–	–	nr	**1811.25**
section T9; 16 m long, 775 mm deep; mass 6.28t	–	–	–	–	nr	**1951.88**
section T10; 18 m long, 815 mm deep; mass 7.43t	–	–	–	–	nr	**2182.50**
'M' beams, flange width 970 mm						
section M2; 17 m long, 720 mm deep; mass 12.95t	–	–	–	–	nr	**4741.88**
section M3; 18 m long, 800 mm deep; mass 15.11t	–	–	–	–	nr	**4460.63**
section M6; 22 m long, 1040 mm deep; mass 20.48t	–	–	–	–	nr	**7155.00**
section M8; 25 m long, 1200 mm deep; mass 23.68t	–	–	–	–	nr	**9292.50**
'U' beams, base width 970 mm						
section U3; 16 m long, 900 mm deep; mass 19.24t	–	–	–	–	nr	**7993.13**
section U5; 20 m long, 1000 mm deep; mass 25.64t	–	–	–	–	nr	**10220.63**
section U8; 24 m long, 1200 mm deep; mass 34.56t	–	–	–	–	nr	**13848.75**
section U12; 30 m long, 1600 mm deep; mass 52.74t	–	–	–	–	nr	**18956.25**

CLASS H: PRECAST CONCRETE

Item	Gang hours	Labour £	Plant £	Material £	Unit	Total rate £
SLABS						
Prestressed precast concrete flooring planks; Bison or similar; cement mortar grout between planks on bearings						
100 mm thick floor						
400 mm wide planks	0.21	11.20	6.58	32.64	m²	**50.42**
1200 mm wide planks	0.12	6.24	3.67	40.15	m²	**50.06**
150 mm thick floor						
400 mm wide planks	0.26	14.00	8.22	34.50	m²	**56.72**
1200 mm wide planks	0.14	7.81	4.58	41.85	m²	**54.24**
SEGMENTAL UNITS						
COPINGS, SILLS AND WEIR BLOCKS						
Concrete mix C30						
Coping; weathered and throated						
178 × 64 mm	1.00	9.16	4.85	8.50	m	**22.51**
305 × 76 mm	1.00	6.88	3.46	15.00	m	**25.34**

CLASS I: PIPEWORK - PIPES

Item	Gang hours	Labour £	Plant £	Material £	Unit	**Total rate £**
NOTES						
The rates assume the most efficient items of plant (excavator) and are optimum rates assuming continuous output with no delays caused by other operations or works.						
Ground conditions are assumed to be good easily worked soil with no abnormal conditions that would affect outputs and consistency of work.						
Multiplier Table for labour and plant for various site conditions for working:						
out of sequence × 2.75 (minimum)						
in hard clay × 1.75 to 2.00						
in running sand × 2.75 (minimum)						
in broken rock × 2.75 to 3.50						
below water table × 2.00 (minimum)						
Variance from CESMM3						
Fittings are included with the pipe concerned, for convenience of reference, rather than in Class J.						
RESOURCES - LABOUR						
Drainage / pipework gang (small bore)						
1 ganger/chargehand (skill rate 4) - 50% of time		7.12				
1 skilled operative (skill rate 4)		13.32				
2 unskilled operatives (general)		24.88				
1 plant operator (skill rate 3)		16.19				
1 plant operator (skill rate 3) - 50% of time		8.10				
Total Gang Rate/Hour		69.60				
Drainage / pipework gang (small bore - not in trenches)						
1 ganger/chargehand (skill rate 4) - 50% of time		7.12				
1 skilled operative (skill rate 4)		13.32				
2 unskilled operatives (general)		24.88				
Total Gang Rate/Hour		45.31				
Drainage / pipework gang (large bore)						
Note: relates to pipes exceeding 700 mm diameter.		–				
1 ganger/chargehand (skill rate 4) - 50% of time		7.12				
1 skilled operative (skill rate 4)		13.32				
2 unskilled operatives (general)		24.88				
1 plant operator (skill rate 3)		16.19				
Total Gang Rate/Hour		61.50				
Drainage / pipework gang (large bore - not in trenches)						
Note: relates to pipes exceeding 700 mm diameter.		–				
1 ganger/chargehand (skill rate 4) - 50% of time		7.12				
1 skilled operative (skill rate 4)		13.32				
2 unskilled operatives (general)		24.88				
Total Gang Rate/Hour		45.31				

CLASS I: PIPEWORK - PIPES

Item	Gang hours	Labour £	Plant £	Material £	Unit	Total rate £
RESOURCES - PLANT						
Field drains						
0.4 m³ hydraulic backacter			22.93			
2t dumper - 30% of time			1.95			
Stihl saw, 12', petrol - 30% of time			0.33			
small pump - 30% of time			0.71			
Total Rate / Hour		£	25.92			
			–			
Add to the above for trench supports appropriate to trench depth (see below).			–			
Field drains (not in trenches)						
2t dumper - 30% of time			1.95			
Stihl saw, 12', petrol - 30% of time			0.33			
Total Rate / Hour		£	2.28			
Drains/sewers (small bore)						
1.0 m³ hydraulic backacter			40.06			
2t dumper - 30% of time			1.95			
2.80 m³/min compressor, 2 tool - 30% of time			2.47			
disc saw - 30% of time			0.41			
extra 50ft / 15m hose - 30% of time			0.09			
small pump - 30% of time			0.71			
sundry tools - 30% of time			1.01			
Total Rate / Hour		£	46.71			
			–			
Add to the above for trench supports appropriate to trench depth (see below).			–			
Drains/sewers (small bore - not in trenches)						
2t dumper - 30% of time			1.95			
2.80 m³/min compressor, 2 tool - 30% of time			2.47			
disc saw - 30% of time			0.41			
extra 50ft / 15m hose - 30% of time			0.09			
sundry tools - 30% of time			1.01			
Total Rate / Hour		£	5.94			
Drains/sewers (large bore)						
1.0 m³ hydraulic backacter			40.06			
20t crawler crane - 50% of time			14.27			
2t dumper (30% of time)			1.95			
2.80 m³/min compressor, 2 tool - 30% of time			2.47			
disc saw - 30% of time			0.41			
extra 50ft / 15m hose - 30% of time			0.09			
small pump - 30% of time			0.71			
sundry tools - 30% of time			1.01			
Total Rate / Hour		£	60.98			
			–			
Add to the above for trench supports appropriate to trench depth (see below).			–			

CLASS I: PIPEWORK - PIPES

Item	Gang hours	Labour £	Plant £	Material £	Unit	Total rate £
RESOURCES - PLANT – cont						
Drains/sewers (large bore - not in trenches)						
2t dumper (30% of time)			1.95			
2.80 m³/min compressor, 2 tool - 30% of time			2.47			
disc saw - 30% of time			0.41			
extra 50ft / 15m hose - 30% of time			0.09			
sundry tools - 30% of time			1.01			
Total Rate / Hour		£	**5.94**			
Trench supports						
In addition to the above, the following allowances for close sheeted trench supports are included in the following unit rates, assuming that the ground conditions warrants it:			–			
ne 1.50 m deep			3.54		m	
1.50–2.00 m deep			4.94		m	
2.00–2.50 m deep			4.72		m	
2.50–3.00 m deep			5.94		m	
3.00–3.50 m deep			7.23		m	
3.50–4.00 m deep			8.13		m	
4.00–4.50 m deep			9.05		m	
4.50–5.00 m deep			15.44		m	
5.00–5.50 m deep			21.26		m	
CLAY PIPES						
Field drains to BS 1196, butt joints, nominal bore; excavation and supports, backfilling						
75 mm pipes; in trench, depth						
not in trenches	0.06	2.70	0.05	5.60	m	**8.35**
ne 1.50 m deep	0.09	6.22	2.35	5.60	m	**14.17**
1.50–2.00 m deep	0.13	8.98	3.40	5.60	m	**17.98**
2.00–2.50 m deep	0.18	12.43	4.72	5.60	m	**22.75**
2.50–3.00 m deep	0.23	15.88	6.03	5.60	m	**27.51**
100 mm pipes; in trench, depth						
not in trenches	0.06	2.70	0.05	8.11	m	**10.86**
ne 1.50 m deep	0.10	6.91	2.61	8.11	m	**17.63**
1.50–2.00 m deep	0.14	9.67	3.67	8.11	m	**21.45**
2.00–2.50 m deep	0.19	13.12	4.97	8.11	m	**26.20**
2.50–3.00 m deep	0.24	16.57	6.29	8.11	m	**30.97**
150 mm pipes; in trench, depth						
not in trenches	0.06	2.70	0.05	14.53	m	**17.28**
ne 1.50 m deep	0.11	7.60	2.86	14.53	m	**24.99**
1.50–2.00 m deep	0.15	10.36	3.92	14.53	m	**28.81**
2.00–2.50 m deep	0.20	13.81	5.23	14.53	m	**33.57**
2.50–3.00 m deep	0.25	17.27	6.56	14.53	m	**38.36**
225 mm pipes; in trench, depth						
not in trenches	0.06	2.70	0.05	34.04	m	**36.79**
ne 1.50 m deep	0.13	8.98	3.39	34.04	m	**46.41**
1.50–2.00 m deep	0.17	11.74	4.45	34.04	m	**50.23**
2.00–2.50 m deep	0.22	15.19	5.76	34.04	m	**54.99**
2.50–3.00 m deep	0.27	18.65	7.08	34.04	m	**59.77**

CLASS I: PIPEWORK - PIPES

Item	Gang hours	Labour £	Plant £	Material £	Unit	Total rate £
Vitrified clay perforated field drains to BS EN295, sleeved joints; excavation and supports, backfilling						
100 mm pipes; in trench, depth						
not in trenches	0.06	2.70	0.05	12.78	m	**15.53**
ne 1.50 m deep	0.15	10.36	3.90	12.78	m	**27.04**
1.50–2.00 m deep	0.19	13.12	4.96	12.78	m	**30.86**
2.00–2.50 m deep	0.24	16.57	6.27	12.78	m	**35.62**
2.50–3.00 m deep	0.29	20.03	7.61	12.78	m	**40.42**
Extra for bend	0.08	5.52	–	10.27	nr	**15.79**
Extra for single junction	0.09	6.22	–	21.17	nr	**27.39**
150 mm pipes; in trench, depth						
not in trenches	0.06	2.70	0.05	23.23	m	**25.98**
ne 1.50 m deep	0.17	11.74	4.43	23.23	m	**39.40**
1.50–2.00 m deep	0.20	13.81	5.23	23.23	m	**42.27**
2.00–2.50 m deep	0.24	16.57	6.27	23.23	m	**46.07**
2.50–3.00 m deep	0.29	20.03	7.61	23.23	m	**50.87**
Extra for bend	0.13	8.63	–	20.19	nr	**28.82**
Extra for single junction	0.13	8.98	–	28.24	nr	**37.22**
225 mm pipes; in trench, depth						
not in trenches	0.08	3.60	0.05	31.54	m	**35.19**
ne 1.50 m deep	0.18	12.43	4.69	31.54	m	**48.66**
1.50–2.00 m deep	0.22	15.19	5.76	31.54	m	**52.49**
2.00–2.50 m deep	0.26	17.96	6.81	31.54	m	**56.31**
2.50–3.00 m deep	0.30	20.72	7.87	31.54	m	**60.13**
Extra for bend	0.12	8.29	–	98.05	nr	**106.34**
Extra for single junction	0.20	13.47	–	136.88	nr	**150.35**
300 mm pipes; in trench, depth						
not in trenches	0.10	4.50	0.23	45.58	m	**50.31**
ne 1.50 m deep	0.19	13.12	4.94	45.58	m	**63.64**
1.50–2.00 m deep	0.24	16.57	6.27	45.58	m	**68.42**
2.00–2.50 m deep	0.28	19.34	7.33	45.58	m	**72.25**
2.50–3.00 m deep	0.32	22.10	8.39	45.58	m	**76.07**
Extra for bend	0.24	16.57	–	186.22	nr	**202.79**
Extra for single junction	0.23	15.88	–	351.35	nr	**367.23**

CLASS I: PIPEWORK - PIPES

Item	Gang hours	Labour £	Plant £	Material £	Unit	Total rate £
CLAY PIPES – cont						
Vitrified clay pipes to BS EN295, plain ends with push-fit polypropylene flexible couplings; excavation and supports, backfilling						
100 mm pipes; in trenches, depth						
not in trenches	0.06	2.70	0.37	7.51	m	10.58
ne 1.50 m deep	0.15	10.36	6.87	7.51	m	24.74
1.50–2.00 m deep	0.19	13.12	8.91	7.51	m	29.54
2.00–2.50 m deep	0.24	16.57	11.27	7.51	m	35.35
2.50–3.00 m deep	0.29	20.03	13.64	7.51	m	41.18
3.00–3.50 m deep	0.36	24.86	16.94	7.51	m	49.31
3.50–4.00 m deep	0.44	30.39	20.74	7.51	m	58.64
4.00–4.50 m deep	0.55	37.98	25.93	7.51	m	71.42
4.50–5.00 m deep	0.70	48.34	33.06	7.51	m	88.91
5.00–5.50 m deep	0.90	62.15	42.55	7.51	m	112.21
Extra for bend	0.05	3.45	–	18.40	nr	21.85
Extra for rest bend	0.06	4.14	–	27.50	nr	31.64
Extra for single junction; equal	0.09	6.22	–	32.57	nr	38.79
Extra for saddle; oblique	0.23	15.88	–	24.09	nr	39.97
150 mm pipes; in trenches, depth						
not in trenches	0.06	2.70	0.37	16.31	m	19.38
ne 1.50 m deep	0.17	11.74	7.97	16.31	m	36.02
1.50–2.00 m deep	0.20	13.81	9.38	16.31	m	39.50
2.00–2.50 m deep	0.24	16.57	11.27	16.31	m	44.15
2.50–3.00 m deep	0.29	20.03	13.64	16.31	m	49.98
3.00–3.50 m deep	0.39	26.93	18.34	16.31	m	61.58
3.50–4.00 m deep	0.45	31.08	21.21	16.31	m	68.60
4.00–4.50 m deep	0.58	40.05	27.37	16.31	m	83.73
4.50–5.00 m deep	0.75	51.80	35.40	16.31	m	103.51
5.00–5.50 m deep	0.96	66.30	45.37	16.31	m	127.98
Extra for bend	0.08	5.52	–	53.55	nr	59.07
Extra for rest bend	0.09	6.22	–	61.20	nr	67.42
Extra for single junction; equal	0.11	7.60	–	80.55	nr	88.15
Extra for taper reducer	0.07	4.83	–	41.33	nr	46.16
Extra for saddle; oblique	0.29	20.03	–	49.05	nr	69.08
225 mm pipes; in trenches, depth						
not in trenches	0.08	3.60	0.47	35.81	m	39.88
ne 1.50 m deep	0.18	12.43	8.43	35.81	m	56.67
1.50–2.00 m deep	0.22	15.19	10.34	35.81	m	61.34
2.00–2.50 m deep	0.26	17.96	12.22	35.81	m	65.99
2.50–3.00 m deep	0.30	20.72	14.12	35.81	m	70.65
3.00–3.50 m deep	0.41	28.31	19.30	35.81	m	83.42
3.50–4.00 m deep	0.47	32.46	22.14	35.81	m	90.41
4.00–4.50 m deep	0.65	44.89	30.66	35.81	m	111.36
4.50–5.00 m deep	0.80	55.25	37.77	35.81	m	128.83
5.00–5.50 m deep	1.02	70.44	48.22	35.81	m	154.47
Extra for bend	0.10	6.91	–	104.09	nr	111.00
Extra for rest bend	0.11	7.60	–	143.06	nr	150.66
Extra for single junction; equal	0.16	11.05	–	176.14	nr	187.19
Extra for taper reducer	0.12	8.29	–	127.48	nr	135.77
Extra for saddle; oblique	0.36	24.86	–	138.54	nr	163.40

CLASS I: PIPEWORK - PIPES

Item	Gang hours	Labour £	Plant £	Material £	Unit	Total rate £
300 mm pipes; in trenches, depth						
not in trenches	0.10	4.50	0.60	47.62	m	52.72
ne 1.50 m deep	0.19	13.12	8.89	47.62	m	69.63
1.50–2.00 m deep	0.24	16.57	11.26	47.62	m	75.45
2.00–2.50 m deep	0.28	19.34	13.14	47.62	m	80.10
2.50–3.00 m deep	0.32	22.10	15.04	47.62	m	84.76
3.00–3.50 m deep	0.42	29.01	19.78	47.62	m	96.41
3.50–4.00 m deep	0.52	35.91	24.51	47.62	m	108.04
4.00–4.50 m deep	0.70	48.34	33.03	47.62	m	128.99
4.50–5.00 m deep	0.90	62.15	42.50	47.62	m	152.27
5.00–5.50 m deep	1.12	77.35	52.93	47.62	m	177.90
Extra for bend	0.15	10.36	–	158.79	nr	169.15
Extra for rest bend	0.16	11.05	–	347.34	nr	358.39
Extra for single junction; equal	0.19	13.12	–	266.58	nr	279.70
Extra for saddle; oblique	0.44	30.39	–	239.49	nr	269.88
Vitrified clay pipes to BS EN295, spigot and socket joints with sealing ring; excavation and supports, backfilling						
100 mm pipes; in trenches, depth						
not in trenches	0.06	2.70	0.40	4.32	m	7.42
ne 1.50 m deep	0.15	10.36	7.15	4.32	m	21.83
1.50–2.00 m deep	0.19	13.12	8.91	4.32	m	26.35
2.00–2.50 m deep	0.24	16.57	11.27	4.32	m	32.16
2.50–3.00 m deep	0.29	20.03	13.64	4.32	m	37.99
3.00–3.50 m deep	0.36	24.86	16.94	4.32	m	46.12
3.50–4.00 m deep	0.44	30.39	20.74	4.32	m	55.45
4.00–4.50 m deep	0.55	37.98	25.93	4.32	m	68.23
4.50–5.00 m deep	0.70	48.34	33.06	4.32	m	85.72
5.00–5.50 m deep	0.90	62.15	42.55	4.32	m	109.02
Extra for bend	0.05	3.45	–	13.71	nr	17.16
Extra for rest bend	0.06	4.14	–	25.63	nr	29.77
Extra for single junction; equal	0.09	6.22	–	18.85	nr	25.07
Extra fro saddle; oblique	0.23	15.88	–	18.86	nr	34.74
150 mm pipes; in trenches, depth						
not in trenches	0.06	2.70	0.37	7.54	m	10.61
ne 1.50 m deep	0.17	11.74	7.97	7.54	m	27.25
1.50–2.00 m deep	0.20	13.81	9.38	7.54	m	30.73
2.00–2.50 m deep	0.24	16.57	11.27	7.54	m	35.38
2.50–3.00 m deep	0.29	20.03	13.64	7.54	m	41.21
3.00–3.50 m deep	0.39	26.93	18.34	7.54	m	52.81
3.50–4.00 m deep	0.45	31.08	21.21	7.54	m	59.83
4.00–4.50 m deep	0.58	40.05	27.37	7.54	m	74.96
4.50–5.00 m deep	0.75	51.80	35.40	7.54	m	94.74
5.00–5.50 m deep	0.96	66.30	45.37	7.54	m	119.21
Extra for bend	0.08	5.52	–	70.18	nr	75.70
Extra for rest bend	0.09	6.22	–	64.06	nr	70.28
Extra for single junction; equal	0.11	7.60	–	30.97	nr	38.57
Extra for double junction; equal	0.13	8.98	–	68.82	nr	77.80
Extra for taper reducer	0.07	4.83	–	23.82	nr	28.65
Extra for saddle; oblique	0.29	20.03	–	33.33	nr	53.36

CLASS I: PIPEWORK - PIPES

Item	Gang hours	Labour £	Plant £	Material £	Unit	Total rate £
CLAY PIPES – cont						
Vitrified clay pipes to BS EN295, spigot and socket joints with sealing ring – cont						
225 mm pipes; in trenches, depth						
not in trenches	0.08	3.60	0.47	20.41	m	24.48
ne 1.50 m deep	0.18	12.43	8.44	20.41	m	41.28
1.50–2.00 m deep	0.22	15.19	10.34	20.41	m	45.94
2.00–2.50 m deep	0.26	17.96	12.22	20.41	m	50.59
2.50–3.00 m deep	0.30	20.72	14.12	20.41	m	55.25
3.00–3.50 m deep	0.41	28.31	19.30	20.41	m	68.02
3.50–4.00 m deep	0.47	32.46	22.11	20.41	m	74.98
4.00–4.50 m deep	0.65	44.89	30.66	20.41	m	95.96
4.50–5.00 m deep	0.80	55.25	37.77	20.41	m	113.43
5.00–5.50 m deep	1.02	70.44	48.22	20.41	m	139.07
Extra for bend	0.10	6.91	–	165.52	nr	172.43
Extra for rest bend	0.11	7.60	–	152.58	nr	160.18
Extra for single junction; equal	0.16	11.05	–	88.60	nr	99.65
Extra for double junction; equal	0.18	12.43	–	160.74	nr	173.17
Extra for taper reducer	0.12	8.29	–	74.27	nr	82.56
Extra for saddle; oblique	0.36	24.86	–	109.58	nr	134.44
300 mm pipes; in trenches, depth						
not in trenches	0.10	4.50	0.60	25.85	m	30.95
ne 1.50 m deep	0.19	13.12	8.89	25.85	m	47.86
1.50–2.00 m deep	0.24	16.57	11.26	25.85	m	53.68
2.00–2.50 m deep	0.28	19.34	13.14	25.85	m	58.33
2.50–3.00 m deep	0.32	22.10	15.04	25.85	m	62.99
3.00–3.50 m deep	0.42	29.01	19.78	25.85	m	74.64
3.50–4.00 m deep	0.52	35.91	24.51	25.85	m	86.27
4.00–4.50 m deep	0.70	48.34	33.03	25.85	m	107.22
4.50–5.00 m deep	0.90	62.15	42.50	25.85	m	130.50
5.00–5.50 m deep	1.12	77.35	52.93	25.85	m	156.13
Extra for bend	0.15	10.36	–	341.21	nr	351.57
Extra for rest bend	0.16	11.05	–	323.72	nr	334.77
Extra for single junction; equal	0.19	13.12	–	210.83	nr	223.95
Extra for double junction; equal	0.21	14.50	–	331.36	nr	345.86
Extra for taper reducer	0.15	10.36	–	205.01	nr	215.37
Extra for saddle; oblique	0.44	30.39	–	190.70	nr	221.09
400 mm pipes; in trenches, depth						
not in trenches	0.10	4.50	0.60	87.03	m	92.13
ne 1.50 m deep	0.23	15.88	10.76	87.03	m	113.67
1.50–2.00 m deep	0.29	20.03	13.61	87.03	m	120.67
2.00–2.50 m deep	0.32	22.10	15.02	87.03	m	124.15
2.50–3.00 m deep	0.38	26.24	17.88	87.03	m	131.15
3.00–3.50 m deep	0.46	31.77	21.66	87.03	m	140.46
3.50–4.00 m deep	0.58	40.05	27.35	87.03	m	154.43
4.00–4.50 m deep	0.75	51.80	35.36	87.03	m	174.19
4.50–5.00 m deep	0.95	65.61	44.84	87.03	m	197.48
5.00–5.50 m deep	1.20	82.87	56.71	87.03	m	226.61
Extra for bend; 90°	0.24	16.57	–	588.18	nr	604.75
Extra for bend; 45°	0.24	16.57	–	588.18	nr	604.75
Extra for bend; 22.5°	0.24	16.57	–	588.18	nr	604.75

CLASS I: PIPEWORK - PIPES

Item	Gang hours	Labour £	Plant £	Material £	Unit	Total rate £
450 mm pipes; in trenches, depth						
not in trenches	0.23	10.34	1.36	121.72	m	**133.42**
ne 1.50 m deep	0.23	15.88	10.76	121.72	m	**148.36**
1.50–2.00 m deep	0.30	20.72	14.09	121.72	m	**156.53**
2.00–2.50 m deep	0.32	22.10	15.02	121.72	m	**158.84**
2.50–3.00 m deep	0.38	26.24	17.88	121.72	m	**165.84**
3.00–3.50 m deep	0.47	32.46	22.11	121.72	m	**176.29**
3.50–4.00 m deep	0.60	41.44	28.28	121.72	m	**191.44**
4.00–4.50 m deep	0.77	53.18	36.32	121.72	m	**211.22**
4.50–5.00 m deep	0.97	66.99	45.80	121.72	m	**234.51**
5.00–5.50 m deep	1.20	82.87	56.71	121.72	m	**261.30**
Extra for bend; 90°	0.29	20.03	–	774.52	nr	**794.55**
Extra for bend; 45°	0.29	20.03	–	774.52	nr	**794.55**
Extra for bend; 22.5°	0.29	20.03	–	774.52	nr	**794.55**
CONCRETE PIPES						
Concrete porous pipes to BS 5911; excavation and supports, backfilling						
150 mm pipes; in trench, depth						
not in trenches	0.06	2.70	0.29	5.85	m	**8.84**
ne 1.50 m deep	0.17	11.74	4.43	5.85	m	**22.02**
1.50–2.00 m deep	0.20	13.81	5.23	5.85	m	**24.89**
2.00–2.50 m deep	0.24	16.57	6.27	5.85	m	**28.69**
2.50–3.00 m deep	0.29	20.03	7.61	5.85	m	**33.49**
3.00–3.50 m deep	0.39	26.93	10.24	5.85	m	**43.02**
3.50–4.00 m deep	0.45	31.08	11.86	5.85	m	**48.79**
4.00–4.50 m deep	0.58	40.05	15.31	5.85	m	**61.21**
4.50–5.00 m deep	0.75	51.80	19.82	5.85	m	**77.47**
5.00–5.50 m deep	0.96	66.30	25.42	5.85	m	**97.57**
225 mm pipes; in trench, depth						
not in trenches	0.08	3.60	0.18	6.30	m	**10.08**
ne 1.50 m deep	0.18	12.43	4.69	6.30	m	**23.42**
1.50–2.00 m deep	0.22	15.19	5.76	6.30	m	**27.25**
2.00–2.50 m deep	0.26	17.96	6.81	6.30	m	**31.07**
2.50–3.00 m deep	0.30	20.72	7.87	6.30	m	**34.89**
3.00–3.50 m deep	0.41	28.31	10.77	6.30	m	**45.38**
3.50–4.00 m deep	0.47	32.46	12.38	6.30	m	**51.14**
4.00–4.50 m deep	0.65	44.89	17.15	6.30	m	**68.34**
4.50–5.00 m deep	0.80	55.25	21.14	6.30	m	**82.69**
5.00–5.50 m deep	1.02	70.44	27.01	6.30	m	**103.75**
300 mm pipes; in trench, depth						
not in trenches	0.10	4.50	0.23	6.94	m	**11.67**
ne 1.50 m deep	0.19	13.12	4.94	6.94	m	**25.00**
1.50–2.00 m deep	0.24	16.57	6.27	6.94	m	**29.78**
2.00–2.50 m deep	0.28	19.34	7.33	6.94	m	**33.61**
2.50–3.00 m deep	0.32	22.10	8.39	6.94	m	**37.43**
3.00–3.50 m deep	0.42	29.01	11.04	6.94	m	**46.99**
3.50–4.00 m deep	0.52	35.91	13.70	6.94	m	**56.55**
4.00–4.50 m deep	0.70	48.34	18.47	6.94	m	**73.75**
4.50–5.00 m deep	0.90	62.15	23.79	6.94	m	**92.88**
5.00–5.50 m deep	1.12	77.35	29.65	6.94	m	**113.94**

CLASS I: PIPEWORK - PIPES

Item	Gang hours	Labour £	Plant £	Material £	Unit	Total rate £
CONCRETE PIPES – cont						
Concrete pipes with rebated flexible joints to BS 5911 Class 120; excavation and supports, backfilling						
300 mm pipes; in trenches, depth						
not in trenches	0.12	5.39	0.37	17.36	m	23.12
ne 1.50 m deep	0.22	15.19	10.31	17.36	m	42.86
1.50–2.00 m deep	0.26	17.96	12.21	17.36	m	47.53
2.00–2.50 m deep	0.30	20.72	14.10	17.36	m	52.18
2.50–3.00 m deep	0.34	23.48	16.00	17.36	m	56.84
3.00–3.50 m deep	0.44	30.39	20.70	17.36	m	68.45
3.50–4.00 m deep	0.56	38.67	26.39	17.36	m	82.42
4.00–4.50 m deep	0.73	50.41	34.44	17.36	m	102.21
4.50–5.00 m deep	0.92	63.54	43.43	17.36	m	124.33
5.00–5.50 m deep	1.19	82.18	56.23	17.36	m	155.77
5.50–6.00 m deep	1.42	98.07	67.19	17.36	m	182.62
Extra for bend	0.08	5.52	3.64	105.00	nr	114.16
375 mm pipes; in trenches, depth						
not in trenches	0.15	6.74	0.89	12.85	m	20.48
ne 1.50 m deep	0.24	16.57	11.24	12.85	m	40.66
1.50–2.00 m deep	0.29	20.03	13.61	12.85	m	46.49
2.00–2.50 m deep	0.33	22.79	15.50	12.85	m	51.14
2.50–3.00 m deep	0.38	26.24	17.88	12.85	m	56.97
3.00–3.50 m deep	0.46	31.77	21.66	12.85	m	66.28
3.50–4.00 m deep	0.58	40.05	27.35	12.85	m	80.25
4.00–4.50 m deep	0.75	51.80	35.36	12.85	m	100.01
4.50–5.00 m deep	0.95	65.61	44.84	12.85	m	123.30
5.00–5.50 m deep	1.20	82.87	56.71	12.85	m	152.43
5.50–6.00 m deep	1.45	100.14	68.60	12.85	m	181.59
Extra for bend	0.10	6.91	5.08	112.30	nr	124.29
450 mm pipes; in trenches, depth						
not in trenches	0.17	7.64	1.01	19.00	m	27.65
ne 1.50 m deep	0.25	17.27	11.71	19.00	m	47.98
1.50–2.00 m deep	0.31	21.41	14.54	19.00	m	54.95
2.00–2.50 m deep	0.35	24.17	16.43	19.00	m	59.60
2.50–3.00 m deep	0.40	27.62	18.80	19.00	m	65.42
3.00–3.50 m deep	0.48	33.15	22.64	19.00	m	74.79
3.50–4.00 m deep	0.63	43.51	29.68	19.00	m	92.19
4.00–4.50 m deep	0.80	55.25	37.73	19.00	m	111.98
4.50–5.00 m deep	1.00	69.06	47.21	19.00	m	135.27
5.00–5.50 m deep	1.25	86.33	59.08	19.00	m	164.41
5.50–6.00 m deep	1.51	104.28	71.43	19.00	m	194.71
Extra for bend	0.13	8.98	7.28	165.00	nr	181.26
525 mm pipes; in trenches, depth						
not in trenches	0.20	8.99	1.19	39.63	m	49.81
ne 1.50 m deep	0.27	18.65	12.63	39.63	m	70.91
1.50–2.00 m deep	0.33	22.79	15.49	39.63	m	77.91
2.00–2.50 m deep	0.37	25.55	17.38	39.63	m	82.56
2.50–3.00 m deep	0.42	29.01	19.76	39.63	m	88.40
3.00–3.50 m deep	0.49	33.84	23.06	39.63	m	96.53
3.50–4.00 m deep	0.65	44.89	30.64	39.63	m	115.16

CLASS I: PIPEWORK - PIPES

Item	Gang hours	Labour £	Plant £	Material £	Unit	Total rate £
4.00–4.50 m deep	0.83	57.32	39.14	39.63	m	**136.09**
4.50–5.00 m deep	1.03	71.13	48.62	39.63	m	**159.38**
5.00–5.50 m deep	1.28	88.40	60.49	39.63	m	**188.52**
5.50–6.00 m deep	1.55	107.04	73.32	39.63	m	**219.99**
Extra for bend	0.16	11.05	9.46	270.00	nr	**290.51**
750 mm pipes; in trenches, depth						
not in trenches	0.22	9.89	1.32	30.85	m	**42.06**
ne 1.50 m deep	0.30	19.75	18.34	30.85	m	**68.94**
1.50–2.00 m deep	0.36	23.71	22.03	30.85	m	**76.59**
2.00–2.50 m deep	0.41	27.00	25.11	30.85	m	**82.96**
2.50–3.00 m deep	0.47	30.95	28.79	30.85	m	**90.59**
3.00–3.50 m deep	0.58	38.19	35.58	30.85	m	**104.62**
3.50–4.00 m deep	0.80	52.68	49.12	30.85	m	**132.65**
4.00–4.50 m deep	1.05	69.14	64.51	30.85	m	**164.50**
4.50–5.00 m deep	1.30	85.61	79.94	30.85	m	**196.40**
5.00–5.50 m deep	1.55	102.07	95.36	30.85	m	**228.28**
5.50–6.00 m deep	1.82	119.85	112.09	30.85	m	**262.79**
Extra for bends	0.24	15.80	21.66	355.00	nr	**392.46**
900 mm pipes; in trenches, depth						
not in trenches	0.25	11.24	1.49	80.00	m	**92.73**
ne 1.50 m deep	0.33	21.73	20.17	80.00	m	**121.90**
1.50–2.00 m deep	0.40	26.34	24.48	80.00	m	**130.82**
2.00–2.50 m deep	0.46	30.29	28.17	80.00	m	**138.46**
2.50–3.00 m deep	0.53	34.90	32.48	80.00	m	**147.38**
3.00–3.50 m deep	0.70	46.09	42.94	80.00	m	**169.03**
3.50–4.00 m deep	0.92	60.58	56.49	80.00	m	**197.07**
4.00–4.50 m deep	1.20	79.02	73.72	80.00	m	**232.74**
4.50–5.00 m deep	1.50	98.78	92.24	80.00	m	**271.02**
5.00–5.50 m deep	1.80	118.53	110.75	80.00	m	**309.28**
5.50–6.00 m deep	2.10	138.28	129.33	80.00	m	**347.61**
Extra for bends	0.39	25.68	35.55	692.00	nr	**753.23**
1200 mm pipes; in trenches, depth						
not in trenches	0.25	11.24	1.49	135.00	m	**147.73**
ne 1.50 m deep	0.46	30.29	28.11	135.00	m	**193.40**
1.50–2.00 m deep	0.53	34.90	32.44	135.00	m	**202.34**
2.00–2.50 m deep	0.60	39.51	36.73	135.00	m	**211.24**
2.50–3.00 m deep	0.70	46.09	42.91	135.00	m	**224.00**
3.00–3.50 m deep	0.85	55.97	52.13	135.00	m	**243.10**
3.50–4.00 m deep	1.12	73.75	68.77	135.00	m	**277.52**
4.00–4.50 m deep	1.45	95.48	89.09	135.00	m	**319.57**
4.50–5.00 m deep	1.75	115.24	107.59	135.00	m	**357.83**
5.00–5.50 m deep	2.05	134.99	126.14	135.00	m	**396.13**
5.50–6.00 m deep	2.36	155.41	145.32	135.00	m	**435.73**
Extra for bends	0.51	33.58	46.70	1190.00	nr	**1270.28**

CLASS I: PIPEWORK - PIPES

Item	Gang hours	Labour £	Plant £	Material £	Unit	Total rate £
CONCRETE PIPES – cont						
Concrete pipes with rebated flexible joints to BS 5911 Class 120 – cont						
1500 mm pipes; in trenches, depth						
not in trenches	0.35	15.73	2.07	225.00	m	242.80
ne 1.50 m deep	0.60	39.51	36.65	225.00	m	301.16
1.50–2.00 m deep	0.70	46.09	42.85	225.00	m	313.94
2.00–2.50 m deep	0.81	53.34	49.59	225.00	m	327.93
2.50–3.00 m deep	0.92	60.58	56.38	225.00	m	341.96
3.00–3.50 m deep	1.05	69.14	64.40	225.00	m	358.54
3.50–4.00 m deep	1.27	83.63	77.97	225.00	m	386.60
4.00–4.50 m deep	1.70	111.94	104.45	225.00	m	441.39
4.50–5.00 m deep	2.05	134.99	126.05	225.00	m	486.04
5.00–5.50 m deep	2.40	158.04	147.67	225.00	m	530.71
5.50–6.00 m deep	2.75	181.09	167.82	226.84	m	575.75
Extra for bends	0.63	41.49	57.68	1329.38	nr	1428.55
1800 mm pipes; in trenches, depth						
not in trenches	0.40	17.98	2.37	405.00	m	425.35
ne 1.50 m deep	0.77	50.70	47.04	405.00	m	502.74
1.50–2.00 m deep	0.91	59.92	55.68	405.00	m	520.60
2.00–2.50 m deep	1.03	67.83	63.05	405.00	m	535.88
2.50–3.00 m deep	1.12	73.75	68.63	405.00	m	547.38
3.00–3.50 m deep	1.20	79.02	73.59	405.00	m	557.61
3.50–4.00 m deep	1.52	100.09	93.33	405.00	m	598.42
4.00–4.50 m deep	2.00	131.70	122.87	405.00	m	659.57
4.50–5.00 m deep	2.40	158.04	147.56	405.00	m	710.60
5.00–5.50 m deep	2.80	184.38	172.28	405.00	m	761.66
5.50–6.00 m deep	3.15	207.43	193.96	405.00	m	806.39
Extra for bends	0.77	50.70	70.52	1848.14	nr	1969.36
2100 mm pipes; in trenches, depth						
not in trenches	0.45	20.23	2.68	525.00	m	547.91
ne 1.50 m deep	0.98	64.53	59.88	525.00	m	649.41
1.50–2.00 m deep	1.13	74.41	69.16	525.00	m	668.57
2.00–2.50 m deep	1.23	81.00	75.14	525.00	m	681.14
2.50–3.00 m deep	1.30	85.61	79.68	525.00	m	690.29
3.00–3.50 m deep	1.50	98.78	92.00	525.00	m	715.78
3.50–4.00 m deep	1.82	119.85	111.76	525.00	m	756.61
4.00–4.50 m deep	2.35	154.75	144.36	525.00	m	824.11
4.50–5.00 m deep	2.80	184.38	172.15	525.00	m	881.53
5.00–5.50 m deep	3.20	210.72	196.90	525.00	m	932.62
5.50–6.00 m deep	3.55	233.77	218.59	525.00	m	977.36
Extra for bends	0.89	58.61	81.62	2377.44	nr	2517.67

CLASS I: PIPEWORK - PIPES

Item	Gang hours	Labour £	Plant £	Material £	Unit	Total rate £
IRON PIPES						
Cast iron pipes to BS 437 plain ended pipe with 'Timesaver' mechanical coupling joints; excavation and supports, backfilling						
75 mm pipes; in trenches, depth						
not in trenches	0.12	5.39	0.71	21.49	m	27.59
ne 1.50 m deep	0.19	13.12	8.70	21.49	m	43.31
1.50–2.00 m deep	0.21	14.50	9.86	21.49	m	45.85
2.00–2.50 m deep	0.28	19.34	13.15	21.49	m	53.98
2.50–3.00 m deep	0.34	23.48	16.00	21.49	m	60.97
3.00–3.50 m deep	0.43	29.70	20.23	21.49	m	71.42
3.50–4.00 m deep	0.55	37.98	25.91	21.49	m	85.38
4.00–4.50 m deep	0.71	49.03	33.48	21.49	m	104.00
4.50–5.00 m deep	0.89	61.46	42.02	21.49	m	124.97
5.00–5.50 m deep	1.15	79.42	54.34	21.49	m	155.25
Extra for bend; 87.5°	0.31	21.41	1.47	30.31	nr	53.19
Extra for bend; 45°	0.31	21.41	1.47	30.31	nr	53.19
Extra for single junction; equal	0.48	33.15	2.91	45.54	nr	81.60
Extra for taper reducer	0.27	18.65	1.47	34.80	nr	54.92
100 mm pipes; in trenches, depth						
not in trenches	0.13	5.84	0.30	32.42	m	38.56
ne 1.50 m deep	0.21	14.50	9.84	32.42	m	56.76
1.50–2.00 m deep	0.23	15.88	10.79	32.42	m	59.09
2.00–2.50 m deep	0.30	20.72	14.10	32.42	m	67.24
2.50–3.00 m deep	0.37	25.55	17.40	32.42	m	75.37
3.00–3.50 m deep	0.48	33.15	22.59	32.42	m	88.16
3.50–4.00 m deep	0.60	41.44	28.28	32.42	m	102.14
4.00–4.50 m deep	0.75	51.80	35.36	32.42	m	119.58
4.50–5.00 m deep	0.95	65.61	44.84	32.42	m	142.87
5.00–5.50 m deep	1.22	84.25	57.67	32.42	m	174.34
Extra for bend; 87.5°	0.38	26.24	2.23	37.78	nr	66.25
Extra for bend; 45°	0.38	26.24	2.23	37.78	nr	66.25
Extra for bend; long radius	0.38	26.24	2.23	99.38	nr	127.85
Extra for single junction; equal	0.59	40.75	3.64	118.53	nr	162.92
Extra for double junction; equal	0.80	55.25	5.82	82.59	nr	143.66
Extra for taper reducer	0.40	27.62	3.64	41.38	nr	72.64
150 mm pipes; in trenches, depth						
not in trenches	0.14	6.29	0.30	64.64	m	71.23
ne 1.50 m deep	0.24	16.57	11.24	64.64	m	92.45
1.50–2.00 m deep	0.28	19.34	13.14	64.64	m	97.12
2.00–2.50 m deep	0.34	23.48	15.98	64.64	m	104.10
2.50–3.00 m deep	0.40	27.62	18.80	64.64	m	111.06
3.00–3.50 m deep	0.54	37.29	25.42	64.64	m	127.35
3.50–4.00 m deep	0.61	42.13	28.75	64.64	m	135.52
4.00–4.50 m deep	0.79	54.56	37.25	64.64	m	156.45
4.50–5.00 m deep	1.05	72.51	49.58	64.64	m	186.73
5.00–5.50 m deep	1.32	91.16	62.38	64.64	m	218.18
Extra for bend; 87.5°	0.56	38.67	2.91	72.21	nr	113.79
Ectra for bend; 45°	0.56	38.67	2.91	72.21	nr	113.79
Extra for bend; long radius	0.56	38.67	2.91	151.47	nr	193.05
Extra for single junction; equal	0.88	60.77	5.82	138.95	nr	205.54
Extra for taper reducer	0.56	38.67	5.82	130.47	nr	174.96

CLASS I: PIPEWORK - PIPES

Item	Gang hours	Labour £	Plant £	Material £	Unit	Total rate £
IRON PIPES – cont						
225 mm pipes; in trenches, depth						
not in trenches	0.15	6.74	0.30	165.65	m	**172.69**
ne 1.50 m deep	0.25	17.27	11.71	165.65	m	**194.63**
1.50–2.00 m deep	0.30	20.72	14.09	165.65	m	**200.46**
2.00–2.50 m deep	0.35	24.17	16.42	165.65	m	**206.24**
2.50–3.00 m deep	0.41	28.31	19.28	165.65	m	**213.24**
3.00–3.50 m deep	0.54	37.29	25.42	165.65	m	**228.36**
3.50–4.00 m deep	0.61	42.13	28.75	165.65	m	**236.53**
4.00–4.50 m deep	0.77	53.18	36.32	165.65	m	**255.15**
4.50–5.00 m deep	1.02	70.44	48.17	165.65	m	**284.26**
5.00–5.50 m deep	1.30	89.78	61.45	165.65	m	**316.88**
Extra for bend; 87.5°	0.77	53.18	2.91	310.16	nr	**366.25**
Extra for bend; 45°	0.77	53.18	2.91	310.16	nr	**366.25**
Extra for single junction; equal	1.21	83.56	6.55	659.63	nr	**749.74**
Extra for taper reducer	0.77	53.18	6.55	259.13	nr	**318.86**
Ductile iron pipes to BS 4772, Tyton joints; excavation and supports, backfilling						
100 mm pipes; in trenches, depth						
not in trenches	0.10	4.50	0.30	55.38	m	**60.18**
ne 1.50 m deep	0.19	13.12	8.44	55.38	m	**76.94**
1.50–2.00 m deep	0.20	13.81	9.38	55.38	m	**78.57**
2.00–2.50 m deep	0.26	17.96	12.22	55.38	m	**85.56**
2.50–3.00 m deep	0.32	22.10	15.04	55.38	m	**92.52**
3.00–3.50 m deep	0.42	29.01	19.78	55.38	m	**104.17**
3.50–4.00 m deep	0.53	36.60	24.98	55.38	m	**116.96**
4.00–4.50 m deep	0.66	45.58	31.14	55.38	m	**132.10**
4.50–5.00 m deep	0.84	58.01	39.66	55.38	m	**153.05**
5.00–5.50 m deep	1.08	74.58	51.04	55.38	m	**181.00**
Extra for bend; 90°	0.38	26.24	12.99	103.94	nr	**143.17**
Extra for single junction; equal	0.60	41.44	4.38	148.30	nr	**194.12**
150 mm pipes; in trenches, depth						
not in trenches	0.11	4.94	0.30	66.28	m	**71.52**
ne 1.50 m deep	0.21	14.50	9.84	66.28	m	**90.62**
1.50–2.00 m deep	0.25	17.27	11.74	66.28	m	**95.29**
2.00–2.50 m deep	0.30	20.72	14.10	66.28	m	**101.10**
2.50–3.00 m deep	0.37	25.55	17.40	66.28	m	**109.23**
3.00–3.50 m deep	0.49	33.84	23.06	66.28	m	**123.18**
3.50–4.00 m deep	0.55	37.98	25.91	66.28	m	**130.17**
4.00–4.50 m deep	0.72	49.72	33.96	66.28	m	**149.96**
4.50–5.00 m deep	0.94	64.92	44.39	66.28	m	**175.59**
5.00–5.50 m deep	1.19	82.18	56.23	66.28	m	**204.69**
Extra for bend; 90°	0.57	39.36	2.91	224.06	nr	**266.33**
Extra for single junction; equal	0.88	60.77	5.82	295.71	nr	**362.30**
250 mm pipes; in trenches, depth						
not in trenches	0.16	7.19	0.30	117.56	m	**125.05**
ne 1.50 m deep	0.24	16.57	11.24	117.56	m	**145.37**
1.50–2.00 m deep	0.30	20.72	14.09	117.56	m	**152.37**

CLASS I: PIPEWORK - PIPES

Item	Gang hours	Labour £	Plant £	Material £	Unit	Total rate £
2.00–2.50 m deep	0.35	24.17	16.43	117.56	m	**158.16**
2.50–3.00 m deep	0.42	29.01	19.76	117.56	m	**166.33**
3.00–3.50 m deep	0.56	38.67	26.35	117.56	m	**182.58**
3.50–4.00 m deep	0.64	44.20	30.16	117.56	m	**191.92**
4.00–4.50 m deep	0.81	55.94	38.21	117.56	m	**211.71**
4.50–5.00 m deep	1.02	70.44	48.17	117.56	m	**236.17**
5.00–5.50 m deep	1.27	87.71	60.01	117.56	m	**265.28**
Extra for bend; 90°	0.96	66.30	4.38	646.86	nr	**717.54**
Extra for single junction; equal	1.32	91.16	7.28	815.52	nr	**913.96**
400 mm pipes; in trenches, depth						
not in trenches	0.24	10.79	0.65	219.78	m	**231.22**
ne 1.50 m deep	0.33	22.79	15.46	219.78	m	**258.03**
1.50–2.00 m deep	0.43	29.70	20.17	219.78	m	**269.65**
2.00–2.50 m deep	0.48	33.15	22.53	219.78	m	**275.46**
2.50–3.00 m deep	0.57	39.36	26.80	219.78	m	**285.94**
3.00–3.50 m deep	0.71	49.03	33.40	219.78	m	**302.21**
3.50–4.00 m deep	0.91	62.84	42.88	219.78	m	**325.50**
4.00–4.50 m deep	1.16	80.11	54.71	219.78	m	**354.60**
4.50–5.00 m deep	1.47	101.52	69.39	219.78	m	**390.69**
5.00–5.50 m deep	1.77	122.24	83.65	219.78	m	**425.67**
600 mm pipes; in trenches, depth						
not in trenches	0.34	15.28	0.89	395.46	m	**411.63**
ne 1.50 m deep	0.47	32.46	22.00	395.46	m	**449.92**
1.50–2.00 m deep	0.55	37.98	25.80	395.46	m	**459.24**
2.00–2.50 m deep	0.66	45.58	31.00	395.46	m	**472.04**
2.50–3.00 m deep	0.78	53.87	36.68	395.46	m	**486.01**
3.00–3.50 m deep	0.89	61.46	41.89	395.46	m	**498.81**
3.50–4.00 m deep	1.09	75.28	51.37	395.46	m	**522.11**
4.00–4.50 m deep	1.37	94.61	64.62	395.46	m	**554.69**
4.50–5.00 m deep	1.70	117.40	80.27	395.46	m	**593.13**
5.00–5.50 m deep	2.03	140.19	95.93	395.46	m	**631.58**
STEEL PIPES						
Carbon steel pipes to BS EN 10126; welded joints;						
(for protection and lining refer to manufacturer);						
excavation and supports, backfilling						
100 mm pipes; in trenches, depth						
not in trenches	0.07	3.15	0.41	31.30	m	**34.86**
ne 1.50 m deep	0.15	10.36	7.02	31.30	m	**48.68**
1.50–2.00 m deep	0.17	11.74	7.98	31.30	m	**51.02**
2.00–2.50 m deep	0.22	15.19	10.34	31.30	m	**56.83**
2.50–3.00 m deep	0.27	18.65	12.68	31.30	m	**62.63**
3.00–3.50 m deep	0.35	24.17	16.46	31.30	m	**71.93**
3.50–4.00 m deep	0.44	30.39	20.74	31.30	m	**82.43**
4.00–4.50 m deep	0.55	37.98	25.93	31.30	m	**95.21**
4.50–5.00 m deep	0.70	48.34	48.14	31.30	m	**127.78**
5.00–5.50 m deep	0.87	60.08	41.11	31.30	m	**132.49**
Extra for bend; 45°	0.07	4.83	2.17	12.81	nr	**19.81**
Extra for bend; 90°	0.07	4.83	2.17	16.18	nr	**23.18**
Extra for single junction; equal	0.11	7.60	3.64	38.03	nr	**49.27**

CLASS I: PIPEWORK - PIPES

Item	Gang hours	Labour £	Plant £	Material £	Unit	Total rate £
STEEL PIPES – cont						
Carbon steel pipes to BS EN 10126 – cont						
150 mm pipes; in trenches, depth						
not in trenches	0.07	3.15	0.41	48.22	m	**51.78**
ne 1.50 m deep	0.17	11.74	7.97	48.22	m	**67.93**
1.50–2.00 m deep	0.20	13.81	9.38	48.22	m	**71.41**
2.00–2.50 m deep	0.24	16.57	11.27	48.22	m	**76.06**
2.50–3.00 m deep	0.29	20.03	13.64	48.22	m	**81.89**
3.00–3.50 m deep	0.39	26.93	18.34	48.22	m	**93.49**
3.50–4.00 m deep	0.44	30.39	20.74	48.22	m	**99.35**
4.00–4.50 m deep	0.57	39.36	26.89	48.22	m	**114.47**
4.50–5.00 m deep	0.72	49.72	49.50	48.22	m	**147.44**
5.00–5.50 m deep	0.90	62.15	42.55	48.22	m	**152.92**
Extra for bend; 45°	0.09	6.22	2.91	26.90	nr	**36.03**
Extra for bend; 90°	0.09	6.22	2.91	36.23	nr	**45.36**
Extra for single junction; equal	0.16	11.05	5.82	67.28	nr	**84.15**
200 mm pipes; in trenches, depth						
not in trenches	0.09	4.05	8.90	50.98	m	**63.93**
ne 1.50 m deep	0.18	12.43	8.44	50.98	m	**71.85**
1.50–2.00 m deep	0.21	14.50	9.86	50.98	m	**75.34**
2.00–2.50 m deep	0.25	17.27	11.75	50.98	m	**80.00**
2.50–3.00 m deep	0.30	20.72	14.12	50.98	m	**85.82**
3.00–3.50 m deep	0.40	27.62	18.82	50.98	m	**97.42**
3.50–4.00 m deep	0.46	31.77	21.69	50.98	m	**104.44**
4.00–4.50 m deep	0.59	40.75	27.82	50.98	m	**119.55**
4.50–5.00 m deep	0.74	51.10	34.95	50.98	m	**137.03**
5.00–5.50 m deep	0.92	63.54	43.48	50.98	m	**158.00**
Extra for bend; 45°	0.12	8.29	2.91	56.86	nr	**68.06**
Extra for bend; 90°	0.12	8.29	2.91	74.20	nr	**85.40**
Extra for single junction; equal	0.21	14.50	5.76	129.24	nr	**149.50**
250 mm pipes; in trenches, depth						
not in trenches	0.10	4.50	0.60	64.11	m	**69.21**
ne 1.50 m deep	0.18	12.43	13.10	64.11	m	**89.64**
1.50–2.00 m deep	0.22	15.19	15.49	64.11	m	**94.79**
2.00–2.50 m deep	0.26	17.96	17.28	64.11	m	**99.35**
2.50–3.00 m deep	0.31	21.41	21.80	64.11	m	**107.32**
3.00–3.50 m deep	0.41	28.31	28.84	64.11	m	**121.26**
3.50–4.00 m deep	0.47	32.46	32.10	64.11	m	**128.67**
4.00–4.50 m deep	0.60	41.44	39.54	64.11	m	**145.09**
4.50–5.00 m deep	0.75	51.80	51.56	64.11	m	**167.47**
5.00–5.50 m deep	0.94	64.92	64.65	64.11	m	**193.68**
Extra for bend; 45°	0.13	8.98	2.91	100.03	nr	**111.92**
Extra for bend; 90°	0.13	8.98	2.91	131.56	nr	**143.45**
Extra for single junction; equal	0.23	15.88	5.82	240.97	nr	**262.67**
300 mm pipes; in trenches, depth						
not in trenches	0.11	4.94	0.65	73.50	m	**79.09**
ne 1.50 m deep	0.20	13.81	9.36	73.50	m	**96.67**
1.50–2.00 m deep	0.25	17.27	11.74	73.50	m	**102.51**

CLASS I: PIPEWORK - PIPES

Item	Gang hours	Labour £	Plant £	Material £	Unit	Total rate £
2.00–2.50m deep	0.28	19.34	13.14	73.50	m	**105.98**
2.50–3.00m deep	0.34	23.48	16.00	73.50	m	**112.98**
3.00–3.50m deep	0.44	30.39	20.70	73.50	m	**124.59**
3.50–4.00m deep	0.53	36.60	24.98	73.50	m	**135.08**
4.00–4.50m deep	0.68	46.96	32.07	73.50	m	**152.53**
4.50–5.00m deep	0.86	59.39	40.61	73.50	m	**173.50**
5.00–5.50m deep	1.06	73.20	50.11	73.50	m	**196.81**
Extra for bend; 45°	0.14	9.67	4.38	133.72	nr	**147.77**
Extra for bend; 90°	0.14	9.67	4.38	180.03	nr	**194.08**
Extra for single junction; equal	0.25	17.27	9.46	372.08	nr	**398.81**
POLYVINYL CHLORIDE PIPES						
Unplasticised pvc perforated pipes; ring seal sockets; excavation and supports, backfilling; 6m pipe lengths unless stated otherwise						
82mm pipes; in trench, depth						
not in trenches; 3.00m pipe lengths	0.06	2.70	0.14	8.62	m	**11.46**
ne 1.50m deep; 3.00m pipe lengths	0.10	6.91	2.61	8.62	m	**18.14**
1.50–2.00m deep	0.13	8.98	3.40	8.62	m	**21.00**
2.00–2.50m deep	0.16	11.05	4.19	8.62	m	**23.86**
2.50–3.00m deep	0.19	13.12	4.98	8.62	m	**26.72**
3.00–3.50m deep	0.22	15.19	5.79	8.62	m	**29.60**
3.50–4.00m deep	0.25	17.27	6.59	8.62	m	**32.48**
4.00–4.50m deep	0.28	19.34	7.39	8.62	m	**35.35**
4.50–5.00m deep	0.32	22.10	8.46	8.62	m	**39.18**
5.00–5.50m deep	0.35	24.17	9.26	8.62	m	**42.05**
110mm pipes; in trench, depth						
not in trenches	0.06	2.70	0.07	12.06	m	**14.83**
ne 1.50m deep	0.10	6.91	2.61	12.06	m	**21.58**
1.50–2.00m deep	0.13	8.98	3.40	12.06	m	**24.44**
2.00–2.50m deep	0.16	11.05	4.19	12.06	m	**27.30**
2.50–3.00m deep	0.19	13.12	4.98	12.06	m	**30.16**
3.00–3.50m deep	0.22	15.19	5.79	12.06	m	**33.04**
3.50–4.00m deep	0.25	17.27	6.59	12.06	m	**35.92**
4.00–4.50m deep	0.28	19.34	7.39	12.06	m	**38.79**
4.50–5.00m deep	0.32	22.10	8.46	12.06	m	**42.62**
5.00–5.50m deep	0.36	24.86	9.53	12.06	m	**46.45**
160mm pipes; in trench, depth						
not in trenches	0.06	2.70	0.14	30.82	m	**33.66**
ne 1.50m deep	0.11	7.60	2.86	30.82	m	**41.28**
1.50–2.00m deep	0.14	9.67	3.67	30.82	m	**44.16**
2.00–2.50m deep	0.18	12.43	4.72	30.82	m	**47.97**
2.50–3.00m deep	0.22	15.19	5.77	30.82	m	**51.78**
3.00–3.50m deep	0.23	15.88	6.04	30.82	m	**52.74**
3.50–4.00m deep	0.26	17.96	6.86	30.82	m	**55.64**
4.00–4.50m deep	0.30	20.72	7.92	30.82	m	**59.46**
4.50–5.00m deep	0.35	24.17	9.25	30.82	m	**64.24**
5.00–5.50m deep	0.40	27.62	10.59	30.82	m	**69.03**

CLASS I: PIPEWORK - PIPES

Item	Gang hours	Labour £	Plant £	Material £	Unit	Total rate £
POLYVINYL CHLORIDE PIPES – cont						
Unplasticised pvc pipes; ring seal sockets; excavation and supports, backfilling; 6 m pipe lengths unless stated otherwise						
82 mm pipes; in trenches, depth						
not in trenches; 3.00 m pipe lengths	0.06	2.70	0.37	7.49	m	**10.56**
ne 1.50 m deep; 3.00 m pipe lengths	0.10	6.91	4.59	7.49	m	**18.99**
1.50–2.00 m deep	0.13	8.98	6.11	7.49	m	**22.58**
2.00–2.50 m deep	0.16	11.05	7.51	7.49	m	**26.05**
2.50–3.00 m deep	0.19	13.12	8.92	7.49	m	**29.53**
3.00–3.50 m deep	0.22	15.19	10.37	7.49	m	**33.05**
3.50–4.00 m deep	0.25	17.27	11.79	7.49	m	**36.55**
4.00–4.50 m deep	0.28	19.34	13.21	7.49	m	**40.04**
4.50–5.00 m deep	0.32	22.10	15.11	7.49	m	**44.70**
5.00–5.50 m deep	0.35	24.17	16.53	7.49	m	**48.19**
Extra for bend; short radius (socket/spigot)	0.05	3.45	–	10.46	nr	**13.91**
Extra for branches; equal (socket/spigot)	0.07	4.83	–	13.22	nr	**18.05**
110 mm pipes; in trenches, depth						
not in trenches	0.06	2.70	0.37	10.49	m	**13.56**
ne 1.50 m deep	0.10	6.91	4.70	10.49	m	**22.10**
1.50–2.00 m deep	0.13	8.98	6.11	10.49	m	**25.58**
2.00–2.50 m deep	0.16	11.05	7.51	10.49	m	**29.05**
2.50–3.00 m deep	0.19	13.12	8.92	10.49	m	**32.53**
3.00–3.50 m deep	0.22	15.19	10.37	10.49	m	**36.05**
3.50–4.00 m deep	0.25	17.27	11.79	10.49	m	**39.55**
4.00–4.50 m deep	0.28	19.34	13.21	10.49	m	**43.04**
4.50–5.00 m deep	0.32	22.10	15.11	10.49	m	**47.70**
5.00–5.50 m deep	0.36	24.86	17.01	10.49	m	**52.36**
Extra for bend; short redius (socket/spigot)	0.05	3.45	–	10.11	nr	**13.56**
Extra for bend; adjustable (socket/spigot)	0.05	3.45	–	10.11	nr	**13.56**
Extra for reducer	0.05	3.45	–	5.87	nr	**9.32**
Extra for branches; equal (socket/spigot)	0.07	4.83	–	20.08	nr	**24.91**
160 mm pipes; in trenches, depth						
not in trenches	0.06	2.70	0.37	26.69	m	**29.76**
ne 1.50 m deep	0.11	7.60	5.14	26.69	m	**39.43**
1.50–2.00 m deep	0.14	9.67	6.58	26.69	m	**42.94**
2.00–2.50 m deep	0.18	12.43	8.47	26.69	m	**47.59**
2.50–3.00 m deep	0.22	15.19	10.36	26.69	m	**52.24**
3.00–3.50 m deep	0.23	15.88	10.82	26.69	m	**53.39**
3.50–4.00 m deep	0.26	17.96	12.27	26.69	m	**56.92**
4.00–4.50 m deep	0.30	20.72	14.16	26.69	m	**61.57**
4.50–5.00 m deep	0.35	24.17	16.52	26.69	m	**67.38**
5.00–5.50 m deep	0.40	27.62	18.90	26.69	m	**73.21**
Extra for bend; short radius (socket/spigot)	0.05	3.45	–	42.66	nr	**46.11**
Extra for branches; equal (socket/spigot)	0.07	4.83	–	68.86	nr	**73.69**

CLASS I: PIPEWORK - PIPES

Item	Gang hours	Labour £	Plant £	Material £	Unit	Total rate £
225 mm pipes; in trenches, depth						
not in trenches	0.07	3.15	0.41	66.38	m	**69.94**
ne 1.50 m deep	0.12	8.29	5.62	66.38	m	**80.29**
1.50–2.00 m deep	0.15	10.36	7.03	66.38	m	**83.77**
2.00–2.50 m deep	0.20	13.81	9.39	66.38	m	**89.58**
2.50–3.00 m deep	0.23	15.88	10.80	66.38	m	**93.06**
3.00–3.50 m deep	0.24	16.57	11.29	66.38	m	**94.24**
3.50–4.00 m deep	0.27	18.65	12.72	66.38	m	**97.75**
4.00–4.50 m deep	0.32	22.10	15.09	66.38	m	**103.57**
4.50–5.00 m deep	0.36	24.86	17.00	66.38	m	**108.24**
5.00–5.50 m deep	0.45	31.08	21.27	66.38	m	**118.73**
Extra for bend; short radius 45° (double socket)	0.07	4.83	–	123.73	nr	**128.56**
Extra for branches; equal (all socket)	0.09	6.22	–	135.97	nr	**142.19**
300 mm pipes; in trenches, depth						
not in trenches	0.08	3.60	0.47	190.39	m	**194.46**
ne 1.50 m deep	0.13	8.98	6.09	190.39	m	**205.46**
1.50–2.00 m deep	0.16	11.05	7.51	190.39	m	**208.95**
2.00–2.50 m deep	0.21	14.50	9.87	190.39	m	**214.76**
2.50–3.00 m deep	0.23	15.88	10.80	190.39	m	**217.07**
3.00–3.50 m deep	0.25	17.27	11.77	190.39	m	**219.43**
3.50–4.00 m deep	0.28	19.34	13.20	190.39	m	**222.93**
4.00–4.50 m deep	0.34	23.48	16.05	190.39	m	**229.92**
4.50–5.00 m deep	0.38	26.24	17.95	190.39	m	**234.58**
5.00–5.50 m deep	0.47	32.46	22.20	190.39	m	**245.05**
Extra for bend; short radius 45° (double socket)	0.07	4.83	–	396.43	nr	**401.26**
Extra for branches; unequal (all socket)	0.09	6.22	–	475.97	nr	**482.19**
Unplasticised pvc pipes; polypropylene couplings; excavation and supports, backfilling; 6 m pipe lengths unless stated otherwise						
110 mm pipes; in trenches, depth						
not in trenches	0.06	2.70	0.37	8.50	m	**11.57**
ne 1.50 m deep	0.10	6.91	4.70	8.50	m	**20.11**
1.50–2.00 m deep	0.13	8.98	6.11	8.50	m	**23.59**
2.00–2.50 m deep	0.16	11.05	7.51	8.50	m	**27.06**
2.50–3.00 m deep	0.19	13.12	8.92	8.50	m	**30.54**
3.00–3.50 m deep	0.22	15.19	10.37	8.50	m	**34.06**
3.50–4.00 m deep	0.25	17.27	11.79	8.50	m	**37.56**
4.00–4.50 m deep	0.28	19.34	13.21	8.50	m	**41.05**
4.50–5.00 m deep	0.32	22.10	15.11	8.50	m	**45.71**
5.00–5.50 m deep	0.36	24.86	17.01	8.50	m	**50.37**
160 mm pipes; in trenches, depth						
not in trenches	0.06	2.70	0.37	28.84	m	**31.91**
ne 1.50 m deep	0.11	7.60	5.14	28.84	m	**41.58**
1.50–2.00 m deep	0.14	9.67	6.58	28.84	m	**45.09**
2.00–2.50 m deep	0.18	12.43	8.47	28.84	m	**49.74**
2.50–3.00 m deep	0.22	15.19	10.36	28.84	m	**54.39**
3.00–3.50 m deep	0.23	15.88	10.82	28.84	m	**55.54**
3.50–4.00 m deep	0.26	17.96	12.27	28.84	m	**59.07**
4.00–4.50 m deep	0.30	20.72	14.16	28.84	m	**63.72**
4.50–5.00 m deep	0.35	24.17	16.52	28.84	m	**69.53**
5.00–5.50 m deep	0.40	27.62	18.90	28.84	m	**75.36**

CLASS I: PIPEWORK - PIPES

Item	Gang hours	Labour £	Plant £	Material £	Unit	Total rate £
POLYVINYL CHLORIDE PIPES – cont						
Unplasticised pvc pipes – cont						
225 mm pipes; in trenches, depth						
not in trenches	0.07	3.15	0.41	57.77	m	**61.33**
ne 1.50 m deep	0.12	8.29	5.62	57.77	m	**71.68**
1.50–2.00 m deep	0.15	10.36	7.03	57.77	m	**75.16**
2.00–2.50 m deep	0.20	13.81	9.39	57.77	m	**80.97**
2.50–3.00 m deep	0.23	15.88	10.80	57.77	m	**84.45**
3.00–3.50 m deep	0.24	16.57	11.29	57.77	m	**85.63**
3.50–4.00 m deep	0.27	18.65	12.72	57.77	m	**89.14**
4.00–4.50 m deep	0.32	22.10	15.09	57.77	m	**94.96**
4.50–5.00 m deep	0.36	24.86	17.00	57.77	m	**99.63**
5.00–5.50 m deep	0.45	31.08	21.27	57.77	m	**110.12**
300 mm pipes; in trenches, depth						
not in trenches	0.08	3.60	0.47	112.03	m	**116.10**
ne 1.50 m deep	0.13	8.98	6.09	112.03	m	**127.10**
1.50–2.00 m deep	0.16	11.05	7.51	112.03	m	**130.59**
2.00–2.50 m deep	0.21	14.50	9.87	112.03	m	**136.40**
2.50–3.00 m deep	0.23	15.88	10.80	112.03	m	**138.71**
3.00–3.50 m deep	0.25	17.27	11.77	112.03	m	**141.07**
3.50–4.00 m deep	0.28	19.34	13.20	112.03	m	**144.57**
4.00–4.50 m deep	0.34	23.48	16.05	112.03	m	**151.56**
4.50–5.00 m deep	0.38	26.24	17.95	112.03	m	**156.22**
5.00–5.50 m deep	0.47	32.46	22.20	112.03	m	**166.69**
Ultrarib unplasticised pvc pipes; ring seal joints;						
excavation and supports, backfilling						
150 mm pipes; in trenches, depth						
not in trenches	0.13	5.84	0.78	14.58	m	**21.20**
ne 1.50 m deep	0.16	11.05	7.49	14.58	m	**33.12**
1.50–2.00 m deep	0.19	13.12	8.91	14.58	m	**36.61**
2.00–2.50 m deep	0.22	15.19	10.35	14.58	m	**40.12**
2.50–3.00 m deep	0.24	16.57	11.28	14.58	m	**42.43**
3.00–3.50 m deep	0.24	16.57	11.29	14.58	m	**42.44**
3.50–4.00 m deep	0.27	18.65	12.72	14.58	m	**45.95**
4.00–4.50 m deep	0.32	22.10	15.09	14.58	m	**51.77**
4.50–5.00 m deep	0.36	24.86	17.00	14.58	m	**56.44**
5.00–5.50 m deep	0.45	31.08	21.27	14.58	m	**66.93**
Extra for 45° bends; short radius (socket/spigot)	0.05	3.45	–	31.38	nr	**34.83**
Extra for branches; equal (socket/spigot)	0.09	6.22	–	75.89	nr	**82.11**

CLASS I: PIPEWORK - PIPES

Item	Gang hours	Labour £	Plant £	Material £	Unit	Total rate £
225 mm pipes; in trenches, depth						
not in trenches	0.08	3.60	0.47	40.79	m	**44.86**
ne 1.50 m deep	0.13	8.98	6.09	40.79	m	**55.86**
1.50–2.00 m deep	0.16	11.05	7.51	40.79	m	**59.35**
2.00–2.50 m deep	0.21	14.50	9.87	40.79	m	**65.16**
2.50–3.00 m deep	0.23	15.88	10.80	40.79	m	**67.47**
3.00–3.50 m deep	0.25	17.27	11.77	40.79	m	**69.83**
3.50–4.00 m deep	0.28	19.34	13.20	40.79	m	**73.33**
4.00–4.50 m deep	0.34	23.48	16.05	40.79	m	**80.32**
4.50–5.00 m deep	0.38	26.24	17.95	40.79	m	**84.98**
5.00–5.50 m deep	0.47	32.46	22.20	40.79	m	**95.45**
Extra for 45° bends; short radius (socket/spigot)	0.05	3.45	–	126.28	nr	**129.73**
Extra for branches; equal (socket/spigot)	0.09	6.22	–	251.81	nr	**258.03**
300 mm pipes; in trenches, depth						
not in trenches	0.08	3.60	0.47	67.28	m	**71.35**
ne 1.50 m deep	0.13	8.98	6.09	67.28	m	**82.35**
1.50–2.00 m deep	0.16	11.05	7.51	67.28	m	**85.84**
2.00–2.50 m deep	0.21	14.50	9.87	67.28	m	**91.65**
2.50–3.00 m deep	0.23	15.88	10.80	67.28	m	**93.96**
3.00–3.50 m deep	0.25	17.27	11.77	67.28	m	**96.32**
3.50–4.00 m deep	0.28	19.34	13.20	67.28	m	**99.82**
4.00–4.50 m deep	0.34	23.48	16.05	67.28	m	**106.81**
4.50–5.00 m deep	0.38	26.24	17.95	67.28	m	**111.47**
5.00–5.50 m deep	0.47	32.46	22.20	67.28	m	**121.94**
EXtra for 45° bends; short radius (socket/spigot)	0.07	4.83	–	229.32	nr	**234.15**
Extra for branches; equal (socket/spigot)	0.09	6.22	–	581.83	nr	**588.05**

CLASS J: PIPEWORK - FITTINGS AND VALVES

Item	Gang hours	Labour £	Plant £	Material £	Unit	Total rate £
NOTES						
Fittings on pipes shown with the appropriate pipe in Class I						
RESOURCES - LABOUR						
Fittings and valves gang						
1 ganger/chargehand (skill rate 4) - 50% of time		7.12				
1 skilled operative (skill rate 4)		13.32				
2 unskilled operatives (general)		24.88				
1 plant operator (skill rate 3)		16.19				
Total Gang Rate / Hour	£	**61.50**				
RESOURCES - PLANT						
Fittings and valves						
1.0 m³ hydraulic backacter			40.06			
disc saw - 30% of time			0.41			
2.80 m³/min compressor, 2 tool - 30% of time			2.47			
2t dumper - 30% of time			1.95			
compressor tools, extra 50ft / 15m hose - 30% of time			0.09			
small pump - 30% of time			0.71			
sundry tools - 30% of time			1.01			
Total Rate / Hour		£	**46.71**			
VALVES AND PENSTOCKS						
Valves						
Non-return valves; cast iron; single door; tidal flap						
250 mm	0.40	26.34	9.34	386.40	nr	**422.08**
350 mm	0.40	27.62	9.34	737.10	nr	**774.06**
450 mm	0.70	48.34	9.34	978.60	nr	**1036.28**
600 mm	0.70	48.34	14.03	1328.25	nr	**1390.62**
800 mm	0.75	51.80	16.34	2361.33	nr	**2429.47**
Penstocks; cast iron; wall mounted; hand operated						
250 mm	1.66	114.64	16.34	488.36	nr	**619.34**
350 mm	2.30	158.84	18.68	689.38	nr	**866.90**
450 mm	2.90	200.27	21.02	851.66	nr	**1072.95**
600 × 600 mm	5.40	372.92	37.36	1375.76	nr	**1786.04**
1000 × 1000 mm	9.00	621.54	70.07	2462.69	nr	**3154.30**

CLASS K: PIPEWORK - MANHOLES AND PIPEWORK ACCESSORIES

Item	Gang hours	Labour £	Plant £	Material £	Unit	Total rate £
NOTES						
The rates assume the most efficient items of plant (excavator) and are optimum rates assuming continuous output with no delays caused by other operations or works.						
Ground conditions are assumed to be good easily worked soil with no abnormal conditions that would affect outputs and consistency of work.						
Multiplier Table for labour and plant for various site conditions for working:						
out of sequence × 2.75 (minimum)						
in hard clay × 1.75 to 2.00						
in running sand × 2.75 (minimum)						
in broken rock × 2.75 to 3.50						
below water table × 2.00 (minimum)						
RESOURCES - LABOUR						
Gullies gang						
1 chargehand pipelayer (skill rate 4) - 50% of time		7.12				
1 skilled operative (skill rate 4)		13.32				
2 unskilled operatives (general)		24.88				
1 plant operator (skill rate 3)		16.19				
Total Gang Rate / Hour		£ **61.50**				
French/rubble drains, ditches and trenches gang; ducts and metal culverts gang						
1 chargehand pipelayer (skill rate 4) - 50% of time		7.12				
1 skilled operative (skill rate 4)		13.32				
2 unskilled operatives (general)		24.88				
1 plant operator (skill rate 3)		16.19				
Total Gang Rate / Hour		£ **61.50**				
RESOURCES - PLANT						
Gullies						
0.4 m³ hydraulic excavator			22.93			
2t dumper (30% of time)			1.95			
2.80 m³/min compressor, 2 tool (30% of time)			2.47			
compaction plate / roller (30% of time)			0.49			
2.40 m³/min road breaker (30% of time)			0.59			
54mm poker vibrator (30% of time)			0.38			
extra 15ft / 50m hose (30% of time)			0.09			
disc saw (30% of time)			0.33			
small pump (30% of time)			0.71			
Total Rate / Hour			£ **29.95**			

CLASS K: PIPEWORK - MANHOLES AND PIPEWORK ACCESSORIES

Item	Gang hours	Labour £	Plant £	Material £	Unit	Total rate £
RESOURCES - PLANT – cont						
French/rubble drains, ditches and trenches;						
ducts and metal culverts						
0.4m³ hydraulic backacter			22.93			
2t dumper (30% of time)			1.95			
disc saw (30% of time)			0.33			
compaction plate / roller (30% of time)			0.86			
2.80 m³/min compressor, 2 tool (30% of time)			2.47			
small pump (30% of time)			0.71			
Total Rate / Hour		£	29.25			
MANHOLES						
Brick construction						
Design criteria used in models:						
* class A engineering bricks						
* 215 thick walls generally; 328 thick to chambers						
exceeding 2.5 m deep						
* 225 mm plain concrete C20/20 base slab						
* 300 mm reinforced concrete C20/20 reducing						
slab						
* 125 mm reinforced concrete C20/20 top slab						
* maximum height of working chamber 2.0 m						
above benching						
* 750 × 750 access shaft						
* plain concrete C15/20 benching, 150 mm clay						
main channel longitudinally and two 100 branch						
channels						
* step irons at 300 mm centres, doubled if depth to						
invert exceeds 3000 mm						
* heavy duty manhole cover and frame						
750 × 700 chamber 500 depth to invert						
excavation, support, backfilling and disposal						44.49
concrete base						142.81
brickwork chamber						248.78
concrete cover slab						159.25
concrete benching, main and branch channels						24.81
step irons						15.93
access cover and frame						387.75
TOTAL					£	1023.82
750 × 700 chamber 1000 depth to invert						
excavation, support, backfilling and disposal						88.99
concrete base						142.81
brickwork chamber						497.56
concrete cover slab						159.25
concrete benching and channels						24.81
step irons						31.86
access cover and frame						387.75
TOTAL					£	1333.03

CLASS K: PIPEWORK - MANHOLES AND PIPEWORK ACCESSORIES

Item	Gang hours	Labour £	Plant £	Material £	Unit	Total rate £
750 × 700 chamber 1500 depth to invert						
excavation, support, backfilling and disposal						133.47
concrete base						142.81
brickwork chamber						746.34
concrete cover slab						159.25
concrete benching and channels						24.81
step irons						49.91
access cover and frame						387.75
TOTAL					£	1644.34
900 × 700 chamber 500 depth to invert						
excavation, support, backfilling and disposal						48.79
concrete base						158.76
brickwork chamber						267.06
concrete cover slab						174.16
concrete benching and channels						29.77
step irons						15.93
access cover and frame						387.75
TOTAL					£	1082.22
900 × 700 chamber 1000 depth to invert						
excavation, support, backfilling and disposal						97.58
concrete base						158.76
brickwork chamber						534.12
concrete cover slab						174.16
concrete benching and channels						29.77
step irons						31.86
access cover and frame						387.75
TOTAL					£	1414.01
900 × 700 chamber 1500 depth to invert						
excavation, support, backfilling and disposal						146.37
concrete base						158.76
brickwork chamber						801.19
concrete cover slab						174.16
concrete benching and channels						29.77
step irons						49.91
access cover and frame						387.75
TOTAL					£	1747.91
1050 × 700 chamber 1500 depth to invert						
excavation, support, backfilling and disposal						159.28
concrete base						174.71
brickwork chamber						856.04
concrete cover slab						189.07
concrete benching and channels						34.73
step irons						49.91
access cover and frame						387.75
TOTAL					£	1851.49
1050 × 700 chamber 2500 depth to invert						
excavation, support, backfilling and disposal						265.45
concrete base						174.71
brickwork chamber						1426.73

CLASS K: PIPEWORK - MANHOLES AND PIPEWORK ACCESSORIES

Item	Gang hours	Labour £	Plant £	Material £	Unit	Total rate £
MANHOLES – cont						
Brick construction – cont						
1050 × 700 chamber 2500 depth to invert – cont						
concrete cover slab						189.07
concrete benching and channels						34.73
step irons						83.19
access cover and frame						387.75
TOTAL					£	2561.63
1050 × 700 chamber 3500 depth to invert						
excavation, support, backfilling and disposal						371.63
concrete base						174.71
brickwork chamber						1997.43
access shaft						2040.08
reducing slab						189.07
concrete cover slab						154.29
concrete benching and channels						34.73
step irons						116.47
access cover and frame						387.75
TOTAL					£	5466.16
1350 × 700 chamber 2500 depth to invert						
excavation, support, backfilling and disposal						308.45
concrete base						206.61
brickwork chamber						1609.55
concrete cover slab						218.88
concrete benching and channels						44.65
step irons						83.19
access cover and frame						387.75
TOTAL					£	2859.08
1350 × 700 chamber 3500 depth to invert						
excavation, support, backfilling and disposal						431.83
concrete base						206.61
brickwork chamber						2253.38
access shaft						2040.08
reducing slab						218.88
concrete cover slab						154.29
concrete benching and channels						44.65
step irons						116.47
access cover and frame						387.75
TOTAL					£	5853.94
1350 × 700 chamber 4500 depth to invert						
excavation, support, backfilling and disposal						555.22
concrete base						206.61
brickwork chamber						2897.21
access shaft						2622.97
reducing slab						218.88
concrete cover slab						154.29
concrete benching and channels						44.65
step irons						149.74
access cover and frame						387.75
TOTAL					£	7237.32

CLASS K: PIPEWORK - MANHOLES AND PIPEWORK ACCESSORIES

Item	Gang hours	Labour £	Plant £	Material £	Unit	Total rate £
Precast concrete construction						
Design criteria used in models:						
* circular shafts						
* 150 mm plain concrete C15/20 surround						
* 225 mm plain concrete C20/20 base slab						
* precast reducing slab						
* precast top slab						
* maximum height of working chamber 2.0 m above benching						
* 750 mm diameter access shaft						
* plain concrete C15/20 benching, 150 mm clay main channel longitudinally and two 100 branch channels						
* step irons at 300 mm centres, doubled if depth to invert exceeds 3000 mm						
* heavy duty manhole cover and frame						
* in manholes over 6 m deep, landings at maximum intervals						
675 diameter × 500 depth to invert						
excavation, support, backfilling and disposal						27.00
concrete base						40.50
main chamber rings						20.25
cover slab						79.88
concrete benching and channels						52.88
concrete surround						41.63
step irons						15.93
access cover and frame						387.75
TOTAL					£	665.81
675 diameter × 750 depth to invert						
excavation, support, backfilling and disposal						36.00
concrete base						40.50
main chamber rings						46.13
cover slab						79.88
concrete benching and channels						52.88
concrete surround						60.75
step irons						–
access cover and frame						387.75
TOTAL					£	727.77
675 diameter × 1000 depth to invert						
excavation, support, backfilling and disposal						43.88
concrete base						40.50
main chamber rings						73.13
cover slab						79.88
concrete benching and channels						52.88
step irons						–
concrete surround						81.00
access cover and frame						387.75
TOTAL					£	790.86

CLASS K: PIPEWORK - MANHOLES AND PIPEWORK ACCESSORIES

Item	Gang hours	Labour £	Plant £	Material £	Unit	Total rate £
MANHOLES – cont						
Precast concrete construction – cont						
675 diameter×1250 depth to invert						
excavation, support, backfilling and disposal						51.75
concrete base						40.50
main chamber rings						100.13
cover slab						79.88
concrete benching and channels						52.88
concrete surround						100.13
step irons						41.59
access cover and frame						387.75
TOTAL					£	854.60
900 diameter×750 depth to invert						
excavation, support, backfilling and disposal						115.35
concrete base						136.54
main chamber rings						64.33
cover slab						94.70
concrete benching and channels						29.77
concrete surround						77.24
step irons						–
access cover and frame						387.75
TOTAL					£	858.30
900 diameter×1000 depth to invert						
excavation, support, backfilling and disposal						153.79
concrete base						136.54
main chamber rings						85.78
cover slab						94.70
concrete benching and channels						29.77
concrete surround						102.99
step irons						–
access cover and frame						387.75
TOTAL					£	962.26
900 diameter×1500 depth to invert						
excavation, support, backfilling and disposal						230.68
concrete base						136.54
main chamber rings						128.67
cover slab						94.70
concrete benching and channels						29.77
concrete surround						154.49
step irons						49.91
access cover and frame						387.75
TOTAL					£	1123.49

CLASS K: PIPEWORK - MANHOLES AND PIPEWORK ACCESSORIES

Item	Gang hours	Labour £	Plant £	Material £	Unit	Total rate £
1200 diameter × 1500 depth to invert						
excavation, support, backfilling and disposal						285.68
concrete base						192.75
main chamber rings						135.42
concrete benching and channels						44.54
cover slab						121.33
concrete surround						199.35
step irons						49.91
access cover and frame						387.75
TOTAL					£	1416.73
1200 diameter × 2000 depth to invert						
excavation, support, backfilling and disposal						380.90
concrete base						192.75
main chamber rings						180.55
cover slab						121.33
concrete benching and channels						44.54
concrete surround						265.80
step irons						66.55
access cover and frame						387.75
TOTAL					£	1640.18
1200 diameter × 2500 depth to invert						
excavation, support, backfilling and disposal						476.13
concrete base						192.75
main chamber rings						225.70
cover slab						121.33
concrete benching and channels						44.54
concrete surround						332.26
step irons						83.19
access cover and frame						387.75
TOTAL					£	1863.65
1200 diameter × 3000 depth to invert						
excavation, support, backfilling and disposal						571.35
concrete base						192.75
main chamber rings						270.83
cover slab						121.33
concrete benching and channels						44.54
concrete surround						398.71
step irons						99.83
access cover and frame						387.75
TOTAL					£	2087.09
1800 diameter × 1500 depth to invert						
excavation, support, backfilling and disposal						415.00
concrete base						332.91
main chamber rings						407.53
cover slab						326.02
concrete benching and channels						100.22
concrete surround						289.09
step irons						49.91
access cover and frame						387.75
TOTAL					£	2308.44

CLASS K: PIPEWORK - MANHOLES AND PIPEWORK ACCESSORIES

Item	Gang hours	Labour £	Plant £	Material £	Unit	Total rate £
MANHOLES – cont						
Precast concrete construction – cont						
1800 diameter × 2000 depth to invert						
excavation, support, backfilling and disposal						553.33
concrete base						332.91
main chamber rings						543.38
cover slab						326.02
concrete benching and channels						100.22
concrete surround						385.45
step irons						66.55
access cover and frame						387.75
TOTAL					£	2695.61
1800 diameter × 2500 depth to invert						
excavation, support, backfilling and disposal						691.66
concrete base						332.91
main chamber rings						679.22
cover slab						326.02
concrete benching and channels						100.22
concrete surround						481.81
step irons						83.19
access cover and frame						387.75
TOTAL					£	3082.78
1800 diameter × 3000 depth to invert						
excavation, suport, backfilling and disposal						829.99
concrete base						332.91
main chamber rings						815.06
cover slab						326.02
concrete surround						578.17
concrete benching and channels						100.22
step irons						99.83
access cover and frame						387.75
TOTAL						3469.95
1800 diameter × 3500 depth to invert						
excavation, support, backfilling and disposal						968.33
concrete base						332.91
access shaft						300.22
main chamber rings						950.91
reducing slab						364.84
cover slab						94.70
concrete benching and channels						100.22
concrete surround						674.54
step irons						116.47
access cover and frame						387.75
TOTAL					£	4426.72

CLASS K: PIPEWORK - MANHOLES AND PIPEWORK ACCESSORIES

Item	Gang hours	Labour £	Plant £	Material £	Unit	Total rate £
1800 diameter×4000 depth to invert						
excavation, support, backfilling and disposal						1106.66
concrete base						332.91
access shaft						343.10
main chamber rings						1086.75
reducing slab						364.84
cover slab						94.70
concrete benching and channels						100.22
concrete surround						770.89
step irons						133.10
access cover and frame						387.75
TOTAL					£	4720.93
2400 diameter×1500 depth to invert						
excavation, support, backfilling and disposal						570.08
concrete base						510.11
main chamber rings						1036.29
cover slab						947.02
concrete benching and channels						178.16
concrete surround						378.82
step irons						49.91
access cover and frame						387.75
TOTAL					£	4058.15
2400 diameter×3000 depth to invert						
excavation, support, backfilling and disposal						1140.16
concrete base						510.11
main chamber rings						2072.59
cover slab						947.02
concrete benching and channels						178.16
concrete surround						757.64
step irons						99.83
access cover and frame						387.75
TOTAL					£	6093.26
2400 diameter×4500 depth to invert						
excavation, support, backfilling and disposal						1710.26
concrete base						510.11
access shaft						385.99
main chamber rings						3108.88
reducing slab						1024.65
cover slab						94.70
concrete benching and channels						178.16
concrete surround						1136.45
step irons						149.74
access cover and frame						387.75
TOTAL					£	8686.69

CLASS K: PIPEWORK - MANHOLES AND PIPEWORK ACCESSORIES

Item	Gang hours	Labour £	Plant £	Material £	Unit	Total rate £
MANHOLES – cont						
Precast concrete construction – cont						
2700 diameter×1500 depth to invert						
excavation, support, backfilling and disposal						657.29
concrete base						612.60
main chamber rings						1199.31
cover slab						1443.83
concrete benching and channels						225.47
concrete surround						423.69
step irons						49.91
access cover and frame						387.75
TOTAL					£	4999.84
2700 diameter×3000 depth to invert						
excavation, support, backfilling and disposal						1314.59
concrete base						612.60
main chamber rings						2398.61
cover slab						1443.83
concrete benching and channels						225.47
concrete surround						847.37
step irons						99.83
access cover and frame						387.75
TOTAL					£	7330.04
2700 diameter×4500 depth to invert						
excavation, support, backfilling and disposal						1971.88
concrete base						612.60
access shaft						385.99
main chamber rings						3597.92
reducing slab						1676.70
cover slab						94.70
concrete benching and channels						225.47
concrete surround						1271.05
step irons						149.74
access cover and frame						387.75
TOTAL					£	10373.79
3000 diameter×3000 depth to invert						
excavation, support, backfilling and disposal						1501.88
concrete base						724.33
main chamber rings						3376.69
cover slab						1863.00
concrete benching and channels						278.36
concrete surround						937.10
step irons						99.83
access cover and frame						387.75
TOTAL					£	9168.93

CLASS K: PIPEWORK - MANHOLES AND PIPEWORK ACCESSORIES

Item	Gang hours	Labour £	Plant £	Material £	Unit	Total rate £
3000 diameter × 4500 depth to invert						
excavation, support, backfilling and disposal						2252.82
concrete base						724.33
access shaft						385.99
main chamber rings						5065.03
reducing slab						2134.69
cover slab						94.70
concrete benching and channels						278.36
concrete surround						1405.65
step irons						149.74
access cover and frame						387.75
TOTAL					£	12879.07
3000 diameter × 6000 depth to invert						
excavation, support, backfilling and disposal						3003.76
concrete base						724.33
access shaft						514.65
main chamber rings						6753.38
reducing slab						2134.69
cover slab						94.70
concrete benching and channels						278.36
concrete surround						1874.20
step irons						199.66
access cover and frame						387.75
TOTAL					£	15965.48
BACKDROPS TO MANHOLES						
Clayware vertical pipe complete with rest bend at base and tumbling bay junction to main drain complete with stopper; concrete grade C20 surround, 150 mm thick; additional excavation and disposal						
100 pipe						
1.15 m to invert	—	—	—	—	nr	109.49
2.15 m to invert	—	—	—	—	nr	139.83
3.15 m to invert	—	—	—	—	nr	168.84
4.15 m to invert	—	—	—	—	nr	200.50
150 pipe						
1.15 m to invert	—	—	—	—	nr	168.84
2.15 m to invert	—	—	—	—	nr	203.14
3.15 m to invert	—	—	—	—	nr	242.71
4.15 m to invert	—	—	—	—	nr	282.28
225 pipe						
1.15 m to invert	—	—	—	—	nr	336.37
2.15 m to invert	—	—	—	—	nr	394.41
3.15 m to invert	—	—	—	—	nr	455.08
4.15 m to invert	—	—	—	—	nr	518.40

CLASS K: PIPEWORK - MANHOLES AND PIPEWORK ACCESSORIES

Item	Gang hours	Labour £	Plant £	Material £	Unit	Total rate £
GULLIES						
Vitrified clay; set in concrete grade C20, 150 mm thick; additional excavation and disposal						
Road gulley						
450 mm diameter × 900 mm deep, 100 mm or 150 mm outlet; PC £126.00.00*; cast iron road gulley grating and frame group 4 434 × 434 mm, PC £49.00*, on Class B engineering brick seating	0.50	22.48	1.14	498.05	nr	521.67
Yard gulley (mud); trapped with rodding eye; galvanized bucket; stopper						
225 mm diameter, 100 mm diameter outlet, PC £89.00*; cast iron hinged grate and frame	0.30	13.48	0.69	236.31	nr	250.48
Grease interceptors; internal access and bucket						
600 × 450 mm, PC £704.00*; metal tray and lid, square hopper with horizontal inlet	0.35	15.73	0.80	1435.81	nr	1452.34
Precast concrete; set in concrete grade C20, 150 mm thick; additional excavation and disposal						
Road gulley; trapped with rodding eye; galvanized bucket; stopper						
450 mm diameter × 750 mm deep, PC £45.00*; cast iron road gulley grating and frame group 4, 434 × 434 mm, PC £55.00*, on Class B engineering brick seating	0.50	22.48	1.14	311.55	nr	335.17
450 mm diameter × 900 mm deep, PC £46.00*; cast iron road gulley grating and frame group 4, 434 × 434 mm, PC £55.00*, on Class B engineering brick seating	0.54	24.27	1.23	314.55	nr	340.05
450 mm diameter × 1050 mm deep, PC £50.00*; cast iron road gulley grating and frame group 4, 434 × 434 mm, PC £55.00*, on Class B engineering brick seating	0.58	26.07	1.32	316.22	nr	343.61
FRENCH DRAINS, RUBBLE DRAINS, DITCHES						
The rates assume the most efficient items of plant (excavator) and are optimum rates assuming continuous output with no delays caused by other operations or works.						
Ground conditions are assumed to be good easily worked soil with no abnormal conditions that would affect outputs and consistency of work.						
Multiplier Table for labour and plant for various site conditions for working:						
out of sequence × 2.75 (minimum)						
in hard clay × 1.75 to 2.00						
in running sand × 2.75 (minimum)						

CLASS K: PIPEWORK - MANHOLES AND PIPEWORK ACCESSORIES

Item	Gang hours	Labour £	Plant £	Material £	Unit	Total rate £
in broken rock × 2.75 to 3.50						
below water table × 2.00 (minimum)						
Excavation of trenches for unpiped rubble drains						
(excluding trench support); cross-sectional area						
0.25–0.50 m²	0.10	6.91	2.54	–	m	**9.45**
0.50–0.75 m²	0.12	8.29	3.04	–	m	**11.33**
0.75–1.00 m²	0.14	9.67	3.55	–	m	**13.22**
1.00–1.50 m²	0.17	11.74	4.31	–	m	**16.05**
1.50–2.00 m²	0.20	13.81	5.07	–	m	**18.88**
Filling French and rubble drains with graded material						
graded material; 20 mm stone aggregate; PC						
£4.45/t	0.30	20.72	7.61	29.50	m³	**57.83**
broken brick/concrete rubble; PC £3.42/t	0.29	20.03	7.36	30.05	m³	**57.44**
Excavation of rectangular section ditches; unlined;						
cross-sectional area						
0.25–0.50 m²	0.11	7.60	2.74	–	m	**10.34**
0.50–0.75 m²	0.13	8.98	3.23	–	m	**12.21**
0.75–1.00 m²	0.16	11.05	3.98	–	m	**15.03**
1.00–1.50 m²	0.20	13.81	4.98	–	m	**18.79**
1.50–2.00 m²	0.25	17.27	6.22	–	m	**23.49**
Excavation of rectangular ditches; lined with precast						
concrete slabs; cross-sectional area						
0.25–0.50 m²	0.15	10.36	4.16	12.99	m	**27.51**
0.50–0.75 m²	0.25	17.27	6.95	21.40	m	**45.62**
0.75–1.00 m²	0.36	24.86	10.00	30.45	m	**65.31**
1.00–1.50 m²	0.40	27.62	11.11	43.02	m	**81.75**
1.50–2.00 m²	0.45	31.08	12.50	60.68	m	**104.26**
Excavation of vee section ditches; unlined;						
cross-sectional area						
0.25–0.50 m²	0.10	6.91	2.95	–	m	**9.86**
0.50–0.75 m²	0.12	8.29	3.52	–	m	**11.81**
0.75–1.00 m²	0.14	9.67	4.12	–	m	**13.79**
1.00–1.50 m²	0.18	12.43	5.29	–	m	**17.72**
1.50–2.00 m²	0.22	15.19	6.47	–	m	**21.66**
DUCTS AND METAL CULVERTS						
Galvanized steel culverts; bitumen coated						
Sectional corrugated metal culverts, nominal internal						
diameter 0.5–1 m; 1000 mm nominal internal						
diameter, 1.6 mm thick						
not in trenches	0.15	9.88	0.71	117.87	m	**128.46**
in trenches, depth not exceeding 1.5 m	0.31	20.41	9.06	117.87	m	**147.34**
in trenches, depth 1.5–2 m	0.43	28.32	12.57	117.87	m	**158.76**
in trenches, depth 2–2.5 m	0.51	33.58	14.91	117.87	m	**166.36**
in trenches, depth 2.5–3 m	0.60	39.51	17.55	117.87	m	**174.93**

CLASS K: PIPEWORK - MANHOLES AND PIPEWORK ACCESSORIES

Item	Gang hours	Labour £	Plant £	Material £	Unit	Total rate £
DUCTS AND METAL CULVERTS – cont						
Galvanized steel culverts – cont						
Sectional corrugated metal culverts, nominal internal diameter exceeding 1.5 m; 1600 mm nominal internal diameter, 1.6 mm thick						
not in trenches	0.21	13.83	1.00	189.10	m	203.93
in trenches, depth 1.5–2 m	0.44	28.97	12.87	189.10	m	230.94
in trenches, depth 2–2.5 m	0.56	36.88	16.38	189.10	m	242.36
in trenches, depth 2.5–3 m	0.68	44.78	19.89	189.10	m	253.77
in trenches, depth 3–3.5 m	0.82	54.00	24.00	189.10	m	267.10
Sectional corrugated metal culverts, nominal internal diameter exceeding 1.5 m; 2000 mm nominal internal diameter, 1.6 mm thick						
not in trenches	0.26	17.12	1.24	400.27	m	418.63
in trenches, depth 2–2.5 m	0.46	30.29	13.47	400.27	m	444.03
in trenches, depth 2.5–3 m	0.60	39.51	17.55	400.27	m	457.33
in trenches, depth 3–3.5 m	0.75	49.39	21.93	400.27	m	471.59
in trenches, depth 3.5–4 m	0.93	61.24	27.21	400.27	m	488.72
Sectional corrugated metal culverts, nominal internal diameter exceeding 1.5 m; 2200 mm nominal internal diameter, 1.6 mm thick						
not in trenches	0.33	21.73	1.57	451.01	m	474.31
in trenches, depth 2.5–3 m	0.64	42.14	18.72	451.01	m	511.87
in trenches, depth 3–3.5 m	0.77	50.70	22.53	451.01	m	524.24
in trenches, depth 3.5–4 m	1.02	67.17	29.85	451.01	m	548.03
in trenches, depth exceeding 4 m	1.32	86.92	38.61	451.01	m	576.54
OTHER PIPEWORK ANCILLARIES						
Notes						
Refer to Section G (Concrete and concrete ancillaries) for costs relevant to the construction of Headwall Structure.						
Build Ends in						
Connections to existing manholes and other chambers, pipe bore						
150 mm diameter	0.60	41.44	3.28	9.86	nr	54.58
225 mm diameter	0.95	65.61	5.18	18.35	nr	89.14
300 mm diameter	1.25	86.33	6.83	27.86	nr	121.02
375 mm diameter	1.45	100.14	7.93	36.70	nr	144.77
450 mm diameter	1.75	115.24	50.54	46.18	nr	211.96

CLASS L: PIPEWORK SUPPORTS AND PIPEWORK ANCILLARIES

Item	Gang hours	Labour £	Plant £	Material £	Unit	Total rate £
NOTES						
The rates assume the most efficient items of plant (excavator) and are optimum rates assuming continuous output with no delays caused by other operations or works.						
Ground conditions are assumed to be good easily worked soil with no abnormal conditions that would affect outputs and consistency of work.						
Multiplier Table for labour and plant for various site conditions for working:						
out of sequence × 2.75 (minimum)						
in hard clay × 1.75 to 2.00						
in running sand × 2.75 (minimum)						
in broken rock × 2.75 to 3.50						
below water table × 2.00 (minimum)						
RESOURCES - LABOUR						
Supports and protection gang						
2 unskilled operatives (general)		24.88				
1 plant operator (skill rate 3)		16.19				
Total Gang Rate / Hour		£ 41.07				
RESOURCES - PLANT						
Supports and protection						
0.40 m³ hydraulic backacter			22.93			
Bomag BW 65S			7.00			
Total Rate / Hour			£ 29.93			
EXTRAS TO EXCAVATION AND BACKFILLING						
Drainage sundries						
Extra over any item of drainage for excavation in						
rock	0.65	26.49	30.04	–	m³	56.53
mass concrete	0.84	34.23	39.07	–	m³	73.30
reinforced concrete	1.18	48.09	54.78	–	m³	102.87
Excavation of soft spots, backfilling						
concrete grade C15P	0.30	12.22	13.87	89.51	m³	115.60

CLASS L: PIPEWORK SUPPORTS AND PIPEWORK ANCILLARIES

Item	Gang hours	Labour £	Plant £	Material £	Unit	Total rate £
SPECIAL PIPE LAYING METHODS						
There are many factors, apart from design consideration, which influence the cost of pipe jacking, so that it is only possible to give guide prices for a sample of the work involved. For more reliable estimates it is advisable to seek the advice of a Specialist Contractor.						
The main cost considerations are :-						
* the nature of the ground						
* length of drive						
* location						
* presence of water						
* depth below surface						
Provision of all plant, equipment and labour establishing						
thrust pit; 6 m × 4 m × 8 m deep	–	–	–	–	item	30000.00
reception pit; 4 m × 4 m × 8 m deep	–	–	–	–	item	22000.00
mobilise and set up pipe jacking equipment	–	–	–	–	item	44000.00
Pipe jacking, excluding the cost of non-drainage materials; concrete pipes BS 5911 Part 1 Class 120 with rebated joints, steel reinforcing band; length not exceeding 50 m; in sand and gravel						
900 mm nominal bore	–	–	–	–	m	1279.26
1200 mm nominal bore	–	–	–	–	m	1651.86
1500 mm nominal bore	–	–	–	–	m	1932.50
1800 mm nominal bore	–	–	–	–	m	2185.50
BEDS						
Imported sand						
100 mm deep bed for pipes nominal bore						
100 mm	0.02	0.81	0.46	2.29	m	3.56
150 mm	0.02	0.81	0.46	2.47	m	3.74
225 mm	0.03	1.22	0.69	2.88	m	4.79
300 mm	0.04	1.63	0.92	3.01	m	5.56
150 mm deep bed for pipes nominal bore						
150 mm	0.06	2.44	1.38	3.73	m	7.55
225 mm	0.07	2.85	1.61	4.31	m	8.77
300 mm	0.09	3.67	2.06	4.57	m	10.30
400 mm	0.12	4.89	2.75	5.11	m	12.75
450 mm	0.14	5.71	3.21	5.98	m	14.90
600 mm	0.17	6.93	3.90	7.39	m	18.22
750 mm	0.19	7.74	4.36	8.26	m	20.36
900 mm	0.21	8.56	4.82	9.67	m	23.05
1200 mm	0.25	10.19	5.73	11.37	m	27.29

CLASS L: PIPEWORK SUPPORTS AND PIPEWORK ANCILLARIES

Item	Gang hours	Labour £	Plant £	Material £	Unit	Total rate £
Imported granular material						
100 mm deep bed for pipes nominal bore						
100 mm	0.02	0.81	0.46	1.78	m	**3.05**
150 mm	0.03	1.22	0.69	1.93	m	**3.84**
225 mm	0.04	1.63	0.92	2.21	m	**4.76**
300 mm	0.05	2.04	1.15	2.36	m	**5.55**
150 mm deep bed for pipes nominal bore						
150 mm	0.06	2.44	1.38	2.88	m	**6.70**
225 mm	0.08	3.26	1.83	3.32	m	**8.41**
300 mm	0.10	4.08	2.29	3.55	m	**9.92**
400 mm	0.13	5.30	2.98	4.00	m	**12.28**
450 mm	0.15	6.11	3.44	4.61	m	**14.16**
600 mm	0.18	7.33	4.13	5.74	m	**17.20**
750 mm	0.20	8.15	4.59	6.40	m	**19.14**
900 mm	0.22	8.96	5.05	7.51	m	**21.52**
1200 mm	0.26	10.60	5.96	8.82	m	**25.38**
Mass concrete						
100 mm deep bed for pipes nominal bore						
100 mm	0.07	2.85	1.61	6.84	m	**11.30**
150 mm	0.08	3.26	1.83	7.42	m	**12.51**
225 mm	0.09	3.67	2.06	8.57	m	**14.30**
300 mm	0.11	4.48	2.52	9.15	m	**16.15**
150 mm deep bed for pipes nominal bore						
100 mm	0.10	4.08	2.29	10.28	m	**16.65**
150 mm	0.12	4.89	2.75	11.13	m	**18.77**
225 mm	0.14	5.71	3.21	12.85	m	**21.77**
300 mm	0.16	6.52	3.67	13.72	m	**23.91**
400 mm	0.19	7.74	4.36	15.39	m	**27.49**
450 mm	0.21	8.56	4.82	17.99	m	**31.37**
600 mm	0.24	9.78	5.50	22.25	m	**37.53**
750 mm	0.26	10.60	5.96	24.82	m	**41.38**
900 mm	0.28	11.41	6.42	29.11	m	**46.94**
1200 mm	0.32	13.04	7.34	34.23	m	**54.61**

CLASS L: PIPEWORK SUPPORTS AND PIPEWORK ANCILLARIES

Item	Gang hours	Labour £	Plant £	Material £	Unit	Total rate £
HAUNCHES						
The following items allow for dressing the haunching material half-way up the pipe barrel for the full width of the bed and then dressing in triangular fashion to the crown of the pipe. The items exclude the drain bed.						
Mass concrete PC £86.20/m³						
Haunches for pipes nominal bore						
150 mm	0.24	9.78	5.50	6.28	m	21.56
225 mm	0.29	11.82	6.65	9.85	m	28.32
300 mm	0.36	14.67	8.26	12.42	m	35.35
400 mm	0.43	17.52	9.86	16.43	m	43.81
450 mm	0.50	20.38	11.47	22.25	m	54.10
600 mm	0.56	22.82	12.84	34.45	m	70.11
750 mm	0.62	25.27	14.22	42.59	m	82.08
900 mm	0.69	28.12	15.82	58.29	m	102.23
1200 mm	0.75	30.56	17.20	76.30	m	124.06
SURROUNDS						
The following items provide for dressing around the pipe above the bed. sand and granular material is quantified on the basis of the full width of the bed to the stated distance above the crown, concrete as an ellipse from the top corners of the bed to a poit at the stated distance above the crown. The items exclude the drain bed.						
Imported sand						
100 mm thick bed for pipes nominal bore						
100 mm	0.04	1.63	0.92	4.15	m	6.70
150 mm	0.05	2.04	1.15	5.95	m	9.14
225 mm	0.06	2.44	1.38	9.53	m	13.35
300 mm	0.08	3.26	1.83	13.28	m	18.37
150 mm thick bed for pipes nominal bore						
100 mm	0.10	4.08	2.29	5.04	m	11.41
150 mm	0.12	4.89	2.75	6.91	m	14.55
225 mm	0.14	5.71	3.21	10.65	m	19.57
300 mm	0.18	7.33	4.13	14.42	m	25.88
400 mm	0.24	9.78	5.50	21.05	m	36.33
450 mm	0.28	11.41	6.42	26.81	m	44.64
600 mm	0.34	13.86	7.80	43.31	m	64.97
750 mm	0.38	15.48	8.71	60.39	m	84.58
900 mm	0.42	17.11	9.63	84.08	m	110.82
1200 mm	0.50	20.38	11.47	134.00	m	165.85

CLASS L: PIPEWORK SUPPORTS AND PIPEWORK ANCILLARIES

Item	Gang hours	Labour £	Plant £	Material £	Unit	Total rate £
Imported granular material						
100 mm thick bed for pipes nominal bore						
100 mm	0.07	4.83	1.61	3.21	m	9.65
150 mm	0.10	6.91	2.29	4.58	m	13.78
225 mm	0.13	8.98	2.98	7.35	m	19.31
300 mm	0.16	11.05	3.67	10.26	m	24.98
150 mm thick bed for pipes nominal bore						
100 mm	0.10	4.08	2.29	3.88	m	10.25
150 mm	0.12	4.89	2.75	5.34	m	12.98
225 mm	0.14	5.71	3.21	8.22	m	17.14
300 mm	0.18	7.33	4.13	11.18	m	22.64
400 mm	0.24	9.78	5.50	16.30	m	31.58
450 mm	0.28	11.41	6.42	20.80	m	38.63
600 mm	0.34	13.86	7.80	33.44	m	55.10
750 mm	0.38	15.48	8.71	46.78	m	70.97
900 mm	0.42	17.11	9.63	65.10	m	91.84
1200 mm	0.50	20.38	11.47	103.79	m	135.64
Mass concrete						
100 mm thick bed for pipes nominal bore						
100 mm	0.14	5.71	3.21	12.37	m	21.29
150 mm	0.16	6.52	3.67	17.77	m	27.96
225 mm	0.18	7.33	4.13	28.45	m	39.91
300 mm	0.22	8.96	5.05	39.69	m	53.70
150 mm thick bed for pipes nominal bore						
150 mm	0.23	9.37	5.27	20.66	m	35.30
225 mm	0.26	10.60	5.96	31.79	m	48.35
300 mm	0.30	12.22	6.88	43.27	m	62.37
400 mm	0.36	14.67	8.26	63.11	m	86.04
450 mm	0.40	16.30	9.17	80.43	m	105.90
600 mm	0.45	18.34	10.32	129.30	m	157.96
750 mm	0.50	20.38	11.47	180.91	m	212.76
900 mm	0.55	22.41	12.61	251.64	m	286.66
1200 mm	0.61	24.86	13.99	401.07	m	439.92
CONCRETE STOOLS AND THRUST BLOCKS						
Mass concrete PC £86.20/m³						
Concrete stools or thrust blocks (nett volume of concrete excluding volume occupied by pipes)						
0.1 m³	0.18	7.33	4.13	10.25	nr	21.71
0.1–0.2 m³	0.32	13.04	7.34	20.51	nr	40.89
0.2–0.5 m³	0.62	25.27	14.22	51.37	nr	90.86
0.5–1.0 m³	0.91	37.08	20.87	102.82	nr	160.77
1.0–2.0 m³	1.29	52.57	29.58	205.83	nr	287.98
2.0–4.0 m³	3.15	128.36	72.24	411.45	nr	612.05

CLASS M: STRUCTURAL METALWORK

Item	Gang hours	Labour £	Plant £	Material £	Unit	Total rate £
NOTES						
The following are guide prices for various structural members commonly found in a Civil Engineering contract. The list is by no means exhaustive and costs are very much dependent on the particular design and will vary greatly according to specific requirements.						
For more detailed prices, reference should be made to Specialist Contractors.						
FABRICATION OF MEMBERS; STEELWORK						
Columns						
universal beams; straight on plan	–	–	–	–	t	1300.00
circular hollow sections; straight on plan	–	–	–	–	t	2500.00
rectangular hollow sections; straight on plan	–	–	–	–	t	2500.00
Beams						
universal beams; straight on plan	–	–	–	–	t	1300.00
universal beams; curved on plan	–	–	–	–	t	2020.00
channels; straight on plan	–	–	–	–	t	1420.00
channels; curved on plan	–	–	–	–	t	2450.00
castellated beams; straight on plan	–	–	–	–	t	1800.00
Portal frames						
straight on plan	–	–	–	–	t	1600.00
Trestles, towers and built-up columns						
straight on plan	–	–	–	–	t	1900.00
Trusses and built-up girders						
straight on plan	–	–	–	–	t	1900.00
curved on plan	–	–	–	–	t	2500.00
Bracings						
angles; straight on plan	–	–	–	–	t	1300.00
circular hollow sections; straight on plan	–	–	–	–	t	2600.00
Purlins and cladding rails						
straight on plan	–	–	–	–	t	1500.00
Cold rolled purlins and rails						
straight on plan	–	–	–	–	t	2500.00
Anchorages and holding down bolt assemblies						
base plate and bolt assemblies complete	–	–	–	–	t	2900.00
ERECTION OF FABRICATED MEMBERS ON SITE						
Trial erection	–	–	–	–	t	650.00
Permanent erection	–	–	–	–	t	650.00
Site bolts						
black	–	–	–	–	t	3200.00
HSFG general grade	–	–	–	–	t	3200.00
HSFG higher	–	–	–	–	t	3600.00
HSFG load indicating or limit types, general grade	–	–	–	–	t	4200.00
HSFG load indicating or limit types, higher grade	–	–	–	–	t	4700.00

CLASS M: STRUCTURAL METALWORK

Item	Gang hours	Labour £	Plant £	Material £	Unit	Total rate £
OFF SITE SURFACE TREATMENT						
Note: The following preparation and painting systems have been calculated on the basis of 20 m² per tonne						
Blast cleaning	–	–	–	–	m²	5.10
Galvanising	–	–	–	–	m²	18.75
Painting						
one coat zinc chromate primer	–	–	–	–	m²	4.89
one coat two pack epoxy zinc phosphate primer (75 microns dry film thickness)	–	–	–	–	m²	7.50
two coats epoxy micaceous iron oxide (100 microns dry film thickness per coat)	–	–	–	–	m²	22.50

CLASS N: MISCELLANEOUS METALWORK

Item	Gang hours	Labour £	Plant £	Material £	Unit	Total rate £
NOTES						
General						
The following are guide prices for various structural members commonly found in a Civil Engineering contract. The list is by no means exhaustive and costs are very much dependent on the particular design and will vary greatly according to specific requirements.						
For more detailed prices, reference should be made to Specialist Contractors.						
Cladding						
CESMM3 N.2.1 requires cladding to be measured in square metres, the item so produced being inclusive of all associated flashings at wall corners and bases, eaves, gables, ridges and around openings. As the relative quantities of these flashings will depend very much on the complexity of the building shape, the guide prices shown below for these items are shown separately to help with the accuracy of the estimate.						
Bridge bearings						
Bridge bearings are manufactured and installed to individual specifications. The following guide prices are for different sizes of simple bridge bearings. If requirements are known, then advice ought to be obtained from specialist manufactureres such as CCL.						
If there is a requirement for testing bridge bearings prior to their being installed then the tests should be enumerated separately.						
Specialist advice should be sought once details are known.						
RESOURCES - LABOUR						
Roofing - cladding gang						
1 ganger/chargehand (skill rate 3) - 50% of time		7.90				
2 skilled operative (skill rate 3)		29.76				
1 unskilled operative (general) - 50% of time		6.22				
Total Gang Rate / Hour	£	**43.88**				
Bridge bearing gang						
1 skilled operative (skill rate 4)		13.32				
2 unskilled operatives (general)		24.88				
Total Gang Rate / Hour	£	**38.20**				

CLASS N: MISCELLANEOUS METALWORK

Item	Gang hours	Labour £	Plant £	Material £	Unit	Total rate £
RESOURCES - PLANT						
Cladding to roofs						
15 m telescopic access platform - 50% of time			10.23			
Total Gang Rate / Hour		£	10.23			
Cladding to walls						
15 m telescopic access platform			20.47			
Total Gang Rate / Hour		£	20.47			
MILD STEEL						
Mild steel						
Stairways and landings	–	–	–	–	t	5375.00
Walkways and platforms	–	–	–	–	t	5125.00
Ladders						
cat ladder; 64 × 13 mm bar strings; 19 mm rungs at 250 mm centres; 450 mm wide with safety hoops	–	–	–	–	m	96.62
Miscellaneous framing						
angle section; 200 × 200 × 16 mm (equal)	–	–	–	–	m	301.44
angle section; 150 × 150 × 10 mm (equal)	–	–	–	–	m	141.30
angle section; 100 × 100 × 12 mm (equal)	–	–	–	–	m	226.08
angle section; 200 × 150 × 15 mm (unequal)	–	–	–	–	m	197.82
angle section; 150 × 75 × 10 mm (unequal)	–	–	–	–	m	105.97
universal beams; 914 × 419 mm	–	–	–	–	m	2059.80
universal beams; 533 × 210 mm	–	–	–	–	m	732.00
universal joists; 127 × 76 mm	–	–	–	–	m	78.00
channel section; 381 × 102 mm	–	–	–	–	m	324.00
channel section; 254 × 76 mm	–	–	–	–	m	165.60
channel section; 152 × 76 mm	–	–	–	–	m	107.40
tubular section; 100 × 100 × 10 mm	–	–	–	–	m	164.40
tubular section; 200 × 200 × 15 mm	–	–	–	–	m	541.80
tubular section; 76.1 × 5.0 mm	–	–	–	–	m	52.62
tubular section; 139.7 × 6.3 mm	–	–	–	–	m	151.20
Mild steel; galvanized						
Handrails						
76 mm diameter tubular handrail, 48 mm diameter standards at 750 mm centres, 48 mm diameter middle rail, 1070 mm high overall	–	–	–	–	m	116.63
Plate flooring						
8 mm (on plain) 'Durbar' pattern floor plates, maximum weight each panel 50 kg	–	–	–	–	m²	300.00
Mild steel; internally and externally acid dipped, rinse and hot dip galvanized, epoxy internal paint						
Uncovered tanks						
1600 litre capacity open top water tank	–	–	–	–	nr	907.27
18180 litre capacity open top water tank	–	–	–	–	nr	7500.00
Covered tanks						
1600 litre capacity open top water tank with loose fitting lid;	–	–	–	–	nr	1230.00
18180 litre capacity open top water tank with loose fitting lid	–	–	–	–	nr	12479.38

CLASS N: MISCELLANEOUS METALWORK

Item	Gang hours	Labour £	Plant £	Material £	Unit	Total rate £
MILD STEEL – cont						
Corrugated steel plates to BS 1449 Pt 1, Gr H4, sealed and bolted; BS729 hot dip galvanized, epoxy internal and external paint						
Uncovered tanks						
713 m³ capacity bolted cylindrical open top tank	–	–	–	–	nr	40546.00
PROPRIETARY WORK						
Galvanized steel troughed sheeting; 0.70 mm metal thickness, 75 mm deep corrugations; colour coating each side; fixing with plastic capped self-tapping screws to steel purlins or rails						
Cladding						
upper surfaces inclined at an angle ne 30° to the horizontal	0.15	6.54	1.53	20.43	m²	28.50
Extra for :						
galvanized steel inner lining sheet, 0.40 mm thick, Plastisol colour coating	0.06	2.61	1.23	7.64	m²	11.48
galvanized steel inner lining sheet, 0.40 mm thick, Plastisol colour coating; insulation, 80 mm thick	0.08	3.49	0.82	13.43	m²	17.74
surfaces inclined at an angle exceeding 60° to the horizontal	0.16	6.97	3.27	16.88	m²	27.12
Extra for :						
galvanized steel inner lining sheet, 0.40 mm thick, Plastisol colour coating	0.11	4.79	2.25	7.64	m²	14.68
galvanized steel inner lining sheet, 0.40 mm thick, Plastisol colour coating; insulation, 80 mm thick	0.13	5.67	2.66	13.43	m²	21.76
Galvanized steel flashings; 0.90 mm metal thickness; bent to profile; fixing with plastic capped self-tapping screws to steel purlins or rails; mastic sealant						
Flashings to cladding						
250 mm girth	0.12	5.23	2.46	12.94	m	20.63
500 mm girth	0.18	7.84	3.68	20.10	m	31.62
750 mm girth	0.22	9.59	4.50	27.27	m	41.36
Aluminium profiled sheeting; 0.90 mm metal thickness, 75 mm deep corrugations; colour coating each side; fixing with plastic capped self-tapping screws to steel purlins or rails						
Cladding						
upper surfaces inclined at an angle exceeding 60° to the horizontal	0.20	8.72	4.09	23.95	m²	36.76
Extra for :						
aluminium inner lining sheet, 0.70 mm thick, Plastisol colour coating	0.13	5.67	2.66	19.08	m²	27.41
aluminium inner lining sheet, 0.70 mm thick, Plastisol colour coating; insulation, 80 mm thick	0.15	6.54	3.07	24.86	m²	34.47

CLASS N: MISCELLANEOUS METALWORK

Item	Gang hours	Labour £	Plant £	Material £	Unit	Total rate £
Aluminium profiled sheeting; 1.00 mm metal thickness, 75 mm deep corrugations; colour coating each side; fixing with plastic capped self-tapping screws to steel purlins or rails						
Cladding						
upper surfaces inclined at an angle ne 30° to the horizontal	0.17	7.41	1.74	27.50	m²	**36.65**
Extra for :						
aluminium inner lining sheet, 0.70 mm thick, Plastisol colour coating	0.09	3.92	0.92	19.08	m²	**23.92**
aluminium inner lining sheet, 0.70 mm thick, Plastisol colour coating; insulation, 80 mm thick	0.11	4.79	1.13	24.86	m²	**30.78**
Aluminium flashings; 0.90 mm metal thickness; bent to profile; fixing with plastic capped self-tapping screws to steel purlins or rails; mastic sealant						
Flashings to cladding						
250 mm girth	0.12	5.23	2.46	14.36	m	**22.05**
500 mm girth	0.18	7.84	3.68	22.95	m	**34.47**
750 mm girth	0.22	9.59	4.50	31.53	m	**45.62**
Flooring; Eurogrid; galvanized mild steel						
Open grid flooring						
type 41/100; 3 × 25 mm bearer bar; 6 mm diameter transverse bar	–	–	–	–	m²	**48.36**
type 41/100; 5 × 25 mm bearer bar; 6 mm diameter transverse bar	–	–	–	–	m²	**62.56**
type 41/100; 3 × 30 mm bearer bar; 6 mm diameter transverse bar	–	–	–	–	m²	**56.13**
Duct covers; Stelduct; galvanized mild steel						
Duct covers; pedestrian duty						
225 mm clear opening	–	–	–	–	m	**89.06**
450 mm clear opening	–	–	–	–	m	**98.51**
750 mm clear opening	–	–	–	–	m	**116.87**
Duct covers; medium duty						
225 mm clear opening	–	–	–	–	m	**138.39**
450 mm clear opening	–	–	–	–	m	**156.43**
750 mm clear opening	–	–	–	–	m	**181.29**
Duct covers; heavy duty						
225 mm clear opening	–	–	–	–	m	**143.34**
450 mm clear opening	–	–	–	–	m	**213.70**
750 mm clear opening	–	–	–	–	m	**304.79**

CLASS N: MISCELLANEOUS METALWORK

Item	Gang hours	Labour £	Plant £	Material £	Unit	Total rate £
PROPRIETARY WORK – cont						
Bridge bearings						
Supply plain rubber bearings (3 m and 5 m lengths)						
150 × 20 mm	0.35	13.59	–	37.63	m	**51.22**
150 × 25 mm	0.35	13.59	–	45.15	m	**58.74**
Supply and place in position laminated elastomeric rubber bearing						
250 × 150 × 19 mm	0.25	9.71	–	27.95	nr	**37.66**
300 × 200 × 19 mm	0.25	9.71	–	30.64	nr	**40.35**
300 × 200 × 30 mm	0.27	10.48	–	38.70	nr	**49.18**
300 × 200 × 41 mm	0.27	10.48	–	58.18	nr	**68.66**
300 × 250 × 41 mm	0.30	11.65	–	72.72	nr	**84.37**
300 × 250 × 63 mm	0.30	11.65	–	101.59	nr	**113.24**
400 × 250 × 19 mm	0.32	12.43	–	51.06	nr	**63.49**
400 × 250 × 52 mm	0.32	12.43	–	117.39	nr	**129.82**
400 × 300 × 19 mm	0.32	12.43	–	61.27	nr	**73.70**
600 × 450 × 24 mm	0.35	13.59	–	188.08	nr	**201.67**
Adhesive fixings to laminated elastomeric rubber bearings						
2 mm thick epoxy adhesive	1.00	38.83	–	46.50	m²	**85.33**
15 mm thick epoxy mortar	1.50	58.24	–	279.35	m²	**337.59**
15 mm thick epoxy pourable grout	2.00	77.66	–	278.65	m²	**356.31**
Supply and install mechanical guides for laminated elastomeric rubber bearings						
500kN SLS design load; FP50 fixed pin Type 1	2.00	77.66	–	834.00	nr	**911.66**
500kN SLS design load; FP50 fixed pin Type 2	2.00	77.66	–	860.00	nr	**937.66**
750kN SLS design load; FP75 fixed pin Type 1	2.10	81.54	–	967.00	nr	**1048.54**
750kN SLS design load; FP75 fixed pin Type 2	2.10	81.54	–	1100.00	nr	**1181.54**
300kN SLS design load; UG300 Uniguide Type 1	2.00	77.66	–	1000.00	nr	**1077.66**
300kN SLS design load; UG300 Uniguide Type 2	2.00	77.66	–	1150.00	nr	**1227.66**
Supply and install fixed pot bearings						
355 × 355; PF200	2.00	77.66	–	894.00	nr	**971.66**
425 × 425; PF300	2.10	81.54	–	1066.00	nr	**1147.54**
Supply and install free sliding pot bearings						
445 × 345; PS200	2.10	81.54	–	1202.00	nr	**1283.54**
520 × 415; PS300	2.20	85.43	–	1533.00	nr	**1618.43**
Supply and install guided sliding pot bearings						
455 × 375; PG200	2.20	85.43	–	1597.00	nr	**1682.43**
545 × 435; PG300	2.30	89.31	–	1901.00	nr	**1990.31**
Testing; laminated elastomeric bearings						
compression test	–	–	–	55.50	nr	**55.50**
shear test	–	–	–	71.75	nr	**71.75**
bond test (Exclusive of cost of bearings as this is a destructive test)	–	–	–	283.50	nr	**283.50**

CLASS O: TIMBER

Item	Gang hours	Labour £	Plant £	Material £	Unit	Total rate £
RESOURCES - LABOUR						
Timber gang						
1 foreman carpenter/joiner (craftsman)		22.32				
1 carpenter/joiner (craftsman)		19.15				
1 unskilled operative (general)		12.44				
1 plant operator (skill rate 3) - 50% of time		8.10				
Total Gang Rate / Hour	£	**62.00**				
Timber fixings gang						
1 carpenter/joiner (craftsman)		22.32				
1 unskilled operative (general)		12.44				
Total Gang Rate / Hour	£	**34.76**				
RESOURCES - PLANT						
Timber						
tractor / trailer			17.28			
10t crawler crane (25% of time)			6.55			
5.6t rough terrain forklift (25% of time)			5.79			
7.5 KVA diesel generator			2.22			
two K637 rotary hammers			1.20			
two electric screwdrivers			1.38			
Total Gang Rate / Hour		£	**34.42**			

RESOURCES - MATERIALS
The timber material prices shown below are averages, actual prices being very much affected by availability of suitably sized forest timbers capable of conversion to the sizes shown. Apart from the practicality of being able to obtain the larger sizes in one timber, normal practice and drive for economy would lead to their neing built up using smaller timbers.

HARDWOOD COMPONENTS

Item	Gang hours	Labour £	Plant £	Material £	Unit	Total rate £
Greenheart						
100 × 75 mm						
length not exceeding 1.5 m	0.15	9.05	5.19	6.10	m	**20.34**
length 1.5–3 m	0.13	7.84	4.50	7.01	m	**19.35**
length 3–5 m	0.12	7.24	4.13	7.01	m	**18.38**
length 5–8 m	0.11	6.64	3.81	7.19	m	**17.64**
150 × 75 mm						
length not exceeding 1.5 m	0.17	10.26	5.98	8.49	m	**24.73**
length 1.5–3 m	0.15	9.05	5.19	8.29	m	**22.53**
length 3–5 m	0.14	8.45	4.82	8.29	m	**21.56**
length 5–8 m	0.12	7.24	5.61	8.49	m	**21.34**

CLASS O: TIMBER

Item	Gang hours	Labour £	Plant £	Material £	Unit	Total rate £
HARDWOOD COMPONENTS – cont						
Greenheart – cont						
200 × 100 mm						
length not exceeding 1.5 m	0.24	14.48	8.26	16.98	m	**39.72**
length 1.5–3 m	0.21	12.67	7.35	16.98	m	**37.00**
length 3–5 m	0.20	11.77	6.73	16.98	m	**35.48**
length 5–8 m	0.18	10.86	6.19	16.98	m	**34.03**
length 8–12 m	0.16	9.65	5.51	20.72	m	**35.88**
200 × 200 mm						
length not exceeding 1.5 m	0.42	25.34	14.45	29.38	m	**69.17**
length 1.5–3 m	0.40	24.14	13.77	29.38	m	**67.29**
length 3–5 m	0.38	22.93	13.08	29.38	m	**65.39**
length 5–8 m	0.34	20.52	11.70	29.38	m	**61.60**
length 8–12 m	0.30	18.10	10.57	31.68	m	**60.35**
225 × 100 mm						
length not exceeding 1.5 m	0.27	16.29	9.42	16.98	m	**42.69**
length 1.5–3 m	0.24	14.48	8.26	16.98	m	**39.72**
length 3–5 m	0.22	13.27	7.82	16.98	m	**38.07**
length 5–8 m	0.20	12.07	9.35	16.98	m	**38.40**
length 8–12 m	0.18	10.86	6.44	20.72	m	**38.02**
300 × 100 mm						
length not exceeding 1.5 m	0.36	21.72	12.39	22.21	m	**56.32**
length 1.5–3 m	0.33	19.91	11.49	22.21	m	**53.61**
length 3–5 m	0.30	18.10	10.57	22.21	m	**50.88**
length 5–8 m	0.27	16.29	9.42	24.74	m	**50.45**
length 8–12 m	0.24	14.48	8.26	3.01	m	**25.75**
300 × 200 mm						
length not exceeding 1.5 m	0.50	30.17	17.21	44.40	m	**91.78**
length 1.5–3 m	0.45	27.15	15.61	44.40	m	**87.16**
length 3–5 m	0.40	24.14	13.77	44.40	m	**82.31**
length 5–8 m	0.35	21.12	12.17	47.13	m	**80.42**
length 8–12 m	0.30	18.10	10.33	47.13	m	**75.56**
300 × 300 mm						
length not exceeding 1.5 m	0.52	31.38	17.90	66.61	m	**115.89**
length 1.5–3 m	0.48	28.96	16.52	66.61	m	**112.09**
length 3–5 m	0.44	26.55	15.14	66.61	m	**108.30**
length 5–8 m	0.40	24.14	13.77	70.83	m	**108.74**
length 8–12 m	0.36	21.72	12.39	70.83	m	**104.94**
450 × 450 mm						
length not exceeding 1.5 m	0.98	59.13	33.98	146.71	m	**239.82**
length 1.5–3 m	0.90	54.31	31.22	146.71	m	**232.24**
length 3–5 m	0.83	50.08	28.69	146.71	m	**225.48**
length 5–8 m	0.75	45.26	25.94	150.37	m	**221.57**
length 8–12 m	0.68	41.03	23.40	157.71	m	**222.14**

CLASS O: TIMBER

Item	Gang hours	Labour £	Plant £	Material £	Unit	Total rate £
SOFTWOOD COMPONENTS						
Softwood; stress graded SC3/4						
100×75mm						
up to 3.00m long	0.10	6.03	3.22	1.94	m	**11.19**
3.00–5.00m long	0.10	6.03	3.22	3.26	m	**12.51**
150×75mm						
up to 3.00m long	0.13	7.84	4.17	4.00	m	**16.01**
3.00–5.00m long	0.11	6.64	3.53	5.91	m	**16.08**
200×100mm						
up to 3.00m long	0.16	9.65	5.13	7.08	m	**21.86**
3.00–5.00m long	0.14	8.45	4.50	12.56	m	**25.51**
200×200mm						
up to 3.00m long	0.25	15.09	8.02	20.47	m	**43.58**
3.00–5.00m long	0.23	13.88	7.38	30.43	m	**51.69**
5.00–8.00m long	0.20	12.07	6.42	23.49	m	**41.98**
300×200mm						
up to 3.00m long	0.27	16.29	8.67	34.72	m	**59.68**
3.00–5.00m long	0.25	15.09	8.02	50.38	m	**73.49**
5.00–8.00m long	0.23	13.88	7.38	34.87	m	**56.13**
300×300mm						
up to 3.00m long	0.30	18.10	9.64	55.83	m	**83.57**
3.00–5.00m long	0.27	16.29	8.67	80.19	m	**105.15**
5.00–8.00m long	0.25	15.09	8.02	56.04	m	**79.15**
450×450mm						
up to 3.00m long	0.35	21.12	11.23	158.79	m	**191.14**
3.00–5.00m long	0.31	18.71	9.95	216.56	m	**245.22**
5.00–8.00m long	0.28	16.90	8.99	157.22	m	**183.11**
600×600mm						
3.00–5.00m long	0.39	23.53	9.66	289.86	m	**323.05**
5.00–8.00m long	0.37	22.33	11.87	301.64	m	**335.84**
ADD to the above prices for vacuum / pressure impregnating to minimum 5.30kg/m³ salt retention	–	–	–	33.83	m³	**33.83**
HARDWOOD DECKING						
Greenheart; wrought finish						
Thickness 25-50mm						
150×50mm	0.58	35.00	18.62	58.57	m²	**112.19**
Thickness 50-75mm						
200×75mm	0.75	45.26	24.07	76.42	m²	**145.75**
Thickness 75-100mm						
250×100mm	0.95	57.32	29.31	96.83	m²	**183.46**

CLASS O: TIMBER

Item	Gang hours	Labour £	Plant £	Material £	Unit	Total rate £
SOFTWOOD DECKING						
Douglas Fir						
Thickness 25-50 mm						
150×50 mm	0.39	23.53	12.52	29.13	m²	**65.18**
Thickness 50-75 mm						
200×75 mm	0.50	30.17	16.06	38.39	m²	**84.62**
Thickness 75-100 mm						
250×100 mm	0.65	39.22	28.09	42.74	m²	**110.05**
FITTINGS AND FASTENINGS						
Metalwork						
Spikes; mild steel material rosehead						
14×14×275 mm long	0.13	4.33	–	0.94	nr	**5.27**
Metric mild steel bolts, nuts and washers						
M6 × 25mm long	0.05	1.66	–	0.18	nr	**1.84**
M6 × 50mm long	0.05	1.66	–	0.20	nr	**1.86**
M6 × 75mm long	0.05	1.66	–	0.23	nr	**1.89**
M6 × 100mm long	0.06	2.00	–	0.25	nr	**2.25**
M6 × 120mm long	0.06	2.00	–	0.37	nr	**2.37**
M6 × 150mm long	0.06	2.00	–	0.44	nr	**2.44**
M8 × 25mm long	0.05	1.66	–	0.20	nr	**1.86**
M8 × 50mm long	0.05	1.66	–	0.24	nr	**1.90**
M8 × 75mm long	0.06	2.00	–	0.28	nr	**2.28**
M8 × 100mm long	0.06	2.00	–	0.32	nr	**2.32**
M8 × 120mm long	0.07	2.33	–	0.49	nr	**2.82**
M8 × 150mm long	0.07	2.33	–	0.58	nr	**2.91**
M10 × 25mm long	0.05	1.66	–	0.37	nr	**2.03**
M10 × 50mm long	0.06	2.00	–	0.41	nr	**2.41**
M10 × 75mm long	0.06	2.00	–	0.46	nr	**2.46**
M10 × 100mm long	0.07	2.33	–	0.53	nr	**2.86**
M10 × 120mm long	0.07	2.33	–	0.55	nr	**2.88**
M10 × 150mm long	0.07	2.33	–	0.90	nr	**3.23**
M10 × 200mm long	0.08	2.66	–	1.49	nr	**4.15**
M12 × 25mm long	0.06	2.00	–	0.47	nr	**2.47**
M12 × 50mm long	0.06	2.00	–	0.51	nr	**2.51**
M12 × 75mm long	0.07	2.33	–	0.58	nr	**2.91**
M12 × 100mm long	0.07	2.33	–	0.66	nr	**2.99**
M12 × 120mm long	0.08	2.66	–	0.72	nr	**3.38**
M12 × 150mm long	0.08	2.66	–	0.93	nr	**3.59**
M12 × 200mm long	0.08	2.66	–	1.30	nr	**3.96**
M12 × 240mm long	0.09	3.00	–	1.96	nr	**4.96**
M12 × 300mm long	0.10	3.33	–	2.10	nr	**5.43**
M16 × 50mm long	0.07	2.33	–	0.86	nr	**3.19**
M16 × 75mm long	0.07	2.33	–	0.95	nr	**3.28**
M16 × 100mm long	0.08	2.66	–	1.07	nr	**3.73**
M16 × 120mm long	0.08	2.66	–	1.24	nr	**3.90**
M16 × 150mm long	0.09	3.00	–	1.39	nr	**4.39**
M16 × 200mm long	0.09	3.00	–	1.91	nr	**4.91**
M16 × 240mm long	0.10	3.33	–	2.71	nr	**6.04**

CLASS O: TIMBER

Item	Gang hours	Labour £	Plant £	Material £	Unit	Total rate £
M16 × 300mm long	0.10	3.33	–	2.96	nr	6.29
M20 × 50mm long	0.07	2.33	–	0.92	nr	3.25
M20 × 75mm long	0.08	2.66	–	1.50	nr	4.16
M20 × 100mm long	0.08	2.66	–	1.69	nr	4.35
M20 × 120mm long	0.09	3.00	–	2.06	nr	5.06
M20 × 150mm long	0.09	3.00	–	2.21	nr	5.21
M20 × 200mm long	0.10	3.33	–	2.85	nr	6.18
M20 × 240mm long	0.10	3.33	–	3.81	nr	7.14
M20 × 300mm long	0.11	3.66	–	4.22	nr	7.88
M24 × 50mm long	0.08	2.66	–	2.99	nr	5.65
M24 × 75mm long	0.08	2.66	–	3.21	nr	5.87
M24 × 100mm long	0.09	3.00	–	3.32	nr	6.32
M24 × 120mm long	0.09	3.00	–	3.55	nr	6.55
M24 × 150mm long	0.10	3.33	–	3.98	nr	7.31
M24 × 200mm long	0.11	3.66	–	4.59	nr	8.25
M24 × 240mm long	0.11	3.66	–	5.91	nr	9.57
M24 × 300mm long	0.12	3.99	–	6.50	nr	10.49
M30 × 100mm long	0.09	3.00	–	7.93	nr	10.93
M30 × 120mm long	0.10	3.33	–	8.25	nr	11.58
M30 × 150mm long	0.11	3.66	–	8.73	nr	12.39
M30 × 200mm long	0.11	3.66	–	9.53	nr	13.19
Carriage bolts, nuts and washer						
M6 × 25mm long	0.05	1.66	–	0.12	nr	1.78
M6 × 50mm long	0.05	1.66	–	0.22	nr	1.88
M6 × 75mm long	0.05	1.66	–	0.26	nr	1.92
M6 × 100mm long	0.06	2.00	–	0.39	nr	2.39
M6 × 150mm long	0.06	2.00	–	0.58	nr	2.58
M8 × 25mm long	0.05	1.66	–	0.26	nr	1.92
M8 × 50mm long	0.05	1.66	–	0.29	nr	1.95
M8 × 75mm long	0.06	2.00	–	0.31	nr	2.31
M8 × 100mm long	0.06	2.00	–	0.39	nr	2.39
M8 × 150mm long	0.07	2.33	–	0.40	nr	2.73
M8 × 200mm long	0.07	2.33	–	0.85	nr	3.18
M10 × 25mm long	0.05	1.66	–	0.42	nr	2.08
M10 × 50mm long	0.06	2.00	–	0.39	nr	2.39
M10 × 75mm long	0.06	2.00	–	0.41	nr	2.41
M10 × 100mm long	0.07	2.33	–	0.44	nr	2.77
M10 × 150mm long	0.07	2.33	–	0.62	nr	2.95
M10 × 200mm long	0.08	2.66	–	1.06	nr	3.72
M10 × 240mm long	0.08	2.66	–	2.36	nr	5.02
M10 × 300mm long	0.09	3.00	–	2.59	nr	5.59
M12 × 25mm long	0.06	2.00	–	0.30	nr	2.30
M12 × 50mm long	0.06	2.00	–	0.61	nr	2.61
M12 × 75mm long	0.07	2.33	–	0.73	nr	3.06
M12 × 100mm long	0.07	2.33	–	0.82	nr	3.15
M12 × 150mm long	0.08	2.66	–	1.17	nr	3.83
M12 × 200mm long	0.08	2.66	–	1.66	nr	4.32
M12 × 240mm long	0.09	3.00	–	2.88	nr	5.88
M12 × 300mm long	0.10	3.33	–	3.14	nr	6.47

CLASS O: TIMBER

Item	Gang hours	Labour £	Plant £	Material £	Unit	Total rate £
FITTINGS AND FASTENINGS – cont						
Galvanized steel						
Straps						
30 × 2.5 × 600 mm girth	0.13	4.33	–	1.86	nr	6.19
30 × 2.5 × 800 mm girth	0.13	4.33	–	2.48	nr	6.81
30 × 2.5 × 1000 mm girth	0.13	4.33	–	3.09	nr	7.42
30 × 2.5 × 1200 mm girth	0.15	4.99	–	3.71	nr	8.70
30 × 2.5 × 1400 mm girth	0.13	4.33	–	4.33	nr	8.66
30 × 2.5 × 1600 mm girth	0.15	4.99	–	4.95	nr	9.94
30 × 2.5 × 1800 mm girth	0.15	4.99	–	5.57	nr	10.56
30 × 5 × 600 mm girth	0.13	4.33	–	3.71	nr	8.04
30 × 5 × 800 mm girth	0.13	4.33	–	4.95	nr	9.28
30 × 5 × 1000 mm girth	0.13	4.33	–	6.18	nr	10.51
30 × 5 × 1200 mm girth	0.15	4.99	–	7.42	nr	12.41
30 × 5 × 1400 mm girth	0.13	4.33	–	8.65	nr	12.98
30 × 5 × 1600 mm girth	0.15	4.99	–	9.89	nr	14.88
30 × 5 × 1800 mm girth	0.15	4.99	–	11.13	nr	16.12
Timber connectors; round toothed plate, single sided for 10 mm or 12 mm bolts						
38 mm diameter	0.01	0.17	–	1.77	nr	1.94
50 mm diameter	0.01	0.17	–	1.81	nr	1.98
63 mm diameter	0.01	0.27	–	1.95	nr	2.22
75 mm diameter	0.01	0.27	–	2.10	nr	2.37
Timber connectors; round toothed plate, double sided for 10 mm or 12 mm bolts						
38 mm diameter	0.01	0.17	–	1.77	nr	1.94
50 mm diameter	0.01	0.17	–	1.81	nr	1.98
63 mm diameter	0.01	0.27	–	1.95	nr	2.22
75 mm diameter	0.01	0.27	–	2.10	nr	2.37
Split ring connectors						
50 mm diameter	0.06	2.00	–	2.60	nr	4.60
63 mm diameter	0.01	0.20	–	2.60	nr	2.80
101 mm diameter	0.01	0.20	–	6.94	nr	7.14
Shear plate connectors						
67 mm diameter	0.01	0.20	–	4.17	nr	4.37
101 mm diameter	0.01	0.20	–	18.05	nr	18.25
Flitch plates						
200 × 75 × 10 mm	0.07	2.33	–	6.18	nr	8.51
300 × 100 × 10 mm	0.09	3.00	–	12.37	nr	15.37
450 × 150 × 12 mm	0.15	4.99	–	33.38	nr	38.37
Stainless steel						
Straps						
30 × 2.5 × 600 mm girth	0.13	4.33	–	4.20	nr	8.53
30 × 2.5 × 800 mm girth	0.13	4.33	–	5.84	nr	10.17
30 × 2.5 × 1000 mm girth	0.13	4.33	–	3.09	nr	7.42
30 × 2.5 × 1200 mm girth	0.15	4.99	–	3.71	nr	8.70
30 × 2.5 × 1400 mm girth	0.13	4.33	–	4.33	nr	8.66
30 × 2.5 × 1600 mm girth	0.15	4.99	–	4.95	nr	9.94
30 × 2.5 × 1800 mm girth	0.15	4.99	–	5.57	nr	10.56

CLASS O: TIMBER

Item	Gang hours	Labour £	Plant £	Material £	Unit	Total rate £
30 × 5 × 600 mm girth	0.13	4.33	–	6.77	nr	11.10
30 × 5 × 800 mm girth	0.13	4.33	–	9.03	nr	13.36
30 × 5 × 1000 mm girth	0.13	4.33	–	6.18	nr	10.51
30 × 5 × 1200 mm girth	0.15	4.99	–	7.42	nr	12.41
30 × 5 × 1400 mm girth	0.13	4.33	–	8.65	nr	12.98
30 × 5 × 1600 mm girth	0.15	4.99	–	9.89	nr	14.88
30 × 5 × 1800 mm girth	0.15	4.99	–	11.13	nr	16.12
Coach screws						
5.0 mm diameter × 75 mm long	0.04	1.33	–	–	nr	1.33
7.0 mm diameter × 105 mm long	0.05	1.66	–	–	nr	1.66
7.0 mm diameter × 140 mm long	0.06	2.00	–	–	nr	2.00
10.0 mm diameter × 95 mm long	0.06	2.00	–	0.01	nr	2.01
10.0 mm diameter × 165 mm long	0.07	2.33	–	0.01	nr	2.34
Timber connectors; round toothed plate, single sided for 10 mm or 12 mm bolts						
38 mm diameter	0.01	0.17	–	2.78	nr	2.95
50 mm diameter	0.01	0.17	–	2.88	nr	3.05
63 mm diameter	0.01	0.27	–	3.22	nr	3.49
75 mm diameter	0.01	0.27	–	3.66	nr	3.93
Timber connectors; round toothed plate, double sided for 10 mm or 12 mm bolts						
38 mm diameter	0.01	0.17	–	2.78	nr	2.95
50 mm diameter	0.01	0.17	–	2.88	nr	3.05
63 mm diameter	0.01	0.27	–	3.22	nr	3.49
75 mm diameter	0.01	0.27	–	3.66	nr	3.93
Split ring connectors						
63 mm diameter	0.06	2.00	–	12.20	nr	14.20
101 mm diameter	0.06	2.00	–	23.95	nr	25.95
Shear plate connectors						
67 mm diameter	0.06	2.00	–	0.44	nr	2.44
101 mm diameter	0.06	2.00	–	0.99	nr	2.99

CLASS P: PILING

Item	Gang hours	Labour £	Plant £	Material £	Unit	Total rate £
GENERALLY						
There are a number of different types of piling which are available for use in differing situations. Selection of the most suitable type of piling for a particular site will depend on a number of factors including the physical conditions likely to be encountered during driving, the loads to be carried, the design of superstructure, etc. The most commonly used systems are included in this section						
It is essential that a thorough and adequate site investigation is carried out to ascertain details of the ground strata and bearing capacities to enable a proper assessment to be made of the most suitable and economical type of piling to be adopted.						
There are so many factors, apart from design considerations, which influence the cost of piling that it is not possible to give more than an approximate indication of costs. To obtain reliable costs for a particular contract advice should be sought from a company specialising in the particular type of piling proposed. Some Specialist Contractors will also provide a design service if required.						
BORED CAST IN PLACE CONCRETE PILES						
Generally						
The items 'number of piles' are calculated based on the following:						
allowance for provision of all plant, equipment and labour including transporting to and from site and establishing and dismantling at £6,000 in total.						
moving the rig to and setting up at each pile position; preparing to commence driving; £40.00, £60.00 and 75.00 per 300 mm, 450 mm and 600 mm diameter piles using the tripod mounted percussion rig .						
Standing time is quoted at £100.00 per hour for tripod rig, £195.00 per hour for mobile rig and £120.00 per hour for continuous flight auger..						

CLASS P: PILING

Item	Gang hours	Labour £	Plant £	Material £	Unit	Total rate £
Disposal of material arising from pile bores						
The disposal of excavated material is shown						
separately, partly as this task is generally carried out						
by the main contractor rather than the piling						
specialist, but also to allow for simple adjustment						
should contaminated ground be envisaged.						
Disposal of material arising from pile bores;						
collection from around piling operations						
storage on site; 100 m maximum distance; using						
22.5 t ADT	0.05	0.90	2.38	–	m³	3.28
removal; 5 km distance; using 20 t tipper	0.12	2.17	6.41	–	m³	8.58
removal; 15 km distance; using 20 t tipper	0.22	3.97	11.74	–	m³	15.71
Add to the above rates where tipping charges apply						
(excluding Landfill Tax):						
non-hazardous waste	–	–	–	–	m³	36.30
hazardous waste	–	–	–	–	m³	74.25
special waste	–	–	–	–	m³	90.75
contaminated liquid	–	–	–	–	m³	99.00
contaminated sludge	–	–	–	–	m³	132.00
Add to the above rates where Landfill Tax applies:						
exempted material	–	–	–	–	m³	–
inactive or inert material	–	–	–	–	m³	3.75
other material	–	–	–	–	m³	84.00
Concrete 35 N/mm², 20 mm aggregate; installed by lorry/crawler-mounted rotary rig						
The following unit costs cover the construction of small						
diameter bored piles using lorry or crawler mounted						
rotary boring rigs. This type of plant is more mobile and						
faster in operation than the tripod rigs and is ideal for						
large contracts in cohesive ground. Construction of						
piles under bentonite suspension can be carried out to						
obviate the use of liners. Standard diameters of 450–						
900 mm diameter can be constructed to depths of 30 m.						
The costs are based on installing 100 piles on a						
clear site with reasonable access.						
Diameter: 300 mm						
number of piles (see above)	–	–	–	–	nr	156.35
concreted length	–	–	–	–	m	10.39
depth bored to 10 m maximum depth	–	–	–	–	m	32.49
depth bored to 15 m maximum depth	–	–	–	–	m	32.49
depth bored to 20 m maximum depth	–	–	–	–	m	32.49
Diameter: 450 mm						
number of piles (see above)	–	–	–	–	nr	156.35
concreted length	–	–	–	–	m	13.00
depth bored to 10 m maximum depth	–	–	–	–	m	40.62
depth bored to 15 m maximum depth	–	–	–	–	m	40.62
depth bored to 20 m maximum depth	–	–	–	–	m	40.62

CLASS P: PILING

Item	Gang hours	Labour £	Plant £	Material £	Unit	Total rate £
BORED CAST IN PLACE CONCRETE PILES – cont						
Concrete 35N/mm², 20mm aggregate – cont						
Diameter: 600mm						
number of piles (see above)	–	–	–	–	nr	156.35
concreted length	–	–	–	–	m	33.51
depth bored to 10m maximum depth	–	–	–	–	m	73.09
depth bored to 15m maximum depth	–	–	–	–	m	73.09
depth bored to 20m maximum depth	–	–	–	–	m	73.09
Concrete 35N/mm², 20mm aggregate; installed by continuous flight auger						
The following unit costs cover the construction of piles by screwing a continuous flight auger into the ground to a design depth (Determined prior to commencement of piling operations and upon which the rates are based and subsequently varied to actual depths). Concrete is then pumped through the hollow stem of the auger to the bottom and the pile formed as the auger is withdrawn. Spoil is removed by the auger as it is withdrawn. This is a fast method of construction without causing disturbance or vibration to adjacent ground. No casing is required even in unsuitable soils. Reinforcement can be placed after grouting is complete.						
The costs are based on installing 100 piles on a clear site with reasonable access.						
Diameter: 300mm						
number of piles	–	–	–	–	nr	36.50
concreted length	–	–	–	–	m	12.00
depth bored to 10m maximum depth	–	–	–	–	nr	12.25
depth bored to 15m maximum depth	–	–	–	–	nr	10.90
depth bored to 20m maximum depth	–	–	–	–	nr	10.50
Diameter: 450mm						
number of piles	–	–	–	–	nr	36.50
concreted length	–	–	–	–	m	22.25
depth bored to 10m maximum depth	–	–	–	–	nr	14.00
depth bored to 15m maximum depth	–	–	–	–	nr	13.40
depth bored to 20m maximum depth	–	–	–	–	nr	12.50
Diameter: 600mm						
number of piles	–	–	–	–	nr	36.50
concreted length	–	–	–	–	m	38.00
depth bored to 10m maximum depth	–	–	–	–	nr	14.30
depth bored to 15m maximum depth	–	–	–	–	nr	13.60
depth bored to 20m maximum depth	–	–	–	–	nr	12.90

CLASS P: PILING

Item	Gang hours	Labour £	Plant £	Material £	Unit	Total rate £
DRIVEN CAST IN PLACE CONCRETE PILES						
Generally						
The items 'number of piles' are calculated based on the following:						
allowance for provision of all plant, equipment and labour including transporting to and from site and establishing and dismantling at £6,250 in total for piles using the Temporary Steel Casing Method and £10,500 in total for piles using the Segmental Casing Method.						
moving the rig to and setting up at each pile position;						
preparing to commence driving; £110.00 per pile.						
For the Temporary Steel Casing Method, obstructions (where within the capabilities of the normal plant) are quoted at £160.00 per hour and standing time at £152.50 per hour.						
For the Segmental Steel Casing Method, obstructions (where within the capabilities of the normal plant) are quoted at £275.00 per hour and standing time at £252.00 per hour.						
Temporary steel casing method; concrete 35 N/ mm²; reinforced for 750 kN						
The following unit costs cover the construction of piles by driving a heavy steel tube into the ground either by using an internal hammer acting on a gravel or concrete plug, as is more usual, or by using an external hammer on a driving helmet at the top of the tube. After driving to the required depth an enlarged base is formed by hammering out sucessive charges of concrete down the tube. The tube is then filled with concrete which is compacted as the tube is vibrated and withdrawn. Piles of 350 to 500 mm diameter can be constructed with rakes up to 1 in 4 to carry working loads up to 120 t per pile. The costs are based on installing 100 piles on a clear site with reasonable access.						
Diameter 430 mm; drive shell and form pile						
number of piles	–	–	–	–	nr	205.78
concreted length	–	–	–	–	m	23.00
depth driven; bottom-driven method	–	–	–	–	m	6.70
depth driven; top-driven method	–	–	–	–	m	3.85

CLASS P: PILING

Item	Gang hours	Labour £	Plant £	Material £	Unit	Total rate £
DRIVEN CAST IN PLACE CONCRETE PILES – cont						
Segmental casing method; concrete 35 N/mm²; nominal reinforcement						
The following unit costs cover the construction of piles by driving into hard material using a serrated thick walled tube. It is oscillated and pressed into the hard material using a hydraulic attachment to the piling rig. The hard material is broken up using chiselling methods and is then removed by mechanical grab. The costs are based on installing 100 piles on a clear site with reasonable access.						
Diameter 620 mm						
number of piles	–	–	–	–	nr	270.75
concreted length	–	–	–	–	m	131.90
depth bored or driven to 15 m maximum depth	–	–	–	–	m	10.00
Diameter 1180 mm						
number of piles	–	–	–	–	nr	270.75
concreted length	–	–	–	–	m	143.15
depth bored or driven to 15 m maximum depth	–	–	–	–	m	15.00
Diameter 1500 mm						
number of piles	–	–	–	–	nr	270.75
concreted length	–	–	–	–	m	192.29
depth bored or driven to 15 m maximum depth	–	–	–	–	m	23.76
PREFORMED CONCRETE PILES						
The following unit costs cover the installation of driven precast concrete piles by using a hammer acting on shoe fitted or cast into the precast concrete pile unit. Single pile lengths are normally a maximum of 13m long, at which point, a mechanical interlocking joint is required to extend the pile. These joints are most economically and practically formed at works. Lengths, sizes of sections, reinforcement details and concrete mixes vary for differing contractors, whose specialist advice should be sought for specific designs. The following unit costs are based on installing 100 piles on a clear site with reasonable access. The items 'number of piles' are calculated based on the following: allowance for provision of all plant, equipment and labour including transporting to and from site and establishing and dismantling at £3,150 in total for piles up to 275 × 275 mm and £3,850 in total for piles 350 × 350 mm and over.						

CLASS P: PILING

Item	Gang hours	Labour £	Plant £	Material £	Unit	Total rate £
moving the rig to and setting up at each pile position; preparing to commence driving; piles up to 275×275mm £35.00 each; piles 350×350mm and over, £55.00 each.						
an allowance for the cost of the driving head and shoe; £35.00 for 235×235mm piles, £45.00 for 275×275mm and £55.00 for 350×350mm.						
cost of providing the pile of the stated length						
Typical allowances for standing time are £138.50 per hour for 235×235mm piles, £157.50 for 275×275mm and £189.50 for 350×350mm.						
Concrete 50 N/mm²; reinforced for 600 kN						
The costs are based on installing 100 piles on a clear site with reasonable access.						
Cross-sectional area: 0.05–0.1 m²; 235×235mm						
number of piles of 10m length	–	–	–	–	nr	301.90
number of piles of 15m length	–	–	–	–	nr	401.90
number of piles of 20m length	–	–	–	–	nr	501.90
number of piles of 25m length	–	–	–	–	nr	601.90
add for mechanical interlocking joint	–	–	–	–	nr	60.00
depth driven	–	–	–	–	m	2.75
Cross-sectional area: 0.05–0.1 m²; 275×275mm						
number of piles of 10m length	–	–	–	–	nr	336.90
number of piles of 15m length	–	–	–	–	nr	454.40
number of piles of 20m length	–	–	–	–	nr	571.90
number of piles of 25m length	–	–	–	–	nr	689.40
add for mechanical interlocking joint	–	–	–	–	nr	62.60
depth driven	–	–	–	–	m	2.90
Cross-sectional area: 0.1–0.15 m²; 350×350mm						
number of piles of 10m length	–	–	–	–	nr	527.60
number of piles of 15m length	–	–	–	–	nr	712.60
number of piles of 20m length	–	–	–	–	nr	897.60
number of piles of 25m length	–	–	–	–	nr	1082.60
number of piles of 30m length	–	–	–	–	nr	1267.60
add for mechanical interlocking joint	–	–	–	–	nr	120.00
depth driven	–	–	–	–	m	4.10

CLASS P: PILING

Item	Gang hours	Labour £	Plant £	Material £	Unit	Total rate £
TIMBER PILES						
The items 'number of piles' are calculated based on the following:						
allowance for provision of all plant, equipment and labour including transporting to and from site and establishing and dismantling at £6,000 in total.						
moving the rig to and setting up at each pile position and preparing to drive at £75.00 per pile.						
allowance for the cost of the driving head and shoe at £40.00 per 225×225 mm pile, £50.00 for 300×300 mm, £60.00 for 350×350 mm and £70.00 for 450×450mm.						
cost of providing the pile of the stated length						
A typical allowance for standing time is £245.50 per hour.						
Douglas Fir; hewn to mean pile size						
The costa are based on installing 100 piles on a clear site with reasonable access.						
Cross-sectional area: 0.05–0.1 m²; 225×225 mm						
number of piles of 10 m length	–	–	–	–	nr	528.68
number of piles of 15 m length	–	–	–	–	nr	696.08
number of piles of 20 m length	–	–	–	–	nr	863.48
depth driven	–	–	–	–	m	3.42
Cross-sectional area: 0.05–0.1 m²; 300×300 mm						
number of piles of 10 m length	–	–	–	–	nr	801.87
number of piles of 15 m length	–	–	–	–	nr	1099.53
number of piles of 20 m length	–	–	–	–	nr	1397.19
depth driven	–	–	–	–	m	4.09
Cross-sectional area: 0.1–0.15 m²; 350×350 mm						
number of piles of 10 m length	–	–	–	–	nr	1029.60
number of piles of 15 m length	–	–	–	–	nr	1434.78
number of piles of 20 m length	–	–	–	–	nr	1839.96
depth driven	–	–	–	–	m	4.69
Cross-sectional area: 0.15–0.25 m²; 450×450 mm						
number of piles of 10 m length	–	–	–	–	nr	1571.47
number of piles of 15 m length	–	–	–	–	nr	2241.25
number of piles of 20 m length	–	–	–	–	nr	2911.03
number of piles of 25 m length	–	–	–	–	nr	3580.81
depth driven	–	–	–	–	m	6.15
Greenheart; hewn to mean pile size						
The costs are based on installing 100 piles on a clear site with reasonable access.						
Cross-sectional area: 0.05–0.1 m²; 225×225 mm						
number of piles of 10 m length	–	–	–	–	nr	595.76
number of piles of 15 m length	–	–	–	–	nr	796.70
number of piles of 20 m length	–	–	–	–	nr	997.64
depth driven	–	–	–	–	m	3.42

CLASS P: PILING

Item	Gang hours	Labour £	Plant £	Material £	Unit	Total rate £
Cross-sectional area: 0.05–0.1 m²; 300 × 300 mm						
number of piles of 10 m length	–	–	–	–	nr	921.03
number of piles of 15 m length	–	–	–	–	nr	1278.27
number of piles of 20 m length	–	–	–	–	nr	1635.51
depth driven	–	–	–	–	m	4.09
Cross-sectional area: 0.1–0.15 m²; 350 × 350 mm						
number of piles of 10 m length	–	–	–	–	nr	1191.60
number of piles of 15 m length	–	–	–	–	nr	1677.78
number of piles of 20 m length	–	–	–	–	nr	2163.96
depth driven	–	–	–	–	m	4.69
Cross-sectional area: 0.15–0.25 m²; 450 × 450 mm						
number of piles of 10 m length	–	–	–	–	nr	1844.35
number of piles of 15 m length	–	–	–	–	nr	2650.57
number of piles of 20 m length	–	–	–	–	nr	3456.79
number of piles of 25 m length	–	–	–	–	nr	4263.01
depth driven	–	–	–	–	m	6.15

ISOLATED STEEL PILES

Steel bearing piles are commonly carried out by a Specialist Contractor and whose advice should be sought to arrive at accurate costing. However the following items can be used to assess a budget cost for such work.

The following unit costs are based upon driving 100 nr steel bearing piles on a clear site with reasonable access.

The items 'number of piles' are calculated based on the following:

allowance for provision of all plant, equipment and labour including transporting to and from site and establishing and dismantling at £6,600 in total up to a maximum 100 miles radius from base and £16,050 up to a maximum 250 miles radius from base.

moving the rig to and setting up at each pile position; preparing to commence driving; £193.15 per pile.

cost of providing the pile of the stated length

A typical allowance for standing time is £283.90 per hour.

Steel EN 10025 grade S275; within 100 miles of steel plant

The costs are based upon installing 100 nr on a clear site with reasonable access.

Item	Gang hours	Labour £	Plant £	Material £	Unit	Total rate £
Mass 45 kg/m; 203 × 203 mm						
number of piles: length 10 m	–	–	–	5205.25	nr	5205.25
number of piles: length 15 m	–	–	–	5529.41	nr	5529.41
number of piles: length 20 m	–	–	–	5851.05	nr	5851.05
depth driven; vertical	–	–	–	–	m	3.23
depth driven; raking	–	–	–	–	m	5.11

CLASS P: PILING

Item	Gang hours	Labour £	Plant £	Material £	Unit	Total rate £
ISOLATED STEEL PILES – cont						
Steel EN 10025 grade S275 – cont						
Mass 54 kg/m; 203 × 203 mm						
number of piles; length 10 m	–	–	–	5333.40	nr	**5333.40**
number of piles; length 15 m	–	–	–	5722.40	nr	**5722.40**
number of piles; length 20 m	–	–	–	6108.36	nr	**6108.36**
depth driven; vertical	–	–	–	–	m	**3.23**
depth driven; raking	–	–	–	–	m	**5.11**
Mass 63 kg/m; 254 × 254 mm						
number of piles; length 10 m	–	–	–	5468.62	nr	**5468.62**
number of piles; length 15 m	–	–	–	5925.98	nr	**5925.98**
number of piles; length 20 m	–	–	–	6379.81	nr	**6379.81**
depth driven; vertical	–	–	–	–	m	**3.49**
depth driven; raking	–	–	–	–	m	**5.38**
Mass 71 kg/m; 254 × 254 mm						
number of piles; length 10 m	–	–	–	5583.43	nr	**5583.43**
number of piles; length 15 m	–	–	–	6098.87	nr	**6098.87**
number of piles; length 20 m	–	–	–	6610.33	nr	**6610.33**
depth driven; vertical	–	–	–	–	m	**3.49**
depth driven; raking	–	–	–	–	m	**5.38**
Mass 85 kg/m; 254 × 254 mm						
number of piles; length 10 m	–	–	–	5784.35	nr	**5784.35**
number of piles; length 15 m	–	–	–	6401.42	nr	**6401.42**
number of piles; length 20 m	–	–	–	7013.73	nr	**7013.73**
depth driven; vertical	–	–	–	–	m	**3.49**
depth driven; raking	–	–	–	–	m	**5.38**
Mass 79 kg/m; 305 × 305 mm						
number of piles; length 10 m	–	–	–	5715.97	nr	**5715.97**
number of piles; length 15 m	–	–	–	6298.35	nr	**6298.35**
number of piles; length 20 m	–	–	–	6876.30	nr	**6876.30**
depth driven; vertical	–	–	–	–	m	**3.49**
depth driven; raking	–	–	–	–	m	**5.38**
Mass 95 kg/m; 305 × 305 mm						
number of piles; length 10 m	–	–	–	5949.17	nr	**5949.17**
number of piles; length 15 m	–	–	–	6649.51	nr	**6649.51**
number of piles; length 20 m	–	–	–	7344.51	nr	**7344.51**
depth driven; vertical	–	–	–	–	m	**3.76**
depth driven; raking	–	–	–	–	m	**5.70**
Mass 110 kg/m; 305 × 305 mm						
number of piles; length 10 m	–	–	–	6167.81	nr	**6167.81**
number of piles; length 15 m	–	–	–	6978.72	nr	**6978.72**
number of piles; length 20 m	–	–	–	7783.46	nr	**7783.46**
depth driven; vertical	–	–	–	–	m	**3.76**
depth driven; raking	–	–	–	–	m	**5.70**
Mass 109 kg/m; 356 × 368 mm						
number of piles; length 10 m	–	–	–	6177.69	nr	**6177.69**
number of piles; length 15 m	–	–	–	6993.46	nr	**6993.46**
number of piles; length 20 m	–	–	–	7803.11	nr	**7803.11**
depth driven; vertical	–	–	–	–	m	**3.76**
depth driven; raking	–	–	–	–	m	**5.70**

CLASS P: PILING

Item	Gang hours	Labour £	Plant £	Material £	Unit	Total rate £
Mass 126 kg/m; 305 × 305 mm						
number of piles; length 10 m	–	–	–	6401.02	nr	6401.02
number of piles; length 15 m	–	–	–	7329.88	nr	7329.88
number of piles; length 20 m	–	–	–	8251.67	nr	8251.67
driving piles; vertical	–	–	–	–	m	4.08
driving piles; raking	–	–	–	–	m	6.02
Mass 149 kg/m; 305 × 305 mm						
number of piles; length 10 m	–	–	–	6736.25	nr	6736.25
number of piles; length 15 m	–	–	–	7834.67	nr	7834.67
number of piles; length 20 m	–	–	–	8924.72	nr	8924.72
driving piles; vertical	–	–	–	–	m	4.08
driving piles; raking	–	–	–	–	m	6.02
Mass 186 kg/m; 305 × 305 mm						
number of piles; length 10 m	–	–	–	7275.55	nr	7275.55
number of piles; length 15 m	–	–	–	8646.72	nr	8646.72
number of piles; length 20 m	–	–	–	10007.46	nr	10007.46
driving piles; vertical	–	–	–	–	m	4.68
drive piles; raking	–	–	–	–	m	6.61
Mass 223 kg/m; 305 × 305 mm						
number of piles; length 10 m	–	–	–	7814.84	nr	7814.84
number of piles; length 15 m	–	–	–	9458.78	nr	9458.78
number of piles; length 20 m	–	–	–	11090.20	nr	11090.20
driving piles; vertical	–	–	–	–	m	4.95
driving piles; raking	–	–	–	–	m	6.88
Mass 133 kg/m; 356 × 368 mm						
number of piles; length 10 m	–	–	–	6532.89	nr	6532.89
number of piles; length 15 m	–	–	–	7528.28	nr	7528.28
number of piles; length 20 m	–	–	–	8516.20	nr	8516.20
driving piles; vertical	–	–	–	–	m	4.08
driving piles; raking	–	–	–	–	m	6.02
Mass 152 kg/m; 356 × 368 mm						
number of piles; length 10 m	–	–	–	6814.09	nr	6814.09
number of piles; length 15 m	–	–	–	7951.67	nr	7951.67
number of piles; length 20 m	–	–	–	9080.73	nr	9080.73
driving piles; vertical	–	–	–	–	m	4.41
driving piles; raking	–	–	–	–	m	6.29
Mass 174 kg/m; 356 × 368 mm						
number of piles; length 10 m	–	–	–	7139.69	nr	7139.69
number of piles; length 15 m	–	–	–	8441.92	nr	8441.92
number of piles; length 20 m	–	–	–	9734.40	nr	9734.40
driving piles; vertical	–	–	–	–	m	4.41
driving piles; raking	–	–	–	–	m	6.29

CLASS P: PILING

Item	Gang hours	Labour £	Plant £	Material £	Unit	Total rate £
INTERLOCKING STEEL PILES						
Sheet steel piling is commonly carried out by a Specialist Contractor, whose advice should be sought to arrive at accurate costings. However the following items can be used to assess a budget for such work.						
Note: area of driven piles will vary from area supplied dependant upon pitch line of piling and provision for such allowance has been made in PC for supply.						
The materials cost below includes the manufacturers tarrifs for a 200 mile delivery radius from works, delivery 10–20t loads and with an allowance of 10% to cover waste / projecting piles etc.						
Arcelor Mittal Z section steel piles; EN 10248 grade S270GP steel						
The following unit costs are based on driving/ extracting 1,500 m² of sheet piling on a clear site with reasonable access.						
Provision of all plant, equipment and labour including transport to and from the site and establishing and dismantling for						
driving of sheet piling	–	–	–	–	sum	7350.00
extraction of sheet piling	–	–	–	–	sum	4987.50
Standing time	–	–	–	–	hr	352.80
Section modulus 800–1200 cm³/m; section reference AZ 12; mass 98.7 kg/m², sectional modulus 1200 cm³/m; EN 10248 grade S270GP steel						
length of welded corner piles	–	–	–	–	m	68.25
length of welded junction piles	–	–	–	–	m	110.20
driven area	–	–	–	–	m²	123.90
area of piles of length not exceeding 14 m	–	–	–	84.76	m²	84.76
area of piles of length 14–24 m	–	–	–	90.90	m²	90.90
area of piles of length exceeding 24 m	–	–	–	90.90	m²	90.90
Section modulus 1200–20000 cm³/m; section reference AZ 17; mass 108.6 kg/m²; sectional modulus 1665 cm³/m; EN 10248 grade S270GP steel						
length of welded corner piles	–	–	–	–	m	68.25
length of welded junction piles	–	–	–	–	m	110.20
driven area	–	–	–	–	m²	147.00
area of piles of length not exceeding 14 m	–	–	–	89.81	m²	89.81
area of piles of length 14–24 m	–	–	–	90.90	m²	90.90
area of piles of length exceeding 24 m	–	–	–	90.90	m²	90.90
Section modulus 2000–3000 cm³/m; section reference AZ 26; mass 155.2 kg/m²; sectional modulus 2600 cm³/m; EN 10248 grade S270GP steel						
driven area	–	–	–	–	m²	178.60
area of piles of length 6–18 m	–	–	–	101.75	m²	101.75
area of piles of length 18–24 m	–	–	–	102.99	m²	102.99

CLASS P: PILING

Item	Gang hours	Labour £	Plant £	Material £	Unit	Total rate £
Section modulus 3000–4000 cm^3/m; section reference AZ 36; mass 194.0 kg/m^2; sectional modulus 3600 cm^3/m; EN 10248 grade S270GP steel						
driven area	–	–	–	–	m^2	245.91
area of piles of length 6–18 m	–	–	–	124.24	m^2	124.24
area of piles of length 18–24 m	–	–	–	125.75	m^2	125.75
Straight section modulus ne 500 cm^3/m; section reference AS 500-12 mass 149 kg/m^2; sectional modulus 51 cm^3/m; EN 10248 grade S270GP steel						
driven area	–	–	–	–	m^2	178.60
area of piles of length 6–18 m	–	–	–	154.81	m^2	154.81
area of piles of length 18–24 m	–	–	–	156.69	m^2	156.69
One coat black tar vinyl (PC1) protective treatment applied all surfaces at shop to minimum dry film thickness up to 150 microns to steel piles						
section reference AZ 12; pile area	–	–	–	11.05	m^2	11.05
section reference AZ 17; pile area	–	–	–	11.30	m^2	11.30
section reference AZ 26; pile area	–	–	–	12.35	m^2	12.35
section reference AZ 36; pile area	–	–	–	13.15	m^2	13.15
section reference AS 500 - 12; pile area	–	–	–	13.73	m^2	13.73
One coat black high build isocyanate cured epoxy pitch (PC2) protective treatment applied all surfaces at shop to minimum dry film thickness up to 450 microns to steel piles						
section reference AZ 12; pile area	–	–	–	17.36	m^2	17.36
section reference AZ 17; pile area	–	–	–	17.75	m^2	17.75
section reference AZ 26; pile area	–	–	–	19.40	m^2	19.40
section reference AZ 36; pile area	–	–	–	20.66	m^2	20.66
section reference AS 500 - 12; pile area	–	–	–	21.58	m^2	21.58
Arcelor Mittal U section steel piles; EN 10248 grade S270GP steel						
The following unit costs are based on driving/ extracting 1,500 m^2 of sheet piling on a clear site with reasonable access.						
Provision of plant, equipment and labour including transport to and from the site and establishing and dismantling						
driving of sheet piling	–	–	–	–	sum	7100.00
extraction of sheet piling	–	–	–	–	sum	4750.00
Standing time	–	–	–	–	hr	330.00
Section modulus 500–800 cm^3/m; section reference PU 6; mass 76.0 kg/m^2; sectional modulus 600 cm^3/m						
driven area	–	–	–	–	m^2	100.00
area of piles of length 6–18 m	–	–	–	72.78	m^2	72.78
area of piles of length 18–24 m	–	–	–	73.72	m^2	73.72
Section modulus 800–1200 cm^3/m; section reference PU 8; mass 90.9 kg/m^2; sectional modulus 830 cm^3/m						
driven area	–	–	–	–	m^2	110.00
area of piles of length 6–18 m	–	–	–	76.97	m^2	76.97
area of piles of length 18–24 m	–	–	–	77.98	m^2	77.98

CLASS P: PILING

Item	Gang hours	Labour £	Plant £	Material £	Unit	Total rate £
INTERLOCKING STEEL PILES – cont						
Arcelor Mittal U section steel piles – cont						
Section modulus 1200–2000 cm^3/m; section reference PU 12; mass 110.1 kg/m^2; sectional modulus 1200 cm^3/m						
driven area	–	–	–	–	m^2	**128.00**
area of piles of length 6–18 m	–	–	–	90.98	m^2	**90.98**
area of piles of length 18–24 m	–	–	–	92.15	m^2	**92.15**
Section modulus 1200–2000 cm^3/m; section reference PU 18; mass 128.2 kg/m^2; sectional modulus 1800 cm^3/m						
driven area	–	–	–	–	m^2	**141.00**
area of piles of length 6–18 m	–	–	–	105.67	m^2	**105.67**
area of piles of length 18–24 m	–	–	–	107.03	m^2	**107.03**
Section modulus 2000–3000 cm^3/m; section reference PU 22; mass 143.6 kg/m^2; sectional modulus 2200 cm^3/m						
driven area	–	–	–	–	m^2	**141.00**
area of piles of length 6–18 m	–	–	–	118.58	m^2	**118.58**
area of piles of length 18–24 m	–	–	–	120.11	m^2	**120.11**
Section modulus 3000–4000 cm^3/m; section reference PU 32; mass 190.2 kg/m^2; sectional modulus 3200 cm^3/m						
driven area	–	–	–	–	m^2	**176.80**
area of piles of length 6–18 m	–	–	–	134.05	m^2	**134.05**
area of piles of length 18–24 m	–	–	–	135.78	m^2	**135.78**
One coat black tar vinyl (PC1) protective treatment applied all surfaces at shop to minimum dry film thickness up to 150 microns to steel piles						
section reference PU 6; pile area	–	–	–	10.15	m^2	**10.15**
section reference PU 8; pile area	–	–	–	9.97	m^2	**9.97**
section reference PU 12; pile area	–	–	–	10.61	m^2	**10.61**
section reference PU 18; pile area	–	–	–	11.38	m^2	**11.38**
section reference PU 22; pile area	–	–	–	11.79	m^2	**11.79**
section reference PU 32; pile area	–	–	–	11.97	m^2	**11.97**
One coat black high build isocyanate cured epoxy pitch (PC2) protective treatment applied all surfaces at shop to minimum dry film thickness up to 450 microns to steel piles						
section reference PU 6; pile area	–	–	–	15.95	m^2	**15.95**
section reference PU 8; pile area	–	–	–	15.68	m^2	**15.68**
section reference PU 12; pile area	–	–	–	16.66	m^2	**16.66**
section reference PU 18; pile area	–	–	–	17.88	m^2	**17.88**
section reference PU 22; pile area	–	–	–	18.54	m^2	**18.54**
section reference PU 32; pile area	–	–	–	18.81	m^2	**18.81**

CLASS Q: PILING ANCILLARIES

Item	Gang hours	Labour £	Plant £	Material £	Unit	Total rate £
CAST IN PLACE CONCRETE PILES						
Bored; lorry/crawler mounted rotary rig						
Backfilling empty bore with selected excavated material						
diameter 500mm	–	–	–	–	m	3.40
Permanent casings; each length not exceeding 13 m						
diameter 500mm	–	–	–	–	m	73.62
Permanent casings; each length exceeding 13 m						
diameter 500mm	–	–	–	–	m	79.44
Enlarged bases						
diameter 1500 mm; to 500mm diameter pile	–	–	–	–	nr	251.23
Cutting off surplus lengths						
diameter 500mm	–	–	–	–	m	27.27
Preparing heads						
500mm diameter	–	–	–	–	nr	40.90
Bored; continuous flight auger						
Backfilling empty bore with selected excavated material						
450 mm diameter piles	–	–	–	–	m	2.73
600 mm diameter piles	–	–	–	–	m	3.75
750 mm diameter piles	–	–	–	–	m	4.08
Permanent casings; each length not exceeding 13 m						
450 mm diameter piles	–	–	–	–	m	66.15
600 mm diameter piles	–	–	–	–	m	89.83
750 mm diameter piles	–	–	–	–	m	112.45
Permanent casings; each length exceeding 13 m						
450 mm diameter piles	–	–	–	–	m	68.78
600 mm diameter piles	–	–	–	–	m	95.55
750 mm diameter piles	–	–	–	–	m	118.23
Enlarged bases						
diameter 1400 mm; to 450 mm diameter piles	–	–	–	–	nr	229.70
diameter 1800 mm; to 600 mm diameter piles	–	–	–	–	nr	279.95
diameter 2100 mm; to 750 mm diameter piles	–	–	–	–	nr	314.94
Cutting off surplus lengths						
450 mm diameter piles	–	–	–	–	m	23.86
600 mm diameter piles	–	–	–	–	m	34.07
750 mm diameter piles	–	–	–	–	m	40.90
Preparing heads						
450 mm diameter piles	–	–	–	–	nr	25.53
600 mm diameter piles	–	–	–	–	nr	40.90
750 mm diameter piles	–	–	–	–	nr	61.33
Collection from around pile heads of spoil accruing from piling operations and depositing in spoil heaps (For final disposal see Class E - Excavation Ancillaries)	–	–	–	–	m³	2.96

Unit Costs – Civil Engineering Works

CLASS Q: PILING ANCILLARIES

Item	Gang hours	Labour £	Plant £	Material £	Unit	Total rate £
CAST IN PLACE CONCRETE PILES – cont						
Reinforcement; mild steel						
Straight bars, nominal size						
6 mm	–	–	–	–	t	619.50
8 mm	–	–	–	–	t	619.50
10 mm	–	–	–	–	t	614.25
12 mm	–	–	–	–	t	603.75
16 mm	–	–	–	–	t	598.50
25 mm	–	–	–	–	t	598.50
32 mm	–	–	–	–	t	603.75
50 mm	–	–	–	–	t	609.00
Helical bars, nominal size						
6 mm	–	–	–	–	t	640.50
8 mm	–	–	–	–	t	640.50
10 mm	–	–	–	–	t	635.25
12 mm	–	–	–	–	t	624.75
Reinforcement; high tensile steel						
Straight bars, nominal size						
6 mm	–	–	–	–	t	640.50
8 mm	–	–	–	–	t	640.50
10 mm	–	–	–	–	t	635.25
12 mm	–	–	–	–	t	624.75
16 mm	–	–	–	–	t	619.50
25 mm	–	–	–	–	t	619.50
32 mm	–	–	–	–	t	624.75
50 mm	–	–	–	–	t	630.00
Helical bars, nominal size						
6 mm	–	–	–	–	t	661.50
8 mm	–	–	–	–	t	661.50
10 mm	–	–	–	–	t	656.25
12 mm	–	–	–	–	t	645.75
Couplers; Lenton type A; threaded ends on reinforcing bars						
12 mm	0.09	10.23	–	6.72	nr	16.95
16 mm	0.09	10.23	–	8.11	nr	18.34
20 mm	0.09	10.23	–	11.84	nr	22.07
25 mm	0.09	10.23	–	16.37	nr	26.60
32 mm	0.09	10.23	–	22.47	nr	32.70
40 mm	0.09	10.23	–	30.87	nr	41.10
Couplers; Lenton type B; threaded ends on reinforcing bars						
12 mm	0.09	10.23	–	21.59	nr	31.82
16 mm	0.09	10.23	–	25.02	nr	35.25
20 mm	0.09	10.23	–	27.75	nr	37.98
25 mm	0.09	10.23	–	32.04	nr	42.27
32 mm	0.09	10.23	–	43.30	nr	53.53
40 mm	0.09	10.23	–	63.48	nr	73.71

CLASS Q: PILING ANCILLARIES

Item	Gang hours	Labour £	Plant £	Material £	Unit	Total rate £
PREFORMED CONCRETE PILES						
General						
Preparing heads						
235×235mm piles	–	–	–	–	nr	33.73
275×275mm piles	–	–	–	–	nr	46.06
350×350mm piles	–	–	–	–	nr	70.83
TIMBER PILES						
Douglas Fir						
Cutting off surplus lengths						
cross-sectional area: 0.025–0.05 m^2	–	–	–	–	nr	2.94
cross-sectional area: 0.05–0.1 m^2	–	–	–	–	nr	5.25
cross-sectional area: 0.1–0.15 m^2	–	–	–	–	nr	6.73
cross-sectional area: 0.15–0.25 m^2	–	–	–	–	nr	12.89
Preparing heads						
cross-sectional area: 0.025–0.05 m^2	–	–	–	–	nr	2.94
cross-sectional area: 0.05–0.1 m^2	–	–	–	–	nr	5.25
cross-sectional area: 0.1–0.15 m^2	–	–	–	–	nr	6.73
cross-sectional area: 0.15–0.25 m^2	–	–	–	–	nr	12.89
Greenheart						
Cutting off surplus lengths						
cross-sectional area: 0.025–0.05 m^2	–	–	–	–	nr	5.81
cross-sectional area: 0.05–0.1 m^2	–	–	–	–	nr	10.44
cross-sectional area: 0.1–0.15 m^2	–	–	–	–	nr	13.52
cross-sectional area: 0.15–0.25 m^2	–	–	–	–	nr	25.05
Preparing heads						
cross-sectional area: 0.025–0.05 m^2	–	–	–	–	nr	5.81
cross-sectional area: 0.05–0.1 m^2	–	–	–	–	nr	10.44
cross-sectional area: 0.1–0.15 m^2	–	–	–	–	nr	13.52
cross-sectional area: 0.15–0.25 m^2	–	–	–	–	nr	25.05

CLASS Q: PILING ANCILLARIES

Item	Gang hours	Labour £	Plant £	Material £	Unit	Total rate £
ISOLATED STEEL PILES						
Steel bearing piles						
Steel bearing piles are commonly carried out by a Specialist Contractor and whose advice should be sought to arrive at accurate costing. However the following items can be used to assess a budget cost for such work.						
The item for number of pile extensions includes for the cost of setting up the rig at the pile position together with welding the extension to the top of the steel bearing pile. The items for length of pile extension cover the material only, the driving cost being included in Class P.						
Number of pile extensions						
at each position	–	–	–	–	nr	64.50
Length of pile extensions, each length not exceeding 3 m; steel EN 10025 grade S275						
mass 45 kg/m	–	–	–	64.33	m	64.33
mass 54 kg/m	–	–	–	77.19	m	77.19
mass 63 kg/m	–	–	–	90.77	m	90.77
mass 71 kg/m	–	–	–	102.29	m	102.29
mass 79 kg/m	–	–	–	115.59	m	115.59
mass 85 kg/m	–	–	–	122.46	m	122.46
mass 95 kg/m	–	–	–	139.00	m	139.00
mass 109 kg/m	–	–	–	161.93	m	161.93
mass 110 kg/m	–	–	–	160.95	m	160.95
mass 126 kg/m	–	–	–	184.36	m	184.36
mass 149 kg/m	–	–	–	218.01	m	218.01
mass 133 kg/m	–	–	–	197.59	m	197.59
mass 152 kg/m	–	–	–	225.81	m	225.81
mass 174 kg/m	–	–	–	258.49	m	258.49
mass 186 kg/m	–	–	–	272.15	m	272.15
mass 223 kg/m	–	–	–	326.29	m	326.29
Length of pile extensions, each length exceeding 3 m; steel EN 10025 grade S275						
mass 45 kg/m	–	–	–	64.33	m	64.33
mass 54 kg/m	–	–	–	77.19	m	77.19
mass 63 kg/m	–	–	–	90.77	m	90.77
mass 71 kg/m	–	–	–	102.29	m	102.29
mass 79 kg/m	–	–	–	115.59	m	115.59
mass 85 kg/m	–	–	–	122.46	m	122.46
mass 95 kg/m	–	–	–	139.00	m	139.00
mass 109 kg/m	–	–	–	161.93	m	161.93
mass 110 kg/m	–	–	–	160.95	m	160.95
mass 126 kg/m	–	–	–	184.36	m	184.36
mass 149 kg/m	–	–	–	218.01	m	218.01
mass 133 kg/m	–	–	–	197.59	m	197.59
mass 152 kg/m	–	–	–	225.81	m	225.81
mass 174 kg/m	–	–	–	258.49	m	258.49
mass 186 kg/m	–	–	–	272.15	m	272.15
mass 223 kg/m	–	–	–	326.29	m	326.29

CLASS Q: PILING ANCILLARIES

Item	Gang hours	Labour £	Plant £	Material £	Unit	Total rate £
Number of pile extensions						
section size 203×203 × any kg/m	–	–	–	–	nr	215.00
section size 254×254 × any kg/m	–	–	–	–	nr	258.00
section size 305×305 × any kg/m	–	–	–	–	nr	295.63
section size 356×368 × any kg/m	–	–	–	–	nr	327.88
Cutting off surplus lengths						
mass 30–60 kg/m	–	–	–	–	nr	75.25
mass 60–120 kg/m	–	–	–	–	nr	107.50
mass 120–250 kg/m	–	–	–	–	nr	145.13
Burning off tops of piles to level						
mass 30–60 kg/m	–	–	–	–	nr	75.25
mass 60–120 kg/m	–	–	–	–	nr	107.50
mass 120–250 kg/m	–	–	–	–	nr	145.13

INTERLOCKING STEEL PILES

Arcelor Mittal Z section steel piles; EN 10248 grade S270GP steel

Steel bearing piles are commonly carried out by a Specialist Contractor and whose advice should be sought to arrive at accurate costing. However the following items can be used to assess a budget cost for such work.
The item for number of pile extensions includes for welding the extension to the top of the steel bearing pile. The items for length of pile extension cover the material only, the driving cost being included in Class P.

Item	Gang hours	Labour £	Plant £	Material £	Unit	Total rate £
Number of pile extensions						
Cutting off surplus lengths						
section modulus 500–800 cm³/m	–	–	–	–	m	23.20
section modulus 800–1200 cm³/m	–	–	–	–	m	23.20
section modulus 1200–2000 cm³/m	–	–	–	–	m	18.50
section modulus 2000–3000 cm³/m	–	–	–	–	m	23.20
section modulus 3000–4000 cm³/m	–	–	–	–	m	18.50
Extract piling and stacking on site						
section modulus 500–800 cm³/m	–	–	–	–	m²	36.93
section modulus 800–1200 cm³/m	–	–	–	–	m²	18.50
section modulus 1200–2000 cm³/m	–	–	–	–	m²	18.50
section modulus 3000–4000 cm³/m	–	–	–	–	m²	18.50
section modulus 2000–3000 cm³/m	–	–	–	–	m²	18.50

CLASS Q: PILING ANCILLARIES

Item	Gang hours	Labour £	Plant £	Material £	Unit	Total rate £
INTERLOCKING STEEL PILES – cont						
Arcelor Mittal U section steel piles; EN 10248 grade S270GP steel						
Steel bearing piles are commonly carried out by a Specialist Contractor and whose advice should be sought to arrive at accurate costing. However the following items can be used to assess a budget cost for such work.						
The item for number of pile extensions includes for welding the extension to the top of the steel bearing pile. The items for length of pile extension cover the material only, the driving cost being included in Class P.						
Number of pile extensions						
Cutting off surplus lengths						
section modulus 500–800 cm³/m	–	–	–	–	m	22.10
section modulus 800–1200 cm³/m	–	–	–	–	m	22.10
section modulus 800–1200 cm³/m	–	–	–	–	m	22.10
section modulus 2000–3000 cm³/m	–	–	–	–	m	22.10
section modulus 3000–4000 cm³/m	–	–	–	–	m	22.10
Extract piling and stack on site						
section modulus 500–800 cm³/m	–	–	–	–	m²	18.50
section modulus 800–1200 cm³/m	–	–	–	–	m²	18.50
section modulus 800–1200 cm³/m PU12	–	–	–	–	m²	18.50
section modulus 800–1200 cm³/m PU18	–	–	–	–	m²	18.50
section modulus 2000–3000 cm³/m	–	–	–	–	m²	18.50
section modulus 3000–4000 cm³/m	–	–	–	–	m²	18.50
OBSTRUCTIONS						
General						
Obstructions	–	–	–	–	hr	94.76
PILE TESTS						
Cast in place						
Pile tests; 500 mm diameter working pile; maximum test load of 600kN using non-working tension piles as reaction tripod						
first pile	–	–	–	–	nr	4725.00
subsequent pile	–	–	–	–	nr	3675.00
Take and test undisturbed soil samples; tripod	–	–	–	–	nr	168.00
Make, cure and test concrete cubes; tripod	–	–	–	–	nr	12.60
Pile tests; working pile; maximum test load of 1½ times working load; first pile						
450 mm / 650kN	–	–	–	–	nr	2500.00
600 mm / 1400kN	–	–	–	–	nr	5250.00
750 mm / 2200kN	–	–	–	–	nr	9500.00

CLASS Q: PILING ANCILLARIES

Item	Gang hours	Labour £	Plant £	Material £	Unit	Total rate £
Pile tests; working pile; maximum test load of 1½ times working load; second and subsequent piles						
450 mm / 650kN	–	–	–	–	nr	2300.00
600 mm / 1400kN	–	–	–	–	nr	5050.00
750 mm / 2200kN	–	–	–	–	nr	9100.00
Pile tests; working pile; electronic integrity testing; each pile (minimum 40 piles per visit)	–	–	–	–	nr	11.50
Make, cure and test concrete cubes	–	–	–	–	nr	22.00
Preformed						
Pile tests; working pile; maximum test load of 1.5 times working load	–	–	–	–	nr	4080.00
Pile tests; working pile; dynamic testing with piling hammer	–	–	–	–	nr	270.00
Steel bearing piles						
Steel bearing piles are commonly carried out by a Specialist Contractor and whose advice should be sought to arrive at accurate costing. However the following items can be used to assess a budget cost for such work.						
The following unit costs are based upon driving 100 nr steel bearing piles 15-24m long on a clear site with reasonable access. Supply is based on delivery 75 miles from works, in loads over 20t.						
Establishment of pile testing equipment on site preliminary to any piling operation	–	–	–	–	sum	1462.00
Carry out pile test on bearing piles irrespective of section using pile testing equipment on site up to 108 t load	–	–	–	–	nr	4676.25
Driven-temporary casing						
Pile tests; 430 mm diameter working pile; maximum test load of 1125kN using non-working tension piles as reaction; first piles						
bottom driven	–	–	–	–	nr	3675.00
top driven	–	–	–	–	nr	2992.50
Pile tests; 430 mm diameter working pile; maximum test load of 1125kN using non-working tension piles as reaction; subsequent piles						
bottom driven	–	–	–	–	nr	1575.00
top driven	–	–	–	–	nr	1575.00
Pile tests; working pile; electronic integrity testing; each pile (minimum 40 piles per visit)	–	–	–	–	nr	14.91
Make cure and test concrete cubes	–	–	–	–	nr	10.50
Driven - segmental casing						
Pile tests; 500 mm diameter working pile; maximum test load of 600kN using non-working tension piles as reaction						
first pile	–	–	–	–	nr	5512.50
subsequent piles	–	–	–	–	nr	4200.00

CLASS R: ROADS AND PAVINGS

Item	Gang hours	Labour £	Plant £	Material £	Unit	Total rate £
GENERAL						
Notes - Labour and Plant						
All outputs are based on clear runs without undue delay to two pavers with 75% utilisation.						
The outputs can be adjusted as follows to take account of space or time influences on the utilisation.						
Factors for varying utilisation of Labour and Plant:						
1 paver @ 75 % utilisation = × 2.00						
1 paver @ 100 % utilisation = × 1.50						
2 paver @ 100 % utilisation = × 0.75						
RESOURCES - LABOUR						
Sub-base laying gang						
1 ganger/chargehand (skill rate 4)		14.23				
1 skilled operative (skill rate 4)		13.32				
2 unskilled operatives (general)		24.88				
1 plant operator (skill rate 2)		18.05				
1 plant operator (skill rate 3)		16.19				
Total Gang Rate / Hour	£	**86.67**				
Flexible paving gang						
1 ganger/chargehand (skill rate 4)		14.23				
2 skilled operatives (skill rate 4)		26.64				
4 unskilled operatives (general)		49.76				
4 plant operators (skill rate 3)		64.76				
Total Gang Rate / Hour	£	**155.39**				
Concrete paving gang						
1 ganger/chargehand (skill rate 4)		14.23				
2 skilled operatives (skill rate 4)		26.64				
4 unskilled operatives (general)		49.76				
1 plant operator (skill rate 2)		18.05				
1 plant operator (skill rate 3)		16.19				
Total Gang Rate / Hour	£	**124.87**				
Road surface spraying gang						
1 plant operator (skill rate 3)		16.19				
Total Gang Rate / Hour	£	**16.19**				
Road chippings gang						
1 ganger/chargehand (skill rate 4) - 50% of time		7.12				
1 skilled operative (skill rate 4)		13.32				
2 unskilled operatives (general)		24.88				
3 plant operators (skill rate 3)		48.57				
Total Gang Rate / Hour	£	**93.89**				
Cutting slabs gang						
1 unskilled operative (generally)		12.44				
Total Gang Rate / Hour	£	**12.44**				
Concrete filled joints gang						
1 ganger/chargehand (skill rate 4) - 50% of time		7.12				
1 skilled operatives (skill rate 4)		13.32				
2 unskilled operatives (general)		24.88				
Total Gang Rate / Hour	£	**45.31**				

CLASS R: ROADS AND PAVINGS

Item	Gang hours	Labour £	Plant £	Material £	Unit	Total rate £
Milling gang						
1 ganger/chargehand (skill rate 4)		14.23				
2 skilled operatives (skill rate 4)		26.64				
4 unskilled operatives (general)		49.76				
1 plant operators (skill rate 3)		16.19				
1 plant operator (skill rate 2)		18.05				
Total Gang Rate / Hour	£	**124.87**				
Rake and compact planed material gang						
1 ganger/chargehand (skill rate 4)		14.23				
1 skilled operatives (skill rate 4)		13.32				
3 unskilled operatives (general)		37.32				
1 plant operator (skill rate 3)		16.19				
1 plant operator (skill rate 4)		14.56				
Total Gang Rate / Hour	£	**95.62**				
Kerb laying gang						
3 skilled operatives (skill rate 4)		39.96				
1 unskilled operative (general)		12.44				
1 plant operator (skill rate 3) - 25% of time		4.05				
Total Gang Rate / Hour	£	**56.45**				
Path sub-base, bitmac and gravel laying gang						
1 ganger/chargehand (skill rate 4)		14.23				
2 unskilled operatives (general)		24.88				
1 plant operator (skill rate 3)		16.19				
Total Gang Rate / Hour	£	**55.30**				
Paviors and flagging gang						
1 skilled operative (skill rate 4)		13.32				
1 unskilled operative (general)		12.44				
Total Gang Rate / Hour	£	**25.76**				
Traffic signs gang						
1 ganger/chargehand (skill rate 3)		15.79				
1 skilled operative (skill rate 3)		14.88				
2 unskilled operatives (general)		24.88				
1 plant operator (skill rate 3) - 25% of time		4.05				
Total Gang Rate / Hour	£	**59.60**				
RESOURCES - PLANT						
Sub-base laying						
93 KW motor grader			29.85			
0.80 m³ tractor loader			23.80			
6t towed roller			10.73			
Total Rate / Hour		£	**64.37**			
Flexible paving						
2 asphalt pavers, 35 kW, 4.0 m			93.57			
2 deadweight rollers, 3 point, 10 t			33.55			
tractor with front bucket and integral 2 tool						
compressor			18.88			
compressor tools: scabbler			1.26			
tar sprayer, 100 litre			6.45			
self propelled chip spreader			9.59			
channel (heat) iron			1.59			
Total Rate / Hour		£	**164.89**			

CLASS R: ROADS AND PAVINGS

Item	Gang hours	Labour £	Plant £	Material £	Unit	Total rate £
RESOURCES - PLANT – cont						
Concrete paving						
wheeled loader, 2.60 m²			56.75			
concrete paver, 6.0 m			75.37			
compaction slipform finisher			18.18			
Total Rate / Hour		£	**150.30**			
Road surface spraying						
tar sprayer, 100 litre			6.45			
Total Rate / Hour		£	**6.45**			
Road chippings						
deadweight rollers, 3 point, 10 t			16.78			
tar sprayer, 100 litre			6.45			
self propelled chip spreader			9.59			
channel (heat) iron			1.59			
Total Rate / Hour		£	**34.04**			
Cutting slabs						
compressor, 65 cfm			6.24			
12' disc cutter			1.38			
Total Rate / Hour		£	**7.62**			
Milling						
cold planer, 2.10 m			53.78			
wheeled loader, 2.60m²			56.75			
Total Rate/ Hour			110.53			
Heat planing						
heat planer, 4.5 m			77.87			
wheeled loader, 2.60 m²			56.75			
Total Rate/ Hour			134.62			
Rake and compact planed material						
deadweight roller, 3 point, 10t			16.78			
tractor with front bucket and integral 2 tool						
compressor			18.88			
channel (heat) iron			1.23			
Total Rate/ Hour			36.89			
Kerb laying						
backhoe JCB 3CX (25% of time)JCB 3CX (25% of time)			4.19		hr	
12' stihl saw			1.11		hr	
road forms			2.16		hr	
TOTAL		£	**7.66**		hr	
Path sub-base, bitmac and gravel laying						
backhoe JCB 3CX			16.77		hr	
2 t dumper			6.49		hr	
pedestrian roller Bomag BW 90S			8.26		hr	
TOTAL		£	**31.53**		hr	
Paviors and flagging						
2 t dumper (33% of time)			2.16			
Total Rate / Hour		£	**2.16**			

CLASS R: ROADS AND PAVINGS

Item	Gang hours	Labour £	Plant £	Material £	Unit	Total rate £
Traffic signs						
JCB 3CX backhoe - 50% of time			8.39			
125 cfm compressor - 50% of time			4.71			
compressor tools: hand held hammer drill - 50% of time			0.47			
compressor tools: clay spade - 50% of time			0.21			
compressor tools: extra 15 m hose - 50% of time			0.16			
8 t lorry with hiab lift - 50% of time			6.39			
Total Rate / Hour		£	**20.31**			
SUB-BASES, FLEXIBLE ROAD BASES AND SURFACING						
Granular material DfT specified type 1						
Sub-base; spread and graded						
75 mm deep	0.04	3.09	2.25	25.73	m³	**31.07**
100 mm deep	0.04	3.54	2.57	25.73	m³	**31.84**
150 mm deep	0.05	3.98	2.90	25.73	m³	**32.61**
200 mm deep	0.05	4.42	3.22	25.73	m³	**33.37**
Lean concrete DfT specified strength mix C20P/ 20 mm aggregate						
Sub-base; spread and graded						
100 mm deep	0.05	3.98	2.90	67.20	m³	**74.08**
200 mm deep	0.05	4.42	3.22	67.20	m³	**74.84**
Hardcore						
Sub-base; spread and graded						
100 mm deep	0.04	3.54	2.57	34.98	m³	**41.09**
150 mm deep	0.05	3.98	2.90	34.98	m³	**41.86**
200 mm deep	0.05	4.42	3.22	34.98	m³	**42.62**
Geotextiles refer to Class E						
Wet mix macadam; DfT clause 808						
Sub-base; spread and graded						
75 mm deep	0.04	3.09	3.41	73.80	m³	**80.30**
100 mm deep	0.04	3.54	3.54	73.80	m³	**80.88**
200 mm deep	0.05	4.42	3.73	73.80	m³	**81.95**
Dense Bitumen Macadam						
Base to DfT clause 903						
100 mm deep	0.02	3.17	3.30	6.20	m²	**12.67**
150 mm deep	0.03	3.96	4.12	9.30	m²	**17.38**
200 mm deep	0.03	4.75	4.95	12.40	m²	**22.10**
Binder Course to DfT clause 906						
50 mm deep	0.02	2.38	2.47	2.61	m²	**7.46**
100 mm deep	0.02	3.17	3.30	5.23	m²	**11.70**

CLASS R: ROADS AND PAVINGS

Item	Gang hours	Labour £	Plant £	Material £	Unit	Total rate £
SUB-BASES, FLEXIBLE ROAD BASES AND SURFACING – cont						
Dense Bitumen Macadam – cont						
Surface Course to DfT clause 912						
30 mm deep	0.01	1.58	1.65	2.23	m^2	**5.46**
50 mm deep	0.02	2.38	2.47	3.71	m^2	**8.56**
Bitumen Macadam						
Binder Course to DfT clause 908						
35 mm deep	0.01	1.58	1.65	2.37	m^2	**5.60**
70 mm deep	0.02	2.38	2.47	4.74	m^2	**9.59**
Dense Tarmacadam						
Base to DfT clause 902						
50 mm deep	0.02	2.38	2.45	3.25	m^2	**8.08**
100 mm deep	0.02	2.38	2.45	6.51	m^2	**11.34**
Binder Course to DfT clause 907						
60 mm deep	0.02	2.38	2.47	4.65	m^2	**9.50**
80 mm deep	0.02	2.38	2.47	6.20	m^2	**11.05**
Dense Tar Surfacing						
Surface Course to DfT clause 913						
30 mm deep	0.01	1.58	1.65	2.38	m^2	**5.61**
50 mm deep	0.02	2.38	2.45	3.97	m^2	**8.80**
Cold Asphalt						
Surface Course to DfT clause 914						
15 mm deep	0.01	1.58	1.65	1.44	m^2	**4.67**
30 mm deep	0.01	1.58	1.65	2.91	m^2	**6.14**
Rolled Asphalt						
Binder Course to DfT clause 905						
60 mm deep	0.02	2.38	2.47	5.76	m^2	**10.61**
80 mm deep	0.02	2.38	2.47	7.68	m^2	**12.53**
Surface Course to DfT clause 911						
40 mm deep	0.02	2.38	2.47	3.84	m^2	**8.69**
60 mm deep	0.02	2.38	2.47	5.76	m^2	**10.61**
Slurry sealing; BS 434 class K3						
Sealing to DfT clause 918						
3 mm deep	0.02	0.25	0.10	2.52	m^2	**2.87**
4 mm deep	0.02	0.25	0.10	2.47	m^2	**2.82**
Coated chippings, 9–11 kg/m^2						
Surface dressing to DfT clause 915						
6 mm nominal size	0.01	0.96	0.34	0.81	m^2	**2.11**
8 mm nominal size	0.01	0.96	0.34	0.83	m^2	**2.13**
10 mm nominal size	0.01	0.96	0.34	0.84	m^2	**2.14**
12 mm nominal size	0.01	0.96	0.34	0.92	m^2	**2.22**

CLASS R: ROADS AND PAVINGS

Item	Gang hours	Labour £	Plant £	Material £	Unit	Total rate £
Bituminous spray; BS 434 K1 - 40						
Tack coat to DfT clause 920						
large areas; over 20m²	0.02	0.25	0.10	0.27	m²	0.62
small areas; under 20m²	0.02	0.25	0.10	0.27	m²	0.62
Removal of flexible surface						
Trimming edges only of existing slabs, floors or similar surfaces (wet or dry); 6mm cutting width						
50mm deep	0.02	0.25	0.15	3.74	m	4.14
100mm deep	0.03	0.38	0.24	3.94	m	4.56
Cutting existing slabs, floors or similar surfaces (wet or dry); 8mm cutting width						
50mm deep	0.03	0.32	0.19	3.74	m	4.25
100mm deep	0.06	0.76	0.46	3.94	m	5.16
150mm deep	0.08	1.02	0.61	4.01	m	5.64
Milling pavement (assumes disposal on site or re-use as fill but excludes transport if required)						
75mm deep	0.03	3.34	2.98	–	m²	6.32
100mm deep	0.04	4.46	3.98	–	m²	8.44
50mm deep; scarifying surface	0.02	2.73	2.43	–	m²	5.16
75mm deep; scarifying surface	0.04	4.58	4.09	–	m²	8.67
25mm deep; heat planing for re-use	0.03	3.96	4.31	–	m²	8.27
50mm deep; heat planing for re-use	0.06	6.94	7.54	–	m²	14.48
Raking over scarified or heat planed material; compacting with 10 t roller						
ne 50mm deep	0.01	0.95	0.37	–	m²	1.32
CONCRETE PAVEMENTS						
The following unit costs are for jointed reinforced concrete slabs, laid in reasonable areas (over 200m²) by paver train/slipformer.						
Designed mix; cement to BS EN 197-1; grade C30, 20mm aggregate						
Carriageway slabs of DfT Specified paving quality						
180mm deep	0.02	1.86	2.25	15.17	m²	19.28
220mm deep	0.02	2.23	2.71	18.54	m²	23.48
260mm deep	0.02	2.73	3.31	21.91	m²	27.95
300mm deep	0.03	3.10	3.76	25.28	m²	32.14
Fabric						
Steel fabric reinforcement to BS 4483						
Ref A142 nominal mass 2.22kg	0.03	3.72	–	2.10	m²	5.82
Ref A252 nominal mass 3.95kg	0.04	4.96	–	3.75	m²	8.71
Ref B385 nominal mass 4.53kg	0.04	4.96	–	4.12	m²	9.08
Ref C636 nominal mass 5.55kg	0.05	6.19	–	3.86	m²	10.05
Ref B503 nominal mass 5.93kg	0.05	6.19	–	3.75	m²	9.94

CLASS R: ROADS AND PAVINGS

Item	Gang hours	Labour £	Plant £	Material £	Unit	Total rate £
CONCRETE PAVEMENTS – cont						
Mild Steel bar reinforcement BS 4449						
Bars; supplied in bent and cut lengths						
6 mm nominal size	8.00	991.04	–	643.88	t	1634.92
8 mm nominal size	6.74	834.95	–	638.72	t	1473.67
10 mm nominal size	6.74	834.95	–	571.76	t	1406.71
12 mm nominal size	6.74	834.95	–	540.86	t	1375.81
16 mm nominal size	6.15	761.86	–	412.08	t	1173.94
High yield steel bar reinforcement BS 4449 or 4461						
Bars; supplied in bent and cut lengths						
6 mm nominal size	8.00	991.04	–	647.03	t	1638.07
8 mm nominal size	6.74	834.95	–	641.86	t	1476.81
10 mm nominal size	6.74	834.95	–	574.56	t	1409.51
12 mm nominal size	6.74	834.95	–	564.21	t	1399.16
16 mm nominal size	6.15	761.86	–	543.51	t	1305.37
Sheeting to prevent moisture loss						
Polyethylene sheeting; lapped joints; horizontal below concrete pavements						
250 micron	0.01	1.24	–	0.34	m^2	1.58
500 micron	0.01	1.24	–	0.96	m^2	2.20
JOINTS IN CONCRETE PAVEMENTS						
General						
Longitudinal joints						
180 mm deep	0.01	1.49	1.80	17.43	m	20.72
220 mm deep	0.01	1.49	1.80	20.50	m	23.79
260 mm deep	0.01	1.49	1.80	25.11	m	28.40
300 mm deep	0.01	1.49	1.80	29.21	m	32.50
Expansion joints						
180 mm deep	0.01	1.49	1.80	34.32	m	37.61
220 mm deep	0.01	1.49	1.80	39.96	m	43.25
260 mm deep	0.01	1.49	1.80	45.60	m	48.89
300 mm deep	0.01	1.49	1.80	46.73	m	50.02
Contraction joints						
180 mm deep	0.01	1.49	1.80	20.02	m	23.31
220 mm deep	0.01	1.49	1.80	21.17	m	24.46
260 mm deep	0.01	1.49	1.80	22.49	m	25.78
300 mm deep	0.01	1.49	1.80	26.80	m	30.09
Construction joints						
180 mm deep	0.01	1.49	1.80	13.20	m	16.49
220 mm deep	0.01	1.49	1.80	14.40	m	17.69
260 mm deep	0.01	1.49	1.80	15.54	m	18.83
300 mm deep	0.01	1.49	1.80	16.65	m	19.94
Open joints with filler						
ne 0.5 m; 10 mm flexcell joint filler	0.11	4.94	–	3.15	m	8.09
0.5–1.00 m; 10 mm flexcell joint filler	0.11	4.94	–	4.52	m	9.46
Joint sealants						
10 × 20 mm cold polysulphide sealant	0.14	6.29	–	3.10	m	9.39
20 × 20 mm cold polysulphide sealant	0.18	8.09	–	6.16	m	14.25

CLASS R: ROADS AND PAVINGS

Item	Gang hours	Labour £	Plant £	Material £	Unit	Total rate £
KERBS, CHANNELS AND EDGINGS						
Foundations to kerbs etc.						
Measurement Note: the following are shown separate from their associated kerb etc. to simplify the presentation of cost alternatives.						
Mass concrete						
200 × 100 mm	0.01	0.56	0.07	1.82	m	**2.45**
300 × 150 mm	0.02	0.84	0.11	4.16	m	**5.11**
450 × 150 mm	0.02	1.12	0.15	6.16	m	**7.43**
100 × 100 mm haunching, one side	0.01	0.28	0.04	0.44	m	**0.76**
Precast concrete kerbs; BS 7263:bedded, jointed and pointed in cement mortar						
Kerbs; bullnosed; splayed or half battered; straight or curved over 12 m radius						
125 × 150 mm	0.06	3.36	0.46	9.80	m	**13.62**
125 × 255 mm	0.07	3.92	0.54	11.00	m	**15.46**
150 × 305 mm	0.07	3.92	0.55	11.85	m	**16.32**
Kerbs; bullnosed; splayed or half battered; curved ne 12 m radius						
125 × 150 mm	0.07	3.64	0.50	9.80	m	**13.94**
125 × 255 mm	0.08	4.20	0.57	11.00	m	**15.77**
150 × 305 mm	0.08	4.20	0.59	11.85	m	**16.64**
Quadrants						
305 × 305 × 150 mm	0.08	4.48	0.63	14.65	nr	**19.76**
455 × 455 × 255 mm	0.10	5.59	0.79	17.84	nr	**24.22**
Drop kerbs						
125 × 255 mm	0.07	3.92	0.55	8.24	m	**12.71**
150 × 305 mm	0.07	3.92	0.55	10.28	m	**14.75**
Channel; straight or curved over 12 m radius						
255 × 125 mm	0.07	3.92	0.55	5.87	m	**10.34**
Channel; curved radius ne 12 m						
255 × 125 mm	0.07	3.92	0.55	5.87	m	**10.34**
Edging; straight or curved over 12 m radius						
50 × 150 mm	0.04	2.24	0.32	7.39	m	**9.95**
Edging; curved ne 12 m radius						
50 × 150 mm	0.05	2.52	0.35	7.39	m	**10.26**
Precast concrete drainage channels; Charcon Safeticurb; channels jointed with plastic rings and bedded; jointed and pointed in cement mortar						
Channel unit; straight; Type DBA/3						
250 × 254 mm; medium duty	0.08	4.20	0.59	48.95	m	**53.74**
305 × 305 mm; heavy duty	0.10	5.32	0.75	103.20	m	**109.27**

CLASS R: ROADS AND PAVINGS

Item	Gang hours	Labour £	Plant £	Material £	Unit	Total rate £
KERBS, CHANNELS AND EDGINGS – cont						
Precast concrete Ellis Trief safety kerb; bedded jointed and pointed in cement mortar						
Kerb; straight or curved over 12 m radius						
415 × 380 mm	0.23	12.59	1.72	84.59	m	98.90
Kerb; curved ne 12 m radius						
415 × 380 mm	0.25	13.99	1.91	84.59	m	100.49
Precast concrete combined kerb and drainage block Beany block system; bedded jointed and pointed in cement mortar						
Kerb; top block, shallow base unit, standard cover plate and frame						
straight or curved over 12 m radius	0.15	8.39	1.15	117.25	m	126.79
curved ne 12 m radius	0.20	11.19	1.52	178.99	m	191.70
Kerb; top block, standard base unit, standard cover plate and frame						
straight or curved over 12 m radius	0.15	8.39	1.16	117.25	m	126.80
curved ne 12 m radius	0.20	11.19	1.52	178.99	m	191.70
Kerb; top block, deep base unit, standard cover plate and frame						
straight or curved over 12 m radius	0.15	8.39	1.14	261.62	m	271.15
curved ne 12 m radius	0.20	11.19	1.51	334.46	m	347.16
base block depth tapers	0.10	5.81	0.77	28.29	m	34.87
Extruded asphalt kerbs to BS 5931; extruded and slip formed						
Kerb; straight or curved over 12 m radius						
75 mm kerb height	–	–	–	7.71	m	7.71
100 mm kerb height	–	–	–	11.47	m	11.47
125 mm kerb height	–	–	–	24.19	m	24.19
Channel; straight or curved over 12 m radius						
300 mm channel width	–	–	–	17.17	m	17.17
250 mm channel width	–	–	–	17.17	m	17.17
Kerb; curved to radius ne 12 m						
75 mm kerb height	–	–	–	14.54	m	14.54
100 mm kerb height	–	–	–	11.47	m	11.47
125 mm kerb height	–	–	–	13.95	m	13.95
Channel; curved to radius ne 12 m						
300 mm channel width	–	–	–	21.94	m	21.94
250 mm channel width	–	–	–	17.10	m	17.10
Extruded concrete; slip formed						
Kerb; straight or curved over 12 m radius						
100 mm kerb height	–	–	–	22.75	m	22.75
125 mm kerb height	–	–	–	28.60	m	28.60
Kerb; curved to radius ne 12 m						
100 mm kerb height	–	–	–	22.75	m	22.75
125 mm kerb height	–	–	–	28.60	m	28.60

CLASS R: ROADS AND PAVINGS

Item	Gang hours	Labour £	Plant £	Material £	Unit	Total rate £
LIGHT DUTY PAVEMENTS						
Sub-bases						
Measurement Note: the following are shown separate from their associated paving to simplify the presentation of cost alternatives.						
To paved area; sloping not exceeding 10° to the horizontal						
100 mm thick sand	0.01	0.49	0.28	3.89	m²	**4.66**
150 mm thick sand	0.01	0.65	0.38	5.83	m²	**6.86**
100 mm thick gravel	0.01	0.49	0.28	2.06	m²	**2.83**
150 mm thick gravel	0.01	0.65	0.38	3.09	m²	**4.12**
100 mm thick hardcore	0.01	0.49	0.28	2.61	m²	**3.38**
150 mm thick hardcore	0.01	0.65	0.38	3.92	m²	**4.95**
100 mm thick concrete grade 20/20	0.02	1.13	0.66	9.05	m²	**10.84**
150 mm thick concrete grade 20/20	0.03	1.73	1.01	13.58	m²	**16.32**
Bitumen macadam surfacing; BS 4987; base course of 20 mm open graded aggregate to clause 2.6.1 tables 5-7; wearing course of 6 mm medium graded aggregate to clause 2.7.6 tables 32-33						
Paved area comprising base course 40 mm thick wearing course 20 mm thick						
sloping not exceeding 10° to the horizontal	0.09	4.59	2.68	6.95	m²	**14.22**
sloping not exceeding 10° to the horizontal; red additives	0.09	4.59	2.68	9.13	m²	**16.40**
sloping not exceeding 10° to the horizontal; green additives	0.09	4.59	2.68	10.62	m²	**17.89**
sloping exceeding 10° to the horizontal	0.10	5.13	3.00	6.95	m²	**15.08**
sloping exceeding 10° to the horizontal; red additives	0.10	5.13	3.00	9.13	m²	**17.26**
sloping exceeding 10° to the horizontal; green additives	0.10	5.13	3.00	10.62	m²	**18.75**
Granular base surfacing; Central Reserve Treatments Limestone, graded 10 mm down; laid and compacted						
Paved area 100 mm thick; surface sprayed twice with two coats of cold bituminous emulsion; blinded with 6 mm quartizite fine gravel						
sloping not exceeding 10° to the horizontal	0.02	1.08	0.63	3.85	m²	**5.56**
Ennstone Johnston Golden gravel; graded 13 mm to fines; rolled wet						
Paved area 50 mm thick; single layer						
sloping not exceeding 10° to the horizontal	0.03	1.62	0.95	8.03	m²	**10.60**

CLASS R: ROADS AND PAVINGS

Item	Gang hours	Labour £	Plant £	Material £	Unit	Total rate £
LIGHT DUTY PAVEMENTS – cont						
Precast concrete flags; BS 7263; grey; bedding in cement mortar						
Paved area; sloping not exceeding 10° to the horizontal						
900×600×63 mm	0.21	5.36	0.45	9.66	m²	**15.47**
900×600×50 mm	0.20	5.11	0.43	8.53	m²	**14.07**
600×600×63 mm	0.25	6.38	0.54	11.80	m²	**18.72**
600×600×50 mm	0.24	6.13	0.52	9.91	m²	**16.56**
600×450×50 mm	0.28	7.15	0.61	12.62	m²	**20.38**
Extra for coloured, 50 mm thick	–	–	–	4.53	m²	**4.53**
Precast concrete rectangular paving blocks; BS 6717; grey; bedding on 50 mm thick dry sharp sand; filling joints; excluding sub-base						
Paved area; sloping not exceeding 10° to the horizontal						
200×100×80 mm thick	0.30	7.66	0.65	14.91	m²	**23.22**
200×100×80 mm thick; coloured blocks	0.30	7.66	0.65	30.51	m²	**38.82**
Brick paviors; bedding on 20 mm thick mortar; excluding sub-base						
Paved area; sloping not exceeding 10° to the horizontal						
215×103×65 mm	0.30	7.66	0.65	25.83	m²	**34.14**
Granite setts; bedding on 25 mm cement mortar; excluding sub-base						
Paved area; sloping not exceeding 10° to the horizontal						
to random pattern	0.90	22.99	1.95	60.04	m²	**84.98**
to specific pattern	1.20	30.65	2.60	60.04	m²	**93.29**
Cobble paving; 50–75 mm; bedding oin 25 mm cement mortar; filling joints; excluding sub-base						
Paved area; sloping not exceeding 10° to the horizontal						
50–75 mm diameter cobbles	1.00	25.54	2.14	35.15	m²	**62.83**

CLASS R: ROADS AND PAVINGS

Item	Gang hours	Labour £	Plant £	Material £	Unit	Total rate £
ANCILLARIES						
Traffic signs						
In this section prices will vary depending upon the diagram configurations. The following are average costs of signs and bollards. Diagram numbers refer to the Traffic Signs Regulations and General Directions 2002 and the figure numbers refer to the Traffic Sign Manual.						
Examples of Prime Costs for Class 1 (High Intensity) traffic and road signs (ex works).						
600 × 450 mm	–	–	–	99.83	nr	**99.83**
600 mm diameter	–	–	–	125.19	nr	**125.19**
600 mm triangular	–	–	–	102.86	nr	**102.86**
500 × 500 mm	–	–	–	88.22	nr	**88.22**
450 × 450 mm	–	–	–	74.83	nr	**74.83**
450 × 300 mm	–	–	–	61.40	nr	**61.40**
1200 × 400 mm (CHEVRONS)	–	–	–	203.19	nr	**203.19**
Examples of Prime Costs for Class 21 (Engineering Grade) traffic and road signs (ex works).						
600 × 450 mm	–	–	–	123.12	nr	**123.12**
600 mm diameter	–	–	–	154.61	nr	**154.61**
600 mm triangular	–	–	–	128.99	nr	**128.99**
500 × 500 mm	–	–	–	108.76	nr	**108.76**
450 × 450 mm	–	–	–	92.85	nr	**92.85**
450 × 300 mm	–	–	–	73.17	nr	**73.17**
1200 × 400 mm (CHEVRONS)	–	–	–	203.97	nr	**203.97**
Standard reflectorized traffic signs						
Note: Unit costs do not include concrete foundations						
Standard one post signs; 600 × 450 mm type C1 signs fixed back to back to another sign (measured separately) with aluminium clips to existing post (measured separately)	0.04	2.37	0.81	104.82	nr	**108.00**
Extra for fixing singly with aluminium clips	0.01	0.59	0.19	6.84	nr	**7.62**
Extra for fixing singly with stainless steel clips	0.01	0.59	0.46	8.76	nr	**9.81**
fixed back to back to another sign (measured separately) with stainless steel clips to one new 76 mm diameter plastic coated steel posts 1.75 m long	0.27	15.98	5.49	167.66	nr	**189.13**
Extra for fixing singly to one face only	0.01	0.59	0.19	–	nr	**0.78**
Extra for 76 mm diameter 1.75 m long aluminium post	0.02	1.18	0.41	11.80	nr	**13.39**
Extra for 76 mm diameter 3.5 m long plastic coated steel post	0.02	1.18	0.41	4.24	nr	**5.83**
Extra for 76 mm diameter 3.5 m long aluminium post	0.02	1.18	0.41	4.24	nr	**5.83**
Extra for excavation for post, in hard material	1.10	65.12	21.93	–	nr	**87.05**
Extra for single external illumination unit with fitted photo cell (excluding trenching and cabling); unit cost per face illuminated	0.33	19.54	6.57	66.24	nr	**92.35**

CLASS R: ROADS AND PAVINGS

Item	Gang hours	Labour £	Plant £	Material £	Unit	Total rate £
ANCILLARIES – cont						
Standard reflectorized traffic signs – cont						
Standard two post signs; 1200 × 400 mm;						
fixed back to back to another sign (measured						
separately) with stainless steel clips to two new						
76 mm diameter plastic coated steel posts	0.51	30.19	10.37	337.63	nr	378.19
Extra for fixing singly to one face only	0.02	1.18	0.41	–	nr	1.59
Extra for two 76 mm diameter 1.75 m long						
aluminium posts	0.04	2.37	0.80	23.60	nr	26.77
Extra for two 76 mm diameter 1.75 m long						
plastic coated steel posts	0.04	2.37	0.80	8.48	nr	11.65
Extra for two 76 mm diameter 3.5 m long						
aluminium posts	0.04	2.37	0.80	8.48	nr	11.65
Extra for excavation for post, in hard material	1.10	65.12	21.93	–	nr	87.05
Extra for single external illumination unit with						
fitted photo cell (including trenching and						
cabling); unite per face illuminated	0.58	34.34	11.57	92.77	nr	138.68
Standard internally illuminated traffic signs						
Bollard with integral mould-in translucent graphics						
(excluding trenching and cabling)						
fixing to concrete base	0.48	28.42	9.75	261.70	nr	299.87
Special traffic signs						
Note: Unit costs do not include concrete foundations						
or trenching and cabling						
Externally illuminated relectorised traffic signs						
manufactured to order						
special signs, surface area 1.50 m² on two						
100 mm diameter steel posts	–	–	–	–	nr	662.50
special signs, surface area 4.00 m² on three						
100 mm diameter steel posts	–	–	–	–	nr	975.00
Internally illuminated traffic signs manufactured to						
order						
special signs, surface area 0.25 m² on one new						
76 mm diameter post	–	–	–	–	nr	164.00
special signs, surface area 0.75 m² on one new						
100 mm diameter steel posts	–	–	–	–	nr	215.25
special signs, surface area 4.00 m² on four new						
120 mm diameter steel posts	–	–	–	–	nr	1076.25
Signs on gantries						
Externally illuminated reflectorized signs						
1.50 m²	1.78	105.67	45.10	186.25	nr	337.02
2.50 m²	2.15	127.28	54.32	310.42	nr	492.02
3.00 m²	3.07	181.74	77.57	371.45	nr	630.76
Internally illuminated sign with translucent optical						
reflective sheeting and remote light source						
0.75 m²	1.56	92.35	39.42	1468.75	nr	1600.52
1.00 m²	1.70	100.64	42.95	1950.00	nr	2093.59
1.50 m²	2.41	142.67	60.89	2931.25	nr	3134.81

CLASS R: ROADS AND PAVINGS

Item	Gang hours	Labour £	Plant £	Material £	Unit	Total rate £
Existing signs						
Take from store and re-erect						
3.0 m high road sign	0.28	16.58	7.07	261.70	nr	285.35
road sign on two posts	0.50	29.60	12.63	523.40	nr	565.63
Surface markings; reflectorized white						
Letters and shapes						
triangles; 1.6 m high	–	–	–	–	nr	8.91
triangles; 2.0 m high	–	–	–	–	nr	12.10
triangles; 3.75 m high	–	–	–	–	nr	15.92
circles with enclosing arrows; 1.6 m diameter	–	–	–	–	nr	63.67
arrows; 4.0 m long; straight	–	–	–	–	nr	25.47
arrows; 4.0 m long; turning	–	–	–	–	nr	25.47
arrows; 6.0 m long; straight	–	–	–	–	nr	31.83
arrows; 6.0 m long; turning	–	–	–	–	nr	31.83
arrows; 6.0 m long; curved	–	–	–	–	nr	31.83
arrows; 6.0 m long; double headed	–	–	–	–	nr	44.57
arrows; 8.0 m long; double headed	–	–	–	–	nr	63.66
arrows; 16.0 m long; double headed	–	–	–	–	nr	95.50
arrows; 32.0 m long; double headed	–	–	–	–	nr	127.33
letters or numerals; 1.6 m high	–	–	–	–	nr	8.29
letters or numerals; 2.0 m high	–	–	–	–	nr	12.10
letters or numerals; 3.75 m high	–	–	–	–	nr	21.02
Continuous lines						
150 mm wide	–	–	–	–	m	0.98
200 mm wide	–	–	–	–	m	1.30
Intermittent lines						
60 mm wide; 0.60 m line and 0.60 m gap	–	–	–	–	m	0.76
100 mm wide; 1.0 m line and 5.0 m gap	–	–	–	–	m	0.76
100 mm wide; 2.0 m line and 7.0 m gap	–	–	–	–	m	0.76
100 mm wide; 4.0 m line and 2.0 m gap	–	–	–	–	m	0.76
100 mm wide; 6.0 m line and 3.0 m gap	–	–	–	–	m	0.76
150 mm wide; 1.0 m line and 5.0 m gap	–	–	–	–	m	1.15
150 mm wide; 6.0 m line and 3.0 m gap	–	–	–	–	m	1.15
150 mm wide; 0.60 m line and 0.30 m gap	–	–	–	–	m	1.15
200 mm wide; 0.60 m line and 0.30 m gap	–	–	–	–	m	1.54
200 mm wide; 1.0 m line and 1.0 m gap	–	–	–	–	m	1.54
Surface markings; reflectorized yellow						
Continuous lines						
100 mm wide	–	–	–	–	m	0.66
150 mm wide	–	–	–	–	m	0.98
Intermittent lines						
kerb marking; 0.25 m long	–	–	–	–	nr	0.65

CLASS R: ROADS AND PAVINGS

Item	Gang hours	Labour £	Plant £	Material £	Unit	Total rate £
ROAD MARKINGS						
Surface markings; thermoplastic screed or spray						
Note: Unit costs based upon new road with clean surface closed to traffic.						
Continuous line in reflectorized white						
150 mm wide	–	–	–	–	m	0.98
200 mm wide	–	–	–	–	m	1.30
Continuous line in reflectorized yellow						
100 mm wide	–	–	–	–	m	0.66
150 mm wide	–	–	–	–	m	0.98
Intermittent line in reflectorized white						
60 mm wide with 0.60 m line and 0.60 m gap	–	–	–	–	m	0.76
100 mm wide with 1.0 m line and 5.0 m gap	–	–	–	–	m	0.76
100 mm wide with 2.0 m line and 7.0 m gap	–	–	–	–	m	0.76
100 mm wide with 4.0 m line and 2.0 m gap	–	–	–	–	m	0.76
100 mm wide with 6.0 m line and 3.0 m gap	–	–	–	–	m	0.76
150 mm wide with 1.0 m line and 5.0 m gap	–	–	–	–	m	1.15
150 mm wide with 6.0 m line and 3.0 m gap	–	–	–	–	m	1.15
150 mm wide with 0.6 m line and 0.3 m gap	–	–	–	–	m	1.15
200 mm wide with 0.6 m line and 0.3 m gap	–	–	–	–	m	1.54
200 mm wide with 1.0 m line and 1.0 m gap	–	–	–	–	m	1.54
Ancillary line in reflectorized white						
150 mm wide in hatched areas	–	–	–	–	m	0.96
200 mm wide in hatched areas	–	–	–	–	m	1.54
Ancillary line in reflectorized yellow						
150 mm wide in hatched areas	–	–	–	–	m	0.96
Triangles in reflectorized white						
1.6 m high	–	–	–	–	nr	8.91
2.0 m high	–	–	–	–	nr	12.10
3.75 m high	–	–	–	–	nr	15.92
Circles with enclosing arrows in reflectorized white						
1.6 m diameter	–	–	–	–	nr	63.67
Arrows in reflectorized white						
4.0 m long straight or turning	–	–	–	–	nr	25.47
6.0 m long straight or turning	–	–	–	–	nr	31.83
6.0 m long curved	–	–	–	–	nr	31.83
6.0 m long double headed	–	–	–	–	nr	44.57
8.0 m long double headed	–	–	–	–	nr	63.66
16.0 m long double headed	–	–	–	–	nr	95.50
32.0 m long double headed	–	–	–	–	nr	127.33
Kerb markings in yellow						
250 mm long	–	–	–	–	nr	0.65
Letters or numerals in reflectorized white						
1.6 m high	–	–	–	–	nr	8.29
2.0 m high	–	–	–	–	nr	12.10
3.75 m high	–	–	–	–	nr	21.02

CLASS R: ROADS AND PAVINGS

Item	Gang hours	Labour £	Plant £	Material £	Unit	Total rate £
Surface markings; Verynyl strip markings						
Note: Unit costs based upon new road with clean surface closed to traffic.						
Verynyl' strip markings (pedestrian crossings and similar locations)						
200 mm wide line	–	–	–	–	m	8.30
600 × 300 mm single stud tile	–	–	–	–	nr	14.18
Removal of thermoplastic screed or spray markings						
Removal of existing reflectorized thermoplastic markings						
100 mm wide line	–	–	–	–	m	1.24
150 mm wide line	–	–	–	–	m	1.86
200 mm wide line	–	–	–	–	m	2.47
arrow or letter ne 6.0 m long	–	–	–	–	nr	19.17
arrow or letter 6.0–16.00 m long	–	–	–	–	nr	80.36
REFLECTING ROAD STUDS						
100 × 100 mm square bi-directional reflecting road studs with amber corner cube reflectors	–	–	–	–	nr	6.33
140 × 254 mm rectangular one way reflecting road studs with red catseye reflectors	–	–	–	–	nr	15.18
140 × 254 mm rectangular one way reflecting road studs with green catseye reflectors	–	–	–	–	nr	15.18
140 × 254 mm rectangular bi-directional reflecting road studs with white catseye reflectors	–	–	–	–	nr	15.18
140 × 254 mm rectangular bi-directional reflecting road studs with amber catseye reflectors	–	–	–	–	nr	15.18
140 × 254 mm rectangular bi-directional reflecting road stud without catseye reflectors	–	–	–	–	nr	9.49
REMOVAL OF ROAD STUDS						
Removal of road studs						
100 × 100 mm corner cube type	–	–	–	–	nr	3.15
140 × 254 mm cateye type	–	–	–	–	nr	7.59
REMOVAL FROM STORE AND REFIX ROAD STUD						
General						
Remove from store and re-install 100 × 100 mm square bi-directional reflecting road stud with corner cube reflectors	–	–	–	–	nr	3.15
Remove from store and re-install 140 × 254 mm rectangular one way reflecting road stud with catseye reflectors	–	–	–	–	nr	7.59

CLASS R: ROADS AND PAVINGS

Item	Gang hours	Labour £	Plant £	Material £	Unit	Total rate £
TRAFFIC SIGNAL INSTALLATIONS						
Traffic signal installation is carried out exclusively by specialist contractors, although certain items are dealt with by the main contractor or a sub-contractor.						
The following of signal pedestals, loop detector unit pedestals, controller unit boxes and cable connection pillars						
Installation of signal pedestals, loop detector unit pedestals, controller unit boxes and cable connection pillars						
signal pedestal	–	–	–	–	nr	**31.74**
loop detector unit pedestral	–	–	–	–	nr	**19.65**
Excavate trench for traffic signal cable, depth ne 1.50 m; supports, backfilling						
450 mm wide	–	–	–	–	m	**5.37**
Extra for excavating in hard material	–	–	–	–	m³	**35.00**
Saw cutting grooves in pavement for detector loops and feeder cables; seal with hot bitumen sealant after installation						
25 mm deep	–	–	–	–	m	**4.81**

CLASS S: RAIL TRACK

Item	Gang hours	Labour £	Plant £	Material £	Unit	Total rate £
NOTES						
Generally						
The following unit costs are for guidance only. For more reliable estimates it is advisable to seek the advice of a Specialist Contractor.						
These rates are for the supply and laying of track other than in connection with the Permanent Way.						
Permanent Way						
The following rates would not reflect work carried out on the existing public track infrastructure (Permanent Way), which tends to be more costly not merely due to differences in technology and the level of specification and control standards, but also due to a number of logistical factors, such as:						
* access to the works for personnel, plant and machinery would be via approved access points to the rail followed by travel along the rail to the work area; this calls for the use of additional and expensive transport plant as well as reducing the effective shift time of the works gang						
* effect of track possession periods will dictate when the work can be carried out and could well force night-time or weekend working and perhaps severely reducing the effective shift hours where coupled to long travel to work distances and the need to clear away before the resumption of traffic.						
* the labour gang will be composed of more highly paid personnel, reflecting the additional training received; in addition there may well be additional gang members acting as look-outs; this could add 30% to the gang rates shown below						
* plant will tend to cost more, especially if the circumstances of the work call for rail/road plant; this could add 20 % to the gang rates shown						
Possession costs						
Where the contractor's work is on, over or poses a risk to the safety of the railway, then the contractor normally applies for possession of the track. During the period for which the contractor is given possession, rail traffic stops .						
Possessions of the Operational Safety Zone may well be fragmented rather than a single continuous period, dependant upon windows in the pattern of rail traffic						
Costs for working alongside an operational rail system are high, the need for safety demanding a high degree of supervision, look-outs, the induction of labour gangs and may involve temporary works such as safety barriers.						

CLASS S: RAIL TRACK

Item	Gang hours	Labour £	Plant £	Material £	Unit	Total rate £
TRACK FOUNDATIONS						
Imported crushed granite						
Bottom ballast	–	–	–	–	m³	60.00
Top ballast	–	–	–	–	m³	60.00
Imported granular material						
Blankets; 150 mm thick	–	–	–	–	m²	50.00
Imported sand						
Blinding; 100 mm thick	–	–	–	–	m²	30.00
Polythene sheeting						
Waterproof membrane; 1200 gauge	–	–	–	–	m²	3.75
LIFTING, PACKING AND SLEWING						
Maximum distance of slew 300 mm; maximum lift 100 mm; no extra ballast allowed						
Bullhead rail track; fishplated; timber sleepers	–	–	–	–	m	128.00
Bullhead rail track; fishplated; concrete sleepers	–	–	–	–	m	128.00
Bullhead rail track with turnout; timber sleepers	–	–	–	–	nr	128.00
Flat bottom rail track; welded; timber sleepers	–	–	–	–	m	128.00
Flat bottom rail track; welded; concrete sleepers	–	–	–	–	m	128.00
Flat bottom rail track with turnout; concrete sleepers	–	–	–	–	nr	80.00
Buffer stops	–	–	–	–	nr	3750.00
TAKING UP						
Taking up; dismantling into individual components; sorting; storing on site where directed						
Bullhead or flat bottom rails						
plain track; fishplated; timber sleepers	–	–	–	–	m	10.67
plain track; fishplated; concrete sleepers	–	–	–	–	m	10.67
plain track; welded; timber sleepers	–	–	–	–	m	10.67
plain track; welded; concrete sleepers	–	–	–	–	m	10.67
turnouts; fishplated; concrete sleepers	–	–	–	–	nr	1500.00
diamond crossings; fishplated; timber sleepers	–	–	–	–	nr	1500.00
Dock and crane rails						
plain track; welded; base plates	–	–	–	–	m	10.67
turnouts; welded; base plates	–	–	–	–	nr	1500.00
diamonds; welded; base plates	–	–	–	–	nr	1500.00
Check and guard rails						
plain track; fishplated	–	–	–	–	m	10.67
Conductor rails						
plain track; fishplated	–	–	–	–	m	5.60

CLASS S: RAIL TRACK

Item	Gang hours	Labour £	Plant £	Material £	Unit	Total rate £
Sundries						
buffer stops	–	–	–	–	nr	3950.00
retarders	–	–	–	–	nr	750.00
wheel stops	–	–	–	–	nr	750.00
lubricators	–	–	–	–	nr	750.00
switch heaters	–	–	–	–	nr	750.00
switch levers	–	–	–	–	nr	750.00
SUPPLYING (STANDARD GAUGE TRACK)						
Supplying						
Bullhead rails; BS 11; delivered in standard 18.288 m lengths						
BS95R section, 47 kg/m; for jointed track	–	–	–	–	t	655.00
BS95R section, 47 kg/m; for welded track	–	–	–	–	t	655.00
Flat bottom rails; BS 11; delivered in standard 18.288 m lengths						
BS113'A' section, 56 kg/m; for jointed track	–	–	–	–	t	655.00
BS113'A' section, 56 kg/m; for welded track	–	–	–	–	t	655.00
Extra for curved rails to form super elevation; radius over 600 m	–	–	–	–	%	20.00
Check and guard rails; BS 11; delivered in standard 18.288 m lengths; flange planed to allow 50 mm free wheel clearance						
BS113'A' section, 56 kg/m; for bolting	–	–	–	–	t	855.00
Conductor rails; BS 11; delivered in standard 18.288 m lengths						
BS113'A' section, 56 kg/m; for bolting	–	–	–	–	t	855.00
Twist rails; BS 11; delivered in standard 18.288 m lengths						
BS113'A' section, 56 kg/m; for bolting	–	–	–	–	t	855.00
Sleepers; bitumen saturated French Maritime pine						
2600 × 250 × 130 mm	–	–	–	–	nr	26.00
Sleepers; bitumen saturated Douglas fir						
2600 × 250 × 130 mm	–	–	–	–	nr	46.15
Sleepers; prestressed concrete						
2525 × 264 × 204 mm; BR type F27, Pandrol inserts	–	–	–	–	nr	62.00
Fittings						
Cast iron chairs complete with chair screws, plastic ferrules, spring steels and keys; BR type S1	–	–	–	–	nr	52.00
Cast iron chairs complete with chair screws, plastic ferrules, spring steels and keys; BR type CC	–	–	–	–	nr	73.00
Cast iron chairs complete with resilient pad, chair screws, ferrules, rail clips and nylon insulators; BR type PAN 6	–	–	–	–	nr	39.75
Cast iron chairs complete with resilient pad, chair screws, ferrules, rail clips and nylon insulators; BR type VN	–	–	–	–	nr	45.00
Cast iron chairs complete with resilient pad, chair screws, ferrules, rail clips and nylon insulators; BR type C	–	–	–	–	nr	34.00

CLASS S: RAIL TRACK

Item	Gang hours	Labour £	Plant £	Material £	Unit	Total rate £
SUPPLYING (STANDARD GAUGE TRACK) – cont						
Supplying – cont						
Fittings – cont						
Pandrol rail clips and nylon insulator	–	–	–	–	nr	2.81
plain fishplates; for BS95R section rail, skirted pattern; sets of two; complete with fishbolts, nuts and washers	–	–	–	–	nr	100.00
plain fishplates; for BS95R section rail, joggled pattern; sets of two; complete with fishbolts, nuts and washers	–	–	–	–	nr	115.00
plain fishplates; for BS 113 'A' section rail, shallow section; sets of two; complete with fishbolts, nuts and washers	–	–	–	–	nr	100.00
insulated fishplates; for BS95R section rail, steel billet pattern; sets of two; complete with high tensile steel bolts, nuts and washers	–	–	–	–	nr	58.00
insulated fishplates; BS95R section rail, steel billet pattern; sets of two; complete with high tensile steel bolts, nuts and washers	–	–	–	–	nr	58.00
cast iron spacer block between running and guard rails; for BS95R section rail; M25 × 220 mm bolt, nut and washers	–	–	–	–	nr	45.00
cast iron spacer block between running and guard rails; for BS 113 'A' section rail; M25 × 220 mm bolt, nut and washers	–	–	–	–	nr	45.00
Turnouts; complete with closures, check rails, fittings, timber sleepers						
Type B8; BS 95R bullhead rail	–	–	–	–	nr	20500.00
Type C10; BS 95R bullhead rail	–	–	–	–	nr	1600.00
Type Bx8; BS 113 'A' section flat bottom rail	–	–	–	–	nr	13500.00
Type Cv9.25; BS 113 'A' section flat bottom rail	–	–	–	–	nr	21600.00
Diamond crossings; complete with closures, check rails, fittings, timber sleepers						
RT standard design, angle 1 in 4; BS95R bullhead rail	–	–	–	–	nr	90215.33
RT standard design, angle 1 in 4; BS 113 'A' section flat bottom rail	–	–	–	–	nr	97519.36
Sundries						
buffer stops; single raker, steel rail and timber; 2 tonnes approximate weight	–	–	–	–	nr	12500.00
buffer stops; double raker, steel rail and timber; 2.5 tonnes approximate weight	–	–	–	–	nr	14000.00
wheel stops; steel; 100 kg approximate weight	–	–	–	–	nr	600.00
lubricators; single rail	–	–	–	–	nr	1100.00
lubricators; double rail	–	–	–	–	nr	2200.00
switch levers; upright pattern	–	–	–	–	nr	750.00
switch levers; flush type	–	–	–	–	nr	750.00

CLASS S: RAIL TRACK

Item	Gang hours	Labour £	Plant £	Material £	Unit	Total rate £
LAYING (STANDARD GAUGE TRACK)						
Laying						
Bullhead rails; jointed with fishplates; softwood sleepers						
plain track	–	–	–	–	m	67.00
form curve in plain track, radius ne 300 m	–	–	–	–	m	67.00
form curve in plain track, radius over 300 m	–	–	–	–	m	67.00
turnouts; standard, type B8	–	–	–	–	nr	13750.00
turnouts; standard, type C10	–	–	–	–	nr	16750.00
diamond crossings; standard	–	–	–	–	nr	35000.00
welded joints; alumino-thermic welding including						
refractory mould	–	–	–	–	nr	650.00
spot re-sleepering	–	–	–	–	nr	97.00
Bullhead rails; jointed with fishplates; concrete sleepers						
plain track	–	–	–	–	m	67.00
form curve in plain track, radius ne 300 m	–	–	–	–	m	67.00
form curve in plain track, radius over 300 m	–	–	–	–	m	67.00
turnouts; standard, type B8	–	–	–	–	nr	13750.00
turnouts; standard, type C10	–	–	–	–	nr	16750.00
diamond crossings; standard	–	–	–	–	nr	35000.00
welded joints; alumino-thermic welding including						
refractory mould	–	–	–	–	nr	650.00
spot re-sleepering	–	–	–	–	nr	97.00
Bullhead rails; welded joints; softwood sleepers						
plain track	–	–	–	–	m	67.00
form curve in plain track, radius ne 300 m	–	–	–	–	m	67.00
form curve in plain track, radius over 300 m	–	–	–	–	m	67.00
turnouts; standard, type B8	–	–	–	–	nr	13750.00
turnouts; standard, type C10	–	–	–	–	nr	16750.00
diamond crossings; standard	–	–	–	–	nr	35000.00
spot re-sleepering	–	–	–	–	nr	97.00
Bullhead rails; welded joints; concrete sleepers						
plain track	–	–	–	–	m	97.00
form curve in plain track, radius ne 300 m	–	–	–	–	m	97.00
form curve in plain track, radius over 300 m	–	–	–	–	m	97.00
turnouts; standard, type B8	–	–	–	–	nr	13750.00
turnouts; standard, type C10	–	–	–	–	nr	16750.00
diamond crossings; standard	–	–	–	–	nr	35000.00
spot re-sleepering	–	–	–	–	nr	97.00
Flat bottom rails; jointed with fishplates; softwood sleepers						
plain track	–	–	–	–	m	67.00
form curve in plain track, radius not exceeding 300 m	–	–	–	–	m	67.00
form curve in plain track, radius exceeding 300 m	–	–	–	–	m	67.00
turnouts; standard, type Bv8	–	–	–	–	nr	13750.00
turnouts; standard, type Cv9.25	–	–	–	–	nr	16750.00
diamond crossings; standard	–	–	–	–	nr	35000.00
welded joints; alumino-thermic welding including						
refractory mould	–	–	–	–	nr	650.00
spot re-sleepering	–	–	–	–	nr	97.00

CLASS S: RAIL TRACK

Item	Gang hours	Labour £	Plant £	Material £	Unit	Total rate £
LAYING (STANDARD GAUGE TRACK) – cont						
Flat bottom rails; jointed with fishplates; concrete sleepers						
plain track	–	–	–	–	m	97.00
form curve in plain track, radius ne 300 m	–	–	–	–	m	97.00
form curve in plain track, radius over 300 m	–	–	–	–	m	97.00
turnouts; standard, type Bv8	–	–	–	–	nr	13750.00
turnouts; standard, type Cv9.25	–	–	–	–	nr	16750.00
diamond crossings; standard	–	–	–	–	nr	35000.00
welded joints; alumino-thermic welding including refractory mould	–	–	–	–	nr	650.00
spot re-sleepering	–	–	–	–	nr	97.00
Flat bottom rails; welded joints; softwood sleepers						
plain track	–	–	–	–	m	97.00
form curve in plain track, radius ne 300 m	–	–	–	–	m	97.00
form curve in plain track, radius over 300 m	–	–	–	–	m	97.00
turnouts; standard, type Bv8	–	–	–	–	nr	13750.00
turnouts; standard, type Cv9.25	–	–	–	–	nr	16750.00
diamond crossings; standard	–	–	–	–	nr	35000.00
spot re-sleepering	–	–	–	–	nr	97.00
Flat bottom rails; welded joints; concrete sleepers						
plain track	–	–	–	–	m	650.00
form curve in plain track, radius ne 300 m	–	–	–	–	m	97.00
form curve in plain track, radius over 300 m	–	–	–	–	m	97.00
turnouts; standard, type Bv8	–	–	–	–	nr	13750.00
turnouts; standard, type Cv9.25	–	–	–	–	nr	16750.00
diamond crossings; standard	–	–	–	–	nr	35000.00
spot re-sleepering	–	–	–	–	nr	97.00
Check rails, flat bottom; jointed with fishplates						
rail	–	–	–	–	m	97.00
welded joints; alumino-thermic welding including refractory mould	–	–	–	–	nr	97.00
Guard rails, bullhead; jointed with fishplates						
rail	–	–	–	–	m	97.00
welded joints; alumino-thermic welding including refractory mould	–	–	–	–	nr	650.00
Guard rails, flat bottom; jointed with fishplates						
rail	–	–	–	–	m	97.00
welded joints; alumino-thermic welding including refractory mould	–	–	–	–	nr	650.00
Conductor rails, bullhead; jointed with fishplates						
rail	–	–	–	–	m	97.00
welded joints; alumino-thermic welding including refractory mould	–	–	–	–	nr	97.00

CLASS S: RAIL TRACK

Item	Gang hours	Labour £	Plant £	Material £	Unit	Total rate £
Sundries						
buffer stops; single raker, steel rail and timber; 2 tonnes approximate weight	–	–	–	–	nr	3750.00
buffer stops; double raker, steel rail and timber; 2.5 tonnes approximate weight	–	–	–	–	nr	3750.00
wheel stops; steel; 100 kg approximate weight	–	–	–	–	nr	3750.00
lubricators; single rail	–	–	–	–	nr	960.00
lubricators; double rail	–	–	–	–	nr	960.00
switch levers; upright pattern	–	–	–	–	nr	1250.00
switch levers; flush type	–	–	–	–	nr	1250.00
conductor rail guard boards	–	–	–	–	m	2.50
DECAUVILLE TRACK						
Supplying						
Dock and crane rails; for welded track						
section 56 crane rail; 12.2 m lengths	–	–	–	–	nr	1477.82
section 101 crane rail; 9.144 m lengths	–	–	–	–	nr	1279.02
Fittings						
20 mm mild steel sole plate 400 mm wide; drilled with two bolt holes at 1200 mm centres	–	–	–	–	m	65.22
M20 × 250 mm holding down bolt, nut and washers	–	–	–	–	nr	2.77
rail clips, spring type, adjustable; complete with M20 × 60 mm stud welded to sole plate	–	–	–	–	nr	16.00
Sundries						
wheel stops; 200 kg each	–	–	–	–	nr	193.76
Laying						
Crane rails, section 56; continuous sole plate; welded						
plain track	–	–	–	–	m	54.79
form curve in plain track, radius ne 300 m	–	–	–	–	m	10.16
Crane rails, section 101; continuous sole plate; welded						
plain track	–	–	–	–	m	69.40
form curve in plain track, radius ne 300 m	–	–	–	–	m	10.16
Sundries						
wheel stops	–	–	–	–	nr	101.78

CLASS T: TUNNELS

Item	Gang hours	Labour £	Plant £	Material £	Unit	Total rate £
NOTES						
There are so many factors, apart from design considerations, which influence the cost of tunnelling that it is only possible to give guide prices for a sample of the work involved. For more reliable estimates it is advisable to seek the advice of a Specialist Contractor.						
The main cost considerations are :						
* the nature of the ground						
* size of tunnel						
* length of drive						
* depth below surface						
* anticipated overbreak						
* support of face and roof of tunnel (rock bolting etc.)						
* necessity for pre-grouting						
* ventilation						
* presence of water						
* use of compressed air working						
The following rates for mass concrete work cast in primary and secondary linings to tunnels and access shafts are based on a 5.0 m depth of shaft and 15.0 m head of tunnel						
Apply the following factors for differing depths and lengths :						
HEAD LENGTH : 15m 30m 60m 90m						
Shaft depth 5m +0% +10% +20% +32½%						
Shaft depth 10m +5% +12½% +27½% +35%						
Shaft depth 15m +10% +17½% +32½% +40%						
Shaft depth 20m +15% +20% +37½% +42½%						
EXCAVATION						
Excavating tunnels in rock						
1.5 m diameter	–	–	–	–	m³	528.00
3.0 m diameter	–	–	–	–	m³	330.00
Excavating tunnels in soft material						
1.5 m diameter	–	–	–	–	m³	250.25
3.0 m diameter	–	–	–	–	m³	137.50
Excavating shafts in rock						
3.0 m diameter	–	–	–	–	m³	203.50
4.5 m diameter	–	–	–	–	m³	170.50
Excavating shafts in soft material						
3.0 m diameter	–	–	–	–	m³	118.25
4.5 m diameter	–	–	–	–	m³	101.75
Excavating other cavities in rock						
1.5 m diameter	–	–	–	–	m³	528.00
3.0 m diameter	–	–	–	–	m³	330.00
Excavating other cavities in soft material						
1.5 m diameter	–	–	–	–	m³	250.25
3.0 m diameter	–	–	–	–	m³	137.50
Excavated surfaces in rock	–	–	–	–	m²	19.25
Excavated surfaces in soft material	–	–	–	–	m²	19.25

CLASS T: TUNNELS

Item	Gang hours	Labour £	Plant £	Material £	Unit	Total rate £
INSITU LINING TO TUNNELS						
Notes						
The following rates for mass concrete work cast in						
primary and secondary linings to tunnels and access						
shafts are based on a 5.0 m depth of shaft and						
15.0 m head of tunnel						
See above for additions for differing shaft depths and						
tunnel lengths.						
Mass concrete; grade C30, 20 mm aggregate						
Cast primary lining to tunnels						
1.5 m diameter	–	–	–	–	m³	333.50
3.0 m diameter	–	–	–	–	m³	287.50
Secondary lining to tunnels						
1.5 m diameter	–	–	–	–	m³	388.13
3.0 m diameter	–	–	–	–	m³	307.63
Formwork; rough finish						
Tunnel lining						
1.5 m diameter	–	–	–	–	m²	54.05
3.0 m diameter	–	–	–	–	m²	54.05
INSITU LINING TO ACCESS SHAFTS						
Notes						
The following rates for mass concrete work cast in						
primary and secondary linings to tunnels and access						
shafts are based on a 5.0 m depth of shaft and						
15.0 m head of tunnel						
See above for additions for differing shaft depths and						
tunnel lengths.						
Mass concrete; grade C30, 20 mm aggregate						
Secondary linings to shafts						
3.0 m int diameter	–	–	–	–	m³	307.63
4.5 m int diameter	–	–	–	–	m³	299.00
Cast primary lining to shafts						
3.0 m int diameter	–	–	–	–	m³	327.75
4.5 m int diameter	–	–	–	–	m³	313.38
Formwork; rough finish						
Shaft lining						
3.0 m int diameter	–	–	–	–	m²	81.65
4.5 m int diameter	–	–	–	–	m²	55.20

CLASS T: TUNNELS

Item	Gang hours	Labour £	Plant £	Material £	Unit	Total rate £
INSITU LINING TO OTHER CAVITIES						
Notes						
The following rates for mass concrete work cast in primary and secondary linings to tunnels and access shafts are based on a 5.0 m depth of shaft and 15.0 m head of tunnel						
See above for additions for differing shaft depths and tunnel lengths.						
Mass concrete; grade C30, 20 mm aggregate						
Cast primary lining to other cavities						
1.5 m int diameter	–	–	–	–	m^3	336.38
3.0 m int diameter	–	–	–	–	m^3	287.50
Secondary linings to other cavities						
1.5 m int diameter	–	–	–	–	m^3	388.13
3.0 m int diameter	–	–	–	–	m^3	307.63
Formwork; rough finish						
Other cavities lining						
1.5 m int diameter	–	–	–	–	m^2	54.05
3.0 m int diameter	–	–	–	–	m^2	54.05
PREFORMED SEGMENTAL LININGS TO TUNNELS						
Precast concrete bolted rings; flanged; including packing; guide price/ring based upon standard bolted concrete segmental rings; ring width 610mm						
Under current market conditions the reader is currently advised to contact specialist contractors directly for costs associated with these works						
PREFORMED SEGMENTAL LININGS TO SHAFTS						
Precast concrete bolted rings; flanged; including packing; guide price/ring based upon standard bolted concrete segmental rings; ring width 610mm						
Under current market conditions the reader is currently advised to contact specialist contractors directly for costs associated with these works						

CLASS T: TUNNELS

Item	Gang hours	Labour £	Plant £	Material £	Unit	Total rate £
PREFORMED SEGMENTAL LININGS TO OTHER						
Precast concrete bolted rings; flanged; including packing; guide price/ring based upon standard bolted concrete segmental rings; ring width 610mm						
Linings to tunnels						
1.5m int diameter; Ring 1.52m ID × 1.77m OD; 6 segments, maximum piece weigth 139 kg	–	–	–	–	nr	589.39
3.0m int diameter; Ring 3.05m ID × 3.35m OD; 7 segments , maximum piece weight 247 kg	–	–	–	–	nr	1055.37
Linings to shafts						
3.0m int diameter; Ring 3.05m ID × 3.35m OD; 7 segments, maximum piece weight 247 kg	–	–	–	–	nr	835.18
Lining ancillaries; bitumen impregnated fibreboard						
Parallel circumferential packing						
1.5m int diameter	–	–	–	–	nr	5.60
3.0m int diameter	–	–	–	–	nr	11.19
Lining ancillaries; PC4AF caulking compound						
Caulking						
1.5m int diameter	–	–	–	–	m	10.11
3.0m int diameter	–	–	–	–	m	10.11
SUPPORT AND STABILIZATION						
Rock bolts						
mechanical	–	–	–	–	m	28.00
mechanical grouted	–	–	–	–	m	43.73
pre-grouted impacted	–	–	–	–	m	42.07
chemical end anchor	–	–	–	–	m	42.07
chemical grouted	–	–	–	–	m	29.92
chemically filled	–	–	–	–	m	47.07
Internal support						
steel arches; supply	–	–	–	–	t	1284.26
steel arches; erection	–	–	–	–	t	485.59
timber supports; supply	–	–	–	–	m^3	358.41
timber supports; erection	–	–	–	–	m^3	309.03
lagging	–	–	–	–	m^2	26.50
sprayed concrete	–	–	–	–	m^2	30.31
mesh or link	–	–	–	–	m^2	9.79
Pressure grouting						
sets of drilling and grouting plant	–	–	–	–	nr	1328.56
face packers	–	–	–	–	nr	58.85
deep packers	–	–	–	–	nr	100.90
drilling and flushing to 40 mm diameter	–	–	–	–	m	22.29
re-drilling and flushing	–	–	–	–	m	17.68
injection of grout materials; chemical grout	–	–	–	–	t	706.33
Forward probing	–	–	–	–	m	22.97

CLASS U: BRICKWORK, BLOCKWORK AND MASONRY

Item	Gang hours	Labour £	Plant £	Material £	Unit	Total rate £
NOTES						
Apply the following multipliers to both labour and plant for rubble walls:-						
height 2 to 5 m 1.21						
height 5 to 10 m 1.37						
wall to radius small 1.75						
wall to radius large 1.50						
wall to rake or batter 1.15						
wall in piers or stanchion 1.50						
wall in butresses 1.15						
RESOURCES - LABOUR						
Brickwork, blockwork and masonry gang						
1 foreman bricklayer (craftsman)		22.32				
4 bricklayers (craftsman)		76.60				
1 unskilled operative (general)		12.44				
Total Gang Rate / Hour	£	**111.36**				
RESOURCES - PLANT						
Brickwork, blockwork and masonry						
2t dumper (50% of time)			3.25			
mixer (50% of time)			0.85			
small power tools			1.96			
scaffold, etc.			1.40			
Total Rate / Hour		£	**7.45**			
COMMON BRICKWORK						
Common bricks in cement mortar designation (ii)						
Thickness 103 mm						
vertical straight walls	0.23	25.90	1.70	28.57	m²	**56.17**
vertical curved walls	0.30	33.62	2.19	28.57	m²	**64.38**
battered straight walls	0.33	37.37	2.45	28.57	m²	**68.39**
battered curved walls	0.37	42.03	2.76	28.57	m²	**73.36**
vertical facing to concrete	0.25	28.06	1.84	29.11	m²	**59.01**
battered facing to concrete	0.37	42.03	2.76	29.13	m²	**73.92**
casing to metal sections	0.32	35.78	2.35	29.13	m²	**67.26**
Thickness 215 mm						
vertical straight walls	0.44	50.21	3.29	57.66	m²	**111.16**
vertical curved walls	0.57	64.74	4.25	57.66	m²	**126.65**
battered straight walls	0.63	68.89	4.70	57.66	m²	**131.25**
battered curved walls	0.71	77.64	5.25	57.66	m²	**140.55**
vertical facing to concrete	0.48	52.49	3.56	58.21	m²	**114.26**
battered facing to concrete	0.71	77.64	5.25	58.21	m²	**141.10**
casing to metal sections	0.61	66.70	4.51	58.21	m²	**129.42**

CLASS U: BRICKWORK, BLOCKWORK AND MASONRY

Item	Gang hours	Labour £	Plant £	Material £	Unit	Total rate £
Thickness 328 mm						
vertical straight walls	0.64	69.98	4.80	87.14	m²	161.92
vertical curved walls	0.82	89.67	6.13	87.14	m²	182.94
battered straight walls	0.91	99.51	6.76	87.14	m²	193.41
battered curved walls	1.01	110.44	7.53	87.14	m²	205.11
vertical facing to concrete	0.69	75.45	5.17	87.69	m²	168.31
battered facing to concrete	1.01	110.44	7.53	87.69	m²	205.66
casing to metal sections	0.87	95.13	6.49	87.69	m²	189.31
Thickness 440 mm						
vertical straight walls	0.84	91.85	6.22	116.47	m²	214.54
vertical curved walls	1.06	115.91	7.88	116.47	m²	240.26
battered straight walls	1.16	126.85	8.66	116.47	m²	251.98
battered curved walls	1.29	141.06	9.60	116.47	m²	267.13
vertical facing to concrete	0.90	98.42	6.69	117.01	m²	222.12
battered facing to concrete	1.29	141.06	9.60	117.02	m²	267.68
casing to metal sections	1.12	122.47	8.33	117.02	m²	247.82
Thickness 890 mm						
vertical straight walls	1.50	164.03	11.18	233.96	m²	409.17
vertical curved walls	1.85	202.30	13.80	233.96	m²	450.06
battered straight walls	2.00	218.70	14.97	233.96	m²	467.63
battered curved walls	2.20	240.57	16.36	233.96	m²	490.89
vertical facing to concrete	1.60	174.96	11.94	234.50	m²	421.40
battered facing to concrete	2.20	240.57	16.36	234.50	m²	491.43
casing to metal sections	1.94	212.14	14.48	234.50	m²	461.12
Thickness exceeding 1 m						
vertical straight walls	1.64	179.33	12.19	261.87	m²	453.39
vertical curved walls	2.00	218.70	14.96	261.87	m²	495.53
battered straight walls	2.17	237.29	16.19	261.87	m²	515.35
battered curved walls	2.37	259.16	17.63	261.87	m²	538.66
vertical facing to concrete	1.74	190.27	12.99	262.42	m²	465.68
battered facing to concrete	2.37	259.16	17.63	262.42	m²	539.21
casing to metal sections	2.10	229.63	15.67	262.42	m²	507.72
Columns and piers						
215 × 215 mm	0.13	14.22	0.97	12.75	m	27.94
440 × 215 mm	0.24	26.24	1.79	25.73	m	53.76
665 × 328 mm	0.44	48.11	3.28	58.17	m	109.56
890 × 890 mm	1.10	120.28	8.20	208.15	m	336.63
Surface features						
copings; standard header-on-edge; 215 mm wide × 103 mm high	0.10	10.94	0.78	6.41	m	18.13
sills; standard header-on-edge; 215 mm wide × 103 mm high	0.13	14.22	0.93	6.26	m	21.41
rebates	0.30	32.80	2.24	–	m	35.04
chases	0.35	38.27	2.61	–	m	40.88
band courses; flush; 215 mm wide	0.05	5.47	0.34	–	m	5.81
band courses; projection 103 mm; 215 mm wide	0.05	5.47	0.37	–	m	5.84
corbels; maximum projection 103 mm; 215 mm wide	0.15	16.40	1.12	6.18	m	23.70
pilasters; 328 mm wide × 103 mm projection	–	–	0.52	12.75	m	13.27
pilasters; 440 mm wide × 215 mm projection	0.12	13.12	0.89	28.74	m	42.75
plinths; projection 103 mm × 900 mm wide	0.19	20.78	1.42	26.10	m	48.30
fair facing	0.06	6.56	0.42	–	m²	6.98

CLASS U: BRICKWORK, BLOCKWORK AND MASONRY

Item	Gang hours	Labour £	Plant £	Material £	Unit	Total rate £
COMMON BRICKWORK – cont						
Common bricks in cement mortar designation (ii) – cont						
Ancillaries						
bonds to existing work; to brickwork	1.50	164.03	11.18	29.48	m²	**204.69**
built-in pipes and ducts, cross-sectional area not exceeding 0.05 m²; excluding supply; brickwork 103 mm thick	0.06	6.56	0.45	0.97	nr	**7.98**
built-in pipes and ducts, cross-sectional area not exceeding 0.05 m²; excluding supply; brickwork 215 mm thick	0.12	13.12	0.86	1.42	nr	**15.40**
built-in pipes and ducts, cross-sectional area 0.05–0.25 m²; excluding supply; brickwork 103 mm thick	0.15	16.40	1.12	1.42	nr	**18.94**
built-in pipes and ducts, cross-sectional area 0.05–0.25 m²; excluding supply; brickwork 215 mm thick	0.29	31.71	2.16	2.10	nr	**35.97**
built-in pipes and ducts, cross-sectional area 0.25-0.5 m²; excluding supply; brickwork 103 mm thick	0.18	19.68	1.30	1.65	nr	**22.63**
built-in pipes and ducts, cross-sectional area 0.25-0.5 m²; excluding supply; brickwork 215 mm thick	0.34	37.18	2.53	2.40	nr	**42.11**
FACING BRICKWORK						
Facing bricks; in plasticised cement mortar designation (iii)						
Thickness 103 mm						
vertical straight walls	0.34	37.18	2.53	43.84	m²	**83.55**
vertical curved walls	0.45	49.21	3.35	43.84	m²	**96.40**
battered straight walls	0.45	49.21	3.35	43.84	m²	**96.40**
battered curved walls	0.56	61.24	4.19	43.84	m²	**109.27**
vertical facing to concrete	0.37	40.46	2.66	44.39	m²	**87.51**
battered facing to concrete	0.56	61.24	4.19	44.39	m²	**109.82**
casing to metal sections	0.49	53.58	3.65	44.39	m²	**101.62**
Thickness 215 mm						
vertical straight walls	0.57	62.33	5.18	88.21	m²	**155.72**
vertical curved walls	0.84	91.85	6.26	88.21	m²	**186.32**
battered straight walls	0.84	91.85	6.26	88.21	m²	**186.32**
battered curved walls	1.02	111.54	7.62	88.21	m²	**207.37**
vertical facing to concrete	0.66	72.17	4.92	88.75	m²	**165.84**
battered facing to concrete	1.02	111.54	7.62	88.75	m²	**207.91**
casing to metal sections	0.82	89.67	6.11	88.75	m²	**184.53**
Thickness 328 mm						
vertical straight walls	0.83	90.76	6.16	132.97	m²	**229.89**
vertical curved walls	1.12	122.47	8.31	132.97	m²	**263.75**
battered straight walls	1.12	122.47	8.31	132.97	m²	**263.75**
battered curved walls	1.36	148.72	10.13	132.97	m²	**291.82**
vertical facing to concrete	0.88	96.23	6.53	133.51	m²	**236.27**
battered facing to concrete	1.36	148.72	10.13	133.51	m²	**292.36**
casing to metal sections	1.18	129.03	8.79	133.51	m²	**271.33**

CLASS U: BRICKWORK, BLOCKWORK AND MASONRY

Item	Gang hours	Labour £	Plant £	Material £	Unit	Total rate £
Thickness 440 mm						
vertical straight walls	1.08	118.10	8.05	177.57	m²	303.72
vertical curved walls	1.39	152.00	10.36	177.57	m²	339.93
battered straight walls	1.39	152.00	10.36	177.57	m²	339.93
vertical facing to concrete	1.09	119.19	8.14	178.11	m²	305.44
battered facing to concrete	1.39	152.00	10.36	178.11	m²	340.47
casing to metal sections	1.39	152.00	10.36	178.11	m²	340.47
Columns and piers						
215×215mm	0.17	18.59	1.24	19.64	m	39.47
440×215mm	0.29	31.71	2.16	39.47	m	73.34
665×328mm	0.58	63.42	4.30	89.10	m	156.82
Surface features						
copings; standard header-on-edge; standard bricks; 215mm wide × 103mm high	0.13	14.22	1.00	9.20	m	24.42
flat arches; standard stretcher-on-end; 215mm wide × 103mm high	0.21	22.96	1.56	8.97	m	33.49
flat arches; standard stretcher-on-end; bullnosed special bricks; 103mm×215mm high	0.22	24.06	1.64	53.19	m	78.89
segmental arches; single ring; standard bricks; 103mm wide × 215mm high	0.37	40.46	2.76	8.97	m	52.19
segmental arches; two ring; standard bricks; 103mm wide × 440mm high	0.49	53.58	3.65	18.10	m	75.33
segmental arches; cut voussoirs; 103mm wide × 215mm high	0.39	42.65	2.91	21.87	m	67.43
rebates	0.33	36.09	2.46	–	m	38.55
chases	0.37	40.46	2.76	–	m	43.22
cornices; maximum projection 103mm; 215mm wide	0.37	40.46	2.76	8.90	m	52.12
band courses; projection 113mm; 215mm wide	0.06	6.56	0.41	–	m	6.97
corbels; maximum projection 113mm; 215mm wide	0.37	40.46	2.76	8.90	m	52.12
pilasters; 328mm wide × 113mm projection	0.05	5.47	0.37	19.63	m	25.47
pilasters; 440mm wide × 215mm projection	0.06	6.56	0.41	44.20	m	51.17
plinths; projection 113mm×900mm wide	0.24	26.24	1.79	39.85	m	67.88
fair facing; pointing as work proceeds	0.06	6.56	0.45	–	m²	7.01
Ancillaries						
bonds to existing work; to brickwork	1.50	164.03	11.18	12.09	m	187.30
built-in pipes and ducts, cross-sectional area not exceeding 0.05m²; excluding supply; brickwork half brick thick	0.10	10.94	0.71	0.95	nr	12.60
built-in pipes and ducts, cross-sectional area not exceeding 0.05m²; excluding supply; brickwork one brick thick	0.15	16.40	0.74	1.67	nr	18.81
built-in pipes and ducts, cross-sectional area 0.05–0.25m²; excluding supply; brickwork half brick thick	0.19	20.78	1.42	1.50	nr	23.70
built-in pipes and ducts, cross-sectional area 0.05–0.25m²; excluding supply; brickwork one brick thick	0.33	36.09	2.46	2.43	nr	40.98
built-in pipes and ducts, cross-sectional area 0.25-0.5m²; excluding supply; brickwork half brick thick	0.23	25.15	1.71	1.73	nr	28.59
built-in pipes and ducts, cross-sectional area 0.25-0.5m²; excluding supply; brickwork one brick thick	0.40	43.74	2.94	2.81	nr	49.49

CLASS U: BRICKWORK, BLOCKWORK AND MASONRY

Item	Gang hours	Labour £	Plant £	Material £	Unit	Total rate £
ENGINEERING BRICKWORK						
Class A engineering bricks, solid; in cement mortar designation (ii)						
Thickness 103 mm						
vertical straight walls	0.27	29.52	2.01	58.43	m²	89.96
vertical curved walls	0.37	40.46	2.76	58.43	m²	101.65
battered straight walls	0.37	40.46	2.76	58.43	m²	101.65
battered curved walls	0.41	44.83	3.07	58.43	m²	106.33
vertical facing to concrete	0.32	34.99	2.35	58.98	m²	96.32
battered facing to concrete	0.46	50.30	3.45	58.98	m²	112.73
casing to metal sections	0.41	44.83	3.07	58.98	m²	106.88
Thickness 215 mm						
vertical straight walls	0.52	56.86	3.88	117.39	m²	178.13
vertical curved walls	0.71	77.64	5.25	117.39	m²	200.28
battered straight walls	0.71	77.64	5.25	117.39	m²	200.28
battered curved walls	0.78	85.29	5.81	117.39	m²	208.49
vertical facing to concrete	0.61	66.70	4.51	117.94	m²	189.15
battered facing to concrete	0.87	95.13	6.49	117.94	m²	219.56
casing to metal sections	0.78	85.29	5.81	117.94	m²	209.04
Thickness 328 mm						
vertical straight walls	0.75	82.01	5.61	176.75	m²	264.37
vertical curved walls	1.01	110.44	7.53	176.75	m²	294.72
battered straight walls	1.01	110.44	7.53	176.75	m²	294.72
battered curved walls	1.11	121.38	8.28	176.75	m²	306.41
vertical facing to concrete	0.87	95.13	6.49	177.29	m²	278.91
battered facing to concrete	1.24	135.59	9.20	177.29	m²	322.08
casing to metal sections	1.11	121.38	8.28	177.29	m²	306.95
Thickness 440 mm						
vertical straight walls	0.97	106.07	7.24	235.94	m²	349.25
vertical curved walls	1.29	141.06	9.60	235.94	m²	386.60
battered straight walls	1.29	141.06	9.60	235.94	m²	386.60
battered curved walls	1.41	154.18	10.51	235.94	m²	400.63
vertical facing to concrete	1.12	122.47	8.33	236.48	m²	367.28
battered facing to concrete	1.56	170.59	11.62	236.48	m²	418.69
casing to metal sections	1.41	154.18	10.51	236.48	m²	401.17
Thickness 890 mm						
vertical straight walls	1.72	188.08	12.80	472.70	m²	673.58
vertical curved walls	2.20	240.57	16.36	472.70	m²	729.63
battered straight walls	2.20	240.57	16.36	472.70	m²	729.63
battered curved walls	2.37	259.16	17.66	472.70	m²	749.52
vertical facing to concrete	1.94	212.14	14.48	473.24	m²	699.86
battered facing to concrete	2.58	282.12	19.20	473.24	m²	774.56
casing to metal sections	2.37	259.16	17.66	473.24	m²	750.06
Thickness exceeding 1 m						
vertical straight walls	1.87	204.48	13.90	529.12	m³	747.50
vertical curved walls	2.37	259.16	17.63	529.12	m³	805.91
battered straight walls	2.37	259.16	17.63	529.12	m³	805.91
battered curved walls	2.55	278.84	18.99	529.12	m³	826.95
vertical facing to concrete	2.10	229.63	15.67	529.59	m³	774.89
battered facing to concrete	2.76	301.81	20.57	529.59	m³	851.97
casing to metal sections	2.55	278.84	18.99	529.59	m³	827.42

CLASS U: BRICKWORK, BLOCKWORK AND MASONRY

Item	Gang hours	Labour £	Plant £	Material £	Unit	Total rate £
Columns and piers						
215×215 mm	0.16	17.50	1.19	26.19	m	**44.88**
440×215 mm	0.28	30.62	2.07	52.60	m	**85.29**
665×328 mm	0.56	61.24	4.14	118.65	m	**184.03**
890×890 mm	1.78	194.64	13.25	423.19	m	**631.08**
Surface features						
copings; standard header-on-edge; 215 mm wide × 103 mm high	0.11	12.03	0.78	13.14	m	**25.95**
sills; standard header-on-edge; 215 mm wide × 103 mm high	0.13	14.22	0.93	12.98	m	**28.13**
rebates	0.33	36.09	2.46	–	m	**38.55**
chases	0.37	40.46	2.76	–	m	**43.22**
band courses; flush; 215 mm wide	0.05	5.47	0.34	–	m	**5.81**
band courses; projection 103 mm; 215 mm wide	0.05	5.47	0.37	–	m	**5.84**
corbels; maximum projection 103 mm; 215 mm wide	0.15	16.40	1.12	12.90	m	**30.42**
pilasters; 328 mm wide × 103 mm projection	0.07	7.65	0.52	26.19	m	**34.36**
pilasters; 440 mm wide × 215 mm projection	0.12	13.12	0.89	58.98	m	**72.99**
plinths; projection 103 mm × 900 mm wide	0.19	20.78	1.42	52.98	m	**75.18**
fair facing	0.06	6.56	0.45	–	m²	**7.01**
Ancillaries						
bonds to existing brickwork; to brickwork	0.29	31.71	2.16	17.01	m²	**50.88**
built-in pipes and ducts, cross-sectional area not exceeding 0.05 m²; excluding supply; brickwork half brick thick	0.10	10.94	0.71	1.61	nr	**13.26**
built-in pipes and ducts, cross-sectional area not exceeding 0.05 m²; excluding supply; brickwork one brick thick	0.15	16.40	1.12	2.19	nr	**19.71**
built-in pipes and ducts, cross-sectional area 0.05–0.25 m²; excluding supply; brickwork half brick thick	0.19	20.78	1.42	2.12	nr	**24.32**
built-in pipes and ducts, cross-sectional area 0.05–0.25 m²; excluding supply; brickwork one brick thick	0.33	36.09	2.46	3.06	nr	**41.61**
built-in pipes and ducts, cross-sectional area 0.025-0.5 m²; excluding supply; brickwork half brick thick	0.23	25.15	1.71	2.37	nr	**29.23**
built-in pipes and ducts, cross-sectional area 0.025-0.5 m²; excluding supply; brickwork one brick thick	0.40	43.74	2.94	3.45	nr	**50.13**

CLASS U: BRICKWORK, BLOCKWORK AND MASONRY

Item	Gang hours	Labour £	Plant £	Material £	Unit	Total rate £
ENGINEERING BRICKWORK – cont						
Class B engineering bricks, perforated; in cement mortar designation (ii)						
Thickness 103 mm						
vertical straight walls	0.27	29.52	2.01	27.94	m²	**59.47**
vertical curved walls	0.37	40.46	2.76	27.94	m²	**71.16**
battered straight walls	0.37	40.46	2.76	27.94	m²	**71.16**
vertical facing to concrete	0.32	34.99	2.35	28.49	m²	**65.83**
casings to metal sections	0.41	44.83	3.07	28.49	m²	**76.39**
Thickness 215 mm						
vertical straight walls	0.52	56.86	3.88	56.41	m²	**117.15**
vertical curved walls	0.71	77.64	5.25	56.41	m²	**139.30**
battered straight walls	0.71	77.64	5.25	56.41	m²	**139.30**
vertical facing to concrete	0.61	66.70	4.51	56.96	m²	**128.17**
casings to metal sections	0.78	85.29	5.81	56.96	m²	**148.06**
Thickness 328 mm						
vertical straight walls	0.76	83.11	5.63	85.28	m²	**174.02**
vertical curved walls	1.01	110.44	7.53	85.28	m²	**203.25**
battered straight walls	1.01	110.44	7.53	85.28	m²	**203.25**
vertical facing to concrete	0.87	95.13	6.49	85.82	m²	**187.44**
casings to metal sections	1.11	121.38	8.28	85.82	m²	**215.48**
Thickness 440 mm						
vertical straight walls	0.97	106.07	7.24	113.98	m²	**227.29**
vertical curved walls	1.29	141.06	9.60	113.98	m²	**264.64**
battered straight walls	1.29	141.06	9.60	113.98	m²	**264.64**
vertical facing to concrete	1.12	122.47	8.33	114.53	m²	**245.33**
casings to metal sections	1.41	154.18	10.51	114.53	m²	**279.22**
Thickness 890 mm						
vertical straight walls	1.72	188.08	12.80	228.79	m²	**429.67**
vertical curved walls	2.20	240.57	16.36	228.79	m²	**485.72**
battered straight walls	2.20	240.57	16.36	228.79	m²	**485.72**
battered curved walls	2.37	259.16	17.66	228.79	m²	**505.61**
vertical facing to concrete	1.94	212.14	14.48	229.34	m²	**455.96**
battered facing to concrete	2.58	282.12	19.20	229.34	m²	**530.66**
casing to metal sections	2.37	259.16	17.66	229.34	m²	**506.16**
Thickness exceeding 1 m						
vertical straight walls	1.87	204.48	13.90	256.44	m³	**474.82**
vertical curved walls	2.37	259.16	17.63	256.44	m³	**533.23**
battered straight walls	2.37	259.16	17.63	256.44	m³	**533.23**
battered curved walls	2.55	278.84	18.99	256.44	m³	**554.27**
vertical facing to concrete	2.10	229.63	15.67	256.91	m³	**502.21**
battered facing to concrete	2.76	301.81	20.57	256.91	m³	**579.29**
casing to metal sections	2.55	278.84	18.99	256.91	m³	**554.74**
Columns and piers						
215 × 215 mm	0.16	17.50	1.19	12.47	m	**31.16**
440 × 215 mm	0.28	30.62	2.07	25.17	m	**57.86**
665 × 328 mm	0.56	61.24	4.14	56.91	m	**122.29**
890 × 890 mm	1.78	194.64	13.25	203.67	m	**411.56**

CLASS U: BRICKWORK, BLOCKWORK AND MASONRY

Item	Gang hours	Labour £	Plant £	Material £	Unit	Total rate £
Surface features						
copings; standard header-on-edge; 215mm wide × 103mm high	0.11	12.03	0.78	6.27	m	**19.08**
sills; standard header-on-edge; 215mm wide × 103mm high	0.13	14.22	0.93	6.12	m	**21.27**
rebates	0.33	36.09	2.46	–	m	**38.55**
chases	0.37	40.46	2.76	–	m	**43.22**
band courses; flush; 215mm wide	0.05	5.47	0.34	–	m	**5.81**
band courses; projection 103mm; 215mm wide	0.05	5.47	0.37	–	m	**5.84**
corbels; maximum projection 103mm; 215mm wide	0.15	16.40	1.12	6.04	m	**23.56**
pilasters; 328mm wide × 103mm projection	0.07	7.65	0.52	12.47	m	**20.64**
pilasters; 440mm wide × 215mm projection	0.12	13.12	0.89	28.11	m	**42.12**
plinths; projection 103mm × 900mm wide	0.19	20.78	1.42	25.54	m	**47.74**
fair facing	0.06	6.56	0.45	–	m²	**7.01**
Ancillaries						
bonds to existing brickwork; to brickwork	0.29	31.71	2.16	16.48	m²	**50.35**
built-in pipes and ducts, cross-sectional area not exceeding 0.05m²; excluding supply; brickwork half brick thick	0.10	10.94	0.71	1.61	nr	**13.26**
built-in pipes and ducts, cross-sectional area not exceeding 0.05m²; excluding supply; brickwork one brick thick	0.15	16.40	1.12	2.19	nr	**19.71**
built-in pipes and ducts, cross-sectional area 0.05–0.25m²; excluding supply; brickwork half brick thick	0.19	20.78	1.42	2.12	nr	**24.32**
built-in pipes and ducts, cross-sectional area 0.05–0.25m²; excluding supply; brickwork one brick thick	0.33	36.09	2.46	3.06	nr	**41.61**
built-in pipes and ducts, cross-sectional area 0.025–0.5m²; excluding supply; brickwork half brick thick	0.23	25.15	1.71	2.37	nr	**29.23**
built-in pipes and ducts, cross-sectional area 0.025–0.5m²; excluding supply; brickwork one brick thick	0.40	43.74	2.94	3.41	nr	**50.09**

CLASS U: BRICKWORK, BLOCKWORK AND MASONRY

Item	Gang hours	Labour £	Plant £	Material £	Unit	Total rate £
LIGHTWEIGHT BLOCKWORK						
Lightweight concrete blocks; 3.5 N/mm²; in cement-lime mortar designation (iii)						
Thickness 100 mm;						
vertical straight walls	0.17	18.59	1.30	13.45	m²	**33.34**
vertical curved walls	0.23	25.15	1.72	13.45	m²	**40.32**
vertical facework to concrete	0.18	19.68	1.33	16.72	m²	**37.73**
casing to metal sections	0.21	22.96	1.56	16.72	m²	**41.24**
Thickness 140 mm;						
vertical straight walls	0.23	25.15	1.68	18.74	m²	**45.57**
vertical curved walls	0.30	32.80	2.23	18.74	m²	**53.77**
vertical facework to concrete	0.23	25.15	1.73	22.04	m²	**48.92**
casing to metal sections	0.27	29.52	2.01	22.04	m²	**53.57**
Thickness 215 mm;						
vertical straight walls	0.28	30.62	2.05	28.83	m²	**61.50**
vertical curved walls	0.37	40.46	2.73	28.83	m²	**72.02**
vertical facework to concrete	0.28	30.62	2.11	32.14	m²	**64.87**
casing to metal sections	0.31	33.90	2.31	32.14	m²	**68.35**
Columns and piers						
440 × 100 mm	0.08	8.75	0.60	5.93	m	**15.28**
890 × 140 mm	0.22	24.06	1.64	16.83	m	**42.53**
Surface features						
fair facing	0.06	6.56	0.45	–	m²	**7.01**
Ancillaries						
built-in pipes and ducts, cross-sectional area not exceeding 0.05 m²; excluding supply; blockwork 100 mm thick	0.04	4.37	0.30	0.33	nr	**5.00**
built-in pipes and ducts, cross-sectional area not exceeding 0.05 m²; excluding supply; blockwork 140 mm thick	0.09	9.84	0.67	1.18	nr	**11.69**
built-in pipes and ducts, cross-sectional area not exceeding 0.05 m²; excluding supply; blockwork 215 mm thick	0.13	14.22	0.94	1.62	nr	**16.78**
built-in pipes and ducts, cross-sectional area 0.05–0.25 m²; excluding supply; blockwork 100 mm thick	0.13	14.22	0.94	1.58	nr	**16.74**
built-in pipes and ducts, cross-sectional area 0.05–0.25 m²; excluding supply; blockwork 140 mm thick	0.18	19.68	1.31	2.59	nr	**23.58**
built-in pipes and ducts, cross-sectional area 0.05–0.25 m²; excluding supply; blockwork 215 mm thick	0.25	27.34	1.86	3.45	nr	**32.65**
built-in pipes and ducts, cross-sectional area 0.25–0.5 m²; excluding supply; blockwork 100 mm thick	0.15	16.40	1.12	2.00	nr	**19.52**
built-in pipes and ducts, cross-sectional area 0.25–0.5 m²; excluding supply; blockwork 140 mm thick	0.21	22.96	1.56	2.59	nr	**27.11**
built-in pipes and ducts, cross-sectional area 0.25–0.5 m²; excluding supply; blockwork 215 mm thick	0.30	32.80	2.20	4.07	nr	**39.07**

CLASS U: BRICKWORK, BLOCKWORK AND MASONRY

Item	Gang hours	Labour £	Plant £	Material £	Unit	Total rate £
DENSE CONCRETE BLOCKWORK						
Dense concrete blocks; 7 N/mm²; in cement mortar designation (iii)						
Walls, built vertical and straight						
100 mm thick	0.19	20.78	1.42	13.88	m²	**36.08**
140 mm thick	0.25	27.34	1.85	17.18	m²	**46.37**
215 mm thick	0.34	37.18	2.53	26.48	m²	**66.19**
Walls, built vertical and curved						
100 mm thick	0.25	27.34	1.89	13.91	m²	**43.14**
140 mm thick	0.33	36.09	2.46	17.18	m²	**55.73**
215 mm thick	0.45	49.21	3.37	26.51	m²	**79.09**
Walls, built vertical in facework to concrete						
100 mm thick	0.20	21.87	1.47	17.30	m²	**40.64**
140 mm thick	0.26	28.43	1.90	20.23	m²	**50.56**
215 mm thick	0.35	38.27	2.61	29.51	m²	**70.39**
Walls, as casings to metal sections, built vertical and straight						
100 mm thick	0.23	25.15	1.71	17.30	m²	**44.16**
140 mm thick	0.30	32.80	2.20	20.23	m²	**55.23**
Columns and piers						
440 × 100 mm	0.08	8.75	0.59	4.95	m	**14.29**
890 × 140 mm thick	0.22	24.06	1.64	15.45	m	**41.15**
Ancillaries						
built-in pipes and ducts, cross-sectional area not exceeding 0.05 m²; excluding supply; blockwork 100 mm thick	0.05	5.47	0.34	0.33	nr	**6.14**
built-in pipes and ducts, cross-sectional area not exceeding 0.05 m²; excluding supply; blockwork 140 mm thick	0.10	10.94	0.75	1.09	nr	**12.78**
built-in pipes and ducts, cross-sectional area not exceeding 0.05 m²; excluding supply; blockwork 215 mm thick	0.14	15.31	1.01	1.53	nr	**17.85**
built-in pipes and ducts, cross-sectional area 0.05–0.25 m²; excluding supply; blockwork 100 mm thick	0.14	15.31	1.01	1.67	nr	**17.99**
built-in pipes and ducts, cross-sectional area 0.05–0.25 m²; excluding supply; blockwork 140 mm thick	0.19	20.78	1.42	2.40	nr	**24.60**
built-in pipes and ducts, cross-sectional area 0.05–0.25 m²; excluding supply; blockwork 215 mm thick	0.27	29.52	2.01	3.22	nr	**34.75**
built-in pipes and ducts, cross-sectional area 0.25–0.5 m²; excluding supply; blockwork 100 mm thick	0.16	17.50	1.21	2.06	nr	**20.77**
built-in pipes and ducts, cross-sectional area 0.25–0.5 m²; excluding supply; blockwork 140 mm thick	0.23	25.15	1.71	2.43	nr	**29.29**
built-in pipes and ducts, cross-sectional area 0.25–0.5 m²; excluding supply; blockwork 215 mm thick	0.32	34.99	2.38	3.80	nr	**41.17**

CLASS U: BRICKWORK, BLOCKWORK AND MASONRY

Item	Gang hours	Labour £	Plant £	Material £	Unit	Total rate £
ARTIFICIAL STONE BLOCKWORK						
Reconstituted stone masonry blocks; Bradstone 100 bed weathered Cotswold or North Cearney rough hewn rockfaced blocks; in coloured cement-lime mortar designation (iii)						
Thickness 100 mm vertical facing						
vertical straight walls	0.30	32.80	2.24	37.19	m²	72.23
vertical curved walls	0.39	42.65	2.91	37.19	m²	82.75
vertical facework to concrete	0.31	33.90	2.33	37.22	m²	73.45
casing to metal sections	0.40	43.74	2.98	37.22	m²	83.94
Ancillaries						
built-in pipes and ducts, cross-sectional area not exceeding 0.05 m²; excluding supply	0.07	7.65	0.52	0.46	nr	8.63
built-in pipes and ducts, cross-sectional area 0.05–0.25 m²; excluding supply	0.22	24.06	1.64	3.96	nr	29.66
built-in pipes and ducts, cross-sectional area 0.25-0.5 m²; excluding supply	0.26	28.43	1.94	4.97	nr	35.34
Reconstituted stone masonry blocks; Bradstone Architectural dressing in weathered Cotswald or North Cerney shades; in coloured cement-lime mortar designation (iii)						
Surface features; Pier Caps						
305 × 305 mm, weathered and throated	0.09	9.84	0.67	19.18	nr	29.69
381 × 381 mm, weathered and throated	0.11	12.03	0.82	27.19	nr	40.04
457 × 457 mm, weathered and throated	0.13	14.22	0.97	37.25	nr	52.44
533 × 533 mm, weathered and throated	0.15	16.40	1.12	51.85	nr	69.37
Surface features; Copings						
152 × 76 mm, twice weathered and throated	0.08	8.75	0.60	8.48	m	17.83
152 × 76 mm, curved on plan, twice weathered and throated	0.11	12.03	0.79	29.02	m	41.84
305 × 76 mm, twice weathered and throated	0.10	10.94	2.98	14.48	m	28.40
305 × 76 mm, curved on plan, twice weathered and throated	0.13	14.22	0.99	76.99	m	92.20
Surface features; Pilasters						
440 × 100 mm	0.14	15.31	1.04	16.38	m	32.73
Surface features; Corbels						
479 × 100 × 215 mm, splayed	0.49	53.58	3.65	128.82	nr	186.05
665 × 100 × 215 mm, splayed	0.55	60.14	4.10	92.94	nr	157.18
Surface features; Lintels						
100 × 140 mm	0.11	12.03	0.82	39.93	m	52.78
100 × 215 mm	0.16	17.50	1.19	52.02	m	70.71

CLASS U: BRICKWORK, BLOCKWORK AND MASONRY

Item	Gang hours	Labour £	Plant £	Material £	Unit	Total rate £
ASHLAR MASONRY						
Portland Whitbed limestone; in cement-lime mortar designation (iv); pointed one side cement-lime mortar designation (iii) incorporating stone dust						
Thickness 50 mm						
vertical facing to concrete	0.85	92.95	6.33	93.29	m^2	**192.57**
Thickness 75 mm						
vertical straight walls	0.95	103.88	7.08	165.06	m^2	**276.02**
vertical curved walls	1.45	158.56	10.81	292.72	m^2	**462.09**
Surface features						
copings; weathered and twice throated; 250 × 150 mm	0.45	49.21	3.35	123.20	m	**175.76**
copings; weathered and twice throated; 250 × 150 mm; curved on plan	0.45	49.21	3.35	147.81	m	**200.37**
copings; weathered and twice throated; 400 × 150 mm	0.49	53.58	3.65	181.09	m	**238.32**
copings; weathered and twice throated; 400 × 150 mm; curved on plan	0.49	53.58	3.65	217.28	m	**274.51**
string courses; shaped and dressed; 75 mm projection × 150 mm high	0.45	49.21	3.35	100.33	m	**152.89**
corbel; shaped and dressed; 500 × 450 × 300 mm	0.55	60.14	4.10	163.27	nr	**227.51**
keystone; shaped and dressed; 750 × 900 × 300 mm (extreme)	1.30	142.16	9.69	516.50	nr	**668.35**
RUBBLE MASONRY						
Rubble masonry; random stones; in cement-lime mortar designation (iii)						
Walls, built vertical and straight; not exceeding 2 m high						
300 mm thick	1.25	136.69	9.31	246.48	m^2	**392.48**
450 mm thick	1.80	196.83	13.41	425.64	m^2	**635.88**
600 mm thick	2.40	262.44	17.88	492.93	m^2	**773.25**
Walls, built vertical, curved on plan; not exceeding 2 m high						
300 mm thick	1.40	153.09	5.27	252.23	m^2	**410.59**
450 mm thick	2.00	218.70	14.90	369.74	m^2	**603.34**
600 mm thick	2.65	289.78	19.75	492.93	m^2	**802.46**
Walls, built with battered face; not exceeding 2 m high						
300 mm thick	1.40	153.09	10.43	246.48	m^2	**410.00**
450 mm thick	2.00	218.70	14.90	369.74	m^2	**603.34**
600 mm thick	2.65	289.78	19.75	492.93	m^2	**802.46**
Rubble masonry; squared stones; in cement-lime mortar designation (iii)						
Walls, built vertical and straight; not exceeding 2 m high						
300 mm thick	1.25	136.69	9.31	444.92	m^2	**590.92**
450 mm thick	1.80	196.83	13.41	667.37	m^2	**877.61**
600 mm thick	2.40	262.44	17.88	889.85	m^2	**1170.17**

CLASS U: BRICKWORK, BLOCKWORK AND MASONRY

Item	Gang hours	Labour £	Plant £	Material £	Unit	Total rate £
RUBBLE MASONRY – cont						
Rubble masonry – cont						
Walls, built vertical, curved on plan; not exceeding 2 m high						
300 mm thick	1.40	153.09	10.43	444.92	m²	**608.44**
450 mm thick	2.00	218.70	14.90	667.37	m²	**900.97**
600 mm thick	2.40	262.44	17.88	936.85	m²	**1217.17**
Walls, built with battered face; not exceeding 2 m high						
300 mm thick	1.40	153.09	10.43	444.92	m²	**608.44**
450 mm thick	2.00	218.70	14.90	667.37	m²	**900.97**
600 mm thick	2.40	262.44	17.88	936.85	m²	**1217.17**
Dry stone walling; random stones						
Average thickness 300 mm						
battered straight walls	1.15	125.75	3.73	227.24	m²	**356.72**
Average thickness 450 mm						
battered straight walls	1.65	180.43	5.36	340.82	m²	**526.61**
Average thickness 600 mm						
battered straight walls	2.15	235.10	6.98	454.45	m²	**696.53**
Surface features						
copings; formed of rough stones 275 × 200 mm (average) high	0.45	49.21	3.35	58.60	m	**111.16**
copings; formed of rough stones 500 × 200 mm (average) high	0.55	60.14	4.10	106.55	m	**170.79**
ANCILLARIES COMMON TO ALL DIVISIONS						
Expamet joint reinforcement						
Ancillaries						
joint reinforcement; 65 mm wide	0.01	1.09	0.05	0.60	m	**1.74**
joint reinforcement; 115 mm wide	0.01	1.09	0.06	0.93	m	**2.08**
joint reinforcement; 175 mm wide	0.01	1.09	0.08	1.13	m	**2.30**
joint reinforcement; 225 mm wide	0.01	1.09	0.10	1.77	m	**2.96**
Hyload pitch polymer damp proof course; lapped joints; in cement mortar						
Ancillaries						
103 mm wide; horizontal	0.01	1.09	0.08	0.54	m	**1.71**
103 mm wide; vertical	0.02	2.19	0.13	0.54	m	**2.86**
215 mm wide; horizontal	0.03	3.28	0.19	1.94	m	**5.41**
215 mm wide; vertical	0.04	4.37	0.30	1.94	m	**6.61**
328 mm wide; horizontal	0.04	4.37	0.28	3.16	m	**7.81**
328 mm wide; vertical	0.06	6.56	0.42	3.16	m	**10.14**

CLASS U: BRICKWORK, BLOCKWORK AND MASONRY

Item	Gang hours	Labour £	Plant £	Material £	Unit	Total rate £
Pre-formed closed cell joint filler; pointing with polysulphide sealant						
Ancillaries						
movement joints; 12 mm filler 90 wide; 12 × 12 sealant one side	0.06	6.56	0.41	0.66	m	**7.63**
movement joints; 12 mm filler 200 wide; 12 × 12 sealant one side	0.07	7.65	0.48	1.00	m	**9.13**
Dritherm cavity insulation						
Infills						
50 mm thick	0.04	4.37	0.30	2.60	m²	**7.27**
75 mm thick	0.05	5.47	0.34	3.28	m²	**9.09**
Concrete						
Infills						
50 mm thick	0.06	6.56	0.42	4.99	m²	**11.97**
Galvanized steel wall ties						
Fixings and ties						
vertical twist strip type; 900 mm horizontal and 450 mm vertical staggered spacings	0.02	2.19	0.11	0.13	m²	**2.43**
Stainless steel wall ties						
Fixings and ties						
vertical twist strip type; 900 mm horizontal and 450 mm vertical staggered spacings	0.02	2.19	0.11	0.29	m²	**2.59**

CLASS V: PAINTING

Item	Gang hours	Labour £	Plant £	Material £	Unit	Total rate £
RESOURCES - LABOUR						
Painting gang						
1 ganger (skill rate 3)		15.79				
3 painters (skill rate 3)		44.64				
1 unskilled operative (general)		12.44				
Total Gang Rate / Hour	£	**72.87**				
RESOURCES - PLANT						
Painting						
1.5KVA diesel generator			1.38			
transformers/cables; junction box			0.64			
4.5' electric grinder			0.63			
transit van (50% of time)			4.98			
ladders			1.41			
Total Rate / Hour		£	**9.03**			
LEAD BASED PRIMER PAINT						
One coat calcium plumbate primer						
Metal						
upper surfaces inclined at an angle ne 30° to the horizontal	0.03	1.81	0.22	0.66	m²	**2.69**
upper surfaces inclined at an angle 30–60° to the horizontal	0.03	1.81	0.22	0.66	m²	**2.69**
surfaces inclined at an angle exceeding 60° to the horizontal	0.03	1.81	0.22	0.66	m²	**2.69**
soffit surfaces and lower surfaces inclined at an angle ne 60° to the horizontal	0.03	2.17	0.28	0.66	m²	**3.11**
surfaces of width ne 300 mm	0.01	0.72	0.09	0.20	m	**1.01**
surfaces of width 300 mm–1 m	0.03	1.81	0.22	0.43	m	**2.46**
Metal sections	0.03	2.17	0.26	0.66	m²	**3.09**
Pipework	0.03	2.17	0.26	0.66	m²	**3.09**
IRON BASED PRIMER PAINT						
One coat iron oxide primer						
Metal						
upper surfaces inclined at an angle ne 30° to the horizontal	0.03	1.81	0.22	0.55	m²	**2.58**
upper surfaces inclined at an angle 30–60° to the horizontal	0.03	1.81	0.22	0.55	m²	**2.58**
surfaces inclined at an angle exceeding 60° to the horizontal	0.03	1.81	0.22	0.55	m²	**2.58**
soffit surfaces and lower surfaces inclined at an angle ne 60° to the horizontal	0.03	2.17	0.28	0.55	m²	**3.00**
surfaces of width ne 300 mm	0.01	0.72	0.09	0.17	m	**0.98**
surfaces of width 300 mm–1 m	0.03	1.81	0.22	0.37	m	**2.40**
Metal sections	0.03	2.17	0.26	0.55	m²	**2.98**
Pipework	0.03	2.17	0.26	0.55	m²	**2.98**

CLASS V: PAINTING

Item	Gang hours	Labour £	Plant £	Material £	Unit	Total rate £
ACRYLIC PRIMER PAINT						
One coat acrylic wood primer						
Timber						
upper surfaces inclined at an angle ne 30° to the horizontal	0.02	1.45	0.18	0.44	m²	2.07
upper surfaces inclined at an angle 30–60° to the horizontal	0.02	1.45	0.18	0.44	m²	2.07
surfaces inclined at an angle exceeding 60° to the horizontal	0.02	1.45	0.17	0.44	m²	2.06
soffit surfaces and lower surfaces inclined at an angle ne 60° to the horizontal	0.03	1.81	0.22	0.44	m²	2.47
surfaces of width ne 300 mm	0.01	0.72	0.08	0.13	m	0.93
surfaces of width 300 mm–1 m	0.03	1.81	0.22	0.29	m	2.32
GLOSS PAINT						
One coat calcium plumbate primer; one undercoat and one finishing coat of gloss paint						
Metal						
upper surfaces inclined at an angle ne 30° to the horizontal	0.07	5.07	0.64	1.63	m²	7.34
upper surfaces inclined at an angle 30–60° to the horizontal	0.08	5.43	0.68	1.63	m²	7.74
surfaces inclined at an angle exceeding 60° to the horizontal	0.08	5.79	0.72	1.63	m²	8.14
soffit surfaces and lower surfaces inclined at an angle ne 60° to the horizontal	0.09	6.15	0.77	1.63	m²	8.55
surfaces of width ne 300 mm	0.03	1.81	0.22	0.49	m	2.52
surfaces of width 300 mm–1 m	0.05	3.62	0.46	1.06	m	5.14
Metal sections	0.09	6.52	0.76	1.63	m²	8.91
Pipework	0.09	6.52	0.76	1.63	m²	8.91
One coat acrylic wood primer; one undercoat and one finishing coat of gloss paint						
Timber						
upper surfaces inclined at an angle ne 30° to the horizontal	0.07	4.71	0.56	1.41	m²	6.68
upper surfaces inclined at an angle 30–60° to the horizontal	0.07	5.07	0.64	1.41	m²	7.12
surfaces inclined at an angle exceeding 60° to the horizontal	0.08	5.43	0.68	1.42	m²	7.53
soffit surfaces and lower surfaces inclined at an angle ne 60° to the horizontal	0.08	5.79	0.72	1.41	m²	7.92
surfaces of width ne 300 mm	0.03	1.81	0.16	0.42	m	2.39
surfaces of width 300 mm–1 m	0.05	3.62	0.46	0.92	m	5.00

CLASS V: PAINTING

Item	Gang hours	Labour £	Plant £	Material £	Unit	Total rate £
GLOSS PAINT – cont						
One coat calcium plumbate primer; two undercoats and one finishing coat of gloss paint						
Metal						
upper surfaces inclined at an angle ne 30° to the horizontal	0.06	4.34	0.54	1.94	m²	**6.82**
upper surfaces inclined at an angle 30–60° to the horizontal	0.07	4.71	0.58	1.94	m²	**7.23**
surfaces inclined at an angle exceeding 60° to the horizontal	0.07	5.07	0.64	1.94	m²	**7.65**
soffit surfaces and lower surfaces inclined at an angle ne 60° to the horizontal	0.08	5.43	0.68	1.89	m²	**8.00**
surfaces of width ne 300 mm	0.03	2.17	0.28	0.58	m	**3.03**
surfaces of width 300 mm–1 m	0.06	4.34	0.54	1.26	m	**6.14**
One coat acrylic wood primer; two undercoats and one finishing coat of gloss paint						
Timber						
upper surfaces inclined at an angle ne 30° to the horizontal	0.07	4.71	0.58	1.72	m²	**7.01**
upper surfaces inclined at an angle 30–60° to the horizontal	0.07	5.07	0.64	1.72	m²	**7.43**
surfaces inclined at an angle exceeding 60° to the horizontal	0.08	5.43	0.68	1.72	m²	**7.83**
soffit surfaces and lower surfaces inclined at an angle ne 60° to the horizontal	0.08	5.79	0.62	1.72	m²	**8.13**
surfaces of width ne 300 mm	0.03	2.17	0.64	0.52	m	**3.33**
surfaces of width 300 mm–1 m	0.06	4.34	0.54	1.12	m	**6.00**
One coat alkali resisting primer; two undercoats and one finishing coat of gloss paint						
Smooth concrete						
upper surfaces inclined at an angle ne 30° to the horizontal	0.04	2.53	0.32	1.71	m²	**4.56**
upper surfaces inclined at an angle 30–60° to the horizontal	0.04	2.90	0.36	1.71	m²	**4.97**
surfaces inclined at an angle exceeding 60° to the horizontal	0.05	3.26	0.40	1.72	m²	**5.38**
soffit surfaces and lower surfaces inclined at an angle ne 60° to the horizontal	0.05	3.62	0.46	1.71	m²	**5.79**
surfaces of width ne 300 mm	0.02	1.09	0.14	0.51	m	**1.74**
surfaces of width 300 mm–1 m	0.03	2.17	0.28	1.11	m	**3.56**

CLASS V: PAINTING

Item	Gang hours	Labour £	Plant £	Material £	Unit	Total rate £
Brickwork and rough concrete						
upper surfaces inclined at an angle ne 30° to the horizontal	0.04	2.90	0.36	1.73	m²	4.99
upper surfaces inclined at an angle 30–60° to the horizontal	0.05	3.26	0.40	1.74	m²	5.40
surfaces inclined at an angle exceeding 60° to the horizontal	0.05	3.62	0.46	1.78	m²	5.86
soffit surfaces and lower surfaces inclined at an angle ne 60° to the horizontal	0.06	3.98	0.50	1.78	m²	6.26
surfaces of width ne 300 mm	0.02	1.45	0.18	0.53	m	2.16
surfaces of width 300 mm–1 m	0.04	2.53	0.32	1.15	m	4.00
Blockwork						
upper surfaces inclined at an angle ne 30° to the horizontal	0.06	3.98	0.50	1.78	m²	6.26
soffit surfaces and lower surfaces inclined at an angle ne 60° to the horizontal	0.06	4.34	0.54	1.78	m²	6.66
surfaces of width ne 300 mm	0.03	1.81	0.22	0.53	m	2.56
surfaces of width 300 mm–1 m	0.04	2.90	0.36	1.16	m	4.42
Two coats anti-condensation paint						
Metal						
upper surfaces inclined at an angle ne 30° to the horizontal	0.08	5.43	0.68	2.05	m²	8.16
upper surfaces inclined at an angle 30–60° to the horizontal	0.08	5.43	0.68	2.05	m²	8.16
surfaces inclined at an angle exceeding 60° to the horizontal	0.08	5.43	0.68	2.05	m²	8.16
soffit surfaces and lower surfaces inclined at an angle ne 60° to the horizontal	0.08	5.79	0.72	2.05	m²	8.56
surfaces of width ne 300 mm	0.03	2.17	0.28	0.61	m	3.06
surfaces of width 300 mm–1 m	0.05	3.62	0.46	1.33	m	5.41
Metal sections	0.09	6.52	0.82	2.05	m²	9.39

CLASS V: PAINTING

Item	Gang hours	Labour £	Plant £	Material £	Unit	Total rate £
EMULSION PAINT						
One thinned coat, two coats vinyl emulsion paint						
Smooth concrete						
upper surfaces inclined at an angle ne 30° to the horizontal	0.07	4.71	0.58	1.15	m²	**6.44**
upper surfaces inclined at an angle 30–60° to the horizontal	0.07	5.07	0.64	1.15	m²	**6.86**
surfaces inclined at an angle exceeding 60° to the horizontal	0.08	5.43	0.68	1.15	m²	**7.26**
soffit surfaces and lower surfaces inclined at an angle ne 60° to the horizontal	0.08	5.79	0.72	1.15	m²	**7.66**
surfaces of width ne 300 mm	0.03	2.17	0.28	0.34	m	**2.79**
surfaces of width 300 mm–1 m	0.07	5.07	0.64	0.75	m	**6.46**
Brickwork and rough concrete						
upper surfaces inclined at an angle ne 30° to the horizontal	0.08	5.43	0.68	1.18	m²	**7.29**
upper surfaces inclined at an angle 30–60° to the horizontal	0.08	5.79	0.72	1.18	m²	**7.69**
surfaces inclined at an angle exceeding 60° to the horizontal	0.09	6.15	0.77	1.18	m²	**8.10**
soffit surfaces and lower surfaces inclined at an angle ne 60° to the horizontal	0.09	6.52	0.82	1.18	m²	**8.52**
surfaces of width ne 300 mm	0.03	2.17	0.28	0.35	m	**2.80**
surfaces of width 300 mm–1 m	0.08	5.79	0.72	0.81	m	**7.32**
Blockwork						
surfaces inclined at an angle exceeding 60° to the horizontal	0.09	6.52	0.82	1.21	m²	**8.55**
soffit surfaces and lower surfaces inclined at an angle ne 60° to the horizontal	0.10	7.24	0.90	1.21	m²	**9.35**
surfaces of width ne 300 mm	0.04	2.90	0.36	0.36	m	**3.62**
surfaces of width 300 mm–1 m	0.08	5.79	0.72	0.79	m	**7.30**

CLASS V: PAINTING

Item	Gang hours	Labour £	Plant £	Material £	Unit	Total rate £
CEMENT PAINT						
Two coats masonry paint						
Smooth concrete						
upper surfaces inclined at an angle ne 30° to the horizontal	0.07	4.71	0.58	0.93	m²	**6.22**
upper surfaces inclined at an angle 30–60° to the horizontal	0.07	5.07	0.64	0.93	m²	**6.64**
surfaces inclined at an angle exceeding 60° to the horizontal	0.08	5.43	0.68	0.93	m²	**7.04**
soffit surfaces and lower surfaces inclined at an angle ne 60° to the horizontal	0.08	5.79	0.72	0.93	m²	**7.44**
surfaces of width ne 300 mm	0.03	2.17	0.28	0.28	m	**2.73**
surfaces of width 300 mm–1 m	0.07	5.07	0.64	0.61	m	**6.32**
Brickwork and rough concrete						
upper surfaces inclined at an angle ne 30° to the horizontal	0.08	5.43	0.68	1.08	m²	**7.19**
upper surfaces inclined at an angle 30–60° to the horizontal	0.08	5.79	0.72	1.08	m²	**7.59**
surfaces inclined at an angle exceeding 60° to the horizontal	0.09	6.15	0.77	1.08	m²	**8.00**
soffit surfaces and lower surfaces inclined at an angle ne 60° to the horizontal	0.09	6.52	0.82	1.08	m²	**8.42**
surfaces of width ne 300 mm	0.03	2.17	0.28	0.32	m	**2.77**
surfaces of width 300 mm–1 m	0.08	5.79	0.72	0.70	m	**7.21**
Blockwork						
surfaces inclined at an angle exceeding 60° to the horizontal	0.09	6.52	0.82	1.40	m²	**8.74**
soffit surfaces and lower surfaces inclined at an angle ne 60° to the horizontal	0.10	7.24	0.90	1.40	m²	**9.54**
surfaces of width ne 300 mm	0.04	2.90	0.36	0.42	m	**3.68**
surfaces of width 300 mm–1 m	0.08	5.79	0.72	0.91	m	**7.42**
One thinned coat, two coats concrete floor paint						
Smooth concrete						
upper surfaces inclined at an angle ne 30° to the horizontal	0.07	4.71	0.58	3.29	m²	**8.58**
upper surfaces inclined at an angle 30–60° to the horizontal	0.07	5.07	0.64	3.29	m²	**9.00**
surfaces inclined at an angle exceeding 60° to the horizontal	0.08	5.43	0.68	3.29	m²	**9.40**
soffit surfaces and lower surfaces inclined at an angle ne 60° to the horizontal	0.08	5.79	0.72	2.87	m²	**9.38**
surfaces of width ne 300 mm	0.03	2.17	0.28	1.02	m	**3.47**
surfaces of width 300 mm–1 m	0.07	5.07	0.64	2.14	m	**7.85**
Additional coats						
width exceeding 1 m	0.03	2.17	0.28	1.32	m²	**3.77**
surfaces of width ne 300 mm	0.01	0.72	0.09	0.40	m	**1.21**
surfaces of width 300 mm–1 m	0.02	1.09	0.22	0.86	m	**2.17**

CLASS V: PAINTING

Item	Gang hours	Labour £	Plant £	Material £	Unit	Total rate £
EPOXY OR POLYURETHANE PAINT						
Blast clean to BS 7079; one coat zinc chromate etch primer, two coats zinc phosphate CR/alkyd undercoat off site; one coat MIO CR undercoat and one coat CR finish on site						
Metal sections	–	–	–	–	m²	28.99
Blast clean to BS 7079; one coat zinc rich 2 pack primer, one coat MIO high build epoxy 2 pack paint off site; one coat polyurethane 2 pack undercoat and one coat polyurethane 2 pack finish on site						
Metal sections	–	–	–	–	m²	34.41
Two coats clear polyurethane varnish						
Timber						
upper surfaces inclined at an angle ne 30° to the horizontal	0.04	2.90	0.36	1.54	m²	4.80
upper surfaces inclined at an angle 30–60° to the horizontal	0.05	3.26	0.40	1.54	m²	5.20
surfaces inclined at an angle exceeding 60° to the horizontal	0.05	3.62	0.46	1.54	m²	5.62
soffit surfaces and lower surfaces inclined at an angle ne 60° to the horizontal	0.06	4.34	0.54	1.54	m²	6.42
surfaces of width ne 300 mm	0.03	1.81	0.22	0.46	m	2.49
surfaces of width 300 mm–1 m	0.05	3.62	0.46	1.00	m	5.08
Two coats colour stained polyurethane varnish						
Timber						
upper surfaces inclined at an angle ne 30° to the horizontal	0.04	2.90	0.36	1.10	m²	4.36
upper surfaces inclined at an angle 30–60° to the horizontal	0.05	3.26	0.40	1.10	m²	4.76
surfaces inclined at an angle exceeding 60° to the horizontal	0.05	3.62	0.46	1.10	m²	5.18
soffit surfaces and lower surfaces inclined at an angle ne 60° to the horizontal	0.06	4.34	0.54	1.10	m²	5.98
surfaces of width ne 300 mm	0.03	1.81	0.22	0.33	m	2.36
surfaces of width 300 mm–1 m	0.05	3.62	0.46	0.71	m	4.79
Three coats colour stained polyurethane varnish						
Timber						
upper surfaces inclined at an angle ne 30° to the horizontal	0.06	3.98	0.50	1.64	m²	6.12
upper surfaces inclined at an angle 30–60° to the horizontal	0.07	4.71	0.58	1.64	m²	6.93
surfaces inclined at an angle exceeding 60° to the horizontal	0.07	5.07	0.64	1.64	m²	7.35
soffit surfaces and lower surfaces inclined at an angle ne 60° to the horizontal	0.09	6.15	0.77	1.64	m²	8.56
surfaces of width ne 300 mm	0.04	2.53	0.32	0.49	m	3.34
surfaces of width 300 mm–1 m	0.07	5.07	0.64	1.07	m	6.78

CLASS V: PAINTING

Item	Gang hours	Labour £	Plant £	Material £	Unit	Total rate £
One coat hardwood stain basecoat and two coats hardwood woodstain						
Timber						
upper surfaces inclined at an angle ne 30° to the horizontal	0.06	4.34	0.54	1.47	m²	**6.35**
upper surfaces inclined at an angle 30–60° to the horizontal	0.07	4.71	0.58	1.47	m²	**6.76**
surfaces inclined at an angle exceeding 60° to the horizontal	0.07	5.07	0.64	1.47	m²	**7.18**
soffit surfaces and lower surfaces inclined at an angle ne 60° to the horizontal	0.08	5.79	0.72	1.58	m²	**8.09**
surfaces of width ne 300 mm	0.04	2.53	0.32	0.44	m	**3.29**
surfaces of width 300 mm–1 m	0.07	5.07	0.64	0.95	m	**6.66**
BITUMINOUS OR PRESERVATIVE PAINT						
Two coats golden brown wood preservative						
Timber						
upper surfaces inclined at an angle ne 30° to the horizontal	0.05	3.26	0.40	1.39	m²	**5.05**
upper surfaces inclined at an angle 30–60° to the horizontal	0.05	3.62	0.46	1.39	m²	**5.47**
surfaces inclined at an angle exceeding 60° to the horizontal	0.06	3.98	0.50	1.39	m²	**5.87**
soffit surfaces and lower surfaces inclined at an angle ne 60° to the horizontal	0.07	4.71	0.58	1.39	m²	**6.68**
surfaces of width ne 300 mm	0.03	2.17	0.28	0.42	m	**2.87**
surfaces of width 300 mm–1 m	0.06	3.98	0.50	0.90	m	**5.38**
Two coats bituminous paint						
Metal sections	0.08	5.79	0.72	0.83	m²	**7.34**
Pipework	0.09	6.52	0.86	0.83	m²	**8.21**

CLASS W: WATERPROOFING

Item	Gang hours	Labour £	Plant £	Material £	Unit	Total rate £
RESOURCES - LABOUR						
Roofing - cladding gang						
1 ganger/chargehand (skill rate 3) - 50% of time		7.90				
2 skilled operative (skill rate 3)		29.76				
1 unskilled operative (general) - 50% of time		6.22				
Total Gang Rate / Hour		£ 43.88				
Damp proofing gang						
1 ganger (skill rate 4)		14.23				
1 skilled operative (skill rate 4)		13.32				
1 unskilled labour (general)		12.44				
Total Gang Rate / Hour		£ 39.99				
Roofing - asphalt gang						
1 ganger/chargehand (skill rate 4) - 50% of time		7.12				
1 skilled operative (skill rate 4)		13.32				
1 unskilled operative (general)		12.44				
Total Gang Rate / Hour		£ 32.88				
Tanking - asphalt gang						
1 ganger/chargehand (skill rate 4) - 50% of time		7.12				
1 skilled operative (skill rate 4)		13.32				
1 unskilled operative (general)		12.44				
Total Gang Rate / Hour		£ 32.88				
Tanking - waterproof sheeting gang						
1 skilled operative (skill rate 4)		13.32				
Total Gang Rate / Hour		£ 13.32				
Tanking - rendering gang						
1 ganger/chargehand (skill rate 4)		14.23				
1 skilled operative (skill rate 4)		13.32				
1 unskilled operative (generally)		12.44				
Total Gang Rate / Hour		£ 39.99				
Potective layers - flexible sheeting, sand and pea gravel coverings gang						
1 unskilled operative (generally)		12.44				
Total Gang Rate / Hour		£ 12.44				
Protective layers - screed gang						
1 ganger/chargehand (skill rate 4)		14.23				
1 skilled operative (skill rate 4)		13.32				
1 unskilled operative (general)		12.44				
Total Gang Rate / Hour		£ 39.99				
Sprayed or brushed waterproofing gang						
1 ganger/chargehand (skill rate 4) - 30% of time		4.27				
1 skilled operative (skill rate 4)		13.32				
Total Gang Rate / Hour		£ 17.59				

CLASS W: WATERPROOFING

Item	Gang hours	Labour £	Plant £	Material £	Unit	Total rate £
RESOURCES - PLANT						
Damp proofing						
2t dumper (50% of time)			3.25			
Total Rate / Hour		£	**3.25**			
Tanking - asphalt						
tar boiler (50% of time)			12.42			
2t dumper (50% of time)			3.25			
Total Rate / Hour		£	**15.66**			
Tanking - waterproof sheeting						
2t dumper (50% of time)			3.25			
Total Rate / Hour		£	**3.25**			
Tanking - rendering						
mixer			23.96			
2t dumper (50% time)			3.25			
Total Rate / Hour		£	**27.21**			
Protective layers - flexible sheeting, sand and pea gravel coverings						
2t dumper (50% of time)			3.25			
Total Rate / Hour		£	**3.25**			
Protective layers - screed						
mixer			23.96			
2t dumper (50% time)			3.25			
Total Rate / Hour		£	**27.21**			
Sprayed or brushed waterproofing						
2t dumper (50% of time)			3.25			
Total Rate / Hour		£	**3.25**			

CLASS W: WATERPROOFING

Item	Gang hours	Labour £	Plant £	Material £	Unit	Total rate £
DAMP PROOFING						
Waterproof sheeting						
0.3 mm polythene sheet						
ne 300 mm wide	0.01	0.12	0.03	–	m	0.15
300 mm–1 m wide	0.01	0.13	0.03	–	m	0.16
on horizontal or included surfaces	0.01	0.16	0.04	–	m²	0.20
on vertical surfaces	0.01	0.13	0.03	–	m²	0.16
NOTES						
Asphalt roofing						
Work has been presented in more detail than						
CESMM3 to allow the user greater freedom to access						
an appropriate rate to suit the complexity of their work						
TANKING						
Asphalt						
13 mm Mastic asphalt to BS 6925, Type T 1097; two						
coats; on concrete surface						
upper surfaces inclined at an angle ne 30° to the						
horizontal	0.45	14.67	7.65	6.62	m²	28.94
upper surfaces inclined at an angle 30–60° to the						
horizontal	0.60	19.57	10.20	6.62	m²	36.39
upper surfaces inclined at an angle exceeding 60°						
to the horizontal	1.00	32.61	16.99	6.62	m²	56.22
curved surfaces	1.20	39.13	20.39	6.62	m²	66.14
domed surfaces	1.50	48.91	25.49	6.62	m²	81.02
ne 300 mm wide	0.20	6.52	3.40	2.00	m	11.92
300 mm–1 m wide	0.45	14.67	7.65	6.62	m	28.94
20 mm Mastic asphalt to BS 6925, Type T 1097; two						
coats; on concrete surface						
upper surfaces inclined at an angle ne 30° to the						
horizontal	0.50	16.30	8.50	10.17	m²	34.97
upper surfaces inclined at an angle 30–60° to the						
horizontal	0.70	22.83	11.90	10.17	m²	44.90
upper surfaces inclined at an angle exceeding 60°						
to the horizontal	1.20	39.13	20.39	10.17	m²	69.69
curved surfaces	1.40	45.65	23.79	10.17	m²	79.61
domed surfaces	1.75	57.07	29.74	10.17	m²	96.98
ne 300 mm wide	0.23	7.50	3.91	3.05	m	14.46
300 mm–1 m wide	0.50	16.30	8.50	10.17	m	34.97
13 mm Mastic asphalt to BS 6925, Type T 1097; two						
coats; on brickwork surface; raking joints to form key						
upper surfaces inclined at an angle 30–60° to the						
horizontal	0.90	29.35	10.23	10.21	m²	49.79
upper surfaces inclined at an angle exceeding 60°						
to the horizontal	1.30	42.39	22.09	10.21	m²	74.69
curved surfaces	1.50	48.91	25.49	10.21	m²	84.61
domed surfaces	1.80	58.70	30.59	10.21	m²	99.50
ne 300 mm wide	0.30	9.78	3.40	3.07	m	16.25
300 mm–1 m wide	0.75	24.46	12.79	10.21	m	47.46

CLASS W: WATERPROOFING

Item	Gang hours	Labour £	Plant £	Material £	Unit	Total rate £
Waterproof sheeting						
Bituthene 3000; lapped joints						
upper surfaces inclined at an angle ne 30° to the horizontal	0.05	1.98	0.16	9.14	m²	**11.28**
upper surfaces inclined at an angle 30–60° to the horizontal	0.05	1.98	0.16	9.14	m²	**11.28**
ne 300 mm wide	0.03	0.99	0.08	2.41	m	**3.48**
300 mm–1 m wide	0.04	1.59	0.13	5.13	m	**6.85**
Bituthene 3000; lapped joints; primer coat						
upper surfaces inclined at an angle exceeding 60° to the horizontal	0.06	2.38	0.19	9.98	m²	**12.55**
Rendering in waterproof cement mortar						
19 mm render in waterproof cement mortar (1:3); two coat work						
upper surfaces inclined at an angle exceeding 60° to the horizontal	0.11	4.37	2.99	9.24	m²	**16.60**
ne 300 mm wide	0.07	2.78	1.91	3.43	m	**8.12**
300 mm–1 m wide	0.11	4.37	2.99	6.35	m	**13.71**
32 mm render in waterproof cement mortar (1:3); one coat work						
upper surfaces inclined at an angle ne 30° to the horizontal	0.11	4.37	2.99	12.17	m²	**19.53**
ne 300 mm wide	0.07	2.78	1.91	3.54	m	**8.23**
300 mm–1 m wide	0.11	4.37	2.99	6.17	m	**13.53**
ROOFING						
Asphalt						
13 mm Mastic asphalt to BS 6925 Type R 988; two coats; on concrete surface						
ne 300 mm wide	0.20	6.52	3.40	2.41	m	**12.33**
20 mm Mastic asphalt to BS 6925 Type R 988; two coats; on concrete surface						
upper surfaces inclined at an angle ne 30° to the horizontal	0.50	16.30	8.50	12.84	m²	**37.64**
upper surfaces inclined at an angle 30–60° to the horizontal	0.70	22.83	11.90	12.84	m²	**47.57**
surfaces inclined at an angle exceeding 60° to the horizontal	1.20	39.13	20.39	12.84	m²	**72.36**
curved surfaces	1.50	48.91	25.49	12.84	m²	**87.24**
domed surfaces	1.75	57.07	29.74	12.84	m²	**99.65**
ne 300 mm wide	0.23	7.50	3.91	3.85	m	**15.26**
300 mm–1 m wide	0.50	16.30	8.50	12.84	m	**37.64**
Extra for :						
10 mm thick limestone chippings bedded in hot bitumen	0.05	1.63	0.40	1.36	m²	**3.39**
dressing with solar reflective paint	0.05	1.63	0.40	3.31	m²	**5.34**
300 × 300 × 8 mm GRP tiles bedded in hot bitumen	0.30	9.78	2.39	29.55	m²	**41.72**

CLASS W: WATERPROOFING

Item	Gang hours	Labour £	Plant £	Material £	Unit	Total rate £
PROTECTIVE LAYERS						
Flexible sheeting						
3 mm Servi-pak protection board to Bituthene						
upper surfaces inclined at an angle ne 30° to the						
horizontal	0.20	2.47	0.65	18.47	m²	**21.59**
ne 300 mm wide	0.10	1.23	0.32	7.89	m	**9.44**
300 mm–1 m wide	0.20	2.47	0.65	18.47	m	**21.59**
3 mm Servi-pak protection board to Bituthene; fixing with adhesive dabs						
upper surfaces inclined at an angle exceeding 60°						
to the horizontal	0.25	3.08	0.81	19.56	m²	**23.45**
ne 300 mm wide	0.12	1.48	0.39	8.31	m	**10.18**
300 mm–1 m wide	0.25	3.08	0.81	19.56	m	**23.45**
6 mm Servi-pak protection board to Bituthene						
upper surfaces inclined at an angle ne 30° to the						
horizontal	0.20	2.47	0.65	30.07	m²	**33.19**
ne 300 mm wide	0.10	1.23	0.32	12.38	m	**13.93**
300 mm 1 m wide	0.20	2.47	0.65	30.07	m	**33.19**
6 mm Servi-pak protection board to Bituthene; fixing with adhesive dabs						
upper surfaces inclined at an angle exceeding 60°						
to the horizontal	0.25	3.08	0.81	31.54	m²	**35.43**
ne 300 mm wide	0.12	1.48	0.39	13.28	m	**15.15**
300 mm 1 m wide	0.25	3.08	0.81	31.54	m	**35.43**
Sand covering						
25 mm thick						
upper surfaces inclined at an angle ne 30° to the						
horizontal	0.02	0.25	0.03	0.68	m²	**0.96**
Pea gravel covering						
50 mm thick						
upper surfaces inclined at an angle ne 30° to the						
horizontal	0.02	0.25	0.03	15.48	m²	**15.76**
Sand and cement screed						
50 mm screed in cement mortar (1:4); one coat work						
upper surfaces inclined at an angle ne 30° to the						
horizontal	0.13	5.16	3.48	4.88	m²	**13.52**
SPRAYED OR BRUSHED WATERPROOFING						
Two coats RIW liquid asphaltic composition						
on horizontal or vertical surfaces	0.06	1.05	0.19	67.20	m²	**68.44**
Two coats Aquaseal						
on horizontal or vertical surfaces	0.06	1.05	0.19	1.03	m²	**2.27**
One coat Ventrot primer; one coat Ventrot hot applied damp proof membrane						
on horizontal or vertical surfaces	0.05	0.87	0.16	5.69	m²	**6.72**

CLASS X: MISCELLANEOUS WORK

Item	Gang hours	Labour £	Plant £	Material £	Unit	Total rate £
RESOURCES - LABOUR						
Fencing / barrier gang						
1 ganger/chargehand (skill rate 4) - 50% of time		7.12				
1 skilled operative (skill rate 4) - 50% of time		6.66				
1 unskilled operative (general)		12.44				
1 plant operator (skill rate 4)		14.56				
Total Gang Rate / Hour	£	**40.77**				
Safety fencing gang						
1 ganger/chargehand (skill rate 4)		14.23				
1 skilled operative (skill rate 4)		13.32				
2 unskilled operatives (general)		24.88				
1 plant operator (skill rate 4)		14.56				
Total Gang Rate / Hour	£	**66.99**				
Guttering gang						
2 skilled operatives (skill rate 4)		26.64				
Total Gang Rate / Hour	£	**26.64**				
Rock filled gabions gang						
1 ganger/chargehand (skill rate 4)		14.23				
4 unskilled operatives (general)		49.76				
1 plant operator (skill rate 3) - 50% of time		8.10				
Total Gang Rate / Hour	£	**72.08**				
RESOURCES - PLANT						
Fencing/barriers						
agricultural type tractor; fencing auger			13.65			
gas oil for ditto			4.41			
drop sided trailer; two axles			0.35			
power tools (fencing)			3.13			
Total Rate / Hour		£	**21.55**			
Guttering						
ladders			2.11			
Total Rate / Hour		£	**2.11**			
Rock filled gabions						
16 tonne crawler backacter (50% of time)			14.94			
Total Rate / Hour		£	**14.94**			

CLASS X: MISCELLANEOUS WORK

Item	Gang hours	Labour £	Plant £	Material £	Unit	Total rate £
FENCES						
Timber fencing						
Timber post and wire						
1.20 m high; DfT type 3; timber posts, driven; cleft chestnut paling	0.07	2.83	1.51	10.36	m	**14.70**
0.90 m high; DfT type 4; galvanized rectangular wire mesh	0.13	5.26	2.80	10.30	m	**18.36**
1.275 m high; DfT type 1; galvanized wire, 2 barbed, 4 plain	0.06	2.43	1.29	2.64	m	**6.36**
1.275 m high; DfT type 2; galvanized wire, 2 barbed, 4 plain	0.06	2.43	1.29	2.37	m	**6.09**
Concrete post and wire						
1.20 m high; DfT type 3; timber posts, driven; cleft chestnut paling	0.07	2.83	1.51	10.36	m	**14.70**
0.90 m high; DfT type 4; galvanized rectangular wire mesh	0.13	5.26	2.80	10.30	m	**18.36**
1.275 m high; DfT type 1; galvanized wire, 2 barbed, 4 plain	0.06	2.43	1.29	2.64	m	**6.36**
1.275 m high; DfT type 2; galvanized wire, 2 barbed, 4 plain	0.06	2.43	1.29	2.37	m	**6.09**
Metal post and wire						
1.20 m high; DfT type 3; timber posts, driven; cleft chestnut paling	0.07	2.83	1.51	10.36	m	**14.70**
0.90 m high; DfT type 4; galvanized rectangular wire mesh	0.13	5.26	2.80	10.30	m	**18.36**
1.275 m high; DfT type 1; galvanized wire, 2 barbed, 4 plain	0.06	2.43	1.29	2.64	m	**6.36**
1.275 m high; DfT type 2; galvanized wire, 2 barbed, 4 plain	0.06	2.43	1.29	2.37	m	**6.09**
Timber post and wire						
1.20 m high; DfT type 3; timber posts, driven; cleft chestnut paling	0.07	2.83	1.51	10.36	m	**14.70**
0.90 m high; DfT type 4; galvanized rectangular wire mesh	0.13	5.26	2.80	10.30	m	**18.36**
1.275 m high; DfT type 1; galvanized wire, 2 barbed, 4 plain	0.06	2.43	1.29	2.64	m	**6.36**
1.275 m high; DfT type 2; galvanized wire, 2 barbed, 4 plain	0.06	2.43	1.29	2.37	m	**6.09**
Timber close boarded; concrete posts						
Timber close boarded						
1.80 m high; 125 × 125 mm posts	0.30	12.13	6.46	36.35	m	**54.94**

CLASS X: MISCELLANEOUS WORK

Item	Gang hours	Labour £	Plant £	Material £	Unit	Total rate £
Wire rope safety fencing to BS 5750; based on 600 m lengths						
Metal crash barriers						
600 mm high; 4 wire ropes; long line posts at 2.40 m general spacings, driven	0.16	10.63	3.45	63.62	m	**77.70**
600 mm high; 4 wire ropes; short line posts at 2.40 m general spacings, 400 × 400 × 600 mm concrete footing	0.27	17.94	5.82	66.06	m	**89.82**
600 mm high; 4 wire ropes; short line posts at 2.40 m general spacings, 400 × 400 × 600 mm concrete footing, socketed	0.32	21.26	6.89	66.06	m	**94.21**
600 mm high; 4 wire ropes; short line posts at 2.40 m general spacings, bolted to structure	0.20	13.29	4.31	62.86	m	**80.46**
Pedestrian guard rails						
Metal guard rails						
1000 mm high; tubular galvanized mild steel to BS 3049, mesh infill (105 swg, 50 × 50 mm mesh; steel posts with concrete footing	0.80	10.57	3.45	169.93	m	**183.95**
Beam safety fencing; based on 600 m lengths						
Metal crash barriers						
600 mm high; untensioned corrugated beam, single sided; long posts at 3.20 m general spacings, driven	0.10	6.64	2.15	67.14	m	**75.93**
600 mm high; untensioned corrugated beam, double sided; long posts at 3.20 m general spacings, driven	0.26	17.28	5.60	123.14	m	**146.02**
600 mm high; untensioned open box beam, single sided; long posts at 3.20 m general spacings, driven	0.10	6.64	2.15	62.81	m	**71.60**
600 mm high; untensioned open box beam, double sided; long posts at 3.20 m general spacings, driven	0.26	17.28	5.60	114.80	m	**137.68**
600 mm high; untensioned open box beam, double height; long posts at 3.20 m general spacings, driven	0.30	19.93	6.46	135.14	m	**161.53**
600 mm high; tensioned corrugated beam, single sided; long posts at 3.20 m general spacings, driven	0.13	8.64	2.80	59.14	m	**70.58**
600 mm high; tensioned corrugated beam, double sided; long posts at 3.20 m general spacings, driven	0.37	24.59	7.97	107.14	m	**139.70**

CLASS X: MISCELLANEOUS WORK

Item	Gang hours	Labour £	Plant £	Material £	Unit	Total rate £
GATES AND STILES						
Notes						
Refer to Part 5 Series 300.						
Timber field gates						
single; 3.00 m wide × 1.27 m high	1.84	74.43	39.01	124.29	nr	**237.73**
single; 3.60 m wide × 1.27 m high	1.90	76.86	40.28	134.33	nr	**251.47**
single; 4.10 m wide × 1.27 m high	2.00	80.90	42.40	144.37	nr	**267.67**
single; 4.71 m wide × 1.27 m high	2.00	80.90	42.40	154.41	nr	**277.71**
Timber wicket gates						
single;1.20 m wide × 1.20 m high; DfT Type 1	1.20	48.54	25.44	60.85	nr	**134.83**
single; 1.20 m wide × 1.02 m high; DfT Type 2	1.20	48.54	25.44	46.16	nr	**120.14**
Metal field gates						
single; steel tubular; 3.60 m wide × 1.175 m high	0.30	12.13	6.36	150.46	nr	**168.95**
single; steel tubular; 4.50 m wide × 1.175 m high	0.30	12.13	6.36	176.08	nr	**194.57**
single; steel tubular; half mesh; 3.60 m wide × 1.175 m high	0.30	12.13	6.36	179.72	nr	**198.21**
single; steel tubular; half mesh; 4.50 m wide × 1.175 m high	0.30	12.13	6.36	213.98	nr	**232.47**
single; steel tubular; extra wide; 4.88 m wide × 1.175 m high	0.30	12.13	6.36	203.96	nr	**222.45**
double; steel tubular; 5.02 m wide × 1.175 m high	0.60	24.27	12.72	209.81	nr	**246.80**
Stiles						
1.00 m wide × 1.45 m high; DfT Type 1	1.50	60.67	31.80	115.37	nr	**207.84**
1.00 m wide × 1.45 m high; DfT Type 2	1.40	56.63	29.68	79.89	nr	**166.20**
DRAINAGE TO STRUCTURES ABOVE GROUND						
Note						
Outputs are based on heights up to 3 m above ground and exclude time spent on erecting access equipment, but include marking, cutting, drilling to wood, brick or concrete and all fixings. testing of finished work is not included.						
Output multipliers for labour and plant for heights over 3 m :						
3–6 m × 1.25						
6–9 m × 1.50						
9–12 m × 1.75						
12–15 m × 2.00						
Cast iron gutters and fittings; BS 460 2002						
100 × 75 mm gutters; support brackets	0.25	6.60	0.53	53.07	m	**60.20**
stop end	0.03	0.71	0.06	41.48	nr	**42.25**
running outlet	0.04	0.92	0.07	41.40	nr	**42.39**
angle	0.04	0.92	0.07	65.91	nr	**66.90**
125 × 75 mm gutters; support brackets	0.30	7.92	0.63	65.73	m	**74.28**
stop end	0.03	0.84	0.07	48.58	nr	**49.49**
running outlet	0.04	1.06	0.08	48.58	nr	**49.72**
angle	0.04	1.06	0.08	80.32	nr	**81.46**

CLASS X: MISCELLANEOUS WORK

Item	Gang hours	Labour £	Plant £	Material £	Unit	Total rate £
Cast iron rainwater pipes and fittings; BS 460 2002						
65 mm diameter; support brackets	0.26	6.86	0.55	34.22	m	41.63
bend	0.28	7.39	0.59	19.56	nr	27.54
offset, 75 mm projection	0.30	7.92	0.63	29.38	nr	37.93
offset, 225 mm projection	0.30	7.92	0.63	34.21	nr	42.76
offset, 455 mm projection	0.30	7.92	0.63	93.60	nr	102.15
shoe	0.18	4.75	0.38	31.36	nr	36.49
75 mm diameter; support brackets	0.28	7.39	0.59	34.09	m	42.07
bend	0.28	7.39	0.59	23.30	nr	31.28
offset, 75 mm projection	0.30	7.92	0.63	29.38	nr	37.93
offset, 225 mm projection	0.30	7.92	0.63	34.21	nr	42.76
offset, 455 mm projection	0.30	7.92	0.63	93.60	nr	102.15
shoe	0.18	4.75	0.38	27.20	nr	32.33
PVC-U gutters and fittings; Marley						
116 × 75 mm gutters; support brackets	0.18	4.75	0.38	10.80	m	15.93
stop end	0.05	1.32	0.11	4.34	nr	5.77
running outlet	0.05	1.32	0.11	15.53	nr	16.96
angle	0.14	3.70	0.30	18.95	nr	22.95
150 mm half round gutters; support brackets	0.18	4.75	0.38	15.62	m	20.75
stop end	0.05	1.32	0.11	6.89	nr	8.32
running outlet	0.05	1.32	0.11	19.12	nr	20.55
angle	0.14	3.70	0.30	17.46	nr	21.46
PVC-U external rainwater pipes and fittings; Marley						
68 mm diameter; support brackets	0.15	3.96	0.32	7.97	m	12.25
bend	0.16	4.22	0.34	8.78	nr	13.34
offset bend	0.18	4.75	0.38	4.86	nr	9.99
shoe	0.10	2.64	0.21	7.59	nr	10.44
110 mm diameter; support brackets	0.15	3.96	0.32	18.17	m	22.45
bend	0.16	4.22	0.34	21.54	nr	26.10
offset bend	0.18	4.75	0.38	20.84	nr	25.97
shoe	0.10	2.64	0.21	17.41	nr	20.26
ROCK FILLED GABIONS						
Gabions						
PVC coated galvanized wire mesh box gabions, wire laced; graded broken stone filling						
1.0 × 1.0 m module sizes	0.65	46.50	9.71	45.25	m³	101.46
1.0 × 0.5 m module sizes	0.80	57.23	11.95	58.09	m³	127.27
Heavily galvanized woven wire mesh box gabions, wire laced; graded broken stone filling						
1.0 × 1.0 m module sizes	0.65	46.50	9.71	40.64	m³	96.85
1.0 × 0.5 m module sizes	0.80	57.23	11.95	50.79	m³	119.97
Reno mattresses						
PVC coated woven wire mesh mattresses, wire tied; graded broken stone filling						
230 mm deep	0.15	10.73	2.24	17.27	m²	30.24
Heavily galvanized woven wire mesh, wire tied; graded broken stone filling						
300 mm deep	0.15	10.73	2.24	19.31	m²	32.28

CLASS Y: SEWER AND WATER MAIN RENOVATION AND ANCILLARY WORKS

Item	Gang hours	Labour £	Plant £	Material £	Unit	Total rate £
RESOURCES - LABOUR						
Drain repair gang						
1 chargehand pipelayer (skill rate 4) - 50% of time		7.12				
1 skilled operative (skill rate 4)		13.32				
2 unskilled operatives (generally)		24.88				
1 plant operator (skill rate 3)		16.19				
Total Gang Rate / Hour		61.50				
RESOURCES - PLANT						
Drainage						
0.40 m^3 hydraulic excavator			22.93			
trench sheets, shores, props etc.			25.99			
2t dumper (30% of time)			1.95			
compaction / plate roller (30% of time)			0.48			
7.30 m^3/min compressor			6.84			
small pump			0.71			
Total Rate / Hour		£	58.90			
PREPARATION OF EXISTING SEWERS						
Cleaning						
eggshape sewer 1300 mm high	–	–	–	–	m	18.31
Removing intrusions						
brickwork	–	–	–	–	m^3	93.05
concrete	–	–	–	–	m^3	161.01
reinforced concrete	–	–	–	–	m^3	204.64
Plugging laterals with concrete plug						
bore not exceeding 300 mm	–	–	–	–	nr	89.78
bore 450 mm	–	–	–	–	nr	138.58
Plugging laterals with brickwork plug						
bore 750 mm	–	–	–	–	nr	417.24
Local internal repairs to brickwork						
area not exceeding 0.1 m²	–	–	–	–	nr	18.49
area 0.1–0.25 m²	–	–	–	–	nr	48.26
area 1m²	–	–	–	–	nr	117.45
area 10 m²	–	–	–	–	nr	541.45
Grouting ends of redundant drains and sewers						
100 mm diameter	0.03	1.83	1.75	5.41	nr	8.99
300 mm diameter	0.13	7.63	7.34	21.62	nr	36.59
450 mm diameter	0.26	15.87	15.35	47.20	nr	78.42
600 mm diameter	0.50	30.51	29.50	95.51	nr	155.52
1200 mm diameter	1.70	103.73	100.16	270.30	nr	471.19
STABILIZATION OF EXISTING SEWERS						
Pointing, cement mortar (1:3)						
faces of brickwork	–	–	–	–	m²	39.46

CLASS Y: SEWER AND WATER MAIN RENOVATION AND ANCILLARY WORKS

Item	Gang hours	Labour £	Plant £	Material £	Unit	Total rate £
RENOVATION OF EXISTING SEWERS						
Sliplining						
GRP one piece unit; eggshape sewer 1300 mm high	–	–	–	–	m	**419.26**
Renovation of existing sewers						
sliplining	–	–	–	–	hr	**101.67**
LATERALS TO RENOVATED SEWERS						
Jointing						
bore not exceeding 150 mm	–	–	–	–	nr	**64.98**
bore 150–300 mm	–	–	–	–	nr	**115.29**
bore 450 mm	–	–	–	–	nr	**161.41**
EXISTING MANHOLES						
Abandonment						
sealing redundant road gullies with grade C15 concrete	0.02	1.40	1.37	17.12	nr	**19.89**
sealing redundant chambers with grade C15 concrete						
ne 1.0 m deep to invert	0.09	5.49	5.32	68.93	nr	**79.74**
1.0–2.0 m deep to invert	0.21	12.81	12.39	111.28	nr	**136.48**
2.0–3.0 m deep to invert	0.55	33.56	185.16	301.12	nr	**251.10**
Alteration						
100 × 100 mm water stop tap boxes on 100 × 100 mm brick chambers						
raising the level by 150 mm or less	0.06	3.66	3.57	21.46	nr	**28.69**
lowering the level by 150 mm or less	0.04	2.44	2.36	12.45	nr	**17.25**
420 × 420 mm cover and frame on 420 × 420 mm in-situ concrete chamber						
raising the level by 150 mm or less	0.10	6.10	5.93	38.34	nr	**50.37**
lowering the level by 150 mm or less	0.06	3.66	3.57	25.35	nr	**32.58**
Raising the level of 700 × 700 mm cover and frame on 700 × 500 mm in-situ concrete chamber						
by 150 mm or less	0.17	10.37	10.03	55.38	nr	**75.78**
600 × 600 mm grade 'A' heavy duty manhole cover and frame on 600 × 600 mm brick chamber						
raising the level by 150 mm or less	0.17	10.37	10.03	55.38	nr	**75.78**
raising the level by 150–300 mm	0.21	12.81	12.39	68.90	nr	**94.10**
lowering the level by 150 mm or less	0.10	6.10	5.93	35.11	nr	**47.14**
INTERRUPTIONS						
Preparation of existing sewers						
cleaning	–	–	–	–	hr	**430.50**
Stabilization of existing sewers						
pointing	–	–	–	–	hr	**87.27**

CLASS Z: SIMPLE BUILDING WORKS INCIDENTAL TO CIVIL ENGINEERING WORKS

Item	Gang hours	Labour £	Plant £	Material £	Unit	Total rate £
NOTES						
The user should refer to Part 6: Unit Costs - Ancillary Building Works for cost guidance on the matters listed in this section.						
RESOURCES - LABOUR						
Carpentry and joinery gang						
1 foreman carpenter (craftsman)		22.32				
5 carpenters (craftsmen)		95.75				
1 unskilled operative (general)		12.44				
Total Gang Rate/Hour		130.51				
Ironmongery gang						
1 carpenter (craftsman)		19.15				
Total Gang Rate/Hour		19.15				
Glazing gang						
1 glazier (craftsman)		19.15				
Total Gang Rate/Hour		19.15				
Finishings gang						
2 plasterers / ceramic tilers (craftsmen)		38.30				
1 unskilled operative (general)		12.44				
Total Gang Rate/Hour		50.74				
Vinyl tiling gang						
1 tiler (craftsman)		19.15				
1 unskilled operative (general)		12.44				
Total Gang Rate/Hour		31.59				
Plumbing gang						
1 plumber (craftsman)		19.15				
1 unskilled operative (general)		12.44				
Total Gang Rate/Hour		31.59				

CLASS Z: SIMPLE BUILDING WORKS INCIDENTAL TO CIVIL ENGINEERING WORKS

Item	Gang hours	Labour £	Plant £	Material £	Unit	Total rate £
CARPENTRY AND JOINERY						
Softwood; structural grade SC3; sawn; tanalised						
Structural and carcassing timber; floors						
38 × 100	0.02	2.82	–	1.35	m	4.17
38 × 125	0.02	2.95	–	1.59	m	4.54
38 × 150	0.03	3.21	–	1.90	m	5.11
50 × 100	0.03	3.21	–	2.90	m	6.11
50 × 125	0.03	3.34	–	3.60	m	6.94
50 × 150	0.03	3.47	–	4.31	m	7.78
50 × 175	0.03	3.47	–	4.10	m	7.57
50 × 200	0.03	3.72	–	8.27	m	11.99
50 × 225	0.03	3.72	–	6.57	m	10.29
75 × 200	0.03	3.98	–	7.89	m	11.87
75 × 225	0.03	3.98	–	9.96	m	13.94
100 × 200	0.04	5.39	–	9.51	m	14.90
100 × 225	0.05	5.78	–	9.96	m	15.74
Structural and carcassing timber; walls and partitions						
38 × 100	0.03	3.47	–	1.35	m	4.82
38 × 125	0.03	3.85	–	1.59	m	5.44
38 × 150	0.03	4.11	–	1.90	m	6.01
50 × 100	0.03	4.24	–	2.90	m	7.14
50 × 125	0.04	4.49	–	3.60	m	8.09
Structural and carcassing timber; flat roofs						
38 × 100	0.03	3.21	–	1.35	m	4.56
38 × 125	0.02	2.95	–	1.59	m	4.54
38 × 150	0.03	3.21	–	1.90	m	5.11
50 × 100	0.03	3.21	–	2.90	m	6.11
50 × 125	0.03	3.34	–	3.60	m	6.94
50 × 150	0.03	3.47	–	4.31	m	7.78
50 × 175	0.03	3.47	–	4.10	m	7.57
50 × 200	0.03	3.72	–	8.27	m	11.99
50 × 225	0.03	3.72	–	6.57	m	10.29
75 × 200	0.03	3.98	–	7.89	m	11.87
75 × 225	0.03	3.98	–	9.96	m	13.94
100 × 200	0.04	5.39	–	9.51	m	14.90
100 × 225	0.05	5.78	–	9.96	m	15.74
Structural and carcassing timber; pitched roofs						
38 × 100	0.03	3.47	–	1.35	m	4.82
38 × 125	0.02	2.95	–	1.59	m	4.54
38 × 150	0.03	3.47	–	1.90	m	5.37
50 × 100	0.03	4.24	–	2.90	m	7.14
50 × 125	0.03	3.34	–	3.60	m	6.94
50 × 150	0.04	4.88	–	4.31	m	9.19
50 × 175	0.04	4.88	–	4.10	m	8.98
50 × 200	0.04	4.88	–	8.27	m	13.15
50 × 225	0.04	4.88	–	6.57	m	11.45
75 × 125	0.05	5.78	–	5.48	m	11.26
75 × 150	0.05	5.78	–	6.57	m	12.35

CLASS Z: SIMPLE BUILDING WORKS INCIDENTAL TO CIVIL ENGINEERING WORKS

Item	Gang hours	Labour £	Plant £	Material £	Unit	Total rate £
CARPENTRY AND JOINERY – cont						
Structural and carcassing timber; plates and bearers						
38 × 100	0.01	1.41	–	1.35	m	2.76
50 × 75	0.01	1.41	–	2.17	m	3.58
50 × 100	0.01	1.67	–	2.90	m	4.57
75 × 100	0.01	1.67	–	4.39	m	6.06
75 × 125	0.02	1.93	–	5.48	m	7.41
75 × 150	0.02	1.93	–	6.57	m	8.50
Structural and carcassing timber; struts						
38 × 100	0.06	7.06	–	1.35	m	8.41
50 × 75	0.06	7.06	–	2.17	m	9.23
50 × 100	0.06	7.06	–	2.90	m	9.96
Structural and carcassing timber; cleats						
225 mm × 100 mm × 75 mm	0.04	4.62	–	1.01	nr	5.63
Structural and carcassing timber; trussed rafters and roof trusses						
Softwood; joinery quality; wrought; tanalised						
Strip boarding; walls						
18 mm nominal thick	0.14	17.97	–	13.73	m²	31.70
18 mm nominal thick; not exceeding 100 mm wide	0.05	5.78	–	1.37	m²	7.15
18 mm nominal thick; 100–200 mm wide	0.05	6.42	–	2.75	m²	9.17
18 mm nominal thick; 200–300 mm wide	0.06	7.70	–	4.12	m²	11.82
Softwood; joinery quality; wrought						
Strip boarding; walls						
18 mm nominal thick	0.14	17.97	–	13.73	m²	31.70
18 mm nominal thick; not exceeding 100 mm wide	0.05	5.78	–	1.37	m²	7.15
18 mm nominal thick; 100–200 mm wide	0.05	6.42	–	2.75	m²	9.17
18 mm nominal thick; 200–300 mm wide	0.06	7.70	–	4.12	m²	11.82
Miscellaneous joinery; skirtings						
19 × 100	0.02	2.57	–	2.56	m	5.13
25 × 150	0.03	3.21	–	3.94	m	7.15
Miscellaneous joinery; architraves						
12 × 50	0.02	3.08	–	1.84	m	4.92
19 × 63	0.02	3.08	–	2.10	m	5.18
Miscellaneous joinery; trims						
12 × 25	0.02	3.08	–	1.44	m	4.52
12 × 50	0.02	3.08	–	1.84	m	4.92
16 × 38	0.02	3.08	–	1.42	m	4.50
19 × 19	0.02	3.08	–	1.35	m	4.43
Plywood; marine quality						
Sheet boarding; walls						
18 mm nominal tihck	0.14	17.97	–	17.26	m²	35.23

CLASS Z: SIMPLE BUILDING WORKS INCIDENTAL TO CIVIL ENGINEERING WORKS

Item	Gang hours	Labour £	Plant £	Material £	Unit	Total rate £
Plywood; external quality						
Sheet boarding; floors						
18 mm nominal thick	0.11	14.12	–	13.73	m²	27.85
18 mm nominal thick; not exceeding 100 mm wide	0.04	4.49	–	1.37	m²	5.86
18 mm nominal thick; 100–200 mm wide	0.04	4.88	–	2.75	m²	7.63
18 mm nominal thick; 200–300 mm wide	0.05	5.78	–	4.12	m²	9.90
Sheet boarding; walls						
18 mm nominal thick	0.14	17.97	–	13.73	m²	31.70
18 mm nominal thick; not exceeding 100 mm wide	0.05	5.78	–	1.37	m²	7.15
18 mm nominal thick; 100–200 mm wide	0.05	6.42	–	2.75	m²	9.17
18 mm nominal thick; 200–300 mm wide	0.06	7.70	–	4.12	m²	11.82
Sheet boarding; soffits						
18 mm nominal thick	0.16	19.90	–	13.73	m²	33.63
18 mm nominal thick; not exceeding 100 mm wide	0.05	6.42	–	1.37	m²	7.79
18 mm nominal thick; 100–200 mm wide	0.06	7.06	–	2.75	m²	9.81
18 mm nominal thick; 200–300 mm wide	0.07	8.47	–	4.12	m²	12.59
INSULATION						
Vapour barrier; Sisalkraft 728 building paper (Class A1F); 150 mm laps; fixed to softwood						
Sheets						
floors	0.01	1.67	–	2.12	m²	3.79
sloping upper surfaces	0.01	1.80	–	2.12	m²	3.92
walls	0.02	2.18	–	2.12	m²	4.30
soffits	0.02	2.57	–	2.12	m²	4.69
Insulation quilt; Isover glass fibre; laid loose between members at 600 mm centres						
Quilts; floors						
60 mm thick	0.03	3.59	–	1.65	m²	5.24
80 mm thick	0.03	3.59	–	2.15	m²	5.74
100 mm thick	0.03	3.59	–	2.55	m²	6.14
150 mm thick	0.03	3.59	–	3.90	m²	7.49
Quilts; sloping upper surfaces						
60 mm thick	0.03	3.98	–	1.65	m²	5.63
80 mm thick	0.03	4.36	–	2.15	m²	6.51
100 mm thick	0.04	4.62	–	2.55	m²	7.17
150 mm thick	0.04	5.13	–	3.90	m²	9.03
Insulation quilt; Isover glass fibre; laid between members at 600 mm centres; fixing with staples						
Quilts; walls						
60 mm thick	0.04	4.49	–	1.83	m²	6.32
80 mm thick	0.04	4.88	–	2.33	m²	7.21
100 mm thick	0.04	5.39	–	2.73	m²	8.12
150 mm thick	0.05	6.42	–	4.07	m²	10.49
Quilts; soffits						
60 mm thick	0.04	5.39	–	1.83	m²	7.22
80 mm thick	0.05	6.03	–	2.33	m²	8.36
100 mm thick	0.05	6.42	–	2.73	m²	9.15
150 mm thick	0.05	6.93	–	4.07	m²	11.00

CLASS Z: SIMPLE BUILDING WORKS INCIDENTAL TO CIVIL ENGINEERING WORKS

Item	Gang hours	Labour £	Plant £	Material £	Unit	Total rate £
INSULATION – cont						
Insulation board; Jablite or similar expanded						
polystyrene standard grade; fixing with adhesive						
Boards; floors						
25 mm thick	0.06	7.19	–	3.83	m²	**11.02**
40 mm thick	0.06	7.19	–	5.56	m²	**12.75**
50 mm thick	0.06	7.19	–	6.99	m²	**14.18**
Boards; sloping upper surfaces						
25 mm thick	0.07	8.34	–	3.83	m²	**12.17**
40 mm thick	0.07	8.34	–	5.56	m²	**13.90**
50 mm thick	0.07	8.34	–	6.99	m²	**15.33**
Boards; walls						
25 mm thick	0.08	9.63	–	3.83	m²	**13.46**
40 mm thick	0.08	9.63	–	5.56	m²	**15.19**
50 mm thick	0.08	9.63	–	6.99	m²	**16.62**
Boards; soffits						
25 mm thick	0.08	10.65	–	3.83	m²	**14.48**
40 mm thick	0.08	10.65	–	5.56	m²	**16.21**
50 mm thick	0.08	10.65	–	6.99	m²	**17.64**
WINDOWS, DOORS AND GLAZING						
Timber windows; treated planed softwood;						
Jeld-Wen; plugged and screwed to masonry						
High performance top hung reversible windows;						
weather stripping; opening panes hung on rustproof						
hinges with aluminized lacquered espagnolette bolts;						
low E 24 mm double glazing						
630 mm × 900 mm; ref LEC109AR	0.16	21.05	–	328.61	nr	**349.66**
630 mm × 1200 mm; ref LEC112AR	0.21	26.83	–	365.34	nr	**392.17**
1200 mm × 1050 mm; ref LEC210AR	0.25	31.45	–	457.73	nr	**489.18**
1770 mm × 1050 mm; ref LEC310AER	0.31	39.66	–	781.50	nr	**821.16**
Metal windows; steel fixed light; factory finished						
polyester powder coating; Crittal Homelight						
range; fixing lugs plugged and screwed to						
masonry						
One piece composites; glazing 4 mm OQ glass;						
easy-glaze beads and weather stripping						
628 mm × 923 mm; ref ZNC5	0.22	27.98	–	193.73	nr	**221.71**
1237 mm × 923 mm; ref ZNC13	0.32	40.82	–	373.98	nr	**414.80**
1237 mm × 1218 mm; ref ZND13	0.32	40.82	–	476.76	nr	**517.58**
1846 mm × 1513 mm; ref ZNDV14	0.32	40.82	–	532.48	nr	**573.30**

CLASS Z: SIMPLE BUILDING WORKS INCIDENTAL TO CIVIL ENGINEERING WORKS

Item	Gang hours	Labour £	Plant £	Material £	Unit	Total rate £
Plastics windows; PVC-U, reinforced where appropriate with aluminium alloy; standard ironmongery; fixing lugs plugged and screwed to masonry						
Casement / fixed light; glazing 4 mm OQ glass; e.p. d.m. glazing gaskets and weather seals						
600 mm × 1200 mm; single glazed	0.32	40.82	–	127.77	nr	**168.59**
600 mm × 1200 mm; double glazed	0.32	40.82	–	132.18	nr	**173.00**
1200 mm × 1200 mm; single glazed	0.36	46.72	–	193.86	nr	**240.58**
1200 mm × 1200 mm; double glazed	0.36	46.72	–	202.67	nr	**249.39**
1800 mm × 1200 mm; single glazed	0.41	52.88	–	286.39	nr	**339.27**
1800 mm × 1200 mm; double glazed	0.41	52.88	–	308.42	nr	**361.30**
Timber doors; treated planed softwood						
Matchboarded, ledged and braced doors; 25 mm thick ledges and braces; 19 mm thick tongued, grooved and v-jointed one side vertical boarding						
762 mm × 1981 mm	0.27	35.04	–	62.33	nr	**97.37**
838 mm × 1981 mm	0.27	35.04	–	62.33	nr	**97.37**
Panelled doors; one open panel for glass; including beads						
762 × 1981 × 44 mm	0.32	41.08	–	105.75	nr	**146.83**
Panelled doors; two open panels for glass; including beads						
762 × 1981 × 44 mm	0.32	41.08	–	137.41	nr	**178.49**
838 × 1981 × 44 mm	0.32	41.08	–	137.41	nr	**178.49**
Timber doors; standard flush pattern						
Flush door; internal quality; skeleton or cellular core; hardboard faced both sides; lipped on two long edges; primed						
626 × 2040 × 40 mm	0.22	28.62	–	22.72	nr	**51.34**
726 × 2040 × 40 mm	0.22	28.62	–	22.72	nr	**51.34**
826 × 2040 × 40 mm	0.22	28.62	–	22.29	nr	**50.91**
Flush door; internal quality; skeleton or cellular core; chipboard faced both sides; lipped all edges; primed						
626 × 2040 × 40 mm	0.22	28.62	–	29.45	nr	**58.07**
726 × 2040 × 40 mm	0.22	28.62	–	29.45	nr	**58.07**
826 × 2040 × 40 mm	0.22	28.62	–	29.45	nr	**58.07**
Flush door; internal quality; skeleton or cellular core; Sapele veneered both sides; lipped all edges; primed						
626 × 2040 × 40 mm	0.34	43.64	–	45.23	nr	**88.87**
726 × 2040 × 40 mm	0.34	43.64	–	45.23	nr	**88.87**
826 × 2040 × 40 mm	0.34	43.64	–	45.23	nr	**88.87**
Flush door; half-hour fire check (30/20); solid core; chipboard faced both sides; lipped all edges; primed						
626 × 2040 × 44 mm	0.32	41.08	–	54.45	nr	**95.53**
726 × 2040 × 44 mm	0.32	41.08	–	54.45	nr	**95.53**
826 × 2040 × 44 mm	0.32	41.08	–	54.45	nr	**95.53**

CLASS Z: SIMPLE BUILDING WORKS INCIDENTAL TO CIVIL ENGINEERING WORKS

Item	Gang hours	Labour £	Plant £	Material £	Unit	Total rate £
WINDOWS, DOORS AND GLAZING – cont						
Timber frames or lining sets; treated planed softwood						
Internal door frame or lining; 30 × 107 mm lining with 12 × 38 mm door stop						
for 726 × 2040 mm door	0.14	18.61	–	39.54	nr	**58.15**
for 826 × 2040 mm door	0.14	18.61	–	39.54	nr	**58.15**
Internal door frame or lining; 30 × 133 mm lining with 12 × 38 mm door stop						
for 726 × 2040 mm door	0.14	18.61	–	44.26	nr	**62.87**
for 826 × 2040 mm door	0.14	18.61	–	44.26	nr	**62.87**
Ironmongery						
Hinges						
100 mm; light steel	0.17	3.33	–	0.88	nr	**4.21**
S/D rising butts; R/L hand; 102 × 67 mm; BRS	0.17	3.33	–	24.63	nr	**27.96**
Door closers						
light duty surface fixed door closer; L/R hand; SIL	1.30	24.71	–	101.15	nr	**125.86**
door selector; face fixing; SAA	0.80	15.21	–	86.57	nr	**101.78**
floor spring; single and double action; ZP	3.33	63.30	–	212.57	nr	**275.87**
Locks						
mortice dead lock; 63 × 108 mm; SSS	1.00	19.01	–	14.01	nr	**33.02**
rim lock; 140 × 73 mm; GYE	0.56	10.65	–	10.65	nr	**21.30**
upright mortice lock; 103 × 82 mm; 3 lever	1.11	21.10	–	14.44	nr	**35.54**
Bolts						
flush; 152 × 25 mm; SCP	0.80	15.21	–	18.99	nr	**34.20**
indicating; 76 × 41 mm; SAA	0.89	16.92	–	13.89	nr	**30.81**
panic; single; SVE	3.33	63.30	–	87.62	nr	**150.92**
panic; double; SVE	4.67	88.78	–	119.81	nr	**208.59**
necked tower; 203 mm; BJ	0.40	7.60	–	5.19	nr	**12.79**
Handles						
pull; 225 mm; back fixing; PAA	0.23	4.37	–	8.41	nr	**12.78**
pull; 225 mm; face fixing with cover rose; PAA	0.44	8.36	–	45.18	nr	**53.54**
lever; PAA	0.44	8.36	–	31.48	nr	**39.84**
Plates						
finger plate; 300 × 75 × 3 mm; SAA	0.23	4.37	–	5.50	nr	**9.87**
kicking plate; 1000 × 150 × 3 mm; PAA	0.44	8.36	–	15.99	nr	**24.35**
letter plate; 330 × 76 mm; aluminium finish	1.77	33.65	–	11.45	nr	**45.10**
Brackets						
head bracket; open; side fixing; bolting to masonry	0.50	9.51	–	9.47	nr	**18.98**
head bracket; open; soffit fixing; bolting to masonry	0.50	9.51	–	4.66	nr	**14.17**
Sundries						
rubber door stop; SAA	0.11	2.09	–	1.60	nr	**3.69**

CLASS Z: SIMPLE BUILDING WORKS INCIDENTAL TO CIVIL ENGINEERING WORKS

Item	Gang hours	Labour £	Plant £	Material £	Unit	Total rate £
Glazing; standard plain glass to BS 952, clear float; glazing with putty or bradded beads						
Glass						
3 mm thick	0.95	18.06	–	20.54	m²	**38.60**
4 mm thick	0.95	18.06	–	21.80	m²	**39.86**
5 mm thick	0.95	18.06	–	26.51	m²	**44.57**
6 mm thick	0.95	18.06	–	29.07	m²	**47.13**
Hermetically sealed units, factory made						
two panes 4 mm thick, 6 mm air space	0.95	18.06	–	–	m²	**18.06**
Glazing; standard plain glass to BS 952, rough cast; glazing with putty or bradded beads						
Glass						
6 mm thick	0.95	18.06	–	29.82	m²	**47.88**
Glazing; standard plain glass to BS 952, Georgian wired cast; glazing with putty or bradded beads						
Glass						
7 mm thick	0.95	18.06	–	26.74	m²	**44.80**
Glazing; standard plain glass to BS 952, Georgian wired polished; glazing with putty or bradded beads						
Glass						
6 mm thick	0.95	18.06	–	41.52	m²	**59.58**
Glazing; special glass to BS 952, toughened clear float; glazing with putty or bradded beads						
Glass						
4 mm thick	0.95	18.06	–	24.82	m²	**42.88**
5 mm thick	0.95	18.06	–	32.92	m²	**50.98**
6 mm thick	0.95	18.06	–	36.25	m²	**54.31**
10 mm thick	1.05	19.96	–	59.51	m²	**79.47**
Glazing; special glass to BS 952, clear laminated safety; glazing with putty or bradded beads						
Glass						
4.4 mm thick	0.95	18.06	–	31.28	m²	**49.34**
5.4 mm thick	0.95	18.06	–	32.10	m²	**50.16**
6.4 mm thick	0.95	18.06	–	37.34	m²	**55.40**
Glazing; standard plain glass to BS 952, clear float; glazing with putty or bradded beads						
Glass						
4.4 mm thick	0.95	18.06	–	31.28	m²	**49.34**
5.4 mm thick	0.95	18.06	–	32.10	m²	**50.16**
6.4 mm thick	0.95	18.06	–	37.34	m²	**55.40**
Patent glazing; aluminium alloy bars 2.55 m long at 622 mm centres; fixed to supports						
Patent glazing						
roofs	–	–	–	393.60	m²	**393.60**

CLASS Z: SIMPLE BUILDING WORKS INCIDENTAL TO CIVIL ENGINEERING WORKS

Item	Gang hours	Labour £	Plant £	Material £	Unit	Total rate £
SURFACE FINISHES, LININGS AND PARTITIONS						
In situ finishes; cement and sand (1:3); steel trowelled						
Floors						
30 mm thick	0.12	6.04	–	2.56	m²	**8.60**
40 mm thick	0.12	6.04	–	3.41	m²	**9.45**
50 mm thick	0.14	7.05	–	4.26	m²	**11.31**
60 mm thick	0.15	7.55	–	5.11	m²	**12.66**
70 mm thick	0.16	8.06	–	5.96	m²	**14.02**
Sloping upper surfaces						
30 mm thick	0.16	8.06	–	2.56	m²	**10.62**
40 mm thick	0.16	8.06	–	3.41	m²	**11.47**
50 mm thick	0.18	9.06	–	4.26	m²	**13.32**
60 mm thick	0.19	9.57	–	5.11	m²	**14.68**
70 mm thick	0.20	10.07	–	5.96	m²	**16.03**
Walls						
12 mm thick	0.31	15.61	–	1.02	m²	**16.63**
15 mm thick	0.34	17.12	–	1.25	m²	**18.37**
20 mm thick	0.38	19.14	–	1.70	m²	**20.84**
Surfaces of width not exceeding 300 mm						
12 mm thick	0.15	7.55	–	1.02	m	**8.57**
15 mm thick	0.13	6.55	–	1.28	m	**7.83**
20 mm thick	0.12	6.04	–	1.70	m	**7.74**
30 mm thick	0.08	4.03	–	2.56	m	**6.59**
40 mm thick	0.08	4.03	–	3.41	m	**7.44**
50 mm thick	0.09	4.53	–	4.26	m	**8.79**
60 mm thick	0.09	4.53	–	5.11	m	**9.64**
70 mm thick	0.10	5.04	–	5.96	m	**11.00**
Surfaces of width 300 mm–1 m						
12 mm thick	0.23	11.58	–	1.02	m	**12.60**
15 mm thick	0.26	13.09	–	1.28	m	**14.37**
20 mm thick	0.26	13.09	–	1.70	m	**14.79**
30 mm thick	0.12	6.04	–	2.56	m	**8.60**
40 mm thick	0.12	6.04	–	3.41	m	**9.45**
50 mm thick	0.13	6.55	–	4.26	m	**10.81**
60 mm thick	0.14	7.05	–	5.11	m	**12.16**
70 mm thick	0.15	7.55	–	5.96	m	**13.51**
In situ finishes; lightweight plaster; Thistle; steel trowelled						
Walls						
12 mm thick	0.22	11.08	–	1.17	m²	**12.25**
Surfaces of width not exceeding 300 mm						
12 mm thick	0.22	11.08	–	0.35	m	**11.43**
Surfaces of width 300 mm–1 m						
12 mm thick	0.16	8.06	–	1.17	m	**9.23**

CLASS Z: SIMPLE BUILDING WORKS INCIDENTAL TO CIVIL ENGINEERING WORKS

Item	Gang hours	Labour £	Plant £	Material £	Unit	Total rate £
Beds and backings; cement and sand (1:3)						
Floors						
50 mm thick	0.14	7.05	–	4.26	m²	**11.31**
70 mm thick	0.16	8.06	–	5.96	m²	**14.02**
Sloping upper surfaces						
30 mm thick	0.16	8.06	–	2.56	m²	**10.62**
40 mm thick	0.16	8.06	–	3.41	m²	**11.47**
Walls						
12 mm thick	0.31	15.61	–	1.02	m²	**16.63**
20 mm thick	0.38	19.14	–	1.70	m²	**20.84**
Surfaces of width not exceeding 300 mm						
12 mm thick	0.15	7.55	–	1.02	m	**8.57**
20 mm thick	0.12	6.04	–	1.70	m	**7.74**
30 mm thick	0.08	4.03	–	2.56	m	**6.59**
50 mm thick	0.09	4.53	–	4.26	m	**8.79**
70 mm thick	0.10	5.04	–	5.96	m	**11.00**
Surfaces of width 300 mm–1 m						
12 mm thick	0.23	11.58	–	1.02	m	**12.60**
20 mm thick	0.26	13.09	–	1.70	m	**14.79**
30 mm thick	0.12	6.04	–	2.56	m	**8.60**
50 mm thick	0.13	6.55	–	4.26	m	**10.81**
70 mm thick	0.15	7.55	–	5.96	m	**13.51**
Tiles; red clay; bedding 10 mm thick and jointing in cement mortar (1:3); grouting with cement mortar (1:1)						
Floors						
150 × 150 × 12.5 mm	0.32	16.12	–	23.94	m²	**40.06**
200 × 200 × 19 mm	0.32	16.12	–	33.27	m²	**49.39**
Surfaces of width not exceeding 300 mm						
150 × 150 × 12.5 mm	0.16	8.06	–	7.18	m	**15.24**
200 × 200 × 19 mm	0.13	6.55	–	9.99	m	**16.54**
Surfaces of width 300 mm–1 m						
150 × 150 × 12.5 mm	0.24	12.09	–	23.94	m	**36.03**
200 × 200 × 19 mm	0.24	12.09	–	33.29	m	**45.38**
Tiles; brown clay; bedding 10 mm thick and jointing in cement mortar (1:3); grouting with cement mortar (1:1)						
Floors						
150 × 150 × 12.5 mm	0.32	16.12	–	27.95	m²	**44.07**
200 × 200 × 19 mm	0.32	16.12	–	38.98	m²	**55.10**
Surfaces of width not exceeding 300 mm						
150 × 150 × 12.5 mm	0.16	8.06	–	8.39	m	**16.45**
200 × 200 × 19 mm	0.13	6.55	–	11.70	m	**18.25**
Surfaces of width 300 mm–1 m						
150 × 150 × 12.5 mm	0.24	12.09	–	27.95	m	**40.04**
200 × 200 × 19 mm	0.24	12.09	–	39.00	m	**51.09**

CLASS Z: SIMPLE BUILDING WORKS INCIDENTAL TO CIVIL ENGINEERING WORKS

Item	Gang hours	Labour £	Plant £	Material £	Unit	Total rate £
SURFACE FINISHES, LININGS AND PARTITIONS – cont						
Tiles; glazed ceramic wall tiles; BS 6431; white; fixing with adhesive; pointing joints with grout						
Sloping upper surfaces						
152 × 152 × 5.5 mm thick	0.28	14.10	–	21.72	m²	**35.82**
200 × 100 × 6.5 mm thick	0.28	14.10	–	21.72	m²	**35.82**
Walls						
152 × 152 × 5.5 mm thick	0.24	12.09	–	21.72	m²	**33.81**
200 × 100 × 6.5 mm thick	0.24	12.09	–	21.72	m²	**33.81**
Soffit						
152 × 152 × 5.5 mm thick	0.28	14.10	–	21.72	m²	**35.82**
200 × 100 × 6.5 mm thick	0.28	14.10	–	21.72	m²	**35.82**
Surfaces of width not exceeding 300 mm						
152 × 152 × 5.5 mm thick	0.12	6.04	–	6.52	m	**12.56**
200 × 100 × 6.5 mm thick	0.12	6.04	–	6.52	m	**12.56**
Surfaces of width 300 mm–1 m						
152 × 152 × 5.5 mm thick	0.18	9.06	–	21.72	m	**30.78**
200 × 100 × 6.5 mm thick	0.18	9.06	–	21.72	m	**30.78**
Tiles; slate; Riven Welsh; bedding 10 mm thick and jointing in cement mortar (1:3); grouting with cement mortar (1:1)						
Floors						
250 × 250 × 12–15 mm	0.24	12.09	–	67.27	m²	**79.36**
Surfaces of width not exceeding 300 mm						
250 × 250 × 12–15 mm	0.12	6.04	–	20.18	m	**26.22**
Surfaces of width 300 mm–1 m						
250 × 250 × 12–15 mm	0.18	9.06	–	67.27	m	**76.33**
Tiles; vinyl; Accoflex; fixing with adhesive						
Floors						
300 × 300 × 2 mm	0.17	5.24	–	7.90	m²	**13.14**
300 × 300 × 2.5 mm	0.17	5.24	–	8.99	m²	**14.23**
Surfaces of width not exceeding 300 mm						
300 × 300 × 2 mm	0.08	2.60	–	2.38	m	**4.98**
300 × 300 × 2.5 mm	0.08	2.60	–	2.70	m	**5.30**
Surfaces of width 300 mm–1 m						
300 × 300 × 2 mm	0.13	3.92	–	7.90	m	**11.82**
300 × 300 × 2.5 mm	0.13	3.92	–	8.99	m	**12.91**
Tiles; vinyl; Marley HD; fixing with adhesive						
Floors						
300 × 300 × 2 mm	0.23	7.30	–	11.02	m²	**18.32**
Surfaces of width not exceeding 300 mm						
300 × 300 × 2 mm	0.12	3.67	–	3.31	m	**6.98**
Surfaces of width 300 mm–1 m						
300 × 300 × 2 mm	0.17	5.49	–	11.02	m	**16.51**

CLASS Z: SIMPLE BUILDING WORKS INCIDENTAL TO CIVIL ENGINEERING WORKS

Item	Gang hours	Labour £	Plant £	Material £	Unit	Total rate £
Tiles; rubber studded; Altro Mondopave; type MRB; black; fixing with adhesive						
Floors						
500 × 500 × 2.5 mm	0.40	12.54	–	30.60	m²	**43.14**
500 × 500 × 4.0 mm	0.40	12.54	–	29.30	m²	**41.84**
Surfaces of width not exceeding 300 mm						
500 × 500 × 2.5 mm	0.20	6.27	–	9.16	m	**15.43**
500 × 500 × 4.0 mm	0.20	6.27	–	8.79	m	**15.06**
Surfaces of width 300 mm–1 m						
500 × 500 × 2.5 mm	0.30	9.40	–	30.60	m	**40.00**
500 × 500 × 4.0 mm	0.30	9.40	–	29.30	m	**38.70**
Tiles; rubber studded; Altro Mondopave; type MRB; colour; fixing with adhesive						
Floors						
500 × 500 × 2.5 mm	0.40	12.54	–	33.97	m²	**46.51**
500 × 500 × 4.0 mm	0.40	12.54	–	33.66	m²	**46.20**
Surfaces of width not exceeding 300 mm						
500 × 500 × 2.5 mm	0.20	6.27	–	10.17	m	**16.44**
500 × 500 × 4.0 mm	0.20	6.27	–	10.10	m	**16.37**
Surfaces of width 300 mm–1 m						
500 × 500 × 2.5 mm	0.30	9.40	–	33.97	m	**43.37**
500 × 500 × 4.0 mm	0.30	9.40	–	33.66	m	**43.06**
Tiles; linoleum; Forbo Nairn; Marmoleum Dual; level; fixing with adhesive						
Floors						
2.50 mm thick; marbled	0.27	8.37	–	16.05	m²	**24.42**
Surfaces of width not exceeding 300 mm						
2.50 mm thick; marbled	0.13	4.17	–	4.82	m	**8.99**
Surfaces of width 300 mm–1 m						
2.50 mm thick; marbled	0.20	6.27	–	1.85	m	**8.12**
Flexible sheer; linoleum; Forbo Nairn; Marmoleum Real; level; fixing with adhesive						
Floors						
2.00 mm thick; marbled	0.33	10.45	–	12.00	m²	**22.45**
2.50 mm thick; marbled	0.33	10.45	–	12.83	m²	**23.28**
3.20 mm thick; marbled	0.33	10.45	–	15.41	m²	**25.86**
4.00 mm thick; marbled	0.33	10.45	–	19.17	m²	**29.62**
Surfaces of width not exceeding 300 mm						
2.00 mm thick; marbled	0.17	5.24	–	3.60	m	**8.84**
2.50 mm thick; marbled	0.17	5.24	–	3.85	m	**9.09**
3.20 mm thick; marbled	0.17	5.24	–	4.62	m	**9.86**
4.00 mm thick; marbled	0.17	5.24	–	5.61	m	**10.85**
Surfaces of width 300 mm–1m						
2.00 mm thick; marbled	0.25	7.84	–	12.00	m	**19.84**
2.50 mm thick; marbled	0.25	7.84	–	12.83	m	**20.67**
3.20 mm thick; marbled	0.25	7.84	–	15.41	m	**23.25**
4.00 mm thick; marbled	0.25	7.84	–	19.17	m	**27.01**

CLASS Z: SIMPLE BUILDING WORKS INCIDENTAL TO CIVIL ENGINEERING WORKS

Item	Gang hours	Labour £	Plant £	Material £	Unit	Total rate £
SURFACE FINISHES, LININGS AND PARTITIONS – cont						
Flexible sheet; vinyl; slip resistant; Forbo Nairn; Surestep PUR; fixing with adhesive						
Floors						
2.00 mm thick	0.33	10.44	–	16.01	m²	26.45
Surfaces of width not exceeding 300 mm						
2.00 mm thick	0.17	5.24	–	4.80	m	10.04
Surfaces of width 300 mm–1 m						
2.00 mm thick	0.25	7.84	–	16.01	m	23.85
Flexible sheet; vinyl; Marley HD; fixing with adhesive						
Floors						
2.00 mm thick	0.30	9.40	–	11.02	m²	20.42
2.50 mm thick	0.33	10.45	–	12.32	m²	22.77
Surfaces of width not exceeding 300 mm						
2.00 mm thick	0.15	4.70	–	3.31	m	8.01
2.50 mm thick	0.17	5.24	–	3.70	m	8.94
Surfaces of width 300 mm–1 m						
2.00 mm thick	0.23	7.05	–	11.02	m	18.07
2.50 mm thick	0.25	7.84	–	12.32	m	20.16
Flexible sheet; vinyl; Armstrong Contract Interior; fixing with adhesive						
Floors						
2.00 mm thick	0.33	10.45	–	13.98	m²	24.43
Surfaces of width not exceeding 300 mm						
2.00 mm thick	0.17	5.24	–	4.20	m	9.44
Surfaces of width 300 mm–1 m						
2.00 mm thick	0.25	7.84	–	13.98	m	21.82
Flexible sheet; carpet; Armstrong Strong; fixing with adhesive						
Floors						
5.00 mm thick	0.17	5.24	–	44.62	m²	49.86
Surfaces of width not exceeding 300 mm						
5.00 mm thick	0.08	2.60	–	13.39	m	15.99
Surfaces of width 300 mm–1 m						
5.00 mm thick	0.13	3.92	–	44.62	m	48.54
Suspended ceilings; mineral fibre tiles in exposed grid; suspension system and wire hangers to structural soffit						
Suspended ceiling						
150–500 mm depth of suspension	–	–	–	–	m²	21.08

CLASS Z: SIMPLE BUILDING WORKS INCIDENTAL TO CIVIL ENGINEERING WORKS

Item	Gang hours	Labour £	Plant £	Material £	Unit	Total rate £
Suspended ceilings; mineral fibre tiles in concealed grid; suspension system and wire hangers to structural soffit						
Suspended ceiling						
150–500 mm depth of suspension	–	–	–	–	m²	34.62
Bulkheads						
250 mm girth	–	–	–	–	m	12.50
500 mm girth	–	–	–	–	m	25.00
PIPED BUILDING SERVICES						
Pipework; copper pipes EN 1057; capillary fittings; pipe clips screwed to background						
Pipes						
15 mm	0.13	3.94	–	2.18	m	6.12
22 mm	0.13	4.10	–	4.31	m	8.41
28 mm	0.14	4.58	–	5.53	m	10.11
35 mm	0.17	5.21	–	13.03	m	18.24
42 mm	0.19	6.00	–	15.71	m	21.71
Fittings						
15 mm made bend	0.08	2.37	–	9.70	nr	12.07
15 mm elbow	0.08	2.37	–	1.25	nr	3.62
15 mm equal tee	0.13	3.94	–	2.32	nr	6.26
15 mm straight coupling	0.09	2.68	–	0.93	nr	3.61
22 mm made bend	0.10	3.16	–	19.07	nr	22.23
22 mm elbow	0.11	3.47	–	2.79	nr	6.26
22 mm equal tee	0.17	5.21	–	6.35	nr	11.56
22 mm straight coupling	0.11	3.47	–	2.10	nr	5.57
28 mm made bend	0.13	3.94	–	24.36	nr	28.30
28 mm elbow	0.14	4.42	–	4.59	nr	9.01
28 mm equal tee	0.20	6.47	–	7.76	nr	14.23
28 mm straight coupling	0.14	4.42	–	3.49	nr	7.91
35 mm made bend	0.15	4.73	–	40.60	nr	45.33
35 mm elbow	0.17	5.21	–	11.68	nr	16.89
35 mm equal tee	0.23	7.26	–	28.81	nr	36.07
35 mm straight coupling	0.17	5.21	–	6.18	nr	11.39
42 mm made bend	0.20	6.31	–	43.87	nr	50.18
42 mm elbow	0.20	6.31	–	45.63	nr	51.94
42 mm equal tee	0.26	8.21	–	30.34	nr	38.55
42 mm straight coupling	0.20	6.31	–	9.78	nr	16.09
Pipework; 19 mm thick rigid mineral glass fibre sectional pipe lagging; plain finish; fixing with aluminium bands						
Insulation						
around 15 mm pipes	0.04	1.10	–	3.29	m	4.39
around 22 mm pipes	0.05	1.58	–	3.50	m	5.08
around 28 mm pipes	0.06	1.74	–	3.76	m	5.50
around 35 mm pipes	0.06	1.89	–	4.20	m	6.09
around 42 mm pipes	0.07	2.05	–	4.48	m	6.53

CLASS Z: SIMPLE BUILDING WORKS INCIDENTAL TO CIVIL ENGINEERING WORKS

Item	Gang hours	Labour £	Plant £	Material £	Unit	Total rate £
PIPED BUILDING SERVICES – cont						
Equipment; polyethylene cold water feed and expansion cistern to BS 4213, with cover; placing in position						
Cisterns and tanks						
68 litres; ref SC15	0.63	19.73	–	30.39	nr	**50.12**
114 litres; ref SC25	0.72	22.88	–	39.33	nr	**62.21**
Equipment; grp cold water storage cistern, with cover; placing in position						
Cisterns and tanks						
68 litres	0.63	19.73	–	88.98	nr	**108.71**
114 litres	0.72	22.88	–	93.05	nr	**115.93**
Equipment; copper single feed coil indirect cylinder to BS 1566 Part 2 grade 3; placing in position						
Cisterns and tanks						
96 litres; ref 2	1.00	31.56	–	161.59	nr	**193.15**
114 litres; ref 3	1.13	35.51	–	181.94	nr	**217.45**
Equipment; combination copper coil direct hot water storage units to BS 3198; placing in position						
Cisterns and tanks						
400 × 900 mm; 65/20 litres	1.40	44.18	–	382.32	nr	**426.50**
450 × 1075 mm; 115/25 litres	2.45	77.32	–	439.45	nr	**516.77**
Sanitary appliances and fittings						
Sink; white glazed fireclay to BS 1206 with pair of cast iron cantilever brackets						
610 × 455 × 205 mm	1.50	47.34	–	124.61	nr	**171.95**
610 × 455 × 205 mm	1.50	47.34	–	201.86	nr	**249.20**
Sink; stainless steel combined bowl and drainer; pair 19 mm chromium plated high neck pillar taps; chain and self colour plug to BS 3380; setting on base unit						
1050 × 500, single drainer, single bowl 420 × 350 × 175 mm	0.88	27.61	–	131.51	nr	**159.12**
1550 × 500, double drainer, single bowl 420 × 350 × 200 mm	0.88	27.61	–	217.67	nr	**245.28**
Lavatory basin; white vitreous china to BS 1188; pair 12 mm chromium plated pillar taps; chain and self colour plug to BS 3380; trap; painted cast iron brackets plugged and screwed to masonry						
560 × 405 mm	1.15	36.29	–	152.59	nr	**188.88**
635 × 455 mm	1.15	36.29	–	159.88	nr	**196.17**
Add for coloured	–	–	–	–	nr	–
Add for pedestal in lieu of brackets	0.10	3.16	–	63.00	nr	**66.16**

CLASS Z: SIMPLE BUILDING WORKS INCIDENTAL TO CIVIL ENGINEERING WORKS

Item	Gang hours	Labour £	Plant £	Material £	Unit	Total rate £
WC suite; low level; white glazed vitreous china pan; black plastic seat; 9 litre white glazed vitreous china cistern and brackets; low pressure ball valve; plastic flush pipe; building bracket into masonry; plugging and screwing pipe brackets and pan; bedding pan in mastic						
P trap outlet	1.50	47.34	–	261.45	nr	**308.79**
Add for coloured	–	–	–	–	nr	–
Bowl type wall urinal; white glazed vitreous china; white glazed vitreous china automatic flushing cistern and brackets; chromium plated flush pipe and spreaders; building brackets into masonry; plugging and screwing pipe brackets						
single; 455×380 × 330mm	2.00	63.12	–	290.67	nr	**353.79**
range of two; 455×380 × 330mm	5.50	173.58	–	448.12	nr	**621.70**
Add for each additional urinal	1.60	50.50	–	173.25	nr	**223.75**
Add for division between urinals	0.38	11.84	–	82.86	nr	**94.70**

PART 7

Unit Costs – Highway Works

INTRODUCTORY NOTES

The Unit Costs in this part represent the net cost to the Contractor of executing the work on site; they are not the prices which would be entered in a tender Bill of Quantities.

It must be emphasised that the unit rates are averages calculated on unit outputs for typical site conditions. Costs can vary considerably from contract to contract depending on individual Contractors, site conditions, working methods and other factors. Reference should be made to Part 1 for a general discussion on Civil Engineering Estimating.

Guidance prices are included for work normally executed by specialists, with a brief description where necessary of the assumptions upon which the costs have been based. Should the actual circumstances differ, it would be prudent to obtain check prices from the specialists concerned, on the basis of actual / likely quantity of the work, nature of site conditions, geographical location, time constraints, etc.

The method of measurement adopted in this section is the **Method of Measurement for Highway Works**, subject to variances where this has been felt to be of advantage to produce more helpful price guidance.

We have structured this Unit Cost section to cover as many aspects of Civil and Highway works as possible.

The Gang hours column shows the output per measured unit in actual time, not the total labour hours; thus for an item involving a gang of 5 men each for 0.3 hours, the total labour hours would naturally be 1.5, whereas the Gang hours shown would be 0.3.

This section is structured in such a manner as to provide the User with adequate background information on how the rates have been calculated, so as to allow them to be readily adjusted to suit other conditions to the example presented:

Alternative gang structures as well as the effect of varying bonus levels, travelling costs etc.

Other types of plant or else different running costs from the medium usage presumed

Other types of materials or else different discount / waste allowances from the levels presumed

Manual for Detailing Reinforced Concrete Structures to EC2

J. Calavera

Detailing 210 structural elements, each with a double page spread, this book is an indispensible guide to designing for reinforced concrete. Drawing on Jose Calavera's many years of experience, each entry provides commentary and recommendations for the element concerned, as 0077ell as summarising the appropriate Eurocode legislation and referring to further standards and literature.

Includes a CD ROM with AutoCad files of all of the models, which you can adapt to suit your own purposes.

Section 1. General Rules for Bending, Placing and Lapping of Reinforcement
Section 2. Constructive Details

1. Foundations 2. Walls 3. Columns and Joints 4. Walls Subjected to Axial Loads. Shear Walls and Cores 5. Beams and Lintels 6. Slabs 7. Flat Slabs 8. Stairs 9. Supports 10. Brackets 11. Pavements and Galleries 12. Cylindrical Chimneys, Towers and Hollow Piles 13. Rectangular Silos, Caissons and Hollow Piles 14. Reservoirs, Tanks and Swimming Pools 15. Special Details for Seismic Areas

October 2011: 512pp
Hb: 978-0-415-66348-9: **£80.00**

To Order: Tel: +44 (0) 1235 400524 **Fax:** +44 (0) 1235 400525
or Post: Taylor and Francis Customer Services,
Bookpoint Ltd, Unit T1, 200 Milton Park, Abingdon, Oxon, OX14 4TA UK
Email: book.orders@tandf.co.uk

**For a complete listing of all our titles visit:
www.tandf.co.uk**

GUIDANCE NOTES

Generally

Adjustments should be made to the rates shown for time, location, local conditions, site constraints and any other factors likely to affect the costs of a specific scheme.

Materials cost

Materials costs within the rates have been calculated using the 'list prices' contained in Part 5: Resources (pages 131 to 202), with an index appearing on page 39), adjusted to allow for delivery charges (if any) and a 'reasonable' level of discount obtainable by the contractor, this will vary very much depending on the contractor's standing, the potential size of the order and the supplier's eagerness and will vary also between raw traded goods such as timber which will attract a low discount of perhaps 3%, if at all, and manufactured goods where the room for bargaining is much greater and can reach levels of 30% to 40%. High demand for a product at the time of pricing can dramatically reduce the potential discount, as can the world economy in the case of imported goods such as timber and copper. Allowance has also been made for wastage on site (generally 2½% to 5%) dependent upon the risk of damage, the actual level should take account of the nature of the material and its method of storage and distribution about the site.

Labour cost

The composition of the labour and type of plant is generally stated at the beginning of each section, more detailed information on the calculation of the labour rates is given in Part 5: Resources. In addition on pages 37 and 38 is a summary of labour grades and responsibilities extracted from the Working Rule Agreement. Within Parts 4 and 5, each section is prefaced by a detailed build-up of the labour gang assumed for each type of work. This should allow the user to see the cost impact of a different gang as well as different levels of bonus payments, allowances for skills and conditions, travelling allowances etc. The user should be aware that the output constants are based on the gangs shown and would also need to be changed.

Plant cost

A rate build-up of suitable plant is generally stated at the beginning of each section, with more detailed information on alternative machines and their average fuel costs being given in Part 5: Resources, pages 131 to 202. Within Parts 6 and 7, each section is prefaced by a detailed build-up of plant assumed for each type of work This should allow the user to see the cost impact of using alternative plant as well as different levels of usage. The user should be aware that the output constants are based on the plant shown and would also need to be changed.
Outputs

The user is directed to Part 10: Outputs (pages 585 to 592), which contains a selection of output constants and in particular a chart of haulage times for various capacities of Tippers.

Method of Measurement

A keynote to bills of quantities for highway works is the brevity of descriptions due to a strong emphasis being placed on the estimator pricing the work described in the Specification and shown on the Drawings.

Although this part of the book is primarily based on MMHW, the specific rules have been varied from in cases where it has been felt that an alternative presentation would be of value to the book's main purpose of providing guidance on prices. This is especially so with a number of specialist contractors but also in the cases of work where a more detailed presentation will enable the user to allow for ancillary items.

LEVEL 1 DIVISION		LEVEL 2 CONSTRUCTION HEADING	LEVEL 3 MMHW SERIES HEADINGS
(i) Preliminaries		Preliminaries	Series 100
(ii) Roadworks		Roadworks General	Series 200 Series 300 Series 400 Series 600
		Main Carriageway	Series 500 Series 700 Series 1100
		Interchanges	Series 500 Series 700 Series 1100
		Side Roads	Series 500 Series 700 Series 1100
		Signs, Motorway Communications and Lighting	Series 1200
			Series 1300 Series 1400 Series 1500
		Landscape and Ecology	Series 3000
(iii) Structures	Structure in form of Bridge or Viaduct; Name or Reference	Special Preliminaries Piling Substructure – End Supports	Series 2700 Series 1600 Series 500 Series 600 Series 1100 Series 1700 Series 1800 Series 1900 Series 2300 Series 2400
		Substructure – Intermediate Supports Substructure – Main Span Substructure – Approach Spans	As for End Supports
		Superstructure – Main Span Superstructure – Approach Spans Superstructure – Arch Ribs	Series 500 Series 1700 Series 1800 Series 1900 Series 2100 Series 2300 Series 2400
		Finishings	Series 400 Series 600 Series 700 Series 1100 Series 2000 Series 2400 Series 5000

LEVEL 1 DIVISION		LEVEL 2 CONSTRUCTION HEADING	LEVEL 3 MMHW SERIES HEADINGS
	Retaining wall, Culvert, Subway, Gantry, Large Headwall, Gabion Wall, Diaphragm wall, Pocket Type Reinforced Brick-work, Retaining Wall and the like; Name or Reference	Special Preliminaries Main Construction	Series 500 Series 600 Series 1100 Series 1600 Series 1700 Series 1800 Series 1900 Series 2300 Series 2400
		Finishings	Series 400 Series 600 Series 700 Series 1100 Series 2000 Series 2400 Series 5000
(iv) Structures where a choice of designs is offered	Structure Designed by the Overseeing Organi-zation; Name or Reference Structure Designed by the Contractor; Name or Reference	To comply with the principles set down above for Structures	
(v) Structures Designed by the Contractor	Structure; Name or Reference		
(vi) Service Areas		Roadworks Structures	To comply with the principles set down above for Roadworks and Structures
(vii) Maintenance Compounds		Roadworks Structures	
(viii) Accommodation Works		Interest; Name or Reference	
(ix) Works for Statutory or Other Bodies		Body; Name or Reference	To comply with the principles set down above for Roadworks and Structures
(x) Daywork		Daywork	
(xi) PC & Provisional Sum		PC & Provisional Sum	

SERIES 100: PRELIMINARIES

Item	Gang hours	Labour £	Plant £	Material £	Unit	Total rate £
NOTES						
General						
Refer also to the example calculation of						
Preliminaries in Part 2 and also to Part 8 Oncosts						
and profit.						
TEMPORARY ACCOMMODATION						
Temporary Accommodation						
Erection of principal offices for the Overseeing						
Organization						
prefabricated unit; connect to services	–	–	540.71	–	nr	**540.71**
Erection of offices and messes for the Contractor						
prefabricated unit; connect to services	–	–	540.71	–	nr	**540.71**
Erection of stores and workshops for the Contractor						
prefabricated unit; connect to services	–	–	540.71	–	nr	**540.71**
Servicing of principal offices for the Overseeing						
Organization						
jack leg hutment; 3.7 m × 2.6 m	–	–	45.00	–	week	**45.00**
jack leg hutment; 7.35 m × 3.1 m	–	–	70.00	–	week	**70.00**
jack leg hutment; 14.7 m × 3.7 m	–	–	250.00	–	week	**250.00**
jack leg toilet unit; 4.9 m × 2.6 m	–	–	46.00	–	week	**46.00**
Servicing of portable offices for the Overseeing						
Organization						
wheeled cabin; 3.7 m × 2.3 m	–	–	42.50	–	week	**42.50**
wheeled cabin; 6.7 m × 2.3 m	–	–	55.00	–	week	**55.00**
Servicing of offices and messes for the Overseeing						
Organization						
jack leg hutment; 3.7 m × 2.6 m	–	–	45.00	–	week	**45.00**
jack leg hutment; 7.35 m × 3.1 m	–	–	70.00	–	week	**70.00**
jack leg hutment; 14.7 m × 3.7 m	–	–	250.00	–	week	**250.00**
wheeled cabin; 3.7 m × 2.3 m	–	–	42.50	–	week	**42.50**
wheeled cabin; 6.7 m × 2.3 m	–	–	55.00	–	week	**55.00**
jack leg toilet unit; 4.9 m × 2.6 m	–	–	46.00	–	week	**46.00**
canteen unit; 9.8 m × 2.6 m	–	–	150.00	–	week	**150.00**
Servicing of stores and workshops for the Contractor						
jack leg hutment; 3.7 m × 2.6 m	–	–	45.00	–	week	**45.00**
jack leg hutment; 7.35 m × 3.1 m	–	–	70.00	–	week	**70.00**
wheeled cabin; 3.7 m × 2.3 m	–	–	42.50	–	week	**42.50**
wheeled cabin; 6.7 m × 2.3 m	–	–	55.00	–	week	**55.00**
pollution decontamination unit; 6.7 m × 2.3 m	–	–	250.00	–	week	**250.00**
Dismantling of principal offices for the Overseeing						
Organization						
prefabricated unit; disconnect from services;						
removing	–	–	540.71	–	nr	**540.71**
Dismantling of offices and messes for the Contractor						
prefabricated unit; disconnect from services;						
removing	–	–	540.71	–	nr	**540.71**
Dismantling of stores and workshops for the Contractor						
prefabricated unit; disconnect from services;						
removing	–	–	540.71	–	nr	**540.71**

SERIES 100: PRELIMINARIES

Item	Gang hours	Labour £	Plant £	Material £	Unit	Total rate £
VEHICLES FOR THE OVERSEEING ORGANIZATION						
Vehicles For The Overseeing Organization						
Vehicles for the Overseeing Organization						
Land Rover or similar, short wheelbase	–	–	470.91	–	week	**470.91**
Land Rover or similar, long wheelbase	–	–	570.51	–	week	**570.51**
For other types of transport vehicles refer to						
Resources-Plant.	–	–	–	–	-	**-**
OPERATIVES FOR THE ENGINEER						
Operatives For The Overseeing Organization						
Chainman for the Overseeing Organization	40.00	539.15	–	–	week	**539.15**
Driver for the Overseeing Organization	40.00	539.15	–	–	week	**539.15**
Laboratory assistant for the Overseeing Organization	40.00	363.67	–	–	week	**363.67**
INFORMATION BOARD						
Information Board						
Roadworks sign ref. 7002	–	–	–	–	nr	**78.55**
Roadworks sign ref. 7003	–	–	–	–	nr	**98.75**
Roadworks sign fref. 7004	–	–	–	–	nr	**86.65**

SERIES 200: SITE CLEARANCE

Item	Gang hours	Labour £	Plant £	Material £	Unit	Total rate £
NOTES						
General						
The prices in this section are to include for the removal of superficial obstructions down to existing ground level.						
Demolition of individual or groups of buildings or structures						
MMHW states that individual structures should be itemized. The following rates are given as £ per m3 to simplify the pricing of different sized structures. (Refer also to Part 4 of this book Class D)						
RESOURCES - LABOUR						
Clearance gang						
1 ganger/chargehand (skill rate 4)		14.58				
1 skilled operative (skill rate 4)		13.65				
2 unskilled operatives (general)		25.50				
1 plant operator (skill rate 3)		16.59				
Total Gang Rate / Hour	£	70.33				
RESOURCES - PLANT						
Clearance						
0.8 m³ tractor loader - 50% of time			11.18			
8 t lorry with hiab - 25% of time			3.19			
4 t dumper - 50% of time			4.83			
20 t mobile crane - 25% of time			9.86			
compressor, 11.3 m³/min (450 cfm), 4 tool			20.39			
compressor tools: two brick hammers / picks - 50% of time			1.38			
compressor tools: chipping hammer			0.66			
compressor tools: medium rock drill 30			0.77			
compressor tools: road breaker			0.97			
compressor tools: two 15 m lengths hose			0.34			
Total Gang Rate / Hour		£	53.56			
SITE CLEARANCE						
Site Clearance						
General site clearance						
open field site	9.91	691.74	525.35	–	ha	**1217.09**
medium density wooded	20.61	1438.63	1092.48	–	ha	**2531.11**
heavy density wooded	32.09	2239.96	1700.67	–	ha	**3940.63**
urban areas (town centre)	30.60	2135.96	1622.06	–	ha	**3758.02**
live dual carriageway	30.60	2135.96	1622.06	–	ha	**3758.02**

SERIES 200: SITE CLEARANCE

Item	Gang hours	Labour £	Plant £	Material £	Unit	Total rate £
Demolition of building or structure						
building; brick construction with timber floor and roof	–	–	–	–	m³	7.22
building; brick construction with concrete floor and roof	–	–	–	–	m³	12.02
building; masonry construction with timber floor and roof	–	–	–	–	m³	9.42
building; reinforced concrete frame construction with brick infill	–	–	–	–	m³	12.54
building; steel frame construction with brick infill	–	–	–	–	m³	6.82
building; steel frame construction with cladding	–	–	–	–	m³	6.50
building; timber	–	–	–	–	m³	5.84
reinforced concrete bridge deck or superstructure	–	–	–	–	m³	15.07
reinforced concrete bridge abutment or bank seat	–	–	–	–	m³	39.60
reinforced concrete retaining wall	–	–	–	–	m³	150.92
brick or masonry retaining wall	–	–	–	–	m³	83.05
brick or masonry boundary wall	–	–	–	–	m³	67.98
dry stone boundary wall	–	–	–	–	m³	81.84
TAKE UP OR TAKE DOWN AND SET ASIDE FOR RE-USE OR REMOVE TO STORE OR TIP OFF SITE						
Take Up Or Down And Set Aside For Re-Use Or Remove To Store Or Tip Off Site						
Take up or down and set aside for re-use						
precast concrete kerbs and channels	0.02	1.19	1.95	–	m	3.14
precast concrete edgings	0.01	0.89	1.56	–	m	2.45
precast concrete drainage and kerb blocks	0.02	1.49	2.37	–	m	3.86
precast concrete drainage channel systems	0.02	1.49	2.37	–	m	3.86
tensioned single sided corrugated beam safety fence	0.14	9.77	7.56	–	m	17.33
timber post and 4 rail fence	0.08	5.58	4.28	–	m	9.86
bench seat	0.13	9.07	7.03	–	nr	16.10
cattle trough	0.16	10.89	8.38	–	nr	19.27
permanent bollard	0.13	9.07	7.03	–	nr	16.10
pedestrian crossing lights; pair	0.26	18.15	13.99	–	nr	32.14
lighting column including bracket arm and lantern; 5m high	0.58	40.49	31.13	–	nr	71.62
lighting column including bracket arm and lantern; 10m high	0.61	42.58	32.64	–	nr	75.22
traffic sign	0.26	18.15	13.99	–	nr	32.14
timber gate	0.13	9.07	7.03	–	nr	16.10
stile	0.13	9.07	7.03	–	nr	16.10
road stud	0.03	1.81	1.36	–	nr	3.17
chamber cover and frame	0.03	1.81	1.36	–	nr	3.17
gully grating and frame	0.03	1.81	1.36	–	nr	3.17
feeder pillars	0.03	1.81	1.36	–	nr	3.17

SERIES 200: SITE CLEARANCE

Item	Gang hours	Labour £	Plant £	Material £	Unit	Total rate £
TAKE UP OR TAKE DOWN AND SET ASIDE FOR RE-USE OR REMOVE TO STORE OR TIP OFF SITE – cont						
Take up or down and remove to store off site						
precast concrete kerbs and channels	0.02	1.19	3.50	–	m	**4.69**
precast concrete edgings	0.01	0.89	2.72	–	m	**3.61**
precast concrete drainage and kerb blocks	0.02	1.49	4.30	–	m	**5.79**
precast concrete drainage channel systems	0.02	1.49	4.30	–	m	**5.79**
tensioned single sided corrugated beam safety fence	0.14	9.77	8.84	–	m	**18.61**
timber post and 4 rail fence	0.08	5.58	5.01	–	m	**10.59**
bench seat	0.13	9.07	11.76	–	nr	**20.83**
cattle trough	0.16	10.89	14.06	–	nr	**24.95**
permanent bollard	0.13	9.07	11.76	–	nr	**20.83**
pedestrian crossing lights; pair	0.26	18.15	16.36	–	nr	**34.51**
lighting column including bracket arm and lantern; 5m high	0.58	40.49	36.41	–	nr	**76.90**
lighting column including bracket arm and lantern; 10m high	0.61	42.58	38.19	–	nr	**80.77**
traffic sign	0.26	18.15	16.36	–	nr	**34.51**
timber gate	0.13	9.07	11.76	–	nr	**20.83**
stile	0.13	9.07	11.76	–	nr	**20.83**
road stud	0.03	1.81	2.30	–	nr	**4.11**
chamber cover and frame	0.03	1.81	2.30	–	nr	**4.11**
gully grating and frame	0.03	1.81	2.30	–	nr	**4.11**
feeder pillars	0.03	1.81	2.30	–	nr	**4.11**
Take up or down and remove to tip off site						
tensioned single sided corrugated beam safety fence	0.17	11.87	9.23	–	m	**21.10**
timber post and 4 rail fence	0.09	6.28	4.87	–	m	**11.15**
low pressure gas mains up to 150mm diameter	0.04	2.79	2.18	–	m	**4.97**
low pressure water mains up to 75mm diameter	0.03	2.09	1.67	–	m	**3.76**
power cable laid singly	0.03	2.09	1.67	–	m	**3.76**
lighting column including bracket arm and lantern; 5m high	0.82	57.24	44.73	–	nr	**101.97**
lighting column including bracket arm and lantern; 10m high	0.85	59.33	46.26	–	nr	**105.59**
traffic sign including posts	0.38	26.52	20.48	–	nr	**47.00**
Removal of existing reflectorised thermoplastic road markings						
100mm wide line	–	–	–	–	m	**1.21**
150mm wide line	–	–	–	–	m	**1.82**
200mm wide line	–	–	–	–	m	**2.42**
arrow or letter ne 6.0m long	–	–	–	–	nr	**18.72**
arrow or letter 6.0 - 16.0m long	–	–	–	–	nr	**78.49**

SERIES 300: FENCING

Item	Gang hours	Labour £	Plant £	Material £	Unit	Total rate £
NOTES						
General						
This section is restricted to those fences and barriers which are most commonly found on Highway Works. Hedges have been included, despite not being specifically catered for in the MMHW.						
The re-erection cost for fencing taken from store assumes that major components are in good condition; the prices below allow a sum of 20% of the value of new materials to cover minor repairs, new fixings and touching up any coatings.						
RESOURCES - LABOUR						
Fencing/barrier gang						
1 ganger/chargehand (skill rate 4) - 50% of time		7.29				
1 skilled operative (skill rate 4) - 50% of time		6.83				
1 unskilled operative (general)		12.75				
1 plant operator (skill rate 4)		14.92				
Total Gang Rate / Hour	£	**41.79**				
Horticultural works gang						
1 skilled operative (skill rate 4)		13.65				
1 unskilled operative (general)		12.75				
Total Gang Rate / Hour	£	**26.40**				
RESOURCES - PLANT						
Fencing/Barriers						
agricultural type tractor; fencing auger			14.34			
gas oil for ditto			4.61			
drop sided trailer, two axles			0.35			
power tools etc. (fencing)			3.13			
Total Gang Rate / Hour		£	**22.44**			
RESOURCES - MATERIALS						
All rates for materials are based on the most economically available materials with a minimum waste allowance of 2.5% and supplier's discount of 15%.						

SERIES 300: FENCING

Item	Gang hours	Labour £	Plant £	Material £	Unit	Total rate £
ENVIRONMENTAL BARRIERS						
Hedges						
Set out, nick out and excavate trench minimum 400 mm deep and break up subsoil to minimum depth 300 mm	0.15	3.93	–	–	m	**3.93**
Supply and plant hedging plants ; backfilling with excavated topsoil						
single row plants at 200 mm centres	0.25	6.54	–	5.33	m	**11.87**
single row plants at 300 mm centres	0.17	4.45	–	3.55	m	**8.00**
single row plants at 400 mm centres	0.13	3.27	–	2.67	m	**5.94**
single row plants at 500 mm centres	0.10	2.62	–	2.13	m	**4.75**
single row plants at 600 mm centres	0.08	2.09	–	1.77	m	**3.86**
double row plants at 200 mm centres	0.50	13.09	–	10.66	m	**23.75**
double row plants at 300 mm centres	0.34	8.90	–	7.10	m	**16.00**
double row plants at 400 mm centres	0.25	6.54	–	5.33	m	**11.87**
double row plants at 500 mm centres	0.20	5.24	–	4.26	m	**9.50**
double row plants at 600 mm centres	0.16	4.19	–	3.54	m	**7.73**
Extra for incorporating manure at $1 m^3 / 30m3$	0.60	7.65	–	0.34	m^3	**7.99**
Noise barriers						
Noise barriers consist of the erection of reflective or absorbent acoustical screening to reduce nuisance from noise. Due to the divergence in performance requirements and specification for various locations it is not practical to state all inclusive unit costs. Therefore advice should be sought from Specialist Contractors in order to obtain accurate costings. However listed below are examples of sample specification together with approximate costings in order to obtain budget prices.						
NB:- The following unit costs are based upon a 2.0 m high barrier						
Noise reflective barriers						
Barrier with architectural precast concrete panels and integral posts	–	–	–	–	m²	**165.49**
Barrier with acoustical timber planks post support system	–	–	–	–	m²	**149.85**
Sound Absorptive barriers						
Barrier with architectural precast wood fibre concrete panels and integral posts	–	–	–	–	m²	**198.13**
Barrier with perforated steel and mineral wool blankets in self-supporting system	–	–	–	–	m²	**219.95**

SERIES 300: FENCING

Item	Gang hours	Labour £	Plant £	Material £	Unit	Total rate £
FENCING, GATES AND STILES						
Fencing, gates and stiles						
Temporary fencing						
Type 1; 1.275 m high, timber posts and two strands of galvanized barbed wire and four strands of galvanized plain wire	0.06	2.49	1.35	2.23	m	**6.07**
Type 2; 1.275 m high, timber posts and two strands of galvanized barbed wire and four strands of galvanized plain wire	0.06	2.49	1.35	2.00	m	**5.84**
Type 3; 1.2 m high, timber posts and cleft chestnut paling	0.07	2.90	1.57	8.72	m	**13.19**
Type 4; 0.9 m high, timber posts and galvanized rectangular wire mesh	0.13	5.39	2.92	4.37	m	**12.68**
Timber rail fencing						
1.4 m high, timber posts and four rails	0.13	5.39	2.92	12.80	m	**21.11**
Plastic coated heavy pattern chain link fencing						
1.40 m high with 125 × 125 mm concrete posts	0.05	2.07	1.12	8.04	m	**11.23**
1.80 m high with 125 × 125 mm concrete posts	0.06	2.49	1.35	15.03	m	**18.87**
Plastic coated strained wire fencing						
1.35 m high, nine strand with 40 × 40 × 3 mm plastic coated RHS posts	0.16	6.63	3.59	14.95	m	**25.17**
1.80 m high, eleven strand with 50 × 50 × 3 mm plastic coated RHS posts	0.20	8.29	4.49	19.91	m	**32.69**
2.10 m high, fifteen strand with 50 × 50 × 3 mm plastic coated RHS posts	0.22	9.12	4.94	23.24	m	**37.30**
Woven wire fencing						
1.23 m high, galvanized wire with 75 × 150 mm timber posts	0.06	2.49	1.35	6.59	m	**10.43**
Close boarded fencing						
1.80 m high with 125 × 125 mm concrete posts	0.30	12.44	6.73	39.71	m	**58.88**
Concrete foundation						
to main posts	0.09	3.73	0.32	5.45	nr	**9.50**
to straining posts	0.09	3.73	0.32	5.45	nr	**9.50**
to struts	0.09	3.73	0.32	5.45	nr	**9.50**
to intermediate posts	0.09	3.73	0.32	5.45	nr	**9.50**
Steel tubular frame single field gates						
1.175 m high 3.60 m wide	0.30	12.44	6.73	143.46	nr	**162.63**
1.175 m high 4.50 m wide	0.30	12.44	6.73	167.89	nr	**187.06**
Steel tubular frame half mesh single field gates						
1.175 m high 3.60 m wide	0.30	12.44	6.73	171.36	nr	**190.53**
1.175 m high 4.50 m wide	0.30	12.44	6.73	204.03	nr	**223.20**
Steel tubular frame extra wide single field gates						
1.175 m high 4.88 m wide	0.30	12.44	6.73	194.47	nr	**213.64**
Steel tubular frame double field gates						
1.175 m high 5.02 m wide	0.60	24.88	13.46	200.05	nr	**238.39**
Timber single field gates						
1.27 m high 3.00 m wide	1.84	76.29	41.28	118.51	nr	**236.08**
1.27 m high 3.60 m wide	1.90	78.78	42.63	128.08	nr	**249.49**
1.27 m high 4.10 m wide	2.00	82.92	44.87	137.65	nr	**265.44**
1.27 m high 4.71 m wide	2.00	82.92	44.87	147.22	nr	**275.01**

SERIES 300: FENCING

Item	Gang hours	Labour £	Plant £	Material £	Unit	Total rate £
FENCING, GATES AND STILES – cont						
Timber Type 1 wicket gates						
1.27 m high 1.20 m wide	1.20	49.75	26.93	57.78	nr	**134.46**
Timber Type 2 wicket gates						
1.27 m high 1.02 m wide	1.20	49.75	26.93	44.01	nr	**120.69**
Timber kissing gates						
1.27 m high 1.77 m wide	2.00	82.92	44.87	262.66	nr	**390.45**
Timber stiles Type 1						
1.45 m high 1.00 m wide	1.50	62.19	33.66	115.37	nr	**211.22**
Timber stiles Type 2						
1.45 m high 1.00 m wide	1.40	58.05	31.41	79.89	nr	**169.35**
Extra for						
sheep netting on post and wire	0.04	1.45	0.12	0.56	m	**2.13**
pig netting on post and wire	0.04	1.45	0.12	0.61	m	**2.18**
REMOVE FROM STORE AND RE-ERECT FENCING, GATES AND STILES						
Remove From Store And Re-Erect Fencing, Gates And Stiles						
Timber rail fencing						
1.4 m high, timber posts and four rails	0.13	5.39	2.92	2.74	m	**11.05**
Plastic coated heavy pattern chain link fencing						
1.40 m high with 125 × 125 mm concrete posts	0.05	2.07	1.12	1.71	m	**4.90**
1.80 m high with 125 × 125 mm concrete posts	0.06	2.49	1.35	2.27	m	**6.11**
Plastic coated strained wire fencing						
1.35 m high, nine strand with 40 × 40 × 3 mm plastic coated RHS posts	0.16	6.63	3.59	3.20	m	**13.42**
1.80 m high, eleven strand with 50 × 50 × 3 mm plastic coated RHS posts	0.20	8.29	4.49	4.27	m	**17.05**
2.10 m high, fifteen strand with 50 × 50 × 3 mm plastic coated RHS posts	0.22	9.12	4.94	4.52	m	**18.58**
Woven wire fencing						
1.23 m high, galvanized wire with 75 × 150 mm timber posts	0.06	2.49	1.35	1.42	m	**5.26**
Close boarded fencing						
1.80 m high with 125 × 125 mm concrete posts	0.30	12.44	6.73	7.63	m	**26.80**
Steel tubular frame single field gates						
1.175 m high 3.60 m wide	0.30	12.44	6.73	12.80	nr	**31.97**
1.175 m high 4.50 m wide	0.30	12.44	6.73	16.44	nr	**35.61**
Steel tubular frame half mesh single field gates						
1.175 m high 3.60 m wide	0.30	12.44	6.73	16.44	nr	**35.61**
1.175 m high 4.50 m wide	0.30	12.44	6.73	20.11	nr	**39.28**
Steel tubular frame extra wide single field gates						
1.175 m high 4.88 m wide	0.30	12.44	6.73	18.28	nr	**37.45**
Steel tubular frame double field gates						
1.175 m high 5.02 m wide	0.60	24.88	13.46	27.43	nr	**65.77**

SERIES 300: FENCING

Item	Gang hours	Labour £	Plant £	Material £	Unit	Total rate £
Timber single field gates						
1.27 m high 3.00 m wide	1.84	76.29	41.28	32.22	nr	**149.79**
1.27 m high 3.60 m wide	1.90	78.78	42.63	37.33	nr	**158.74**
1.27 m high 4.10 m wide	2.00	82.92	44.87	45.13	nr	**172.92**
1.27 m high 4.71 m wide	2.00	82.92	44.87	46.49	nr	**174.28**
Timber Type 1 wicket gates						
1.27 m high 1.20 m wide	1.20	49.75	26.93	10.18	nr	**86.86**
Timber Type 2 wicket gates						
1.27 m high 1.02 m wide	1.20	49.75	26.93	11.62	nr	**88.30**
Timber kissing gates						
1.27 m high 1.77 m wide	2.00	82.92	44.87	32.22	nr	**160.01**
Timber stiles Type 1						
1.45 m high 1.00 m wide	1.50	62.19	33.66	26.63	nr	**122.48**
Timber stiles Type 2						
1.45 m high 1.00 m wide	1.40	58.05	31.41	20.02	nr	**109.48**
EXCAVATION IN HARD MATERIAL						
Extra over excavation for excavation in Hard Material in fencing works	0.50	20.73	–	–	m^3	**20.73**

SERIES 400: ROAD RESTRAINT SYSTEMS (VEHICLE AND PEDESTRIAN)

Item	Gang hours	Labour £	Plant £	Material £	Unit	Total rate £
NOTES						
General						
The re-erection cost for safety fencing taken from store assumes that major components are in good condition; the prices below allow a sum of 20% of the value of new materials to cover minor repairs, new fixings and touching up any coatings.						
The heights of the following parapets are in accordance with the Standard Designs and DfT requirements. The rates include for all anchorages and fixings and in the case of steel, galvanising at works and painting four coat paint system on site together with etching the galvanized surface, as necessary.						
RESOURCES - LABOUR						
Safety barrier gang						
1 ganger/chargehand (skill rate 4)		14.58				
1 skilled operative (skill rate 4)		13.65				
2 unskilled operatives (general)		25.50				
1 plant operator (skill rate 4)		14.92				
Total Gang Rate / Hour	£	**68.66**				
Parapet gang						
1 ganger/chargehand (skill rate 4)		14.58				
1 skilled operative (skill rate 4)		13.65				
2 unskilled operatives (general)		25.50				
1 plant operator (skill rate 4)		14.92				
Total Gang Rate / Hour	£	**68.66**				
RESOURCES - PLANT						
Safety barriers						
agricultural type tractor; fencing auger			14.34			
gas oil for ditto			4.61			
drop sided trailer, two axles			0.35			
small tools (part time)			2.06			
Total Gang Rate / Hour		£	21.36			
Parapets						
agricultural type tractor; front bucket - 50% of time			7.21			
gas oil for ditto			2.31			
2.80 m^3/min (100 cfm) compressor; two tool			3.38			
gas oil for ditto			5.11			
compressor tools: heavy rock drill 33, 84 cfm			2.10			
compressor tools: rotary drill, 10 cfm			1.06			
8 t lorry with 1 t hiab - 50% of time			5.00			
gas oil for ditto			1.39			
Total Gang Rate / Hour		£	27.56			

SERIES 400: ROAD RESTRAINT SYSTEMS (VEHICLE AND PEDESTRIAN)

Item	Gang hours	Labour £	Plant £	Material £	Unit	Total rate £
BEAM SAFETY BARRIERS						
Prices generally are for beams 'straight or curved exceeding 120m radius', for work to a tighter radius 50–120m radius Add 15%						
not exceeding 50 m radius Add 40%						
Untensioned beams						
single sided corrugated beam	0.07	4.77	1.49	19.16	m	**25.42**
double sided corrugated beam	0.22	14.98	4.70	38.29	m	**57.97**
single sided open box beam	0.07	4.77	1.49	38.87	m	**45.13**
single sided double rail open box beam	0.22	14.98	4.70	77.73	m	**97.41**
double height open box beam	0.24	16.35	5.13	80.16	m	**101.64**
Tensioned beams						
single sided corrugated beam	0.10	6.81	2.13	20.24	m	**29.18**
double sided corrugated beam	0.12	8.17	2.57	41.74	m	**52.48**
Long driven post						
for single sided tensioned corrugated beam	0.06	3.75	1.17	45.38	nr	**50.30**
for double sided tensioned corrugated beam	0.06	3.75	1.17	45.38	nr	**50.30**
for single sided open box beam	0.06	3.75	1.17	45.38	nr	**50.30**
for double sided open box beam	0.06	3.75	1.17	45.38	nr	**50.30**
for double height open box beam	0.06	3.75	1.17	45.38	nr	**50.30**
Short post for setting in concrete or socket						
for single sided tensioned corrugated beam	0.06	4.09	1.28	17.50	nr	**22.87**
for double sided tensioned corrugated beam	0.06	4.09	1.28	17.82	nr	**23.19**
for single sided open box beam	0.06	4.09	1.28	21.24	nr	**26.61**
for double sided open box beam	0.06	4.09	1.28	25.32	nr	**30.69**
Mounting bracket fixed to structure						
for single sided open box beam	0.16	10.90	3.42	68.19	nr	**82.51**
Terminal section						
for untensioned single sided corrugated beam	0.71	48.36	15.17	490.88	nr	**554.41**
for untensioned double sided corrugated beam	1.25	85.14	26.70	519.75	nr	**631.59**
for untensioned single sided open box beam	1.01	68.79	21.58	525.52	nr	**615.89**
for untensioned double sided open box beam	1.78	121.24	38.02	883.58	nr	**1042.84**
for tensioned single sided corrugated beam	0.96	65.39	20.50	342.46	nr	**428.35**
for tensioned double sided corrugated beam	1.70	115.79	36.31	524.09	nr	**676.19**
Full height anchorage						
for single sided tensioned corrugated beam	3.95	269.04	84.38	756.82	nr	**1110.24**
for double sided tensioned corrugated beam	4.35	296.28	92.92	882.42	nr	**1271.62**
for single sided open box beam	3.80	258.82	81.17	652.00	nr	**991.99**
for double sided open box beam	4.20	286.07	89.72	768.08	nr	**1143.87**
Expansion joint anchorage						
for single sided open box beam	4.52	307.86	96.55	1342.95	nr	**1747.36**
for double sided open box beam	5.15	350.77	110.01	1627.97	nr	**2088.75**
Type 048 connection to bridge parapet						
for single sided open box beam	0.70	47.68	14.95	179.81	nr	**242.44**
Connection piece for single sided open box beam to single sided corrugated beam	0.78	53.13	16.66	176.43	nr	**246.22**

SERIES 400: ROAD RESTRAINT SYSTEMS (VEHICLE AND PEDESTRIAN)

Item	Gang hours	Labour £	Plant £	Material £	Unit	Total rate £
BEAM SAFETY BARRIERS – cont						
Standard concrete foundation						
for post for corrugated beam	0.23	15.67	4.83	7.11	nr	27.61
for post for open box beam	0.23	15.67	4.91	8.85	nr	29.43
Concrete foundation Type 1 spanning filter drain						
for post for corrugated beam	0.25	17.03	5.34	10.58	nr	32.95
for post for open box beam	0.25	17.03	5.34	9.36	nr	31.73
Standard socketed foundation for post for open box beam	0.30	20.43	6.41	10.70	nr	37.54
CONCRETE SAFETY BARRIERS						
Permanent vertical concrete safety barrier; TRL Design - DfT Approved						
Intermediate units Type V01 & V02; 3 m long						
straight or curved exceeding 50 m radius	0.16	10.90	3.42	433.98	nr	448.30
curved not exceeding 50 m radius	0.20	13.62	4.27	433.98	nr	451.87
Make up units Type V05 & V06; 1 m long						
straight or curved exceeding 50 m radius	0.30	20.43	6.41	548.33	nr	575.17
Termination units Type V03 & V04; 3 m long	0.50	34.06	10.68	538.21	nr	582.95
Transition to single sided open box beam unit Type V08 & V09; 1.5 m long	0.37	25.20	7.90	639.85	nr	672.95
Transition to rectangular hollow section beam unit Type V10, V11 & V12; 1.5 m long	0.37	25.20	7.90	550.91	nr	584.01
Transition to double sided open box beam; 1.5 m long unit Type V07	0.37	25.20	7.90	615.59	nr	648.69
Anchor plate sets (normally two plates per first and last three units in any run)	0.15	10.22	6.37	42.64	nr	59.23
Temporary concrete safety barrier; TRL Design - DfT Approved						
Intermediate units Type V28; 3 m long						
straight or curved exceeding 50 m radius	0.16	10.90	3.42	387.37	nr	401.69
curved not exceeding 50 m radius	0.20	13.62	4.27	387.37	nr	405.26
Termination units Type V29; 3 m long	0.50	34.06	10.68	434.14	nr	478.88
WIRE ROPE SAFETY FENCES						
Brifen wire rope safety fencing DfT approved; based on 600 m lengths; 4 rope system; posts at 2.40 m general spacings.						
Wire rope	0.03	2.04	0.64	14.62	m	17.30
Long driven line posts	0.05	3.41	1.07	26.98	nr	31.46
Long driven deflection posts	0.05	3.41	1.07	26.98	nr	31.46
Long driven height restraining posts	0.05	3.41	1.07	26.98	nr	31.46
Short line post for setting in concrete or socket	0.06	4.09	1.28	24.97	nr	30.34
Short deflection post for setting in concrete or socket	0.06	4.09	1.28	24.97	nr	30.34
Short height restraining post for setting in concrete or socket	0.06	4.09	1.28	24.97	nr	30.34

SERIES 400: ROAD RESTRAINT SYSTEMS (VEHICLE AND PEDESTRIAN)

Item	Gang hours	Labour £	Plant £	Material £	Unit	Total rate £
Fixed height surface mounted post fixed to structure or foundation	0.09	6.13	1.92	28.35	nr	**36.40**
Standard intermediate anchorage	2.00	136.22	42.72	327.60	nr	**506.54**
Standard end anchorage	2.00	136.22	42.72	118.13	nr	**297.07**
In situ standard concrete foundation for post	0.23	15.67	4.91	8.42	nr	**29.00**
In situ standard socketed foundation for post	0.33	22.48	7.05	8.42	nr	**37.95**
Concrete foundation Type 1 spanning filter drain for post	0.37	25.20	7.90	10.13	nr	**43.23**

VEHICLE PARAPETS

Vehicle Parapets
The heights of the following parapets are in accordance with the Standard Designs and DfT requirements. The rates include for all anchorages and fixings and in the case of steel, galvanising at works and painting four coat paint system on site, together with etching the galvanized surface as necessary

Steel Parapets

Item	Gang hours	Labour £	Plant £	Material £	Unit	Total rate £
Metal parapet Group P1; 1.0m high; comprising steel yielding posts and steel horizontal rails						
straight or curved exceeding 50m radius	–	–	–	163.75	m	**163.75**
curved not exceeding 50m radius	–	–	–	177.39	m	**177.39**
Metal parapet Group P2 (48 Kph); 1.0m high; comprising steel yielding posts and steel horizontal rails with vertical infill bars						
straight or curved exceeding 50m radius	–	–	–	231.98	m	**231.98**
curved not exceeding 50m radius	–	–	–	245.62	m	**245.62**
Metal parapet Group P2 (80 Kph); 1.0m high; comprising steel yielding posts and steel horizontal rails						
straight or curved exceeding 50m radius	–	–	–	191.04	m	**191.04**
curved not exceeding 50m radius	–	–	–	204.69	m	**204.69**
Metal parapet Group P2 (113 Kph); 1.0m high; comprising steel yielding posts and steel horizontal rails						
straight or curved exceeding 50m radius	–	–	–	191.04	m	**191.04**
curved not exceeding 50m radius	–	–	–	204.69	m	**204.69**
Metal parapet Group P4; 1.15m high; comprising steel yielding posts and steel horizontal rails						
straight or curved exceeding 50m radius	–	–	–	163.75	m	**163.75**
curved not exceeding 50m radius	–	–	–	177.39	m	**177.39**
Metal parapet Group P5; 1.25m high; comprising steel yielding posts and steel horizontal rails						
straight or curved exceeding 50m radius	–	–	–	207.80	m	**207.80**
curved not exceeding 50m radius	–	–	–	231.98	m	**231.98**

SERIES 400: ROAD RESTRAINT SYSTEMS (VEHICLE AND PEDESTRIAN)

Item	Gang hours	Labour £	Plant £	Material £	Unit	Total rate £
VEHICLE PARAPETS – cont						
Steel Parapets – cont						
Metal parapet Group P5; 1.50 m high; comprising steel yielding posts and steel horizontal rails						
straight or curved exceeding 50 m radius	–	–	–	231.98	m	**231.98**
curved not exceeding 50 m radius	–	–	–	245.62	m	**245.62**
Metal parapet Group P6; 1.50 m high; comprising steel yielding posts and steel horizontal rails						
straight or curved exceeding 50 m radius	–	–	–	1084.82	m	**1084.82**
curved not exceeding 50 m radius	–	–	–	1173.51	m	**1173.51**
Aluminium Parapets						
Metal parapet Group P1; 1.0 m high; comprising aluminium yielding / frangible posts and aluminium horizontal rails						
straight or curved exceeding 50 m radius	–	–	–	163.75	m	**163.75**
curved not exceeding 50 m radius	–	–	–	177.39	m	**177.39**
Metal parapet Group P2 (80 Kph); 1.0 m high; comprising aluminium yielding / frangible posts and aluminium horizontal rails with mesh infill						
straight or curved exceeding 50 m radius	–	–	–	177.39	m	**177.39**
curved not exceeding 50 m radius	–	–	–	191.00	m	**191.00**
Metal parapet Group P2 (113 Kph); 1.0 m high; comprising aluminium yielding / frangible posts and aluminium horizontal rails with mesh infill						
straight or curved exceeding 50 m radius	–	–	–	177.39	m	**177.39**
curved not exceeding 50 m radius	–	–	–	186.38	m	**186.38**
Metal parapet Group P4; 1.15 m high; comprising aluminium yielding / frangible posts and aluminium horizontal rails						
straight or curved exceeding 50 m radius	–	–	–	170.56	m	**170.56**
curved not exceeding 50 m radius	–	–	–	184.22	m	**184.22**
Metal parapet Group P5; 1.25 m high; comprising aluminium yielding / frangible posts and aluminium horizontal rails with solid sheet infill, anti-access panels						
straight or curved exceeding 50 m radius	–	–	–	204.69	m	**204.69**
curved not exceeding 50 m radius	–	–	–	218.32	m	**218.32**
Metal parapet Group P5; 1.50 m high; comprising aluminium yielding / frangible posts and aluminium horizontal rails with solid sheet infill, anti-access panels						
straight or curved exceeding 50 m radius	–	–	–	231.99	m	**231.99**
curved not exceeding 50 m radius	–	–	–	231.99	m	**231.99**
CRASH CUSHIONS						
Static crash cushion system to BS EN1317-3, Class 110km/hr	–	–	–	–	nr	**38000.00**

SERIES 400: ROAD RESTRAINT SYSTEMS (VEHICLE AND PEDESTRIAN)

Item	Gang hours	Labour £	Plant £	Material £	Unit	Total rate £
REMOVE FROM STORE AND RE-ERECT BEAM SAFETY BARRIERS						
Remove From Store And Re-Erect Beam Safety Barriers						
Prices generally are for beams 'straight or curved exceeding 120m radius', for work to a tighter radius						
50–120m radius Add 15%						
not exceeding 50m radius Add 25%						
Untensioned beams						
single sided corrugated beam	0.07	4.77	1.49	3.82	m	**10.08**
double sided corrugated beam	0.22	14.98	4.70	7.67	m	**27.35**
single sided open box beam	0.07	4.77	1.49	7.78	m	**14.04**
single sided double rail open box beam	0.22	14.98	4.70	15.81	m	**35.49**
double height open box beam	0.24	16.35	5.13	16.03	m	**37.51**
Tensioned beams						
single sided corrugated beam	0.10	6.81	2.13	4.04	m	**12.98**
double sided corrugated beam	0.12	8.17	2.57	8.35	m	**19.09**
Long driven post						
for single sided tensioned corrugated beam	0.06	3.75	1.17	9.07	nr	**13.99**
for double sided tensioned corrugated beam	0.06	3.75	1.17	9.08	nr	**14.00**
for single sided open box beam	0.06	3.75	1.17	9.15	nr	**14.07**
for double sided open box beam	0.06	3.75	1.17	10.64	nr	**15.56**
for double height open box beam	0.06	3.75	1.17	27.07	nr	**31.99**
Short post for setting in concrete or socket						
for single sided tensioned corrugated beam	0.06	4.09	1.28	17.65	nr	**23.02**
for double sided tensioned corrugated beam	0.06	4.09	1.28	18.00	nr	**23.37**
for single sided open box beam	0.06	4.09	1.28	21.44	nr	**26.81**
for double sided open box beam	0.06	4.09	1.28	25.55	nr	**30.92**
Mounting bracket fixed to structure						
for single sided open box beam	0.16	10.90	3.42	6.76	nr	**21.08**
Terminal section						
for untensioned single sided corrugated beam	0.71	48.36	15.17	98.17	nr	**161.70**
for untensioned double sided corrugated beam	1.25	85.14	26.70	104.07	nr	**215.91**
for untensioned single sided open box beam	1.01	68.79	21.58	105.11	nr	**195.48**
for untensioned double sided open box beam	1.78	121.24	38.02	171.52	nr	**330.78**
for tensioned single sided corrugated beam	0.96	65.39	20.50	65.55	nr	**151.44**
for tensioned double sided corrugated beam	1.70	115.79	36.31	102.17	nr	**254.27**
Full height anchorage						
for single sided tensioned corrugated beam	3.95	269.04	84.38	129.65	nr	**483.07**
for double sided tensioned corrugated beam	4.35	296.28	92.92	165.46	nr	**554.66**
for single sided open box beam	3.80	258.82	81.17	130.40	nr	**470.39**
for double sided open box beam	4.20	286.07	89.72	153.73	nr	**529.52**
Expansion joint anchorage						
for single sided open box beam	4.52	307.86	96.55	268.83	nr	**673.24**
for double sided open box beam	5.15	350.77	110.01	325.71	nr	**786.49**
Type 048 connection to bridge parapet						
for single sided open box beam	0.70	47.68	14.95	7.53	nr	**70.16**
Connection piece for single sided open box beam to single sided corrugated beam	0.78	53.13	16.66	176.43	nr	**246.22**

SERIES 400: ROAD RESTRAINT SYSTEMS (VEHICLE AND PEDESTRIAN)

Item	Gang hours	Labour £	Plant £	Material £	Unit	Total rate £
REMOVE FROM STORE AND RE-ERECT BEAM SAFETY BARRIERS – cont						
Standard concrete foundation						
for post for corrugated beam	0.23	15.67	4.83	7.11	nr	**27.61**
for post for open box beam	0.23	15.67	4.91	8.85	nr	**29.43**
Concrete foundation Type 1 spanning filter drain						
for post for corrugated beam	0.25	17.03	5.34	10.58	nr	**32.95**
for post for open box beam	0.25	17.03	5.34	9.36	nr	**31.73**
Standard socketed foundation for post for open box beam	0.30	20.43	6.41	10.70	nr	**37.54**
PEDESTRIAN GUARD RAILS AND HANDRAILS						
New work						
Tubular galvanized mild steel pedestrian guard rails to BS 7818 with mesh infill (105 swg, 50×50 mm mesh); 1.0 m high						
mounted on posts with concrete footing	0.16	10.90	3.42	176.93	m	**191.25**
mounted on posts bolted to structure or ground beam	0.14	9.54	2.99	176.93	m	**189.46**
Solid section galvanized steel pedestrian guard rails with vertical rails (group P4 parapet); 1.0 m high						
mounted on posts with concrete footing	0.19	12.94	4.06	23.28	m	**40.28**
mounted on posts bolted to structure or ground beam	0.17	11.58	3.63	161.07	m	**176.28**
Tubular double ball galvanized steel handrail						
50 mm diameter; 1.20 m high posts	0.15	10.22	3.20	148.57	m	**161.99**
63 mm diameter; 1.20 m high posts	0.15	10.22	3.20	210.69	m	**224.11**
Extra for concrete footings for handrail support posts	0.05	3.41	–	4.51	m	**7.92**
Existing guard rails						
Take from store and re-erect						
pedestrian guard railing, 3.0 m long × 1.0 m high panels	0.15	10.22	3.20	5.17	nr	**18.59**

SERIES 500: DRAINAGE AND SERVICE DUCTS

Item	Gang hours	Labour £	Plant £	Material £	Unit	Total rate £
NOTES						
General						
The re-erection cost for covers and grating complete with frames taken from store assumes that major components are in good condition; the prices below allow a sum of 10% of the value of new materials to cover minor repairs, new fixings and touching up any coatings.						
RESOURCES - LABOUR						
Drains/sewers/culverts gang (small bore)						
1 ganger/chargehand (skill rate 4) - 50% of time		7.29				
1 skilled operative (skill rate 4)		13.65				
2 unskilled operatives (general)		25.50				
1 plant operator (skill rate 3)		16.59				
Total Gang Rate / Hour	£	63.04				
Drains/sewers/culverts gang (large bore)						
1 ganger/chargehand (skill rate 4) - 50% of time		7.29				
1 skilled operative (skill rate 4)		13.65				
2 unskilled operatives (general)		25.50				
1 plant operator (skill rate 3)		16.59				
1 plant operator (skill rate 3) - 30% of time		4.98				
Total Gang Rate / Hour	£	68.02				
Gullies gang						
1 ganger/chargehand (skill rate 4) - 50% of time		7.29				
1 skilled operative (skill rate 4)		13.65				
2 unskilled operatives (general)		25.50				
Total Gang Rate / Hour	£	46.45				
RESOURCES - PLANT						
Drains/sewers/culverts gang (small bore)						
0.4 m³ hydraulic excavator			23.33			
2t dumper - 30% of time			1.95			
360mm compaction plate - 30% of time			0.62			
2.80m3/min compressor, 2 tool - 30% of time			2.55			
disc saw - 30% of time			0.43			
extra 15ft/50m hose - 30% of time			0.10			
small pump - 30% of time			0.71			
Total Gang Rate / Hour		£	29.69			
Note: in addition to the above are the following allowances for trench struts/props/sheeting, assuming the need for close boarded earth support:			–			
average 1.00 m deep			1.35		m	
average 1.50 m deep			1.51		m	
average 2.00 m deep			1.74		m	
average 2.50 m deep			2.07		m	
average 3.00 m deep			2.61		m	
average 3.50 m deep			3.23		m	

SERIES 500: DRAINAGE AND SERVICE DUCTS

Item	Gang hours	Labour £	Plant £	Material £	Unit	Total rate £
RESOURCES - PLANT – cont						
Drains/sewers/culverts gang (small bore) – cont						
Note: excavation in hard materials as above but with the addition of breaker attachments to the excavator as follows:			–			
generally : BRH91 (141kg)			3.75			
reinforced concrete; BRH125 (310kg)			5.31			
Drains/sewers/culverts gang (large bore)] i.e. greater than 700mm bore						
0.4 m³ hydraulic excavator			23.33			
2t dumper - 30% of time			1.95			
2.80 m³/min compressor, 2 tool - 30% of time			2.55			
compaction plate / roller - 30% of time			0.62			
disc saw - 30% of time			0.33			
small pump - 30% of time			0.71			
10 t crawler crane - 30% of time			8.22			
Total Gang Rate / Hour		£	37.71			
Gullies						
2t dumper - 30% of time			1.95			
Stihl disc saw - 30% of time			0.33			
Total Gang Rate / Hour		£	2.28			
RESOURCES - MATERIALS						
For the purposes of bedding widths for pipe bedding materials, trenches have been taken as exceeding 1.50 m in depth; trenches to lesser depths are generally 150 mm narrower than those given here so that the rates need to be reduced proportionately.						
DRAINS AND SERVICE DUCTS (EXCLUDING FILTER, NARROW FILTER AND FIN DRAINS)						
Vitrified clay pipes to Bs 65, plain ends with push-fit polypropylene flexible couplings						
150 mm diameter drain or sewer in trench, depth to invert						
average 1.00 m deep	0.17	10.63	9.37	19.89	m	**39.89**
average 1.50 m deep	0.19	11.89	10.34	19.89	m	**42.12**
average 2.00 m deep	0.22	13.76	12.12	19.89	m	**45.77**
average 2.50 m deep	0.26	16.26	14.79	19.89	m	**50.94**
average 3.00 m deep	0.34	21.27	18.65	19.89	m	**59.81**
average 3.50 m deep	0.42	26.27	23.03	19.89	m	**69.19**
Extra for						
Type N sand bed 650 × 100 mm	0.04	2.50	2.20	1.25	m	**5.95**
Type T sand surround 650 wide × 100 mm	0.08	5.00	4.40	3.00	m	**12.40**
Type F granular bed 650 × 100 mm	0.05	3.13	2.76	1.82	m	**7.71**
Type S granular surround 650 wide × 100 mm	0.16	10.01	8.81	4.34	m	**23.16**
Type A concrete bed 650 × 100 mm	0.11	6.88	5.65	7.76	m	**20.29**
Type B 100 mm concrete bed and haunch	0.24	15.01	12.32	6.63	m	**33.96**
Type Z concrete surround 650 wide × 100 mm	0.22	13.76	11.30	18.60	m	**43.66**

SERIES 500: DRAINAGE AND SERVICE DUCTS

Item	Gang hours	Labour £	Plant £	Material £	Unit	Total rate £
225 mm diameter drain or sewer in trench, depth to invert						
average 1.50 m deep	0.20	12.51	6.94	33.43	m	52.88
average 2.00 m deep	0.24	15.01	13.21	33.43	m	61.65
average 2.50 m deep	0.28	17.52	15.34	33.43	m	66.29
average 3.00 m deep	0.35	21.89	19.17	33.43	m	74.49
average 3.50 m deep	0.44	27.52	24.11	33.43	m	85.06
average 4.00 m deep	0.56	35.03	30.69	33.43	m	99.15
Extra for						
Type N sand bed 750 × 150 mm	0.12	7.51	6.16	2.17	m	15.84
Type T sand surround 750 wide × 150 mm	0.24	15.01	12.32	5.35	m	32.68
Type F granular bed 750 × 150 mm	0.13	8.13	6.68	3.12	m	17.93
Type S granular surround 750 wide × 150 mm	0.24	15.01	12.32	7.74	m	35.07
Type A concrete bed 750 × 150 mm	0.19	11.89	9.75	13.45	m	35.09
Type B 150 mm concrete bed and haunch	0.29	18.14	14.89	10.35	m	43.38
Type Z concrete surround 750 wide × 150 mm	0.36	22.52	18.49	33.33	m	74.34
300 mm diameter drain or sewer in trench, depth to invert						
average 1.50 m deep	0.21	13.14	11.57	44.46	m	69.17
average 2.00 m deep	0.26	16.26	14.32	44.46	m	75.04
average 2.50 m deep	0.30	18.77	16.46	44.46	m	79.69
average 3.00 m deep	0.37	23.15	20.32	44.46	m	87.93
average 4.00 m deep	0.61	38.16	33.45	44.46	m	116.07
average 5.00 m deep	1.01	63.18	55.30	44.46	m	162.94
Extra for						
Type N sand bed 800 × 150 mm	0.13	8.13	6.68	2.31	m	17.12
Type T sand surround 800 × 150 mm	0.26	16.26	13.02	7.44	m	36.72
Type F granular bed 800 × 150 mm	0.14	8.76	7.19	3.32	m	19.27
Type S granular surround 800 wide × 150 mm	0.26	16.26	13.35	10.54	m	40.15
Type A concrete bed 800 × 150 mm	0.20	12.51	10.27	14.35	m	37.13
Type B 150 mm concrete bed and haunch	0.36	22.52	18.49	13.07	m	54.08
Type Z concrete surround 800 wide × 150 mm	0.22	13.76	33.99	42.80	m	90.55
Vitrified clay pipes to Bs 65, spigot and socket joints with sealing ring						
400 mm diameter drain or sewer in trench, depth to invert						
average 2.00 m deep	0.31	19.39	17.05	81.26	m	117.70
average 2.50 m deep	0.35	21.89	19.17	81.26	m	122.32
average 3.00 m deep	0.42	26.27	23.03	81.26	m	130.56
average 4.00 m deep	0.66	41.29	36.19	81.26	m	158.74
average 5.00 m deep	1.07	66.93	58.64	81.26	m	206.83
average 6.00 m deep	1.85	115.73	101.42	81.26	m	298.41
Extra for						
Type N sand bed 900 × 150 mm	0.14	8.76	7.19	2.60	m	18.55
Type T sand surround 900 wide × 150 mm	0.28	17.52	14.38	10.64	m	42.54
Type F granular bed 900 × 150 mm	0.15	9.38	7.70	3.77	m	20.85
Type S granular surround 900 × 150 mm	0.28	17.52	14.38	15.39	m	47.29
Type A concrete bed 900 × 150 mm	0.21	13.14	10.79	16.12	m	40.05
Type B, 150 mm concrete bed and haunch	0.43	26.90	22.08	17.22	m	66.20
Type Z concrete surround 900 wide × 150 mm	0.40	25.02	20.54	66.14	m	111.70

SERIES 500: DRAINAGE AND SERVICE DUCTS

Item	Gang hours	Labour £	Plant £	Material £	Unit	Total rate £
DRAINS AND SERVICE DUCTS (EXCLUDING FILTER, NARROW FILTER AND FIN DRAINS) – cont						
Concrete pipes with rebated flexible joints to Bs 5911-1						
450 mm diameter piped culvert in trench, depth to invert						
average 2.00 m deep	0.33	20.64	18.17	23.66	m	**62.47**
average 3.00 m deep	0.44	27.52	23.92	23.66	m	**75.10**
average 4.00 m deep	0.72	45.04	39.14	23.66	m	**107.84**
average 6.00 m deep	1.65	103.22	90.22	23.66	m	**217.10**
Extra for						
Type A concrete bed 1050 × 150 mm	0.24	15.01	12.32	18.80	m	**46.13**
Type Z concrete surround 1050 wide × 150 mm	0.45	28.15	23.11	84.27	m	**135.53**
750 mm diameter piped culvert in trench, depth to invert						
average 2.00 m deep	0.38	25.65	26.59	70.10	m	**122.34**
average 3.00 m deep	0.52	35.10	36.26	70.10	m	**141.46**
average 4.00 m deep	0.92	62.10	64.18	70.10	m	**196.38**
average 6.00 m deep	2.05	138.37	142.73	70.10	m	**351.20**
Extra for						
Type A concrete bed 1250 × 150 mm	0.26	17.55	13.35	26.04	m	**56.94**
Type Z concrete surround 1250 wide × 150 mm	0.50	33.75	25.68	189.38	m	**248.81**
900 mm diameter piped culvert in trench, depth to invert						
average 2.00 m deep	0.43	29.02	30.06	94.16	m	**153.24**
average 3.00 m deep	0.62	41.85	43.33	94.16	m	**179.34**
average 4.00 m deep	1.06	71.55	73.93	94.16	m	**239.64**
average 6.00 m deep	2.25	151.87	156.63	94.16	m	**402.66**
Extra for						
Type A concrete bed 1500 × 150 mm	0.28	18.90	14.38	30.49	m	**63.77**
Type Z concrete surround 1500 wide × 150 mm	0.55	37.12	28.24	263.33	m	**328.69**
Corrugated steel pipes galvanized, hot dip bitumen coated (Armco type)						
1000 mm diameter piped culvert in trench, Type S granular surround, depth to invert						
average 2.00 m deep	0.99	66.82	69.15	136.29	m	**272.26**
1600 mm diameter piped culvert in trench, Type S granular surround, depth to invert						
average 2.00 m deep	2.13	143.77	148.82	224.44	m	**517.03**
2000 mm diameter piped culvert in trench, Type S granular surround, depth to invert						
average 3.00 m deep	2.97	200.46	207.49	455.70	m	**863.65**
2200 mm diameter piped culvert in trench, Type S granular surround, depth to invert						
average 3.00 m deep	3.29	222.06	229.79	514.58	m	**966.43**

SERIES 500: DRAINAGE AND SERVICE DUCTS

Item	Gang hours	Labour £	Plant £	Material £	Unit	Total rate £
Clay cable ducts; Hepduct						
100 mm diameter service duct in trench, Type S granular surround, depth to invert						
average 1.00 m deep	0.14	8.76	7.51	13.74	m	**30.01**
Two 100 mm diameter service ducts in trench, Type S granular surround, depth to invert						
average 1.00 m deep	0.24	15.01	12.92	27.26	m	**55.19**
Three 100 mm diameter service ducts in trench, Type S granular surround, depth to invert						
average 1.00 m deep	0.32	20.02	17.21	39.48	m	**76.71**
Four 100 mm diameter service ducts in trench, Type S granular surround, depth to invert						
average 1.00 m deep	0.40	25.02	21.55	54.20	m	**100.77**
Six 100 mm diameter service ducts in trench, Type S granular surround, depth to invert						
average 1.00 m deep	0.60	37.53	32.24	75.51	m	**145.28**
Extra for						
Type Z concrete surround on single duct	0.08	5.00	4.11	11.04	m	**20.15**
Type Z concrete surround on additional ways	0.08	5.00	4.11	11.04	m	**20.15**
150 mm diameter conduit, per way	0.01	0.63	0.52	14.81	m	**15.96**
225 mm diameter conduit, per way	0.01	0.63	0.52	41.05	m	**42.20**
FILTER DRAINS						
Vitrified clay perforated pipes to Bs 65, sleeved joints						
150 mm diameter filter drain in trench with Type A bed and Type A fill filter material,						
average 1.00 m deep	0.26	16.26	13.52	22.14	m	**51.92**
average 2.00 m deep	0.30	18.77	15.55	29.65	m	**63.97**
average 3.00 m deep	0.35	21.89	18.26	37.18	m	**77.33**
150 mm pipes with Type A bed and Type B fill, depth						
average 1.00 m deep	0.26	16.26	13.52	31.09	m	**60.87**
average 2.00 m deep	0.30	18.77	15.55	46.99	m	**81.31**
average 3.00 m deep	0.35	21.89	18.26	62.82	m	**102.97**
225 mm pipes with Type A bed and Type A fill, depth						
average 1.00 m deep	0.27	16.89	13.98	38.52	m	**69.39**
average 2.00 m deep	0.31	19.39	15.99	47.74	m	**83.12**
average 3.00 m deep	0.36	22.52	18.76	56.27	m	**97.55**
225 mm pipes with Type A bed and Type B fill, depth						
average 1.00 m deep	0.27	16.89	13.98	47.85	m	**78.72**
average 2.00 m deep	0.31	19.39	15.99	66.86	m	**102.24**
average 3.00 m deep	0.36	22.52	18.76	85.03	m	**126.31**
300 mm pipes with Type A bed and Type A fill, depth						
average 1.00 m deep	0.28	17.52	14.48	51.87	m	**83.87**
average 2.00 m deep	0.32	20.02	16.59	62.46	m	**99.07**
average 3.00 m deep	0.37	23.15	19.22	72.55	m	**114.92**
300 mm pipes with Type A bed and Type B fill, depth						
average 1.00 m deep	0.28	17.52	14.48	59.97	m	**91.97**
average 2.00 m deep	0.32	20.02	16.59	81.30	m	**117.91**
average 3.00 m deep	0.37	23.15	19.22	102.42	m	**144.79**

SERIES 500: DRAINAGE AND SERVICE DUCTS

Item	Gang hours	Labour £	Plant £	Material £	Unit	Total rate £
FILTER DRAINS – cont						
Conc. porous pipe BS 5911-114, sleeved joints						
150 mm pipes with Type A bed and Type A fill, depth						
average 1.00 m deep	0.26	16.26	13.52	16.61	m	**46.39**
average 2.00 m deep	0.30	18.77	15.55	24.12	m	**58.44**
average 3.00 m deep	0.35	21.89	18.26	31.65	m	**71.80**
150 mm pipes with Type A bed and Type B fill, depth						
average 1.00 m deep	0.26	16.26	13.52	25.56	m	**55.34**
average 2.00 m deep	0.30	18.77	15.55	41.47	m	**75.79**
average 3.00 m deep	0.35	21.89	18.26	57.29	m	**97.44**
225 mm pipes with Type A bed and Type A fill, depth						
average 1.00 m deep	0.27	16.89	13.98	19.09	m	**49.96**
average 2.00 m deep	0.31	19.39	15.99	28.30	m	**63.68**
average 3.00 m deep	0.36	22.52	18.76	36.83	m	**78.11**
225 mm pipes with Type A bed and Type B fill, depth						
average 1.00 m deep	0.27	16.89	13.98	28.41	m	**59.28**
average 2.00 m deep	0.31	19.39	15.99	47.42	m	**82.80**
average 3.00 m deep	0.36	22.52	18.76	65.59	m	**106.87**
300 mm pipes with Type A bed and Type A fill, depth						
average 1.00 m deep	0.28	17.52	14.48	22.44	m	**54.44**
average 2.00 m deep	0.32	20.02	16.59	33.03	m	**69.64**
average 3.00 m deep	0.37	23.15	19.22	43.12	m	**85.49**
300 mm pipes with Type A bed and Type B fill, depth						
average 1.00 m deep	0.28	17.52	14.48	30.53	m	**62.53**
average 2.00 m deep	0.32	20.02	16.59	51.87	m	**88.48**
average 3.00 m deep	0.37	23.15	19.22	72.99	m	**115.36**
Filter material contiguous with filter drains, sub-base material and lightweight aggregate infill						
Type A	0.07	4.38	3.31	16.97	m³	**24.66**
Type B	0.07	4.38	3.44	36.21	m³	**44.03**
Excavate and replace filter material contiguous with filter drain						
Type A	0.47	14.55	15.70	16.97	m³	**47.22**
Type B	0.47	14.55	15.70	36.21	m³	**66.46**
FIN DRAINS AND NARROW FILTER DRAINS						
Fin Drain DfT type 6 using 'Trammel' drainage fabrics and perforated clay drain; surrounding pipe with sand and granular fill						
100 mm clay perforated pipes, depth						
average 1.00 m deep	0.17	10.63	9.17	13.94	m	**33.74**
average 2.00 m deep	0.23	14.39	12.37	20.20	m	**46.96**

SERIES 500: DRAINAGE AND SERVICE DUCTS

Item	Gang hours	Labour £	Plant £	Material £	Unit	Total rate £
Fin Drain DfT type 7 using 'Trammel' drainage fabrics and slotted UPVC drain; surrounding pipe with sand and granular fill, backfilling with selected suitable material						
100 mm UPVC slotted pipes, depth						
average 1.00 m deep	0.17	10.63	9.17	23.62	m	**43.42**
average 2.00 m deep	0.23	14.39	12.37	29.87	m	**56.63**
Narrow Filter Drain DfT type 8 using 'Trammel' drainage fabrics and perforated UPVC drain; surrounding pipe with sand and granular fill, backfilling with granular material						
110 mm UPVC perforated pipes, depth						
average 1.00 m deep	0.17	10.63	9.17	22.41	m	**42.21**
average 2.00 m deep	0.22	13.76	11.81	24.32	m	**49.89**
Narrow Filter Drain DfT type 9 using 'Trammel' drainage fabrics and perforated UPVC drain; surrounding pipe with sand and granular fill, backfilling with granular material						
110 mm UPVC perforated pipes, depth						
average 1.00 m deep	0.17	10.63	9.17	24.65	m	**44.45**
average 2.00 m deep	0.23	14.39	12.37	28.49	m	**55.25**
CONNECTIONS						
Note: excavation presumed covered by new trench						
Connection of pipe to existing drain, sewer or piped culvert						
150 mm	0.42	26.27	15.82	26.78	nr	**68.87**
225 mm	0.60	37.53	26.82	99.82	nr	**164.17**
300 mm	1.15	71.94	53.68	173.71	nr	**299.33**
Connection of pipes to existing chambers						
150 mm to one brick	1.20	75.07	64.60	8.51	nr	**148.18**
150 mm to precast	0.60	37.53	32.29	8.51	nr	**78.33**
300 mm to one and a half brick	2.40	150.13	129.14	13.87	nr	**293.14**

SERIES 500: DRAINAGE AND SERVICE DUCTS

Item	Gang hours	Labour £	Plant £	Material £	Unit	Total rate £
CHAMBERS AND GULLIES						
Notes						
The rates assume the most efficient items of plant (excavator) and are optimum rates, assuming continuous output with no delays caused by other operations or works. Ground conditions are assumed to be good soil with no abnormal conditions that would affect outputs and consistency of work. Multiplier Table for Labour and Plant for various site conditions and for working:						
out of sequence with other works × 2.75 minimum						
in hard clay × 1.75–2.00						
in running sand × 2.75 minimum						
in broken rock × 2.75–3.50						
below water table × 2.00 minimum						
Brick construction						
Design criteria used in models:						
* class A engineering bricks						
* 215 thick walls generally; 328 thick to chambers exceeding 2.5 m deep						
* 225 plain concrete C20/20 base slab						
* 300 reinforced concrete C20/20 reducing slab						
* 125 reinforced concrete C20/20 top slab						
* maximum height of working chamber 2.0 m above benching						
* 750 × 750 access shaft						
* plain concrete C15/20 benching, 150 clay main channel longitudinally and two 100 branch channels						
* step irons at 300 mm centres, doubled if depth to invert exceeds 3000 mm						
* heavy duty manhole cover and frame						
750 × 700 chamber 500 depth to invert						
excavation, support, backfilling and disposal						40.34
concrete base						129.48
brickwork chamber						236.05
concrete cover slab						144.39
concrete benching, main and branch channels						22.49
step irons						15.93
access cover and frame						382.50
TOTAL					£	**971.18**
750 × 700 chamber 1000 depth to invert						
excavation, support, backfilling and disposal						80.68
concrete base						129.48
brickwork chamber						472.11
concrete cover slab						144.39
concrete benching; main and branch channels						22.49
step irons						31.86
access cover and frame						382.50
TOTAL					£	**1263.51**

SERIES 500: DRAINAGE AND SERVICE DUCTS

Item	Gang hours	Labour £	Plant £	Material £	Unit	Total rate £
750 × 700 chamber 1500 depth to invert						
excavation, support, backfilling and disposal						121.01
concrete base						129.48
brickwork chamber						708.16
concrete cover slab						144.39
concrete benching; main and branch channels						22.49
step irons						47.79
access cover and frame						382.50
TOTAL					£	**1555.82**
900 × 700 chamber 500 depth to invert						
excavation, support, backfilling and disposal						44.24
concrete base						143.94
brickwork chamber						253.40
concrete cover slab						157.91
concrete benching; main and branch channels						26.99
step irons						15.93
access cover and frame						382.50
TOTAL					£	**1024.90**
900 × 700 chamber 1000 depth to invert						
excavation, support, backfilling and disposal						88.47
concrete base						143.94
brickwork chamber						506.80
concrete cover slab						157.91
concrete benching; main and branch channels						26.99
step irons						31.86
access cover and frame						382.50
TOTAL					£	**1338.47**
900 × 700 chamber 1500 depth to invert						
excavation, support, backfilling and disposal						132.71
concrete base						143.94
brickwork chamber						760.20
concrete cover slab						157.91
concrete benching; main and branch channels						26.99
step irons						47.79
access cover and frame						382.50
TOTAL					£	**1652.04**
1050 × 700 chamber 1500 depth to invert						
excavation, support, backfilling and disposal						144.41
concrete base						158.41
brickwork chamber						812.25
concrete cover slab						171.42
concrete benching; main and branch channels						31.49
step irons						47.79
access cover and frame						382.50
TOTAL					£	**1748.26**

SERIES 500: DRAINAGE AND SERVICE DUCTS

Item	Gang hours	Labour £	Plant £	Material £	Unit	Total rate £
CHAMBERS AND GULLIES – cont						
Brick construction – cont						
1050 × 700 chamber 2500 depth to invert						
excavation, support, backfilling and disposal						240.68
concrete base						158.41
brickwork chamber						1353.73
concrete cover slab						171.42
concrete benching; main and branch channels						31.49
step irons						79.65
access cover and frame						382.50
TOTAL					£	**2417.88**
1050 × 700 chamber 3500 depth to invert						
excavation, support, backfilling and disposal						336.95
concrete base						158.41
brickwork chamber						1895.23
concrete reducing slab						171.42
brickwork access shaft						1935.70
concrete cover slab						139.89
concrete benching; main and branch channels						31.49
step irons						111.51
access cover and frame						382.50
TOTAL					£	**5163.10**
1350 × 700 chamber 2500 depth to invert						
excavation, support, backfilling and disposal						279.66
concrete base						187.32
brickwork chamber						1527.21
concrete cover slab						198.45
concrete benching; main and branch channels						40.48
step irons						79.65
access cover and frame						382.50
TOTAL					£	**2695.28**
1350 × 700 chamber 3500 depth to invert						
excavation, support, backfilling and disposal						391.53
concrete base						187.32
brickwork chamber						2138.09
concrete reducing slab						198.45
brickwork access shaft						1935.70
concrete cover slab						139.89
concrete benching; main and branch channels						40.48
step irons						111.51
access cover and frame						382.50
TOTAL					£	**5525.49**

SERIES 500: DRAINAGE AND SERVICE DUCTS

Item	Gang hours	Labour £	Plant £	Material £	Unit	Total rate £
1350 × 700 chamber 4500 depth to invert						
excavation, support, backfilling and disposal						503.40
concrete base						187.32
brickwork chamber						2748.98
concrete reducing slab						198.45
brickwork access shaft						2488.77
concrete cover slab						139.89
concrete benching; main and branch channels						40.48
step irons						143.37
access cover and frame						382.50
TOTAL					£	**6833.17**
Precast concrete construction						
Design criteria used in models:						
* circular shafts						
* 150 plain concrete surround						
* 225 plain concrete C20/20 base slab						
* precast reducing slab						
* precast top slab						
* maximum height of working chamber 2.0 m above benching						
* 750 diameter access shaft						
* plain concrete C15/20 benching, 150 clay main channel longitudinally and two 100 branch channels						
* step irons at 300 mm centres, doubled if depth to invert exceeds 3000 mm						
* heavy duty manhole cover and frame						
* in manholes over 6 m deep, landings at maximum intervals						
675 diameter x 500 depth to invert						
excavation, support, backfilling and disposal						28.34
concrete base						44.06
main chamber rings						21.57
cover slab						72.42
concrete surround						45.29
concrete benching, main and branch channels						57.53
step irons						15.93
access cover and frame						382.50
TOTAL					£	**667.64**
675 diameter x 750 depth to invert						
excavation, support, backfilling and disposal						37.78
concrete base						44.06
main chamber rings						49.14
cover slab						72.42
concrete surround						66.10
concrete benching, main and branch channels						57.53
step irons						23.89
access cover and frame						382.50
TOTAL					£	**733.43**

SERIES 500: DRAINAGE AND SERVICE DUCTS

Item	Gang hours	Labour £	Plant £	Material £	Unit	Total rate £
CHAMBERS AND GULLIES – cont						
Precast concrete construction – cont						
675 diameter × 1000 depth to invert						
excavation, support, backfilling and disposal						45.75
concrete base						44.06
main chamber rings						77.91
cover slab						72.42
concrete surround						88.13
concrete benching, main and branch channels						57.53
step irons						31.86
access cover and frame						382.50
TOTAL					£	**800.15**
675 diameter × 1250 depth to invert						
excavation, support, backfilling and disposal						53.96
concrete base						44.06
main chamber rings						106.67
cover slab						72.42
concrete surround						108.94
concrete benching, main and branch channels						57.53
step irons						39.83
access cover and frame						382.50
TOTAL					£	**865.90**
900 diameter × 750 depth to invert						
excavation, support, backfilling and disposal						104.58
concrete base						123.80
main chamber rings						58.32
cover slab						83.53
concrete surround						70.03
concrete benching, main and branch channels						22.72
step irons						23.89
access cover and frame						382.50
TOTAL					£	**869.38**
900 diameter × 1000 depth to invert						
excavation, support, backfilling and disposal						139.43
concrete base						123.80
main chamber rings						77.78
cover slab						83.53
concrete surround						93.38
concrete benching, main and branch channels						22.72
step irons						31.86
access cover and frame						382.50
TOTAL					£	**954.99**

SERIES 500: DRAINAGE AND SERVICE DUCTS

Item	Gang hours	Labour £	Plant £	Material £	Unit	Total rate £
900 diameter × 1500 depth to invert						
excavation, support, backfilling and disposal						209.15
concrete base						123.80
main chamber rings						116.66
cover slab						83.53
concrete surround						140.07
concrete benching, main and branch channels						22.72
step irons						47.79
access cover and frame						382.50
TOTAL					£	**1126.21**
1200 diameter × 1500 depth to invert						
excavation, support, backfilling and disposal						259.02
concrete base						174.76
main chamber rings						122.78
cover slab						109.00
concrete surround						180.74
concrete benching, main and branch channels						40.38
step irons						47.79
access cover and frame						382.50
TOTAL					£	**1316.97**
1200 diameter × 2000 depth to invert						
excavation, support, backfilling and disposal						345.35
concrete base						174.76
main chamber rings						163.70
cover slab						109.00
concrete surround						241.00
concrete benching, main and branch channels						40.38
step irons						63.72
access cover and frame						382.50
TOTAL					£	**1520.41**
1200 diameter × 2500 depth to invert						
excavation, support, backfilling and disposal						431.69
concrete base						174.76
main chamber rings						204.63
cover slab						109.00
concrete surround						301.25
concrete benching, main and branch channels						40.38
step irons						79.65
access cover and frame						382.50
TOTAL					£	**1723.86**
1200 diameter × 3000 depth to invert						
excavation, support, backfilling and disposal						518.03
concrete base						174.76
main chamber rings						245.55
cover slab						109.00
concrete surround						361.50
concrete benching, main and branch channels						40.38
step irons						95.58
access cover and frame						382.50
TOTAL					£	**1927.30**

SERIES 500: DRAINAGE AND SERVICE DUCTS

Item	Gang hours	Labour £	Plant £	Material £	Unit	Total rate £
CHAMBERS AND GULLIES – cont						
Precast concrete construction – cont						
1800 diameter × 1500 depth to invert						
excavation, support, backfilling and disposal						376.27
concrete base						301.84
main chamber rings						369.50
cover slab						330.18
concrete surround						262.11
concrete benching, main and branch channels						90.86
step irons						47.79
access cover and frame						382.50
TOTAL					£	**2161.04**
1800 diameter × 2000 depth to invert						
excavation, support, backfilling and disposal						501.69
concrete base						301.84
main chamber rings						492.66
cover slab						330.18
concrete surround						349.47
concrete benching, main and branch channels						90.86
step irons						63.72
access cover and frame						382.50
TOTAL					£	**2512.92**
1800 diameter × 2500 depth to invert						
excavation, support, backfilling and disposal						627.11
concrete base						301.84
main chamber rings						615.83
cover slab						330.18
concrete surround						436.85
concrete benching, main and branch channels						90.86
step irons						79.65
access cover and frame						382.50
TOTAL					£	**2864.81**
1800 diameter × 3000 depth to invert						
excavation, support, backfilling and disposal						752.53
concrete base						301.84
main chamber rings						738.48
cover slab						330.18
concrete surround						524.21
concrete benching, main and branch channels						90.86
step irons						95.58
access cover and frame						382.50
TOTAL					£	**3216.17**

SERIES 500: DRAINAGE AND SERVICE DUCTS

Item	Gang hours	Labour £	Plant £	Material £	Unit	Total rate £
1800 diameter × 3500 depth to invert						
excavation, support, backfilling and disposal						877.95
concrete base						301.84
main chamber rings						862.15
reducing slab						348.08
access shaft						272.20
cover slab						83.53
concrete surround						611.58
concrete benching, main and branch channels						90.86
step irons						111.51
access cover and frame						382.50
TOTAL					£	**3942.21**
1800 diameter × 4000 depth to invert						
excavation, support, backfilling and disposal						1003.37
concrete base						301.84
main chamber rings						985.32
reducing slab						348.08
access shaft						311.08
cover slab						83.53
concrete surround						698.94
concrete benching, main and branch channels						90.86
step irons						127.44
access cover and frame						382.50
TOTAL					£	**4332.97**
2400 diameter × 1500 depth to invert						
excavation, support, backfilling and disposal						516.87
concrete base						462.50
main chamber rings						939.57
cover slab						955.53
concrete surround						343.46
concrete benching, main and branch channels						161.53
step irons						47.79
access cover and frame						382.50
TOTAL					£	**3809.76**
2400 diameter × 3000 depth to invert						
excavation, support, backfilling and disposal						1033.75
concrete base						462.50
main chamber rings						1879.15
cover slab						955.53
concrete surround						686.93
concrete benching, main and branch channels						161.53
step irons						95.58
access cover and frame						382.50
TOTAL					£	**5657.46**

SERIES 500: DRAINAGE AND SERVICE DUCTS

Item	Gang hours	Labour £	Plant £	Material £	Unit	Total rate £
CHAMBERS AND GULLIES – cont						
Precast concrete construction – cont						
2400 diameter × 4500 depth to invert						
excavation, support, backfilling and disposal						1550.63
concrete base						462.50
main chamber rings						2818.72
reducing slab						978.08
access shaft						349.96
cover slab						83.53
concrete surround						1030.38
concrete benching, main and branch channels						161.53
step irons						143.37
access cover and frame						382.50
TOTAL					£	**7961.21**
2700 diameter × 1500 depth to invert						
excavation, support, backfilling and disposal						595.95
concrete base						555.42
main chamber rings						1087.37
cover slab						1466.29
concrete surround						384.14
concrete benching, main and branch channels						204.43
step irons						47.79
access cover and frame						382.50
TOTAL					£	**4723.89**
2700 diameter × 3000 depth to invert						
excavation, support, backfilling and disposal						1191.89
concrete base						555.42
main chamber rings						2174.74
cover slab						1466.29
concrete surround						768.28
concrete benching, main and branch channels						204.43
step irons						95.58
access cover and frame						382.50
TOTAL					£	**6839.14**
2700 diameter × 4500 depth to invert						
excavation, support, backfilling and disposal						1787.75
concrete base						555.42
main chamber rings						3262.11
reducing slab						1595.18
access shaft						349.96
cover slab						83.53
concrete surround						1152.42
concrete benching, main and branch channels						204.43
step irons						143.37
access cover and frame						382.50
TOTAL					£	**9516.67**

SERIES 500: DRAINAGE AND SERVICE DUCTS

Item	Gang hours	Labour £	Plant £	Material £	Unit	Total rate £
3000 diameter × 3000 depth to invert						
excavation, support, backfilling and disposal						1361.70
concrete base						656.73
main chamber rings						3061.53
cover slab						1869.32
concrete surround						849.64
concrete benching, main and branch channels						252.38
step irons						95.58
access cover and frame						382.50
TOTAL						**8529.38**
3000 diameter × 4500 depth to invert						
excavation, support, backfilling and disposal						2042.56
concrete base						656.73
main chamber rings						4592.30
reducing slab						2035.65
access shaft						349.96
cover slab						83.53
concrete surround						1274.46
concrete benching, main and branch channels						252.38
step irons						143.37
access cover and frame						382.50
TOTAL						**11813.43**
3000 diameter × 6000 depth to invert						
excavation, support, backfilling and disposal						2723.41
concrete base						656.73
main chamber rings						6123.06
reducing slab						2035.65
access shaft						466.62
cover slab						83.53
concrete surround						1699.28
concrete benching, main and branch channels						252.38
step irons						191.16
access cover and frame						382.50
TOTAL						**14614.31**
Clayware vertical pipe complete with rest bend at base and tumbling bay junction to main drain complete with stopper; concrete grade C20 surround, 150 mm thick; additional excavation and disposal						
100 pipe						
1.15 m to invert						nr	103.89
2.15 m to invert						nr	132.67
3.15 m to invert						nr	160.20
4.15 m to invert						nr	190.24
150 pipe						
1.15 m to invert						nr	160.20
2.15 m to invert						nr	192.75
3.15 m to invert						nr	230.30
4.15 m to invert						nr	267.84

SERIES 500: DRAINAGE AND SERVICE DUCTS

Item	Gang hours	Labour £	Plant £	Material £	Unit	Total rate £
CHAMBERS AND GULLIES – cont						
Clayware vertical pipe complete with rest bend at base and tumbling bay junction to main drain complete with stopper – cont						
225 pipe						
1.15m to invert					nr	319.16
2.15m to invert					nr	374.23
3.15m to invert					nr	431.80
4.15m to invert					nr	491.87
Vitrified clay; set in concrete grade C20, 150mm thick; additional excavation and disposal						
Road gulley						
450mm diameter × 900mm deep, 100mm or 150mm outlet; cast iron road gulley grating and frame group 4, 434×434mm, on Class B engineering brick seating	0.50	23.04	1.14	393.38	nr	417.56
Yard gulley (mud); trapped with rodding eye; galvanized bucket; stopper						
225mm diameter, 100mm diameter outlet, cast iron hinged grate and frame	0.30	13.82	0.72	292.48	nr	307.02
Grease interceptors; internal access and bucket						
600×450mm, metal tray and lid, square hopper with horizontal inlet	0.35	16.13	0.80	1089.82	nr	1106.75
Precast concrete; set in concrete grade C20P, 150mm thick; additional excavation and disposal						
Road gulley; trapped with rodding eye; galvanized bucket; stopper						
450mm diameter × 900mm deep, cast iron road gulley grating and frame group 4, 434×434mm, on Class B engineering brick seating	0.54	24.88	1.23	249.42	nr	275.53
SOFT SPOTS AND OTHER VOIDS						
Excavation of soft spots and other voids in bottom of trenches, chambers and gullies	0.07	4.38	3.79	–	m³	**8.17**
Filling of soft spots and other voids in bottom of trenches, chambers and gullies with imported selected sand	0.09	5.63	4.52	20.38	m³	**30.53**
Filling of soft spots and other voids in bottom of trenches, chambers and gullies with imported natural gravel	0.09	5.63	4.52	30.30	m³	**40.45**
Filling of soft spots and other voids in bottom of trenches, chambers and gullies with concrete Grade C15, 40mm aggregate	0.09	5.63	4.47	103.24	m³	**113.34**
Filling of soft spots and other voids in bottom of trenches, chambers and gullies with concrete Grade C20, 20mm aggregate	0.09	5.63	4.47	91.43	m³	**101.53**

SERIES 500: DRAINAGE AND SERVICE DUCTS

Item	Gang hours	Labour £	Plant £	Material £	Unit	Total rate £
SUPPORTS LEFT IN EXCAVATION						
Timber close boarded supports left in						
trench	–	–	8.67	–	m²	8.67
pits	–	–	8.67	–	m²	8.67
Steel trench sheeting supports left in						
trench	–	–	36.38	–	m²	36.38
pits	–	–	36.38	–	m²	36.38
RENEWAL, RAISING OR LOWERING OF COVERS AND GRATINGS ON EXISTING CHAMBERS						
Raising the level of 100×100 mm water stop tap boxes on 100×100 mm brick chambers						
by 150 mm or less	0.06	2.76	0.44	13.68	nr	16.88
Lowering the level of 100×100 mm water stop tap boxes on 100×100 mm brick chambers						
by 150 mm or less	0.04	1.84	0.29	27.99	nr	30.12
Raising the level of 420×420 mm cover and frame on 420×420 mm in-situ concrete chamber						
by 150 mm or less	0.10	4.61	0.73	35.35	nr	40.69
Lowering the level of 420×420 mm British Telecom cover and frame on 420×420 mm in-situ concrete chamber						
by 150 mm or less	0.08	3.46	2.14	16.96	nr	22.56
Raising the level of 700×500 mm cover and frame on 700×500 mm in-situ concrete chamber						
by 150 mm or less	0.17	7.83	1.22	13.68	nr	22.73
Raising the level of 600×600 mm grade A heavy duty manhole cover and frame on 600×600 mm brick chamber						
by 150 mm or less	0.17	7.83	1.22	27.99	nr	37.04
by 150–300 mm	0.21	9.68	1.51	35.35	nr	46.54
Lowering the level of 600×600 mm grade A heavy duty manhole cover and frame on 600×600 mm brick chamber						
by 150 mm or less	0.10	4.61	0.73	16.96	nr	22.30
REMOVE FROM STORE AND REINSTALL COVERS						
Remove from store and reinstall cover and frame						
600×600 mm; Group 5; super heavy duty E600 cast iron	0.25	11.52	0.65	367.55	nr	379.72
600×600 mm; Group 4; heavy duty triangular D400 cast iron	0.25	11.52	0.65	224.91	nr	237.08
600×600 × 75 mm; Group 2; light duty single seal B125 cast iron	0.25	11.52	0.65	120.12	nr	132.29
600×600 × 100mm; Group 2; medium duty single seal B125 cast iron	0.25	11.52	0.65	131.30	nr	143.47

SERIES 500: DRAINAGE AND SERVICE DUCTS

Item	Gang hours	Labour £	Plant £	Material £	Unit	Total rate £
GROUTING UP OF EXISTING DRAINS AND SERVICE DUCTS						
Concrete Grade C15						
Sealing redundant road gullies	0.02	0.92	0.13	14.11	m³	**15.16**
Filling redundant chambers with						
ne 1.0 m deep to invert	0.09	5.63	4.20	56.61	nr	**66.44**
1.0–2.0 m deep to invert	0.21	13.14	9.80	91.49	nr	**114.43**
2.0–3.0 m deep to invert	0.55	34.41	25.64	152.45	nr	**212.50**
Grouting up of existing drains and service ducts						
100 mm diameter	0.03	1.88	1.41	3.38	m	**6.67**
300 mm diameter	0.13	8.13	6.16	14.30	m	**28.59**
450 mm diameter	0.26	16.26	12.32	29.70	m	**58.28**
600 mm diameter	0.50	31.28	23.68	60.31	m	**115.27**
1200 mm diameter	1.70	106.34	80.48	168.60	m	**355.42**
EXCAVATION IN HARD MATERIAL						
Extra over excavation for excavation in Hard Material in drainage:						
existing pavement, brickwork, concrete, masonry and the like	0.15	9.38	8.61	–	m³	**17.99**
rock	0.35	21.89	20.09	–	m³	**41.98**
reinforced concrete	0.60	37.53	35.38	–	m³	**72.91**
Reinstatement of pavement construction; extra over excavation for breaking up and subsequently reinstating 150 mm flexible surfacing and 280 mm sub-base						
100 mm diameter sewer, drain or service duct	0.09	5.63	5.03	14.51	m	**25.17**
150 mm diameter sewer, drain or service duct	0.10	6.26	5.60	15.47	m	**27.33**
225 mm diameter sewer, drain or service duct	0.10	6.26	5.60	16.92	m	**28.78**
300 mm diameter sewer, drain or service duct	0.10	6.26	5.60	17.41	m	**29.27**
375 mm diameter sewer, drain or service duct	0.10	6.26	5.60	17.89	m	**29.75**
450 mm diameter sewer, drain or service duct	0.12	7.51	6.13	18.85	m	**32.49**
2 way 100mm diameter service ducts	0.10	6.26	5.60	14.99	m	**26.85**
3 way 100mm diameter service ducts	0.10	6.26	5.60	15.47	m	**27.33**
4 way 100mm diameter service ducts	0.10	6.26	5.60	16.92	m	**28.78**

SERIES 500: DRAINAGE AND SERVICE DUCTS

Item	Gang hours	Labour £	Plant £	Material £	Unit	Total rate £
MANHOLES						
Materials						
In order to arrive at an all in rate for enumerated manholes the various constituent elements have been broken down and unit rates given against each item.						
Earthworks						
Excavation by machine						
ne 2.0 m deep	0.22	13.76	5.86	–	m³	19.62
2.0–3.0 m deep	0.24	15.01	6.39	–	m³	21.40
3.0–4.0 m deep	0.26	16.26	6.93	–	m³	23.19
Backfiling						
ne 2.0 m deep	0.11	6.88	2.92	–	m³	9.80
2.0–3.0 m deep	0.11	6.88	2.92	–	m³	9.80
3.0–4.0 m deep	0.11	6.88	2.92	–	m³	9.80
Earthwork support						
ne 2.0 m deep	0.01	0.44	1.17	–	m³	1.61
2.0–3.0 m deep	0.01	0.69	2.05	–	m³	2.74
3.0–4.0 m deep	0.01	0.88	3.43	–	m³	4.31
Disposal of excavated material						
off site	0.55	22.38	7.65	–	m³	30.03
to spoil heap average 50 m distant	0.19	2.84	1.23	–	m³	4.07
Mass concrete grade C15						
Blinding; thickness						
ne 150 mm deep	1.05	13.77	4.35	91.01	m³	109.13
Surrounds to manholes; thickness						
ne 150 mm deep	2.30	30.16	9.50	91.01	m³	130.67
150 - 500 mm deep	1.70	22.29	7.02	91.01	m³	120.32
Benching to bottom of manhole						
650 mm diameter	1.00	13.11	3.74	7.56	nr	24.41
900 mm diameter	2.00	26.22	7.59	14.48	nr	48.29
1050 mm diameter	3.60	47.20	13.46	19.70	nr	80.36
1200 mm diameter	4.75	62.28	17.77	25.74	nr	105.79
1500 mm diameter	5.75	75.39	21.51	40.23	nr	137.13
rectangular	3.50	45.89	13.09	22.75	m²	81.73
Reinforced concrete grade C20						
Bases; thickness						
150–300 mm deep	2.30	30.16	8.60	91.43	m³	130.19
300–500 mm deep	2.50	32.78	9.35	91.43	m³	133.56
Suspended slabs; thickness						
ne 150 mm deep	3.45	45.24	12.91	91.43	m³	149.58
150–300 mm deep	2.70	35.40	10.10	91.43	m³	136.93
Formwork; GRP forms						
Plain curved, vertical	–	–	–	–	m²	-
to base and benching	0.60	7.87	2.24	9.35	m²	19.46
to surrounds	0.40	5.24	1.50	6.87	hr	13.61

SERIES 500: DRAINAGE AND SERVICE DUCTS

Item	Gang hours	Labour £	Plant £	Material £	Unit	Total rate £
MANHOLES – cont						
Bar reinforcement						
See Series 1700						
Precast concrete circular manhole rings; BS 5911 Part 1						
Shaft rings						
675 mm; plain/reinforced	3.20	45.32	23.42	57.69	m	**126.43**
900 mm; plain/reinforced	3.80	53.82	27.80	76.86	m	**158.48**
1050 mm; plain/reinforced	4.20	59.48	40.36	83.34	m	**183.18**
1200 mm; plain/reinforced	4.68	66.28	34.25	101.77	m	**202.30**
1500 mm; reinforced	5.28	74.78	38.60	143.70	m	**257.08**
1800 mm; reinforced	6.00	84.98	43.92	225.09	m	**353.99**
2100 mm; reinforced	6.60	101.07	73.11	427.79	m	**601.97**
2400 mm; reinforced	7.20	110.26	81.68	540.96	m	**732.90**
2700 mm; reinforced	8.00	122.51	90.76	626.80	m	**840.07**
3000 mm; reinforced	8.60	131.70	97.56	854.53	m	**1083.79**
Reducing slabs + landing slabs; 900 mm diameter access						
1050 mm; plain/reinforced	2.52	35.69	18.43	101.29	nr	**155.41**
1200 mm; plain/reinforced	2.80	39.66	20.50	121.97	nr	**182.13**
1500 mm; reinforced	3.16	44.75	23.12	201.23	nr	**269.10**
1800 mm; reinforced	3.60	50.99	26.36	282.36	nr	**359.71**
2100 mm; reinforced	4.00	61.25	45.38	591.99	nr	**698.62**
2400 mm; reinforced	4.40	67.38	49.93	770.65	nr	**887.96**
2700 mm; reinforced	4.80	73.50	54.46	1234.35	nr	**1362.31**
3000 mm; reinforced	5.20	79.63	59.01	1564.82	nr	**1703.46**
Cover slabs						
675 mm plain/reinforced	1.98	27.85	13.48	54.12	nr	**95.45**
900 mm; plain/reinforced	2.16	30.59	15.82	73.28	nr	**119.69**
1050 mm; plain/reinforced	2.52	35.69	18.43	79.62	nr	**133.74**
1200 mm; plain/reinforced	2.80	39.66	20.50	96.78	nr	**156.94**
1500 mm; reinforced	3.16	44.75	23.12	159.43	nr	**227.30**
1800 mm; reinforced	3.60	50.99	26.36	243.06	nr	**320.41**
2100 mm; reinforced	4.00	61.25	45.38	502.90	nr	**609.53**
2400 mm; reinforced	4.40	67.38	49.93	675.24	nr	**792.55**
2700 mm; reinforced	4.80	73.50	54.46	1116.42	nr	**1244.38**
3000 mm; reinforced	5.20	79.63	59.01	1304.90	nr	**1443.54**
Engineering bricks, Class B; PC £330.00/1000; in cement mortar (1:3)						
Walls, built vertical and straight						
one brick thick; depth ne 1.0 m	2.63	47.59	4.45	11.40	m²	**63.44**
one brick thick; depth 1.0–2.0 m	2.80	50.76	4.75	11.40	m²	**66.91**
one brick thick; depth 2.0–3.0 m	3.04	55.20	5.16	11.40	m²	**71.76**
one brick thick; depth 3.0–4.0 m	3.29	59.64	5.58	11.40	m²	**76.62**
one brick thick; depth 4.0–5.0 m	3.68	66.62	6.23	11.40	m²	**84.25**
one and a half bricks thick; depth ne 1.0 m	3.60	65.35	6.11	15.55	m²	**87.01**
one and a half bricks thick; depth 1.0–2.0 m	4.02	72.96	6.83	15.55	m²	**95.34**

SERIES 500: DRAINAGE AND SERVICE DUCTS

Item	Gang hours	Labour £	Plant £	Material £	Unit	Total rate £
one and a half bricks thick; depth 2.0–3.0 m	4.41	79.94	7.48	15.55	m²	102.97
one and a half bricks thick; depth 3.0–4.0 m	4.79	86.92	8.13	15.55	m²	110.60
one and a half bricks thick; depth 4.0–5.0 m	5.18	93.90	8.78	15.55	m²	118.23
Step irons; BS 1247; malleable; galvanized						
Built into joints; at depths						
ne 1.0 m	0.11	1.90	0.18	15.02	nr	17.10
1.0–2.0 m	0.11	1.90	0.18	15.02	nr	17.10
2.0–3.0 m	0.14	2.54	0.24	15.02	nr	17.80
3.0–4.0 m	0.18	3.17	0.30	15.02	nr	18.49
4.0–5.0 m	0.21	3.81	0.63	15.02	nr	19.46
Cast into concrete; at depths						
ne 1.0 m	0.09	1.20	0.18	15.02	nr	16.40
1.0–2.0 m	0.09	1.20	0.18	15.02	nr	16.40
2.0–3.0 m	0.12	1.60	0.24	15.02	nr	16.86
3.0–4.0 m	0.15	2.00	0.30	15.02	nr	17.32
4.0–5.0 m	0.18	2.40	0.63	15.02	nr	18.05
Vitrified clayware channels; bedding in cement mortar in bottom of manhole						
Half section straight						
100 mm diameter	0.54	7.21	0.90	7.66	nr	15.77
150 mm diameter	0.69	9.21	1.15	13.62	nr	23.98
225 mm diameter	0.96	12.82	1.60	31.73	nr	46.15
300 mm diameter	1.29	17.22	2.16	64.28	nr	83.66
Channel bend						
100 mm diameter	0.42	5.61	0.70	7.22	nr	13.53
150 mm diameter	0.57	7.61	0.95	11.88	nr	20.44
225 mm diameter	0.84	11.22	1.40	39.59	nr	52.21
300 mm diameter	1.20	16.02	2.00	80.69	nr	98.71
Channel branch						
100 mm diameter	0.48	6.45	0.80	14.51	nr	21.76
150 mm diameter	0.78	10.41	1.30	23.75	nr	35.46
225 mm diameter	2.01	26.84	3.36	67.90	nr	98.10
300 mm diameter	3.21	42.86	5.36	161.35	nr	209.57
Half section straight taper						
150 mm diameter	0.66	8.81	1.10	40.66	nr	50.57
225 mm diameter	0.99	13.22	1.65	90.01	nr	104.88
300 mm diameter	1.44	19.23	2.40	177.51	nr	199.14
Half section branch channel bend						
100 mm diameter	0.42	5.61	0.70	9.59	nr	15.90
150 mm diameter	0.57	7.61	0.95	15.82	nr	24.38
225 mm diameter	0.84	11.22	1.40	52.88	nr	65.50
300 mm diameter	1.20	16.02	2.00	107.72	nr	125.74
Three quarter section branch channel bend						
100 mm diameter	0.42	5.61	0.70	9.59	nr	15.90
150 mm diameter	0.57	7.61	0.95	15.82	nr	24.38
225 mm diameter	0.84	11.22	1.40	52.88	nr	65.50
300 mm diameter	1.20	16.02	2.00	107.72	nr	125.74

SERIES 500: DRAINAGE AND SERVICE DUCTS

Item	Gang hours	Labour £	Plant £	Material £	Unit	Total rate £
MANHOLES – cont						
Spun concrete channels; bedding in cement mortar in bottom of manhole						
Half section straight						
300 mm diameter	1.14	15.22	2.26	9.02	nr	**26.50**
450 mm diameter	2.25	30.04	4.45	15.34	nr	**49.83**
600 mm diameter	3.15	42.06	5.86	27.85	nr	**75.77**
900 mm diameter	5.91	78.91	11.69	53.41	nr	**144.01**
1200 mm diameter	9.42	125.78	18.63	93.44	nr	**237.85**
Labours						
Build in pipes; to brickwork						
300 mm diameter	0.28	5.08	0.45	2.00	m²	**7.53**
450 mm diameter	0.39	6.98	0.62	2.81	m²	**10.41**
600 mm diameter	0.52	9.52	0.85	3.77	m²	**14.14**
900 mm diameter	0.70	12.69	1.12	6.66	m²	**20.47**
1200 mm diameter	0.77	13.96	1.24	13.39	m²	**28.59**
Build in pipes; to precast concrete						
300 mm diameter	1.35	18.03	2.53	2.81	nr	**23.37**
450 mm diameter	2.01	26.84	3.77	4.02	nr	**34.63**
600 mm diameter	2.61	34.85	4.89	5.62	nr	**45.36**
900 mm diameter	3.84	51.27	7.19	12.11	nr	**70.57**
1200 mm diameter	4.83	64.49	9.05	21.24	nr	**94.78**
Access covers and frames; coated; BS EN 124; bed frame in cement mortar; set cover in grease and sand						
Group 5; super heavy duty; solid top						
600 × 600 mm; E600	1.80	24.03	2.64	333.36	nr	**360.03**
Group 4; heavy duty; solid top						
600 × 600 mm; double trianglar, hinged; D400	1.44	19.23	2.11	204.18	nr	**225.52**
Group 2; medium duty						
600 × 600 mm; single seal; B125	1.17	15.62	1.72	128.86	nr	**146.20**
Group 2; light duty						
600 × 600 mm; single seal; B125	0.87	11.62	1.28	109.37	nr	**122.27**
Manhole ancillaries						
Manhole interceptors						
100 mm diameter	0.54	7.21	1.01	106.95	nr	**115.17**
150 mm diameter	0.60	8.01	1.12	154.20	nr	**163.33**
225 mm diameter	0.90	12.02	1.69	481.28	nr	**494.99**
300 mm diameter	1.20	16.02	2.25	973.97	nr	**992.24**
Rodding eye points (excluding pipework)						
100 mm diameter	1.20	16.02	2.25	37.92	nr	**56.19**
150 mm diameter	1.35	18.03	2.25	64.77	nr	**85.05**

SERIES 600: EARTHWORKS

Item	Gang hours	Labour £	Plant £	Material £	Unit	Total rate £
GENERAL						
Notes						
The cost of earth moving and other associated works is dependent on matching the overall quantities and production rates called for by the programme of works with the most appropriate plant and assessing the most suitable version of that plant that will:						
* deal with the site conditions (e.g. type of ground, type of excavation, length of haul, prevailing weather, etc.);						
* comply with the specification requirements (e.g. compaction, separation of materials, surface tolerances, etc.);						
* complete the work economically (e.g. provide surface tolerances which will avoid undue excessive thickness of expensive imported materials).						
Excavation rates						
Unless stated the units for excavation in material other than topsoil rock or artificial hard materials are based on excavation in firm gravel soils.						
Factors for alternative types of soil:						
Multiply the rates by:-						
Scrapers Tractor Backacters						
Dozers (minimum						
and bucket size						
Loaders 0.5m³)						
Stiff clay 1.5 2.0 1.7						
Chalk 2.5 3.0 2.0						
Soft rock 3.5 2.5 2.0						
Broken rock 3.7 2.5 1.7						
Disposal rates						
The other important consideration in excavation of material is bulkage of material after it is dug and loaded onto transport.						
All pricing and estimating for disposal is based on the volume of solid material excavated and rates for disposal should be adjusted by the following factors for bulkage:						
Sand bulkage × 1.10						
Gravel bulkage × 1.20						
Compacted soil bulkage × 1.30						
Compacted sub-base, acceptable fill etc. bulkage × 1.30						

SERIES 600: EARTHWORKS

Item	Gang hours	Labour £	Plant £	Material £	Unit	Total rate £
GENERAL – cont						
Disposal rates – cont						
All pricing and estimating for disposal is based on the volume of solid material excavated and rates for disposal should be adjusted by the following factors for bulkage: – cont						
Stiff clay bulkage x 1.20						
Fill rates						
The price for filling presume the material being supplied in loose state.						
RESOURCES - LABOUR						
Scraper gang						
1 plant operator (skill rate 2)		18.50				
Total Gang Rate / Hour	£	18.50				
Scraper and ripper bulldozer gang (hard material)						
2 plant operator (skill rate 2)		37.00				
Total Gang Rate / Hour	£	37.00				
General excavation gang						
1 plant operator (skill rate 3)		16.59				
1 plant operator (skill rate 3) - 25% of time		4.15				
1 banksman (skill rate 4)		13.65				
Total Gang Rate / Hour	£	34.40				
Shore defences (armour stones) gang						
1 plant operator (skill rate 2)		18.50				
1 banksman (skill rate 4)		13.65				
Total Gang Rate / Hour	£	32.15				
Filling gang						
1 plant operator (skill rate 4)		14.92				
2 unskilled operatives (general)		25.50				
Total Gang Rate / Hour	£	40.43				
Treatment of filled surfaces gang						
1 plant operator (skill rate 2)		18.50				
Total Gang Rate / Hour	£	18.50				
Geotextiles (light sheets) gang						
1 ganger/chargehand (skill rate 4) - 50% of time		7.29				
2 unskilled operatives (general)		25.50				
Total Gang Rate / Hour	£	32.79				

SERIES 600: EARTHWORKS

Item	Gang hours	Labour £	Plant £	Material £	Unit	Total rate £
Geotextiles (medium sheets) gang						
1 ganger/chargehand (skill rate 4) - 20% of time		2.92				
3 unskilled operatives (general)		38.25				
Total Gang Rate / Hour	£	**41.17**				
Geotextiles (heavy sheets) gang						
1 ganger/chargehand (skill rate 4) - 50% of time		7.29				
2 unskilled operatives (general)		25.50				
1 plant operator (skill rate 4)		14.92				
Total Gang Rate / Hour	£	**47.72**				
Horticultural works gang						
1 skilled operative (skill rate 4)		13.65				
1 unskilled operative (general)		12.75				
Total Gang Rate / Hour	£	**26.40**				
RESOURCES - PLANT						
Scraper excavation						
motor scraper, 16.80 m³, elevating			94.42			
Total Gang Rate / Hour		£	**94.42**			
Scraper and ripper bulldozer excavation						
motor scraper, 16.80 m³, elevating			94.42			
D8 tractor dozer			98.47			
dozer attachment: triple shank ripper			6.83			
Total Gang Rate / Hour		£	**199.72**			
General excavation						
hydraulic crawler backacter, 0.40 m³			23.33			
backacter attachments: percussion breaker (25% of time)			1.34			
tractor loader, 1.50 m³ (25% of time)			9.00			
loader attachments: ripper (25% of time)			1.79			
Total Gang Rate / Hour		£	**35.46**			
Shore defences (armour stones)						
hydraulic crawler backacter, 1.20 m³			55.36			
backacter attachments: rock bucket			2.93			
backacter attachments: clamshell grab			3.93			
Total Gang Rate / Hour		£	**62.21**			
Geotextiles (heavy sheets)						
tractor loader, 1.5 m³			36.00			
Total Gang Rate / Hour		£	**36.00**			
Filling						
tractor loader, 1.5 m³			36.00			
Total Gang Rate / Hour		£	**36.00**			

SERIES 600: EARTHWORKS

Item	Gang hours	Labour £	Plant £	Material £	Unit	Total rate £
RESOURCES - PLANT – cont						
Treatment of filled surfaces						
tractor loader, 1.50 m³			36.00			
3 wheel deadweight roller, 10 tonne			16.78			
Total Gang Rate / Hour		£	**52.78**			
EXCAVATION						
Typical motorway cutting generally using motorised scrapers and/or dozers on an average haul of 2000 m (one way)						
Excavation of acceptable material Class 5A	0.01	0.15	0.76	–	m³	**0.91**
Excavation of acceptable material excluding Class 5A in cutting and other excavation	0.01	0.23	1.93	–	m³	**2.16**
General excavation using backacters						
Excavation of acceptable material excluding Class 5A in new watercourses	0.06	2.05	2.41	–	m³	**4.46**
Excavation of unacceptable material Class U1 / U2 in new watercourses	0.07	2.22	2.48	–	m³	**4.70**
General excavation using backacters						
Excavation of unacceptable material Class U1/U2 in clearing abandoned watercourses	0.06	1.88	2.11	–	m³	**3.99**
General excavation using backacters and tractor loaders						
Excavation of acceptable material Class 5A	0.04	1.37	1.48	–	m³	**2.85**
Excavation of acceptable material excluding Class 5A in structural foundations						
ne 3.0 m deep	0.08	2.56	2.86	–	m³	**5.42**
ne 6.0 m deep	0.20	6.83	7.48	–	m³	**14.31**
Excavation of acceptable material excluding Class 5A in foundations for corrugated steel buried structures and the like						
ne 3.0 m deep	0.08	2.56	2.86	–	m³	**5.42**
ne 6.0 m deep	0.20	6.83	7.48	–	m³	**14.31**
Excavation of unacceptable material Class U1 / U2 in structural foundations						
ne 3.0 m deep	0.09	2.90	3.26	–	m³	**6.16**
ne 6.0 m deep	0.21	7.17	7.90	–	m³	**15.07**
Excavation of unacceptable material Class U1 / U2 in foundations for corrugated steel buried structures and the like						
ne 3.0 m deep	0.09	2.90	3.26	–	m³	**6.16**
ne 6.0 m deep	0.21	7.17	7.90	–	m³	**15.07**

SERIES 600: EARTHWORKS

Item	Gang hours	Labour £	Plant £	Material £	Unit	Total rate £
General excavation using backacters and tractor loader with ripper						
Excavation of acceptable material Class 5A	0.03	1.02	1.11	–	m³	2.13
Excavation of acceptable material excluding Class 5A in cutting and other excavation	0.04	1.37	1.48	–	m³	2.85
Excavate unacceptable material Class U1 / U2 in cutting and other excavation	0.04	1.37	1.48	–	m³	2.85
EXCAVATION IN HARD MATERIAL						
Typical motorway cutting generally using motorised scrapers and/or dozers on an average haul of 2000 m (one way)						
Excavate in hard material; using scraper and ripper bulldozer						
mass concrete/medium hard rock	0.09	3.31	17.97	–	m³	21.28
reinforced concrete/hard rock	0.15	5.51	29.96	–	m³	35.47
tarmacadam	0.16	5.88	3.20	–	m³	9.08
General excavation using backacters						
Extra over excavation in new watercourses for excavation in hard material						
rock	0.74	25.26	27.19	–	m³	52.45
pavements, brickwork, concrete and masonry	0.69	23.55	25.33	–	m³	48.88
reinforced concrete	1.12	38.23	41.10	–	m³	79.33
General excavation using backacters and tractor loaders						
Extra over excavation in structural foundations for excavation in hard material						
rock	0.74	25.26	27.19	–	m³	52.45
pavements, brickwork, concrete and masonry	0.69	23.55	25.33	–	m³	48.88
reinforced concrete	1.12	38.23	41.10	–	m³	79.33
Extra over excavation for excavation in foundations for corrugated steel buried structures and the like for excavation in hard material						
rock	0.74	25.26	27.19	–	m³	52.45
pavements, brickwork, concrete and masonry	0.69	23.55	25.33	–	m³	48.88
reinforced concrete	1.12	38.23	41.10	–	m³	79.33
General excavation using backacters and tractor loader with ripper						
Extra over excavation for excavation in cutting and other excavation for excavation in hard material						
rock	0.71	24.23	26.06	–	m³	50.29
pavements, brickwork, concrete and masonry	0.66	22.53	24.23	–	m³	46.76
reinforced concrete	1.04	35.50	38.18	–	m³	73.68

SERIES 600: EARTHWORKS

Item	Gang hours	Labour £	Plant £	Material £	Unit	Total rate £
DEPOSITION OF FILL						
Deposition of acceptable material Class 1C/6B in						
embankments and other areas of fill	0.01	0.44	0.40	–	m³	**0.84**
strengthened embankments	0.01	0.44	0.40	–	m³	**0.84**
reinforced earth structures	0.01	0.44	0.40	–	m³	**0.84**
anchored earth structures	0.01	0.44	0.40	–	m³	**0.84**
landscaped areas	0.01	0.44	0.40	–	m³	**0.84**
environmental bunds	0.01	0.44	0.40	–	m³	**0.84**
fill to structures	0.01	0.44	0.40	–	m³	**0.84**
fill above structural concrete foundations	0.01	0.44	0.40	–	m³	**0.84**
DISPOSAL OF MATERIAL						
Disposal of acceptable material excluding Class 5A						
using 10 tonnes capacity tipping lorry for on-site or						
off-site use; haul distance to tip not exceeding 1 Km	0.06	1.05	1.63	–	m³	**2.68**
ADD per further Km haul	0.03	0.53	0.81	–	m³	**1.34**
using 10–15 tonnes capacity tipping lorry for						
on-site or off-site use; haul distance to tip not						
exceeding 1 Km	0.05	0.88	1.36	–	m³	**2.24**
ADD per further Km haul	0.03	0.44	0.68	–	m³	**1.12**
using 15–25 tonnes capacity tipping lorry for						
on-site or off-site use; haul distance to tip not						
exceeding 1 Km	0.05	0.78	1.22	–	m³	**2.00**
ADD per further Km haul	0.02	0.38	0.60	–	m³	**0.98**
Disposal of acceptable material Class 5A (excluding						
resale value of soil)						
using 10 tonnes capacity tipping lorry for on-site or						
off-site use; haul distance to tip not exceeding 1 Km	0.06	1.05	1.63	–	m³	**2.68**
ADD per further Km haul	0.03	0.53	0.81	–	m³	**1.34**
using 10–15 tonnes capacity tipping lorry for						
on-site or off-site use; haul distance to tip not						
exceeding 1 Km	0.05	0.88	1.36	–	m³	**2.24**
ADD per further Km haul	0.03	0.44	0.68	–	m³	**1.12**
using 15–25 tonnes capacity tipping lorry for						
on-site or off-site use; haul distance to tip not						
exceeding 1 Km	0.05	0.78	1.22	–	m³	**2.00**
ADD per further Km haul	0.02	0.38	0.60	–	m³	**0.98**
Disposal of unacceptable material Class U1						
using 10 tonnes capacity tipping lorry for on-site						
or off-site use; haul distance to tip not exceeding						
1 Km	0.06	1.05	1.63	–	m³	**2.68**
ADD per further Km haul	0.03	0.53	0.81	–	m³	**1.34**
using 10–15 tonnes capacity tipping lorry for						
on-site or off-site use; haul distance to tip not						
exceeding 1 Km	0.05	0.88	1.36	–	m³	**2.24**
ADD per further Km haul	0.03	0.44	0.68	–	m³	**1.12**
using 15–25 tonnes capacity tipping lorry for						
on-site or off-site use; haul distance to tip not						
exceeding 1 Km	0.05	0.78	1.22	–	m³	**2.00**
ADD per further Km haul	0.02	0.38	0.60	–	m³	**0.98**

SERIES 600: EARTHWORKS

Item	Gang hours	Labour £	Plant £	Material £	Unit	Total rate £
Disposal of unacceptable material Class U2						
using 10 tonnes capacity tipping lorry for on-site or off-site use; haul distance to tip not exceeding						
1 Km	0.06	1.05	1.63	–	m³	**2.68**
ADD per further Km haul	0.03	0.53	0.81	–	m³	**1.34**
using 10–15 tonnes capacity tipping lorry for on-site or off-site use; haul distance to tip not						
exceeding 1 Km	0.05	0.88	1.36	–	m³	**2.24**
ADD per further Km haul	0.03	0.44	0.68	–	m³	**1.12**
using 15–25 tonnes capacity tipping lorry for on-site or off-site use; haul distance to tip not						
exceeding 1 Km	0.05	0.78	1.22	–	m³	**2.00**
ADD per further Km haul	0.02	0.38	0.60	–	m³	**0.98**
Add to the above rates where tipping charges apply:						
non-hazardous waste	–	–	–	–	m³	**23.51**
hazardous waste	–	–	–	–	m³	**92.60**
special waste	–	–	–	–	m³	**98.50**
contaminated liquid	–	–	–	–	m³	**112.50**
contaminated sludge	–	–	–	–	m³	**141.75**
Add to the above rates where Landfill Tax applies:						
exempted material	–	–	–	–	m³	**-**
inactive or inert material	–	–	–	–	m³	**2.50**
other material	–	–	–	–	m³	**56.00**
IMPORTED FILL						
Imported graded granular fill, natural gravels DfT Class 1A/B/C [1.9 t/m³]						
Imported acceptable material in						
embankments and other areas of fill	0.02	0.72	0.65	19.89	m³	**21.26**
extra for Aggregate Tax	–	–	–	4.75	m³	**4.75**
Imported graded granular fill, crushed gravels or rock DfT Class 1A/B/C [1.9 t/m³]; using tractor loader						
Imported acceptable material in						
embankments and other areas of fill	0.02	0.72	0.65	22.40	m³	**23.77**
extra for Aggregate Tax	–	–	–	4.75	m³	**4.75**
Cohesive material DfT Class 2A/B/C/D [1.8 t/m³]; using tractor loader						
Imported acceptable material in						
embankments and other areas of fill	0.02	0.72	0.65	20.83	m³	**22.20**
landscaped areas	0.02	0.72	0.65	20.83	m³	**22.20**
environmental bunds	0.02	0.84	0.76	20.83	m³	**22.43**
fill to structures	0.02	0.96	0.86	20.83	m³	**22.65**
fill above structural concrete foundations	0.02	0.76	0.68	20.83	m³	**22.27**

SERIES 600: EARTHWORKS

Item	Gang hours	Labour £	Plant £	Material £	Unit	Total rate £
IMPORTED FILL – cont						
Reclaimed pulverized fuel ash DfT Class 2E [2.8 t/m3]; using tractor loader						
Imported acceptable material in						
embankments and other areas of fill	0.02	0.72	0.65	22.57	m³	**23.94**
Reclaimed quarry waste DfT Class 2E [1.8 t/m3]; using tractor loader						
Imported acceptable material in						
embankments and other areas of fill	0.02	0.72	0.65	14.51	m³	**15.88**
extra for Aggregate Tax (fill other than from						
acceptable process)	–	–	–	4.50	m³	**4.50**
Imported topsoil DfT Class 5B [1.44 t/m3]; using tractor loader						
Imported topsoil Class 5B	0.02	0.72	0.65	23.22	m³	**24.59**
Imported selected well graded granular fill DfT Class 6A (1.90 t/m3); using tractor loader						
Imported acceptable material						
embankments and other areas of fill	–	–	0.76	27.64	m³	**28.40**
landscape areas	0.02	0.84	0.76	27.64	m³	**29.24**
environmental bunds	0.02	0.72	0.65	27.64	m³	**29.01**
fill to structures	0.02	0.96	0.86	27.64	m³	**29.46**
fill above structural concrete foundations	0.02	0.76	0.68	27.64	m³	**29.08**
extra for Aggregate Tax	–	–	–	4.75	m³	**4.75**
Imported selected granular fill, DfT Class 6F (1.9 t/m3); using tractor loader						
Imported acceptable material in						
embankments and other areas of fill	0.10	3.81	3.42	25.53	m³	**32.76**
extra for Aggregate Tax	–	–	–	4.75	m³	**4.75**
Imported selected well graded fill DfT Class 6I (1.90 t/m3); using tractor loader						
Imported acceptable material						
reinforced earth structures	0.02	0.88	0.79	21.98	m³	**23.65**
extra for Aggregate Tax	–	–	–	4.75	m³	**4.75**
Imported rock fill (1.90 t/m3); using tractor loader						
Imported acceptable material						
in embankments and other areas of fill	0.02	0.80	0.72	32.35	m³	**33.87**
extra for Aggregate Tax	–	–	–	4.75	m³	**4.75**
Imported well graded granular material (1.90 t/ m³); (bedding/free draining materials under shore protection) using tractor loader						
Imported acceptable material						
in embankments and other areas of fill	0.02	0.80	0.72	26.17	m³	**27.69**

SERIES 600: EARTHWORKS

Item	Gang hours	Labour £	Plant £	Material £	Unit	Total rate £
Imported rock fill (as a core embankment) (1.90 t/ m³); using tractor loader						
Imported acceptable material						
in embankments and other areas of fill	0.03	1.32	1.19	32.35	m³	34.86
extra for Aggregate Tax	–	–	–	4.75	m³	4.75
Imported Armour Stones (1.90 t/m³) (shore protection of individual rocks up to 0.5 t each); using backacter						
Imported acceptable material						
in embankments and other areas of fill	0.06	1.79	3.48	46.06	m³	51.33
extra for Aggregate Tax	–	–	–	4.75	m³	4.75
Imported Armour Stones (1.90 t/m³) (shore protection of individual rocks up to 1.0 t each); using backacter						
Imported acceptable material						
in embankments and other areas of fill	0.03	1.02	1.99	64.90	m³	67.91
extra for Aggregate Tax	–	–	–	4.75	m³	4.75
Imported Armour Stones (1.90 t/m³) (shore protection of individual rocks up to 3.0 t each); using backacter						
Imported acceptable material						
in embankments and other areas of fill	0.03	0.83	1.63	72.23	m³	74.69
extra for Aggregate Tax	–	–	–	4.75	m³	4.75
COMPACTION OF FILL						
Compaction of granular fill material						
in embankments and other areas of fill	0.01	0.06	0.26	–	m³	0.32
adjacent to structures	0.01	0.09	0.42	–	m³	0.51
above structural concrete foundations	0.01	0.13	0.63	–	m³	0.76
Compaction of fill material						
in sub-base or capping layers under verges, central reserves or side slopes	0.02	0.22	1.06	–	m³	1.28
adjacent to structures	0.04	0.39	1.85	–	m³	2.24
above structural concrete foundations	0.04	0.45	2.11	–	m³	2.56
Compaction of graded fill material						
in embankments and other areas of fill	0.01	0.07	0.32	–	m³	0.39
adjacent to structures	0.01	0.11	0.53	–	m³	0.64
above structural concrete foundations	0.02	0.17	0.79	–	m³	0.96
Compaction of rock fill materials						
in embankments and other areas of fill	0.01	0.10	0.47	–	m³	0.57
adjacent to structures	0.01	0.16	0.74	–	m³	0.90
above structural concrete foundations	0.02	0.19	0.90	–	m³	1.09
Compaction of clay fill material						
in embankments and other areas of fill	0.03	0.33	1.58	–	m³	1.91
adjacent to structures	0.05	0.53	2.53	–	m³	3.06
above structural concrete foundations	0.05	0.53	2.53	–	m³	3.06

SERIES 600: EARTHWORKS

Item	Gang hours	Labour £	Plant £	Material £	Unit	Total rate £
GEOTEXTILES						
Notes						
The geotextile products mentioned below are not specifically confined to the individual uses stated but are examples of one of many scenarios to which they may be applied. Conversely, the scenarios are not limited to the geotextile used as an example. The heavier grades of sheeting will need to be manipulated into place by machine and cutting would be by hacksaw rather than knife. Care should be taken in assessing the wastage of the more expensive sheeting. The prices include for preparing surfaces, overlaps and turnups, jointing and sealing, fixing material in place if required and reasonable waste between 5% and 10%.						
Stabilization applications for reinforcement of granular sub-bases and capping layers placed over weak and variable soils.						
For use in weak soils with moderate traffic intensities e.g. light access roads: Tensar SS20 Polypropylene Geogrid						
horizontal	0.04	1.37	–	2.18	m²	**3.55**
inclined at an angle 10-45° to the horizontal	0.05	1.72	–	2.18	m²	**3.90**
For use in weak soils with high traffic intensities and/ or high axle loadings: Tensar SS30 Polypropylene Geogrid						
horizontal	0.05	1.84	–	5.39	m²	**7.23**
inclined at an angle 10-45° to the horizontal	0.06	2.29	–	5.39	m²	**7.68**
For construction over very weak soils e.g. alluvium, marsh or peat, or firmer soil subject to exceptionally high axle loadings: Tensar SS40 Polypropylene Geogrid						
horizontal	0.05	2.13	1.62	6.22	m²	**9.97**
inclined at an angle 10-45° to the horizontal	0.06	2.65	2.02	6.22	m²	**10.89**
For trafficked areas where fill comprises of aggregate exceeding 100 mm: Tensar SSLA20 Polypropylene Geogrid						
horizontal	0.04	1.37	–	–	m²	**1.37**
inclined at an angle 10-45° to the horizontal	0.05	1.72	–	–	m²	**1.72**
For stabilisation and separation of granular fill from soft sub grade to prevent intermixing: Terram 1000						
horizontal	0.05	2.08	–	0.57	m²	**2.65**
inclined at an angle 10-45° to the horizontal	0.06	2.61	–	0.57	m²	**3.18**
For stabilisation and separation of granular fill from soft sub grade to prevent intermixing: Terram 2000						
horizontal	0.04	1.99	1.51	1.64	m²	**5.14**
inclined at an angle 10-45° to the horizontal	0.05	2.51	1.91	1.64	m²	**6.06**

SERIES 600: EARTHWORKS

Item	Gang hours	Labour £	Plant £	Material £	Unit	Total rate £
Reinforcement applications for asphalt pavements						
For roads, hardstandings and airfield pavements: Tensar AR-G grid bonded to a geotextile						
horizontal	0.05	1.84	–	3.90	m²	**5.74**
inclined at an angle 10-45° to the horizontal	0.06	2.29	–	3.90	m²	**6.19**
Slope Reinforcement and Embankment Support; For use where soils can only withstand limited shear stresses, therefore steep slopes require external support						
Paragrid 30/155; 330g/m2						
horizontal	0.04	1.37	–	2.07	m²	**3.44**
inclined at an angle 10-45° to the horizontal	0.05	1.72	–	2.07	m²	**3.79**
Paragrid 100/255; 330g/m2						
horizontal	0.04	1.37	–	2.87	m²	**4.24**
inclined at an angle 10-45° to the horizontal	0.05	1.72	–	2.87	m²	**4.59**
Paralink 200; 1120g/m2						
horizontal	0.05	2.56	1.94	5.70	m²	**10.20**
inclined at an angle 10-45° to the horizontal	0.07	3.22	2.45	5.70	m²	**11.37**
Paralink 600; 2040g/m2						
horizontal	0.06	2.98	2.27	12.04	m²	**17.29**
inclined at an angle 10-45° to the horizontal	0.08	3.74	2.84	12.04	m²	**18.62**
TerramGrid 3/3 W						
horizontal	0.06	2.70	2.05	2.89	m²	**7.64**
inclined at an angle 10-45° to the horizontal	0.07	3.36	2.56	2.89	m²	**8.81**
Scour and Erosion Protection						
For use where erosion protection is required to the surface of a slope once its geotechnical stability has been achieved, and to allow grass establishment: Tensar 'Mat' Polyethylene mesh, fixed with Tensar pegs						
horizontal	0.04	1.67	–	4.80	m²	**6.47**
inclined at an angle 10-45° from the horizontal	0.05	2.08	–	5.11	m²	**7.19**
For use where hydraulic action exists, such as coastline protection from pressures exerted by waves, currents and tides: Typar SF56						
horizontal	0.05	2.41	1.84	0.49	m²	**4.74**
inclined at an angle 10-45° from the horizontal	0.06	3.03	2.30	0.49	m²	**5.82**
For protection against puncturing to reservoir liner: Typar SF56						
horizontal	0.05	2.41	1.84	0.49	m²	**4.74**
inclined at an angle 10-45° from the horizontal	0.06	3.03	2.30	0.49	m²	**5.82**
Temporary parking areas						
For reinforcement of grassed areas subject to wear from excessive pedestrian and light motor vehicle traffic: Netlon CE131 high density polyethylene geogrid, including fixing pegs						
sheeting	0.04	1.43	–	4.68	m²	**6.11**

SERIES 600: EARTHWORKS

Item	Gang hours	Labour £	Plant £	Material £	Unit	Total rate £
GEOTEXTILES – cont						
Landscaping applications						
For prevention of weed growth in planted areas by incorporating a geotextile over topsoil: Typar SF20, including pegs						
horizontal	0.08	3.06	–	0.60	m²	**3.66**
inclined at an angle 10-45° to the horizontal	0.09	3.84	–	0.60	m²	**4.44**
For root growth control-Prevention of lateral spread of roots and mixing of road base and humus: Typar SF20						
horizontal	0.08	3.06	–	0.29	m²	**3.35**
inclined at an angle 10-45° to the horizontal	0.09	3.84	–	0.29	m²	**4.13**
SOFT SPOTS AND OTHER VOIDS						
Excavation of soft spots and other voids using motorised scrapers and/or dozers						
Excavate below cuttings or under embankments	0.05	0.93	4.72	–	m³	**5.65**
Excavate in side slopes	0.06	1.11	5.66	–	m³	**6.77**
Excavation of soft spots and other voids using backacters and tractor loader						
Excavate below structural foundations and foundations for corrugated steel buried structures	0.09	1.49	1.48	–	m³	**2.97**
Imported graded granular fill; deposition using tractor loader and towed roller						
Filling of soft spots and other voids below cuttings or under embankments	0.04	0.58	0.95	20.10	m³	**21.63**
Filling of soft spots and other voids in side slopes	0.04	0.71	1.16	20.10	m³	**21.97**
Filling of soft spots and other voids below structural foundations and foundations for corrugated steel buried structures	0.04	0.58	0.95	20.10	m³	**21.63**
Imported rock fill; 1.9 t/m3; using tractor loader						
Deposition into soft areas (rock punching)	0.03	1.00	0.90	32.35	m³	**34.25**
DISUSED SEWERS, DRAINS, CABLES, DUCTS, PIPELINES AND THE LIKE						
Removal of disused sewer or drain						
100 mm internal diameter; with less than 1 m of cover to formation level	0.16	2.42	3.20	–	m	**5.62**
150 mm internal diameter; with 1 to 2 m of cover to formation level	0.20	2.96	3.93	–	m	**6.89**
Backfilling with acceptable material of disused sewer or drain						
100 mm internal diameter; with less than 1 m of cover to formation level	0.23	3.48	4.61	–	m³	**8.09**
150 mm internal diameter; with 1 to 2 m of cover to formation level	0.23	3.48	4.61	–	m³	**8.09**

SERIES 600: EARTHWORKS

Item	Gang hours	Labour £	Plant £	Material £	Unit	Total rate £
Backfilling of disused basements, cellars and the like with acceptable material	0.26	3.93	5.21	–	m³	**9.14**
Backfilling of disused gullies with concrete Grade C15	0.10	1.51	2.00	10.92	nr	**14.43**
SUPPORTS LEFT IN EXCAVATION						
Timber close boarded supports left in excavation	–	–	8.67	–	m²	**8.67**
Steel trench sheeting supports left in excavation	–	–	36.38	–	m²	**36.38**
TOPSOILING AND STORAGE OF TOPSOIL						
Topsoiling 150 mm thick to surfaces						
at 10° or less to horizontal	0.04	0.65	3.30	–	m²	**3.95**
more than 10° to horizontal	0.04	0.74	3.78	–	m²	**4.52**
Topsoiling 350 mm thick to surfaces						
at 10° or less to horizontal	0.05	0.83	4.25	–	m²	**5.08**
more than 10° to horizontal	0.05	0.93	4.72	–	m²	**5.65**
Topsoiling 450 mm thick to surfaces						
at 10° or less to horizontal	0.05	0.93	4.72	–	m²	**5.65**
more than 10° to horizontal	0.06	1.02	5.19	–	m²	**6.21**
Topsoiling 600 mm thick to surfaces						
at 10° or less to horizontal	0.05	0.93	4.72	–	m²	**5.65**
more than 10° to horizontal	0.06	1.02	5.19	–	m²	**6.21**
Permanent storage of topsoil	0.06	1.11	5.66	–	m³	**6.77**
COMPLETION OF FORMATION AND SUB-FORM						
Completion of sub-formation on material other than Class 1C, 6B or rock in cuttings	0.01	0.20	0.60	–	m²	**0.80**
Completion of formation on material other than Class 1C, 6B or rock in cuttings	0.01	0.21	0.47	–	m²	**0.68**
LINING OF WATERCOURSES						
Lining new watercourse invert with precast concrete units 63 mm thick	0.21	5.50	0.45	16.64	m²	**22.59**
Lining new watercourse side slopes with precast concrete units 63 mm thick	0.25	6.54	0.50	15.64	m²	**22.68**
Lining enlarged watercourse invert with precast concrete units 63 mm thick	0.24	6.28	0.52	16.64	m²	**23.44**
Lining enlarged watercourse side slopes with precast concrete units 63 mm thick	0.29	7.59	0.63	15.64	m²	**23.86**

478   *Unit Costs – Highway Works*

SERIES 600: EARTHWORKS

Item	Gang hours	Labour £	Plant £	Material £	Unit	Total rate £
GROUND IMPROVEMENT - DYNAMIC COMPACTION						
Ground consolidation by dynamic compaction is a technique which involves the dropping of a steel or concrete pounder several times in each location in a grid pattern that covers the whole site. For a ground compaction of up to 10 m, a 15 t pounder with a free fall of 20 m would be typical. Several passes over the site are normally required to achieve full compaction. The process is recommended for naturally cohesive soils and is usually uneconomic for areas of less than 4000 m², for sites with granular or mixed granular cohesive soils and 6000 m² for a site with weak cohesive soils. The main considerations to be taken into account when using this method of consolidation are:						
* sufficient area to be viable						
* proximity and condition of adjacent property and services						
* need for blanket layer of granular material for a working surface and as backfill to offset induced settlement						
* water table level						
The granular blanket layer performs the dual functions of working surface and backfill material. Generally 300 mm thickness is required.						
The final bearing capacity and settlement criteria that can be achieved depends on the nature of the material being compacted. Allowable bearing capacity may be increased by up to twice the pre-treated value for the same settlement. Control testing can be by crater volume measurements, site levelling between passes, penetration tests or plate loading tests.						
Dynamic Compaction						
Dynamic compaction in main compaction with a 15 t pounder	–	–	–	–	m²	1.50
Dynamic compaction plant standing time	–	–	–	–	hr	15.00
Free-draining granular material in granular blanket	–	–	–	–	t	8.25
Control testing including levelling, piezometers and penetrameter testing	–	–	–	–	m²	5.00
Kentledge load test	–	–	–	–	nr	8000.00
Establishment Of Plant						
Establishment of dynamic compaction plant	–	–	–	–	sum	21750.00
Vibrated Stone Columns						
Refer to part 4, Class C: Ground Consolidation-vibro-replacement, for an explanation and items for vibrated stone columns						

SERIES 600: EARTHWORKS

Item	Gang hours	Labour £	Plant £	Material £	Unit	Total rate £
GABION WALLING AND MATTRESSES						
Gabion walling with plastic coated galvanized wire mesh, wire laced; filled with 50 mm Class 6G material						
2.0 x 1.0 × 1.0 m module sizes	3.98	46.88	9.88	76.40	m³	**133.16**
2.0 x 1.0 × 0.5 m module sizes	5.20	59.11	24.32	191.40	m³	**274.83**
Gabion walling with heavily galvanized woven wire mesh, wire laced; filled with 50 mm Class 6G material						
2.0 x 1.0 × 1.0 m module sizes	4.22	48.02	19.76	57.38	m³	**125.16**
2.0 x 1.0 × 0.5 m module sizes	5.20	59.11	24.32	188.01	m³	**271.44**
Mattress with plastic coated galvanized wire mesh; filled with 50 mm Class 6G material installed at 10° or less to the horizontal						
6.0 x 2.0 × 0.23 m module sizes	4.22	48.02	19.76	48.84	m³	**116.62**
CRIB WALLING						
Notes						
There are a number of products specially designated for large scale earth control. Crib walling consists of a rigid unit built of rectangular interlocking timber or precast concrete members forming a skeleton of cells laid on top of each other and filled with earth or rock. Prices for these items depend on quantity, difficulty of access to the site and availability of suitable filling material; estimates should be obtained from the manufacturer when the site conditions have been determined.						
Crib walling of timber components laid with a battering face; stone infill						
ne 1.5 m high	5.28	63.30	13.23	94.68	m²	**171.21**
ne 3.7 m high	7.11	85.24	15.95	99.17	m²	**200.36**
ne 4.2 m high	8.13	97.47	17.46	103.79	m²	**218.72**
ne 5.9 m high	10.71	128.40	21.30	116.26	m²	**265.96**
ne 7.4 m high	17.16	205.72	30.88	150.40	m²	**387.00**
Crib walling of precast concrete crib units, laid dry jointed with a battered face; stone infill						
1.0 m high; no dowels	6.75	80.92	29.40	104.93	m	**215.25**
1.5 m high; no dowels	9.36	112.21	40.76	157.39	m	**310.36**
2.5 m high; no dowels	22.11	265.06	99.02	312.63	m	**676.71**
4.0 m high; no dowels	34.11	408.92	152.76	500.20	m	**1061.88**

SERIES 600: EARTHWORKS

Item	Gang hours	Labour £	Plant £	Material £	Unit	Total rate £
GROUND ANCHORAGES						
Ground anchorages consist of the installation of a cable or solid bar tendon fixed in the ground by grouting and tensioned to exceed the working load to be carried. Ground anchors may be of a permanent or temporary nature and can be used in conjunction with diaphragm walling or sheet piling to eliminate the use of strutting etc. The following costs are based on the installation of 50 nr ground anchors						
Ground Anchorage Plant						
Establishment of ground anchorage plant	–	–	–	–	sum	10500.00
Ground anchorage plant standing time	–	–	–	–	hr	168.00
Ground Anchorages						
Ground anchorages; temporary or permanent						
15.0 m maximum depth; in rock, alluvial or clay; 0 - 50 t load	–	–	–	–	nr	86.26
15.0 m maximum depth; in rock or alluvial; 50 - 90 t load	–	–	–	–	nr	103.21
15.0 m maximum depth; in rock only; 90 - 150 t load	–	–	–	–	nr	138.57
Temporary tendons						
in rock, alluvial or clay; 0 - 50 t load	–	–	–	–	nr	66.04
in rock or alluvial; 50 - 90 t load	–	–	–	–	nr	100.75
in rock only; 90 - 150 t load	–	–	–	–	nr	135.27
Permanent tendons						
in rock, alluvial or clay; 0 - 50 t load	–	–	–	–	nr	101.42
in rock or alluvial; 50 - 90 t load	–	–	–	–	nr	138.80
in rock only; 90 - 150 t load	–	–	–	–	nr	173.36
GROUND WATER LOWERING						
The following unit costs are for dewatering pervious ground only and are for sets of equipment comprising:						
* hire of 1 nr diesel driven pump (WP 150/60 or similar) complete with allowance of £50 for fuel	–	–	–	–	day	109.25
* hire of 50 m of 150 mm diameter header pipe	–	–	–	–	day	19.00
* purchase of 35 nr of disposable well points	–	–	–	–	sum	9143.75
* hire of 18 m of delivery pipe	–	–	–	–	day	7.36
* hire of 1 nr diesel driven standby pump	–	–	–	–	day	33.01
* hire of 1 nr jetting pump with hoses (for installation of wellpoints only)	–	–	–	–	day	57.00
* cost of attendant labour and plant (2 hrs per day) inclusive of small dumper and bowser	–	–	–	–	day	95.00
Costs are based on 24 hr operation in 12 hr shifts with attendant operators (specialist advice)						

SERIES 600: EARTHWORKS

Item	Gang hours·	Labour £	Plant £	Material £	Unit	Total rate £
Guide price for single set of equipment comprising pump, 150 mm diameter header pipe, 35 nr well points, delivery pipe and attendant labour and plant						
Bring to site equipment and remove upon completion	–	–	–	–	sum	**2500.00**
Installation: 3 day hire of jetting pump	–	–	–	–	sum	**171.00**
Operating costs						
purchase of well points	–	–	–	–	nr	**261.25**
hire of pump, header pipe, delivery pipe and standby pump complete with fuel etc. and attendant labour and plant	–	–	–	–	day	**263.63**
TRIAL PITS						
The following costs assume the use of mechanical plant and excavating and backfilling on the same day						
Trial pit						
ne 1.0 m deep	1.44	17.26	10.76	–	nr	**28.02**
1.0 - 2.0 m deep	2.62	31.26	21.08	–	nr	**52.34**
over 2.0 m deep	3.18	38.12	23.76	–	nr	**61.88**
BREAKING UP AND PERFORATION OF REDUNDANT PAVEMENTS						
Using scraper and ripper bulldozer						
Breaking up of redundant concrete slab						
ne 100 mm deep	0.05	0.93	9.99	–	m²	**10.92**
100 to 200 mm deep	0.10	1.85	19.97	–	m²	**21.82**
Breaking up of redundant flexible pavement						
ne 100 mm deep	0.02	0.37	3.99	–	m²	**4.36**
100 to 200 mm deep	0.03	0.56	5.99	–	m²	**6.55**
Using backacters and tractor loader with ripper						
Breaking up of redundant reinforced concrete pavement						
ne 100 mm deep	0.28	4.29	4.26	–	m²	**8.55**
100 to 200 mm deep	0.49	7.51	7.46	–	m²	**14.97**
200 to 300 mm deep	0.70	10.73	10.68	–	m²	**21.41**
Using scraper and ripper bulldozer						
Breaking up of redundant flexible pavement using scraper and ripper bulldozer						
ne 100 mm deep	0.02	0.37	3.99	–	m²	**4.36**
100 to 200 mm deep	0.03	0.56	5.99	–	m²	**6.55**

SERIES 600: EARTHWORKS

Item	Gang hours	Labour £	Plant £	Material £	Unit	Total rate £
BREAKING UP AND PERFORATION OF REDUNDANT PAVEMENTS – cont						
Using backacters and breakers						
Perforation of redundant reinforced concrete pavement						
ne 100 mm deep	0.02	0.33	0.49	–	m²	**0.82**
100 to 200 mm deep	0.03	0.50	0.74	–	m²	**1.24**
200 to 300 mm deep	0.05	0.83	1.23	–	m²	**2.06**
Perforation of redundant flexible pavement						
ne 100 mm deep	0.01	0.17	0.25	–	m²	**0.42**
100 to 200 mm deep	0.02	0.25	0.37	–	m²	**0.62**
PERFORATION OF REDUNDANT SLABS, BASEMENTS AND THE LIKE						
Perforation Of Redundant Slabs, BASEMENTS AND THE LIKE						
Using backacters and breakers						
Perforation of redundant reinforced concrete slab						
ne 100 mm deep	0.02	0.33	0.49	–	m²	**0.82**
100 to 200 mm deep	0.03	0.50	0.74	–	m²	**1.24**
200 to 300 mm deep	0.05	0.83	1.23	–	m²	**2.06**
Perforation of redundant reinforced concrete basement						
ne 100 mm deep	0.02	0.35	0.51	–	m²	**0.86**
100 to 200 mm deep	0.03	0.52	0.78	–	m²	**1.30**
200 to 300 mm deep	0.05	0.87	1.29	–	m²	**2.16**
REINFORCED AND ANCHORED EARTH STRUCTURES						
Specialist Advice						
As each structure is different, it is virtually impossible to give accurate unit cost prices, as they will vary with the following parameters:						
* Type of structure						
* Where located (in water, dry condition)						
* Where geographically in the country						
* Type of fill						
* Duration of structure						
* Size of structure, etc.						
To arrive at the unit costs below assumptions have been made for a structure with the following characteristics :						
* Structure - retaining wall 6 m high x 150 m in length						
* Construction - as DfT Specification BE 3/78						
* Site conditions - good foundations						
* Fill - 5 m³ per m² of wall face						
* Fill costs - DfT Specification average £9.50 / tonne						

SERIES 600: EARTHWORKS

Item	Gang hours	Labour £	Plant £	Material £	Unit	Total rate £
Therefore specialist advice should be sought in order to give accurate budget costings for individual projects						
Retaining wall (per m^2 of face)						
concrete faced using ribbed strip	1.00	12.98	4.49	181.73	m^2	**199.20**
concrete faced using flat strip	1.00	12.98	4.49	231.29	m^2	**248.76**
concrete faced using polyester strip	1.00	34.46	9.25	198.24	m^2	**241.95**
concrete faced using geogrid reinforcement	1.00	27.59	10.18	206.53	m^2	**244.30**
preformed mesh using ribbed strip	1.00	15.29	6.99	148.68	m^2	**170.96**

SERIES 700: PAVEMENTS

Item	Gang hours	Labour £	Plant £	Material £	Unit	Total rate £
RESOURCES - LABOUR						
Sub-base laying gang						
1 ganger/chargehand (skill rate 4)		14.58				
1 skilled operative (skill rate 4)		13.65				
2 unskilled operatives (general)		25.50				
1 plant operator (skill rate 2)		18.50				
1 plant operator (skill rate 3)		16.59				
Total Gang Rate / Hour	£	**88.83**				
Flexible paving gang						
1 ganger/chargehand (skill rate 4)		14.58				
2 skilled operatives (skill rate 4)		27.31				
4 unskilled operatives (general)		51.00				
4 plant operators (skill rate 3)		66.38				
Total Gang Rate / Hour	£	**159.27**				
[Concrete paving gang						
1 ganger/chargehand (skill rate 4)		14.58				
2 skilled operatives (skill rate 4)		27.31				
4 unskilled operatives (general)		51.00				
1 plant operator (skill rate 2)		18.50				
1 plant operator (skill rate 3)		16.59				
Total Rate / Hour	£	**127.99**				
Road surface spraying gang						
1 plant operator (skill rate 3)		16.59				
Total Gang Cost / Hour		16.59				
Road chippings gang						
1 ganger/chargehand (skill rate 4) - 50% of time		7.29				
1 skilled operative (skill rate 4)		13.65				
2 unskilled operatives (general)		25.50				
3 plant operators (skill rate 3)		49.78				
Total Gang Rate / Hour	£	**96.23**				
Cutting slabs gang						
1 unskilled operative (general)		12.75				
Total Gang Rate / Hour	£	**12.75**				
Concrete filled joints gang						
1 ganger/chargehand - 50% of time		7.29				
1 skilled operative (skill rate 4)		13.65				
2 unskilled operatives (general)		25.50				
Total Gang Rate / Hour	£	**46.45**				
Milling gang						
1 ganger/chargehand (skill rate 4)		14.58				
2 skilled operatives (skill rate 4)		27.31				
4 unskilled operatives (general)		51.00				
1 plant operator (skill rate 2)		18.50				
1 plant operator (skill rate 3)		16.59				
Total Gang Rate / Hour	£	**127.99**				

SERIES 700: PAVEMENTS

Item	Gang hours	Labour £	Plant £	Material £	Unit	Total rate £
Rake and compact planed material gang						
1 ganger/chargehand (skill rate 4)		14.58				
1 skilled operative (skill rate 4)		13.65				
3 unskilled operatives (general)		38.25				
1 plant operator (skill rate 4)		14.92				
1 plant operator (skill rate 3)		16.59				
Total Gang Rate / Hour	£	**98.01**				
RESOURCES - PLANT						
Sub-base laying						
93 kW motor grader			27.58			
0.8 m³ tractor loader			22.35			
6 t towed vibratory roller			10.73			
Total Gang Rate / Hour		£	**60.65**			
Flexible paving						
2 asphalt pavers, 35 kW, 4.0 m			99.59			
2 deadweight rollers, 3 point, 10 t			33.55			
tractor with front bucket and integral 2 tool						
compressor			21.16			
compressor tools : rammer			0.45			
compressor tools : poker vibrator			2.04			
compressor tools : extra 15 m hose			0.34			
tar sprayer, 100 litre			7.81			
self propelled chip spreader			10.75			
channel (heat) iron			1.21			
Total Gang Rate / Hour		£	**176.91**			
Concrete paving						
wheeled loader, 2.60 m³			58.81			
concrete paver, 6.0 m			80.52			
concrete slipform finisher			20.02			
Total Gang Rate / Hour		£	**159.35**			
Road surface spraying						
tar sprayer; 100 litre			7.81			
Total Gang Rate / Hour		£	**7.81**			
Road chippings						
deadweight roller, 3 point, 10 t			16.78			
tar sprayer, 100 litre			7.81			
self propelled chip spreader			10.75			
channel (heat) iron			1.21			
Total Gang Rate / Hour		£	**36.56**			
Cutting slabs						
compressor, 65 cfm			6.41			
12' disc cutter			1.44			
Total Gang Rate / Hour		£	**7.85**			

SERIES 700: PAVEMENTS

Item	Gang hours	Labour £	Plant £	Material £	Unit	Total rate £
RESOURCES - PLANT – cont						
Cold milling						
cold planer, 2.10 m			57.44			
wheeled loader, 2.60 m³			58.81			
Total Gang Rate / Hour		£	**116.24**			
Heat planing						
heat planer, 4.5 m			82.63			
wheeled loader, 2.60 m³			58.81			
Total Gang Rate / Hour		£	**141.44**			
Rake and compact planed material						
deadweight roller, 3 point, 10 t			16.78			
tractor with front bucket and integral 2 tool						
compressor			21.16			
channel (heat) iron			1.21			
Total Gang Rate / Hour		£	**39.15**			
SUB-BASE						
The following unit costs are generally based on the Highways Agency Specification for Highway Works and reference is made throughout this Section to clauses within that specification.						
Granular material DfT Type 1;						
Sub-base in carriageway, hardshoulder and hardstrip						
75 mm deep	0.04	3.09	2.12	25.73	m³	**30.94**
100 mm deep	0.04	3.53	2.43	25.73	m³	**31.69**
150 mm deep	0.05	3.97	2.73	25.73	m³	**32.43**
200 mm deep	0.05	4.41	3.03	25.73	m³	**33.17**
Granular material DfT Type 2;						
Sub-base; spread and graded						
75 mm deep	0.04	3.09	2.12	23.06	m³	**28.27**
100 mm deep	0.04	3.53	2.43	23.06	m³	**29.02**
150 mm deep	0.05	3.97	2.73	23.06	m³	**29.76**
200 mm deep	0.05	4.41	3.03	23.06	m³	**30.50**
Wet lean concrete DfT specified strength mix C20, 20 mm aggregate;						
Sub-base; spread and graded						
100 mm deep	0.05	3.97	2.73	90.76	m³	**97.46**
200 mm deep	0.05	4.41	3.03	90.76	m³	**98.20**
Hardcore;						
Sub-base; spread and graded						
100 mm deep	0.04	3.53	2.43	38.93	m³	**44.89**
150 mm deep	0.05	3.97	2.73	38.93	m³	**45.63**
200 mm deep	0.05	4.41	3.03	38.93	m³	**46.37**

SERIES 700: PAVEMENTS

Item	Gang hours	Labour £	Plant £	Material £	Unit	Total rate £
Wet mix macadam; DfT Series 900;						
Sub-base; spread and graded						
75 mm deep	0.04	3.09	3.21	25.48	m³	**31.78**
100 mm deep	0.04	3.53	3.34	25.48	m³	**32.35**
200 mm deep	0.05	4.41	3.52	25.48	m³	**33.41**
PAVEMENT (FLEXIBLE)						
Notes - Labour and Plant						
All outputs are based on clear runs without undue						
delay to two pavers with a 75% utilisation						
The outputs can be adjusted as follows to take						
account of space or time influences on the utilisation						
Factors for varying utilisation of Labour and Plant:						
1 paver @ 75% utilisation = x 2.00						
1 paver @ 100% utilisation = x 1.50						
2 pavers @ 100% utilisation = x 0.75						
Dense Bitumen Macadam						
Base to DfT Clause 903						
100 mm deep	0.02	3.16	3.54	6.00	m²	**12.70**
150 mm deep	0.03	3.95	4.42	9.01	m²	**17.38**
200 mm deep	0.03	4.74	5.31	12.01	m²	**22.06**
Binder Course to DfT Clause 904						
50 mm deep	0.02	2.37	2.66	2.61	m²	**7.64**
100 mm deep	0.02	3.16	3.54	5.23	m²	**11.93**
Surface Course to DfT Clause 909						
30 mm deep	0.01	1.58	1.77	2.23	m²	**5.58**
50 mm deep	0.02	2.37	2.66	3.71	m²	**8.74**
Bitumen Macadam						
Binder Course to DfT Clause 901						
35 mm deep	0.01	1.58	1.77	2.37	m²	**5.72**
70 mm deep	0.02	2.37	2.66	4.74	m²	**9.77**
Dense Tarmacadam						
Base to DfT Clause 903						
50 mm deep	0.02	2.37	2.66	3.25	m²	**8.28**
100 mm deep	0.02	2.37	2.66	6.51	m²	**11.54**
Binder Course to DfT Clause 907						
60 mm deep	0.02	2.37	2.66	4.24	m²	**9.27**
80 mm deep	0.02	2.37	2.66	5.65	m²	**10.68**
Dense Tar Surfacing						
Surface Course to DfT Series 900						
30 mm deep	0.01	1.58	1.77	2.27	m²	**5.62**
50 mm deep	0.02	2.37	2.66	3.78	m²	**8.81**

SERIES 700: PAVEMENTS

Item	Gang hours	Labour £	Plant £	Material £	Unit	Total rate £
PAVEMENT (FLEXIBLE) – cont						
Cold Asphalt						
Surface Course to DfT Series 900						
15 mm deep	0.01	1.58	1.77	1.15	m²	**4.50**
30 mm deep	0.01	1.58	1.77	2.30	m²	**5.65**
Rolled Asphalt						
Base to DfT Clause 904						
60 mm deep	0.02	2.37	2.66	3.99	m²	**9.02**
80 mm deep	0.02	2.37	2.66	5.32	m²	**10.35**
Surface Course to DfT Clause 905						
40 mm deep	0.02	2.37	2.66	3.61	m²	**8.64**
60 mm deep	0.02	2.37	2.66	5.42	m²	**10.45**
PAVEMENT (CONCRETE)						
The following unit costs are for jointed reinforced concrete slabs, laid in reasonable areas (over 200 m²) by paver train/slipformer						
Designed mix; cement to BS EN 197-1; grade C30, 20 mm aggregate						
Slab, runway, access roads or similar						
180 mm deep	0.02	1.90	2.39	15.17	m²	**19.46**
220 mm deep	0.02	2.29	2.87	18.54	m²	**23.70**
260 mm deep	0.02	2.79	3.51	21.91	m²	**28.21**
300 mm deep	0.03	3.17	3.98	25.28	m²	**32.43**
Fabric reinforcement						
Steel fabric reinforcement to BS4483						
Ref A142 nominal mass 2.22 kg	0.03	3.81	–	1.96	m²	**5.77**
Ref A252 nominal mass 3.95 kg	0.04	5.08	–	2.50	m²	**7.58**
Ref B385 nominal mass 4.53 kg	0.04	5.08	–	3.75	m²	**8.83**
Ref C636 nominal mass 5.55 kg	0.05	6.35	–	4.03	m²	**10.38**
Ref B503 nominal mass 5.93 kg	0.05	6.35	–	5.52	m²	**11.87**
Mild Steel bar reinforcement Bs 4449						
Bars; supplied in bent and cut lengths						
6 mm nominal size	8.00	1015.90	–	588.83	t	**1604.73**
8 mm nominal size	6.74	855.89	–	588.83	t	**1444.72**
10 mm nominal size	6.74	855.89	–	583.27	t	**1439.16**
12 mm nominal size	6.74	855.89	–	572.16	t	**1428.05**
16 mm nominal size	6.15	780.97	–	566.05	t	**1347.02**
High yield steel bar reinforcement Bs 4449 or 4461						
Bars; supplied in bent and cut lengths						
6 mm nominal size	8.00	1015.90	–	638.83	t	**1654.73**
8 mm nominal size	6.74	855.89	–	633.02	t	**1488.91**
10 mm nominal size	6.74	855.89	–	633.02	t	**1488.91**
12 mm nominal size	6.74	855.89	–	621.40	t	**1477.29**
16 mm nominal size	6.15	780.97	–	615.60	t	**1396.57**

SERIES 700: PAVEMENTS

Item	Gang hours	Labour £	Plant £	Material £	Unit	Total rate £
Sheeting to prevent moisture loss						
Polyethelene sheeting; lapped joints; horizontal below concrete pavements						
1000 gauge	0.01	1.27	–	0.88	m²	**2.15**
2000 gauge	0.01	1.27	–	1.60	m²	**2.87**
Joints in concrete slabs						
Longitudinal joints						
180 mm deep concrete	0.01	1.52	1.91	20.07	m	**23.50**
220 mm deep concrete	0.01	1.52	1.91	34.32	m	**37.75**
260 mm deep concrete	0.01	1.52	1.91	34.83	m	**38.26**
300 mm deep concrete	0.01	1.52	1.91	45.59	m	**49.02**
Expansion joints						
180 mm deep concrete	0.01	1.52	1.91	46.73	m	**50.16**
220 mm deep concrete	0.01	1.52	1.91	20.02	m	**23.45**
260 mm deep concrete	0.01	1.52	1.91	21.18	m	**24.61**
300 mm deep concrete	0.01	1.52	1.91	22.49	m	**25.92**
Contraction joints						
180 mm deep concrete	0.01	1.52	1.91	26.80	m	**30.23**
220 mm deep concrete	0.01	1.52	1.91	13.20	m	**16.63**
260 mm deep concrete	0.01	1.52	1.91	14.40	m	**17.83**
300 mm deep concrete	0.01	1.52	1.91	15.54	m	**18.97**
Construction joints						
180 mm deep concrete	0.01	1.52	1.91	16.65	m	**20.08**
220 mm deep concrete	0.01	1.52	1.91	11.14	m	**14.57**
260 mm deep concrete	0.01	1.52	1.91	12.26	m	**15.69**
300 mm deep concrete	0.01	1.52	1.91	13.36	m	**16.79**
Open joints with filler						
ne 0.5 m; 10 mm flexcell joint filler	0.11	5.07	–	3.15	m	**8.22**
0.5 - 1 m; 10 mm flexcell joint filler	0.11	5.07	–	4.52	m	**9.59**
Joint sealants						
10 × 20 mm hot bitumen sealant	0.14	6.45	–	3.10	m	**9.55**
20 × 20 mm cold polysulphide sealant	0.18	8.29	–	6.16	m	**14.45**
Trimming edges only of existing slabs, floors or similar surfaces (wet or dry); 6 mm cutting width						
50 mm deep	0.02	0.25	0.16	3.28	m	**3.69**
100 mm deep	0.03	0.38	0.24	3.44	m	**4.06**
Cutting existing slabs, floors or similar surfaces (wet or dry); 8 mm cutting width						
50 mm deep	0.03	0.32	0.19	3.49	m	**4.00**
100 mm deep	0.06	0.77	0.47	3.67	m	**4.91**
150 mm deep	0.08	1.02	0.63	3.74	m	**5.39**

SERIES 700: PAVEMENTS

Item	Gang hours	Labour £	Plant £	Material £	Unit	Total rate £
SURFACE TREATMENT						
Slurry sealing; BS 434 class K3						
Slurry sealing to DfT Clause 918						
3 mm deep	0.02	0.25	0.12	1.28	m²	**1.65**
4 mm deep	0.02	0.25	0.12	1.77	m²	**2.14**
Coated chippings, 9 - 11 kg/m²						
Surface dressing to DfT Clause 915						
6 mm nominal size	0.01	0.96	0.37	0.80	m²	**2.13**
8 mm nominal size	0.01	0.96	0.37	0.82	m²	**2.15**
10 mm nominal size	0.01	0.96	0.37	0.83	m²	**2.16**
12 mm nominal size	0.01	0.96	0.37	0.91	m²	**2.24**
Anti Skid Surfacing System						
High friction surfacing to DfT Clause 924						
Proprietary resin bonded surfacing system,						
colours (Buff, Grey, Red, Green)	–	–	–	–	m²	**19.25**
TACK COAT						
Bituminous spray; BS 434 K1 - 40						
Tack coat to DfT Clause 920						
large areas; over 20 m²	0.02	0.25	0.12	0.26	m²	**0.63**
COLD MILLING (PLANING)						
Milling pavement (assumes disposal on site or						
re-use as fill but excludes transport if required)						
75 mm deep	0.03	3.43	3.14	–	m²	**6.57**
100 mm deep	0.04	4.57	4.18	–	m²	**8.75**
50 mm deep; scarifying surface	0.02	2.79	2.56	–	m²	**5.35**
75 mm deep; scarifying surface	0.04	4.70	4.30	–	m²	**9.00**
25 mm deep; heat planing for re-use	0.03	4.06	4.53	–	m²	**8.59**
50 mm deep; heat planing for re-use	0.06	7.11	7.92	–	m²	**15.03**
INSITU RECYCLING						
Raking over scarified or heat planed material;						
compacting with 10 t roller						
50 mm deep	0.01	0.97	0.39	–	m²	**1.36**

SERIES 1100: KERBS,FOOTWAYS AND PAVED AREAS

Item	Gang hours	Labour £	Plant £	Material £	Unit	Total rate £
NOTES						
Measurement Note: sub-bases are shown separate from their associated paving to simplify the presentation of cost alternatives.						
Measurement Note: bases are shown separate from their associated kerb etc. to simplify the presentation of cost alternatives.						
Kerb quadrants, droppers are shown separately						
The re-erection cost for kerbs, channels and edgings etc. taken from store assumes that major components are in good condition; the prices below allow a sum of 20% of the value of new materials to cover minor repairs together with an allowance for replacing a proportion of units.						
RESOURCES - LABOUR						
Kerb laying gang						
3 skilled operatives (skill rate 4)		40.96				
1 unskilled operative (general)		12.75				
1 plant operator (skill rate 3) - 25% of time		4.15				
Total Gang Rate / Hour	£	57.86				
Path sub-base, bitmac and gravel laying gang						
1 skilled operative (skill rate 4)		13.65				
2 unskilled operatives (general)		25.50				
1 plant operator (skill rate 3)		16.59				
Total Gang Rate / Hour	£	55.75				
Paviors and flagging gang						
1 skilled operative (skill rate 4)		13.65				
1 unskilled operative (general)		12.75				
Total Gang Rate / Hour	£	26.40				
RESOURCES - PLANT						
Kerb laying						
backhoe JCB 3CX (25% of time)			4.49			
12' Stihl saw			1.11			
road forms			2.16			
Total Gang Rate / Hour		£	7.76			
Path sub-base, bitmac and gravel laying						
backhoe JCB3CX			17.96			
2 t dumper			6.49			
pedestrian roller `Bomag BW90S			4.76			
Total Gang Rate / Hour		£	29.21			
Paviors and flagging						
2 t dumper (33% of time)			2.16			
Total Gang Rate / Hour		£	2.16			

SERIES 1100: KERBS,FOOTWAYS AND PAVED AREAS

Item	Gang hours	Labour £	Plant £	Material £	Unit	Total rate £
KERBS, CHANNELS, EDGINGS, COMBINED DRAINAGE AND KERB BLOCKS AND LINEAR DRAINAGE						
Kerbs, Channels, Edgings, Combined Drainage And Kerb Blocks And Linear Drainage Channel Systems						
Foundations to kerbs etc.						
Mass concrete						
200 × 100 mm	0.01	0.57	0.07	1.84	m	**2.48**
300 × 150 mm	0.02	0.86	0.12	4.21	m	**5.19**
450 × 150 mm	0.02	1.15	0.16	6.22	m	**7.53**
100 × 100 mm haunching, per side	0.01	0.29	0.04	0.44	m	**0.77**
Precast concrete units; BS 7263; bedded jointed and pointed in cement mortar						
Kerbs; bullnosed, splayed or half battered; laid straight or curved exceeding 12 m radius						
125 × 150 mm	0.06	3.44	0.47	4.93	m	**8.84**
125 × 255 mm	0.07	4.01	0.54	7.39	m	**11.94**
150 × 305 mm	0.07	4.01	0.54	7.64	m	**12.19**
Kerbs; bullnosed, splayed or half battered; laid to curves not exceeding 12 m radius						
125 × 150 mm	0.07	3.73	0.50	3.99	m	**8.22**
125 × 255 mm	0.08	4.30	0.58	5.67	m	**10.55**
150 × 305 mm	0.08	4.30	0.58	8.44	m	**13.32**
Quadrants (normally included in general rate for kerbs; shown separately for estimating purposes)						
305 × 305 × 150 mm	0.08	4.59	0.62	10.71	nr	**15.92**
455 × 455 × 255 mm	0.10	5.73	0.78	13.67	nr	**20.18**
Drop kerbs (normally included in general rate for kerbs; shown separately for estimating purposes)						
125 × 255 mm	0.07	4.01	0.54	5.70	nr	**10.25**
150 × 305 mm	0.07	4.01	0.54	10.47	nr	**15.02**
Channels; laid straight or curved exceeding 12 m radius						
125 × 255 mm	0.07	4.01	0.54	6.98	m	**11.53**
Channels; laid to curves not exceeding 12 m radius						
255 × 125 mm	0.07	4.01	0.54	6.98	m	**11.53**
Edgings; laid straight or curved exceeding 12 m radius						
150 × 50 mm	0.04	2.29	0.31	3.09	m	**5.69**
Edgings; laid to curves not exceeding 12 m radius						
150 × 50 mm	0.05	2.58	0.35	3.94	m	**6.87**
Precast concrete drainage channels;Charcon 'Safeticurb';channels jointed with plastic rings and bedded, jointed and pointed in cement mortar						
Channel unit; Type DBA/3; laid straight or curved exceeding 12 m radius						
250 × 250 mm; medium duty	0.08	4.30	0.58	47.64	m	**52.52**
305 × 305 mm; heavy duty	0.10	5.45	0.74	103.42	m	**109.61**

SERIES 1100: KERBS,FOOTWAYS AND PAVED AREAS

Item	Gang hours	Labour £	Plant £	Material £	Unit	Total rate £
Precast concrete Ellis Trief safety kerb; bedded jointed and pointed in cement mortar						
Kerbs; laid straight or curved exceeding 12m radius						
415×380mm	0.23	12.90	1.75	60.86	m	**75.51**
Kerbs; laid to curves not exceeding 12m radius						
415×380mm	0.25	14.34	1.94	64.40	m	**80.68**
Precast concrete combined kerb and drainage block Beany Block System; bedded jointed and pointed in cement mortar						
Kerb; top block, shallow base unit, standard cover plate and frame						
laid straight or curved exceeding 12m radius	0.15	8.60	1.17	117.82	m	**127.59**
laid to curves not exceeding 12m radius	0.20	11.47	1.55	179.56	m	**192.58**
Kerb; top block, standard base unit, standard cover plate and frame						
laid straight or curved exceeding 12m radius	0.15	8.60	1.17	117.82	m	**127.59**
laid to curves not exceeding 12m radius	0.20	11.47	1.55	179.56	m	**192.58**
Kerb; top block, deep base unit, standard cover plate and frame						
Straight or curved over 12m radius	0.15	8.60	1.17	262.19	m	**271.96**
laid to curves not exceeding 12m radius	0.20	11.47	1.55	335.03	m	**348.05**
Base block depth tapers	0.10	5.73	0.78	28.86	m	**35.37**
Extruded asphalt edgings to pavings; slip formed BS 5931						
Kerb; laid straight or curved exceeding 12m radius						
75mm kerb height	–	–	–	5.88	m	**5.88**
100mm kerb height	–	–	–	10.77	m	**10.77**
125mm kerb height	–	–	–	12.75	m	**12.75**
Channel; laid straight or curved exceeding 12m radius						
300mm channel width	–	–	–	13.91	m	**13.91**
250mm channel width	–	–	–	13.91	m	**13.91**
Kerb; laid to curves not exceeding 12m radius						
75mm kerb height	–	–	–	12.92	m	**12.92**
100mm kerb height	–	–	–	10.77	m	**10.77**
125mm kerb height	–	–	–	11.27	m	**11.27**
Channel; laid to curves not exceeding 12m radius						
300mm channel width	–	–	–	18.89	m	**18.89**
250mm channel width	–	–	–	14.39	m	**14.39**
Extruded concrete; slip formed						
Kerb; laid straight or curved exceeding 12m radius						
100mm kerb height	–	–	–	12.42	m	**12.42**
125mm kerb height	–	–	–	13.44	m	**13.44**
Kerb; laid to curves not exceeding 12m radius						
100mm kerb height	–	–	–	11.88	m	**11.88**
125mm kerb height	–	–	–	12.69	m	**12.69**

SERIES 1100: KERBS,FOOTWAYS AND PAVED AREAS

Item	Gang hours	Labour £	Plant £	Material £	Unit	Total rate £
KERBS, CHANNELS, EDGINGS, COMBINED DRAINAGE AND KERB BLOCKS AND LINEAR DRAINAGE – cont						
Additional Concrete For Kerbs, Channels, Edgings, Combined Drainage And Kerb Blocks And Linear Drainage Channel Systems						
Additional in situ concrete concrete						
kerbs	0.50	28.67	3.43	91.43	m³	**123.53**
channels	0.50	28.67	3.43	91.43	m³	**123.53**
edgings	0.50	28.67	3.43	91.43	m³	**123.53**
Remove From Store And Relay Kerbs, Channels, Edgings, Combined Drainage and Kerb Blocks And Linear Drainage Channel Systems						
Remove from store and relay precast concrete units; bedded jointed and pointed in cement mortar						
Kerbs; laid straight or curved exceeding 12 m radius						
125 × 150 mm	0.06	3.44	0.47	2.07	m	**5.98**
125 × 255 mm	0.07	4.01	0.54	2.56	m	**7.11**
150 × 305 mm	0.07	4.01	0.54	2.61	m	**7.16**
Kerbs; laid to curves not exceeding 12 m radius						
125 × 150 mm	0.07	3.73	0.50	1.88	m	**6.11**
125 × 255 mm	0.08	4.30	0.58	2.22	m	**7.10**
150 × 305 mm	0.08	4.30	0.58	2.77	m	**7.65**
Quadrants (normally included in general rate for kerbs; shown separately for estimating purposes)						
305 × 305 × 150 mm	0.08	4.59	0.62	3.23	nr	**8.44**
455 × 455 × 255 mm	0.10	5.73	0.78	3.82	nr	**10.33**
Drop kerbs (normally included in general rate for kerbs; shown separately for estimating purposes)						
125 × 255 mm	0.07	4.01	0.54	2.22	nr	**6.77**
150 × 305 mm	0.07	4.01	0.54	3.18	nr	**7.73**
Channels; laid straight or curved exceeding 12 m radius						
125 × 255 mm	0.07	4.01	0.54	2.48	m	**7.03**
Channels; laid to curves not exceeding 12 m radius						
255 × 125 mm	0.07	4.01	0.54	2.48	m	**7.03**
Edgings; laid straight or curved exceeding 12 m radius						
150 × 50 mm	0.04	2.29	0.31	1.70	m	**4.30**
Edgings; laid to curves not exceeding 12 m radius						
150 × 50 mm	0.05	2.58	0.35	1.70	m	**4.63**

SERIES 1100: KERBS,FOOTWAYS AND PAVED AREAS

Item	Gang hours	Labour £	Plant £	Material £	Unit	Total rate £
Remove from store and relay precast concrete drainage channels;Charcon Safeticurb; channels jointed with plastic rings and bedded, jointed and pointed in cement mortar						
Channel unit; Type DBA/3; laid straight or curved exceeding 12m radius						
250 × 254mm; medium duty	0.08	4.30	0.58	10.61	m	**15.49**
305 × 305mm; heavy duty	0.10	5.45	0.74	21.77	m	**27.96**
Remove from store and relay precast concrete Ellis Trief safety kerb; bedded jointed and pointed in cement mortar						
Kerbs; laid straight or curved exceeding 12m radius						
415 × 380mm	0.23	12.90	1.75	13.26	m	**27.91**
Kerbs; laid to curves not exceeding 12m radius						
415 × 380mm	0.25	14.34	1.94	13.97	m	**30.25**
Remove from store and relay precast concrete combined kerb and drainage block Beany Block System; bedded jointed and pointed in cement mortar						
Kerb; top block, shallow base unit, standard cover plate and frame						
laid straight or curved exceeding 12m radius	0.15	8.60	1.17	24.65	m	**34.42**
laid to curves not exceeding 12m radius	0.20	11.47	1.55	37.00	m	**50.02**
Kerb; top block, standard base unit, standard cover plate and frame						
laid straight or curved exceeding 12m radius	0.15	8.60	1.17	24.65	m	**34.42**
laid to curves not exceeding 12m radius	0.20	11.47	1.55	37.00	m	**50.02**
Kerb; top block, deep base unit, standard cover plate and frame						
Straight or curved over 12m radius	0.15	8.60	1.17	53.52	m	**63.29**
laid to curves not exceeding 12m radius	0.20	11.47	1.55	68.09	m	**81.11**
Base block depth tapers	0.10	5.73	0.78	6.86	m	**13.37**

SERIES 1100: KERBS,FOOTWAYS AND PAVED AREAS

Item	Gang hours	Labour £	Plant £	Material £	Unit	Total rate £
FOOTWAYS AND PAVED AREAS						
Sub-bases						
To paved area; sloping not exceeding 10° to the horizontal						
100 mm thick sand	0.01	0.50	0.26	3.72	m²	**4.48**
150 mm thick sand	0.01	0.66	0.35	5.59	m²	**6.60**
100 mm thick gravel	0.01	0.50	0.26	2.06	m²	**2.82**
150 mm thick gravel	0.01	0.66	0.35	3.09	m²	**4.10**
100 mm thick hardcore	0.01	0.50	0.26	1.21	m²	**1.97**
150 mm thick hardcore	0.01	0.66	0.35	1.82	m²	**2.83**
100 mm thick concrete grade 20/20	0.02	1.16	0.61	9.16	m²	**10.93**
150 mm thick concrete grade 20/20	0.03	1.77	0.94	13.74	m²	**16.45**
Bitumen macadam surfacing; BS 4987; binder course of 20 mm open graded aggregate to clause 2.6.1 tables 5 - 7; surface course of 6 mm medium graded aggregate to clause 2.7.6 tables 32 - 33; excluding sub-base						
Paved area 60 mm thick; comprising binder course 40 mm thick surface course 20 mm thick						
sloping at 10° or less to the horizontal	0.09	4.70	2.48	10.78	m²	**17.96**
sloping at more than 10° to the horizontal	0.10	5.25	2.78	10.78	m²	**18.81**
Bitumen macadam surfacing; red additives; BS 4987; binder course of 20 mm open graded aggregate to clause 2.6.1 tables 5 - 7; surface course of 6 mm medium graded aggregate to clause 2.7.6 tables 32 - 33; excluding sub-base						
Paved area 60 mm thick;comprising binder course 40 mm thick surface course 20 mm thick						
sloping at 10° or less to the horizontal	0.09	4.70	2.48	13.06	m²	**20.24**
sloping at more than 10° to the horizontal	0.10	5.25	2.78	13.06	m²	**21.09**
Bitumen macadam surfacing; green additives; BS 4987; binder course of 20 mm open graded aggregate to clause 2.6.1 tables 5 - 7; surface course of 6 mm medium graded aggregate to clause 2.7.6 tables 32 - 33; excluding sub-base						
Paved area 60 mm thick; comprising binder course 40 mm thick surface course 20 mm thick						
sloping at 10° or less to the horizontal	0.09	4.70	2.48	14.62	m²	**21.80**
sloping at more than 10° to the horizontal	0.10	5.25	2.78	14.62	m²	**22.65**
Granular base surfacing; Central Reserve Treatments Limestone, graded 10 mm down laid and compacted; excluding sub-base						
Paved area 100 mm thick; surface sprayed with two coats of cold bituminous emulsion; blinded with 6 mm quarzite fine gravel						
sloping not exceeding 10° to the horizontal	0.02	1.11	0.58	6.19	m²	**7.88**

SERIES 1100: KERBS,FOOTWAYS AND PAVED AREAS

Item	Gang hours	Labour £	Plant £	Material £	Unit	Total rate £
Ennstone Johnston Golden gravel; graded 13mm to fines; rolled wet						
Paved area 50mm thick; single layer						
sloping not exceeding 10° to the horizontal	0.03	1.66	0.88	8.06	m²	**10.60**
Precast concrete slabs; BS 7263; grey; 5 point bedding and pointing joints in cement mortar; excluding sub-base						
Paved area 50mm thick; comprising 600×450 x 50mm units						
sloping at 10° or less to the horizontal	0.28	7.33	0.60	11.65	m²	**19.58**
Paved area 50mm thick; comprising 600×600 x 50mm units						
sloping at 10° or less to the horizontal	0.24	6.28	0.52	9.43	m²	**16.23**
Paved area 50mm thick; comprising 900×600 x 50mm units						
sloping at 10° or less to the horizontal	0.20	5.24	0.43	8.33	m²	**14.00**
Extra for coloured, 50mm thick	–	–	–	5.32	m²	**5.32**
Paved area 63mm thick; comprising 600×600 x 63mm units						
sloping at 10° or less to the horizontal	0.25	6.54	0.54	11.53	m²	**18.61**
Paved area 63mm thick; comprising 900×600 x 63mm units						
sloping at 10° or less to the horizontal	0.21	5.50	0.45	9.43	m²	**15.38**
Precast concrete rectangular paving blocks; BS 6717; grey; bedding on 50mm thick dry sharp sand; filling joints; excluding sub-base						
Paved area 80mm thick; comprising 200×100 x 80mm units						
sloping at 10° or less to the horizontal	0.30	7.85	0.65	16.31	m²	**24.81**
Precast concrete rectangular paving blocks; BS 6717; coloured; bedding on 50mm thick dry sharp sand; filling joints; excluding sub-base						
Paved area 80mm thick; comprising 200×100 x 80mm units						
sloping at 10° or less to the horizontal	0.30	7.85	0.65	17.23	m²	**25.73**
Brick paviors delivered to site; bedding on 20mm thick mortar; excluding sub-base						
Paved area 85mm thick; comprising 215×103 x 65mm units						
sloping at 10° or less to the horizontal	0.30	7.85	0.65	26.93	m²	**35.43**
Granite setts (2.88kg/mm thickness/m2); bedding on 25mm cement mortar; excluding sub-base						
Paved area 100mm thick; comprising 100×100 x 100mm units; laid to random pattern						
sloping at 10° or less to the horizontal	0.90	23.56	1.95	63.59	m²	**89.10**

SERIES 1100: KERBS,FOOTWAYS AND PAVED AREAS

Item	Gang hours	Labour £	Plant £	Material £	Unit	Total rate £
FOOTWAYS AND PAVED AREAS – cont						
Granite setts (2.88 kg/mm thickness/m2) – cont						
Paved area 100 mm thick; comprising 100 × 100 x 100 mm units; laid to specific pattern						
sloping at 10° or less to the horizontal	1.20	31.41	2.60	63.59	m²	**97.60**
Cobble paving; 50–75 mm stones; bedding on 25 mm cement mortar; filling joints; excluding sub-base						
Paved area; comprising 50–75 mm stones; laid to random pattern						
sloping at 10° or less to the horizontal	1.00	26.18	2.14	18.57	m²	**46.89**
Remove From Store And Relay Paving Flags, SLABS AND BLOCKS						
Note						
An allowance of 20% of the cost of providing new pavings has been included in the rates below to allow for units which have to be replaced through unacceptable damage.						
Remove from store and relay precast concrete units; bedded jointed and pointed in cement mortar						
Remove from store and relay precast concrete slabs; 5 point bedding and pointing joints in cement mortar; excluding sub-base						
Paved area 50 mm thick; comprising 600 × 450 x 50 mm units						
sloping at 10° or less to the horizontal	0.28	7.33	0.60	2.95	m²	**10.88**
Paved area 50 mm thick; comprising 600 × 600 x 50 mm units						
sloping at 10° or less to the horizontal	0.24	6.28	0.52	2.51	m²	**9.31**
Paved area 50 mm thick; comprising 900 × 600 x 50 mm units						
sloping at 10° or less to the horizontal	0.20	5.24	0.43	2.29	m²	**7.96**
Paved area 63 mm thick; comprising 600 × 600 x 63 mm units						
sloping at 10° or less to the horizontal	0.25	6.54	0.54	2.92	m²	**10.00**
Paved area 63 mm thick; comprising 900 × 600 x 63 mm units						
sloping at 10° or less to the horizontal	0.21	5.50	0.45	2.51	m²	**8.46**

SERIES 1100: KERBS,FOOTWAYS AND PAVED AREAS

Item	Gang hours	Labour £	Plant £	Material £	Unit	Total rate £
Remove from store and relay precast concrete rectangular paving blocks; bedding on 50 mm thick dry sharp sand; filling joints; excluding sub-base Paved area 80 mm thick; comprising 200 × 100 x 80 mm units						
sloping at 10° or less to the horizontal	0.30	7.85	0.65	4.14	m²	**12.64**
Remove from store and relay brick paviors; bedding on 20 mm thick mortar; excluding sub-base Paved area 85 mm thick; comprising 215 × 103 x 65 mm units						
sloping at 10° or less to the horizontal	0.30	7.85	0.65	7.17	m²	**15.67**

SERIES 1200: TRAFFIC SIGNS AND ROAD MARKINGS

Item	Gang hours	Labour £	Plant £	Material £	Unit	Total rate £
NOTES						
The re-erection cost for traffic signs taken from store assumes that major components are in good condition; the prices below allow a sum of 20% of the value of new materials to cover minor repairs, new fixings and touching up any coatings.						
RESOURCES - LABOUR						
Traffic signs gang						
1 ganger/chargehand (skill rate 3)		16.18				
1 skilled operative (skill rate 3)		15.25				
2 unskilled operatives (general)		25.50				
1 plant operator (skill rate 3) - 25% of time		4.15				
Total Gang Rate / Hour	£	**61.08**				
Bollards, furniture gang						
1 ganger/chargehand (skill rate 4)		14.58				
1 skilled operative (skill rate 4)		13.65				
2 unskilled operatives (general)		25.50				
Total Gang Rate / Hour	£	**53.74**				
RESOURCES - PLANT						
Traffic signs						
JCB 3CX backhoe - 50% of time			8.98			
125 cfm compressor - 50% of time			4.85			
compressor tools: hand held hammer drill - 50% of time			0.49			
compressor tools: clay spade - 50% of time			0.22			
compressor tools: extra 15 m hose - 50% of time			0.17			
8 t lorry with hiab lift - 50% of time			6.39			
Total Rate / Hour		£	**21.10**			
Bollards, furniture						
125 cfm compressor - 50% of time			4.85			
compressor tools: hand held hammer drill - 50% of time			0.49			
compressor tools:clay spade - 50% of time			0.22			
compressor tools: extra 15 m hose - 50% of time			0.17			
8 t lorry with hiab lift - 25% of time			3.19			
Total Rate / Hour		£	**8.92**			

SERIES 1200: TRAFFIC SIGNS AND ROAD MARKINGS

Item	Gang hours	Labour £	Plant £	Material £	Unit	Total rate £
TRAFFIC SIGNS						
In this section prices will vary depending upon the diagram configurations. The following are average costs of signs and Bollards. Diagram numbers refer to theTraffic Signs Regulations and General Directions 2002 and the figure numbers refer to the Traffic Signs Manual.						
Examples of Prime Costs for Class 1 (High Intensity) traffic and road signs (ex works) for orders exceeding £1,000.						
Permanent traffic sign as non-lit unit on						
600 × 450 mm	–	–	–	91.22	nr	91.22
600 mm diameter	–	–	–	114.39	nr	114.39
600 mm triangular	–	–	–	93.99	nr	93.99
500 × 500 mm	–	–	–	80.61	nr	80.61
450 × 450 mm	–	–	–	68.38	nr	68.38
450 × 300 mm	–	–	–	56.11	nr	56.11
1200 × 400 mm (CHEVRONS)	–	–	–	185.67	nr	185.67
Examples of Prime Costs for Class 2 (Engineering Grade) traffic and road signs (ex works) for orders exceeding £1,000.						
600 × 450 mm	–	–	–	112.51	nr	112.51
600 mm diameter	–	–	–	141.28	nr	141.28
600 mm triangular	–	–	–	117.87	nr	117.87
500 × 500 mm	–	–	–	99.38	nr	99.38
450 × 450 mm	–	–	–	84.84	nr	84.84
450 × 300 mm	–	–	–	66.86	nr	66.86
1200 × 400 mm (CHEVRONS)	–	–	–	186.38	nr	186.38
Standard reflectorised traffic signs						
Note: Unit costs do not include concrete foundations (see Series 1700)						
Standard one post signs; 600 × 450 mm type C1 signs fixed back to back to another sign (measured separately) with aluminium clips to existing post						
(measured separately)	0.04	2.43	0.84	95.78	nr	99.05
Extra for fixing singly with aluminium clips	0.01	0.61	0.20	5.76	nr	6.57
Extra for fixing singly with stainless steel clips	0.01	0.61	0.48	8.01	nr	9.10

SERIES 1200: TRAFFIC SIGNS AND ROAD MARKINGS

Item	Gang hours	Labour £	Plant £	Material £	Unit	Total rate £
TRAFFIC SIGNS – cont						
Standard reflectorised traffic signs – cont						
Standard one post signs – cont						
fixed back to back to another sign (measured separately) with stainless steel clips to one new 76 mm diameter plastic coated steel posts 1.75 m long	0.27	16.38	5.70	148.70	nr	**170.78**
Extra for fixing singly to one face only	0.01	0.61	0.20	–	nr	**0.81**
Extra for 76 mm diameter 1.75 m long aluminium post	0.02	1.21	0.43	32.64	nr	**34.28**
Extra for 76 mm diameter 3.5 m long plastic coated steel post	0.02	1.21	0.43	25.50	nr	**27.14**
Extra for 76 mm diameter 3.5 m long aluminium post	0.02	1.21	0.43	40.80	nr	**42.44**
Extra for excavation for post, in hard material	1.10	66.75	22.87	–	nr	**89.62**
Extra for single external illumination unit with fitted photo cell (excluding trenching and cabling - see Series 1400); unit cost per face illuminated	0.33	20.02	6.85	69.95	nr	**96.82**
Standard two post signs; 1200 × 400 mm, signs fixed back to back to another sign (measured separately) with stainless steel clips to two new 76 mm diameter plastic coated steel posts 1.75 m long	0.51	30.95	10.76	299.51	nr	**341.22**
Extra for fixing singly to one face only	0.02	1.21	0.43	–	nr	**1.64**
Extra for two 76 mm diameter 1.75 m long aluminium posts	0.04	2.43	0.83	65.28	nr	**68.54**
Extra for two 76 mm diameter 3.5 m long plastic coated steel posts	0.04	2.43	0.83	51.00	nr	**54.26**
Extra for two 76 mm diameter 3.5 m long aluminium post	0.04	2.43	0.83	81.60	nr	**84.86**
Extra for excavation for post, in hard material	1.10	66.75	22.87	–	nr	**89.62**
Extra for single external illumination unit with fitted photo cell (excluding trenching and cabling - see Series 1400); unit cost per face illuminated	0.58	35.19	12.06	97.97	nr	**145.22**
Standard internally illuminated traffic signs						
Bollard with integral mould-in translucent graphics (excluding trenching and cabling)						
fixing to concrete base	0.48	29.13	10.13	211.20	nr	**250.46**

SERIES 1200: TRAFFIC SIGNS AND ROAD MARKINGS

Item	Gang hours	Labour £	Plant £	Material £	Unit	Total rate £
Special traffic signs						
Note: Unit costs do not include concrete foundations (see Series 1700) or trenching and cabling (see Series 1400)						
Externally illuminated reflectorised traffic signs manufactured to order						
special signs, surface area 1.50 m² on two 100 mm diameter steel posts	–	–	–	–	nr	556.50
special signs, surface area 4.00 m² on three 100 mm diameter steel posts	–	–	–	–	nr	787.50
Internally illuminated traffic signs manufactured to order						
special signs, surface area 0.25 m² on one new 76 mm diameter steel post	–	–	–	–	nr	163.20
special signs, surface area 0.75 m² on one new 100 mm diameter steel post	–	–	–	–	nr	214.20
special signs, surface area 4.00 m² on four new 120 mm diameter steel posts	–	–	–	–	nr	1071.00
Signs on gantries						
Externally illuminated reflectorised signs						
1.50 m²	1.78	108.31	46.60	194.35	nr	349.26
2.50 m²	2.15	130.46	56.13	323.92	nr	510.51
3.00 m²	3.07	186.29	80.15	387.60	nr	654.04
Internally illuminated sign with translucent optical reflective sheeting and remote light source						
0.75 m²	1.56	94.66	40.73	1410.00	nr	1545.39
1.00 m²	1.70	103.16	44.38	1872.00	nr	2019.54
1.50 m²	2.41	146.24	62.92	2814.00	nr	3023.16
Remove from Store and Re-erect Traffic Signs						
Take from store and re-erect						
3.0 m high road sign	0.28	16.99	7.31	88.00	nr	112.30
road sign on two posts	0.50	30.34	13.05	175.99	nr	219.38
ROAD MARKINGS						
Thermoplastic screed or spray						
Note: Unit costs based upon new road with clean surface closed to traffic)						
Continuous line in reflectorised white						
150 mm wide	–	–	–	–	m	0.93
200 mm wide	–	–	–	–	m	1.24
Continuous line in reflectorised yellow						
100 mm wide	–	–	–	–	m	0.63
150 mm wide	–	–	–	–	m	0.93
Intermittent line in reflectorised white						
60 mm wide with 0.60 m line and 0.60 m gap	–	–	–	–	m	0.75
100 mm wide with 1.0 m line and 5.0 m gap	–	–	–	–	m	0.75
100 mm wide with 2.0 m line and 7.0 m gap	–	–	–	–	m	0.75

SERIES 1200: TRAFFIC SIGNS AND ROAD MARKINGS

Item	Gang hours	Labour £	Plant £	Material £	Unit	Total rate £
ROAD MARKINGS – cont						
Thermoplastic screed or spray – cont						
Intermittent line in reflectorised white – cont						
100 mm wide with 4.0 m line and 2.0 m gap	–	–	–	–	m	0.75
100 mm wide with 6.0 m line and 3.0 m gap	–	–	–	–	m	0.75
150 mm wide with 1.0 m line and 5.0 m gap	–	–	–	–	m	1.12
150 mm wide with 6.0 m line and 3.0 m gap	–	–	–	–	m	1.12
150 mm wide with 0.60 m line and 0.30 m gap	–	–	–	–	m	1.12
200 mm wide with 0.60 m line and 0.30 m gap	–	–	–	–	m	1.50
200 mm wide with 1.0 m line and 1.0 m gap	–	–	–	–	m	1.50
Ancillary line in reflectorised white						
150 mm wide in hatched areas	–	–	–	–	m	0.93
200 mm wide in hatched areas	–	–	–	–	m	1.50
Ancillary line in yellow						
150 mm wide in hatched areas	–	–	–	–	m	0.93
Triangles in reflectorised white						
1.6 m high	–	–	–	–	nr	8.70
2.0 m high	–	–	–	–	nr	11.82
3.75 m high	–	–	–	–	nr	15.55
Circles with enclosing arrows in reflectorised white						
1.6 m diameter	–	–	–	–	nr	62.19
Arrows in reflectorised white						
4.0 m long straight or turning	–	–	–	–	nr	24.87
6.0 m long straight or turning	–	–	–	–	nr	31.09
6.0 m long curved	–	–	–	–	nr	31.09
6.0 m long double headed	–	–	–	–	nr	43.53
8.0 m long double headed	–	–	–	–	nr	62.18
16.0 m long double headed	–	–	–	–	nr	93.28
32.0 m long double headed	–	–	–	–	nr	124.37
Kerb markings in yellow						
250 mm long	–	–	–	–	nr	0.63
Letters or numerals in reflectorised white						
1.6 m high	–	–	–	–	nr	8.10
2.0 m high	–	–	–	–	nr	11.82
3.75 m high	–	–	–	–	nr	20.53
Verynyl strip markings						
Note: Unit costs based upon new road with clean surface closed to traffic						
'Verynyl' strip markings (pedestrian crossings and similar locations)						
200 mm wide line	–	–	–	–	m	8.12
600 × 300 mm single stud tile	–	–	–	–	nr	13.86

SERIES 1200: TRAFFIC SIGNS AND ROAD MARKINGS

Item	Gang hours	Labour £	Plant £	Material £	Unit	Total rate £
REFLECTING ROAD STUDS						
Road Studs						
100 × 100 mm square bi-directional reflecting road stud with amber corner cube reflectors	–	–	–	–	nr	6.18
140 × 254 mm rectangular one way reflecting road stud with red catseye reflectors	–	–	–	–	nr	14.84
140 × 254 mm rectangular one way reflecting road stud with green catseye reflectors	–	–	–	–	nr	14.84
140 × 254 mm rectangular bi-directional reflecting road stud with white catseye reflectors	–	–	–	–	nr	14.84
140 × 254 mm rectangular bi-directional reflecting road stud with amber catseye reflectors	–	–	–	–	nr	14.84
140 × 254 mm rectangular bi-directional reflecting road stud without catseye reflectors	–	–	–	–	nr	9.28
REMOVE FROM STORE AND RE-INSTALL STUDS						
Remove From Store And Re-Install Road Studs						
Remove from store and re-install 100 × 100 mm square bi-directional reflecting road stud with corner cube reflectors	–	–	–	–	nr	3.07
Remove from store and re-install 140 × 254 mm rectangular one way reflecting road stud with catseye reflectors	–	–	–	–	nr	7.42
TRAFFIC SIGNAL INSTALLATIONS						
Traffic signal installation is carried out exclusively by specialist contractors, although certain items are dealt with by the main contractor or a sub-contractor. The following detailed prices are given to assist in the calculation of the total installation cost.						
Installation of signal pedestals, loop detector unit pedestals, controller unit boxes and cable connection pillars						
signal pedestal	–	–	–	–	nr	47.25
loop detector unit pedestal	–	–	–	–	nr	20.96
controller unit box	–	–	–	–	nr	51.59
Excavate trench for traffic signal cable, depth ne 1.50 m; supports, backfilling						
450 mm wide	–	–	–	–	m	6.44
Extra for excavating in hard material	–	–	–	–	m³	35.47
Saw cutting grooves in pavement for detector loops and feeder cables; seal with hot bitumen sealant after installation						
25 mm deep	–	–	–	–	m	4.20

SERIES 1200: TRAFFIC SIGNS AND ROAD MARKINGS

Item	Gang hours	Labour £	Plant £	Material £	Unit	Total rate £
MARKER POSTS						
Glass reinforced plastic marker posts						
types 1,2,3 or 4	–	–	–	–	nr	**13.30**
types 5,6,7 or 8	–	–	–	–	nr	**12.60**
Line posts for emergency crossing	–	–	–	–	nr	**7.82**
Standard reflectorized traffic cylinder 1000 mm high 125 mm diameter; mounted in cats eye base (deliniator)	–	–	–	–	nr	**24.41**
PERMANENT BOLLARDS						
Permanent bollard; non-illuminated; precast concrete						
150 mm minimum diameter 750 mm high	0.80	42.66	6.87	140.61	nr	**190.14**
300 mm minimum diameter 750 mm high	0.80	42.66	6.87	164.81	nr	**214.34**
Extra for exposed aggregate finish	–	–	–	11.67	nr	**11.67**
Permanent bollard; non-illuminated; galvanized steel						
removable and lockable pattern	0.80	42.66	10.46	124.11	nr	**177.23**
MISCELLANEOUS FURNITURE						
Galvanised steel lifting traffic barrier						
4.0 m wide	2.40	127.99	20.60	883.21	nr	**1031.80**
Precast concrete seats						
bench seat 2.0 m long	0.75	40.00	6.44	226.69	nr	**273.13**
bench seat with concrete ends and timber slats						
2.0 m long	0.75	40.00	6.44	254.21	nr	**300.65**
Timber seat fixed to concrete base						
bench seat 2.0 m long	0.45	24.00	2.75	325.69	nr	**352.44**
Metal seat						
bench seat 2.0 m long	0.75	40.00	6.44	624.84	nr	**671.28**

SERIES 1300: ROAD LIGHTING COLUMNS AND BRACKETS, CCTV MASTS AND CANTILEVER MASTS

Item	Gang hours	Labour £	Plant £	Material £	Unit	Total rate £
NOTES						
For convenience in pricing this section departs from Series 1300 requirements and shows lighting column costs broken down into main components of columns, brackets, and lamps and cabling.						
The outputs assume operations are continuous and are based on at least 10 complete units and do not include any allowance for on site remedial works after erection of the columns.						
Painting and protection of the columns apart from galvanising is not included in the following prices or outputs.						
The re-erection cost for lighting columns taken from store assumes that major components are in good condition; the prices below allow a sum of 20% of the value of new materials to cover minor repairs, new fixings and touching up any coatings.						
RESOURCES - LABOUR						
Column erection gang						
1 ganger/chargehand (skill rate 4)		14.58				
1 skilled operative (skill rate 4)		13.65				
1 unskilled operative (general)		12.75				
1 plant operator (craftsman) - 50% of time		10.60				
Total Gang Rate / Hour	£	**51.58**				
Bracket erection gang						
1 ganger/chargehand (skill rate 4)		14.58				
1 skilled operative (skill rate 4)		13.65				
1 plant operator (skill rate 4)		14.92				
1 plant operator (craftsman)		21.20				
Total Gang Rate / Hour	£	**64.36**				
Lanterns gang						
1 skilled operative (skill rate 3)		15.25				
1 skilled operative (skill rate 4)		13.65				
1 plant operator (skill rate 4)		14.92				
1 plant operator (craftsman)		21.20				
Total Gang Rate / Hour	£	**65.03**				
RESOURCES - PLANT						
Columns and bracket arms						
15 t mobile crane - 50% of time			16.51			
125 cfm compressor - 50% of time			4.85			
compressor tools: 2 single head scabbler - 50% of time			1.32			
2 t dumper - 50% of time			3.25			
Total Rate / Hour		£	**25.93**			

SERIES 1300: ROAD LIGHTING COLUMNS AND BRACKETS, CCTV MASTS AND CANTILEVER MASTS

Item	Gang hours	Labour £	Plant £	Material £	Unit	Total rate £
RESOURCES - PLANT – cont						
Bracket arms						
15 t mobile crane			33.02			
access platform, Simon hoist (50 ft)			35.74			
Total rate / Hour			68.76			
Lanterns						
15 t mobile crane			33.02			
access platform, Simon hoist (50 ft)			35.74			
Total Rate / Hour		£	**68.76**			
ROAD LIGHTING COLUMNS, BRACKETS, WALL MOUNTINGS, CCTV MASTS AND CANTILEVER MASTS						
Galvanised steel road lighting columns to BS EN 40 with flange plate base (including all control gear, switching, fuses and internal wiring)						
4.0 m nominal height	0.75	38.40	19.45	226.01	nr	**283.86**
6.0 m nominal height	0.80	40.96	27.92	339.02	nr	**407.90**
8.0 m nominal height	0.96	49.15	33.51	423.77	nr	**506.43**
10.0 m nominal height	1.28	65.53	44.68	508.53	nr	**618.74**
12.0 m nominal height	1.44	73.73	50.26	734.54	nr	**858.53**
15.0 m nominal height	1.76	90.11	61.43	1017.06	nr	**1168.60**
3.0 m cast iron column (pedestrian / landscape area)	0.75	38.40	26.19	1457.24	nr	**1521.83**
Precast concrete lighting columns to BS EN 40 with flange plate base (including all control gear, switching, fuses and internal wiring)						
5.0 m nominal height	0.75	38.40	26.19	238.50	nr	**303.09**
10.0 m nominal height	1.28	65.53	44.71	600.97	nr	**711.21**
Galvanised steel bracket arm to BS EN 40; with 5° uplift						
0.5 m projection, single arm	0.16	10.22	20.34	35.13	nr	**65.69**
1.0 m projection, single arm	0.19	12.13	13.06	75.09	nr	**100.28**
1.5 m projection, single arm	0.21	10.75	14.20	82.32	nr	**107.27**
2.0 m projection, single arm	0.27	17.24	18.56	99.63	nr	**135.43**
1.0 m projection, double arm	0.29	18.52	19.94	124.59	nr	**163.05**
2.0 m projection, double arm	0.32	20.44	22.00	150.24	nr	**192.68**
Precast concrete bracket arm to BS EN 40; with 5° uplift						
1.0 m projection, single arm	0.24	15.33	16.50	145.15	nr	**176.98**
2.0 m projection, single arm	0.32	20.44	22.00	174.16	nr	**216.60**
1.0 m projection, double arm	0.35	22.35	24.06	168.53	nr	**214.94**
2.0 m projection, doube arm	0.37	23.63	25.44	204.81	nr	**253.88**
Lantern unit with photo-electric control set to switch on at 100 lux; lamps						
55W SON (P226); to suit 4 m and 5 m columns	0.40	25.80	20.90	284.16	nr	**330.86**
70W SON (P236); to suit 5 m and 6 m columns	0.40	25.80	20.90	287.15	nr	**333.85**
250W SON (P426); to suit 8 m, 10 m and 12 m columns	0.50	32.25	26.12	430.72	nr	**489.09**
Sphere 70W SON; to suit 3 m columns (P456)	0.50	32.25	26.12	515.98	nr	**574.35**
400W SON High pressure sodium; to suit 12 m and 15 m columns	0.50	32.25	26.12	598.21	nr	**656.58**

SERIES 1300: ROAD LIGHTING COLUMNS AND BRACKETS, CCTV MASTS AND CANTILEVER MASTS

Item	Gang hours	Labour £	Plant £	Material £	Unit	Total rate £
REMOVE FROM STORE AND RE-ERECT ROAD LIGHTING COLUMNS AND BRACKETS AND MASTS						
Remove From Store And Re-Erect Road Lighting Columns, Brackets And Wall Mountings						
Re-erection of galvanized steel road lighting columns with flange plate base; including all control gear, switching, fuses and internal wiring						
4.0 m nominal height	0.75	38.40	19.46	18.70	nr	**76.56**
6.0 m nominal height	0.80	40.96	20.77	24.46	nr	**86.19**
8.0 m nominal height	0.96	49.15	24.89	30.22	nr	**104.26**
10.0 m nominal height	1.28	65.53	33.21	35.97	nr	**134.71**
12.0 m nominal height	1.44	73.73	37.36	41.73	nr	**152.82**
15.0 m nominal height	1.76	90.11	45.66	50.36	nr	**186.13**
3.0 m cast iron column (pedestrian / landscape area)	0.75	38.40	19.48	15.82	nr	**73.70**
Re-erection of precast concrete lighting columns with flange plate base; including all control gear, switching, fuses and internal wiring						
5.0 m nominal height	0.75	38.40	19.46	21.59	nr	**79.45**
10.0 m nominal height	1.28	65.53	33.21	35.97	nr	**134.71**
Re-erection of galvanized steel bracket arms						
0.5 m projection, single arm	0.16	10.22	11.00	30.75	nr	**51.97**
1.0 m projection, single arm	0.19	12.13	13.06	30.75	nr	**55.94**
1.5 m projection, single arm	0.21	13.41	14.44	30.75	nr	**58.60**
2.0 m projection, single arm	0.27	17.24	18.56	30.75	nr	**66.55**
1.0 m projection, double arm	0.29	18.52	19.94	30.75	nr	**69.21**
2.0 m projection, double arm	0.32	20.44	22.00	30.75	nr	**73.19**
Re-erection of precast concrete bracket arms						
1.0 m projection, single arm	0.24	15.33	16.50	30.75	nr	**62.58**
2.0 m projection, single arm	0.32	20.44	22.00	30.75	nr	**73.19**
1.0 m projection, double arm	0.35	22.35	24.87	30.75	nr	**77.97**
2.0 m projection, double arm	0.37	23.63	24.06	30.75	nr	**78.44**
Re-installing lantern unit with photo-electric control set to switch on at 100 lux; lamps						
55W SON (P226); to suit 4 m and 5 m columns	0.40	25.80	20.90	–	nr	**46.70**
70W SON (P236); to suit 5 m and 6 m columns	0.22	14.19	20.90	–	nr	**35.09**
250W SON (P426); to suit 8 m, 10 m and 12 m columns	0.50	32.25	26.12	–	nr	**58.37**
Sphere 70W SON; to suit 3 m (P456)	0.50	32.25	26.12	–	nr	**58.37**
400W SON High pressure sodium; to suit 12 m and 15 m columns	0.50	32.25	26.12	–	nr	**58.37**

SERIES 1400: ELECTRICAL WORK FOR ROAD LIGHTING AND TRAFFIC SIGNS

Item	Gang hours	Labour £	Plant £	Material £	Unit	Total rate £
RESOURCES - LABOUR						
Trenching gang						
1 ganger/chargehand (skill rate 4)		13.65				
1 skilled operative (skill rate 4)		13.65				
2 unskilled operatives (general)		25.50				
1 plant operator (skill rate 3) - 75% of time		12.45				
Total Gang Rate / Hour	£	**65.25**				
Cable laying gang						
1 ganger/chargehand (skill rate 4)		13.65				
2 skilled operatives (skill rate 3)		30.50				
1 skilled operative (skill rate 4)		13.65				
Total Gang Rate / Hour	£	**57.81**				
RESOURCES - PLANT						
Service trenching						
JCB 3CX backhoe - 50% of time			8.98			
125 cfm compressor - 50% of time			4.85			
compressor tools: 2 single head scabbler - 50% of time			1.32			
2 t dumper - 50% of time			3.25			
trench excavator - 25% of time			17.06			
Total Rate / Hour		£	**35.45**			
Cable laying						
8 t IVECO chassis or similar - 50% of time			6.07			
Total Rate / Hour		£	**6.07**			
LOCATING BURIED ROAD LIGHTING AND TRAFFIC SIGNS CABLES						
Locating Buried Road Lighting And Traffic Signs Cable						
Locating buried road lighting and traffic signs cable						
in carriageways, footways, bridge decks and paved areas	0.25	16.18	8.86	–	m	**25.04**
in verges and central reserves	0.20	12.94	7.09	–	m	**20.03**
in side slopes of cuttings or side slopes of embankments	0.15	9.71	5.32	–	m	**15.03**

SERIES 1400: ELECTRICAL WORK FOR ROAD LIGHTING AND TRAFFIC SIGNS

Item	Gang hours	Labour £	Plant £	Material £	Unit	Total rate £
TRENCH FOR CABLE OR DUCT						
Trench for cable						
300 to 450 mm wide; depth not exceeding 1.5 m	0.15	9.71	5.32	–	m	**15.03**
450 to 600 mm wide; depth not exceeding 1.5 m	0.20	12.94	7.09	–	m	**20.03**
Extra for excavating rock or reinforced concrete in trench	0.50	32.35	17.74	–	m³	**50.09**
Extra for excavating brickwork or mass concrete in trench	0.40	25.88	14.18	–	m³	**40.06**
Extra for backfilling with pea gravel	0.02	1.29	0.72	19.81	m³	**21.82**
Extra for 450 × 100 mm sand cable bedding and covering	0.02	1.29	0.72	1.45	m	**3.46**
Extra for PVC marker tape	0.01	0.32	0.19	2.17	m	**2.68**
Extra for 150 × 300 clay cable tiles	0.05	3.24	1.77	4.94	m	**9.95**
Extra for 150 × 900 concrete cable tiles	0.03	1.94	1.07	8.39	m	**11.40**
CABLE AND DUCT						
600/1000V 2 core, PVC/SWA/PVC cable with copper conductors						
Cable; in trench not exceeding 1.5 m deep						
2.5 mm²	0.01	0.57	0.06	2.42	m	**3.05**
4 mm²	0.02	1.15	0.12	3.24	m	**4.51**
6 mm²	0.02	1.15	0.12	3.83	m	**5.10**
10 mm²	0.02	1.15	0.12	5.28	m	**6.55**
16 mm²	0.02	1.15	0.12	6.65	m	**7.92**
25 mm²	0.03	1.72	0.18	8.48	m	**10.38**
600/1000V 4 core, PVC/SWA/PVC cable with copper conductors						
Cable; in trench not exceeding 1.5 m deep						
16. mm²	0.03	1.72	0.18	9.75	m	**11.65**
35 mm²	0.13	7.45	0.79	11.72	m	**19.96**
70 mm²	0.15	8.60	0.91	15.87	m	**25.38**
600/1000V 2 core, PVC/SWA/PVC cable drawn into ducts, pipe bays or troughs						
Cable; in trench not exceeding 1.5 m deep						
2.5 mm²	0.01	0.57	0.06	2.42	m	**3.05**
4 mm²	0.02	1.15	0.12	3.24	m	**4.51**
6 mm²	0.02	1.15	0.12	3.83	m	**5.10**
10 mm²	0.02	1.15	0.12	5.28	m	**6.55**
16 mm²	0.03	1.72	0.18	6.65	m	**8.55**
35 mm²	0.12	6.88	0.73	8.48	m	**16.09**
70 mm²	0.14	8.03	0.85	12.85	m	**21.73**

SERIES 1400: ELECTRICAL WORK FOR ROAD LIGHTING AND TRAFFIC SIGNS

Item	Gang hours	Labour £	Plant £	Material £	Unit	Total rate £
CABLE JOINTS AND CABLE TERMINATIONS						
Straight joint in 2 core PVC/SWA/PVC cable						
2.5 mm²	0.30	17.20	1.82	50.91	nr	69.93
4.0 mm²	0.30	17.20	1.82	50.91	nr	69.93
6.0 mm²	0.32	18.35	1.94	50.91	nr	71.20
10.0 mm²	0.42	24.08	2.55	51.96	nr	78.59
16.0 mm²	0.50	28.67	3.04	54.50	nr	86.21
35.0 mm²	0.80	45.87	4.86	69.32	nr	120.05
70.0 mm²	1.15	65.94	6.98	76.70	nr	149.62
Tee joint in 2 core PVC/SWA/PVC cable						
2.5 mm²	0.46	26.38	2.79	70.63	nr	99.80
4.0 mm²	0.46	26.38	2.79	70.63	nr	99.80
6.0 mm²	0.48	27.52	2.91	70.63	nr	101.06
10.0 mm²	0.62	35.55	3.76	72.45	nr	111.76
16.0 mm²	0.74	42.43	4.49	75.63	nr	122.55
25.0 mm²	0.91	52.18	5.53	81.01	nr	138.72
Tee joint in 4 core PVC/SWA/PVC cable						
16.0 mm²	1.10	63.07	6.68	85.22	nr	154.97
35.0 mm²	1.30	74.54	7.89	95.42	nr	177.85
70.0 mm²	1.60	91.74	9.71	137.03	nr	238.48
Looped terminations of 2 core PVC/SWA/PVC cable in lit sign units, traffic signals installation control unit, pedestrian crossing control unit, road lighting column, wall mounting, subway distribution box, gantry distribution box or feeder pillar.						
2.5 mm²	0.15	8.60	0.91	6.93	nr	16.44
6.0 mm²	0.15	8.60	0.91	9.38	nr	18.89
10.0 mm²	0.16	9.17	0.97	12.20	nr	22.34
16.0 mm²	0.25	14.33	1.52	12.45	nr	28.30
25.0 mm²	0.30	17.20	1.82	17.37	nr	36.39
Terminations of 4 core PVC/SWA/PVC cable in lit sign units, traffic signals installation control unit, pedestrian crossing control unit, road lighting column, wall mounting, subway distribution box, gantry distribution box or feeder pillar.						
16.0 mm²	0.35	20.07	2.13	13.39	nr	35.59
35.0 mm²	0.55	31.54	3.34	24.42	nr	59.30
70.0 mm²	0.68	38.99	4.13	46.34	nr	89.46
FEEDER PILLARS						
Galvanised steel feeder pillars						
411 × 610 mm	4.64	266.05	28.17	246.00	nr	540.22
611 × 810 mm	4.24	243.12	25.74	362.85	nr	631.71
811 × 1110 mm	4.96	284.40	30.12	479.70	nr	794.22
1111 × 1203 mm	4.19	240.25	25.50	553.50	nr	819.25

SERIES 1400: ELECTRICAL WORK FOR ROAD LIGHTING AND TRAFFIC SIGNS

Item	Gang hours	Labour £	Plant £	Material £	Unit	Total rate £
EARTH ELECTRODES						
Earth electrodes providing minimal protection using earth rods, plates or stops and protective tape and joint						
to suit columns ne 12.0 m	0.40	22.94	–	263.30	nr	**286.24**
to suit columns ne 15.0 m	0.40	22.94	–	349.20	nr	**372.14**
to suit Superstructure or Buildings using copper lead conductor (per 23 m height)	1.00	57.34	–	593.89	nr	**651.23**
CHAMBERS						
Brick chamber with galvanized steel cover and frame; depth to uppermost surface of base slab						
ne 1.0 m deep	–	–	–	–	nr	**892.42**

SERIES 1500: MOTORWAY COMMUNICATIONS

Item	Gang hours	Labour £	Plant £	Material £	Unit	Total rate £
RESOURCES - LABOUR						
Service trenching gang						
1 ganger/chargehand (skill rate 4)		14.58				
1 skilled operative (skill rate 4)		13.65				
2 unskilled operatives (general)		25.50				
1 plant operator (skill rate 3) - 75% of time		12.45				
Total Gang Rate / Hour	£	**66.18**				
RESOURCES - PLANT						
Trenching						
JCB 3CX backhoe (50% of time)			8.98			
125 cfm compressor (50% of time)			4.85			
compressor tools: 2 single head scabbler (50% of time)			1.32			
2 t dumper (50% of time)			3.25			
trench excavator (25% of time)			17.06			
Total Gang Rate / Hour		£	**35.45**			
LOCATING BURIED COMMUNICATIONS CABLING						
Locating buried road lighting and traffic signs cable						
in carriageways, footways, bridge decks and paved areas	0.25	16.18	8.86	–	m	**25.04**
in verges and central reserves	0.20	12.94	7.09	–	m	**20.03**
in side slopes of cuttings or side slopes of embankments	0.15	9.71	5.32	–	m	**15.03**
TRENCH FOR COMMUNICATIONS CABLE OR DUCT						
Trench for cable						
300 to 450 mm wide; depth not exceeding 1.5 m	0.15	9.71	5.32	–	m	**15.03**
450 to 600 mm wide; depth not exceeding 1.5 m	0.20	12.94	7.09	–	m	**20.03**
Extra for excavating rock or reinforced concrete in trench	0.50	32.35	17.74	–	m³	**50.09**
Extra for excavating brickwork or mass concrete in trench	0.40	25.88	14.18	–	m³	**40.06**
Extra for backfilling with pea gravel	0.02	1.29	0.72	19.81	m³	**21.82**
Extra for 450 × 100 mm sand cable bedding and covering	0.02	1.29	0.72	1.45	m	**3.46**
Extra for PVC marker tape	0.01	0.32	0.19	2.17	m	**2.68**
Extra for 150 × 300 clay cable tiles	0.05	3.24	1.77	4.94	m	**9.95**
Extra for 150 × 900 concrete cable tiles	0.03	1.94	1.07	8.39	m	**11.40**

SERIES 1500: MOTORWAY COMMUNICATIONS

Item	Gang hours	Labour £	Plant £	Material £	Unit	Total rate £
COMMUNICATIONS CABLING AND DUCT						
Communication cables laid in trench						
type A1 2 pair 0.9mm² armoured multi-pair	–	–	–	–	m	4.62
type A2 20 pair 0.9mm² armoured multi-pair	–	–	–	–	m	8.29
type A3 30 pair 0.9mm² armoured multi-pair	–	–	–	–	m	10.19
Power cables laid in trench						
type A4 10.0mm² armoured split concentric	–	–	–	–	m	7.07
Detector feeder cables laid in trench						
type A6 50/0.25m² single core detector feeder cable	–	–	–	–	m	8.93
type A7 50/0.25m² single core detector feeder cable	–	–	–	–	m	8.93
COMMUNICATIONS CABLE JOINTS AND TERMINATIONS						
Cable terminations						
of type 1 cable	–	–	–	–	nr	65.94
of type 2 cable	–	–	–	–	nr	294.63
of type 3 cable	–	–	–	–	nr	422.63
of type 4 cable	–	–	–	–	nr	70.67
of type 6 cable	–	–	–	–	nr	42.10
of type 7 cable	–	–	–	–	nr	42.10
COMMUNICATIONS EQUIPMENT						
Cabinet bases						
foundation plinth	–	–	–	–	nr	133.46
Matrix signal post bases						
foundation plinth	–	–	–	–	nr	138.18
CCTV camera bases						
foundation plinth	–	–	–	–	nr	133.46
Wall mounted brackets						
at maximum 15.0m height	–	–	–	–	nr	44.52
Fix only the following equipment						
communication equipment cabinet, 600 type series	–	–	–	–	nr	140.32
terminator type II	–	–	–	–	nr	148.73
emergency telephone post	–	–	–	–	nr	74.52
telephone housing	–	–	–	–	nr	23.27
matrix signal post	–	–	–	–	nr	74.52
Motorwarn / fogwarn	–	–	–	–	nr	148.93
distributor on gantry	–	–	–	–	nr	189.63
isolator switch for gantry	–	–	–	–	nr	412.97
heater unit mounted on gantry; Henleys' 65 W type 22501	–	–	–	–	nr	28.29

SERIES 1500: MOTORWAY COMMUNICATIONS

Item	Gang hours	Labour £	Plant £	Material £	Unit	Total rate £
COMMUNICATIONS EQUIPMENT – cont						
Terminal blocks						
Klippon BK6	–	–	–	–	nr	7.67
Klippon BK12	–	–	–	–	nr	8.61
Work to pavement for loop detection circuits						
cut or form grooves in pavement for detector						
loops and feeders	–	–	–	–	m	5.00
additional cost for sealing with hot bitumen						
sealant	–	–	–	–	m	0.51
CHAMBERS						
Brick chamber with galvanized steel cover and						
frame; depth to uppermost surface of base slab						
ne 1.0 m deep	–	–	–	–	nr	892.42

SERIES 1600: PILING AND EMBEDDED RETAINING WALLS

Item	Gang hours	Labour £	Plant £	Material £	Unit	Total rate £
GENERAL						
Notes						
There are a number of different types of piling which are available for use in differing situations. Selection of the most suitable type of piling for a particular site will depend on a number of factors including the physical conditions likely to be encountered during driving, the loads to be carried, the design of superstructure, etc.						
The most commonly used systems are included in this section.						
It is essential that a thorough and adequate site investigation is carried out to ascertain details of the ground strata and bearing cpacities to enable a proper assessment to be made of the most suitable and economical type of piling to be adopted.						
There are so many factors, apart from design considerations, which influence the cost of piling that it is not possible to give more than an approximate indication of costs. To obtain reliable costs for a particular contract advice should be sought from a company specialising in the particular type of piling proposed. Some Specialist Contractors will also provide a design service if required.						
PILING PLANT						
Driven precast concrete reinforced piles						
Establishment of piling plant for						
235 × 235 mm precast reinforced and prestressed concrete piles in main piling	–	–	–	–	item	4490.00
275 × 275 mm precast reinforced and prestressed concrete piles in main piling	–	–	–	–	item	4490.00
350 × 350 mm precast reinforced and prestressed concrete piles in main piling	–	–	–	–	item	4760.00
Moving piling plant for						
235 × 235 mm precast reinforced and prestressed concrete piles in main piling	–	–	–	–	nr	50.00
275 × 275 mm precast reinforced and prestressed concrete piles in main piling	–	–	–	–	nr	50.00
350 × 350 mm precast reinforced and prestressed concrete piles in main piling	–	–	–	–	nr	55.00
Bored in-situ reinforced concrete piling (tripod rig)						
Establishment of piling plant for 500 mm diameter cast-in-place concrete piles (tripod rig) in main piling	–	–	–	–	item	6175.00
Moving piling plant for 500 mm diameter cast-in-place concrete piles (tripod rig) in main piling	–	–	–	–	nr	158.09

SERIES 1600: PILING AND EMBEDDED RETAINING WALLS

Item	Gang hours	Labour £	Plant £	Material £	Unit	Total rate £
PILING PLANT – cont						
Bored in-situ reinforced concrete piling (mobile rig)						
Establishment of piling plant for 500 mm diameter cast-in-place concrete piles (mobile rig) in main piling	–	–	–	–	item	14250.00
Moving piling plant for 500 mm diameter cast-in-place concrete piles in (mobile rig) main piling	–	–	–	–	nr	130.24
Concrete injected piles (continuous flight augered)						
Establishment of piling plant for cast-in-place concrete piles (CFA) in main piling						
450 mm diameter; 650kN	–	–	–	–	item	8200.00
600 mm diameter; 1400kN	–	–	–	–	item	8200.00
750 mm diameter; 2200kN	–	–	–	–	item	8200.00
Moving piling plant for cast-in-place concrete piles (CFA) in main piling						
450 mm diameter; 650kN	–	–	–	–	nr	41.00
600 mm diameter 1400kN	–	–	–	–	nr	41.00
750 mm diameter 2200kN	–	–	–	–	nr	41.00
Driven cast in place piles; segmental casing method						
Establishment of piling plant for cast-in-place concrete piles in main piling	–	–	–	–	item	11500.00
Moving piling plant for cast-in-place concrete piles in main piling	–	–	–	–	nr	25.00
Establishment of piling plant for cast-in-place concrete piles in main piling						
bottom driven	–	–	–	–	item	6000.00
top driven	–	–	–	–	item	6000.00
Moving piling plant for 430 mm diameter cast-in-place concrete piles in main piling						
bottom driven	–	–	–	–	nr	90.76
top driven	–	–	–	–	nr	62.80
Steel bearing piles						
Establishment of piling plant for steel bearing piles in main piling						
maximum 100 miles radius from base	–	–	–	–	item	4500.00
maximum 250 miles radius from base	–	–	–	–	item	10000.00
Moving piling plant for steel bearing piles in main piling	–	–	–	–	nr	60.00

SERIES 1600: PILING AND EMBEDDED RETAINING WALLS

Item	Gang hours	Labour £	Plant £	Material £	Unit	Total rate £
Z section sheet steel piles						
Provision of all plant, equipment and labour including transport to and from the site and establishing and dismantling for						
driving of sheet piling	–	–	–	–	item	6820.00
extraction of sheet piling	–	–	–	–	item	6050.00
U section sheet steel piles						
Provision of plant, equipment and labour including transport to and from the site and establishing and dismantling for						
driving of sheet piling	–	–	–	–	item	6200.00
extraction of sheet piling	–	–	–	–	item	5500.00
Establishment of piling plant for steel tubular piles in main piling						
maximum 100 miles radius from base	–	–	–	–	item	6750.00
maximum 250 miles radius from base	–	–	–	–	item	16750.00
Moving piling plant for steel tubular piles in main piling	–	–	–	–	nr	248.06
PRECAST CONCRETE PILES						
Driven precast reinforced concrete piles						
The following unit costs cover the installation of driven precast concrete piles by using a hammer acting on a shoe fitted onto or cast into the pile unit. The costs are based installing 100 piles of nominal sizes stated, and a concrete strength of 50N/mm² suitably reinforced for a working load not exceeding 600kN, with piles average 15m long, on a clear site with reasonable access.						
Single pile lengths are normally a maximum of 13m long, at which point, a mechanical interlocking joint is required to extend the pile. These joints are most economically and practically formed at works.						
Lengths, sizes of sections, reinforcement details and concrete mixes vary for differing contractors, whose specialist advice should be sought for specific designs.						
Precast concrete piles; concrete 50N/mm²						
235 × 235 mm; 5–10 m in length; main piling	–	–	–	–	m	20.00
275 × 275 mm; 5–10 m in length; main piling	–	–	–	–	m	23.50
350 × 350 mm; 5–10 m in length; main piling	–	–	–	–	m	37.00
Mechanical Interlocking joint						
235 × 235 mm	–	–	–	–	nr	60.00
275 × 275 mm	–	–	–	–	nr	62.60
350 × 350 mm	–	–	–	–	nr	120.00

SERIES 1600: PILING AND EMBEDDED RETAINING WALLS

Item	Gang hours	Labour £	Plant £	Material £	Unit	Total rate £
PRECAST CONCRETE PILES – cont						
Driven precast reinforced concrete piles – cont						
Driving vertical precast piles						
235×235mm; 5–10m in length; in main piling	–	–	–	–	m	2.75
275×275mm; 5–10m in length; in main piling	–	–	–	–	m	2.90
350×350mm; 5–10m in length; in main piling	–	–	–	–	m	4.10
Stripping vertical precast concrete pile heads						
235×235mm piles in main piling	–	–	–	–	nr	33.14
275×275mm piles in main piling	–	–	–	–	nr	43.11
350×350mm piles in main piling	–	–	–	–	nr	66.30
Standing time						
275×275mm	–	–	–	–	hr	167.82
350×350mm	–	–	–	–	hr	200.24
CAST IN PLACE PILES						
Bored in-situ reinforced concrete piling (tripod rig)						
The following unit costs cover the construction of small diameter bored piling using light and compact tripod rigs requiring no expensive site levelling or access ways. Piling can be constructed in very restricted headroom or on confined and difficult sites. Standard diameters are between 400 and 600mm with a normal maximum depth of 30m.						
The costs are based on installing 100 piles of 500mm nominal diameter, a concrete strength of 20N/mm² with nominal reinforcement, on a clear site with reasonable access. Disposal of excavated material is included separately.						
Vertical 500mm diameter cast-in-place piles; 20N/mm² concrete; nominal reinforcement; in main piling	–	–	–	–	m	178.32
Vertical 500mm diameter empty bores in main piling	–	–	–	–	m	35.76
Add for boring through obstructions	–	–	–	–	hr	108.03
Standing time	–	–	–	–	hr	103.53

SERIES 1600: PILING AND EMBEDDED RETAINING WALLS

Item	Gang hours	Labour £	Plant £	Material £	Unit	Total rate £
Bored in-situ reinforced concrete piling (mobile rig)						
The following unit costs cover the construction of small diameter bored piles using lorry or crawler mounted rotary boring rigs. This type of plant is more mobile and faster in operation than the tripod rigs and is ideal for large contracts in cohesive ground. Construction of piles under bentonite suspension can be carried out to obviate the use of liners. Standard diameters of 450 to 900 mm diameter can be constructed to depths of 30 m.						
The costs are based on installing 100 piles of 500 mm nominal diameter, a concrete strength of 20N/mm² with nominal reinforcement, on a clear site with reasonable access. Disposal of excavated material is included separately.						
Vertical 500 mm diameter cast-in-place piles; 20N/mm² concrete; nominal reinforcement	–	–	–	–	m	108.39
Vertical 500 mm diameter empty bores	–	–	–	–	m	49.69
Add for boring through obstructions	–	–	–	–	hr	427.50
Standing time	–	–	–	–	hr	332.50
Concrete injected piles (continuous flight augered)						
The following unit costs cover the construction of piles by screwing a continuous flight auger into the ground to a design depth (Determined prior to commencement of piling operations and upon which the rates are based and subsequently varied to actual depths). Concrete is then pumped through the hollow stem of the auger to the bottom and the pile formed as the auger is withdrawn. Spoil is removed by the auger as it is withdrawn. This is a fast method of construction without causing disturbance or vibration to adjacent ground. No casing is required even in unsuitable soils. Reinforcement can be placed after grouting is complete.						
The costs are based on installing 100 piles on a clear site with reasonable access. Disposal of excavated material is included separately.						
Vertical cast-in-place piles; 20N/mm² concrete						
450 mm diameter; 650kN; 10–15m in length; main piling	–	–	–	–	m	20.87
600 mm diameter; 1400kN; 10–15m in length; main piling	–	–	–	–	m	50.47
750 mm diameter; 2200kN; 10–15m in length; main piling	–	–	–	–	m	54.69

SERIES 1600: PILING AND EMBEDDED RETAINING WALLS

Item	Gang hours	Labour £	Plant £	Material £	Unit	Total rate £
CAST IN PLACE PILES – cont						
Concrete injected piles (continuous flight augered) – cont						
Vertical empty bores						
450 mm diameter	–	–	–	–	m	25.87
600 mm diameter	–	–	–	–	m	2.21
750 mm diameter	–	–	–	–	m	30.65
Standing time / Boring through obstructions time						
450 mm diameter	–	–	–	–	hr	410.00
600 mm diameter	–	–	–	–	hr	410.00
750 mm diameter	–	–	–	–	hr	410.00
DRIVEN CAST-IN-PLACE PILES						
Driven cast-in-place piles; segmental casing method						
The following unit costs cover the construction of piles by driving into hard material using a serrated thick wall tube. It is oscillated and pressed into the hard material using a hydraulic attachment to the piling rig. The hard material is broken up using chiselling methods and is then removed by mechanical grab.						
Vertical cast-in-place piles; 20N/mm2 concrete						
620 mm diameter; 10–15m in length; main piling	–	–	–	–	m	166.48
1180 mm diameter; 10–15m in length; main piling	–	–	–	–	m	185.45
1500 mm diameter; 10–15m in length; main piling	–	–	–	–	m	253.26
Standing time	–	–	–	–	hr	275.63
Add for driving through obstructions	–	–	–	–	hr	293.00
Driven in-situ reinforced concrete piling						
The following unit costs cover the construction of piles by driving a tube into the ground either by using an internal hammer acting on a gravel or concrete plug or, as is more usual, by using an external hammer on a driving helmet at the top of the tube. After driving to the required depth an enlarged base is formed by hammering out sucessive charges of concrete down the tube. The tube is then filled with concrete which is compacted as the tube is vibrated and withdrawn. Piles of 350 to 500 mm diameter can be constructed with rakes up to 1 in 4 to carry working loads up to 120t per pile.						
The costs are based on installing 100 piles of 430 mm nominal diameter, a concrete strength of 20N/mm² suitably reinforced for a working load not exceeding 750kN, on a clear site with reasonable access.						

SERIES 1600: PILING AND EMBEDDED RETAINING WALLS

Item	Gang hours	Labour £	Plant £	Material £	Unit	Total rate £
SEE SECTION PILING PLANT Establishment of piling plant for cast-in-place concrete piles in main piling						
bottom driven	–	–	–	–	item	8000.00
top driven	–	–	–	–	item	8000.00
SEE SECTION PILING PLANT Moving piling plant for 430 mm diameter cast-in-place concrete piles in main piling						
bottom driven	–	–	–	–	nr	40.00
top driven	–	–	–	–	nr	40.00
Vertical 430 mm diameter cast-in-place piles 20N/mm² concrete; reinforcement for 750kN maximum load						
bottom driven	–	–	–	–	m	20.36
top driven	–	–	–	–	m	20.36
Standing time	–	–	–	–	hr	175.00
Add for driving through obstructions where within the capabilities of the normal plant	–	–	–	–	hr	185.00
Stripping vertical concrete pile heads						
430 mm diameter heads	–	–	–	–	nr	75.00
REINFORCEMENT FOR CAST-IN-PLACE PILES						
Mild steel						
Steel bar reinforcement						
6 mm nominal size; not exceeding 12 in length	–	–	–	–	t	918.00
8 mm nominal size; not exceeding 12 in length	–	–	–	–	t	918.00
10 mm nominal size; not exceeding 12 in length	–	–	–	–	t	918.00
12 mm nominal size; not exceeding 12 in length	–	–	–	–	t	918.00
16 mm nominal size; not exceeding 12 in length	–	–	–	–	t	918.00
25 mm nominal size; not exceeding 12 in length	–	–	–	–	t	918.00
32 mm nominal size; not exceeding 12 in length	–	–	–	–	t	918.00
40 mm nominal size; not exceeding 12 in length	–	–	–	–	t	918.00
Steel helical reinforcement						
6 mm nominal size; not exceeding 12 in length	–	–	–	–	t	918.00
8 mm nominal size; not exceeding 12 in length	–	–	–	–	t	918.00
10 mm nominal size; not exceeding 12 in length	–	–	–	–	t	918.00
12 mm nominal size; not exceeding 12 in length	–	–	–	–	t	918.00
High tensile steel						
Steel bar reinforcement						
6 mm nominal size; not exceeding 12 in length	–	–	–	–	t	1020.00
8 mm nominal size; not exceeding 12 in length	–	–	–	–	t	1020.00
10 mm nominal size; not exceeding 12 in length	–	–	–	–	t	1020.00
12 mm nominal size; not exceeding 12 in length	–	–	–	–	t	1020.00
16 mm nominal size; not exceeding 12 in length	–	–	–	–	t	918.00
25 mm nominal size; not exceeding 12 in length	–	–	–	–	t	918.00
32 mm nominal size; not exceeding 12 in length	–	–	–	–	t	1020.00
40 mm nominal size; not exceeding 12 in length	–	–	–	–	t	1020.00

SERIES 1600: PILING AND EMBEDDED RETAINING WALLS

Item	Gang hours	Labour £	Plant £	Material £	Unit	Total rate £
REINFORCEMENT FOR CAST-IN-PLACE PILES – cont						
High tensile steel – cont						
Steel helical reinforcement						
6 mm nominal size; not exceeding 12 in length	–	–	–	–	t	**1020.00**
8 mm nominal size; not exceeding 12 in length	–	–	–	–	t	**1020.00**
10 mm nominal size; not exceeding 12 in length	–	–	–	–	t	**1020.00**
12 mm nominal size; not exceeding 12 in length	–	–	–	–	t	**1020.00**
STEEL BEARING PILES						
Steel bearing piles are commonly carried out by a Specialist Contractor and whose advice should be sought to arrive at accurate costing. However the following items can be used to assess a budget cost for such work.						
The following unit costs are based upon driving 100nr steel bearing piles on a clear site with reasonable access. Supply is based on delivery 75 miles from works, in loads over 20t.						
Steel bearing piles						
Standing time	–	–	–	–	hr	**350.00**
203 × 203 × 45 kg/m steel bearing piles; Grade S275						
not exceeding 5m in length in main piling	–	–	–	66.45	m	**66.45**
5–10m in length in main piling	–	–	–	65.67	m	**65.67**
10–15m in length in main piling	–	–	–	65.67	m	**65.67**
15–20m in length in main piling	–	–	–	66.06	m	**66.06**
203 × 203 × 54 kg/m steel bearing piles; Grade S275						
not exceeding 5m in length in main piling	–	–	–	79.74	m	**79.74**
5–10m in length in main piling	–	–	–	78.81	m	**78.81**
10–15m in length in main piling	–	–	–	78.81	m	**78.81**
15–20m in length in main piling	–	–	–	79.27	m	**79.27**
254 × 254 × 63 kg/m steel bearing piles; Grade S275						
ne 5m in length	–	–	–	93.75	m	**93.75**
5–10m in length	–	–	–	92.66	m	**92.66**
10–15m in length	–	–	–	92.66	m	**92.66**
15–20m in length	–	–	–	93.21	m	**93.21**
254 × 254 × 71 kg/m steel bearing piles; Grade S275						
ne 5m in length	–	–	–	105.65	m	**105.65**
5–10m in length	–	–	–	104.43	m	**104.43**
10–15m in length	–	–	–	104.43	m	**104.43**
15–20m in length	–	–	–	105.04	m	**105.04**
254 × 254 × 85 kg/m steel bearing piles; Grade S275						
ne 5m in length	–	–	–	126.49	m	**126.49**
5–10m in length	–	–	–	125.02	m	**125.02**
10–15m in length	–	–	–	125.02	m	**125.02**
15–20m in length	–	–	–	125.75	m	**125.75**

SERIES 1600: PILING AND EMBEDDED RETAINING WALLS

Item	Gang hours	Labour £	Plant £	Material £	Unit	Total rate £
305 × 305 × 79 kg/m steel bearing piles; Grade S275						
ne 5m in length	–	–	–	119.38	m	**119.38**
5–10m in length	–	–	–	118.01	m	**118.01**
10–15m in length	–	–	–	118.01	m	**118.01**
15–20m in length	–	–	–	118.69	m	**118.69**
305 × 305 × 95kg/m steel bearing piles; Grade S275						
ne 5m in length	–	–	–	143.55	m	**143.55**
5–10m in length	–	–	–	141.91	m	**141.91**
10–15m in length	–	–	–	141.91	m	**141.91**
15–20m in length	–	–	–	142.73	m	**142.73**
305 × 305 × 110 kg/m steel bearing piles; Grade S275						
ne 5m in length	–	–	–	166.22	m	**166.22**
5–10m in length	–	–	–	164.32	m	**164.32**
10–15m in length	–	–	–	164.32	m	**164.32**
15–20m in length	–	–	–	165.27	m	**165.27**
305 × 305 × 126 kg/m steel bearing piles; Grade S275						
ne 5m in length	–	–	–	190.40	m	**190.40**
5–10m in length	–	–	–	188.22	m	**188.22**
10–15m in length	–	–	–	188.22	m	**188.22**
15–20m in length	–	–	–	189.31	m	**189.31**
305 × 305 × 149 kg/m steel bearing piles; Grade S275						
ne 5m in length	–	–	–	225.15	m	**225.15**
5–10m in length	–	–	–	222.58	m	**222.58**
10–15m in length	–	–	–	222.58	m	**222.58**
15–20m in length	–	–	–	223.87	m	**223.87**
305 × 305 × 186 kg/m steel bearing piles; Grade S275						
ne 5m in length	–	–	–	281.06	m	**281.06**
5–10m in length	–	–	–	277.85	m	**277.85**
10–15m in length	–	–	–	277.85	m	**277.85**
15–20m in length	–	–	–	279.46	m	**279.46**
305 × 305 × 233 kg/m steel bearing piles; Grade S275						
ne 5m in length	–	–	–	336.97	m	**336.97**
5–10m in length	–	–	–	333.13	m	**333.13**
10–15m in length	–	–	–	333.13	m	**333.13**
15–20m in length	–	–	–	335.05	m	**335.05**
356 × 368 × 109 kg/m steel bearing piles; Grade S275						
ne 5m in length	–	–	–	167.21	m	**167.21**
5–10m in length	–	–	–	165.33	m	**165.33**
10–15m in length	–	–	–	165.33	m	**165.33**
15–20m in length	–	–	–	166.27	m	**166.27**
356 × 368 × 133 kg/m steel bearing piles; Grade S275						
ne 5m in length	–	–	–	204.03	m	**204.03**
5–10m in length	–	–	–	201.74	m	**201.74**
10–15m in length	–	–	–	201.74	m	**201.74**
15–20m in length	–	–	–	202.88	m	**202.88**
356 × 368 × 152kg/m steel bearing piles; Grade S275						
ne 5m in length	–	–	–	233.18	m	**233.18**
5–10m in length	–	–	–	230.56	m	**230.56**
10–15m in length	–	–	–	230.56	m	**230.56**
15–20m in length	–	–	–	231.87	m	**231.87**

SERIES 1600: PILING AND EMBEDDED RETAINING WALLS

Item	Gang hours	Labour £	Plant £	Material £	Unit	Total rate £
STEEL BEARING PILES – cont						
Steel bearing piles – cont						
356×368×174 kg/m steel bearing piles; Grade S275						
ne 5m in length	–	–	–	266.93	m	266.93
5–10m in length	–	–	–	263.93	m	263.93
10–15m in length	–	–	–	263.93	m	263.93
15–20m in length	–	–	–	265.43	m	265.43
Driving vertical steel bearing piles						
section weight not exceeding 70 kg/m	–	–	–	–	m	3.00
section weight 70–90 kg/m	–	–	–	–	m	3.25
section weight 90–110 kg/m	–	–	–	–	m	3.50
section weight 90–110 kg/m	–	–	–	–	m	3.80
section weight 110–130 kg/m	–	–	–	–	m	3.80
section weight 150–170 kg/m	–	–	–	–	m	4.10
Driving raking steel bearing piles						
section weight not exceeding 70 kg/m	–	–	–	–	m	4.75
section weight 70–90 kg/m	–	–	–	–	m	5.00
section weight 90–110 kg/m	–	–	–	–	m	5.30
section weight 110–130 kg/m	–	–	–	–	m	5.30
section weight 130–150 kg/m	–	–	–	–	m	3.80
section weight 150–170 kg/m	–	–	–	–	m	4.10
section weight 170–190 kg/m	–	–	–	–	m	4.35
section weight 190–210 kg/m	–	–	–	–	m	4.60
allow 30% of the respective item above for the lengthened section only						
Welding on lengthening pieces to vertical steel bearing piles						
203×203× any kg/m	–	–	–	–	nr	200.00
254×254× any kg/m	–	–	–	–	nr	240.00
305×305× any kg/m	–	–	–	–	nr	275.00
356×368× any kg/m	–	–	–	–	nr	305.00
Cutting or burning off surplus length of vertical steel bearing piles						
203×203× any kg/m	–	–	–	–	nr	70.00
254×254× any kg/m	–	–	–	–	nr	100.00
305×305× any kg/m	–	–	–	–	nr	120.00
356×368× any kg/m	–	–	–	–	nr	135.00

SERIES 1600: PILING AND EMBEDDED RETAINING WALLS

Item	Gang hours	Labour £	Plant £	Material £	Unit	Total rate £
STEEL TUBULAR PILES						
Steel tubular piles are commonly carried out by a						
Specialist Contractor and whose advice should be						
sought to arrive at accurate costings. However the						
following items can be used to assess a budget cost						
for each work.						
The following unit costs are based upon driving						
100nr steel tubular piles on a clear site with						
reasonable access.						
Standing time	–	–	–	–	hr	309.19
Steel Grade 275; delivered in 10–20 t loads; mass						
60–120 kg/m						
section 508 mm × 8 mm × 98.6 kg/m	–	–	–	130.15	m	130.15
section 559 mm × 8 mm × 109 kg/m	–	–	–	143.88	m	143.88
Steel Grade 275; delivered in 10–20 t loads; mass						
120–250 kg/m						
section 508 mm × 10 mm × 123 kg/m	–	–	–	162.36	m	162.36
section 508 mm × 12.5 mm × 153 kg/m	–	–	–	201.96	m	201.96
section 508 mm × 16 mm × 194 kg/m	–	–	–	256.08	m	256.08
section 508 mm × 20 mm × 241 kg/m	–	–	–	318.12	m	318.12
section 559 mm × 10 mm × 135 kg/m	–	–	–	178.20	m	178.20
section 559 mm × 12.5 mm × 168 kg/m	–	–	–	221.76	m	221.76
section 559 mm × 16 mm × 214 kg/m	–	–	–	282.48	m	282.48
section 610 mm × 8 mm × 119 kg/m	–	–	–	157.08	m	157.08
section 610 mm × 10 mm × 148 kg/m	–	–	–	195.36	m	195.36
section 610 mm × 12.5 mm × 184 kg/m	–	–	–	242.88	m	242.88
section 610 mm × 16 mm × 234 kg/m	–	–	–	308.88	m	308.88
section 660 mm × 8 mm × 129 kg/m	–	–	–	170.28	m	170.28
section 660 mm × 10 mm × 160 kg/m	–	–	–	211.20	m	211.20
section 660 mm × 12.5 mm × 200 kg/m	–	–	–	264.00	m	264.00
section 711 mm × 8 mm × 134 kg/m	–	–	–	176.88	m	176.88
section 711 mm × 10 mm × 173 kg/m	–	–	–	228.36	m	228.36
section 711 mm × 12 mm × 215 kg/m	–	–	–	283.80	m	283.80
section 762 mm × 8 mm × 149 kg/m	–	–	–	196.68	m	196.68
section 762 mm × 10 mm × 185 kg/m	–	–	–	244.20	m	244.20
section 762 mm × 12.5 mm × 231 kg/m	–	–	–	304.92	m	304.92
Steel Grade 275; delivered in 10–20 t loads; mass						
250–500 kg/m						
section 559 mm × 20 mm × 266 kg/m	–	–	–	351.12	m	351.12
section 610 mm × 20 mm × 291 kg/m	–	–	–	384.12	m	384.12
section 660 mm × 16 mm × 254 kg/m	–	–	–	335.28	m	335.28
section 660 mm × 20 mm × 316 kg/m	–	–	–	417.12	m	417.12
section 660 mm × 25 mm × 392 kg/m	–	–	–	517.44	m	517.44
section 711 mm × 16 mm × 274 kg/m	–	–	–	361.68	m	361.68
section 711 mm × 20 mm × 341 kg/m	–	–	–	450.12	m	450.12
section 711 mm × 25 mm × 423 kg/m	–	–	–	558.36	m	558.36
section 762 mm × 16 mm × 294 kg/m	–	–	–	388.08	m	388.08
section 762 mm × 20 mm × 366 kg/m	–	–	–	483.12	m	483.12
section 762 mm × 25 mm × 454 kg/m	–	–	–	599.28	m	599.28

SERIES 1600: PILING AND EMBEDDED RETAINING WALLS

Item	Gang hours	Labour £	Plant £	Material £	Unit	Total rate £
STEEL TUBULAR PILES – cont						
Driving vertical steel tubular piles						
mass 60–120 kg/m	–	–	–	–	m	**7.08**
mass 120–150 kg/m	–	–	–	–	m	**7.20**
mass 150–160 kg/m	–	–	–	–	m	**7.31**
mass 160–190 kg/m	–	–	–	–	m	**7.42**
mass 190–220 kg/m	–	–	–	–	m	**7.53**
mass 220–250 kg/m	–	–	–	–	m	**7.64**
mass 250–280 kg/m	–	–	–	–	m	**7.75**
mass 280–310 kg/m	–	–	–	–	m	**7.88**
mass 310–340 kg/m	–	–	–	–	m	**7.98**
mass 340–370 kg/m	–	–	–	–	m	**8.10**
mass 370–400 kg/m	–	–	–	–	m	**8.32**
mass 400–430 kg/m	–	–	–	–	m	**8.43**
mass 430–460 kg/m	–	–	–	–	m	**8.66**
Driving raking steel tubular piles						
mass 60–120 kg/m	–	–	–	–	m	**7.46**
mass 120–150 kg/m	–	–	–	–	m	**7.56**
mass 150–160 kg/m	–	–	–	–	m	**7.69**
mass 160–190 kg/m	–	–	–	–	m	**7.80**
mass 190–220 kg/m	–	–	–	–	m	**7.92**
mass 220–250 kg/m	–	–	–	–	m	**8.04**
mass 250–280 kg/m	–	–	–	–	m	**8.16**
mass 280–310 kg/m	–	–	–	–	m	**8.27**
mass 310–340 kg/m	–	–	–	–	m	**8.39**
mass 340–370 kg/m	–	–	–	–	m	**8.52**
mass 370–400 kg/m	–	–	–	–	m	**8.75**
mass 400–430 kg/m	–	–	–	–	m	**8.87**
mass 430–460 kg/m	–	–	–	–	m	**9.10**
Driving lengthened vertical steel tubular piles						
mass 60–120 kg/m	–	–	–	–	m	**9.20**
mass 120–150 kg/m	–	–	–	–	m	**9.35**
mass 150–160 kg/m	–	–	–	–	m	**9.50**
mass 160–190 kg/m	–	–	–	–	m	**9.65**
mass 190–220 kg/m	–	–	–	–	m	**9.80**
mass 220–250 kg/m	–	–	–	–	m	**9.93**
mass 250–280 kg/m	–	–	–	–	m	**10.08**
mass 280–310 kg/m	–	–	–	–	m	**10.24**
mass 310–340 kg/m	–	–	–	–	m	**10.37**
mass 340–370 kg/m	–	–	–	–	m	**10.52**
mass 370–400 kg/m	–	–	–	–	m	**10.82**
mass 400–430 kg/m	–	–	–	–	m	**10.96**
mass 430–460 kg/m	–	–	–	–	m	**11.26**

SERIES 1600: PILING AND EMBEDDED RETAINING WALLS

Item	Gang hours	Labour £	Plant £	Material £	Unit	Total rate £
Driving lengthened raking steel tubular piles						
mass 60–120 kg/m	–	–	–	–	m	9.69
mass 120–150 kg/m	–	–	–	–	m	9.84
mass 150–160 kg/m	–	–	–	–	m	10.00
mass 160–190 kg/m	–	–	–	–	m	10.14
mass 190–220 kg/m	–	–	–	–	m	10.30
mass 220–250 kg/m	–	–	–	–	m	10.46
mass 250–280 kg/m	–	–	–	–	m	10.61
mass 280–310 kg/m	–	–	–	–	m	10.76
mass 310–340 kg/m	–	–	–	–	m	10.92
mass 340–370 kg/m	–	–	–	–	m	11.07
mass 370–400 kg/m	–	–	–	–	m	11.37
mass 400–430 kg/m	–	–	–	–	m	11.53
mass 430–460 kg/m	–	–	–	–	m	11.83
Welding on lengthening piece to steel tubular piles						
section diameter 508 × any thickness	–	–	–	–	nr	157.41
section diameter 559 × any thickness	–	–	–	–	nr	166.40
section diameter 610 × any thickness	–	–	–	–	nr	175.40
section diameter 660 × any thickness	–	–	–	–	nr	184.39
section diameter 711 × any thickness	–	–	–	–	nr	193.39
section diameter 762 × any thickness	–	–	–	–	nr	202.38
Cutting or burning off surplus length of steel tubular piles						
section diameter 508 × any thickness	–	–	–	–	nr	10.29
section diameter 559 × any thickness	–	–	–	–	nr	10.34
section diameter 610 × any thickness	–	–	–	–	nr	10.40
section diameter 660 × any thickness	–	–	–	–	nr	10.45
section diameter 711 × any thickness	–	–	–	–	nr	10.53
section diameter 762 × any thickness	–	–	–	–	nr	10.59
PROOF LOADING OF PILES						
Driven precast concrete piles						
Establishment of proof loading equipment; proof loading of vertical precast concrete piles with maintained load to 900 kN	–	–	–	–	item	2760.00
Establishment of proof loading equipment; proof loading of vertical precast concrete piles by dynamic testing with piling hammer	–	–	–	–	nr	690.00
Bored in-situ reinforced concrete piling (tripod rig)						
Establishment of proof loading equipment for bored cast-in-place piles	–	–	–	–	item	950.00
Proof loading of vertical cast-in-place piles with maximum test load of 600kN on a working pile 500mm diameter using tension piles as reaction	–	–	–	–	nr	3325.00

SERIES 1600: PILING AND EMBEDDED RETAINING WALLS

Item	Gang hours	Labour £	Plant £	Material £	Unit	Total rate £
PROOF LOADING OF PILES – cont						
Bored in-situ reinforced concrete piling (mobile rig)						
Establishment of proof loading equipment for bored cast-in-place piles	–	–	–	–	item	1330.00
Proof loading of vertical cast-in-place piles with maximum test load of 600kN on a working pile 500mm diameter using tension piles as reaction	–	–	–	–	nr	3325.00
Concrete injected piles (continuous flight augered)						
Establishment of proof loading equipment for bored cast-in-place piles						
450 mm diameter; 650kN	–	–	–	–	item	1537.50
600 mm diameter; 1400kN	–	–	–	–	item	1537.50
750 mm diameter; 2200kN	–	–	–	–	item	1537.50
Proof loading of vertical cast-in-place piles to 1.5 times working load						
450 mm diameter; 650kN	–	–	–	–	nr	2500.00
600 mm diameter; 1400kN	–	–	–	–	nr	5250.00
750 mm diameter; 2200kN	–	–	–	–	nr	9500.00
Electronic integrity testing						
cost per pile (minimum 40 piles per visit)	–	–	–	–	nr	11.50
Segmental casing method piles						
Establishment of proof loading equipment for driven cast-in-place piles	–	–	–	–	item	1250.00
Proof loading of vertical cast-in-place piles with maximum test load of 600kN on a working pile 500 mm diameter using non-working tension piles as reaction	–	–	–	–	nr	4000.00
Driven in-situ reinforced concrete piling						
Establishment of proof loading equipment for driven cast-in-place piles						
bottom driven	–	–	–	–	item	1000.00
top driven	–	–	–	–	item	1000.00
Proof loading of vertical cast-in-place piles with maximum test load of 1125kN on a working pile 430 mm diameter using non-working tension piles as reaction						
bottom driven	–	–	–	–	nr	3500.00
top driven	–	–	–	–	nr	3500.00
Electronic integrity testing						
Cost per pile (minimum 40 piles per visit)	–	–	–	–	nr	25.00
Steel bearing piles						
Establishment of proof loading equipment for steel bearing piles in main piling	–	–	–	–	item	1360.00
Proof loading of vertical steel bearing piles with maximum test load of 108 t load on a working pile using non-working tension piles as reaction	–	–	–	–	nr	4350.00

SERIES 1600: PILING AND EMBEDDED RETAINING WALLS

Item	Gang hours	Labour £	Plant £	Material £	Unit	Total rate £
Steel tubular piles						
Establishment of proof loading equipment for steel tubular piles	–	–	–	–	item	7000.00
Proof loading of steel tubular piles with maximum test load of 108 t load on a working pile using non-working tension piles as reaction	–	–	–	–	nr	6500.00
STEEL SHEET PILES						
Sheet steel piling is commonly carried out by a Specialist Contractor, whose advice should be sought to arrive at accurate costings. However, the following items can be used to assess a budget for such work. The following unit costs are based on driving/ extracting 1,500m² of sheet piling on a clear site with reasonable access.						
Note: area of driven piles will vary from area supplied dependant upon pitch line of piling and provision for such allowance has been made in PC for supply. The materials cost below includes the manufacturers tariffs for a 200 mile delivery radius from works, delivery in 5-10t loads and with an allowance of 10% to cover waste / projecting piles etc.						
Arcelor Mittal Z section steel piles; EN 10248 grade S270GP steel						
The following unit costs are based on driving/ extracting 1,500m² of sheet piling on a clear site with reasonable access.						
Provision of all plant, equipment and labour including transport to and from the site and establishing and dismantling for						
driving of sheet piling	–	–	–	–	sum	6820.00
extraction of sheet piling	–	–	–	–	sum	6050.00
Standing time	–	–	–	–	hr	422.22
Section modulus 800–1200 cm³/m; section reference AZ 12; mass 98.7 kg/m², sectional modulus 1200 cm³/m; EN 10248 grade S270GP steel						
length of welded corner piles	–	–	–	–	m	94.57
length of welded junction piles	–	–	–	–	m	132.39
driven area	–	–	–	–	m²	46.71
area of piles of length not exceeding 14 m	–	–	–	–	m²	78.49
length 14- 24 m	–	–	–	–	m²	84.36
area of piles of length exceeding 24 m	–	–	–	–	m²	87.18
Section modulus 1200–2000 cm³/m; section reference AZ 17; mass 108.6 kg/m²; sectional modulus 1665 cm³/m; EN 10248 grade S270GP steel						
length of welded corner piles	–	–	–	–	m	94.57
length of welded junction piles	–	–	–	–	m	132.39
driven area	–	–	–	–	m²	41.62
area of piles of length not exceeding 14 m	–	–	–	–	m²	82.96
length 14- 24 m	–	–	–	–	m²	84.36
area of piles of length exceeding 24 m	–	–	–	–	m²	87.18

SERIES 1600: PILING AND EMBEDDED RETAINING WALLS

Item	Gang hours	Labour £	Plant £	Material £	Unit	Total rate £
STEEL SHEET PILES – cont						
Arcelor Mittal Z section steel piles – cont						
Section modulus 2000–3000 cm³/m; section reference AZ 26; mass 155.2 kg/m²; sectional modulus 2600 cm³/m; EN 10248 grade S270GP steel						
driven area	–	–	–	–	m²	37.04
area of piles of length 6–18 m	–	–	–	–	m²	94.00
area of piles of length 18–24 m	–	–	–	–	m²	95.59
Section modulus 3000–4000 cm³/m; section reference AZ 36; mass 194.0 kg/m²; sectional modulus 3600 cm³/m; EN 10248 grade S270GP steel						
driven area	–	–	–	–	m²	29.43
area of piles of length 6–18 m	–	–	–	–	m²	114.77
area of piles of length 18–24 m	–	–	–	–	m²	116.72
Straight section modulus ne 500 cm³/m; section reference AS 500-12 mass 149 kg/m²; sectional modulus 51 cm³/m; EN 10248 grade S270GP steel						
driven area	–	–	–	–	m²	30.24
area of piles of length 6–18 m	–	–	–	–	m²	155.49
area of piles of length 18–24 m	–	–	–	–	m²	159.12
One coat black tar vinyl (PC1) protective treatment applied all surfaces at shop to minimum dry film thickness up to 150 microns to steel piles						
section reference AZ 12; pile area	–	–	–	–	m²	8.49
section reference AZ 17; pile area	–	–	–	–	m²	8.67
section reference AZ 26; pile area	–	–	–	–	m²	9.49
section reference AZ 36; pile area	–	–	–	–	m²	10.10
section reference AS 500 - 12; pile area	–	–	–	–	m²	10.55
One coat black high build isocyanate cured epoxy pitch (PC2) protective treatment applied all surfaces at shop to minimum dry film thickness up to 450 microns to steel piles						
section reference AZ 12; pile area	–	–	–	–	m²	16.96
section reference AZ 17; pile area	–	–	–	–	m²	17.36
section reference AZ 26; pile area	–	–	–	–	m²	18.97
section reference AZ 36; pile area	–	–	–	–	m²	20.20
section reference AS 500 - 12; pile area	–	–	–	–	m²	21.09
Arcelor Mittal U section steel piles; EN 10248 grade S270GP steel						
The following unit costs are based on driving/extracting 1,500 m² of sheet piling on a clear site with reasonable access.						
Provision of plant, equipment and labour including transport to and from the site and establishing and dismantling						
driving of sheet piling	–	–	–	–	sum	6200.00
extraction of sheet piling	–	–	–	–	sum	5500.00
Standing time	–	–	–	–	hr	290.70

SERIES 1600: PILING AND EMBEDDED RETAINING WALLS

Item	Gang hours	Labour £	Plant £	Material £	Unit	Total rate £
Section modulus 500–800 cm³/m; section reference PU 6; mass 76.0 kg/m²; sectional modulus 600 cm³/m						
driven area	–	–	–	–	m²	50.82
area of piles of length 6–18 m	–	–	–	–	m²	65.65
area of piles of length 18–24 m	–	–	–	–	m²	66.76
Section modulus 800–1200 cm³/m; section reference PU 8; mass 90.9 kg/m²; sectional modulus 830 cm³/m						
driven area	–	–	–	–	m²	43.31
area of piles of length 6–18 m	–	–	–	–	m²	65.65
area of piles of length 18–24 m	–	–	–	–	m²	66.83
Section modulus 1200–2000 cm³/m; section reference PU 12; mass 110.1 kg/m²; sectional modulus 1200 cm³/m						
driven area	–	–	–	–	m²	38.60
area of piles of length 6–18 m	–	–	–	–	m²	79.76
area of piles of length 18–24 m	–	–	–	–	m²	81.14
Section modulus 1200–2000 cm³/m; section reference PU 18; mass 128.2 kg/m²; sectional modulus 1800 cm³/m						
driven area	–	–	–	–	m²	34.34
area of piles of length 6–18 m	–	–	–	–	m²	92.65
area of piles of length 18–24 m	–	–	–	–	m²	94.25
Section modulus 2000–3000 cm³/m; section reference PU 22; mass 143.6 kg/m²; sectional modulus 2200 cm³/m						
driven area	–	–	–	–	m²	31.53
area of piles of length 6–18 m	–	–	–	–	m²	112.13
area of piles of length 18–24 m	–	–	–	–	m²	114.84
Section modulus 3000–4000 cm³/m; section reference PU 32; mass 190.2 kg/m²; sectional modulus 3200 cm³/m						
driven area	–	–	–	–	m²	27.29
area of piles of length 6–18 m	–	–	–	–	m²	126.77
area of piles of length 18–24 m	–	–	–	–	m²	129.82
One coat black tar vinyl (PC1) protective treatment applied all surfaces at shop to minimum dry film thickness up to 150 microns to steel piles						
section reference PU 6; pile area	–	–	–	–	m²	7.14
section reference PU 8; pile area	–	–	–	–	m²	7.26
section reference PU 12; pile area	–	–	–	–	m²	7.59
section reference PU 18; pile area	–	–	–	–	m²	8.14
section reference PU 22; pile area	–	–	–	–	m²	8.44
section reference PU 32; pile area	–	–	–	–	m²	8.56
One coat black high build isocyanate cured epoxy pitch (PC2) protective treatment applied all surfaces at shop to minimum dry film thickness up to 450 microns to steel piles						
section reference PU 6; pile area	–	–	–	–	m²	14.27
section reference PU 8; pile area	–	–	–	–	m²	14.51
section reference PU 12; pile area	–	–	–	–	m²	15.17

SERIES 1600: PILING AND EMBEDDED RETAINING WALLS

Item	Gang hours	Labour £	Plant £	Material £	Unit	Total rate £
STEEL SHEET PILES – cont						
Arcelor Mittal U section steel piles – cont						
One coat black high build isocyanate cured epoxy pitch (PC2) protective treatment applied all surfaces at shop to minimum dry film thickness up to 450 microns to steel piles – cont						
section reference PU 18; pile area	–	–	–	–	m²	16.27
section reference PU 22; pile area	–	–	–	–	m²	16.87
section reference PU 32; pile area	–	–	–	–	m²	17.13
EMBEDDED RETAINING WALL PLANT						
Diaphgram walls are the construction of vertical walls, cast in place in a trench excavation. They can be formed in reinforced concrete to provide structural elements for temporary or permanent retaining walls. Wall thicknesses of 500 to 1,500mm up to 40m deep may be constructed. Special equipment such as the Hydrofraise can construct walls up to 100 m deep. Restricted urban sites will significantly increase the costs.						
The following costs are based on constructing a diaphgram wall with an excavated volume of 4000 m³ using a grab. Typical progress would be up to 500 m per week.						
Establishment of standard diaphragm walling plant, including bentonite storage tanks.	–	–	–	–	item	76875.00
Standing time	–	–	–	–	hr	1076.25
Guide walls (twin)	–	–	–	–	m	358.75
Waterproofed joints	–	–	–	–	m	6.50
DIAPHRAGM WALLS						
Excavation for walls 1000mm thick, disposal of soil and placing of concrete	–	–	–	–	m³	440.00
Provide and place reinforcement cages	–	–	–	–	t	750.00
Excavate/chisel in hard materials/rock	–	–	–	–	hr	975.00

SERIES 1700: STRUCTURAL CONCRETE

Item	Gang hours	Labour £	Plant £	Material £	Unit	Total rate £
GENERAL						
Notes						
Refer also to Civil Engineering - Concrete, Formwork, Reinforcement and Precast Concrete, although this section is fundamentally different in that the provision of concrete of different classes and its placement is combined in the unit costs.						
RESOURCES - LABOUR						
Concreting gang						
1 ganger/chargehand (skill rate 4)		14.58				
2 skilled operatives (skill rate 4)		27.31				
4 unskilled operatives (general)		51.00				
1 plant operator (skill rate 3) - 25% of time		4.15				
Total Gang Rate / Hour	£	**97.04**				
Formwork gang						
1 foreman (craftsman)		22.88				
2 joiners (craftsman)		39.26				
1 unskilled operative (general)		12.75				
1 plant operator (craftsman) - 25% of time		5.30				
Total Gang Rate / Hour	£	**80.19**				
Reinforcement gang						
1 foreman (craftsman)		22.88				
4 steel fixers (craftsman)		78.52				
1 unskilled operative (general)		12.75				
1 plant operator (craftsman) - 25% of time		5.30				
Total Gang Rate / Hour	£	**119.44**				
RESOURCES - PLANT						
Concreting						
10 t crane- 25% of time			6.57			
gas oil for ditto			0.28			
0.76 m³ concrete skip - 25% of time			0.63			
11.3 m³/min compressor, 4 tool			28.00			
gas oil for ditto			12.78			
4 poker vibrators P5475 mm or less in thickness			5.37			
Total Gang Rate / Hour		£	**53.62**			
Formwork						
20 t crawler crane - 25% of time			7.13			
gas oil for ditto			0.32			
22' saw bench			1.48			
gas oil for ditto			0.37			
small power tools (formwork)			2.25			
Total Gang Rate / Hour		£	**11.55**			

SERIES 1700: STRUCTURAL CONCRETE

Item	Gang hours	Labour £	Plant £	Material £	Unit	Total rate £
RESOURCES - PLANT – cont						
Reinforcement						
30 t crawler crane (25% of time)			7.82			
gas oil for ditto			0.32			
bar cropper			2.16			
small power tools (reinforcement)			1.65			
tirfors, kentledge etc.			0.89			
Total Gang Rate / Hour		£	12.84			
RESOURCES - MATERIALS						
Wastage allowances 5% onto delivered ready mixed concrete						
INSITU CONCRETE						
In-situ concrete Grade C10						
Blinding						
75 mm or less in thickness	0.18	17.33	9.65	94.03	m³	**121.01**
Blinding; in narrow widths up to 1.0 m wide or in bottoms of trenches up to 2.5 m wide; excluding formwork						
75 mm or less in thickness	0.20	19.25	10.73	94.03	m³	**124.01**
In-situ concrete Grade C15						
Blinding; excluding formwork						
75 mm or less in thickness	0.16	15.40	8.58	94.03	m³	**118.01**
Blinding; in narrow widths up to 1.0 m wide or in bottoms of trenches up to 2.5 m wide; excluding formwork						
75 mm or less in thickness	0.18	17.33	9.65	94.03	m³	**121.01**
In-situ concrete Grade C20/20						
Bases, footings, pile caps and ground beams; thickness						
ne 150 mm	0.20	19.25	10.73	96.11	m³	**126.09**
150–300 mm	0.17	16.37	9.17	96.11	m³	**121.65**
300–500 mm	0.15	14.44	8.10	96.11	m³	**118.65**
exceeding 500 mm	0.14	13.48	7.51	96.11	m³	**117.10**
Walls; thickness						
ne 150 mm	0.21	20.22	11.31	96.11	m³	**127.64**
150–300 mm	0.15	14.44	8.10	96.11	m³	**118.65**
300–500 mm	0.13	12.51	7.03	96.11	m³	**115.65**
exceeding 500 mm	0.12	11.55	6.44	96.11	m³	**114.10**
Suspended slabs; thickness						
ne 150 mm	0.27	25.99	14.53	96.11	m³	**136.63**
150–300 mm	0.21	20.22	11.31	96.11	m³	**127.64**
300–500 mm	0.19	18.29	10.24	96.11	m³	**124.64**
exceeding 500 mm	0.19	18.29	10.24	96.11	m³	**124.64**

SERIES 1700: STRUCTURAL CONCRETE

Item	Gang hours	Labour £	Plant £	Material £	Unit	Total rate £
Columns, piers and beams; cross-sectional area						
ne 0.03 m²	0.50	48.13	26.81	96.11	m³	**171.05**
0.03–0.10 m²	0.40	38.51	21.45	96.11	m³	**156.07**
0.10–0.25 m²	0.35	33.69	18.82	96.11	m³	**148.62**
0.12–1.00 m²	0.35	33.69	18.82	96.11	m³	**148.62**
exceeding 1 m²	0.28	26.95	15.02	96.11	m³	**138.08**
ADD to the above prices for						
sulphate resisting cement	–	–	–	9.59	m³	**9.59**
air entrained concrete	–	–	–	4.95	m³	**4.95**
water repellant concrete	–	–	–	4.99	m³	**4.99**
In-situ concrete Grade C30/20						
Bases, footings, pile caps and ground beams; thickness						
ne 150 mm	0.21	20.22	11.31	98.28	m³	**129.81**
150–300 mm	0.18	17.33	9.65	98.28	m³	**125.26**
300–500 mm	0.15	14.44	8.10	98.28	m³	**120.82**
exceeding 500 mm	0.14	13.48	7.51	98.28	m³	**119.27**
Walls; thickness						
ne 150 mm	0.22	21.18	11.80	98.28	m³	**131.26**
150–300 mm	0.16	15.40	9.77	98.28	m³	**123.45**
300–500 mm	0.13	12.51	8.00	98.28	m³	**118.79**
exceeding 500 mm	0.12	11.55	7.33	98.28	m³	**117.16**
Suspended slabs; thickness						
ne 150 mm	0.28	26.95	17.11	98.28	m³	**142.34**
150–300 mm	0.22	21.18	13.45	98.28	m³	**132.91**
300–500 mm	0.19	18.29	11.67	98.28	m³	**128.24**
exceeding 500 mm	0.18	17.33	11.00	98.28	m³	**126.61**
Columns, piers and beams; cross-sectional area						
ne 0.03 m²	0.53	51.02	32.43	98.28	m³	**181.73**
0.03–0.10 m²	0.42	40.43	25.66	98.28	m³	**164.37**
0.10–0.25 m²	0.36	34.66	21.99	98.28	m³	**154.93**
0.25–1.00 m²	0.35	33.69	21.44	98.28	m³	**153.41**
exceeding 1 m²	0.28	26.95	17.11	98.28	m³	**142.34**
ADD to the above prices for						
sulphate resisting cement	–	–	–	9.59	m³	**9.59**
air entrained concrete	–	–	–	4.95	m³	**4.95**
water repellant concrete	–	–	–	4.99	m³	**4.99**

SERIES 1700: STRUCTURAL CONCRETE

Item	Gang hours	Labour £	Plant £	Material £	Unit	Total rate £
PRECAST CONCRETE						
The cost of precast concrete item is very much dependant on the complexity of the moulds, the number of units to be cast from each mould and the size and the weight of the unit to be handled. The unit rates below are for standard precast items that are often to be found on a Civil Engineering project. It would be misleading to quote for indicative costs for tailor-made precast concrete units and it is advisable to contact specialist maunfacturers for guide prices.						
Pretensioned prestressed beams; concrete Grade C20						
Beams						
100 × 65 × 1050 mm long	1.00	3.40	1.63	6.87	nr	**11.90**
265 × 65 × 1800 mm long	1.00	4.26	3.42	25.97	nr	**33.65**
Inverted 'T' beams, flange width 495 mm						
section T1; 8 m long, 380 mm deep; mass 1.88t	–	–	–	–	nr	**714.00**
section T2; 9 m long, 420 mm deep; mass 2.29t	–	–	–	–	nr	**816.00**
section T3; 11 m long, 535 mm deep; mass 3.02t	–	–	–	–	nr	**918.00**
section T4; 12 m long, 575 mm deep; mass 3.54t	–	–	–	–	nr	**1020.00**
section T5; 13 m long, 615 mm deep; mass 4.08t	–	–	–	–	nr	**1224.00**
section T6; 13 m long, 655 mm deep; mass 4.33t	–	–	–	–	nr	**1326.00**
section T7; 12 m long, 695 mm deep; mass 4.95t	–	–	–	–	nr	**1530.00**
section T8; 15 m long, 735 mm deep; mass 5.60t	–	–	–	–	nr	**1734.00**
section T9; 16 m long, 775 mm deep; mass 6.28t	–	–	–	–	nr	**1836.00**
section T10; 18 m long, 815 mm deep; mass 7.43t	–	–	–	–	nr	**2040.00**
'M' beams, flange width 970 mm						
section M2; 17 m long, 720 mm deep; mass 12.95t	–	–	–	–	nr	**4692.00**
section M3; 18 m long, 800 mm deep; mass 15.11t	–	–	–	–	nr	**5304.00**
section M6; 22 m long, 1040 mm deep; mass 20.48t	–	–	–	–	nr	**7344.00**
section M8; 25 m long, 1200 mm deep; mass 23.68t	–	–	–	–	nr	**8364.00**
'U' beams, base width 970 mm						
section U3; 16 m long, 900 mm deep; mass 19.24t	–	–	–	–	nr	**9792.00**
section U5; 20 m long, 1000 mm deep; mass 25.64t	–	–	–	–	nr	**11424.00**
section U8; 24 m long, 1200 mm deep; mass 34.56t	–	–	–	–	nr	**15912.00**
section U12 ; 30 m long, 1600 mm deep; mass 52.74t	–	–	–	–	nr	**23868.00**
Precast concrete culverts, cattle creeps and subway units; rebated joints						
Rectangular cross section						
500 mm high × 1000 mm wide	1.00	4.62	4.89	290.70	m	**300.21**
1000 mm high × 1500 mm wide	1.00	10.29	10.90	474.30	m	**495.49**
1500 mm high × 1500 mm wide	1.00	18.66	20.16	550.80	m	**589.62**
2000 mm high × 2750 mm wide	1.00	35.01	32.07	1616.70	m	**1683.78**
2750 mm high × 3000 mm wide	1.00	46.34	64.37	1785.00	m	**1895.71**
Extra for units curved on plan to less than 20 m radius	–	–	–	295.80	m	**295.80**

SERIES 1700: STRUCTURAL CONCRETE

Item	Gang hours	Labour £	Plant £	Material £	Unit	Total rate £
SURFACE FINISH OF CONCRETE - FORMWORK						
Materials						
Formwork materials include for shutter, bracing, ties,						
support, kentledge and all consumables.						
These unit costs are based upon those outputs and						
prices detailed in Civil Engineering - Concrete						
Formwork but are referenced to The Specification for						
Highway Works, clause 1708.						
The following unit rates do not include for formwork						
outside the payline and are based on an optimum of						
a minimum 8 uses with 10% per use towards the						
cost of repairs / replacement of components						
damaged during disassembly.						
ADJUST formwork material costs generally						
depending on the number of uses:						

Nr of Uses	% Adjustment	Inclusion for waste
1	Add 90–170%	7%
2	Add 50–180%	7%
3	Add 15–30%	6%
6	Add 5–10%	5%
8	No change	5%
10	Deduct 5–7%	5%

Definitions
'Class F1' formwork is rough finish
'Class F2' formwork is fair finish
'Class F3' formwork is extra smooth finish

Item	Gang hours	Labour £	Plant £	Material £	Unit	Total rate £
Formwork Class F1						
Horizontal more than 300 mm wide	0.52	40.34	6.00	7.54	m²	**53.88**
Inclined more than 300 mm wide	0.55	42.67	6.35	10.62	m²	**59.64**
Vertical more than 300 mm wide	0.61	47.33	7.05	10.58	m²	**64.96**
300 mm wide or less at any inclination	0.72	55.86	8.31	10.58	m²	**74.75**
Curved of both girth and width more than 300 mm at any inclination	0.95	73.70	10.97	12.10	m²	**96.77**
Curved of girth or width of 300 mm or less at any inclination	0.72	55.86	8.31	12.10	m²	**76.27**
Domed	1.20	93.10	13.86	14.89	m²	**121.85**
Void former cross-section 100 × 100 mm	0.07	5.43	0.29	4.60	m	**10.32**
Void former cross-section 250 × 250 mm	0.12	9.31	0.49	10.87	m	**20.67**
Void former cross-section 500 × 500 mm	0.30	23.27	1.23	21.66	m	**46.16**
Formwork Class F2						
Horizontal more than 300 mm wide	0.54	41.89	6.24	13.46	m²	**61.59**
Inclined more than 300 mm wide	0.57	44.22	6.59	22.01	m²	**72.82**
Vertical more than 300 mm wide	0.63	48.88	7.28	22.01	m²	**78.17**

SERIES 1700: STRUCTURAL CONCRETE

Item	Gang hours	Labour £	Plant £	Material £	Unit	Total rate £
SURFACE FINISH OF CONCRETE - FORMWORK – cont						
Formwork Class F2 – cont						
300 mm wide or less at any inclination	0.74	57.41	8.55	22.01	m²	**87.97**
Curved of both girth and width more than 300 mm at any inclination	0.98	76.03	11.32	26.28	m²	**113.63**
Curved of girth or width of 300 mm or less at any inclination	0.75	58.19	8.66	26.28	m²	**93.13**
Domed	1.40	108.62	16.17	34.13	m²	**158.92**
Formwork Class F3						
Horizontal more than 300 mm wide	0.56	43.45	6.47	15.79	m²	**65.71**
Inclined more than 300 mm wide	0.59	45.77	6.82	24.34	m²	**76.93**
Vertical more than 300 mm wide	0.65	50.43	7.51	24.34	m²	**82.28**
300 mm wide or less at any inclination	0.76	58.96	8.78	24.34	m²	**92.08**
Curved of both girth and width more than 300 mm at any inclination	0.99	76.81	11.44	28.61	m²	**116.86**
Curved of girth or width of 300 mm or less at any inclination	0.77	59.74	8.89	28.61	m²	**97.24**
Domed	1.45	112.49	16.75	36.46	m²	**165.70**
Formwork ancillaries						
Allowance for additional craneage and rub up where required	0.13	10.09	1.50	0.16	m²	**11.75**
PATTERNED PROFILE FORMWORK						
Patterned Profile Formwork						
Extra over formwork for patterned profile formliners, INSITEX or similar	–	–	–	–	m²	**34.85**
STEEL REINFORCEMENT FOR STRUCTURES						
Stainless steel bars						
Bar reinforcement nominal size 16 mm and under not exceeding 12 m in length	6.74	785.70	86.54	3015.00	t	**3887.24**
Bar reinforcement nominal size 20 mm and over not exceeding 12 m in length						
20 mm nominal size	4.44	517.59	57.01	2880.00	t	**3454.60**
25 mm nominal size	4.44	517.59	57.01	2790.00	t	**3364.60**
32 mm nominal size	4.44	517.59	57.01	2790.00	t	**3364.60**
ADD to the above for bars						
12–13.5 m long	–	–	–	20.70	t	**20.70**
13.5–15 m long	–	–	–	20.70	t	**20.70**
over 15 m long; per 500 mm increment	–	–	–	4.60	t	**4.60**
High yield steel bars Bs 4449; deformed, Grade 500C						
Bar reinforcement nominal size 16 mm and under not exceeding 12 m in length	6.74	785.70	86.54	563.50	t	**1435.74**

SERIES 1700: STRUCTURAL CONCRETE

Item	Gang hours	Labour £	Plant £	Material £	Unit	Total rate £
Bar reinforcement nominal size 20 mm and over not exceeding 12 m in length						
20 mm nominal size	4.44	517.59	57.01	609.50	t	1184.10
25 mm nominal size	4.44	517.59	57.01	609.50	t	1184.10
32 mm nominal size	4.44	517.59	57.01	615.25	t	1189.85
40 mm nominal size	4.44	517.59	57.01	621.00	t	1195.60
ADD to the above for bars						
12–13.5 m long	–	–	–	20.70	t	20.70
13.5–15 m long	–	–	–	20.70	t	20.70
over 15 m long; per 500 mm increment	–	–	–	4.60	t	4.60
Helical reinforcement nominal size 16 mm and under						
ne 12 m in length	6.74	785.70	86.54	634.73	t	1506.97
Helical reinforcement nominal size 20 mm and over						
20 mm nominal size	4.44	517.59	57.01	615.60	t	1190.20
25 mm nominal size	4.44	517.59	57.01	615.60	t	1190.20
32 mm nominal size	4.44	517.59	57.01	621.40	t	1196.00
40 mm nominal size	4.44	517.59	57.01	627.21	t	1201.81
Dowels						
16 mm diameter × 600 mm long	0.10	11.66	1.28	1.97	nr	14.91
20 mm diameter × 600 mm long	0.10	11.66	1.28	2.64	nr	15.58
25 mm diameter × 600 mm long	0.10	11.66	1.28	3.76	nr	16.70
32 mm diameter × 600 mm long	0.10	11.66	1.28	5.78	nr	18.72
Mild steel bars Bs 4449; Grade 250						
Bar reinforcement nominal size 16 mm and under						
ne 12 m in length	6.74	785.70	86.54	589.43	t	1461.67
Bar reinforcement nominal size 20 mm and over; not exceeding 12 m in length						
20 mm nominal size	4.44	517.59	57.01	566.05	t	1140.65
25 mm nominal size	4.44	517.59	57.01	566.05	t	1140.65
32 mm nominal size	4.44	517.59	57.01	572.16	t	1146.76
40 mm nominal size	4.44	517.59	57.01	581.61	t	1156.21
ADD to the above for bars						
12–13.5 m long	–	–	–	20.70	t	20.70
13.5–15 m long	–	–	–	20.70	t	20.70
over 15 m long, per 500 mm increment	–	–	–	4.60	t	4.60
ADD for cutting, bending, tagging and baling reinforcement on site						
6 mm nominal size	4.87	180.55	62.53	1.65	t	244.73
8 mm nominal size	4.58	169.80	58.81	1.65	t	230.26
10 mm nominal size	3.42	126.79	43.91	1.65	t	172.35
12 mm nominal size	2.55	94.54	32.75	1.65	t	128.94
16 mm nominal size	2.03	75.26	26.07	1.65	t	102.98
20 mm nominal size	1.68	62.28	21.57	1.65	t	85.50
25 mm nominal size	1.68	62.28	21.57	1.65	t	85.50
32 mm nominal size	1.39	51.53	17.85	1.65	t	71.03
40 mm nominal size	1.39	51.53	17.85	1.65	t	71.03

SERIES 1700: STRUCTURAL CONCRETE

Item	Gang hours	Labour £	Plant £	Material £	Unit	Total rate £
STEEL REINFORCEMENT FOR STRUCTURES – cont						
Fabric reinforcement; high yield steel BS 4483						
Fabric reinforcement						
BS ref A98; nominal mass 1.54 kg/m^2	0.03	3.50	0.39	1.68	m^2	**5.57**
BS ref A142; nominal mass 2.22 kg/m^2	0.03	3.50	0.39	1.96	m^2	**5.85**
BS ref A193; nominal mass 3.02 kg/m^2	0.04	4.66	0.51	2.14	m^2	**7.31**
BS ref A252; nominal mass 3.95 kg/m^2	0.04	4.66	0.51	2.50	m^2	**7.67**
BS ref A393; nominal mass 6.16 kg/m^2	0.07	8.16	0.90	5.05	m^2	**14.11**
BS ref B196; nominal mass 3.05 kg/m^2	0.04	4.66	0.51	4.62	m^2	**9.79**
BS ref B283; nominal mass 3.73 kg/m^2	0.04	4.66	0.51	3.08	m^2	**8.25**
BS ref B385; nominal mass 4.53 kg/m^2	0.05	5.83	0.65	3.75	m^2	**10.23**
BS ref B503; nominal mass 5.93 kg/m^2	0.05	5.83	0.65	5.52	m^2	**12.00**
BS ref B785; nominal mass 8.14 kg/m^2	0.08	9.33	1.03	6.30	m^2	**16.66**
BS ref B1131; nominal mass 10.90 kg/m^2	0.09	10.49	1.16	8.68	m^2	**20.33**
BS ref C282; nominal mass 2.61 kg/m^2	0.03	3.50	0.39	2.01	m^2	**5.90**
BS ref C385; nominal mass 3.41 kg/m^2	0.04	4.66	0.51	2.61	m^2	**7.78**
BS ref C503; nominal mass 4.34 kg/m^2	0.05	5.83	0.65	3.28	m^2	**9.76**
BS ref C636; nominal mass 5.55 kg/m^2	0.05	5.83	0.65	4.03	m^2	**10.51**
BS ref C785; nominal mass 6.72 kg/m^2	0.07	8.16	0.90	4.89	m^2	**13.95**
BS ref D49; nominal mass 0.77 kg/m^2	0.02	2.33	0.26	0.66	m^2	**3.25**
BS ref D98; nominal mass 1.54 kg/m^2	0.02	2.33	0.26	1.12	m^2	**3.71**

SERIES 1800: STEELWORK FOR STRUCTURES

Item	Gang hours	Labour £	Plant £	Material £	Unit	Total rate £
FABRICATION OF STEELWORK						
Steelwork to Bs En10025; Grade S275						
Fabrication of main members						
rolled sections	–	–	–	–	t	1492.43
plated rolled sections	–	–	–	–	t	3847.38
plated girders	–	–	–	–	t	3207.76
box girders	–	–	–	–	t	1521.69
Fabrication of deck panels						
rolled sections	–	–	–	–	t	2568.15
plated rolled sections	–	–	–	–	t	2345.26
plated girders	–	–	–	–	t	2345.26
Fabrication of subsiduary steelwork						
rolled sections	–	–	–	–	t	2345.26
plated rolled sections	–	–	–	–	t	2345.26
plated girders	–	–	–	–	t	2451.86
ERECTION OF STEELWORK						
Trial erection at the place of fabrication	–	–	–	–	t	304.00
Permanent erection of steelwork; substructure	–	–	–	–	t	237.50
Permanent erection of steelwork; superstructure	–	–	–	–	t	237.50
MISCELLANEOUS METALWORK						
Mild steel						
Ladders						
Cat ladder; 64 × 13mm bar strings; 19mm rungs at 250mm centres; 450mm wide with safety hoops	–	–	–	–	m	92.02
Handrails						
Galvanised tubular metal; 76mm diameter handrail, 48mm diameter standards at 750mm centres, 48mm diameter rail; 1070mm high overall	–	–	–	–	m	139.96
Metal access cover and frame						
Group 4, ductile iron, single seal 610 × 610 x 100mm depth; D400	–	–	–	–	nr	190.14
Group 2, ductile iron, double seal single piece cover 600 × 450mm; B125	–	–	–	–	nr	154.15

SERIES 1900: PROTECTION OF STEELWORK AGAINST CORROSION

Item	Gang hours	Labour £	Plant £	Material £	Unit	Total rate £
RESOURCES - LABOUR						
Protective painting gang						
1 ganger/chargehand (skill rate 4)		14.58				
2 skilled operatives (skill rate 4)		27.31				
2 unskilled operatives (general)		25.50				
Total Gang rate / Hour	£	**67.39**				
RESOURCES - PLANT						
Protective painting						
power tools (protection of steelwork)			3.21			
access scaffolding, trestles and ladders			3.15			
5 t transit van (50% of time)			3.75			
gas oil for ditto			1.30			
Total Gang Rate / Hour		£	**11.41**			
RESOURCES - MATERIALS						
All coats applied off site except as noted						
The external environment has been taken as Inland 'B' Exposed.						
PROTECTIVE SYSTEM						
Galvanising to Bs En Iso 1461; apply protective coatings comprising: 1st coat: Mordant T wash; 2nd coat: Zinc rich epoxy primer; 3rd coat: Zinc phosphate, CR/Alkyd Undercoat; 4th coat; MIO CR Undercoat-on site externally; 5th coat: CR coloured finish-on site externally						
To metal parapets and fencing, lighting columns, brackets						
by brush or airless spray to dry film thickness 200 microns	0.20	13.37	2.28	20.22	m²	**35.87**
Blast clean to Bs 7079 (surface preparation); apply protective coatings comprising: 1st coat: Zinc Chromate,Red Oxide Blast Primer; 2nd coat: Zinc Phosphate, Epoxy Ester Undercoat; 3rd coat: MIO Undercoat; 4th coat: MIO coloured finish-on site externally						
To subsiduary steelwork, interior finishes						
By brush or airless spray to dry film thickness 175 microns	0.15	10.03	1.72	14.59	m²	**26.34**

SERIES 1900: PROTECTION OF STEELWORK AGAINST CORROSION

Item	Gang hours	Labour £	Plant £	Material £	Unit	Total rate £

Blast clean to Bs 7079 (surface preparation); apply metal coating of aluminium spray at works; apply protective coatings comprising: 1st coat: Zinc Chromate Etch Primer (2 pack); 2nd coat: Zinc Phosphate, CR/Aalkyd Undercoat; 3rd coat: Zinc Phosphate, CR/alkyd Undercoat; 4th coat: MIO CR Undercoat; 5th coat: CR finish-on site externally
To main steel members
  By brush or airless spray to dry film thickness
  250 microns
	0.18	12.03	2.05	16.53	m^2	**30.61**

Blast clean to Bs 7079 (second quality surface preparation); remove all surface defects to BS EN10025;apply protective coatings comprising: 1st coat: Zinc rich primer (2 pack); 2nd coat: Epoxy High Build M10 (2 pack); 3rd coat: Polyurethane Undercoat (2 pack)-on site internally; 4th coat: Finish coat polyurethane (2 pack)-on site externally To internal steel members By brush or airless spray	0.15	10.03	1.72	22.88	m^2	**34.63**
ALTERNATIVE SURFACE TREATMENTS						
Galvanising (Hot dip) to BS EN ISO 1461, assuming average depth 20 m^2 per tonne of steel	–	–	–	–	m^2	**15.19**
Shot blasting (at works)	–	–	–	–	m^2	**3.05**
Grit blasting (at works)	–	–	–	–	m^2	**4.35**
Sand blasting (at works)	–	–	–	–	m^2	**6.55**
Shot blasting (on site)	–	–	–	–	m^2	**4.28**
Grit blasting (on site)	–	–	–	–	m^2	**6.10**
Sand blasting (on site)	–	–	–	–	m^2	**9.17**

SERIES 2000: WATERPROOFING FOR STRUCTURES

Item	Gang hours	Labour £	Plant £	Material £	Unit	Total rate £
GENERAL						
Notes						
This section is based around the installation of proprietary systems to new/recently completed works as part of major scheme, for minor works/ repairs outputs will be many times more. Outputs are also based on use of skilled labour, therefore effieciency is high and wastage low (5-7% only allowed) excepting laps where required.						
RESOURCES - LABOUR						
Asphalting gang						
1 ganger/chargehand (skill rate 4)		14.58				
2 unskilled operative (general)		25.50				
Total Gang rate / Hour	£	**40.08**				
Damp proofing gang						
1 ganger/chargehand (skill rate 4)		14.58				
1 skilled operative (skill rate 4)		13.65				
1 unskilled operative (general)		12.75				
Total Gang rate / Hour	£	**40.99**				
Sprayed/brushed waterproofing gang						
1 ganger/chargehand (skill rate 4) - 30% of time		4.37				
1 skilled operative (skill rate 4)		13.65				
Total Gang rate / Hour	£	**18.03**				
Protective layers - screed gang						
1 ganger/chargehand (skill rate 4)		14.58				
1 skilled operative (skill rate 4)		13.65				
1 unskilled operative (general)		12.75				
Total Gang rate / Hour	£	**40.99**				
RESOURCES - PLANT						
Asphalting						
45 litre portable tar boiler including sprayer (50% of time)			0.38			
2 t dumper (50% of time)			2.69			
gas oil for ditto			0.56			
Total Gang Rate / Hour		£	3.81			
Damp proofing						
2 t dumper (50% of time)			2.69			
gas oil for ditto			0.56			
Total Gang Rate / Hour		£	3.25			

SERIES 2000: WATERPROOFING FOR STRUCTURES

Item	Gang hours	Labour £	Plant £	Material £	Unit	Total rate £
WATERPROOFING						
Mastic asphalt; BS 6925 Type T 1097; 20 mm thick; two coats						
over 300 mm wide; ne 30° to horizontal	0.33	13.13	1.90	11.07	m^2	**26.10**
over 300 mm wide; 30–90° to horizontal	0.50	19.90	1.26	11.07	m^2	**32.23**
ne 300 mm wide; at any inclination	0.60	23.88	1.26	11.07	m^2	**36.21**
to domed surfaces	0.75	29.85	1.26	11.07	m^2	**42.18**
'Bituthene 1000'; lapped joints						
over 300 mm wide; ne 30° to horizontal	0.05	2.03	0.16	8.10	m^2	**10.29**
over 300 mm wide; 30–90° to horizontal	0.06	2.44	0.19	10.33	m^2	**12.96**
ne 300 mm wide; at any inclination	0.08	3.25	0.26	10.33	m^2	**13.84**
Extra; one coat primer on vertical surfaces	0.03	1.22	0.10	0.03	m^2	**1.35**
'Bituthene 4000' ; lapped joints						
over 300 mm wide; ne 30° to horizontal	0.05	2.03	0.16	9.63	m^2	**11.82**
over 300 mm wide; 30–90° to horizontal	0.06	2.44	1.07	12.84	m^2	**16.35**
ne 300 mm wide; at any inclination	0.08	3.25	1.23	12.84	m^2	**17.32**
'Famguard' (hot applied) with Fam-primer						
over 300 mm wide; ne 30° to horizontal	0.32	13.02	1.04	17.53	m^2	**31.59**
over 300 mm wide; 30–90° to horizontal	0.35	14.24	1.14	17.87	m^2	**33.25**
ne 300 mm wide; at any inclination	0.40	16.27	1.30	20.64	m^2	**38.21**
'Famflex' (hot applied) with Fam-primer						
over 300 mm wide; ne 30° to horizontal	0.32	13.02	1.04	13.58	m^2	**27.64**
over 300 mm wide; 30–90° to horizontal	0.34	13.83	1.10	13.84	m^2	**28.77**
ne 300 mm wide; at any inclination	0.38	15.46	1.23	15.97	m^2	**32.66**
Two coats of RIW liquid asphaltic composition sprayed or brushed on						
over 300 mm wide; ne 30° to horizontal	0.03	0.54	0.10	7.82	m^2	**8.46**
over 300 mm wide; 30–90° to horizontal	0.03	0.54	0.10	7.82	m^2	**8.46**
ne 300 mm wide; at any inclination	0.04	0.71	0.13	7.82	m^2	**8.66**
Two coats of 'Mulseal' sprayed or brushed on						
any inclination	0.07	1.25	0.23	15.59	m^2	**17.07**
20 mm thick red tinted sand asphalt layer						
onto bridge deck	0.02	0.81	0.39	10.89	m^2	**12.09**
SURFACE IMPREGNATION OF CONCRETE						
Silane waterproofing						
Surface impregnation to plain surfaces	–	–	–	–	m^2	**3.60**
REMOVALOF EXISTING WATERPROOFING						
Removal of existing asphalt waterproofing						
over 300 mm wide; ne 30° to horizontal	0.13	5.17	–	–	m^2	**5.17**
over 300 mm wide; 30–90° to horizontal	0.18	7.16	–	–	m^2	**7.16**
over 300 mm wide; at any inclination	0.20	7.96	–	–	m^2	**7.96**
to domed surfaces	0.25	9.95	–	–	m^2	**9.95**

SERIES 2100: BRIDGE BEARINGS

Item	Gang hours	Labour £	Plant £	Material £	Unit	Total rate £
GENERAL						
Notes						
Bridge bearings are manufactured and installed to individual specifications. The following guide prices are for different sizes of simple bridge bearings. If requirements are known, then advice ought to be obtained from specialist maunfacturers such as CCL.						
RESOURCES - LABOUR						
Bridge bearing gang						
1 ganger/chargehand (skill rate 4)		14.58				
2 unskilled operatives (general)		25.50				
Total Gang Rate / Hour	£	**40.08**				
BEARINGS						
Supply plain rubber bearings (3 m and 5 m lengths)						
150 × 20 mm	0.35	13.93	–	38.57	m	**52.50**
150 × 25 mm	0.35	13.93	–	46.28	m	**60.21**
Supply and place in position laminated elastomeric rubber bearing						
250 × 150 × 19 mm	0.25	9.95	–	28.65	nr	**38.60**
300 × 200 × 19 mm	0.25	9.95	–	31.40	nr	**41.35**
300 × 200 × 30 mm	0.27	10.75	–	39.67	nr	**50.42**
300 × 200 × 41 mm	0.27	10.75	–	59.63	nr	**70.38**
300 × 250 × 41 mm	0.30	11.94	–	74.54	nr	**86.48**
300 × 250 × 63 mm	0.30	11.94	–	104.13	nr	**116.07**
400 × 250 × 19 mm	0.32	12.74	–	52.34	nr	**65.08**
400 × 250 × 52 mm	0.32	12.74	–	120.32	nr	**133.06**
400 × 300 × 19 mm	0.32	12.74	–	62.81	nr	**75.55**
600 × 450 × 24 mm	0.35	13.93	–	192.78	nr	**206.71**
Adhesive fixings to laminated elastomeric rubber bearings						
2 mm thick epoxy adhesive	1.00	39.80	–	47.67	m²	**87.47**
15 mm thick epoxy mortar	1.50	59.70	–	266.36	m²	**326.06**
15 mm thick epoxy pourable grout	2.00	79.60	–	285.61	m²	**365.21**
Supply and install mechanical guides for laminated elastomeric rubber bearings						
500kN SLS design load; FP50 fixed pin Type 1	2.00	79.60	–	854.85	nr	**934.45**
500kN SLS design load; FP50 fixed pin Type 2	2.00	79.60	–	881.50	nr	**961.10**
750kN SLS design load; FP75 fixed pin Type 1	2.10	83.58	–	991.17	nr	**1074.75**
750kN SLS design load; FP75 fixed pin Type 2	2.10	83.58	–	1127.50	nr	**1211.08**
300kN SLS design load; UG300 Uniguide Type 1	2.00	79.60	–	1025.00	nr	**1104.60**
300kN SLS design load; UG300 Uniguide Type 2	2.00	79.60	–	1178.75	nr	**1258.35**
Supply and install fixed pot bearings						
355 × 355; PF200	2.00	79.60	–	916.35	nr	**995.95**
425 × 425; PF300	2.10	83.58	–	1092.65	nr	**1176.23**

SERIES 2100: BRIDGE BEARINGS

Item	Gang hours	Labour £	Plant £	Material £	Unit	Total rate £
Supply and install free sliding pot bearings						
445×345; PS200	2.10	83.58	–	1232.05	nr	**1315.63**
520×415; PS300	2.20	87.56	–	1571.33	nr	**1658.89**
Supply and install guided sliding pot bearings						
455×375; PG200	2.20	87.56	–	1636.92	nr	**1724.48**
545×435; PG300	2.30	91.54	–	1948.53	nr	**2040.07**
Testing Bearings						
If there is a requirement for testing bridge bearings prior to their being installed then the tests should be enumerated separately. Specialist advice should be sought once details are known.						
Compression test for laminated elastomeric bearings						
generally	–	–	–	–	nr	**59.66**
Shear test for laminated elastomeric bearings						
generally	–	–	–	–	nr	**77.13**

SERIES 2300: BRIDGE EXPANSION JOINTS AND SELAING OF GAPS

Item	Gang hours	Labour £	Plant £	Material £	Unit	Total rate £
GENERAL						
Notes						
Major movement joints to bridge and viaduct decks are manufactured and installed to individual specifications determined by the type of structure location in the deck, amount of movement to be expected, and many other variables. The following unit rates for other types of movement joints found in structures.						
RESOURCES - LABOUR						
Bridge jointing gang						
1 ganger/chargehand (skill rate 4)		14.58				
1 skilled operative (skill rate 4)		13.65				
1 unskilled operative		12.75				
TOTAL	£	**40.99**				
Jointing gang						
1 ganger/chargehand (skill rate 3)		16.18				
1 skilled operative (skill rate 3)		15.25				
1 unskilled operative		12.75				
TOTAL	£	**44.18**				
SEALING OF GAPS						
Flexcell joint filler board						
10 mm thick	0.10	4.07	–	3.54	m²	**7.61**
19 mm thick	0.16	6.51	–	5.96	m²	**12.47**
25 mm thick	0.16	6.51	–	7.46	m²	**13.97**
Building paper slip joint to abutment toe	0.01	0.45	–	21.70	m²	**22.15**
Bond breaking agent	0.03	1.02	–	4.04	m²	**5.06**
Hot poured rubber bitumen joint sealant						
10 × 20 mm	0.03	1.34	–	7.31	m	**8.65**
20 × 20 mm	0.04	1.63	–	3.11	m	**4.74**
25 × 15 mm	0.07	2.64	–	2.97	m	**5.61**
25 × 25 mm	0.07	2.97	–	4.86	m	**7.83**
Cold applied polysulphide joint sealant						
20 × 20 mm	0.07	2.64	–	3.91	m	**6.55**
Gun grade cold applied elastomeric joint sealant						
25 × 25 mm on 3 mm foam strip	0.07	2.64	–	10.09	m	**12.73**
50 × 25 mm on 3 mm foam strip	0.09	3.66	–	19.39	m	**23.05**
PVC centre bulb waterstop						
160 mm wide	0.08	3.25	–	8.33	m	**11.58**
210 mm wide	0.09	3.66	–	8.68	m	**12.34**
260 mm wide	0.11	4.47	–	13.45	m	**17.92**

SERIES 2300: BRIDGE EXPANSION JOINTS AND SELAING OF GAPS

Item	Gang hours	Labour £	Plant £	Material £	Unit	Total rate £
PVC flat dumbell waterstop						
170 mm wide	0.08	3.25	–	36.90	m	**40.15**
210 mm wide	0.10	4.07	–	43.62	m	**47.69**
250 mm wide	0.12	4.88	–	90.54	m	**95.42**
Dowels, plain or greased						
12 mm mild steel 450 mm long	0.04	1.63	–	1.67	nr	**3.30**
16 mm mild steel 750 mm long	0.05	1.83	–	3.17	nr	**5.00**
16 mm mild steel 750 mm long with debonding agent for 375 mm	0.05	2.16	–	3.79	nr	**5.95**
Dowels, sleeved or capped						
12 mm mild steel 450 mm long with debonding agent for 225 mm and PVC dowel cap	0.05	1.83	–	1.00	nr	**2.83**

SERIES 2400: BRICKWORK, BLOCKWORK AND STONEWORK

Item	Gang hours	Labour £	Plant £	Material £	Unit	Total rate £
RESOURCES - LABOUR						
Masonry gang						
1 foreman bricklayer (craftsman)		22.88				
4 bricklayers (craftsman)		78.52				
1 unskilled operative (general)		12.75				
Total Gang Rate / Hour	£	**114.14**				
RESOURCES - PLANT						
Masonry						
dumper 2 t (50% of time)			2.69			
gas oil for ditto			0.56			
cement mixer 4/3 (50% of time)			0.36			
petrol for ditto			0.64			
small power tools (masonry)			1.96			
minor scaffolding (masonry)			1.40			
Total Rate / Hour		£	**7.60**			
RESOURCES - MATERIALS						
Half brick thick walls are in stretcher bond, thicker than this in English bond (3 stretchers : 1 header) unless otherwise stated.						
DfT Table 24/1: Mortar Proportions by Volume :-						

Mortar Type	Cement: Lime:sand	Masonry Cement:sand	Cement: sand
(i)	1.0 to 1/4.3	-	-
(ii)	1:½:4 to 4½	1:2½ to 3½	1:3 to 4
(iii)	1:1:5 to 6	1:4½	1:5 to 6

Item	Gang hours	Labour £	Plant £	Material £	Unit	Total rate £
BRICKWORK						
Common bricks; in stretcher bond; in cement mortar designation (ii)						
Walls						
half brick thick	0.23	25.38	1.73	5.14	m²	**32.25**
one brick thick	0.44	49.20	3.36	11.40	m²	**63.96**
one and a half bricks thick	0.64	71.68	4.90	18.49	m²	**95.07**
two bricks thick	0.83	92.77	6.34	25.23	m²	**124.34**
Walls, curved on plan						
half brick thick	0.30	32.95	2.23	5.14	m²	**40.32**
one brick thick	0.57	63.44	4.33	11.40	m²	**79.17**
one and a half bricks thick	0.82	91.27	6.26	18.49	m²	**116.02**
two bricks thick	1.06	117.76	8.04	25.23	m²	**151.03**
Walls, with a battered face						
half brick thick	0.33	36.62	2.50	5.14	m²	**44.26**
one brick thick	0.63	70.12	4.79	11.40	m²	**86.31**
one and a half bricks thick	0.91	100.95	6.89	18.49	m²	**126.33**
two bricks thick	1.16	129.34	8.83	25.23	m²	**163.40**

SERIES 2400: BRICKWORK, BLOCKWORK AND STONEWORK

Item	Gang hours	Labour £	Plant £	Material £	Unit	Total rate £
Facework to concrete						
half brick thick	0.25	27.49	1.88	6.63	m²	**36.00**
one brick thick	0.48	53.20	3.63	12.89	m²	**69.72**
one and a half bricks thick	0.69	77.25	5.28	19.97	m²	**102.50**
two bricks thick	0.90	99.95	6.83	26.70	m²	**133.48**
In alteration work						
half brick thick	0.30	32.95	2.25	6.63	m²	**41.83**
one brick thick	0.57	63.44	4.33	12.89	m²	**80.66**
one and a half bricks thick	0.82	91.60	6.26	19.97	m²	**117.83**
two bricks thick	1.06	117.76	8.04	26.70	m²	**152.50**
ADD or DEDUCT to materials costs for variation of £10.00/1000 in PC of common bricks						
half brick thick	–	–	–	0.66	m²	**0.66**
one brick thick	–	–	–	1.32	m²	**1.32**
one and a half bricks thick	–	–	–	1.98	m²	**1.98**
two bricks thick	–	–	–	2.65	m²	**2.65**
Copings; standard header-on-edge;						
215mm wide x 103mm high	0.11	11.69	0.80	1.13	m	**13.62**
ADD or DEDUCT to copings for variation of £1.00/100 in PC of common bricks	–	–	–	0.14	m	**0.14**
Class A engineering bricks, perforated; in cement mortar designation (ii)						
Walls						
half brick thick	0.27	29.94	2.05	5.15	m²	**37.14**
one brick thick	0.52	57.88	3.95	11.42	m²	**73.25**
one and a half bricks thick	0.75	83.81	5.72	18.51	m²	**108.04**
two bricks thick	0.97	108.08	7.38	25.26	m²	**140.72**
Walls, curved on plan						
half brick thick	0.37	41.18	2.81	5.15	m²	**49.14**
one brick thick	0.70	78.47	5.36	11.42	m²	**95.25**
one and a half bricks thick	1.01	112.42	7.68	18.51	m²	**138.61**
two bricks thick	1.29	143.36	9.79	25.26	m²	**178.41**
Walls, with a battered face						
half brick thick	0.37	41.18	2.81	5.15	m²	**49.14**
one brick thick	0.70	78.47	5.36	11.42	m²	**95.25**
one and a half bricks thick	1.01	112.42	7.68	18.51	m²	**138.61**
two bricks thick	1.29	143.36	9.79	25.26	m²	**178.41**
Facework to concrete						
half brick thick	0.32	35.06	2.39	6.64	m²	**44.09**
one brick thick	0.60	67.34	4.60	12.90	m²	**84.84**
ADD or DEDUCT to materials costs for variation of £10.00/1000 in PC of engineering bricks						
half brick thick	–	–	–	0.66	m²	**0.66**
one brick thick	–	–	–	1.32	m²	**1.32**
one and a half bricks thick	–	–	–	1.98	m²	**1.98**
two bricks thick	–	–	–	2.65	m²	**2.65**
Brick coping in standard bricks in headers on edge;						
215mm wide x 103mm high	0.29	32.28	2.20	5.14	m	**39.62**
ADD or DEDUCT to copings for variation of £1.00/100 in PC of Class A enginreering bricks	–	–	–	0.14	m	**0.14**

SERIES 2400: BRICKWORK, BLOCKWORK AND STONEWORK

Item	Gang hours	Labour £	Plant £	Material £	Unit	Total rate £
BRICKWORK – cont						
Class B engineering bricks,perforated; in cement mortar designation (ii)						
Walls						
half brick thick	0.29	32.28	2.20	5.15	m²	**39.63**
one brick thick	0.48	53.98	3.69	11.41	m²	**69.08**
one and a half bricks thick	0.69	77.36	5.28	18.49	m²	**101.13**
two bricks thick	0.91	100.73	6.88	25.24	m²	**132.85**
Walls, curved on plan						
half brick thick	0.42	46.30	3.16	5.15	m²	**54.61**
one brick thick	0.78	86.82	5.93	11.41	m²	**104.16**
one and a half bricks thick	1.04	115.76	7.91	18.49	m²	**142.16**
two bricks thick	1.30	144.70	9.88	25.24	m²	**179.82**
Walls, with battered face						
half brick thick	0.42	46.30	3.16	5.15	m²	**54.61**
one brick thick	0.78	86.82	5.93	11.41	m²	**104.16**
one and a half bricks thick	1.04	115.76	7.91	18.49	m²	**142.16**
two bricks thick	1.30	144.70	9.88	25.24	m²	**179.82**
Facework to concrete						
half brick thick	0.32	35.39	2.42	6.63	m²	**44.44**
one brick thick	0.56	62.33	4.26	12.89	m²	**79.48**
ADD or DEDUCT to materials costs for variation of £10.00/1000 in PC of engineering bricks						
half brick thick	–	–	–	0.66	m²	**0.66**
one brick thick	–	–	–	1.32	m²	**1.32**
one and a half bricks thick	–	–	–	1.98	m²	**1.98**
two bricks thick	–	–	–	2.65	m²	**2.65**
Brick coping in standard bricks in headers on edge; 215mm wide x 103mm high	0.13	14.91	1.02	2.17	m	**18.10**
ADD or DEDUCT to copings for variation of £1.00/ 100 in PC of Class B engineering bricks	–	–	–	0.14	m	**0.14**
Facing bricks; in lime mortar designation (ii)						
Walls						
half brick thick	0.34	37.84	2.59	5.16	m²	**45.59**
one brick thick	0.57	63.89	5.26	11.44	m²	**80.59**
one and a half bricks thick	0.83	92.05	6.29	18.54	m²	**116.88**
two bricks thick	1.08	120.21	8.21	25.31	m²	**153.73**
Walls, curved on plan						
half brick thick	0.45	50.09	3.42	5.16	m²	**58.67**
one brick thick	0.84	93.50	6.39	11.44	m²	**111.33**
one and a half bricks thick	1.11	124.10	8.47	18.54	m²	**151.11**
two bricks thick	1.39	154.71	10.57	25.31	m²	**190.59**
Walls, with a battered face						
half brick thick	0.45	50.09	3.42	5.16	m²	**58.67**
one brick thick	0.84	93.50	6.39	11.44	m²	**111.33**
one and a half bricks thick	1.11	124.10	8.47	18.54	m²	**151.11**
two bricks thick	1.39	154.71	10.57	25.31	m²	**190.59**

SERIES 2400: BRICKWORK, BLOCKWORK AND STONEWORK

Item	Gang hours	Labour £	Plant £	Material £	Unit	Total rate £
Facework to concrete						
half brick thick	0.37	40.96	2.71	6.65	m²	**50.32**
one brick thick	0.66	73.46	5.02	12.93	m²	**91.41**
Extra over common brickwork in English bond for facing with facing bricks in lime mortar designation (ii)	0.11	12.47	0.85	0.02	m²	**13.34**
ADD or DEDUCT to materials costs for variation of £10.00/1000 in PC of facing bricks						
half brick thick	–	–	–	0.66	m²	**0.66**
one brick thick	–	–	–	1.32	m²	**1.32**
one and a half bricks thick	–	–	–	1.98	m²	**1.98**
two bricks thick	–	–	–	2.65	m²	**2.65**
Brick coping in standard bricks in headers on edge;						
215mm wide x 103mm high	0.13	14.91	1.02	1.19	m	**17.12**
Flat arches in standard stretchers on end;						
103mm wide x 215mm high	0.21	23.37	1.60	0.70	m	**25.67**
Flat arches in bullnose stretchers on end;						
103mm × 215mm high	0.22	24.49	1.67	0.79	m	**26.95**
Segmental arches in single ring stretchers on end;						
103mm wide x 215mm high	0.37	41.18	2.81	0.70	m	**44.69**
Segmental arches in double ring stretchers on end;						
103mm wide x 440mm high	0.49	54.54	3.72	1.73	m	**59.99**
Segmental arches; cut voussoirs						
103mm wide x 215mm high	0.39	43.41	2.97	0.93	m	**47.31**
ADD or DEDUCT to copings and arches for variation of £1.00/100 in PC of facing bricks						
header-on-edge	–	–	–	0.14	m	**0.14**
stretcher-on-end	–	–	–	0.14	m	**0.14**
stretcher-on-end bullnose specials	–	–	–	0.14	m	**0.14**
single ring	–	–	–	0.14	m	**0.14**
two ring	–	–	–	0.28	m	**0.28**
BLOCKWORK AND STONEWORK						
Lightweight concrete blocks; solid; 3.5N/mm²; in cement-lime mortar						
Walls						
100mm thick;	0.17	19.37	1.32	8.10	m²	**28.79**
140mm thick;	0.23	25.04	1.71	13.52	m²	**40.27**
215mm thick;	0.28	30.61	2.09	25.96	m²	**58.66**
Walls, curved on plan						
100mm thick;	0.23	25.71	1.75	8.10	m²	**35.56**
140mm thick;	0.30	33.28	2.27	13.52	m²	**49.07**
215mm thick;	0.37	40.74	2.78	25.96	m²	**69.48**
Facework to concrete						
100mm thick;	0.18	19.92	1.36	9.58	m²	**30.86**
140mm thick;	0.23	25.82	1.76	15.00	m²	**42.58**
215mm thick;	0.28	31.50	2.15	27.45	m²	**61.10**
In alteration work						
100mm thick;	0.17	19.37	1.32	8.10	m²	**28.79**
140mm thick;	0.23	25.04	1.71	13.52	m²	**40.27**
215mm thick;	0.28	30.61	2.09	25.96	m²	**58.66**

SERIES 2400: BRICKWORK, BLOCKWORK AND STONEWORK

Item	Gang hours	Labour £	Plant £	Material £	Unit	Total rate £
BLOCKWORK AND STONEWORK – cont						
Dense concrete blocks; solid; 3.5 or 7 N/mm²; in cement-lime mortar						
Walls						
100 mm thick;	0.17	18.92	1.29	10.80	m²	**31.01**
140 mm thick;	0.20	22.26	1.52	15.00	m²	**38.78**
215 mm thick;	0.24	26.71	1.82	23.95	m²	**52.48**
Walls, curved on plan						
100 mm thick;	0.23	25.15	1.72	10.80	m²	**37.67**
140 mm thick;	0.23	25.15	2.02	15.00	m²	**42.17**
215 mm thick;	0.32	35.51	2.43	23.21	m²	**61.15**
Facework to concrete						
100 mm thick;	0.17	19.48	1.33	12.28	m²	**33.09**
140 mm thick;	0.21	22.93	1.57	16.48	m²	**40.98**
215 mm thick;	0.35	38.96	2.22	24.70	m²	**65.88**
In alteration work						
100 mm thick;	0.27	29.61	1.72	10.80	m²	**42.13**
140 mm thick;	0.26	28.94	2.02	15.00	m²	**45.96**
215 mm thick;	0.32	35.51	2.43	23.21	m²	**61.15**
Reconstituted stone; Bradstone 100 bed weathered Cotswold or North Cerney masonary blocks; rough hewn rockfaced blocks; in coloured cement-lime mortar designation (1:2:9) (iii)						
Walls, thickness 100mm						
vertical and straight	0.30	33.39	2.28	54.63	m²	**90.30**
curved on plan	0.39	43.41	2.97	54.63	m²	**101.01**
with a battered face	0.34	38.40	2.62	54.63	m²	**95.65**
in arches	0.57	64.00	4.37	54.63	m²	**123.00**
Facing to concrete; wall ties						
vertical and straight	0.24	26.38	1.80	62.14	m²	**90.32**
curved on plan	0.32	35.06	2.39	62.14	m²	**99.59**
with a battered face	0.36	39.62	2.71	62.14	m²	**104.47**
Reconstituted stone; Bradstone Architectural dressings in weathered Cotswold or North Cerney shades; in coloured cement-lime mortar designation (1:2:9) (iii)						
Copings; twice weathered and throated						
152 × 76 mm ;	0.08	8.90	0.61	16.67	m	**26.18**
152 × 76 mm; curved on plan;	0.11	11.80	0.80	56.54	m	**69.14**
305 × 76 mm;	0.10	11.13	0.76	39.00	m	**50.89**
305 × 76 mm; curved on plan;	0.13	14.80	1.01	77.85	m	**93.66**
Corbels						
479 × 100 × 215 mm, splayed	0.49	54.54	3.72	131.77	nr	**190.03**
665 × 100 × 215 mm, splayed	0.55	61.22	4.18	180.36	nr	**245.76**
Pier caps						
305 × 305 mm	0.09	10.02	0.68	20.01	nr	**30.71**
381 × 381 mm	0.11	12.24	0.84	28.80	nr	**41.88**
457 × 457 mm	0.13	14.47	0.99	39.38	nr	**54.84**
533 × 533 mm	0.15	16.70	1.14	54.84	nr	**72.68**

SERIES 2400: BRICKWORK, BLOCKWORK AND STONEWORK

Item	Gang hours	Labour £	Plant £	Material £	Unit	Total rate £
Lintels						
100 × 140 mm	0.11	12.24	0.84	49.47	m	**62.55**
100 × 215 mm	0.16	17.81	1.22	54.85	m	**73.88**
Natural stone ashlar; Portland Whitbed limestone; in cement-lime mortar designation (iii)						
Walls						
vertical and straight	12.70	1413.57	84.11	2460.99	m³	**3958.67**
curved on plan	19.30	2148.18	146.71	3669.21	m³	**5964.10**
with a battered face	19.30	2148.18	146.71	3669.21	m³	**5964.10**
Facing to concrete; wall ties						
vertical and straight	17.00	1892.18	129.23	2468.91	m³	**4490.32**
curved on plan	25.80	2871.66	196.12	3677.13	m³	**6744.91**
with a battered face	25.80	2871.66	196.12	3677.13	m³	**6744.91**
Copings; twice weathered and throated						
250 × 150 mm	0.45	50.09	3.42	133.10	m	**186.61**
250 × 150 mm; curved on plan	0.45	50.09	3.42	159.65	m	**213.16**
400 × 150 mm	0.49	54.54	3.72	195.56	m	**253.82**
400 × 150 mm; curved on plan	0.49	54.54	3.72	234.61	m	**292.87**
Shaped and dressed string courses						
75 mm projection x 150 mm high	0.45	50.09	3.42	124.48	m	**177.99**
Corbel						
500 × 450 × 300 mm	0.55	61.22	4.18	178.56	nr	**243.96**
Keystone						
750 × 900 × 300 mm (extreme)	1.30	144.70	9.88	559.54	nr	**714.12**
Random rubble uncoursed , weighing 2.0 t/m3 of wall; in cement-lime mortar designation (iii)						
Walls						
vertical and straight	4.17	464.14	31.70	277.40	m³	**773.24**
curved on plan	4.67	519.79	35.50	359.09	m³	**914.38**
with a battered face	4.67	519.79	35.50	359.09	m³	**914.38**
in arches	8.53	949.43	64.84	359.09	m³	**1373.36**
Facework to concrete						
vertical and straight	4.17	464.14	31.70	360.57	m³	**856.41**
curved on plan	4.67	519.79	35.50	360.57	m³	**915.86**
with a battered face	4.67	519.79	35.50	360.57	m³	**915.86**
in arches	8.53	949.43	64.84	360.57	m³	**1374.84**
Copings						
500 × 125 mm	0.49	54.54	3.72	227.88	m	**286.14**
Squared random rubble uncoursed , weighing 2.0 t/m3 of wall; in cement-lime mortar designation (iii)						
Walls						
vertical and straight	4.17	464.14	31.70	482.96	m³	**978.80**
curved on plan	4.67	519.79	35.50	482.96	m³	**1038.25**
with a battered face	4.67	519.79	35.50	482.96	m³	**1038.25**
in arches	8.53	949.43	64.84	482.96	m³	**1497.23**

SERIES 2400: BRICKWORK, BLOCKWORK AND STONEWORK

Item	Gang hours	Labour £	Plant £	Material £	Unit	Total rate £
BLOCKWORK AND STONEWORK – cont						
Squared random rubble uncoursed , weighing						
2.0 t/m3 of wall – cont						
Facework to concrete						
vertical and straight	4.17	464.14	31.70	484.44	m³	**980.28**
curved on plan	4.67	519.79	35.50	484.44	m³	**1039.73**
with a battered face	4.67	519.79	35.50	484.44	m³	**1039.73**
in arches	8.53	949.43	64.84	484.44	m³	**1498.71**
Copings						
500 × 125 mm	0.49	54.54	3.72	227.88	m	**286.14**
Dry rubble , weighing 2.0 t/m3 of wall						
Walls						
vertical and straight	3.83	426.30	29.11	371.17	m³	**826.58**
curved on plan	4.33	481.95	32.91	371.17	m³	**886.03**
with a battered face	4.33	481.95	32.91	371.17	m³	**886.03**
Copings formed of rough stones						
275 × 200 mm (average) high	0.45	50.09	3.42	31.74	m	**85.25**
500 × 200 mm	0.55	61.22	4.18	53.25	m	**118.65**

SERIES 2500: SPECIAL STRUCTURES

Item	Gang hours	Labour £	Plant £	Material £	Unit	Total rate £
SPECIAL STRUCTURES DESIGNED BY CONTRACTOR						
Notes						
This section envisages the following types of structure which may be required to be designed by the Contractor based on stipulated performance criteria:						
* Buried structures						
* Earth retaining structures						
* Environmental barriers						
* Underbridges up to 8 m span						
* Footbridges						
* Piped culverts						
* Box culverts						
* Drainage exceeding 900 mm diameter						
* Other structures						
Naturally, this work cannot be catered for directly in this section and will require the preparation of a sketch solution and approximate quantities to allow pricing using the various other Unit Costs sections as well as the Approximate Estimates section.						
An allowance must be added to such an estimate to cover the Contractor's design fee(s) and expenses.						

SERIES 2700: ACCOMMODATION WORKS, WORKS FOR STATUTORY UNDERTAKERS

Item	Gang hours	Labour £	Plant £	Material £	Unit	Total rate £
ACCOMMODATION WORKS, WORKS FOR STATUTORY UNDERTAKERS						
GENERAL						
Cost items in this series will be specific to individual contract agreements and it is felt that inclusion of prices in this publication would not provide useful guidance.						

SERIES 3000: LANDSCAPING AND ECOLOGY

Item	Gang hours	Labour £	Plant £	Material £	Unit	Total rate £
GROUND PREPARATION AND CULTIVATION						
Supply and apply granular cultivation treatments by hand						
35 grammes/m^2	0.50	13.09	–	33.68	100m2	**46.77**
50 grammes/m^2	0.65	17.02	–	48.12	100m2	**65.14**
75 grammes/m^2	0.85	22.25	–	72.17	100m2	**94.42**
100 grammes/m^2	1.00	26.18	–	96.23	100m2	**122.41**
150 grammes/m^2	1.20	31.41	–	144.35	100m2	**175.76**
Supply and apply granular cultivation treatments by machine in suitable economically large areas						
100 grammes/m^2	–	–	4.25	96.23	100m2	**100.48**
ADD to above for:						
granular treatments per £0.10/kg PC variation +10%						
selective weedkiller +171%						
herbicide +567%						
fertilizer +100%						
Supply and incorporate cultivation additives into top 150 mm topsoil by hand						
1 m^3/10m^2	20.00	523.57	–	12.34	100m2	**535.91**
1 m^3/13m^2	20.00	523.57	–	9.49	100m2	**533.06**
1 m^3/20m^2	19.00	497.39	–	6.17	100m2	**503.56**
1 m^3/40m^2	17.00	445.03	–	3.08	100m2	**448.11**
Supply and incorporate cultivation additives into top 150 mm topsoil by machine in suitable economically large areas						
1 m^3/10m^2	–	–	153.03	12.34	100m2	**165.37**
1 m^3/13m^2	–	–	141.26	9.49	100m2	**150.75**
1 m^3/20m^2	–	–	125.56	6.17	100m2	**131.73**
1 m^3/40m^2	–	–	115.75	3.08	100m2	**118.83**
ADD to above for						
cultavation additives per £0.10/m^2 PC variation + 10%						
compost + 300%						
manure +166%						
peat +1800%						

SERIES 3000: LANDSCAPING AND ECOLOGY

Item	Gang hours	Labour £	Plant £	Material £	Unit	Total rate £
SEEDING AND TURFING						
Selected grass seed ; at the rate of 0.050 kg/m² in two operations						
Grass seeding; by conventional sowing						
at 10° or less to the horizontal	0.01	0.26	–	0.40	m²	**0.66**
more than 10° to horizontal	0.02	0.39	–	0.40	m²	**0.79**
Wild Flora mixture Bsh ref Wf1 combined with Low maintenance conservation grass BSH ref A4 (20%:80%); at the rate of 80 g/m²						
Wildflower seeding; by conventional sowing						
at 10° or less to the horizontal	0.50	13.09	–	1312.32	100 m²	**1325.41**
more than 10° to horizontal	0.02	0.39	–	0.40	m²	**0.79**
Hydraulic mulch grass seed						
Grass seeding by hydraulic seeding						
at 10° or less to the horizontal	0.01	0.13	0.10	1.96	m²	**2.19**
at more than 10° to the horizontal	0.01	0.18	0.10	1.96	m²	**2.24**
Imported turf						
Turfing to surfaces						
at 10° or less to the horizontal	0.12	3.14	–	7.22	m²	**10.36**
more than 10 to the horizontal; pegging down	0.17	4.45	–	7.22	m²	**11.67**
PLANTING						
Trees						
The cost of planting semi-mature trees will depend on the size and species, and on the access to the site for tree handling machines. Prices should be obtained for individual trees and planting.						
Break up subsoil to a depth of 200mm						
in treepit	0.05	1.31	–	–	nr	**1.31**
Supply and plant tree in prepared pit; backfill with excavated topsoil minimum 600 mm deep						
light standard; in pits	0.25	6.54	–	79.71	nr	**86.25**
standard tree	0.45	11.78	–	107.63	nr	**119.41**
selected standard tree	0.75	19.63	–	121.53	nr	**141.16**
heavy standard tree	0.85	22.25	–	166.04	nr	**188.29**
extra heavy standard tree	1.50	39.27	–	206.86	nr	**246.13**
extra for filling with topsoil from spoil heap ne 100 m distant	0.15	3.93	0.97	–	m³	**4.90**
extra for filling with imported topsoil	0.08	2.09	0.52	26.91	m³	**29.52**
extra for incorporating manure or compost into top soil at the rate of 1 m³ per 5 m³ +60%						
Supply tree stake and drive 500 mm into firm ground and trim to approved height, including two tree ties to approved pattern						
one stake; 2.4 m long, 100 mm diameter	0.16	4.19	–	6.88	nr	**11.07**
one stake; 3.0 m long, 100 mm diameter	0.20	5.24	–	6.98	nr	**12.22**
two stakes; 2.4 m long, 100 mm diameter	0.24	6.28	–	11.09	nr	**17.37**
two stakes; 3.0 m long, 100 mm diameter	0.30	7.85	–	13.97	nr	**21.82**

SERIES 3000: LANDSCAPING AND ECOLOGY

Item	Gang hours	Labour £	Plant £	Material £	Unit	Total rate £
Supply and fit tree support comprising three collars and wire guys; including pickets						
galvanized steel 50×600 mm	1.50	39.27	–	23.23	nr	**62.50**
hardwood 75×600 mm	1.50	39.27	–	23.76	nr	**63.03**
Supply and fix standard steel tree guard	0.30	7.85	–	24.08	nr	**31.93**
Hedge plants						
Excavate trench by hand for hedge and deposit soil alongside trench						
300 wide x 300 mm deep	0.10	2.62	–	–	m	**2.62**
450 wide x 300 mm deep	0.13	3.40	–	–	m	**3.40**
Excavate trench by machine for hedge and deposit soil alongside trench						
300 wide x 300 mm deep	0.02	0.30	0.36	–	m	**0.66**
450 wide x 300 mm deep	0.02	0.30	0.36	–	m	**0.66**
Set out, nick out and excavate trench and break up subsoil to minimum depth of 300 mm						
400 mm minimum deep	0.15	3.93	–	–	m	**3.93**
Supply and plant hedging plants; backfill with excavated topsoil						
single row plants at 200 mm centres	0.25	6.54	–	5.33	m	**11.87**
single row plants at 300 mm centres	0.17	4.45	–	3.55	m	**8.00**
single row plants at 400 mm centres	0.13	3.27	–	2.67	m	**5.94**
single row plants at 500 mm centres	0.10	2.62	–	2.13	m	**4.75**
single row plants at 600 mm centres	0.08	2.09	–	1.77	m	**3.86**
double row plants at 200 mm centres	0.50	13.09	–	10.66	m	**23.75**
double row plants at 300 mm centres	0.34	8.90	–	7.10	m	**16.00**
double row plants at 400 mm centres	0.25	6.54	–	5.33	m	**11.87**
double row plants at 500 mm centres	0.20	5.24	–	4.26	m	**9.50**
double row plants at 600 mm centres	0.16	4.19	–	3.54	m	**7.73**
Extra for incorporating manure at 1 m³ / 30m3	0.60	7.65	–	0.34	-	**7.99**
Shrubs						
Form planting hole in previously cultivated area, supply and plant specified shrub and backfill with excavated material						
shrub 300 mm high	0.10	2.62	–	2.73	each	**5.35**
shrub 600 mm high	0.10	2.62	–	4.04	each	**6.66**
shrub 900 mm high	0.10	2.62	–	4.75	each	**7.37**
shrub 1.0 m high and over	0.10	2.62	–	6.13	each	**8.75**
Supply and fix shrub stake including two ties						
one stake; 1.5 m long, 75 mm diameter	0.12	3.14	–	5.67	each	**8.81**
Extra for the above items for planting in prefabricated or in-situ planters +20%						
Herbaceous plants						
Form planting hole in previously cultivated area, supply and plant specified herbaceous plants and backfill with excavated material						
5 plants/m²	0.05	1.31	–	9.30	m²	**10.61**
10 plants/m²	0.16	4.19	–	18.60	m²	**22.79**
25 plants/m²	0.42	10.99	–	46.49	m²	**57.48**

SERIES 3000: LANDSCAPING AND ECOLOGY

Item	Gang hours	Labour £	Plant £	Material £	Unit	Total rate £
PLANTING – cont						
Herbaceous plants – cont						
Form planting hole in previously cultivated area, supply and plant specified herbaceous plants and backfill with excavated material – cont						
35 plants/m²	0.58	15.18	–	65.08	m²	**80.26**
50 plants/m²	0.83	21.73	–	92.98	m²	**114.71**
Supply and fix plant support netting on 50 mm diameter stakes 750 mm long driven into the ground at 1.5 m centres						
1.15 m high green extruded plastic mesh, 125 mm square mesh,	0.06	1.57	–	43.08	m²	**44.65**
Extra to the above items for planting in prefabricated or in-situ planters +20%						
Form planting hole in previously cultivated area; supply and plant bulbs and backfill with excavated material						
small	0.01	0.26	–	0.17	each	**0.43**
medium	0.01	0.26	–	0.27	each	**0.53**
large	0.01	0.26	–	0.32	each	**0.58**
Supply and plant bulb in grassed area using bulb planter and backfill with screened topsoil or peat and cut turf plug						
small	0.01	0.26	–	0.17	each	**0.43**
medium	0.01	0.26	–	0.27	each	**0.53**
large	0.01	0.26	–	0.32	each	**0.58**
Extra to the above items for planting in prefabricated or in-situ planters +15%						
MULCHING						
Organic mulching of medium bark mulch to a depth of 50 mm in planting areas to surfaces						
at 10° or less to the horizontal	0.01	0.26	–	1.75	m²	**2.01**
more than 10° to horizontal	0.02	0.39	–	1.75	m²	**2.14**
Organic mulching of timber mulch to a depth of 75 mm in planting areas to surfaces						
at 10° or less to the horizontal	0.02	0.39	–	3.50	m²	**3.89**
more than 10° to horizontal	0.02	0.52	–	3.50	m²	**4.02**
WEED CONTROL						
Weed and handfork planted areas including removing and dumping weed and debris on site	0.07	1.83	–	–	m²	**1.83**
Supply and apply selective weed killer						
35 grammes/m²	0.01	0.13	–	0.13	m²	**0.26**
70 grammes/m²	0.01	0.18	–	0.26	m²	**0.44**
100 grammes/m²	0.01	0.26	–	0.37	m²	**0.63**

SERIES 3000: LANDSCAPING AND ECOLOGY

Item	Gang hours	Labour £	Plant £	Material £	Unit	Total rate £
MAINTENANCE OF ESTABLISHED TREES AND SHRUBS						
Initial cut back to shrubs and hedge plants, clear away all cuttings	0.04	1.05	–	–	m	**1.05**
Protect planted areas with windbreak fencing fixed to stakes 1.5 m × 50 mm diameter at 2.0 m centres and clear away on completion						
1.0 m high	0.06	1.57	–	4.93	m	**6.50**
Cut out dead and diseased wood, prune, trim and cut to shape; treat wounds with sealant; clear away cuttings						
shrubs ne 1.0 m high	0.20	5.24	–	–	each	**5.24**
shrubs 1.0–2.0 m high	0.30	7.85	–	–	each	**7.85**
shrubs 2.0–3.0 m high	0.40	10.47	–	–	each	**10.47**
Cut and trim ornamental hedge to specified profile; clear away cuttings						
ornamental hedge 2.0 m high	0.60	15.71	–	–	m	**15.71**
ornamental hedge 4.0 m high	0.90	23.56	–	–	m	**23.56**
Trim field hedge to specified height and shape; clear away cuttings						
using flail	0.05	1.31	–	–	m	**1.31**
using cutting bar	0.07	1.83	–	–	m	**1.83**
MAINTENANCE ESTABLISHED GRASSED AREAS						
Grass cutting at medium frequency						
on central reserves	–	–	–	–	m²	**6.63**

SERIES 5000 MAINTENANCE PAINTING OF STEELWORK

Item	Gang hours	Labour £	Plant £	Material £	Unit	Total rate £
SURFACE PREPARATION						
Surface Preparation						
Surface preparation to general surfaces by dry blast cleaning to DfT Clause 5003 to remove unsound paint down to sound paint	–	–	–	–	m²	**9.25**
Protective System						
Protective system Type I (M) to DfT Table 50/2 to general surfaces prepared down to sound paint	–	–	–	–	m²	**26.60**

Spon's Asia-Pacific Construction Costs Handbook

4th Edition

Edited by **Davis Langdon & Seah**

Spon's Asia Pacific Construction Costs Handbook includes construction cost data for 20 countries. This new edition has been extended to include Pakistan and Cambodia. Australia, UK and America are also included, to facilitate comparison with construction costs elsewhere. Information is presented for each country in the same way, as follows:

• key data on the main economic and construction indicators.

• an outline of the national construction industry, covering structure, tendering and contract procedures, materials cost data, regulations and standards

• labour and materials cost data

• Measured rates for a range of standard construction work items

• Approximate estimating costs per unit area for a range of building types

• price index data and exchange rate movements against £ sterling, $US and Japanese Yen

The book also includes a Comparative Data section to facilitate country-to-country comparisons. Figures from the national sections are grouped in tables according to national indicators, construction output, input costs and costs per square metre for factories, offices, warehouses, hospitals, schools, theatres, sports halls, hotels and housing.

This unique handbook will be an essential reference for all construction professionals involved in work outside their own country and for all developers or multinational companies assessing comparative development costs.

April 2010: 234x156: 480pp
Hb: 978-0-415-46565-6: **£120.00**

To Order: Tel: +44 (0) 1235 400524 **Fax:** +44 (0) 1235 400525
or Post: Taylor and Francis Customer Services,
Bookpoint Ltd, Unit T1, 200 Milton Park, Abingdon, Oxon, OX14 4TA UK
Email: book.orders@tandf.co.uk

For a complete listing of all our titles visit:
www.tandf.co.uk

Handbook of Road Technology

4th Edition

M. Lay

"Widely considered to be the single most comprehensive and informative road technology reference text." – *Highway Engineering in Australia*

"A fully revised edition..." – *Transport and Road Update*

"This handbook has its origins in Max Lay's work over many decades in Australia, but reflects a breadth of knowledge far beyond that continent. In that sense it is perhaps the most international in its coverage of practice of such works." – *Mike Chrimes, Librarian, Institution of Civil Engineers, MyICE newsletter*

This fully revised fourth edition of Max Lay's well-established reference work covers all aspects of the technology of roads and road transport, and urban and rural road technology. It forms a comprehensive but accessible reference for all professionals and students interested in roads, road transport and the wide range of disciplines involved with roads.

International in scope, it begins with the preliminary construction procedures; from road planning policies and design considerations to the selection of materials and the building of roads and bridges. It then explores road operating environments that include driver behaviour, traffic flow, lighting and maintenance, and assesses the cost, economics, transport implications and environmental impact of road use.

June 2009: 944pp
Hb: 978-0-415-47265-4: **£150.00**

To Order: Tel: +44 (0) 1235 400524 **Fax:** +44 (0) 1235 400525
or Post: Taylor and Francis Customer Services,
Bookpoint Ltd, Unit T1, 200 Milton Park, Abingdon, Oxon, OX14 4TA UK
Email: book.orders@tandf.co.uk

For a complete listing of all our titles visit:
www.tandf.co.uk

PART 8

Daywork

ιNTRODUCTION

'Dayworks' relates to work which is carried out incidental to a contract but where no other rates have been agreed. In the case of the ICE Conditions of Contract It is ordered by the Engineer pursuant to clause 52(6):

> '(6) The Engineer may if in his opinion it is necessary or desirable order in writing that any additional or substituted work shall be carried out on a daywork basis in accordance with the provisions of Clause 56(4).'

Clause 56(4) states that:

> '(4) Where any work is carried out on a daywork basis the Contractor shall be paid for such work under the conditions and at the rates and prices set out in the daywork schedule included in the Contract or failing the inclusion of a daywork schedule he shall be paid at the rates and prices and under the conditions contained in the 'Schedules of Dayworks carried out incidental to Contract Work' issued by The Civil Engineering Contractor's Association, current at the date of carrying out of the daywork.
>
> The contractor shall furnish to the Engineer such records receipts and other documentation as may be necessary to prove amounts paid and/or costs incurred. Such returns shall be in the form and delivered at the times the Engineer shall direct and shall be agreed within a reasonable time.
>
> Before ordering materials the Contractor shall if so required submit to the Engineer quotations for the same for his approval.'

The most recent Schedule is dated July 2008 and is published by the Civil Engineering Contractors Association. Copies may be obtained from:

CIVIL ENGINEERING CONTRACTORS ASSOCIATION
55 Tufton Street
London
SW1P 3QL
Tel: 020 7227 4620
Fax: 020 7227 4621

These schedules identify a number of items that are excluded from the rates and percentages quoted, but should be recovered by the Contractor when valuing his Daywork account. Under normal circumstances it is impractical to accurately value these items for each Daywork event, although certain items can be allowed for by means of a percentage addition. Suggested methods of calculating such additions are included in this section.

In Summary the Daywork schedule allows for an addition:

> of 148% to wages paid to workmen.
> of 88% to labour only subcontractors and hired plant drivers
> of 12.5% to subsistence allowances and travel paid to workmen.
> of 12.5% to materials used on dayworks
> of 12.5% to full cost of plant hired for dayworks
> of 12.5% to cost of operating welfare facilities
> of 12.5% to cost of additional insurance premiums for abnormal contract work or special conditions.

Part 3 of this book (Resources – Plant) includes detailed references to the plant section of the 2008 Daywork Schedule.

The text of the document is as follows:

SCHEDULES OF DAYWORKS CARRIED OUT INCIDENTAL TO CONTRACT WORK

The General clauses in the Schedule are repeated in full as follows:

1   Labour

Add to the amount of wages paid to operatives – 148%

1   'Amount of wages' means:
Actual wages and bonus paid, daily travelling allowances (fare and/or time), tool allowance and all prescribed payments including those in respect of time lost due to inclement weather paid to operatives at plain time rates and/or at overtime rates.

2   The percentage addition provides for all statutory charges at the date of publication and other charges including:

- National Insurance and Surcharge.
- Normal Contract Works, Third Party & Employer's Liability Insurances
- Annual and Public Holidays with pay.
- Statutory and industry sick pay.
- Welfare benefits.
- Industrial Training Levy.
- Redundancy Payments.
- Employment Rights Act 1996.
- Employment Relations Act 1999.
- Employment Act 2002.
- Site Supervision and staff including foremen and walking gangers, but the time of the gangers or charge hands' working with their gangs is to be paid for as for operatives.
- Small tools – such as picks, shovels, barrows, trowels, hand saws, buckets, trestles, hammers, chisels and all items of a like nature.
- Protective clothing.
- Head Office charges and profit.

3   The time spent in training, mobilization, demobilization etc. for the Dayworks operation is chargeable.
4   All hired plant drivers and labour subcontractor's accounts to be charged in full (without deduction of any cash discounts not exceeding 2.5%) plus 88%.
5   Subsistence or lodging allowances and periodic travel allowances (fare and/or time) paid to or incurred on behalf of operatives are chargeable at cost plus 12.5%.

2   Materials

'Add to the cost of materials – 12.5%'

1   The percentage addition provides for Head office charges and Profit
2   The cost of materials means the invoiced price of materials including delivery to site without deduction of any cash discounts not exceeding 2.5%
3   Unloading of materials:
> The percentage added to the cost of materials excludes the cost of handling which shall be charged in addition. An allowance for unloading into site stock or storage including wastage should be added where materials are taken from existing stock.

3   Supplementary Charges

1   Transport provided by contractors for operatives to, from, in and around the site to be charged at the appropriate Schedule rates.
2   Any other charges incurred in respect of any Dayworks operation including, tipping charges, professional fees, subcontractor's accounts and the like shall be paid for in full plus 12.5% (without deduction of cash discounts not exceeding 2.5%). Labour subcontractors being dealt with in Section 1.
3   The cost of operating welfare facilities to be charged by the contractor at cost plus 12.5%.
4   The cost of additional insurance premiums for abnormal contract work or special site conditions to be charged at cost plus 12.5%.
5   The cost of watching and lighting specially necessitated by Dayworks is to be paid for separately at Schedule rates.

4   Plant

1   These rates apply only to plant already on site, exclusive of drivers and attendants, but inclusive of fuel and consumable stores unless stated to be charged in addition, repairs and maintenance, insurance of plant but excluding time spent on general servicing.
2   Where plant is hired specifically for Dayworks: plant hire (exclusive of drivers and attendants), fuel, oil and grease, insurance, transport etc., to be charged at full amount of invoice (without deduction of any cash discount not exceeding 2.5%) to which should be added consumables where supplied by the contractor, all plus 12.5%.
3   Fuel distribution, mobilization and demobilization are not included in the rates quoted which shall be an additional charge.
4   Metric capacities are adopted and these are not necessarily exact conversions from their imperial equivalents, but cater for the variations arising from comparison of plant manufacturing firms' ratings
5   SAE rated capacities of plant means rated in accordance with the standards specified by the Society of Automotive Engineers.
6   Minimum hire charge will be for the period quoted
7   Hire rates for plant not included below shall be settled at prices reasonably related to the rates quoted.
8   The rates provide for Head Office charges and Profit

The Schedule then gives twenty-one pages of hire rates for a wide range of plant and equipment.

APPLICATION OF DAYWORKS

Generally

A check should be made on the accuracy of the recorded resources and times.

Tender documents generally allow the contractor to tender percentage variations to the figure calculated using the published percentage additions. These vary widely but a reasonable average indication can be along the lines of:

Labour	20% less
Materials	10% less
Plant	30% less

Labour

The Contractor should provide substantiation of the hourly rates he wishes to be paid for the various classes of labour and should demonstrate that the basic 'amount of wages' does not include any of the items actually covered by the percentage addition.

The wage bill is intended to reflect the cost to the Contractor. The time involved is not restricted to the duration of the task, but also includes mobilization and demobilization, together with any training needed – which would include induction courses required for Health & Safety requirements. The rate paid is the actual value of wages and bonuses paid (not simply the basic rate promulgated for the labour grade involved, and includes overtime rates if applicable, tool money, time lost due to inclement weather. In addition, daily travelling allowances are included, as are periodic travel allowances and also subsistence or lodging allowances.

Care should be taken that the matters deemed included in the percentage addition are not duplicated in the amount of wages. For example, it should be noted that foremen, gangers and other supervisory staff are covered by the percentage addition, unless they work in which case they are paid for at the correct rate for the task involved. The proportion of the time they spend working rather than supervising must be agreed. Refer to the amplification of labour categories in Part 3.

Hired or subcontracted labour is paid at invoiced cost, adjusted only where any cash discount exceeded 2½%, in which case the excess percentage is deducted.

Materials

The cost of materials delivered to the site is simply the invoiced price of the materials plus any delivery charges.

Should the cash discount exceed 2½%, the excess percentage is deducted from the amount to be paid.

The percentage addition simply covers the cost of Head Office charges and profit. It does not include for unloading or temporarily storing the materials nor for distributing them on site to the work place. Such cost can be charged, even in cases where the materials may already be in the site stock.

Material waste should be added direct to the cost of materials used for each particular Daywork items as an appropriate percentage.

Handling and offloading materials

Schedule 2.3 states that an allowance for handling materials, and an allowance for unloading or storage including wastage should be an additional charge.
Example

For a 12 month, £10.0m Civil Engineering scheme where the total cost of materials that require handling (excluding Ready Mix concrete, imported fills, fuels and similar items) is £2,500,000.
The following gang is employed (part time) throughout the contract for offloading and handling of materials.

	Net cost (£)
2 labourers	27.18
Lorry (8T) with Hiab lift	23.94
Rate per hour (£)	51.12

Allow an average of 5 hours per week over 50 weeks

250 hours @ £51.12 / hour =   £ 12,780

This cost as a percentage of the materials element of the contract.

(£12,780/£2,500,000) × 100   = 0.51 %

Supplementary charges

1   Transport

Schedule 3.1 states that transport provided by contractors to take operatives to and from the site as well as in and around the site shall be charged at the appropriate Schedule Rate. This would entail the driver and vehicle being included with the labour and plant parts of the Daywork calculation.

2   Any other charges

This relates to any other charges incurred in respect of any Dayworks operation and includes tipping charges, professional fees, subcontractor's accounts and the like. Schedule 3.2 provides for full payment of such charges in full plus the addition of 12½% for Head Office charges and Profit.

Should the cash discount exceed 2½%, the excess percentage is deducted from the amount to be paid.

Welfare Facilities

Schedule 3.3 allows for the net cost of operating these facilities plus 12.5%.

Example

How the costs of operating welfare facilities may be charged to the Daywork account on a 12 month, £10.0m Civil Engineering scheme, where the total labour element is £1,400,000.

Facility	Weekly Cost £
Toilet unit (4 nr)	160.00
Jack leg hutments 24' (2 nr)	120.00
Jack leg hutment 12' (1 nr)	42.00
Labour to clean, maintain and make tea, etc.	
= 1 man, 2 hours per day, 6 days × 24.88	149.28
Consumables (heat, light, soap, disinfectant, etc) say	50.00
Rates, insurance, taxes, etc (add 2%)	10.43
Total weekly cost	531.71

Multiply by 50 weeks (construction period) plus 6 weeks (maintenance period) Total cost to contract = £ 531.71 × 56	29,775.76
Thus cost as a percentage of the Labour element of the contract	2.13 %

=(£29,776/£1,400,000) × 100 = £1,400,000

Add, as schedule 3.3 12.5%	0.27 %
Percentage addition for facilities	2.40 %

Insurances

Schedule 3.4 allows for the cost of additional insurance premiums for abnormal work or special site conditions to be charged at cost plus 12.5%.

Watching and lighting

Schedule 3.5 allows for all such costs necessitated by Dayworks to be paid for separately at Schedule rates.

Plant

The cost of the driver(s) and any required attendants such as banksmen should be covered in the labour section of the dayworks calculation.

The Schedule rates include fuel and consumable stores, repairs and maintenance (but not the time spent on general servicing) and insurance.

The Schedule rates only apply to machinery which is on site at the time of the work – where plant is specifically hired for the task, the Contractor is entitled to be paid the invoiced value. If the invoice excludes consumables used (fuel, oil and grease) then the Contractor is entitled to add the cost – together with insurance and any transport costs incurred in getting the equipment on site all subject to a 12½% addition for Head Office costs and profit.

Head Office charges and Profit allowances are included in the Schedule rates, 12½% being added to the charged value of hired plant.

1   General servicing of plant

Schedule 4.1 specifically excludes time spent on general servicing from the Hire Rates.

General servicing in this context can be assumed to mean:

Checking, replenishing (or changing, if applicable)
   i.e.   engine lubrication
           transmission lubrications
           general greasing
           coolants
           hydraulic oils and brake systems
           filters
           tyres

Inspecting special items
   e.g.   buckets
           hoses / airlines
           shank protectors
           Cables / ropes / hawsers
           rippers
           blades, steels, etc

Example

Assuming an 8 hour working day these operations could take a plant operator on average:

large machine	20 minutes per day	(equating to 1 hour for each 24 worked)
medium machine	10 minutes per day	(equating to 1 hour for each 48 worked)
small machine	5 minutes per day	(equating to 1 hour for each 96 worked)

The cost of the servicing labour would be as follows

large machine	£26.54/hr / 24hrs	= £1.11
medium machine	£17.58/hr / 48 hrs	= £0.37
small machine	£14.68/hr / 96 hrs	= £0.15

These labour costs can be expressed as a percentage of the schedule hire rates:

D8 Dozer	£1.11 / £121.54/hr × 100	= 0.9 %
JCB 3CX	£0.37 / £18.16/hr × 100	= 2.0%
2 tonne dumper	£0.15 / £6.62/hr × 100	= 2.3 %

Taking into account the range of these sizes of plant which are normally deployed on site, the following would provide a reasonable average percentage addition:

0.9 % × 2	= 1.8 %
2.0 % × 3	= 6.0 %
2.3 % × 6	= 13.8 %

Total = 21.6 % for 11 items of plant   i.e. average = 2.0 %

2 Fuel distribution

Schedule 4.3 allows for charging for fuel distribution.

Assuming this is done with a towed fuel bowser behind a farm type tractor with driver/labourer in attendance, the operation cycle would involve visiting, service, and return or continue on to the next machine.

The attendance cost based on the hourly rate tractor/bowser/driver would be: = £15.96 + £1.38 + £13.59 = £30.93

Example:

A heavy item of plant, for example a Cat D8R Tractor Bulldozer with a 212 kW engine
  Fuel consumption is 30.4 litres/hr (38 litres/hr × 80% site utilization factor)
  Fuel capacity is 200 litres
  Requires filling after 6.5 working hours operation (200 litres divided by 30.40 l/hr)
  Tractor / bowser service taking 30 minutes

The machine cost during this 6.5 hr period would be £121.54 × 6.5, i.e. £ 790.01
The attendance cost for the 30 min cycle would be £30.93 × 0.5, i.e. £15.47
The percentage addition for fuelling the machine would be: (£15.47/£790.01) × 100 = 2.0%

Example:

A medium sized item of plant, for example a JCB 3CX
  Fuel consumption is 5.6 litres/hr (7.5 litres/hr × 75% site utilization factor)
  Fuel capacity is 90 litres
  Requires filling after 16 working hours operation (90 litres divided by 5.60 l/hr)
  Tractor / bowser service taking 10 minutes

The machine cost during this 16 hr period would be £18.16 × 16 hrs, i.e. £290.56
The attendance cost for the 10 min cycle would be £30.93 × 0.17 hrs, i.e. £5.26
The percentage addition for fuelling the machine would be: (£5.26/£290.56) × 100 = 1.8%

Example:

A medium sized item of plant, for example a 2 tonne dumper
  Fuel consumption is 2.4 litres/hr (3 litres/hr × 80% site utilization factor)
  Fuel capacity is 35 litres
  Requires filling after 14.5 working hours operation (35 litres divided by 2.40 l/hr)
  Tractor / bowser service taking 5 minutes

The machine cost during this 14.5 hr period would be £6.62 × 14.5 hrs, i.e. £95.99
The attendance cost for the 5 min cycle would be £30.93 × 0.08 hrs, i.e. £2.47
 The percentage addition for fuelling the machine would be: (£2.47/£95.99) × 100 = 2.6%

Considering the range of these categories of plant which are normally deployed on site, the following would provide a reasonable average percentage addition for the above:

2.0% × 2	= 4.0 %
1.8% × 3	= 5.4 %
2.6% × 6	= 15.6 %
Total	= 25.0 % for 11 items of plant – average = 2.3 %

▪ Mobilization

Schedule 4.3 allows for charging for mobilization and demobilization.

Construction Contracts Questions and Answers

2nd Edition

By **David Chappell**

What they said about the first edition: "A fascinating concept, full of knowledgeable gems put in the most frank of styles... A book to sample when the time is right and to come back to when another time is right, maybe again and again." – *David A Simmonds,* Building Engineer *magazine*

• Is there a difference between inspecting and supervising?

• What does 'time-barred' mean?

• Is the contractor entitled to take possession of a section of the work even though it is the contractor's fault that possession is not practicable?

Construction law can be a minefield. Professionals need answers which are pithy and straightforward, as well as legally rigorous. The two hundred questions in the book are real questions, picked from the thousands of telephone enquiries David Chappell has received as a Specialist Adviser to the Royal Institute of British Architects. Although the enquiries were originally from architects, the answers to most of them are of interest to project managers, contractors, QSs, employers and others involved in construction.

The material is considerably updated from the first edition – weeded, extended and almost doubled in coverage. The questions range in content from extensions of time, liquidated damages and loss and/or expense to issues of warranties, bonds, novation, practical completion, defects, valuation, certificates and payment, architects' instructions, adjudication and fees. Brief footnotes and a table of cases will be retained for those who may wish to investigate further.

August 2010: 216x138: 352pp
Pb: 978-0-415-56650-6: **£34.99**

To Order: Tel: +44 (0) 1235 400524 **Fax:** +44 (0) 1235 400525
or Post: Taylor and Francis Customer Services,
Bookpoint Ltd, Unit T1, 200 Milton Park, Abingdon, Oxon, OX14 4TA UK
Email: book.orders@tandf.co.uk

For a complete listing of all our titles visit:
www.tandf.co.uk

PART 9

Professional Fees

CONSULTING ENGINEER CONDITIONS OF APPOINTMENT

Introduction

A scale of professional charges for consulting engineering services in connection with civil engineering works is published by the Association for Consultancy and Engineering (ACE)

Copies of the document can be obtained direct from:

> Association for Consultancy and Engineering
> Alliance House
> 12 Caxton Street
> London SW1H OQL
> Tel 020 7222 6557
> Fax 020 7222 0750

Comparisons

Instead of the previous arrangement of having different agreements designed for each major discipline of engineering, the current agreements have been developed primarily to suit the different roles that Consulting Engineers may be required to perform, with variants of some of them for different disciplines. The agreements have been standardised as far as possible whilst retaining essential differences.

Greater attention is required than with previous agreements to ensure the documents are completed properly. This is because of the perceived need to allow for a wider choice of arrangements, particularly of methods of payment.

The agreements are not intended to be used as unsigned reference material with the details of an engagement being covered in an exchange of letters, although much of their content could be used as a basis for drafting such correspondence.

For 2009 the ACE has published a new suite of Agreements with a broader set of services, these are listed below.

Forms of Agreement

> ACE Agreement 1: Design
>
> ACE Agreement 2: Advise and Report
>
> ACE Agreement 3: Design and Construct
>
> ACE Agreement 4: Sub-Consultancy
>
> ACE Agreement 5: Homeowner
>
> ACE Agreement 6: Expert Witness (Sole Practitioner)
>
> ACE Agreement 7: Expert Witness (Firm)
>
> ACE Agreement 8: Adjudicator

To a number of the the above ACE Agreements Schedules of Services are appended and currently these are:

For use with ACE Agreement 1: Design

ACE Schedule of Services – Part G(a):
Civil and Structural Engineer – single consultant or non-lead consultant

ACE Schedule of Services – Part G(b):
Mechanical and Electrical Engineering (detailed design in buildings)

ACE Schedule of Services – Part G(c):
Mechanical and Electrical Engineering (performance design in buildings)

ACE Schedule of Services – Part G(d):
Civil and Structural Engineer – Lead consultant

ACE Schedule of Services – Part G(e):
Mechanical and Electrical Engineering Design in buildings – Lead consultant

For use with ACE Agreement 3: Design and Construct

ACE Schedule of Services – Part G(f):
Civil and Structural Engineer

ACE Schedule of Services – Part G(g):
Mechanical and Electrical Engineering (detailed design in buildings)

ACE Schedule of Services – Part G(h):
Mechanical and Electrical Engineering (performance design in buildings)

ACE Agreement 1: Design
Design for the appointment of a consultant by a client to undertake detailed design and/or specification of permanent works to be undertaken or installed by a contractor including any studies, appraisals, investigations, contract administration or construction monitoring leading to or resulting from such detailed design and/or specification.

ACE Agreement 2: Advise and Report
Advise and report for the appointment of a consultant by a client to provide any type of advisory, research, checking, reviewing, investigatory, monitoring, reporting or technical services in the built and natural environments where such services do not consist of detailed design or specification of permanent works to be constructed or installed by a contractor.

ACE Agreement 3: Design and Construct
For the appointment of a consultant by a contractor in circumstances where the contractor is to construct permanent works designed by the consultant.

ACE Agreement 4: Sub-consultancy
For the appointment of a sub-consultant by a consultant in circumstances where the consultant is appointed on the terms of an ACE Agreement by its client

ACE Agreement 5: Homeowner
Model letter for the appointment of a consultant by a homeowner

ACE Agreement 6: Expert Witness (Sole Practitioner)
For the appointment of an individual to act as an expert witness

ACE Agreement 7: Expert Witness (Firm)
    For the appointment of a firm to provide an expert witness

ACE Agreement 8: Adjudicator
    For the appointment of an adjudicator

Collateral Warranties

The association is convinced that collateral warranties are generally unnecessary and should only be used in exceptional circumstances. The interests of clients, employers and others are better protected by taking out project or BUILD type latent defects insurance. Nevertheless, in response to observations raised when the pilot editions excluded any mention of warranties, references and arrangements have been included in the Memorandum and elsewhere by which Consulting Engineers may agree to enter into collateral warranty agreements; these should however only be given when the format and requirements thereof have been properly defined and recorded in advance of undertaking the commission.

Requirements for the provision of collateral warranties will be justified even less with commissions under Agreement D than with those under the other ACE agreements. Occasional calls may be made for them, such as when a client intends to dispose of property and needs evidence of a duty of care being owed to specific third parties, but these will be few and far between.

Remuneration

Guidance on appropriate levels of fees to be charged is given at the end of each agreement. Firms and their clients may use this or other sources, including their own records, to determine suitable fee arrangements.

Need for formal documentation

The Association for Consultancy and Engineering recommends that formal written documentation should be executed to record the details of each commission awarded to a Consulting Engineer. These Conditions are published as model forms of agreement suitable for the purpose. However, even if these particular Conditions are not used, it is strongly recommended that, whenever a Consulting Engineer is appointed, there should be at least an exchange of letters defining the duties to be performed and the terms of payment.

Appointments outside the United Kingdom

These conditions of Engagement are designed for use within the UK. For work overseas it is impracticable to give definite recommendations; circumstances differ too widely between countries. There are added complications in documentation relating to local legislation, import customs, conditions of payment, insurance, freight, etc. Furthermore, it is often necessary to arrange for visits to be made by principals and senior staff whose absence abroad during such periods represents a serious reduction of their earning power. The additional duties, responsibilities and non-recoverable costs involved, and the extra work on general co-ordination, should be reflected in the levels of fees. Special arrangements are also necessary to cover travelling and other out-of-pocket expenses in excess of those normally incurred on similar work in the UK, including such matters as local cost-of-living allowances and the cost of providing home-leave facilities for expatriate staff.

CONDITIONS OF ENGAGEMENT

Obligations of the Consulting Engineer

The responsibilities of the Consultant Engineer for the works are as set out in the actual agreement The various standard clauses in the Conditions relate to such matters as differentiating between Normal and Additional services, the duty to exercise skill and care, the need for Client's written consent to the assignment or transfer of any benefit or obligation of the agreement, the rendering of advice if requested on the appointment of other consultants and specialist sub- consultants, any recommendations for design of any part of the Works by Contractors or

Sub-contractors (with the proviso that the Consulting Engineer is not responsible for detailed design of contractors or for defects or omissions in such design), the designation of a Project Leader, the need for timeliness in requests to the Client for information etc., freezing the design once it has been given Client approval and the specific exclusion of any duty to advise on the actual or possible presence of pollution or contamination or its consequences.

Obligations of the Client

The Consultant Engineer shall be supplied with all necessary data and information in good time. The Client shall designate a Representative authorised to make decisions on his behalf and ensure that all decisions, instructions, and approvals are given in time so as not to delay or disrupt the Consultant Engineer.

Site Staff

The Consulting Engineer may employ site staff he feels are required to perform the task, subject to the prior written agreement of the Client. The Client shall bear the cost of local office accommodation, equipment and running costs.

Commencement, Determination, Postponement, Disruption and Delay

The Consulting Engineer's appointment commences at the date of the execution of the Memorandum of Agreement or such earlier date when the Consulting Engineer first commenced the performance of the Services, subject to the right of the Client to determine or postpone all or any of the Services at any time by Notice.

The Client or the Consulting Engineer may determine the appointment in the event of a breach of the Agreement by the other party after two weeks notice. In addition, the Consulting Engineer may determine his appointment after two weeks notice in the event of the Client failing to make proper payment.

The Consulting Engineer may suspend the performance of all or any of the Services for up to twenty-six weeks if he is prevented or significantly impeded from performance by circumstances outside his control. The appointment may be determined by either party in the event of insolvency subject to the issue of notice of determination.

Payments

The Client shall pay fees for the performance of the agreed service(s) together with all fees and charges to the local or other authorities for seeking and obtaining statutory permissions, for all site staff on a time basis, together with additional payments for any variation or the disruption of the Consulting Engineer's work due to the Client varying the task list or brief or to delay caused by the Client, others or unforeseeable events.

If any part of any invoice submitted by the Consulting Engineer is contested, payment shall be made in full of all that is not contested.

Payments shall be made within 28 days of the date of the Consulting Engineer's invoice; interest shall be added to all amounts remaining unpaid thereafter.

Ownership of Documents and Copyright

The Consulting Engineer retains the copyright in all drawings, reports, specifications, calculations etc. prepared in connection with the Task; with the agreement of the Consulting Engineer and subject to certain conditions, the Client may have a licence to copy and use such intellectual property solely for his own purpose on the Task in hand, subject to reservations.

The Consulting Engineer must obtain the client's permission before he publishes any articles, photographs or other illustrations relating to the Task, nor shall he disclose to any person any information provided by the Client as private and confidential unless so authorised by the Client.

Liability, Insurance and Warranties

The liability of the Consulting Engineer is defined, together with the duty of the Client to indemnify the Consulting Engineer against all claims etc. in excess of the agreed liability limit.

The Consulting Engineer shall maintain Professional Indemnity Insurance for an agreed amount and period at commercially reasonable rates, together with Public Liability Insurance and shall produce the brokers' certificates for inspection to show that the required cover is being maintained as and when requested by the Client.

The Consulting Engineer shall enter into and provide collateral warranties for the benefit of other parties if so agreed.

Disputes and Differences

Provision is made for mediation to solve disputes, subject to a time limit of six weeks of the appointment of the mediator at which point it should be referred to an independent adjudicator. Further action could be by referring the dispute to an arbitrator.

QUANTITY SURVEYOR CONDITIONS OF APPOINTMENT

Introduction

Authors' Note:
The Royal Institution of Chartered Surveyors formally abolished standard Quantity Surveyor's fee scales with effect from 31st December 1998. However, in the absence of any alternative guidance and for the benefit of readers, extracts from relevant fee scales have been reproduced in part with the permission of the Royal Institution of Chartered Surveyors, which owns the copyright.

Summary of Scale of Professional Charges

Scale No 38. issued by The Royal Institution of Chartered Surveyors provides an itemised scale of professional charges for Quantity Surveying Services for Civil Engineering Works which is summarised as follows :-

1.1	Generally
1.2	The Scale of professional charges is applicable where the contract provides for the bills of quantities and final account to be based on measurements prepared in accordance with or based on the principles of the Standard Method of Measurement of Civil Engineering Quantities issued by the Institution of Civil Engineers.
1.3	The fees are in all cases exclusive of travelling and other expenses (for which the actual disbursement is recoverable unless there is some prior arrangement for such charges) and of the cost of reproduction of bills of quantities and other documents, which are chargeable in addition at net cost.
1.4	The fees are in all cases exclusive of services in connection with the allocation of the cost of the works for purposes of calculating value added tax for which there shall be an additional fee based on the time involved.
1.5	If any of the materials used in the works are supplied by the employer or charged at a preferential rate, then the actual or estimated market value thereof shall be included in the amounts upon which fees are to be calculated.
1.6	The fees are in all cases exclusive of preparing a specification of the materials to be used and the works to be done.
1.7	If the quantity surveyor incurs additional costs due to exceptional delays in construction operations or any other cause beyond the control of the quantity surveyor then the fees may be adjusted by agreement between the employer and the quantity surveyor.
1.8	If the works are substantially varied at any stage or if the quantity surveyor is involved in abortive work there shall be an additional fee based on the time involved.
1.9	The fees and charges are in all cases exclusive of value added tax which will be applied in accordance with legislation.
1.10	The scale is not intended to apply to works of a civil engineering nature which form a subsidiary part of a building contract or to buildings which are ancillary to a civil engineering contract. In these cases the fees to be charged for quantity surveying services shall be in accordance with the scales applicable to building works.
1.11	When works of both categories I* and II** are included in one contract the fee to be charged shall be calculated by taking the total value of the sections of work in each of the categories and applying the appropriate scale from the beginning in each case. General items such as preliminaries (and in the case of post contract fees contract price fluctuations) or sections of works which cannot be specifically allocated to either category shall be apportioned pro-rata to the values of the other sections of the works and added thereto in order to ascertain the total value of works in each category.
1.12	When a project is the subject of a number of contracts then, for the purpose of calculating fees, the value of such contracts shall not be aggregated but each contract shall be taken separately and the scale of charges applied as appropriate.
1.13	Roads, railways, earthworks and dredging which are ancillary only to any Category II** work shall be regarded as Category II** work. Works or sections of works of Category I* which incorporate piled construction shall be regarded as Category II** works.
1.14	No addition to the fees given hereunder shall be made in respect of works of alteration or repair where such works are incidental to the new works. If the work covered by a single contract is mainly one of alteration or repair then an additional fee shall be negotiated.

1.15   In the absence of agreement to the contrary, payments to the quantity surveyor shall be made by instal-ments by arrangement between the employer and the quantity surveyor.

1.16   Copyright in bills of quantities and other documents prepared by the quantity surveyor is reserved to the quantity surveyor.

*Category I	Works or sections of works such as monolithic walls for quays, jetties dams and reservoirs; caissons; tunnels; airport runways and tracks roads; railways; and earthworks and dredging
** Category II	Works or sections of works such as piled quay walls; suspended jetties and quays; bridges and their abutments; culverts; sewers; pipe-lines; electric mains; storage and treatment tanks; water cooling towers and structures for housing heavy industrial and public utility furnace houses and rolling mills to steel works; and boiler houses,plant, e.g. reactor blocks and turbine halls to electricity generating stations.

Scale of charges

For the full Scale of Fees for Professional Charges for Quantity Surveying Services together with a detailed description of the full service provided the appropriate RICS Fee Scale should be consulted.

PART 10

Outputs

This part lists a selection of OUTPUT CONSTANTS for use within various areas of Civil Engineering Work.

DISPOSAL OF EXCAVATED MATERIALS

Outputs Per Hundred Cubic Metres

Tipper Capacity and Length of Haul	Driver (hours)	Attendant Labour (hours)	Number of cycles	Average Speed (km/hr)	Cycle time (minutes)				
					Loading	Haul	Discharge	Return	Total
3 m³ tipper:									
1 km haul	8.8	0.0	43.3	10	0.70	6.0	0.7	5.5	12.2
5 km haul	26.3	0.0	43.3	16	0.70	18.8	0.7	17.0	36.5
9 km haul	41.9	0.0	43.3	18	0.70	30.0	0.7	27.3	58.0
12 km haul	52.7	0.0	43.3	19	0.70	37.9	0.7	34.4	73.0
15 km haul	65.8	0.0	43.3	19	0.70	47.4	0.7	43.1	91.2
8 m³ tipper:									
1 km haul	10.6	0.8	16.3	3	2.0	20.0	0.9	18.2	39.1
5 km haul	19.7	0.8	16.3	8	2.0	37.5	0.9	34.1	72.5
9 km haul	25.7	0.8	16.3	11	2.0	49.1	0.9	44.6	94.6
12 km haul	26.9	0.8	16.3	14	2.0	51.4	0.9	46.8	99.1
15 km haul	31.4	0.8	16.3	15	2.0	60.0	0.9	54.5	115.4
12 m³ tipper:									
1 km haul	7.1	0.7	10.8	3	2.90	20.0	1.1	18.2	39.3
5 km haul	13.1	0.7	10.8	8	2.90	37.5	1.1	34.1	72.7
9 km haul	17.1	0.7	10.8	11	2.90	49.1	1.1	44.6	94.8
12 km haul	17.9	0.7	10.8	14	2.90	51.4	1.1	46.8	99.3
15 km haul	20.8	0.7	10.8	15	2.90	60.0	1.1	54.5	115.6
15 m³ tipper:									
1 km haul	5.8	0.8	8.7	3	3.70	20.0	1.5	18.2	39.7
5 km haul	10.6	0.8	8.7	8	3.70	37.5	1.5	34.1	73.1
9 km haul	13.8	0.8	8.7	11	3.70	49.1	1.5	44.6	95.2
12 km haul	14.5	0.8	8.7	14	3.70	51.4	1.5	46.8	99.7
15 km haul	16.8	0.8	8.7	15	3.70	60.0	1.5	54.5	116.0

Man hours include round trip for tipper and driver together with attendant labour for positioning during loading and unloading.

The number of cycles are based on the stated heaped capacity of the tipper divided into the total volume of 100 m³ to be moved, being multiplied by a bulking factor of x 1.30 to the loose soil volume.

The average speeds are calculated assuming that the vehicles run on roads or on reasonably level firm surfaces and allow for acceleration and deceleration with the return empty journey being say 10% faster than the haul.

The cycle time shows in detail the time spent being loaded (calculated using a 1.5 m³ loader with a cycle time of 22 seconds), haul journey, discharge (turning / manoeuvring / tipping) and return journey.

BREAKING OUT OBSTRUCTIONS BY HAND

Breaking out pavements, brickwork, concrete and masonry by hand and pneumatic breaker

Description	Unit	By hand using picks, shovels & points Labour	7 m³ Compressor (2 Tool)		10 m³ Compressor (3 Tool)	
			Compressor	Labour	Compressor	Labour
Break out bitmac surfaces on sub-base or hardcore						
75 mm thick	m²/hr	1.00	25	13	50	17
100 mm thick	m²/hr	1.00	20	10	33	11
Break out asphalt roads on hardcore:						
150 mm thick	m²/hr	0.60	8	4	11	4
225 mm thick	m²/hr	0.50	6	3	8	3
300 mm thick	m²/hr	0.40	4	2	7	2
Remove existing set paving	m²/hr	0.80	10	5	17	6
Break out brickwork in cement mortar: 215 mm thick	m²/hr	0.20	3	2	5	2
Break out concrete in areas						
100 mm thick	m²/hr	0.90	9	5	14	5
150 mm thick	m²/hr	0.50	7	4	10	4
225 mm thick	m²/hr	0.30	4	2	6	2
300 mm thick	m²/hr	0.20	3	1	4	1
Break out reinforced concrete	m³/hr	0.02	0.40	0.20	0.60	0.20
Break out sandstone	m³/hr	0.03	0.50	0.30	0.80	0.30

Loading loose materials and items by hand

Material	Unit	Loading into Vehicles	
		Tonne	m³
Bricks	hr	1.7	2.9
Concrete, batches	hr	1.4	1.2
Gulley grates and frames	hr	1.0	
Kerb	hr	1.1	
Paving slabs	hr	0.9	
Pipes, concrete and clayware	hr	0.9	
Precast concrete items	hr	0.9	
Soil	hr	1.4	2.0
Steel reinforcement	hr	0.8	
Steel sections, etc	hr	1.0	
Stone and aggregates:			
bedding material	hr	1.5	1.8
filter/subbase	hr	1.3	1.4
rock fill (6" down)	hr	1.3	1.3
Trench planking and shoring	hr	0.9	2.9

CONCRETE WORK

Placing ready mixed concrete in the works

Description	Labour Gang (m³ per hour)
MASS CONCRETE	
Blinding	
150 mm thick	5.50
150 - 300 mm thick	6.25
300 - 500 mm thick	7.00
Bases and oversite concrete	
not exceeding 150 mm thick	5.00
not exceeding 300 mm thick	5.75
not exceeding 500 mm thick	6.75
exceeding 500 mm thick	7.00
REINFORCED CONCRETE	
Bases	
not exceeding 300 mm thick	5.50
not exceeding 500 mm thick	6.25
exceeding 500 mm thick	6.75
Suspended slabs (not exceeding 3m above pavement level)	
not exceeding 150 mm thick	3.75
not exceeding 300 mm thick	4.75
exceeding 300 mm thick	5.75
Walls and stems (not exceeding 3m above pavement):	
not exceeding 150 mm thick	3.50
not exceeding 300 mm thick	4.50
exceeding 300 mm thick	5.00
Beams, columns and piers (not exceeding 3m above pavement):	
sectional area not exceeding 0.03 m²	2.00
sectional area not exceeding 0.03 - 1.0 m²	2.50
sectional area exceeding 1.0 m²	3.50

Fixing bar reinforcement

All bars delivered to site cut and bent and marked, including craneage and hoisting (maximum height 5 m).

Description (fix only)	Unit	up to 6 mm		7 to 12 mm		13 to 19 mm		over 19 mm	
		steelfixer	labourer	steelfixer	labourer	steelfixer	labourer	steelfixer	labourer
Straight round bars to beams, floors, roofs and walls	t / hr	0.03	0.03	0.04	0.04	0.06	0.06	0.08	0.08
to braces, columns, sloping roofs and battered walls	t / hr	0.01	0.01	0.02	0.02	0.03	0.03	0.05	0.05
Bent round bars to beams, floors, roofs and walls	t / hr	0.02	0.02	0.03	0.03	0.03	0.03	0.04	0.04
to braces, columns, sloping roofs and battered walls	t / hr	0.01	0.01	0.01	0.01	0.02	0.02	0.03	0.03
Straight, indented or square bars to beams, floors, roofs and walls	t / hr	0.02	0.02	0.04	0.04	0.05	0.05	0.07	0.07
to braces, columns, sloping roofs and battered walls	t / hr	0.01	0.01	0.02	0.02	0.02	0.02	0.04	0.04
Bent, indented or square bars to beams, floors, roofs and walls	t / hr	0.02	0.02	0.02	0.02	0.03	0.03	0.03	0.03
to braces, columns, sloping roofs and battered walls	t / hr	0.01	0.01	0.01	0.01	0.02	0.02	0.02	0.02
(Average based on Gang D)		0.125		0.148		0.163		0.225	

Erecting formwork to beams and walls

Erect and strike formwork	Unit	Joiner	Labourer
Walls - vertical face (first fix)			
up to 1.5 m	m²/hr	1.7	0.8
1.5 to 3.0 m	m²/hr	1.4	0.7
3.0 to 4.5 m	m²/hr	1.2	0.6
4.5 to 6.0 m	m²/hr	1.0	0.5

Erecting formwork to slabs

Erect and strike formwork	Unit	Joiner	Labourer
Horizontal flat formwork at heights (first fix)			
up to 3.0 m	m²/hr	1.11	1.11
3.0 to 3.6 m	m²/hr	1.05	1.05
3.6 to 4.2 m	m²/hr	1.00	1.00
4.2 to 4.8 m	m²/hr	0.95	0.95
4.8 to 5.4 m	m²/hr	0.90	0.90
5.4 to 6.0 m	m²/hr	0.83	0.83

Multipliers for formwork

Description	Multiplier
Walls built to batter	1.20
Walls built circular to large radius	1.70
Walls built circular to small radius	2.10
Formwork used once	1.00
Formwork used twice, per use	0.85
Formwork used three times, per use	0.75
Formwork used four times, per use	0.72
Formwork used five times, per use	0.68
Formwork used six times, per use	0.66
Formwork used seven or more times, per use	0.63
Formwork to slope not exceeding 45° from horizontal	1.25

DRAINAGE

Laying and jointing flexible-jointed clayware pipes

Diameter of pipe in mm	Drainage Gang	In trench not exceeding 1.5 m	In trench not exceeding 3 m	In trench 3 - 4.5 m
Pipework				
100 mm	m/hr	10	8	7
150 mm	m/hr	7	6	5
225 mm	m/hr	5	4	3
300 mm	m/hr	4	3	2
375 mm	m/hr	3	2	2
450 mm	m/hr	2	2	1
Bends				
100 mm	nr/hr	20	17	14
150 mm	nr/hr	17	14	13
225 mm	nr/hr	13	10	8
300 mm	nr/hr	10	8	7
375 mm	nr/hr	7	6	5
450 mm	nr/hr	3	3	2
Single junctions				
100 mm	nr/hr	13	10	8
150 mm	nr/hr	7	6	5
225 mm	nr/hr	6	5	4
300 mm	nr/hr	4	3	3
375 mm	nr/hr	3	2	2
450 mm	nr/hr	2	2	1

Precast concrete manholes in sections

Description	Unit	Pipelayer	Labourer
Place 675 mm dia shaft rings	m/hr	1.00	0.30
Place 900 mm manhole rings	m/hr	0.60	0.20
Place 1200 mm manhole rings	m/hr	0.40	0.10
Place 1500 mm manhole rings	m/hr	0.30	0.10
Place 900/675 mm tapers	nr/hr	1.00	0.30
Place 1200/675 mm tapers	nr/hr	0.60	0.20
Place 1500/675 mm tapers	nr/hr	0.50	0.20
Place cover slabs to 675 mm rings	nr/hr	2.50	0.80
Place cover slabs to 900 mm rings	nr/hr	2.00	0.70
Place cover slabs to 1200 mm rings	nr/hr	1.40	0.50
Place cover slabs to 1500 mm rings	nr/hr	0.90	0.30
Build in pipes and make good base:			
150 mm diameter	nr/hr	4.00	-
300 mm diameter	nr/hr	2.00	-
450 mm diameter	nr/hr	1.00	-
Benching 150 mm thick	m²/hr	1.20	1.20
Benching 300 mm thick	m²/hr	0.60	0.60
Render benching 25 mm thick	m²/hr	1.10	1.10
Fix manhole covers frame	nr/hr	1.30	1.30

Part 11

Useful addresses for further information

ACOUSTICAL INVESTIGATION AND
RESEARCH ORGANISATION LTD
Duxon's Turn, Maylands Avenue,
Hemel Hempstead, Herts HP2 4SB
Tel: 01442 247 146
Fax: 01442 256 749
Website: www.airo.co.uk

ASBESTOS REMOVAL CONTRACTORS
ASSOCIATION
ARCA House, 237 Branston Road,
Burton upon Trent, Staffordshire DE14 3BT
Tel: 01283 513 126
Fax: 01283 568 228
Website: www.arca.org.uk

AGGREGATE CONCRETE BLOCK ASSOCIATION
See BRITISH PRECAST CONCRETE FEDERATION

ASSOCIATION OF BUILDERS HARDWARE
MANUFACTURERS
42 Heath Street, Tamworth, Staffs B79 7JH
Tel: 01827 52337
Fax: 01827 310 827
Website: www.abhm.org.uk

ALUMINIUM FEDERATION LTD
Broadway House, Calthorpe Road, Five Ways,
Birmingham B15 1TN
Tel: 0121 456 1103
Fax: 0121 456 2274
Website: www.alfed.org.uk

ASSOCIATION FOR CONSULTANCY AND
ENGINEERING
Alliance House, 12 Caxton Street, London
SW1H 0QL
Tel: 0207 222 6557
Fax: 0207 222 0750
Website: www.acenet.co.uk

AMERICAN HARDWOOD EXPORT COUNCIL
3 St Michael's Alley, London EC3V 9DS
Tel: 0207 626 4111
Fax: 0207 626 4222
Website: www.ahec-europe.org

ASSOCIATION OF LOADING AND ELEVATING
EQUIPMENT MANUFACTURERS
Orbital House, 85 Croydon Road, Caterham, Surrey
CR3 6PD
Tel: 01883 334494
Fax: 01883 334490
Website: www.alem.org.uk

ARBORICULTURAL ADVISORY &
INFORMATION SERVICE
Tel: 0897 161 147

ASSOCIATION OF PROJECT MANAGEMENT
150 West Wycombe Road, High Wycombe,
Bucks HP12 3AE
Tel: 0845 458 1944
Fax: 01494 528 937
Website: www.apm.org.uk

ARBORICULTURAL ASSOCIATION
Ampfield House, Ampfield, Nr Romsey, Hants
SO51 9PA
Tel: 01794 368 717
Fax: 01794 368 978
Website: www.trees.org.uk

BRITISH AGGREGATE CONSTRUCTION
MATERIALS INDUSTRIES LTD
See QUARRY PRODUCTS ASSOCIATION

ARCHITECTURAL ASSOCIATION SCHOOL OF
ARCHITECTURE
34 - 36 Bedford Square, London WC1B 3ES
Tel: 020 7887 4000
Fax: 020 7414 0782
Website: www.aaschool.ac.uk

BRITISH AIRPORTS AUTHORITY PLC
Corporate Office, 130 Wilton Road, London
SW1V 1LQ
Tel: 0207 834 9449
Fax: 0207 932 6699

ASBESTOS INFORMATION CENTRE LTD
ATSS House, Station Road East,
Stowmarket, Suffolk IP14 1RQ
Tel: 01449 676900
Fax: 01449 770028
Website:

BRITISH ANODISING ASSOCIATION
See ALUMINIUM FEDERATION

BRITISH ARCHITECTURAL LIBRARY
RIBA, 66 Portland Place, London W1B 1AD
Tel: 0207 580 5533
Fax: 0207 251 1541
Website: www.architecture.com

BRITISH FIRE PROTECTION SYSTEMS
ASSOCIATION LTD
55 Eden Street, Kingston-upon-Thames,
Surrey KT1 1BW
Tel: 0208 549 5855
Fax: 0208 547 1564
Website: www.bfpsa.org.uk

BRITISH ASSOCIATION OF LANDSCAPE
INDUSTRIES
Landscape House, Stoneleigh Park, Warwickshire
CV8 2LG
Tel: 0870 770 4971
Fax: 0870 770 4972
Website: www.bali.org.uk

BRITISH FIRE SERVICES ASSOCIATION
86 London Road, Leicester LR2 0QR
Tel + Fax: 0116 254 2879

BRITISH BOARD OF AGRéMENT
PO Box 195, Bucknalls Lane, Garston,
Watford, Herts WA25 9BA
Tel: 01923 665 300
Fax: 01923 665 301
Website: www.bbacerts.co.uk

BRITISH FLUE & CHIMNEY
MANUFACTURERS ASSOCIATION
See FEDERATION OF ENVIRONMENTAL TRADE
ASSOCIATIONS

BRITISH CABLE ASSOCIATION
37a Walton Rd, East Molesey,
Surrey KT8 9DW
Tel: 0208 941 4079
Fax: 0208 783 0104

BRITISH GEOLOGICAL SURVEY
Kingsley Dunham Centre
Keyworth, Nottingham NG12 5GG
Tel: 0115 936 3100
Fax: 0115 936 3200
Website: www.bgs.ac.uk

BRITISH CEMENT ASSOCIATION
Riverside House, 4 Meadows Business Park
Station Approach, Blackwater,
Camberley, Surrey GU17 9AB
Tel: 01276 608700
Fax: 01276 608701
Website: www.bca.org.uk

BRITISH INSTITUTE OF ARCHITECTURAL
TECHNOLOGISTS
397 City Road, London EC1V 1NH
Tel: 0207 278 2206
Fax: 0207 837 3194
Website: www.biat.org.uk

BRITISH CERAMIC RESEARCH
Queens Road, Penkhull, Stoke-on-Trent, Staffs
ST4 7LQ
Tel: 01782 764 444
Fax: 01782 412 331
Website: www.ceram.com

BRITISH LIBRARY LENDING DIVISION
Thorpe Arch, Boston Spa, Wetherby, West Yorks
LS23 7BQ
Tel: 01937 546 000

BRITISH CONSTRUCTIONAL STEELWORK
ASSOCIATION LTD
4 Whitehall Court, Westminster, London
SW1A 2ES
Tel: 0207 839 8566
Fax: 0207 976 1634
Website: www.bcsa.org.uk

BRITISH LIBRARY, SCIENCE REFERENCE
AND INFORMATION LIBRARY
The British Library, St Pancras, 96 Euston Road,
London NW1 2DB
Tel: 0870 444 1500
Website: www.bl.uk

BRITISH ELECTROTECHNICAL AND ALLIED
MANUFACTURERS ASSOCIATION
Westminster Tower, 3 Albert Embankment,
London SE1 7SL
Tel: 0207 793 3000
Fax: 0207 793 3003
Website: www.beama.org.uk

BRITISH NON-FERROUS METALS
FEDERATION
Broadway House, 60 Calthorpe Road, Edgbaston,
Birmingham B15 1TN
Tel: 0121 456 6110
Fax: 0121 456 2274

BRITISH PLASTICS FEDERATION
6 Bath Place, Rivington Street, London EC2A 3JE
Tel: 0207 457 5000
Fax: 0207 457 5045
Website: www.bpf.co.uk

BRITISH WOODWORKING FEDERATION
55 Tufton Street, London SW1 3QL
Tel: 0870 458 6939
Fax: 0870 458 6949
Website: www.bwf.org.uk

BRITISH PRECAST CONCRETE FEDERATION
60 Charles St, Leicester LE1 1FB
Tel: 0116 253 6161
Fax: 0116 251 4568
Website: www.britishprecast.org.uk

THE BUILDING CENTRE GROUP
26 Store Street, London WC1E 7BT
Tel: 0207 692 4000
Fax: 0207 580 9641
Website: www.buildingcentre.co.uk

BRITISH REINFORCEMENT MANUFACTURERS
ASSOCIATION (BRMA)
See UK STEEL ASSOCIATION

BUILDING MAINTENANCE INFORMATION (BMI)
3 Cadogan gate, London SW1X 0AS
Tel: 0207 695 1500
Fax: 0207 695 1501

BRITISH RUBBER MANUFACTURERS
ASSOCIATION LTD
90 Tottenham Court Road, London W1P 0BR
Tel: 0207 457 5040
Fax: 0207 631 5471
Website: www.brma.co.uk

BUILDING RESEARCH ESTABLISHMENT (BRE)
Bucknalls Lane, Garston, Watford, Herts
WD25 9XX
Tel: 01923 664 000
Fax: 01923 664 010
Website: www.bre.co.uk

BRITISH STAINLESS STEEL ASSOCIATION
Broomgrove, 59 Clarkhouse Road, Sheffield
S10 2LE
Tel: 0114 267 1260
Fax: 0114 266 1252
Website: www.bssa.org.uk

BUILDING RESEARCH ESTABLISHMENT:
SCOTLAND (BRE)
Kelvin Rd, East Kilbride, Glasgow G75 0RZ
Tel: 01355 576 200
Fax: 01355 576 210

BRITISH STANDARDS INSTITUTION
389 Chiswick High Road, Chiswick W4 4AL
Tel: 0208 996 9000
Fax: 0208 996 7001
Website: www.bsi-global.com.uk

BUILDING SERVICES RESEARCH and
INFORMATION ASSOCIATION
Old Bracknell Lane West, Bracknell, Berks
RG12 7AH
Tel: 01344 465 600
Fax: 01344 465 626
Website: www.bsria.co.uk

BRITISH WATER
1 Queen Anne's Gate, London SW1 9BT
Tel: 0207 957 4554
Fax: 0207 957 4565
Website: www.britishwater.co.uk

CASTINGS technology international
7 East Bank Road, Sheffield S2 3PT
Tel: 0114 272 8647
Fax: 0114 273 0854
Website: www.castingsdev.com

BRITISH WOOD PRESERVING & DAMP PROOFING
ASSOCIATION
1 Gleneagles House, Vernon, Gate, South Street,
Derby DE1 1UP
Tel: 01332 225 100
Fax: 01332 225 101
Website: www.bwpda.co.uk

CEMENT ADMIXTURES ASSOCIATION
38 Tilehouse, Green Lane, Knowle,
West Midlands B93 9EY
Tel: + Fax: 01564 776 362
Website: www.admixtures.org.uk

CHARTERED INSTITUTE OF ARBITRATORS
International Arbitration Centre
12 Bloomsbury Square, London WC1A 2LP
Tel: 0207 421 7444
Fax: 0207 404 4023
Website: www.arbitrators.org.uk

CONSTRUCTION CONFEDERATION
55 Tufton Street, Westminster, London
SW1P 3QL
Tel: 0870 8989090
Fax: 0207 8989095
Website: www.thecc.org.uk

CHARTERED INSTITUTE OF BUILDING (CIOB)
Englemere, Kings Ride, Ascot, Berks SL5 7TB
Tel: 01344 630 700
Fax: 01344 630 777
Website: www.ciob.org.uk

CONSTRUCTION EMPLOYERS FEDERATION 143
Malone Rd, Belfast BT9 6SU
Tel: 02890 877 143
Fax: 02890 877 155
Website: www.cefni.co.uk

CHARTERED INSTITUTION OF WATER AND
ENVIRONMENTAL MANAGEMENT
15 John Street, London WC1N 2EB
Tel: 0207 831 3110
Fax: 0207 405 4967
Website: www.ciwem.org.uk

CONSTRUCTION INDUSTRY RESEARCH
& INFORMATION ASSOCIATION (CIRIA)
Classic House, 174-180 Old Street,
London EC1V 9BP
Tel: 020 7549 3000
Fax: 020 7253 0523
Website: www.ciria.org.uk

CIVIL ENGINEERING CONTRACTORS
ASSOCIATION
55 Tufton Street, Westminster, London
SW1P 3QL
Tel: 0207 227 4620
Fax: 0207 227 4621
Website: www.ceca.co.uk

CORUS CONSTRUCTION & INDUSTRIAL
PO Box L, Brigg Road, Scunthorpe, North
Lincolnshire, DN16 1BP
Tel: 01724 404040
Fax: 01724 402191
Website: www.corusgroup.com

CLAY PIPE DEVELOPMENT ASSOCIATION
Copsham House, 53 Broad Street, Chesham,
Bucks HP5 3EA
Tel: 01494 791 456
Fax: 01494 792 378
Website: www.cpda.co.uk

DEPARTMENT OF TRADE AND INDUSTRY
1 Victoria Street, London SW1H 0ET
Tel: 0207 215 5000
Fax: 0207 828 3258

COLD ROLLED SECTIONS ASSOCIATIONS
National Metal Forming Centre, 47 Birmingham Road,
West Bromwich, B70 6PY
Tel: 0121 601 6350
Fax: 0121 601 6373
Website: www.crsauk.com

DEPARTMENT FOR TRANSPORT
Ashdown House, 123 Victoria Street, London
SW1E 6DE
Tel: 020 7944 3000
Website: www.dft.gov.uk

CONCRETE PIPELINE SYSTEMS ASSOCIATION
60 Charles St, Leicester LE1 1FB
Tel: 0116 253 6161
Fax: 0116 251 4568
Website: www.concretepipes.co.uk

ELECTRICAL CONTRACTORS ASSOCIATION
(ECA)
ESCA House, 34 Palace Court, Bayswater,
London W2 4HY
Tel: 0207 313 4800
Fax: 0207 221 7344
Website: www.eca.co.uk

CONFEDERATION OF BRITISH INDUSTRY
Centre Point, 103 New Oxford St, London
WC1A 1DU
Tel: 0207 379 7400
Fax: 0207 240 1578
Website: www.cbi.org.uk

ELECTRICAL CONTRACTORS ASSOCIATION OF
SCOTLAND (SELECT)
The Walled Garden, Bush Estate, Midlothian
EH26 0SB
Tel: 0131 445 5577
Fax: 0131 445 5548
Website: www.select.org.uk

FEDERATION OF ENVIRONMENTAL TRADE
ASSOCIATIONS
2 Waltham Court, Milley Lane, Hare Hatch, Reading
RG10 9TH
Tel: 0118 940 3416
Fax: 0118 940 6258
Website: www.feta.co.uk

HEATING AND VENTILATORS CONTRACTORS
ASSOCIATION
ESCA House, 34 Palace Court,
London W2 4JG
Tel: 0207 313 4900
Fax: 0207 727 9268
Website: www.hvca.org.uk

FEDERATION OF MASTER BUILDERS
Gordon Fisher House, 14 – 15 Gt James St,
London WC1N 3DP
Tel: 0207 242 7583
Fax: 0207 404 0296
Website: www.fmb.org.uk

HM LAND REGISTRY (HQ)
32 Lincolns Inn Fields, London WC2A 3PH
Tel: 0207 917 8888
Fax: 0207 955 0110
Website: www.landreg.gov.uk

FEDERATION OF PILING SPECIALISTS
Forum Court, 83 Copers Cope Road, Beckenham,
Kent BR3 1NR
Tel: 0208 663 0947
Fax: 0208 663 0949
Website: www.fps.org.uk

ICOM ENERGY ASSOCIATION
6th Floor, The MacLaren Building, 35 Dale End,
Birmingham BI4 7LN
Tel: 0121 200 2100
Fax: 0121 200 1306

FEDERATION OF PLASTERING AND
DRYWALL CONTRACTORS
The Building Centre, 26 Store Street,
London WC1E 7BT
Tel: 020 7580 3545
Fax: 020 7580 3288
Website: www.fpdc.org.uk

INSTITUTE OF CONCRETE TECHNOLOGY
4 Meadows Business Park, Station Approach,
Blackwater, Camberley GU17 9AB
Tel/Fax: 01276 37831
Website: www.ictech.org

FENCING CONTRACTORS ASSOCIATION
Warren Road, Trelleck, Monmouthshire
NP25 4PQ
Tel: 07000 560 722
Fax: 01600 860 614
Website: www.fencingcontractors.org

INSTITUTE OF MATERIALS, MINERALS & MINING
1 Carlton House Terrace, London SW1Y 5DB
Tel: 0207 451 7300
Fax: 0207 839 1702
Website: www.materials.org.uk

FLAT ROOFING ALLIANCE
Fields House, Gower Road, Haywards Heath, West
Sussex RH16 4PL
Tel: 01444 440 027
Fax: 01444 415 616
Website: www.fra.org.uk

INSTITUTE OF QUALITY ASSURANCE
12 Grosvenor Crescent, London SW1X 7EE
Tel: 0207 245 6722
Fax: 0207 245 6755
Website: www.iqa.org.uk

HEALTH & SAFETY LABORATORY
Business Development Unit Health & Safety
Laboratory, Harpur Hill, Buxton, Derbyshire
SK17 9JN
Tel: 01298 218 218
Fax: 01298 218 822
Website: www.hsl.gov.uk

INSTITUTION OF BRITISH ENGINEERS
Clifford Hill Court, Clifford Chambers,
Stratford Upon-Avon,
Warwickshire CU37 8AA
Tel: 01789 298 739
Fax: 01789 294 442
Website: www.britishengineers.com

INSTITUTION OF CIVIL ENGINEERS
1 Great George St, London SW1P 3AA
Tel: 0207 222 7722
Fax: 0207 222 7500
Website: www.ice.org.uk

THE JOINT CONTRACTS TRIBUNAL
9 Cavendish Place, London W1G 0QD
Tel: 0207 637 8650
Fax: 0207 637 8670
Website: www.jctltd.co.uk

INSTITUTION OF ELECTRICAL ENGINEERS
Savoy Place, London WC2R 0BL
Tel: 0207 240 1871
Fax: 0207 240 7735
Website: www.iee.org.uk

LANDSCAPE INSTITUTE
33 Great Portland Street, London W1W 8QG
Tel: 0207 299 4500
Fax: 0207 299 4501
Website:

INSTITUTION OF INCORPORATED ENGINEERS
Savoy Hill House, Savoy Hill, London WC2R 0BS
Tel: 0207 836 3357
Fax: 0207 497 9006
Website: www.iie.org.uk

LEAD DEVELOPMENT ASSOCIATION
42 Weymouth Street, London W1G 6NP
Tel: 0207 499 8422
Fax: 0207 493 1555
Website: www.ldaint.org

INSTITUTION OF MECHANICAL ENGINEERS
1 Birdcage Walk, London SW1H 9JJ
Tel: 0207 222 7899
Fax: 0207 222 4557
Website: www.imeche.org.uk

MASTIC ASPHALT COUNCIL LTD
PO Box 77, Hastings,
Kent TN35 4WL
Tel: 01424 814 400
Fax: 01424 814 446
Website: www.masticasphaltcouncil.co.uk

INSTITUTION OF STRUCTURAL ENGINEERS
11 Upper Belgrave Street, London SW1X 8BH
Tel: 0207 235 4535
Fax: 0207 235 4294
Website: www.istructe.org.uk

MET OFFICE
Fitzroy Road, Exeter, Devon EX1 3PB
Tel: 0870 900 0100
Fax: 0870 900 5050
Website: www.met-office.gov.uk

INSTITUTION OF WATER AND ENVIRONMENTAL
MANAGEMENT
See CHARTERED INSTITUTION OF WATER AND
ENVIRONMENTAL MANAGEMENT

NATIONAL ASSOCIATION OF STEEL
STOCKHOLDERS
6th Floor, McLaren Building, 35 Dale End, Birmingham
B4 7LN
Tel: 0121 200 2288
Fax: 0121 236 7444
Website: www.nass.org.uk

INSTITUTION OF WASTES MANAGEMENT
9 Saxon Court, St Peter's Gardens, Marefair,
Northampton NN1 1SX
Tel: 01604 620 426
Fax: 01604 621 339
Website: www.ciwm.co.uk

ORDNANCE SURVEY
Romsey Road, Maybush, Southampton
SO16 4GU
Tel: 0845 605 0505
Fax: 02380 792 615
Website: www.ordsvy.gov.uk

INTERNATIONAL CONCRETE BRICK ASSOCIATION
See BRITISH PRECAST CONCRETE FEDERATION

PIPELINE INDUSTRIES GUILD
14/15 Belgrave Square, London SW1X 8PS
Tel: 0207 235 7938
Fax: 0207 235 0074
Website: www.pipeguild.co.uk

INTERPAVE
See BRITISH PRECAST CONCRETE FEDERATION

PLASTIC PIPE MANUFACTURERS SOCIETY
9 Cornwall Street, Birmingham B3 3BY
Tel: 0121 236 1866
Fax: 0121 200 1389

PUBLIC RECORDS OFFICE
(the national archives)
Ruskin Avenue, Q Richmond, Surrey TW9 4DU
Tel: 0208 876 3444
Fax: 0208 392 5286
Website: www.nationalarchives.gov.uk

SCOTTISH ENTERPRISE
5 Atlantic Quay, 150 Broomielaw,
Glasgow G2 8LU
Tel: 0141 248 2700
Fax: 0141 221 3217
Website: www.scottish-enterprise.com

QUARRY PRODUCTS ASSOCIATION
Gillingham House, 38-44 Gillingham Street,
London SW1V 1HU
Tel: 0207 963 8000
Fax: 0207 963 8001
Website: www.qpa.org

SOCIETY OF GLASS TECHNOLOGY
Don Valley House, Saville Street East,
Sheffield S4 7UQ
Tel: 0114 263 4455
Fax: 0114 263 4411
Website: www.societyofglasstechnology.org

REINFORCED CONCRETE COUNCIL
Riverside House, 4 Meadows Business Park, Station
Approach, Camberley GU17 9AB
Tel: 01276 607140
Fax: 01276 607141
Website: www.rcc-info.org.uk

SPECIALISED ACCESS ENGINEERING AND
MAINTENANCE ASSOCIATION
Carthusian Court, 12 Carthusian Street,
London EC1M 6EZ
Tel: 020 7397 8122
Fax: 020 7397 8121
Website: www.saema.org

THE RESIN FLOORING FEDERATION
Association House, 99 West Street,
Farnham, Surrey GU9 7EN
Tel: 01252 739 149
Fax: 01252 739 140
Website: www.ferfa.org.uk

STEEL CONSTRUCTION INSTITUTE
Silwood Park, Ascot, Berks SL5 7QN
Tel: 01344 623 345
Fax: 01344 622 944
Website: www.steel-sci.org

ROYAL INSTITUTE OF BRITISH ARCHITECTS
66 Portland Place, London W1B 1AD
Tel: 0207 580 5533
Fax: 0207 255 1541
Website: www.riba.org.uk

SWIMMING POOL AND ALLIED TRADES
ASSOCIATION (SPATA)
SPATA House, 1A Junction Road, Andover, Hants
SP10 3QT
Tel: 01264 356 210
Fax: 01264 332 628
Website: www.spata.co.uk

ROYAL INSTITUTION OF CHARTERED
SURVEYORS (RICS)
12 Great George Street, Parliament Square London
SW1P 3AD
Tel: 0870 333 1600
Fax: 0207 222 9430
Website: www.rics.org.uk

THERMAL INSULATION CONTRACTORS
ASSOCIATION
TICA House, Allington Way, Yarm Road Business
Park, Darlington, County Durham DL1 4QB
Tel: 01325 466 704
Fax: 01325 487 691
Website: www.tica-acad.co.uk

SCOTTISH BUILDING EMPLOYERS FEDERATION
Carron Grange, Carrongrange Avenue,
Stenhousemuir, FK5 3BQ
Tel: 01324 555 550
Fax: 01324 555 551
Website: www.scottish-building.co.uk

THERMAL INSULATION MANUFACTURERS
AND SUPPLIERS ASSOCIATION
Association House, 99 West Street, Farnham,
Surrey GU9 7EN
Tel: 01252 739 154
Fax: 01252 739 140
Website: www.timsa.org.uk

TIMBER TRADE FEDERATION LTD
Clareville House, 26-27 Oxenden Street,
London SW1Y 4EL
Tel: 0207 839 1891
Fax: 0207 930 0094
Website: www.ttf.co.uk

TWI - THE WELDING INSTITUTE
Granta Park, Great Abington, Cambridge
CB1 6AL
Tel: 01123 891 162
Fax: 01223 892 588
Website: www.twi.co.uk

TOWN AND COUNTRY PLANNING
ASSOCIATION
17 Carlton House Terrace, London SW1Y 5AS
Tel: 0207 930 8903
Fax: 0207 930 3280
Website: www.tcpa.org.uk

UK STEEL ASSOCIATION
Broadway House, Tothill Street, London
SW1H 9NQ
Tel: 0207 222 7777
Fax: 0207 222 2782
Website: www.uksteel.org.uk

TRADA TECHNOLOGY LTD
Stocking Lane, Hughenden Valley,
High Wycombe, Bucks HP14 4ND
Tel: 01494 569 602
Fax: 01494 565 487
Website: www.trada.co.uk

Handbook of Road Technology

4th Edition

M. Lay

"Widely considered to be the single most comprehensive and informative road technology reference text." – *Highway Engineering in Australia*

"A fully revised edition..." – *Transport and Road Update*

"This handbook has its origins in Max Lay's work over many decades in Australia, but reflects a breadth of knowledge far beyond that continent. In that sense it is perhaps the most international in its coverage of practice of such works."
– *Mike Chrimes, Librarian, Institution of Civil Engineers, MyICE newsletter*

This fully revised fourth edition of Max Lay's well-established reference work covers all aspects of the technology of roads and road transport, and urban and rural road technology. It forms a comprehensive but accessible reference for all professionals and students interested in roads, road transport and the wide range of disciplines involved with roads.

International in scope, it begins with the preliminary construction procedures; from road planning policies and design considerations to the selection of materials and the building of roads and bridges. It then explores road operating environments that include driver behaviour, traffic flow, lighting and maintenance, and assesses the cost, economics, transport implications and environmental impact of road use.

June 2009: 944pp
Hb: 978-0-415-47265-4: **£150.00**

To Order: Tel: +44 (0) 1235 400524 **Fax:** +44 (0) 1235 400525
or Post: Taylor and Francis Customer Services,
Bookpoint Ltd, Unit T1, 200 Milton Park, Abingdon, Oxon, OX14 4TA UK
Email: book.orders@tandf.co.uk

For a complete listing of all our titles visit:
www.tandf.co.uk

PART 8

Tables and Memoranda

This part of the book contains the following sections:

Reynolds's Reinforced Concrete Designer Handbook

11th Edition

A. Threlfall et al.

This classic and essential work has been thoroughly revised and updated in line with the requirements of new codes and standards which have been introduced in recent years, including the new Eurocode as well as up-to-date British Standards.

It provides a general introduction along with details of analysis and design of a wide range of structures and examination of design according to British and then European Codes.

Highly illustrated with numerous line diagrams, tables and worked examples, *Reynolds's Reinforced Concrete Designer's Handbook* is a unique resource providing comprehensive guidance that enables the engineer to analyze and design reinforced concrete buildings, bridges, retaining walls, and containment structures.

Written for structural engineers, contractors, consulting engineers, local and health authorities, and utilities, this is also excellent for civil and architecture departments in universities and FE colleges.

July 2007: 416pp
Hb: 978-0-419-25820-9: **£99.99**
Pb: 978-0-419-25830-8: **£41.99**

To Order: Tel: +44 (0) 1235 400524 **Fax:** +44 (0) 1235 400525
or Post: Taylor and Francis Customer Services,
Bookpoint Ltd, Unit T1, 200 Milton Park, Abingdon, Oxon, OX14 4TA UK
Email: book.orders@tandf.co.uk

For a complete listing of all our titles visit:
www.tandf.co.uk

CONVERSION TABLES

CONVERSION TABLES

Length	Unit	Conversion factors			
Millimetre	mm	1 in	= 25.4 mm	1 mm	= 0.0394 in
Centimetre	cm	1 in	= 2.54 cm	1 cm	= 0.3937 in
Metre	m	1 ft	= 0.3048 m	1 m	= 3.2808 ft
		1 yd	= 0.9144 m		= 1.0936 yd
Kilometre	km	1 mile	= 1.6093 km	1km	= 0.6214 mile

Note: 1 cm = 10 mm   1 ft = 12 in
         1 m = 1 000 mm   1 yd = 3 ft
         1 km = 1 000 m   1 mile = 1 760 yd

Area	Unit	Conversion factors			
Square Millimetre	mm^2	$1\ in^2$	$= 645.2\ mm^2$	$1\ mm^2$	$= 0.0016\ in^2$
Square Centimetre	cm^2	$1\ in^2$	$= 6.4516\ cm^2$	$1\ cm^2$	$= 1.1550\ in^2$
Square Metre	m^2	$1\ ft^2$	$= 0.0929\ m^2$	$1\ m^2$	$= 10.764\ ft^2$
		$1\ yd^2$	$= 0.8361\ m^2$	$1\ m^2$	$= 1.1960\ yd^2$
Square Kilometre	km^2	$1\ mile^2$	$= 2.590\ km^2$	$1\ km^2$	$= 0.3861\ mile^2$

Note: $1\ cm^2$ = $100\ mm^2$ $1\ ft^2$ = $144\ in^2$
         $1\ m^2$ = $10\ 000\ cm^2$ $1\ yd^2$ = $9\ ft^2$
         $1\ km^2$ = 100 hectares   1 acre = $4\ 840\ yd^2$
                                $1\ mile^2$ = 640 acres

Volume	Unit	Conversion factors			
Cubic Centimetre	cm^3	$1\ cm^3$	$= 0.0610\ in^3$	$1\ in^3$	$= 16.387\ cm^3$
Cubic Decimetre	dm^3	$1\ dm^3$	$= 0.0353\ ft^3$	$1\ ft^3$	$= 28.329\ dm^3$
Cubic Metre	m^3	$1\ m^3$	$= 35.3147\ ft^3$	$1\ ft^3$	$= 0.0283\ m^3$
		$1\ m^3$	$= 1.3080\ yd^3$	$1\ yd^3$	$= 0.7646\ m^3$
Litre	l	1 l	= 1.76 pint	1 pint	= 0.5683 l
			= 2.113 US pt		= 0.4733 US l

Note: $1\ dm^3$ = $1\ 000\ cm^3$ $1\ ft^3$ = $1\ 728\ in^3$ 1 pint = 20 fl oz
         $1\ m^3$ = $1\ 000\ dm^3$ $1\ yd^3$ = $27\ ft^3$ 1 gal = 8 pints
         1 l = $1\ dm^3$

Neither the Centimetre nor Decimetre are SI units, and as such their use, particularly that of the Decimetre, is not widespread outside educational circles.

Mass	Unit	Conversion factors			
Milligram	mg	1 mg	= 0.0154 grain	1 grain	= 64.935 mg
Gram	g	1 g	= 0.0353 oz	1 oz	= 28.35 g
Kilogram	kg	1 kg	= 2.2046 lb	1 lb	= 0.4536 kg
Tonne	t	1 t	= 0.9842 ton	1 ton	= 1.016 t

Note: 1 g = 1000 mg   1 oz = 437.5 grains   1 cwt = 112 lb
         1 kg = 1000 g   1 lb = 16 oz   1 ton = 20 cwt
         1 t = 1000 kg   1 stone = 14 lb

Force	Unit	Conversion factors			
Newton	N	1 lbf	= 4.448 N	1 kgf	= 9.807 N
Kilonewton	kN	1 lbf	= 0.004448 kN	1 ton f	= 9.964 kN
Meganewton	MN	100 tonf	= 0.9964 MN		

Tables and Memoranda

CONVERSION TABLES

Pressure and stress	Unit	Conversion factors
Kilonewton per square metre	kN/m^2	1 lbf/in^2 = 6.895 kN/m^2
		1 bar = 100 kN/m^2
Meganewton per square metre	MN/m^2	1 tonf/ft^2 = 107.3 kN/m^2 = 0.1073 MN/m^2
		1 kgf/cm^2 = 98.07 kN/m^2
		1 lbf/ft^2 = 0.04788 kN/m^2

Coefficient of consolidation (Cv) or swelling	Unit	Conversion factors
Square metre per year	m^2/year	1 cm^2/s = 3 154 m^2/year
		1 ft^2/year = 0.0929 m^2/year

Coefficient of permeability	Unit	Conversion factors
Metre per second	m/s	1 cm/s = 0.01 m/s
Metre per year	m/year	1 ft/year = 0.3048 m/year
		= 0.9651 × (10)^{8}m/s

Temperature	Unit	Conversion factors	
Degree Celsius	°C	°C = 5/9 × (°F − 32)	°F = (9 × °C)/ 5 + 32

Power	Unit	Conversion factors
Kilowatt	kW	1 kW = 1.341 HP
Horsepower	HP	1 HP = 0.746 kW

CONVERSION TABLES

SPEED CONVERSION

km/h	m/min	mph	fpm
1	16.7	0.6	54.7
2	33.3	1.2	109.4
3	50.0	1.9	164.0
4	66.7	2.5	218.7
5	83.3	3.1	273.4
6	100.0	3.7	328.1
7	116.7	4.3	382.8
8	133.3	5.0	437.4
9	150.0	5.6	492.1
10	166.7	6.2	546.8
11	183.3	6.8	601.5
12	200.0	7.5	656.2
13	216.7	8.1	710.8
14	233.3	8.7	765.5
15	250.0	9.3	820.2
16	266.7	9.9	874.9
17	283.3	10.6	929.6
18	300.0	11.2	984.3
19	316.7	11.8	1038.9
20	333.3	12.4	1093.6
21	350.0	13.0	1148.3
22	366.7	13.7	1203.0
23	383.3	14.3	1257.7
24	400.0	14.9	1312.3
25	416.7	15.5	1367.0
26	433.3	16.2	1421.7
27	450.0	16.8	1476.4
28	466.7	17.4	1531.1
29	483.3	18.0	1585.7
30	500.0	18.6	1640.4
31	516.7	19.3	1695.1
32	533.3	19.9	1749.8
33	550.0	20.5	1804.5
34	566.7	21.1	1859.1
35	583.3	21.7	1913.8
36	600.0	22.4	1968.5
37	616.7	23.0	2023.2
38	633.3	23.6	2077.9
39	650.0	24.2	2132.5
40	666.7	24.9	2187.2
41	683.3	25.5	2241.9
42	700.0	26.1	2296.6
43	716.7	26.7	2351.3
44	733.3	27.3	2405.9
45	750.0	28.0	2460.6

Tables and Memoranda

CONVERSION TABLES

km/h	m/min	mph	fpm
46	766.7	28.6	2515.3
47	783.3	29.2	2570.0
48	800.0	29.8	2624.7
49	816.7	30.4	2679.4
50	833.3	31.1	2734.0

GEOMETRY

Two dimensional figures

Figure	Diagram of figure	Surface area	Perimeter
Square		a^2	$4a$
Rectangle		ab	$2(a+b)$
Triangle		$\frac{1}{2}ch$	$a+b+c$
Circle		πr^2 $\frac{1}{4}\pi d^2$ where $2r = d$	$2\pi r$ πd
Parallelogram		ah	$2(a+b)$
Trapezium		$\frac{1}{2}h(a+b)$	$a+b+c+d$
Ellipse		Approximately πab	$\pi(a+b)$
Hexagon		$2.6 \times a^2$	

GEOMETRY

Figure	Diagram of figure	Surface area	Perimeter
Octagon		$4.83 \times a^2$	6a
Sector of a circle		$\frac{1}{2}rb$ or $\frac{q}{360}\pi r^2$ note b = angle $\frac{q}{360} \times \pi 2r$	
Segment of a circle		S − T where S = area of sector, T = area of triangle	
Bellmouth		$\frac{3}{14} \times r^2$	

GEOMETRY

Three dimensional figures

Figure	Diagram of figure	Surface area	Volume
Cube		$6a^2$	a^3
Cuboid/ rectangular block		$2(ab + ac + bc)$	abc
Prism/ triangular block		$bd + hc + dc + ad$	$\frac{1}{2}hcd$
Cylinder		$2\pi r^2 + 2\pi h$	$\pi r^2 h$ $\frac{1}{4}\pi d^2 h$
Sphere		$4\pi r^2$	$\frac{4}{3}\pi r^3$
Segment of sphere		$2\pi Rh$	$\frac{1}{6}\pi h(3r^2 + h^2)$ $\frac{1}{3}\pi h^2(3R - H)$
Pyramid		$(a + b)l + ab$	$\frac{1}{3}abh$

GEOMETRY

Figure	Diagram of figure	Surface area	Volume
Frustum of a pyramid		$l(a + b + c + d) + \sqrt{(ab + cd)}$ [rectangular figure only]	$\frac{h}{3}(ab + cd + \sqrt{abcd})$
Cone		πrl (excluding base) $\pi rl + \pi r^2$ (including base)	$\frac{1}{3}\pi r^2 h$ $\frac{1}{12}\pi d^2 h$
Frustum of a cone		$\pi r^2 + \pi R^2 + \pi l(R + r)$	$\frac{1}{3}\pi(R^2 + Rr + r^2)$

FORMULAE

FORMULAE

Formula	Description
Pythagoras theorem	$A^2 = B^2 + C^2$ where A is the hypotenuse of a right-angled triangle and B and C are the two adjacent sides
Simpsons Rule	The Area is divided into an even number of strips of equal width, and therefore has an odd number of ordinates at the division points $$\text{area} = \frac{S(A + 2B + 4C)}{3}$$ where S = common interval (strip width) A = sum of first and last ordinates B = sum of remaining odd ordinates C = sum of the even ordinates The Volume can be calculated by the same formula, but by substituting the area of each coordinate rather than its length
Trapezoidal Rule	A given trench is divided into two equal sections, giving three ordinates, the first, the middle and the last $$\text{volume} = \frac{S \times (A + B + 2C)}{2}$$ where S = width of the strips A = area of the first section B = area of the last section C = area of the rest of the sections
Prismoidal Rule	A given trench is divided into two equal sections, giving three ordinates, the first, the middle and the last $$\text{volume} = \frac{L \times (A + 4B + C)}{6}$$ where L = total length of trench A = area of the first section B = area of the middle section C = area of the last section

EARTHWORK

Weights of Typical Materials Handled by Excavators

The weight of the material is that of the state in its natural bed and includes moisture
Adjustments should be made to allow for loose or compacted states

Material	kg/m³	lb/cu yd
Adobe	1914	3230
Ashes	610	1030
Asphalt, rock	2400	4050
Basalt	2933	4950
Bauxite: alum ore	2619	4420
Borax	1730	2920
Caliche	1440	2430
Carnotite	2459	4150
Cement	1600	2700
Chalk (hard)	2406	4060
Cinders	759	1280
Clay: dry	1908	3220
Clay: wet	1985	3350
Coal: bituminous	1351	2280
Coke	510	860
Conglomerate	2204	3720
Dolomite	2886	4870
Earth: dry	1796	3030
Earth: moist	1997	3370
Earth: wet	1742	2940
Feldspar	2613	4410
Felsite	2495	4210
Fluorite	3093	5220
Gabbro	3093	5220
Gneiss	2696	4550
Granite	2690	4540
Gravel, dry	1790	3020
Gypsum	2418	4080
Hardcore (consolidated)	1928	120
Lignite broken	1244	2100
Limestone	2596	4380
Magnesite, magnesium ore	2993	5050
Marble	2679	4520
Marl	2216	3740
Peat	700	1180
Potash	2193	3700
Pumice	640	1080
Quarry waste	1438	90
Quartz	2584	4360
Rhyolite	2400	4050

EARTHWORK

Material	kg/m³	lb/cu yd
Sand: dry	1707	2880
Sand: wet	1831	3090
Sand and gravel – dry	1790	3020
– wet	2092	3530
Sandstone	2412	4070
Schist	2684	4530
Shale	2637	4450
Slag (blast)	2868	4840
Slate	2667	4500
Snow – dry	130	220
– wet	510	860
Taconite	3182	5370
Topsoil	1440	2430
Trachyte	2400	4050
Traprock	2791	4710
Water	1000	62

Transport Capacities

Type of vehicle	Capacity of vehicle	
	Payload	Heaped capacity
Wheelbarrow	150	0.10
1 tonne dumper	1250	1.00
2.5 tonne dumper	4000	2.50
Articulated dump truck (Volvo A20 6 × 4)	18500	11.00
Articulated dump truck (Volvo A35 6 × 6)	32000	19.00
Large capacity rear dumper (Euclid R35)	35000	22.00
Large capacity rear dumper (Euclid R85)	85000	50.00

EARTHWORK

Machine Volumes for Excavating and Filling

Machine type	Cycles per minute	Volume per minute (m^3)
1.5 tonne excavator	1	0.04
	2	0.08
	3	0.12
3 tonne excavator	1	0.13
	2	0.26
	3	0.39
5 tonne excavator	1	0.28
	2	0.56
	3	0.84
7 tonne excavator	1	0.28
	2	0.56
	3	0.84
21 tonne excavator	1	1.21
	2	2.42
	3	3.63
Backhoe loader JCB3CX excavator Rear bucket capacity 0.28 m^3	1	0.28
	2	0.56
	3	0.84
Backhoe loader JCB3CX loading Front bucket capacity 1.00 m^3	1	1.00
	2	2.00

Machine Volumes for Excavating and Filling

Machine type	Loads per hour	Volume per hour (m^3)
1 tonne high tip skip loader Volume 0.485 m^3	5	2.43
	7	3.40
	10	4.85
3 tonne dumper Max volume 2.40 m^3 Available volume 1.9 m^3	4	7.60
	5	9.50
	7	13.30
	10	19.00
6 tonne dumper Max volume 3.40 m^3 Available volume 3.77 m^3	4	15.08
	5	18.85
	7	26.39
	10	37.70

EARTHWORK

Bulkage of Soils (after excavation)

Type of soil	Approximate bulking of 1 m³ after excavation
Vegetable soil and loam	25–30%
Soft clay	30–40%
Stiff clay	10–20%
Gravel	20–25%
Sand	40–50%
Chalk	40–50%
Rock, weathered	30–40%
Rock, unweathered	50–60%

Shrinkage of Materials (on being deposited)

Type of soil	Approximate bulking of 1 m³ after excavation
Clay	10%
Gravel	8%
Gravel and sand	9%
Loam and light sandy soils	12%
Loose vegetable soils	15%

Voids in Material Used as Subbases or Beddings

Material	m³ of voids/m³
Alluvium	0.37
River grit	0.29
Quarry sand	0.24
Shingle	0.37
Gravel	0.39
Broken stone	0.45
Broken bricks	0.42

Angles of Repose

Type of soil		Degrees
Clay	– dry	30
	– damp, well drained	45
	– wet	15–20
Earth	– dry	30
	– damp	45
Gravel	– moist	48
Sand	– dry or moist	35
	– wet	25
Loam		40

EARTHWORK

Slopes and Angles

Ratio of base to height	Angle in degrees
5:1	11
4:1	14
3:1	18
2:1	27
1½:1	34
1:1	45
1:1½	56
1:2	63
1:3	72
1:4	76
1:5	79

Grades (in degrees and percents)

Degrees	Percent	Degrees	Percent
1	1.8	24	44.5
2	3.5	25	46.6
3	5.2	26	48.8
4	7.0	27	51.0
5	8.8	28	53.2
6	10.5	29	55.4
7	12.3	30	57.7
8	14.0	31	60.0
9	15.8	32	62.5
10	17.6	33	64.9
11	19.4	34	67.4
12	21.3	35	70.0
13	23.1	36	72.7
14	24.9	37	75.4
15	26.8	38	78.1
16	28.7	39	81.0
17	30.6	40	83.9
18	32.5	41	86.9
19	34.4	42	90.0
20	36.4	43	93.3
21	38.4	44	96.6
22	40.4	45	100.0

EARTHWORK

Bearing Powers

Ground conditions		Bearing power		
		kg/m^2	lb/in^2	Metric t/m^2
Rock,	broken	483	70	50
	solid	2415	350	240
Clay,	dry or hard	380	55	40
	medium dry	190	27	20
	soft or wet	100	14	10
Gravel,	cemented	760	110	80
Sand,	compacted	380	55	40
	clean dry	190	27	20
Swamp and alluvial soils		48	7	5

Earthwork Support

Maximum depth of excavation in various soils without the use of earthwork support

Ground conditions	Feet (ft)	Metres (m)
Compact soil	12	3.66
Drained loam	6	1.83
Dry sand	1	0.3
Gravelly earth	2	0.61
Ordinary earth	3	0.91
Stiff clay	10	3.05

It is important to note that the above table should only be used as a guide. Each case must be taken on its merits and, as the limited distances given above are approached, careful watch must be kept for the slightest signs of caving in

CONCRETE WORK

Weights of Concrete and Concrete Elements

Type of material		kg/m³	lb/cu ft
Ordinary concrete (dense aggregates)			
Non-reinforced plain or mass concrete			
Nominal weight		2305	144
Aggregate	– limestone	2162 to 2407	135 to 150
	– gravel	2244 to 2407	140 to 150
	– broken brick	2000 (av)	125 (av)
	– other crushed stone	2326 to 2489	145 to 155
Reinforced concrete			
Nominal weight		2407	150
Reinforcement	– 1%	2305 to 2468	144 to 154
	– 2%	2356 to 2519	147 to 157
	– 4%	2448 to 2703	153 to 163
Special concretes			
Heavy concrete			
Aggregates	– barytes, magnetite	3210 (min)	200 (min)
	– steel shot, punchings	5280	330
Lean mixes			
Dry-lean (gravel aggregate)		2244	140
Soil-cement (normal mix)		1601	100

CONCRETE WORK

Type of material		kg/m² per mm thick	lb/sq ft per inch thick
Ordinary concrete (dense aggregates)			
Solid slabs (floors, walls etc.)			
Thickness:	75 mm or 3 in	184	37.5
	100 mm or 4 in	245	50
	150 mm or 6 in	378	75
	250 mm or 10 in	612	125
	300 mm or 12 in	734	150
Ribbed slabs			
Thickness:	125 mm or 5 in	204	42
	150 mm or 6 in	219	45
	225 mm or 9 in	281	57
	300 mm or 12 in	342	70
Special concretes			
Finishes etc.			
	Rendering, screed etc. Granolithic, terrazzo	1928 to 2401	10 to 12.5
	Glass-block (hollow) concrete	1734 (approx)	9 (approx)
Prestressed concrete		Weights as for reinforced concrete (upper limits)	
Air-entrained concrete		Weights as for plain or reinforced concrete	

CONCRETE WORK

Average Weight of Aggregates

Materials	Voids %	Weight kg/m³
Sand	39	1660
Gravel 10–20 mm	45	1440
Gravel 35–75 mm	42	1555
Crushed stone	50	1330
Crushed granite (over 15 mm)	50	1345
(n.e. 15 mm)	47	1440
'All-in' ballast	32	1800–2000

Material	kg/m³	lb/cu yd
Vermiculite (aggregate)	64–80	108–135
All-in aggregate	1999	125

Applications and Mix Design

Site mixed concrete

Recommended mix	Class of work suitable for	Cement (kg)	Sand (kg)	Coarse aggregate (kg)	Nr 25 kg bags cement per m³ of combined aggregate
1:3:6	Roughest type of mass concrete such as footings, road haunching over 300 mm thick	208	905	1509	8.30
1:2.5:5	Mass concrete of better class than 1:3:6 such as bases for machinery, walls below ground etc.	249	881	1474	10.00
1:2:4	Most ordinary uses of concrete, such as mass walls above ground, road slabs etc. and general reinforced concrete work	304	889	1431	12.20
1:1.5:3	Watertight floors, pavements and walls, tanks, pits, steps, paths, surface of 2 course roads, reinforced concrete where extra strength is required	371	801	1336	14.90
1:1:2	Works of thin section such as fence posts and small precast work	511	720	1206	20.40

CONCRETE WORK

Ready mixed concrete

Application	Designated concrete	Standardized prescribed concrete	Recommended consistence (nominal slump class)
Foundations			
Mass concrete fill or blinding	GEN 1	ST2	S3
Strip footings	GEN 1	ST2	S3
Mass concrete foundations			
Single storey buildings	GEN 1	ST2	S3
Double storey buildings	GEN 3	ST4	S3
Trench fill foundations			
Single storey buildings	GEN 1	ST2	S4
Double storey buildings	GEN 3	ST4	S4
General applications			
Kerb bedding and haunching	GEN 0	ST1	S1
Drainage works – immediate support	GEN 1	ST2	S1
Other drainage works	GEN 1	ST2	S3
Oversite below suspended slabs	GEN 1	ST2	S3
Floors			
Garage and house floors with no embedded steel	GEN 3	ST4	S2
Wearing surface: Light foot and trolley traffic	RC30	ST4	S2
Wearing surface: General industrial	RC40	N/A	S2
Wearing surface: Heavy industrial	RC50	N/A	S2
Paving			
House drives, domestic parking and external parking	PAV 1	N/A	S2
Heavy-duty external paving	PAV 2	N/A	S2

CONCRETE WORK

Prescribed Mixes for Ordinary Structural Concrete

Weights of cement and total dry aggregates in kg to produce approximately one cubic metre of fully compacted concrete together with the percentages by weight of fine aggregate in total dry aggregates

Conc. grade	Nominal max size of aggregate (mm)	40		20		14		10	
	Workability	**Med.**	**High**	**Med.**	**High**	**Med.**	**High**	**Med.**	**High**
	Limits to slump that may be expected (mm)	**50–100**	**100–150**	**25–75**	**75–125**	**10–50**	**50–100**	**10–25**	**25–50**
7	Cement (kg)	180	200	210	230	–	–	–	–
	Total aggregate (kg)	1950	1850	1900	1800	–	–	–	–
	Fine aggregate (%)	30–45	30–45	35–50	35–50	–	–	–	–
10	Cement (kg)	210	230	240	260	–	–	–	–
	Total aggregate (kg)	1900	1850	1850	1800	–	–	–	–
	Fine aggregate (%)	30–45	30–45	35–50	35–50	–	–	–	–
15	Cement (kg)	250	270	280	310	–	–	–	–
	Total aggregate (kg)	1850	1800	1800	1750	–	–	–	–
	Fine aggregate (%)	30–45	30–45	35–50	35–50	–	–	–	–
20	Cement (kg)	300	320	320	350	340	380	360	410
	Total aggregate (kg)	1850	1750	1800	1750	1750	1700	1750	1650
	Sand								
	Zone 1 (%)	35	40	40	45	45	50	50	55
	Zone 2 (%)	30	35	35	40	40	45	45	50
	Zone 3 (%)	30	30	30	35	35	40	40	45
25	Cement (kg)	340	360	360	390	380	420	400	450
	Total aggregate (kg)	1800	1750	1750	1700	1700	1650	1700	1600
	Sand								
	Zone 1 (%)	35	40	40	45	45	50	50	55
	Zone 2 (%)	30	35	35	40	40	45	45	50
	Zone 3 (%)	30	30	30	35	35	40	40	45
30	Cement (kg)	370	390	400	430	430	470	460	510
	Total aggregate (kg)	1750	1700	1700	1650	1700	1600	1650	1550
	Sand								
	Zone 1 (%)	35	40	40	45	45	50	50	55
	Zone 2 (%)	30	35	35	40	40	45	45	50
	Zone 3 (%)	30	30	30	35	35	40	40	45

REINFORCEMENT

Weights of Bar Reinforcement

Nominal sizes (mm)	Cross-sectional area (mm²)	Mass (kg/m)	Length of bar (m/tonne)
6	28.27	0.222	4505
8	50.27	0.395	2534
10	78.54	0.617	1622
12	113.10	0.888	1126
16	201.06	1.578	634
20	314.16	2.466	405
25	490.87	3.853	260
32	804.25	6.313	158
40	1265.64	9.865	101
50	1963.50	15.413	65

Weights of Bars (at specific spacings)

Weights of metric bars in kilogrammes per square metre

Size (mm)	Spacing of bars in millimetres									
	75	100	125	150	175	200	225	250	275	300
6	2.96	2.220	1.776	1.480	1.27	1.110	0.99	0.89	0.81	0.74
8	5.26	3.95	3.16	2.63	2.26	1.97	1.75	1.58	1.44	1.32
10	8.22	6.17	4.93	4.11	3.52	3.08	2.74	2.47	2.24	2.06
12	11.84	8.88	7.10	5.92	5.07	4.44	3.95	3.55	3.23	2.96
16	21.04	15.78	12.63	10.52	9.02	7.89	7.02	6.31	5.74	5.26
20	32.88	24.66	19.73	16.44	14.09	12.33	10.96	9.87	8.97	8.22
25	51.38	38.53	30.83	25.69	22.02	19.27	17.13	15.41	14.01	12.84
32	84.18	63.13	50.51	42.09	36.08	31.57	28.06	25.25	22.96	21.04
40	131.53	98.65	78.92	65.76	56.37	49.32	43.84	39.46	35.87	32.88
50	205.51	154.13	123.31	102.76	88.08	77.07	68.50	61.65	56.05	51.38

Basic weight of steelwork taken as 7850 kg/m³
Basic weight of bar reinforcement per metre run = 0.00785 kg/mm²
The value of π has been taken as 3.141592654

REINFORCEMENT

Fabric Reinforcement

Preferred range of designated fabric types and stock sheet sizes

Fabric reference	Longitudinal wires			Cross wires			
	Nominal wire size (mm)	Pitch (mm)	Area (mm²/m)	Nominal wire size (mm)	Pitch (mm)	Area (mm²/m)	Mass (kg/m²)
Square mesh							
A393	10	200	393	10	200	393	6.16
A252	8	200	252	8	200	252	3.95
A193	7	200	193	7	200	193	3.02
A142	6	200	142	6	200	142	2.22
A98	5	200	98	5	200	98	1.54
Structural mesh							
B1131	12	100	1131	8	200	252	10.90
B785	10	100	785	8	200	252	8.14
B503	8	100	503	8	200	252	5.93
B385	7	100	385	7	200	193	4.53
B283	6	100	283	7	200	193	3.73
B196	5	100	196	7	200	193	3.05
Long mesh							
C785	10	100	785	6	400	70.8	6.72
C636	9	100	636	6	400	70.8	5.55
C503	8	100	503	5	400	49.0	4.34
C385	7	100	385	5	400	49.0	3.41
C283	6	100	283	5	400	49.0	2.61
Wrapping mesh							
D98	5	200	98	5	200	98	1.54
D49	2.5	100	49	2.5	100	49	0.77

Stock sheet size 4.8 m × 2.4 m, Area 11.52 m²

Average weight kg/m³ of steelwork reinforcement in concrete for various building elements

Substructure	kg/m³ concrete	Substructure	kg/m³ concrete
Pile caps	110–150	Plate slab	150–220
Tie beams	130–170	Cant slab	145–210
Ground beams	230–330	Ribbed floors	130–200
Bases	125–180	Topping to block floor	30–40
Footings	100–150	Columns	210–310
Retaining walls	150–210	Beams	250–350
Raft	60–70	Stairs	130–170
Slabs – one way	120–200	Walls – normal	40–100
Slabs – two way	110–220	Walls – wind	70–125

Note: For exposed elements add the following %:
Walls 50%, Beams 100%, Columns 15%

FORMWORK

Formwork Stripping Times – Normal Curing Periods

Conditions under which concrete is maturing	Minimum periods of protection for different types of cement					
	Number of days (where the average surface temperature of the concrete exceeds 10°C during the whole period)			Equivalent maturity (degree hours) calculated as the age of the concrete in hours multiplied by the number of degrees Celsius by which the average surface temperature of the concrete exceeds 10°C		
	Other	SRPC	OPC or RHPC	Other	SRPC	OPC or RHPC
1. Hot weather or drying winds	7	4	3	3500	2000	1500
2. Conditions not covered by 1	4	3	2	2000	1500	1000

KEY
OPC – Ordinary Portland Cement
RHPC – Rapid-hardening Portland Cement
SRPC – Sulphate-resisting Portland Cement

Minimum Period before Striking Formwork

	Minimum period before striking		
	Surface temperature of concrete		
	16°C	17°C	t°C (0–25)
Vertical formwork to columns, walls and large beams	12 hours	18 hours	300 hours t+10
Soffit formwork to slabs	4 days	6 days	100 days t+10
Props to slabs	10 days	15 days	250 days t+10
Soffit formwork to beams	9 days	14 days	230 days t+10
Props to beams	14 days	21 days	360 days t+10

MASONRY

Number of Bricks required for Various Types of Work per m² of Walling

Description	Brick size	
	215 × 102.5 × 50 mm	215 × 102.5 × 65 mm
Half brick thick		
Stretcher bond	74	59
English bond	108	86
English garden wall bond	90	72
Flemish bond	96	79
Flemish garden wall bond	83	66
One brick thick and cavity wall of two half brick skins		
Stretcher bond	148	119

Quantities of Bricks and Mortar required per m² of Walling

	Unit	No of bricks required	Mortar required (cubic metres)		
Standard bricks			**No frogs**	**Single frogs**	**Double frogs**
Brick size 215 × 102.5 × 50 mm					
half brick wall (103 mm)	m²	72	0.022	0.027	0.032
2 × half brick cavity wall (270 mm)	m²	144	0.044	0.054	0.064
one brick wall (215 mm)	m²	144	0.052	0.064	0.076
one and a half brick wall (322 mm)	m²	216	0.073	0.091	0.108
Mass brickwork	m³	576	0.347	0.413	0.480
Brick size 215 × 102.5 × 65 mm					
half brick wall (103 mm)	m²	58	0.019	0.022	0.026
2 × half brick cavity wall (270 mm)	m²	116	0.038	0.045	0.055
one brick wall (215 mm)	m²	116	0.046	0.055	0.064
one and a half brick wall (322 mm)	m²	174	0.063	0.074	0.088
Mass brickwork	m³	464	0.307	0.360	0.413
Metric modular bricks			**Perforated**		
Brick size 200 × 100 × 75 mm					
90 mm thick	m²	67	0.016	0.019	
190 mm thick	m²	133	0.042	0.048	
290 mm thick	m²	200	0.068	0.078	
Brick size 200 × 100 × 100 mm					
90 mm thick	m²	50	0.013	0.016	
190 mm thick	m²	100	0.036	0.041	
290 mm thick	m²	150	0.059	0.067	
Brick size 300 × 100 × 75 mm					
90 mm thick	m²	33	–	0.015	
Brick size 300 × 100 × 100 mm					
90 mm thick	m²	44	0.015	0.018	

Note: Assuming 10 mm thick joints

MASONRY

Mortar required per m² Blockwork (9.88 blocks/m²)

Wall thickness	75	90	100	125	140	190	215
Mortar m³/m²	0.005	0.006	0.007	0.008	0.009	0.013	0.014

Mortar Group	Cement: lime: sand	Masonry cement: sand	Cement: sand with plasticizer
1	1:0–0.25:3		
2	1:0.5:4–4.5	1:2.5-3.5	1:3–4
3	1:1:5–6	1:4–5	1:5–6
4	1:2:8–9	1:5.5–6.5	1:7–8
5	1:3:10–12	1:6.5–7	1:8

Group 1: strong inflexible mortar
Group 5: weak but flexible

All mixes within a group are of approximately similar strength
Frost resistance increases with the use of plasticizers
Cement: lime: sand mixes give the strongest bond and greatest resistance to rain penetration
Masonry cement equals ordinary Portland cement plus a fine neutral mineral filler and an air entraining agent

Calcium Silicate Bricks

Type	Strength	Location
Class 2 crushing strength	14.0 N/mm²	not suitable for walls
Class 3	20.5 N/mm²	walls above dpc
Class 4	27.5 N/mm²	cappings and copings
Class 5	34.5 N/mm²	retaining walls
Class 6	41.5 N/mm²	walls below ground
Class 7	48.5 N/mm²	walls below ground

The Class 7 calcium silicate bricks are therefore equal in strength to Class B bricks
Calcium silicate bricks are not suitable for DPCs

Durability of Bricks	
FL	Frost resistant with low salt content
FN	Frost resistant with normal salt content
ML	Moderately frost resistant with low salt content
MN	Moderately frost resistant with normal salt content

Tables and Memoranda

MASONRY

Brickwork Dimensions

No. of horizontal bricks	Dimensions (mm)	No. of vertical courses	Height of vertical courses (mm)
½	112.5	1	75
1	225.0	2	150
1½	337.5	3	225
2	450.0	4	300
2½	562.5	5	375
3	675.0	6	450
3½	787.5	7	525
4	900.0	8	600
4½	1012.5	9	675
5	1125.0	10	750
5½	1237.5	11	825
6	1350.0	12	900
6½	1462.5	13	975
7	1575.0	14	1050
7½	1687.5	15	1125
8	1800.0	16	1200
8½	1912.5	17	1275
9	2025.0	18	1350
9½	2137.5	19	1425
10	2250.0	20	1500
20	4500.0	24	1575
40	9000.0	28	2100
50	11250.0	32	2400
60	13500.0	36	2700
75	16875.0	40	3000

TIMBER

Weights of Timber

Material	kg/m^3	lb/cu ft
General	806 (avg)	50 (avg)
Douglas fir	479	30
Yellow pine, spruce	479	30
Pitch pine	673	42
Larch, elm	561	35
Oak (English)	724 to 959	45 to 60
Teak	643 to 877	40 to 55
Jarrah	959	60
Greenheart	1040 to 1204	65 to 75
Quebracho	1285	80
Material	**kg/m^2 per mm thickness**	**lb/sq ft per inch thickness**
Wooden boarding and blocks		
Softwood	0.48	2.5
Hardwood	0.76	4
Hardboard	1.06	5.5
Chipboard	0.76	4
Plywood	0.62	3.25
Blockboard	0.48	2.5
Fibreboard	0.29	1.5
Wood-wool	0.58	3
Plasterboard	0.96	5
Weather boarding	0.35	1.8

TIMBER

Conversion Tables (for timber only)

Inches	Millimetres	Feet	Metres
1	25	1	0.300
2	50	2	0.600
3	75	3	0.900
4	100	4	1.200
5	125	5	1.500
6	150	6	1.800
7	175	7	2.100
8	200	8	2.400
9	225	9	2.700
10	250	10	3.000
11	275	11	3.300
12	300	12	3.600
13	325	13	3.900
14	350	14	4.200
15	375	15	4.500
16	400	16	4.800
17	425	17	5.100
18	450	18	5.400
19	475	19	5.700
20	500	20	6.000
21	525	21	6.300
22	550	22	6.600
23	575	23	6.900
24	600	24	7.200

Planed Softwood
The finished end section size of planed timber is usually 3/16" less than the original size from which it is produced. This however varies slightly depending upon availability of material and origin of the species used.

Standards (timber) to cubic metres and cubic metres to standards (timber)

Cubic metres	Cubic metres standards	Standards
4.672	1	0.214
9.344	2	0.428
14.017	3	0.642
18.689	4	0.856
23.361	5	1.070
28.033	6	1.284
32.706	7	1.498
37.378	8	1.712
42.050	9	1.926
46.722	10	2.140
93.445	20	4.281
140.167	30	6.421
186.890	40	8.561
233.612	50	10.702
280.335	60	12.842
327.057	70	14.982
373.779	80	17.122

TIMBER

1 cu metre = 35.3148 cu ft = 0.21403 std

1 cu ft = 0.028317 cu metres

1 std = 4.67227 cu metres

Basic sizes of sawn softwood available (cross-sectional areas)

Thickness (mm)	Width (mm)								
	75	100	125	150	175	200	225	250	300
16	X	X	X	X					
19	X	X	X	X					
22	X	X	X	X					
25	X	X	X	X	X	X	X	X	X
32	X	X	X	X	X	X	X	X	X
36	X	X	X	X					
38	X	X	X	X	X	X	X		
44	X	X	X	X	X	X	X	X	X
47*	X	X	X	X	X	X	X	X	X
50	X	X	X	X	X	X	X	X	X
63	X	X	X	X	X	X	X		
75	X	X	X	X	X	X	X	X	
100		X		X		X		X	X
150				X		X			X
200						X			
250								X	
300									X

* This range of widths for 47 mm thickness will usually be found to be available in construction quality only

Note: The smaller sizes below 100 mm thick and 250 mm width are normally but not exclusively of European origin. Sizes beyond this are usually of North and South American origin

Basic lengths of sawn softwood available (metres)

1.80	2.10	3.00	4.20	5.10	6.00	7.20
	2.40	3.30	4.50	5.40	6.30	
	2.70	3.60	4.80	5.70	6.60	
		3.90			6.90	

Note: Lengths of 6.00 m and over will generally only be available from North American species and may have to be recut from larger sizes

TIMBER

Reductions from basic size to finished size by planning of two opposed faces

Purpose	Reductions from basic sizes for timber			
	15–35 mm	36–100 mm	101–150 mm	over 150 mm
a) Constructional timber	3 mm	3 mm	5 mm	6 mm
b) Matching interlocking boards	4 mm	4 mm	6 mm	6 mm
c) Wood trim not specified in BS 584	5 mm	7 mm	7 mm	9 mm
d) Joinery and cabinet work	7 mm	9 mm	11 mm	13 mm

Note: The reduction of width or depth is overall the extreme size and is exclusive of any reduction of the face by the machining of a tongue or lap joints

Maximum Spans for Various Roof Trusses

Maximum permissible spans for rafters for Fink trussed rafters

Basic size (mm)	Actual size (mm)	Pitch (degrees)								
		15 (m)	17.5 (m)	20 (m)	22.5 (m)	25 (m)	27.5 (m)	30 (m)	32.5 (m)	35 (m)
38 × 75	35 × 72	6.03	6.16	6.29	6.41	6.51	6.60	6.70	6.80	6.90
38 × 100	35 × 97	7.48	7.67	7.83	7.97	8.10	8.22	8.34	8.47	8.61
38 × 125	35 × 120	8.80	9.00	9.20	9.37	9.54	9.68	9.82	9.98	10.16
44 × 75	41 × 72	6.45	6.59	6.71	6.83	6.93	7.03	7.14	7.24	7.35
44 × 100	41 × 97	8.05	8.23	8.40	8.55	8.68	8.81	8.93	9.09	9.22
44 × 125	41 × 120	9.38	9.60	9.81	9.99	10.15	10.31	10.45	10.64	10.81
50 × 75	47 × 72	6.87	7.01	7.13	7.25	7.35	7.45	7.53	7.67	7.78
50 × 100	47 × 97	8.62	8.80	8.97	9.12	9.25	9.38	9.50	9.66	9.80
50 × 125	47 × 120	10.01	10.24	10.44	10.62	10.77	10.94	11.00	11.00	11.00

TIMBER

Sizes of Internal and External Doorsets

Description	Internal size (mm)	Permissible deviation	External size (mm)	Permissible deviation
Coordinating dimension: height of door leaf height sets	2100		2100	
Coordinating dimension: height of ceiling height set	2300 2350 2400 2700 3000		2300 2350 2400 2700 3000	
Coordinating dimension: width of all door sets S = Single leaf set D = Double leaf set	600 S 700 S 800 S&D 900 S&D 1000 S&D 1200 D 1500 D 1800 D 2100 D		900 S 1000 S 1200 D 1800 D 2100 D	
Work size: height of door leaf height set	2090	± 2.0	2095	± 2.0
Work size: height of ceiling height set	2285 2335 2385 2685 2985	± 2.0	2295 2345 2395 2695 2995	± 2.0
Work size: width of all door sets S = Single leaf set D = Double leaf set	590 S 690 S 790 S&D 890 S&D 990 S&D 1190 D 1490 D 1790 D 2090 D	± 2.0	895 S 995 S 1195 D 1495 D 1795 D 2095 D	± 2.0
Width of door leaf in single leaf sets F = Flush leaf P = Panel leaf	526 F 626 F 726 F&P 826 F&P 926 F&P	± 1.5	806 F&P 906 F&P	± 1.5
Width of door leaf in double leaf sets F = Flush leaf P = Panel leaf	362 F 412 F 426 F 562 F&P 712 F&P 826 F&P 1012 F&P	± 1.5	552 F&P 702 F&P 852 F&P 1002 F&P	± 1.5
Door leaf height for all door sets	2040	± 1.5	1994	± 1.5

ROOFING

Total Roof Loadings for Various Types of Tiles/Slates

	Roof load (slope) kg/m²		
	Slate/Tile	Roofing underlay and battens²	Total dead load kg/m
Asbestos cement slate (600 × 300)	21.50	3.14	24.64
Clay tile interlocking	67.00	5.50	72.50
plain	43.50	2.87	46.37
Concrete tile interlocking	47.20	2.69	49.89
plain	78.20	5.50	83.70
Natural slate (18" × 10")	35.40	3.40	38.80
	Roof load (plan) kg/m²		
Asbestos cement slate (600 × 300)	28.45	76.50	104.95
Clay tile interlocking	53.54	76.50	130.04
plain	83.71	76.50	60.21
Concrete tile interlocking	57.60	76.50	134.10
plain	96.64	76.50	173.14

ROOFING

Tiling Data

Product		Lap (mm)	Gauge of battens	No. slates per m²	Battens (m/m²)	Weight as laid (kg/m²)
CEMENT SLATES						
Eternit slates	600 × 300 mm	100	250	13.4	4.00	19.50
(Duracem)		90	255	13.1	3.92	19.20
		80	260	12.9	3.85	19.00
		70	265	12.7	3.77	18.60
	600 × 350 mm	100	250	11.5	4.00	19.50
		90	255	11.2	3.92	19.20
	500 × 250 mm	100	200	20.0	5.00	20.00
		90	205	19.5	4.88	19.50
		80	210	19.1	4.76	19.00
		70	215	18.6	4.65	18.60
	400 × 200 mm	90	155	32.3	6.45	20.80
		80	160	31.3	6.25	20.20
		70	165	30.3	6.06	19.60
CONCRETE TILES/SLATES						
Redland Roofing						
Stonewold slate	430 × 380 mm	75	355	8.2	2.82	51.20
Double Roman tile	418 × 330 mm	75	355	8.2	2.91	45.50
Grovebury pantile	418 × 332 mm	75	343	9.7	2.91	47.90
Norfolk pantile	381 × 227 mm	75	306	16.3	3.26	44.01
		100	281	17.8	3.56	48.06
Renown interlocking tile	418 × 330 mm	75	343	9.7	2.91	46.40
'49' tile	381 × 227 mm	75	306	16.3	3.26	44.80
		100	281	17.8	3.56	48.95
Plain, vertical tiling	265 × 165 mm	35	115	52.7	8.70	62.20
Marley Roofing						
Bold roll tile	420 × 330 mm	75	344	9.7	2.90	47.00
		100	–	10.5	3.20	51.00
Modern roof tile	420 × 330 mm	75	338	10.2	3.00	54.00
		100	–	11.0	3.20	58.00
Ludlow major	420 × 330 mm	75	338	10.2	3.00	45.00
		100	–	11.0	3.20	49.00
Ludlow plus	387 × 229 mm	75	305	16.1	3.30	47.00
		100	–	17.5	3.60	51.00
Mendip tile	420 × 330 mm	75	338	10.2	3.00	47.00
		100	–	11.0	3.20	51.00
Wessex	413 × 330 mm	75	338	10.2	3.00	54.00
		100	–	11.0	3.20	58.00
Plain tile	267 × 165 mm	65	100	60.0	10.00	76.00
		75	95	64.0	10.50	81.00
		85	90	68.0	11.30	86.00
Plain vertical tiles (feature)	267 × 165 mm	35	110	53.0	8.70	67.00
		34	115	56.0	9.10	71.00

ROOFING

Slate Nails, Quantity per Kilogram

Length	Type			
	Plain wire	Galvanized wire	Copper nail	Zinc nail
28.5 mm	325	305	325	415
34.4 mm	286	256	254	292
50.8 mm	242	224	194	200

Metal Sheet Coverings

Thicknesses and weights of sheet metal coverings								
Lead to BS 1178								
BS Code No	3	4	5	6	7	8		
Colour code	Green	Blue	Red	Black	White	Orange		
Thickness (mm)	1.25	1.80	2.24	2.50	3.15	3.55		
Density (kg/m²)	14.18	20.41	25.40	30.05	35.72	40.26		
Copper to BS 2870								
Thickness (mm)		0.60	0.70					
Bay width								
Roll (mm)		500	650					
Seam (mm)		525	600					
Standard width to form bay	600	750						
Normal length of sheet	1.80	1.80						
Zinc to BS 849								
Zinc Gauge (Nr)	9	10	11	12	13	14	15	16
Thickness (mm)	0.43	0.48	0.56	0.64	0.71	0.79	0.91	1.04
Density (kg/m²)	3.1	3.2	3.8	4.3	4.8	5.3	6.2	7.0
Aluminium to BS 4868								
Thickness (mm)	0.5	0.6	0.7	0.8	0.9	1.0	1.2	
Density (kg/m²)	12.8	15.4	17.9	20.5	23.0	25.6	30.7	

ROOFING

Type of felt	Nominal mass per unit area (kg/10m)	Nominal mass per unit area of fibre base (g/m²)	Nominal length of roll (m)
Class 1			
1B fine granule	14	220	10 or 20
surfaced bitumen	18	330	10 or 20
	25	470	10
1E mineral surfaced bitumen	38	470	10
1F reinforced bitumen	15	160 (fibre) 110 (hessian)	15
1F reinforced bitumen, aluminium faced	13	160 (fibre) 110 (hessian)	15
Class 2			
2B fine granule surfaced bitumen asbestos	18	500	10 or 20
2E mineral surfaced bitumen asbestos	38	600	10
Class 3			
3B fine granule surfaced bitumen glass fibre	18	60	20
3E mineral surfaced bitumen glass fibre	28	60	10
3E venting base layer bitumen glass fibre	32	60*	10
3H venting base layer bitumen glass fibre	17	60*	20

* Excluding effect of perforations

GLAZING

GLAZING

Nominal thickness (mm)	Tolerance on thickness (mm)	Approximate weight (kg/m²)	Normal maximum size (mm)
Float and polished plate glass			
3	+ 0.2	7.50	2140 × 1220
4	+ 0.2	10.00	2760 × 1220
5	+ 0.2	12.50	3180 × 2100
6	+ 0.2	15.00	4600 × 3180
10	+ 0.3	25.00)	6000 × 3300
12	+ 0.3	30.00)	
15	+ 0.5	37.50	3050 × 3000
19	+ 1.0	47.50)	3000 × 2900
25	+ 1.0	63.50)	
Clear sheet glass			
2 *	+ 0.2	5.00	1920 × 1220
3	+ 0.3	7.50	2130 × 1320
4	+ 0.3	10.00	2760 × 1220
5 *	+ 0.3	12.50)	2130 × 2400
6 *	+ 0.3	15.00)	
Cast glass			
3	+ 0.4		
	− 0.2	6.00)	2140 × 1280
4	+ 0.5	7.50)	
5	+ 0.5	9.50	2140 × 1320
6	+ 0.5	11.50)	3700 × 1280
10	+ 0.8	21.50)	
Wired glass			
(Cast wired glass)			
6	+ 0.3	−)	
	− 0.7	)	3700 × 1840
7	+ 0.7	−)	
(Polished wire glass)			
6	+ 1.0	−	330 × 1830

* The 5 mm and 6 mm thickness are known as *thick drawn sheet*. Although 2 mm sheet glass is available it is not recommended for general glazing purposes

METAL

METAL

Weights of Metals

Material	kg/m^3	lb/cu ft
Metals, steel construction, etc.		
Iron		
– cast	7207	450
– wrought	7687	480
– ore – general	2407	150
– (crushed) Swedish	3682	230
Steel	7854	490
Copper		
– cast	8731	545
– wrought	8945	558
Brass	8497	530
Bronze	8945	558
Aluminium	2774	173
Lead	11322	707
Zinc (rolled)	7140	446

	g/mm^2 per metre	lb/sq ft per foot
Steel bars	7.85	3.4

Structural steelwork	Net weight of member @ 7854 kg/m^3	
riveted	+ 10% for cleats, rivets, bolts, etc.	
welded	+ 1.25% to 2.5% for welds, etc.	
Rolled sections		
beams	+ 2.5%	
stanchions	+ 5% (extra for caps and bases)	
Plate		
web girders	+ 10% for rivets or welds, stiffeners, etc.	

	kg/m	lb/ft
Steel stairs: industrial type		
1 m or 3 ft wide	84	56
Steel tubes		
50 mm or 2 in bore	5 to 6	3 to 4
Gas piping		
20 mm or ¾ in	2	1¼

METAL

Universal Beams BS 4: Part 1: 2005

Designation	Mass (kg/m)	Depth of section (mm)	Width of section (mm)	Thickness		Surface area (m²/m)
				Web (mm)	Flange (mm)	
1016 × 305 × 487	487.0	1036.1	308.5	30.0	54.1	3.20
1016 × 305 × 438	438.0	1025.9	305.4	26.9	49.0	3.17
1016 × 305 × 393	393.0	1016.0	303.0	24.4	43.9	3.15
1016 × 305 × 349	349.0	1008.1	302.0	21.1	40.0	3.13
1016 × 305 × 314	314.0	1000.0	300.0	19.1	35.9	3.11
1016 × 305 × 272	272.0	990.1	300.0	16.5	31.0	3.10
1016 × 305 × 249	249.0	980.2	300.0	16.5	26.0	3.08
1016 × 305 × 222	222.0	970.3	300.0	16.0	21.1	3.06
914 × 419 × 388	388.0	921.0	420.5	21.4	36.6	3.44
914 × 419 × 343	343.3	911.8	418.5	19.4	32.0	3.42
914 × 305 × 289	289.1	926.6	307.7	19.5	32.0	3.01
914 × 305 × 253	253.4	918.4	305.5	17.3	27.9	2.99
914 × 305 × 224	224.2	910.4	304.1	15.9	23.9	2.97
914 × 305 × 201	200.9	903.0	303.3	15.1	20.2	2.96
838 × 292 × 226	226.5	850.9	293.8	16.1	26.8	2.81
838 × 292 × 194	193.8	840.7	292.4	14.7	21.7	2.79
838 × 292 × 176	175.9	834.9	291.7	14.0	18.8	2.78
762 × 267 × 197	196.8	769.8	268.0	15.6	25.4	2.55
762 × 267 × 173	173.0	762.2	266.7	14.3	21.6	2.53
762 × 267 × 147	146.9	754.0	265.2	12.8	17.5	2.51
762 × 267 × 134	133.9	750.0	264.4	12.0	15.5	2.51
686 × 254 × 170	170.2	692.9	255.8	14.5	23.7	2.35
686 × 254 × 152	152.4	687.5	254.5	13.2	21.0	2.34
686 × 254 × 140	140.1	383.5	253.7	12.4	19.0	2.33
686 × 254 × 125	125.2	677.9	253.0	11.7	16.2	2.32
610 × 305 × 238	238.1	635.8	311.4	18.4	31.4	2.45
610 × 305 × 179	179.0	620.2	307.1	14.1	23.6	2.41
610 × 305 × 149	149.1	612.4	304.8	11.8	19.7	2.39
610 × 229 × 140	139.9	617.2	230.2	13.1	22.1	2.11
610 × 229 × 125	125.1	612.2	229.0	11.9	19.6	2.09
610 × 229 × 113	113.0	607.6	228.2	11.1	17.3	2.08
610 × 229 × 101	101.2	602.6	227.6	10.5	14.8	2.07
533 × 210 × 122	122.0	544.5	211.9	12.7	21.3	1.89
533 × 210 × 109	109.0	539.5	210.8	11.6	18.8	1.88
533 × 210 × 101	101.0	536.7	210.0	10.8	17.4	1.87
533 × 210 × 92	92.1	533.1	209.3	10.1	15.6	1.86
533 × 210 × 82	82.2	528.3	208.8	9.6	13.2	1.85
457 × 191 × 98	98.3	467.2	192.8	11.4	19.6	1.67
457 × 191 × 89	89.3	463.4	191.9	10.5	17.7	1.66
457 × 191 × 82	82.0	460.0	191.3	9.9	16.0	1.65
457 × 191 × 74	74.3	457.0	190.4	9.0	14.5	1.64
457 × 191 × 67	67.1	453.4	189.9	8.5	12.7	1.63
457 × 152 × 82	82.1	465.8	155.3	10.5	18.9	1.51
457 × 152 × 74	74.2	462.0	154.4	9.6	17.0	1.50
457 × 152 × 67	67.2	458.0	153.8	9.0	15.0	1.50
457 × 152 × 60	59.8	454.6	152.9	8.1	13.3	1.50
457 × 152 × 52	52.3	449.8	152.4	7.6	10.9	1.48
406 × 178 × 74	74.2	412.8	179.5	9.5	16.0	1.51
406 × 178 × 67	67.1	409.4	178.8	8.8	14.3	1.50
406 × 178 × 60	60.1	406.4	177.9	7.9	12.8	1.49

METAL

Designation	Mass (kg/m)	Depth of section (mm)	Width of section (mm)	Thickness		Surface area (m²/m)
				Web (mm)	Flange (mm)	
406 × 178 × 50	54.1	402.6	177.7	7.7	10.9	1.48
406 × 140 × 46	46.0	403.2	142.2	6.8	11.2	1.34
406 × 140 × 39	39.0	398.0	141.8	6.4	8.6	1.33
356 × 171 × 67	67.1	363.4	173.2	9.1	15.7	1.38
356 × 171 × 57	57.0	358.0	172.2	8.1	13.0	1.37
356 × 171 × 51	51.0	355.0	171.5	7.4	11.5	1.36
356 × 171 × 45	45.0	351.4	171.1	7.0	9.7	1.36
356 × 127 × 39	39.1	353.4	126.0	6.6	10.7	1.18
356 × 127 × 33	33.1	349.0	125.4	6.0	8.5	1.17
305 × 165 × 54	54.0	310.4	166.9	7.9	13.7	1.26
305 × 165 × 46	46.1	306.6	165.7	6.7	11.8	1.25
305 × 165 × 40	40.3	303.4	165.0	6.0	10.2	1.24
305 × 127 × 48	48.1	311.0	125.3	9.0	14.0	1.09
305 × 127 × 42	41.9	307.2	124.3	8.0	12.1	1.08
305 × 127 × 37	37.0	304.4	123.3	7.1	10.7	1.07
305 × 102 × 33	32.8	312.7	102.4	6.6	10.8	1.01
305 × 102 × 28	28.2	308.7	101.8	6.0	8.8	1.00
305 × 102 × 25	24.8	305.1	101.6	5.8	7.0	0.992
254 × 146 × 43	43.0	259.6	147.3	7.2	12.7	1.08
254 × 146 × 37	37.0	256.0	146.4	6.3	10.9	1.07
254 × 146 × 31	31.1	251.4	146.1	6.0	8.6	1.06
254 × 102 × 28	28.3	260.4	102.2	6.3	10.0	0.904
254 × 102 × 25	25.2	257.2	101.9	6.0	8.4	0.897
254 × 102 × 22	22.0	254.0	101.6	5.7	6.8	0.890
203 × 133 × 30	30.0	206.8	133.9	6.4	9.6	0.923
203 × 133 × 25	25.1	203.2	133.2	5.7	7.8	0.915
203 × 102 × 23	23.1	203.2	101.8	5.4	9.3	0.790
178 × 102 × 19	19.0	177.8	101.2	4.8	7.9	0.738
152 × 89 × 16	16.0	152.4	88.7	4.5	7.7	0.638
127 × 76 × 13	13.0	127.0	76.0	4.0	7.6	0.537

METAL

Universal Columns BS 4: Part 1: 2005

Designation	Mass (kg/m)	Depth of section (mm)	Width of section (mm)	Thickness Web (mm)	Flange (mm)	Surface area (m²/m)
356 × 406 × 634	633.9	474.7	424.0	47.6	77.0	2.52
356 × 406 × 551	551.0	455.6	418.5	42.1	67.5	2.47
356 × 406 × 467	467.0	436.6	412.2	35.8	58.0	2.42
356 × 406 × 393	393.0	419.0	407.0	30.6	49.2	2.38
356 × 406 × 340	339.9	406.4	403.0	26.6	42.9	2.35
356 × 406 × 287	287.1	393.6	399.0	22.6	36.5	2.31
356 × 406 × 235	235.1	381.0	384.8	18.4	30.2	2.28
356 × 368 × 202	201.9	374.6	374.7	16.5	27.0	2.19
356 × 368 × 177	177.0	368.2	372.6	14.4	23.8	2.17
356 × 368 × 153	152.9	362.0	370.5	12.3	20.7	2.16
356 × 368 × 129	129.0	355.6	368.6	10.4	17.5	2.14
305 × 305 × 283	282.9	365.3	322.2	26.8	44.1	1.94
305 × 305 × 240	240.0	352.5	318.4	23.0	37.7	1.91
305 × 305 × 198	198.1	339.9	314.5	19.1	31.4	1.87
305 × 305 × 158	158.1	327.1	311.2	15.8	25.0	1.84
305 × 305 × 137	136.9	320.5	309.2	13.8	21.7	1.82
305 × 305 × 118	117.9	314.5	307.4	12.0	18.7	1.81
305 × 305 × 97	96.9	307.9	305.3	9.9	15.4	1.79
254 × 254 × 167	167.1	289.1	265.2	19.2	31.7	1.58
254 × 254 × 132	132.0	276.3	261.3	15.3	25.3	1.55
254 × 254 × 107	107.1	266.7	258.8	12.8	20.5	1.52
254 × 254 × 89	88.9	260.3	256.3	10.3	17.3	1.50
254 × 254 × 73	73.1	254.1	254.6	8.6	14.2	1.49
203 × 203 × 86	86.1	222.2	209.1	12.7	20.5	1.24
203 × 203 × 71	71.0	215.8	206.4	10.0	17.3	1.22
203 × 203 × 60	60.0	209.6	205.8	9.4	14.2	1.21
203 × 203 × 52	52.0	206.2	204.3	7.9	12.5	1.20
203 × 203 × 46	46.1	203.2	203.6	7.2	11.0	1.19
152 × 152 × 37	37.0	161.8	154.4	8.0	11.5	0.912
152 × 152 × 30	30.0	157.6	152.9	6.5	9.4	0.901
152 × 152 × 23	23.0	152.4	152.2	5.8	6.8	0.889

METAL

Joists BS 4: Part 1: 2005 (retained for reference, Corus have ceased manufacture in UK)

Designation	Mass	Depth of section	Width of section	Thickness		Surface area
	(kg/m)	(mm)	(mm)	Web (mm)	Flange (mm)	(m²/m)
254 × 203 × 82	82.0	254.0	203.2	10.2	19.9	1.210
203 × 152 × 52	52.3	203.2	152.4	8.9	16.5	0.932
152 × 127 × 37	37.3	152.4	127.0	10.4	13.2	0.737
127 × 114 × 29	29.3	127.0	114.3	10.2	11.5	0.646
127 × 114 × 27	26.9	127.0	114.3	7.4	11.4	0.650
102 × 102 × 23	23.0	101.6	101.6	9.5	10.3	0.549
102 × 44 × 7	7.5	101.6	44.5	4.3	6.1	0.350
89 × 89 × 19	19.5	88.9	88.9	9.5	9.9	0.476
76 × 76 × 13	12.8	76.2	76.2	5.1	8.4	0.411

Parallel Flange Channels

Designation	Mass	Depth of section	Width of section	Thickness		Surface area
	(kg/m)	(mm)	(mm)	Web (mm)	Flange (mm)	(m²/m)
430 × 100 × 64	64.4	430	100	11.0	19.0	1.23
380 × 100 × 54	54.0	380	100	9.5	17.5	1.13
300 × 100 × 46	45.5	300	100	9.0	16.5	0.969
300 × 90 × 41	41.4	300	90	9.0	15.5	0.932
260 × 90 × 35	34.8	260	90	8.0	14.0	0.854
260 × 75 × 28	27.6	260	75	7.0	12.0	0.79
230 × 90 × 32	32.2	230	90	7.5	14.0	0.795
230 × 75 × 26	25.7	230	75	6.5	12.5	0.737
200 × 90 × 30	29.7	200	90	7.0	14.0	0.736
200 × 75 × 23	23.4	200	75	6.0	12.5	0.678
180 × 90 × 26	26.1	180	90	6.5	12.5	0.697
180 × 75 × 20	20.3	180	75	6.0	10.5	0.638
150 × 90 × 24	23.9	150	90	6.5	12.0	0.637
150 × 75 × 18	17.9	150	75	5.5	10.0	0.579
125 × 65 × 15	14.8	125	65	5.5	9.5	0.489
100 × 50 × 10	10.2	100	50	5.0	8.5	0.382

Tables and Memoranda

METAL

Equal Angles BS EN 10056-1

Designation	Mass (kg/m)	Surface area (m²/m)
200 × 200 × 24	71.1	0.790
200 × 200 × 20	59.9	0.790
200 × 200 × 18	54.2	0.790
200 × 200 × 16	48.5	0.790
150 × 150 × 18	40.1	0.59
150 × 150 × 15	33.8	0.59
150 × 150 × 12	27.3	0.59
150 × 150 × 10	23.0	0.59
120 × 120 × 15	26.6	0.47
120 × 120 × 12	21.6	0.47
120 × 120 × 10	18.2	0.47
120 × 120 × 8	14.7	0.47
100 × 100 × 15	21.9	0.39
100 × 100 × 12	17.8	0.39
100 × 100 × 10	15.0	0.39
100 × 100 × 8	12.2	0.39
90 × 90 × 12	15.9	0.35
90 × 90 × 10	13.4	0.35
90 × 90 × 8	10.9	0.35
90 × 90 × 7	9.61	0.35
90 × 90 × 6	8.30	0.35

Unequal Angles BS EN 10056-1

Designation	Mass (kg/m)	Surface area (m²/m)
200 × 150 × 18	47.1	0.69
200 × 150 × 15	39.6	0.69
200 × 150 × 12	32.0	0.69
200 × 100 × 15	33.7	0.59
200 × 100 × 12	27.3	0.59
200 × 100 × 10	23.0	0.59
150 × 90 × 15	26.6	0.47
150 × 90 × 12	21.6	0.47
150 × 90 × 10	18.2	0.47
150 × 75 × 15	24.8	0.44
150 × 75 × 12	20.2	0.44
150 × 75 × 10	17.0	0.44
125 × 75 × 12	17.8	0.40
125 × 75 × 10	15.0	0.40
125 × 75 × 8	12.2	0.40
100 × 75 × 12	15.4	0.34
100 × 75 × 10	13.0	0.34
100 × 75 × 8	10.6	0.34
100 × 65 × 10	12.3	0.32
100 × 65 × 8	9.94	0.32
100 × 65 × 7	8.77	0.32

METAL

Structural Tees Split from Universal Beams BS 4: Part 1: 2005

Designation	Mass (kg/m)	Surface area (m²/m)
305 × 305 × 90	89.5	1.22
305 × 305 × 75	74.6	1.22
254 × 343 × 63	62.6	1.19
229 × 305 × 70	69.9	1.07
229 × 305 × 63	62.5	1.07
229 × 305 × 57	56.5	1.07
229 × 305 × 51	50.6	1.07
210 × 267 × 61	61.0	0.95
210 × 267 × 55	54.5	0.95
210 × 267 × 51	50.5	0.95
210 × 267 × 46	46.1	0.95
210 × 267 × 41	41.1	0.95
191 × 229 × 49	49.2	0.84
191 × 229 × 45	44.6	0.84
191 × 229 × 41	41.0	0.84
191 × 229 × 37	37.1	0.84
191 × 229 × 34	33.6	0.84
152 × 229 × 41	41.0	0.76
152 × 229 × 37	37.1	0.76
152 × 229 × 34	33.6	0.76
152 × 229 × 30	29.9	0.76
152 × 229 × 26	26.2	0.76

Universal Bearing Piles BS 4: Part 1: 2005

Designation	Mass (kg/m)	Depth of Section (mm)	Width of Section (mm)	Thickness	
				Web (mm)	Flange (mm)
356 × 368 × 174	173.9	361.4	378.5	20.3	20.4
356 × 368 × 152	152.0	356.4	376.0	17.8	17.9
356 × 368 × 133	133.0	352.0	373.8	15.6	15.7
356 × 368 × 109	108.9	346.4	371.0	12.8	12.9
305 × 305 × 223	222.9	337.9	325.7	30.3	30.4
305 × 305 × 186	186.0	328.3	320.9	25.5	25.6
305 × 305 × 149	149.1	318.5	316.0	20.6	20.7
305 × 305 × 126	126.1	312.3	312.9	17.5	17.6
305 × 305 × 110	110.0	307.9	310.7	15.3	15.4
305 × 305 × 95	94.9	303.7	308.7	13.3	13.3
305 × 305 × 88	88.0	301.7	307.8	12.4	12.3
305 × 305 × 79	78.9	299.3	306.4	11.0	11.1
254 × 254 × 85	85.1	254.3	260.4	14.4	14.3
254 × 254 × 71	71.0	249.7	258.0	12.0	12.0
254 × 254 × 63	63.0	247.1	256.6	10.6	10.7
203 × 203 × 54	53.9	204.0	207.7	11.3	11.4
203 × 203 × 45	44.9	200.2	205.9	9.5	9.5

Tables and Memoranda

METAL

Hot Formed Square Hollow Sections EN 10210 S275J2H & S355J2H

Size (mm)	Wall thickness (mm)	Mass (kg/m)	Superficial area (m²/m)
40 × 40	2.5	2.89	0.154
	3.0	3.41	0.152
	3.2	3.61	0.152
	3.6	4.01	0.151
	4.0	4.39	0.150
	5.0	5.28	0.147
50 × 50	2.5	3.68	0.194
	3.0	4.35	0.192
	3.2	4.62	0.192
	3.6	5.14	0.191
	4.0	5.64	0.190
	5.0	6.85	0.187
	6.0	7.99	0.185
	6.3	8.31	0.184
60 × 60	3.0	5.29	0.232
	3.2	5.62	0.232
	3.6	6.27	0.231
	4.0	6.90	0.230
	5.0	8.42	0.227
	6.0	9.87	0.225
	6.3	10.30	0.224
	8.0	12.50	0.219
70 × 70	3.0	6.24	0.272
	3.2	6.63	0.272
	3.6	7.40	0.271
	4.0	8.15	0.270
	5.0	9.99	0.267
	6.0	11.80	0.265
	6.3	12.30	0.264
	8.0	15.00	0.259
80 × 80	3.2	7.63	0.312
	3.6	8.53	0.311
	4.0	9.41	0.310
	5.0	11.60	0.307
	6.0	13.60	0.305
	6.3	14.20	0.304
	8.0	17.50	0.299
90 × 90	3.6	9.66	0.351
	4.0	10.70	0.350
	5.0	13.10	0.347
	6.0	15.50	0.345
	6.3	16.20	0.344
	8.0	20.10	0.339
100 × 100	3.6	10.80	0.391
	4.0	11.90	0.390
	5.0	14.70	0.387
	6.0	17.40	0.385
	6.3	18.20	0.384
	8.0	22.60	0.379
	10.0	27.40	0.374
120 × 120	4.0	14.40	0.470

METAL

Size (mm)	Wall thickness (mm)	Mass (kg/m)	Superficial area (m²/m)
	5.0	17.80	0.467
	6.0	21.20	0.465
	6.3	22.20	0.464
	8.0	27.60	0.459
	10.0	33.70	0.454
	12.0	39.50	0.449
	12.5	40.90	0.448
140 × 140	5.0	21.00	0.547
	6.0	24.90	0.545
	6.3	26.10	0.544
	8.0	32.60	0.539
	10.0	40.00	0.534
	12.0	47.00	0.529
	12.5	48.70	0.528
150 × 150	5.0	22.60	0.587
	6.0	26.80	0.585
	6.3	28.10	0.584
	8.0	35.10	0.579
	10.0	43.10	0.574
	12.0	50.80	0.569
	12.5	52.70	0.568
Hot formed from seamless hollow	16.0	65.2	0.559
160 × 160	5.0	24.10	0.627
	6.0	28.70	0.625
	6.3	30.10	0.624
	8.0	37.60	0.619
	10.0	46.30	0.614
	12.0	54.60	0.609
	12.5	56.60	0.608
	16.0	70.20	0.599
180 × 180	5.0	27.30	0.707
	6.0	32.50	0.705
	6.3	34.00	0.704
	8.0	42.70	0.699
	10.0	52.50	0.694
	12.0	62.10	0.689
	12.5	64.40	0.688
	16.0	80.20	0.679
200 × 200	5.0	30.40	0.787
	6.0	36.20	0.785
	6.3	38.00	0.784
	8.0	47.70	0.779
	10.0	58.80	0.774
	12.0	69.60	0.769
	12.5	72.30	0.768
	16.0	90.30	0.759
250 × 250	5.0	38.30	0.987

METAL

Size (mm)	Wall thickness (mm)	Mass (kg/m)	Superficial area (m²/m)
	6.0	45.70	0.985
	6.3	47.90	0.984
	8.0	60.30	0.979
	10.0	74.50	0.974
	12.0	88.50	0.969
	12.5	91.90	0.968
	16.0	115.00	0.959
300 × 300	6.0	55.10	1.18
	6.3	57.80	1.18
	8.0	72.80	1.18
	10.0	90.20	1.17
	12.0	107.00	1.17
	12.5	112.00	1.17
	16.0	141.00	1.16
350 × 350	8.0	85.40	1.38
	10.0	106.00	1.37
	12.0	126.00	1.37
	12.5	131.00	1.37
	16.0	166.00	1.36
400 × 400	8.0	97.90	1.58
	10.0	122.00	1.57
	12.0	145.00	1.57
	12.5	151.00	1.57
	16.0	191.00	1.56
(Grade S355J2H only)	20.00*	235.00	1.55

Note: * SAW process

METAL

Hot Formed Square Hollow Sections JUMBO RHS: JIS G3136

Size (mm)	Wall thickness (mm)	Mass (kg/m)	Superficial area (m²/m)
350 × 350	19.0	190.00	1.33
	22.0	217.00	1.32
	25.0	242.00	1.31
400 × 400	22.0	251.00	1.52
	25.0	282.00	1.51
450 × 450	12.0	162.00	1.76
	16.0	213.00	1.75
	19.0	250.00	1.73
	22.0	286.00	1.72
	25.0	321.00	1.71
	28.0 *	355.00	1.70
	32.0 *	399.00	1.69
500 × 500	12.0	181.00	1.96
	16.0	238.00	1.95
	19.0	280.00	1.93
	22.0	320.00	1.92
	25.0	360.00	1.91
	28.0 *	399.00	1.90
	32.0 *	450.00	1.89
	36.0 *	498.00	1.88
550 × 550	16.0	263.00	2.15
	19.0	309.00	2.13
	22.0	355.00	2.12
	25.0	399.00	2.11
	28.0 *	443.00	2.10
	32.0 *	500.00	2.09
	36.0 *	555.00	2.08
	40.0 *	608.00	2.06
600 × 600	25.0 *	439.00	2.31
	28.0 *	487.00	2.30
	32.0 *	550.00	2.29
	36.0 *	611.00	2.28
	40.0 *	671.00	2.26
700 × 700	25.0 *	517.00	2.71
	28.0 *	575.00	2.70
	32.0 *	651.00	2.69
	36.0 *	724.00	2.68
	40.0 *	797.00	2.68

Note: * SAW process

METAL

Hot Formed Rectangular Hollow Sections: EN10210 S275J2h & S355J2H

Size (mm)	Wall thickness (mm)	Mass (kg/m)	Superficial area (m²/m)
50 × 30	2.5	2.89	0.154
	3.0	3.41	0.152
	3.2	3.61	0.152
	3.6	4.01	0.151
	4.0	4.39	0.150
	5.0	5.28	0.147
60 × 40	2.5	3.68	0.194
	3.0	4.35	0.192
	3.2	4.62	0.192
	3.6	5.14	0.191
	4.0	5.64	0.190
	5.0	6.85	0.187
	6.0	7.99	0.185
	6.3	8.31	0.184
80 × 40	3.0	5.29	0.232
	3.2	5.62	0.232
	3.6	6.27	0.231
	4.0	6.90	0.230
	5.0	8.42	0.227
	6.0	9.87	0.225
	6.3	10.30	0.224
	8.0	12.50	0.219
76.2 × 50.8	3.0	5.62	0.246
	3.2	5.97	0.246
	3.6	6.66	0.245
	4.0	7.34	0.244
	5.0	8.97	0.241
	6.0	10.50	0.239
	6.3	11.00	0.238
	8.0	13.40	0.233
90 × 50	3.0	6.24	0.272
	3.2	6.63	0.272
	3.6	7.40	0.271
	4.0	8.15	0.270
	5.0	9.99	0.267
	6.0	11.80	0.265
	6.3	12.30	0.264
	8.0	15.00	0.259
100 × 50	3.0	6.71	0.292
	3.2	7.13	0.292
	3.6	7.96	0.291
	4.0	8.78	0.290
	5.0	10.80	0.287
	6.0	12.70	0.285
	6.3	13.30	0.284
	8.0	16.30	0.279

METAL

Size (mm)	Wall thickness (mm)	Mass (kg/m)	Superficial area (m²/m)
100 × 60	3.0	7.18	0.312
	3.2	7.63	0.312
	3.6	8.53	0.311
	4.0	9.41	0.310
	5.0	11.60	0.307
	6.0	13.60	0.305
	6.3	14.20	0.304
	8.0	17.50	0.299
120 × 60	3.6	9.70	0.351
	4.0	10.70	0.350
	5.0	13.10	0.347
	6.0	15.50	0.345
	6.3	16.20	0.344
	8.0	20.10	0.339
120 × 80	3.6	10.80	0.391
	4.0	11.90	0.390
	5.0	14.70	0.387
	6.0	17.40	0.385
	6.3	18.20	0.384
	8.0	22.60	0.379
	10.0	27.40	0.374
150 × 100	4.0	15.10	0.490
	5.0	18.60	0.487
	6.0	22.10	0.485
	6.3	23.10	0.484
	8.0	28.90	0.479
	10.0	35.30	0.474
	12.0	41.40	0.469
	12.5	42.80	0.468
160 × 80	4.0	14.40	0.470
	5.0	17.80	0.467
	6.0	21.20	0.465
	6.3	22.20	0.464
	8.0	27.60	0.459
	10.0	33.70	0.454
	12.0	39.50	0.449
	12.5	40.90	0.448
200 × 100	5.0	22.60	0.587
	6.0	26.80	0.585
	6.3	28.10	0.584
	8.0	35.10	0.579
	10.0	43.10	0.574
	12.0	50.80	0.569
	12.5	52.70	0.568
	16.0	65.20	0.559
250 × 150	5.0	30.40	0.787
	6.0	36.20	0.785
	6.3	38.00	0.784
	8.0	47.70	0.779
	10.0	58.80	0.774
	12.0	69.60	0.769
	12.5	72.30	0.768
	16.0	90.30	0.759

METAL

Size (mm)	Wall thickness (mm)	Mass (kg/m)	Superficial area (m²/m)
300 × 200	5.0	38.30	0.987
	6.0	45.70	0.985
	6.3	47.90	0.984
	8.0	60.30	0.979
	10.0	74.50	0.974
	12.0	88.50	0.969
	12.5	91.90	0.968
	16.0	115.00	0.959
400 × 200	6.0	55.10	1.18
	6.3	57.80	1.18
	8.0	72.80	1.18
	10.0	90.20	1.17
	12.0	107.00	1.17
	12.5	112.00	1.17
	16.0	141.00	1.16
450 × 250	8.0	85.40	1.38
	10.0	106.00	1.37
	12.0	126.00	1.37
	12.5	131.00	1.37
	16.0	166.00	1.36
500 × 300	8.0	98.00	1.58
	10.0	122.00	1.57
	12.0	145.00	1.57
	12.5	151.00	1.57
	16.0	191.00	1.56
	20.0	235.00	1.55

METAL

Hot Formed Circular Hollow Sections EN 10210 S275J2H & S355J2H

Outside diameter (mm)	Wall thickness (mm)	Mass (kg/m)	Superficial area (m²/m)
21.3	3.2	1.43	0.067
26.9	3.2	1.87	0.085
33.7	3.0	2.27	0.106
	3.2	2.41	0.106
	3.6	2.67	0.106
	4.0	2.93	0.106
42.4	3.0	2.91	0.133
	3.2	3.09	0.133
	3.6	3.44	0.133
	4.0	3.79	0.133
48.3	2.5	2.82	0.152
	3.0	3.35	0.152
	3.2	3.56	0.152
	3.6	3.97	0.152
	4.0	4.37	0.152
	5.0	5.34	0.152
60.3	2.5	3.56	0.189
	3.0	4.24	0.189
	3.2	4.51	0.189
	3.6	5.03	0.189
	4.0	5.55	0.189
	5.0	6.82	0.189
76.1	2.5	4.54	0.239
	3.0	5.41	0.239
	3.2	5.75	0.239
	3.6	6.44	0.239
	4.0	7.11	0.239
	5.0	8.77	0.239
	6.0	10.40	0.239
	6.3	10.80	0.239
88.9	2.5	5.33	0.279
	3.0	6.36	0.279
	3.2	6.76	0.27
	3.6	7.57	0.279
88.9	4.0	8.38	0.279
	5.0	10.30	0.279
	6.0	12.30	0.279
	6.3	12.80	0.279
114.3	3.0	8.23	0.359
	3.2	8.77	0.359
	3.6	9.83	0.359
	4.0	10.09	0.359
	5.0	13.50	0.359
	6.0	16.00	0.359
	6.3	16.80	0.359

Tables and Memoranda

METAL

Outside diameter (mm)	Wall thickness (mm)	Mass (kg/m)	Superficial area (m²/m)
139.7	3.2	10.80	0.439
	3.6	12.10	0.439
	4.0	13.40	0.439
	5.0	16.60	0.439
	6.0	19.80	0.439
	6.3	20.70	0.439
	8.0	26.00	0.439
	10.0	32.00	0.439
168.3	3.2	13.00	0.529
	3.6	14.60	0.529
	4.0	16.20	0.529
	5.0	20.10	0.529
	6.0	24.00	0.529
	6.3	25.20	0.529
	8.0	31.60	0.529
	10.0	39.00	0.529
	12.0	46.30	0.529
	12.5	48.00	0.529
193.7	5.0	23.30	0.609
	6.0	27.80	0.609
	6.3	29.10	0.609
	8.0	36.60	0.609
	10.0	45.30	0.609
193.7	12.0	53.80	0.609
	12.5	55.90	0.609
219.1	5.0	26.40	0.688
	6.0	31.50	0.688
	6.3	33.10	0.688
	8.0	41.60	0.688
	10.0	51.60	0.688
	12.0	61.30	0.688
	12.5	63.70	0.688
	16.0	80.10	0.688
244.5	5.0	29.50	0.768
	6.0	35.30	0.768
	6.3	37.00	0.768
	8.0	46.70	0.768
	10.0	57.80	0.768
	12.0	68.80	0.768
	12.5	71.50	0.768
	16.0	90.20	0.768
273.0	5.0	33.00	0.858
	6.0	39.50	0.858
	6.3	41.40	0.858
	8.0	52.30	0.858
	10.0	64.90	0.858
	12.0	77.20	0.858
	12.5	80.30	0.858
	16.0	101.00	0.858

METAL

Outside diameter (mm)	Wall thickness (mm)	Mass (kg/m)	Superficial area (m²/m)
323.9	5.0	39.30	1.02
	6.0	47.00	1.02
	6.3	49.30	1.02
	8.0	62.30	1.02
	10.0	77.40	1.02
	12.0	92.30	1.02
	12.5	96.00	1.02
	16.0	121.00	1.02
355.6	6.3	54.30	1.12
	8.0	68.60	1.12
	10.0	85.30	1.12
	12.0	102.00	1.12
	12.5	106.00	1.12
	16.0	134.00	1.12
406.4	6.3	62.20	1.28
	8.0	79.60	1.28
	10.0	97.80	1.28
	12.0	117.00	1.28
	12.5	121.00	1.28
	16.0	154.00	1.28
457.0	6.3	70.00	1.44
	8.0	88.60	1.44
	10.0	110.00	1.44
	12.0	132.00	1.44
	12.5	137.00	1.44
	16.0	174.00	1.44
508.0	6.3	77.90	1.60
	8.0	98.60	1.60
	10.0	123.00	1.60
	12.0	147.00	1.60
	12.5	153.00	1.60
	16.0	194.00	1.60

METAL

Spacing of Holes in Angles

Nominal leg length (mm)	Spacing of holes						Maximum diameter of bolt or rivet		
	A	B	C	D	E	F	A	B and C	D, E and F
200		75	75	55	55	55		30	20
150		55	55					20	
125		45	60					20	
120									
100	55						24		
90	50						24		
80	45						20		
75	45						20		
70	40						20		
65	35						20		
60	35						16		
50	28						12		
45	25								
40	23								
30	20								
25	15								

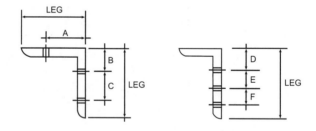

KERBS, PAVING, ETC.

KERBS/EDGINGS/CHANNELS

Precast Concrete Kerbs to BS 7263

Straight kerb units: length from 450 to 915 mm

150 mm high × 125 mm thick		
bullnosed	type BN	
half battered	type HB3	
255 mm high × 125 mm thick		
45° splayed	type SP	
half battered	type HB2	
305 mm high × 150 mm thick		
half battered	type HB1	
Quadrant kerb units		
150 mm high × 305 and 455 mm radius to match	type BN	type QBN
150 mm high × 305 and 455 mm radius to match	type HB2, HB3	type QHB
150 mm high × 305 and 455 mm radius to match	type SP	type QSP
255 mm high × 305 and 455 mm radius to match	type BN	type QBN
255 mm high × 305 and 455 mm radius to match	type HB2, HB3	type QHB
225 mm high × 305 and 455 mm radius to match	type SP	type QSP
Angle kerb units		
305 × 305 × 225 mm high × 125 mm thick		
bullnosed external angle	type XA	
splayed external angle to match type SP	type XA	
bullnosed internal angle	type IA	
splayed internal angle to match type SP	type IA	
Channels		
255 mm wide × 125 mm high flat	type CS1	
150 mm wide × 125 mm high flat type	CS2	
255 mm wide × 125 mm high dished	type CD	

KERBS, PAVING, ETC.

Transition kerb units			
from kerb type SP to HB	left handed	type TL	
	right handed	type TR	
from kerb type BN to HB	left handed	type DL1	
	right handed	type DR1	
from kerb type BN to SP	left handed	type DL2	
	right handed	type DR2	

Number of kerbs required per quarter circle (780mm kerb lengths)

Radius (m)	Number in quarter circle
12	24
10	20
8	16
6	12
5	10
4	8
3	6
2	4
1	2

Precast Concrete Edgings

Round top type ER	Flat top type EF	Bullnosed top type EBN
150 × 50 mm	150 × 50 mm	150 × 50 mm
200 × 50 mm	200 × 50 mm	200 × 50 mm
250 × 50 mm	250 × 50 mm	250 × 50 mm

KERBS, PAVING, ETC.

BASES

Cement Bound Material for Bases and Subbases

CBM1:	very carefully graded aggregate from 37.5–75 ym, with a 7-day strength of 4.5 N/mm^2
CBM2:	same range of aggregate as CBM1 but with more tolerance in each size of aggregate with a 7-day strength of 7.0 N/mm^2
CBM3:	crushed natural aggregate or blast furnace slag, graded from 37.5 mm – 150 ym for 40 mm aggregate, and from 20–75 ym for 20 mm aggregate, with a 7-day strength of 10 N/mm^2
CBM4:	crushed natural aggregate or blast furnace slag, graded from 37.5 mm – 150 ym for 40 mm aggregate, and from 20–75 ym for 20 mm aggregate, with a 7-day strength of 15 N/mm^2

INTERLOCKING BRICK/BLOCK ROADS/PAVINGS

Sizes of Precast Concrete Paving Blocks

Type R blocks
200 × 100 × 60 mm
200 × 100 × 65 mm
200 × 100 × 80 mm
200 × 100 × 100 mm

Type S
Any shape within a 295 mm space

Sizes of clay brick pavers
200 × 100 × 50 mm
200 × 100 × 65 mm
210 × 105 × 50 mm
210 × 105 × 65 mm
215 × 102.5 × 50 mm
215 × 102.5 × 65 mm

Type PA: 3 kN
Footpaths and pedestrian areas, private driveways, car parks, light vehicle traffic and over-run

Type PB: 7 kN
Residential roads, lorry parks, factory yards, docks, petrol station forecourts, hardstandings, bus stations

KERBS, PAVING, ETC.

PAVING AND SURFACING

Weights and Sizes of Paving and Surfacing

Description of item	Size	Quantity per tonne
Paving 50 mm thick	900 × 600 mm	15
Paving 50 mm thick	750 × 600 mm	18
Paving 50 mm thick	600 × 600 mm	23
Paving 50 mm thick	450 × 600 mm	30
Paving 38 mm thick	600 × 600 mm	30
Path edging	914 × 50 × 150 mm	60
Kerb (including radius and tapers)	125 × 254 × 914 mm	15
Kerb (including radius and tapers)	125 × 150 × 914 mm	25
Square channel	125 × 254 × 914 mm	15
Dished channel	125 × 254 × 914 mm	15
Quadrants	300 × 300 × 254 mm	19
Quadrants	450 × 450 × 254 mm	12
Quadrants	300 × 300 × 150 mm	30
Internal angles	300 × 300 × 254 mm	30
Fluted pavement channel	255 × 75 × 914 mm	25
Corner stones	300 × 300 mm	80
Corner stones	360 × 360 mm	60
Cable covers	914 × 175 mm	55
Gulley kerbs	220 × 220 × 150 mm	60
Gulley kerbs	220 × 200 × 75 mm	120

KERBS, PAVING, ETC.

Weights and Sizes of Paving and Surfacing

Material	kg/m³	lb/cu yd
Tarmacadam	2306	3891
Macadam (waterbound)	2563	4325
Vermiculite (aggregate)	64–80	108–135
Terracotta	2114	3568
Cork – compressed	388	24
	kg/m²	lb/sq ft
Clay floor tiles, 12.7 mm	27.3	5.6
Pavement lights	122	25
Damp-proof course	5	1
	kg/m² per mm thickness	lb/sq ft per inch thickness
Paving Slabs (stone)	2.3	12
Granite setts	2.88	15
Asphalt	2.30	12
Rubber flooring	1.68	9
Polyvinyl chloride	1.94 (avg)	10 (avg)

Coverage (m²) Per Cubic Metre of Materials Used as Subbases or Capping Layers

Consolidated thickness laid in (mm)	Square metre coverage		
	Gravel	Sand	Hardcore
50	15.80	16.50	–
75	10.50	11.00	–
100	7.92	8.20	7.42
125	6.34	6.60	5.90
150	5.28	5.50	4.95
175	–	–	4.23
200	–	–	3.71
225	–	–	3.30
300	–	–	2.47

KERBS, PAVING, ETC.

Approximate Rate of Spreads

Average thickness of course (mm)	Description	Approximate rate of spread			
		Open Textured		Dense, Medium & Fine Textured	
		(kg/m²)	(m²/t)	(kg/m²)	(m²/t)
35	14 mm open textured or dense wearing course	60–75	13–17	70–85	12–14
40	20 mm open textured or dense base course	70–85	12–14	80–100	10–12
45	20 mm open textured or dense base course	80–100	10–12	95–100	9–10
50	20 mm open textured or dense, or 28 mm dense base course	85–110	9–12	110–120	8–9
60	28 mm dense base course, 40 mm open textured of dense base course or 40 mm single course as base course		8–10	130–150	7–8
65	28 mm dense base course, 40 mm open textured or dense base course or 40 mm single course	100–135	7–10	140–160	6–7
75	40 mm single course, 40 mm open textured or dense base course, 40 mm dense roadbase	120–150	7–8	165–185	5–6
100	40 mm dense base course or roadbase	–	–	220–240	4–4.5

KERBS, PAVING, ETC.

Surface Dressing Roads: Coverage (m²) per Tonne of Material

Size in mm	Sand	Granite chips	Gravel	Limestone chips
Sand	168	–	–	–
3	–	148	152	165
6	–	130	133	144
9	–	111	114	123
13	–	85	87	95
19	–	68	71	78

Sizes of Flags

Reference	Nominal size (mm)	Thickness (mm)
A	600 × 450	50 and 63
B	600 × 600	50 and 63
C	600 × 750	50 and 63
D	600 × 900	50 and 63
E	450 × 450	50 and 70 chamfered top surface
F	400 × 400	50 and 65 chamfered top surface
G	300 × 300	50 and 60 chamfered top surface

Sizes of Natural Stone Setts

Width (mm)		Length (mm)		Depth (mm)
100	×	100	×	100
75	×	150 to 250	×	125
75	×	150 to 250	×	150
100	×	150 to 250	×	100
100	×	150 to 250	×	150

SEEDING/TURFING AND PLANTING

Topsoil Quality

Topsoil grade	Properties
Premium	Natural topsoil, high fertility, loamy texture, good soil structure, suitable for intensive cultivation.
General purpose	Natural or manufactured topsoil of lesser quality than Premium, suitable for agriculture or amenity landscape, may need fertilizer or soil structure improvement.
Economy	Selected subsoil, natural mineral deposit such as river silt or greensand. The grade comprises two subgrades; 'Low clay' and 'High clay' which is more liable to compaction in handling. This grade is suitable for low-production agricultural land and amenity woodland or conservation planting areas.

Forms of Trees

Standards:	Shall be clear with substantially straight stems. Grafted and budded trees shall have no more than a slight bend at the union. Standards shall be designated as Half, Extra light, Light, Standard, Selected standard, Heavy, and Extra heavy.
Sizes of Standards	
Heavy standard	12–14 cm girth × 3.50 to 5.00 m high
Extra Heavy standard	14–16 cm girth × 4.25 to 5.00 m high
Extra Heavy standard	16–18 cm girth × 4.25 to 6.00 m high
Extra Heavy standard	18–20 cm girth × 5.00 to 6.00 m high
Semi-mature trees:	Between 6.0 m and 12.0 m tall with a girth of 20 to 75 cm at 1.0 m above ground.
Feathered trees:	Shall have a defined upright central leader, with stem furnished with evenly spread and balanced lateral shoots down to or near the ground.
Whips:	Shall be without significant feather growth as determined by visual inspection.
Multi-stemmed trees:	Shall have two or more main stems at, near, above or below ground.

Seedlings grown from seed and not transplanted shall be specified when ordered for sale as:

1+0	one year old seedling
2+0	two year old seedling
1+1	one year seed bed, one year transplanted = two year old seedling
1+2	one year seed bed, two years transplanted = three year old seedling
2+1	two year seed bed, one year transplanted = three year old seedling
1u1	two years seed bed, undercut after 1 year = two year old seedling
2u2	four years seed bed, undercut after 2 years = four year old seedling

SEEDING/TURFING AND PLANTING

Cuttings

The age of cuttings (plants grown from shoots, stems, or roots of the mother plant) shall be specified when ordered for sale. The height of transplants and undercut seedlings/cuttings (which have been transplanted or undercut at least once) shall be stated in centimetres. The number of growing seasons before and after transplanting or undercutting shall be stated.

0 + 1	one year cutting
0 + 2	two year cutting
0 + 1 + 1	one year cutting bed, one year transplanted = two year old seedling
0 + 1 + 2	one year cutting bed, two years transplanted = three year old seedling

Grass Cutting Capacities in m² per hour

Speed mph	Width of cut in metres												
	0.5	0.7	1.0	1.2	1.5	1.7	2.0	2.0	2.1	2.5	2.8	3.0	3.4
1.0	724	1127	1529	1931	2334	2736	3138	3219	3380	4023	4506	4828	5472
1.5	1086	1690	2293	2897	3500	4104	4707	4828	5069	6035	6759	7242	8208
2.0	1448	2253	3058	3862	4667	5472	6276	6437	6759	8047	9012	9656	10944
2.5	1811	2816	3822	4828	5834	6840	7846	8047	8449	10058	11265	12070	13679
3.0	2173	3380	4587	5794	7001	8208	9415	9656	10139	12070	13518	14484	16415
3.5	2535	3943	5351	6759	8167	9576	10984	11265	11829	14082	15772	16898	19151
4.0	2897	4506	6115	7725	9334	10944	12553	12875	13518	16093	18025	19312	21887
4.5	3259	5069	6880	8690	10501	12311	14122	14484	15208	18105	20278	21726	24623
5.0	3621	5633	7644	9656	11668	13679	15691	16093	16898	20117	22531	24140	27359
5.5	3983	6196	8409	10622	12834	15047	17260	17703	18588	22128	24784	26554	30095
6.0	4345	6759	9173	11587	14001	16415	18829	19312	20278	24140	27037	28968	32831
6.5	4707	7322	9938	12553	15168	17783	20398	20921	21967	26152	29290	31382	35566
7.0	5069	7886	10702	13518	16335	19151	21967	22531	23657	28163	31543	33796	38302

Number of Plants per m²: For Plants Planted on an Evenly Spaced Grid

Planting distances

mm	0.10	0.15	0.20	0.25	0.35	0.40	0.45	0.50	0.60	0.75	0.90	1.00	1.20	1.50
0.10	100.00	66.67	50.00	40.00	28.57	25.00	22.22	20.00	16.67	13.33	11.11	10.00	8.33	6.67
0.15	66.67	44.44	33.33	26.67	19.05	16.67	14.81	13.33	11.11	8.89	7.41	6.67	5.56	4.44
0.20	50.00	33.33	25.00	20.00	14.29	12.50	11.11	10.00	8.33	6.67	5.56	5.00	4.17	3.33
0.25	40.00	26.67	20.00	16.00	11.43	10.00	8.89	8.00	6.67	5.33	4.44	4.00	3.33	2.67
0.35	28.57	19.05	14.29	11.43	8.16	7.14	6.35	5.71	4.76	3.81	3.17	2.86	2.38	1.90
0.40	25.00	16.67	12.50	10.00	7.14	6.25	5.56	5.00	4.17	3.33	2.78	2.50	2.08	1.67
0.45	22.22	14.81	11.11	8.89	6.35	5.56	4.94	4.44	3.70	2.96	2.47	2.22	1.85	1.48
0.50	20.00	13.33	10.00	8.00	5.71	5.00	4.44	4.00	3.33	2.67	2.22	2.00	1.67	1.33
0.60	16.67	11.11	8.33	6.67	4.76	4.17	3.70	3.33	2.78	2.22	1.85	1.67	1.39	1.11
0.75	13.33	8.89	6.67	5.33	3.81	3.33	2.96	2.67	2.22	1.78	1.48	1.33	1.11	0.89
0.90	11.11	7.41	5.56	4.44	3.17	2.78	2.47	2.22	1.85	1.48	1.23	1.11	0.93	0.74
1.00	10.00	6.67	5.00	4.00	2.86	2.50	2.22	2.00	1.67	1.33	1.11	1.00	0.83	0.67
1.20	8.33	5.56	4.17	3.33	2.38	2.08	1.85	1.67	1.39	1.11	0.93	0.83	0.69	0.56
1.50	6.67	4.44	3.33	2.67	1.90	1.67	1.48	1.33	1.11	0.89	0.74	0.67	0.56	0.44

SEEDING/TURFING AND PLANTING

Grass Clippings Wet: Based on 3.5 m³/tonne

Annual kg/100 m²		Average 20 cuts kg/100m²			m²/tonne	m²/m³
32.0		1.6			61162.1	214067.3

Nr of cuts	22	20	18	16	12	4
kg/cut	1.45	1.60	1.78	2.00	2.67	8.00
Area capacity of 3 tonne vehicle per load						
m²	206250	187500	168750	150000	112500	37500
Load m³	100 m² units/m³ of vehicle space					
1	196.4	178.6	160.7	142.9	107.1	35.7
2	392.9	357.1	321.4	285.7	214.3	71.4
3	589.3	535.7	482.1	428.6	321.4	107.1
4	785.7	714.3	642.9	571.4	428.6	142.9
5	982.1	892.9	803.6	714.3	535.7	178.6

Transportation of Trees

To unload large trees a machine with the necessary lifting strength is required. The weight of the trees must therefore be known in advance. The following table gives a rough overview. The additional columns with root ball dimensions and the number of plants per trailer provide additional information for example about preparing planting holes and calculating unloading times.

Girth in cm	Rootball diameter in cm	Ball height in cm	Weight in kg	Numbers of trees per trailer
16–18	50–60	40	150	100–120
18–20	60–70	40–50	200	80–100
20–25	60–70	40–50	270	50–70
25–30	80	50–60	350	50
30–35	90–100	60–70	500	12–18
35–40	100–110	60–70	650	10–15
40–45	110–120	60–70	850	8–12
45–50	110–120	60–70	1100	5–7
50–60	130–140	60–70	1600	1–3
60–70	150–160	60–70	2500	1
70–80	180–200	70	4000	1
80–90	200–220	70–80	5500	1
90–100	230–250	80–90	7500	1
100–120	250–270	80–90	9500	1

Data supplied by Lorenz von Ehren GmbH
The information in the table is approximate; deviations depend on soil type, genus and weather

FENCING AND GATES

Types of Preservative

Creosote (tar oil) can be 'factory' applied	by pressure to BS 144: pts 1&2 by immersion to BS 144: pt 1 by hot and cold open tank to BS 144: pts 1&2
Copper/chromium/arsenic (CCA)	by full cell process to BS 4072 pts 1&2
Organic solvent (OS)	by double vacuum (vacvac) to BS 5707 pts 1&3 by immersion to BS 5057 pts 1&3
Pentachlorophenol (PCP)	by heavy oil double vacuum to BS 5705 pts 2&3
Boron diffusion process (treated with disodium octaborate to BWPA Manual 1986)	

Note: Boron is used on green timber at source and the timber is supplied dry

Cleft Chestnut Pale Fences

Pales	Pale spacing	Wire lines	
900 mm	75 mm	2	temporary protection
1050 mm	75 or 100 mm	2	light protective fences
1200 mm	75 mm	3	perimeter fences
1350 mm	75 mm	3	perimeter fences
1500 mm	50 mm	3	narrow perimeter fences
1800 mm	50 mm	3	light security fences

Close-Boarded Fences

Close-boarded fences 1.05 to 1.8 m high
Type BCR (recessed) or BCM (morticed) with concrete posts 140 × 115 mm tapered and Type BW with timber posts

Palisade Fences

Wooden palisade fences
Type WPC with concrete posts 140 × 115 mm tapered and Type WPW with timber posts

For both types of fence:
Height of fence 1050 mm: two rails
Height of fence 1200 mm: two rails
Height of fence 1500 mm: three rails
Height of fence 1650 mm: three rails
Height of fence 1800 mm: three rails

FENCING AND GATES

Post and Rail Fences

Wooden post and rail fences
Type MPR 11/3 morticed rails and Type SPR 11/3 nailed rails
Height to top of rail 1100 mm
Rails: three rails 87 mm, 38 mm

Type MPR 11/4 morticed rails and Type SPR 11/4 nailed rails
Height to top of rail 1100 mm
Rails: four rails 87 mm, 38 mm

Type MPR 13/4 morticed rails and Type SPR 13/4 nailed rails
Height to top of rail 1300 mm
Rail spacing 250 mm, 250 mm, and 225 mm from top
Rails: four rails 87 mm, 38 mm

Steel Posts

Rolled steel angle iron posts for chain link fencing

Posts	Fence height	Strut	Straining post
1500 × 40 × 40 × 5 mm	900 mm	1500 × 40 × 40 × 5 mm	1500 × 50 × 50 × 6 mm
1800 × 40 × 40 × 5 mm	1200 mm	1800 × 40 × 40 × 5 mm	1800 × 50 × 50 × 6 mm
2000 × 45 × 45 × 5 mm	1400 mm	2000 × 45 × 45 × 5 mm	2000 × 60 × 60 × 6 mm
2600 × 45 × 45 × 5 mm	1800 mm	2600 × 45 × 45 × 5 mm	2600 × 60 × 60 × 6 mm
3000 × 50 × 50 × 6 mm	1800 mm	2600 × 45 × 45 × 5 mm	3000 × 60 × 60 × 6 mm
with arms			

Concrete Posts

Concrete posts for chain link fencing

Posts and straining posts	Fence height	Strut
1570 mm 100 × 100 mm	900 mm	1500 mm × 75 × 75 mm
1870 mm 125 × 125 mm	1200 mm	1830 mm × 100 × 75 mm
2070 mm 125 × 125 mm	1400 mm	1980 mm × 100 × 75 mm
2620 mm 125 × 125 mm	1800 mm	2590 mm × 100 × 85 mm
3040 mm 125 × 125 mm	1800 mm	2590 mm × 100 × 85 mm (with arms)

FENCING AND GATES

Rolled Steel Angle Posts

Rolled steel angle posts for rectangular wire mesh (field) fencing

Posts	Fence height	Strut	Straining post
1200 × 40 × 40 × 5 mm	600 mm	1200 × 75 × 75 mm	1350 × 100 × 100 mm
1400 × 40 × 40 × 5 mm	800 mm	1400 × 75 × 75 mm	1550 × 100 × 100 mm
1500 × 40 × 40 × 5 mm	900 mm	1500 × 75 × 75 mm	1650 × 100 × 100 mm
1600 × 40 × 40 × 5 mm	1000 mm	1600 × 75 × 75 mm	1750 × 100 × 100 mm
1750 × 40 × 40 × 5 mm	1150 mm	1750 × 75 × 100 mm	1900 × 125 × 125 mm

Concrete Posts

Concrete posts for rectangular wire mesh (field) fencing

Posts	Fence height	Strut	Straining post
1270 × 100 × 100 mm	600 mm	1200 × 75 × 75 mm	1420 × 100 × 100 mm
1470 × 100 × 100 mm	800 mm	1350 × 75 × 75 mm	1620 × 100 × 100 mm
1570 × 100 × 100 mm	900 mm	1500 × 75 × 75 mm	1720 × 100 × 100 mm
1670 × 100 × 100 mm	600 mm	1650 × 75 × 75 mm	1820 × 100 × 100 mm
1820 × 125 × 125 mm	1150 mm	1830 × 75 × 100 mm	1970 × 125 × 125 mm

Cleft Chestnut Pale Fences

Timber Posts

Timber posts for wire mesh and hexagonal wire netting fences

Round timber for general fences

Posts	Fence height	Strut	Straining post
1300 × 65 mm dia.	600 mm	1200 × 80 mm dia.	1450 × 100 mm dia.
1500 × 65 mm dia.	800 mm	1400 × 80 mm dia.	1650 × 100 mm dia.
1600 × 65 mm dia.	900 mm	1500 × 80 mm dia.	1750 × 100 mm dia.
1700 × 65 mm dia.	1050 mm	1600 × 80 mm dia.	1850 × 100 mm dia.
1800 × 65 mm dia.	1150 mm	1750 × 80 mm dia.	2000 × 120 mm dia.

Squared timber for general fences

Posts	Fence height	Strut	Straining post
1300 × 75 × 75 mm	600 mm	1200 × 75 × 75 mm	1450 × 100 × 100 mm
1500 × 75 × 75 mm	800 mm	1400 × 75 × 75 mm	1650 × 100 × 100 mm
1600 × 75 × 75 mm	900 mm	1500 × 75 × 75 mm	1750 × 100 × 100 mm
1700 × 75 × 75 mm	1050 mm	1600 × 75 × 75 mm	1850 × 100 × 100 mm
1800 × 75 × 75 mm	1150 mm	1750 × 75 × 75 mm	2000 × 125 × 100 mm

FENCING AND GATES

Steel Fences to BS 1722: Part 9: 1992

	Fence height	Top/bottom rails and flat posts	Vertical bars
Light	1000 mm	40 × 10 mm 450 mm in ground	12 mm dia. at 115 mm cs
	1200 mm	40 × 10 mm 550 mm in ground	12 mm dia. at 115 mm cs
	1400 mm	40 × 10 mm 550 mm in ground	12 mm dia. at 115 mm cs
Light	1000 mm	40 × 10 mm 450 mm in ground	16 mm dia. at 120 mm cs
	1200 mm	40 × 10 mm 550 mm in ground	16 mm dia. at 120 mm cs
	1400 mm	40 × 10 mm 550 mm in ground	16 mm dia. at 120 mm cs
Medium	1200 mm	50 × 10 mm 550 mm in ground	20 mm dia. at 125 mm cs
	1400 mm	50 × 10 mm 550 mm in ground	20 mm dia. at 125 mm cs
	1600 mm	50 × 10 mm 600 mm in ground	22 mm dia. at 145 mm cs
	1800 mm	50 × 10 mm 600 mm in ground	22 mm dia. at 145 mm cs
Heavy	1600 mm	50 × 10 mm 600 mm in ground	22 mm dia. at 145 mm cs
	1800 mm	50 × 10 mm 600 mm in ground	22 mm dia. at 145 mm cs
	2000 mm	50 × 10 mm 600 mm in ground	22 mm dia. at 145 mm cs
	2200 mm	50 × 10 mm 600 mm in ground	22 mm dia. at 145 mm cs

Notes: Mild steel fences: round or square verticals; flat standards and horizontals. Tops of vertical bars may be bow-top, blunt, or pointed. Round or square bar railings

Timber Field Gates to BS 3470: 1975

Gates made to this standard are designed to open one way only
All timber gates are 1100 mm high
Width over stiles 2400, 2700, 3000, 3300, 3600, and 4200 mm
Gates over 4200 mm should be made in two leaves

Steel Field Gates to BS 3470: 1975

All steel gates are 1100 mm high
Heavy duty: width over stiles 2400, 3000, 3600 and 4500 mm
Light duty: width over stiles 2400, 3000, and 3600 mm

FENCING AND GATES

Domestic Front Entrance Gates to BS 4092: Part 1: 1966

Metal gates:	Single gates are 900 mm high minimum, 900 mm, 1000 mm and 1100 mm wide

Domestic Front Entrance Gates to BS 4092: Part 2: 1966

Wooden gates:	All rails shall be tenoned into the stiles Single gates are 840 mm high minimum, 801 mm and 1020 mm wide Double gates are 840 mm high minimum, 2130, 2340 and 2640 mm wide

Timber Bridle Gates to BS 5709:1979 (Horse or Hunting Gates)

Gates open one way only	
Minimum width between posts	1525 mm
Minimum height	1100 mm

Timber Kissing Gates to BS 5709:1979

Minimum width	700 mm
Minimum height	1000 mm
Minimum distance between shutting posts	600 mm
Minimum clearance at mid-point	600 mm

Metal Kissing Gates to BS 5709:1979

Sizes are the same as those for timber kissing gates Maximum gaps between rails 120 mm

Categories of Pedestrian Guard Rail to BS 3049:1976

Class A for normal use Class B where vandalism is expected Class C where crowd pressure is likely

DRAINAGE

Width required for Trenches for Various Diameters of Pipes

Pipe diameter (mm)	Trench n.e. 1.50 m deep	Trench over 1.50 m deep
n.e. 100 mm	450 mm	600 mm
100–150 mm	500 mm	650 mm
150–225 mm	600 mm	750 mm
225–300 mm	650 mm	800 mm
300–400 mm	750 mm	900 mm
400–450 mm	900 mm	1050 mm
450–600 mm	1100 mm	1300 mm

Weights and Dimensions – Vitrified Clay Pipes

Product	Nominal diameter (mm)	Effective length (mm)	BS 65 limits of tolerance min (mm)	max (mm)	Crushing strength (kN/m)	Weight (kg/pipe)	(kg/m)
Supersleve	100	1600	96	105	35.00	14.71	9.19
	150	1750	146	158	35.00	29.24	16.71
Hepsleve	225	1850	221	236	28.00	84.03	45.42
	300	2500	295	313	34.00	193.05	77.22
	150	1500	146	158	22.00	37.04	24.69
Hepseal	225	1750	221	236	28.00	85.47	48.84
	300	2500	295	313	34.00	204.08	81.63
	400	2500	394	414	44.00	357.14	142.86
	450	2500	444	464	44.00	454.55	181.63
	500	2500	494	514	48.00	555.56	222.22
	600	2500	591	615	57.00	796.23	307.69
	700	3000	689	719	67.00	1111.11	370.45
	800	3000	788	822	72.00	1351.35	450.45
Hepline	100	1600	95	107	22.00	14.71	9.19
	150	1750	145	160	22.00	29.24	16.71
	225	1850	219	239	28.00	84.03	45.42
	300	1850	292	317	34.00	142.86	77.22
Hepduct (conduit)	90	1500	–	–	28.00	12.05	8.03
	100	1600	–	–	28.00	14.71	9.19
	125	1750	–	–	28.00	20.73	11.84
	150	1750	–	–	28.00	29.24	16.71
	225	1850	–	–	28.00	84.03	45.42
	300	1850	–	–	34.00	142.86	77.22

Weights and Dimensions – Vitrified Clay Pipes

Nominal internal diameter (mm)	Nominal wall thickness (mm)	Approximate weight (kg/m)
150	25	45
225	29	71
300	32	122
375	35	162
450	38	191
600	48	317
750	54	454
900	60	616
1200	76	912
1500	89	1458
1800	102	1884
2100	127	2619

Wall thickness, weights and pipe lengths vary, depending on type of pipe required
The particulars shown above represent a selection of available diameters and are applicable to strength class 1 pipes with flexible rubber ring joints
Tubes with Ogee joints are also available

DRAINAGE

Weights and Dimensions – PVC-u Pipes

	Nominal size	Mean outside diameter (mm)		Wall thickness	Weight
		min	max	(mm)	(kg/m)
Standard pipes	82.4	82.4	82.7	3.2	1.2
	110.0	110.0	110.4	3.2	1.6
	160.0	160.0	160.6	4.1	3.0
	200.0	200.0	200.6	4.9	4.6
	250.0	250.0	250.7	6.1	7.2
Perforated pipes heavy grade	As above	As above	As above	As above	As above
thin wall	82.4	82.4	82.7	1.7	–
	110.0	110.0	110.4	2.2	–
	160.0	160.0	160.6	3.2	–

Width of Trenches Required for Various Diameters of Pipes

Pipe diameter (mm)	Trench n.e. 1.5 m deep (mm)	Trench over 1.5 m deep (mm)
n.e. 100	450	600
100–150	500	650
150–225	600	750
225–300	650	800
300–400	750	900
400–450	900	1050
450–600	1100	1300

DRAINAGE

DRAINAGE BELOW GROUND AND LAND DRAINAGE

Flow of Water Which Can Be Carried by Various Sizes of Pipe

Clay or concrete pipes

	Gradient of pipeline							
	1:10	1:20	1:30	1:40	1:50	1:60	1:80	1:100
Pipe size	Flow in litres per second							
DN 100 15.0	8.5	6.8	5.8	5.2	4.7	4.0	3.5	
DN 150 28.0	19.0	16.0	14.0	12.0	11.0	9.1	8.0	
DN 225 140.0	95.0	76.0	66.0	58.0	53.0	46.0	40.0	

Plastic pipes

	Gradient of pipeline							
	1:10	1:20	1:30	1:40	1:50	1:60	1:80	1:100
Pipe size	Flow in litres per second							
82.4 mm i/dia.	12.0	8.5	6.8	5.8	5.2	4.7	4.0	3.5
110 mm i/dia.	28.0	19.0	16.0	14.0	12.0	11.0	9.1	8.0
160 mm i/dia.	76.0	53.0	43.0	37.0	33.0	29.0	25.0	22.0
200 mm i/dia.	140.0	95.0	76.0	66.0	58.0	53.0	46.0	40.0

Vitrified (Perforated) Clay Pipes and Fittings to BS En 295-5 1994

Length not specified		
75 mm bore	250 mm bore	600 mm bore
100	300	700
125	350	800
150	400	1000
200	450	1200
225	500	

Precast Concrete Pipes: Prestressed Non-pressure Pipes and Fittings: Flexible Joints to BS 5911: Pt. 103: 1994

Rationalized metric nominal sizes: 450, 500	
Length:	500–1000 by 100 increments 1000–2200 by 200 increments 2200–2800 by 300 increments
Angles: length:	450–600 angles 45, 22.5,11.25° 600 or more angles 22.5, 11.25°

DRAINAGE

Precast Concrete Pipes: Un-reinforced and Circular Manholes and Soakaways to BS 5911: Pt. 200: 1994

Nominal sizes:	
Shafts:	675, 900 mm
Chambers:	900, 1050, 1200, 1350, 1500, 1800, 2100, 2400, 2700, 3000 mm
Large chambers:	To have either tapered reducing rings or a flat reducing slab in order to accept the standard cover
Ring depths:	1. 300–1200 mm by 300 mm increments except for bottom slab and rings below cover slab, these are by 150 mm increments
	2. 250–1000 mm by 250 mm increments except for bottom slab and rings below cover slab, these are by 125 mm increments
Access hole:	750 × 750 mm for DN 1050 chamber
	1200 × 675 mm for DN 1350 chamber

Calculation of Soakaway Depth
The following formula determines the depth of concrete ring soakaway that would be required for draining given amounts of water.

$$h = \frac{4ar}{3\pi D^2}$$

h = depth of the chamber below the invert pipe
a = The area to be drained
r = The hourly rate of rainfall (50 mm per hour)
π = pi
D = internal diameter of the soakaway

This table shows the depth of chambers in each ring size which would be required to contain the volume of water specified. These allow a recommended storage capacity of ⅓ (one third of the hourly rainfall figure).

Table Showing Required Depth of Concrete Ring Chambers in Metres

Area m²	50	100	150	200	300	400	500
Ring size							
0.9	1.31	2.62	3.93	5.24	7.86	10.48	13.10
1.1	0.96	1.92	2.89	3.85	5.77	7.70	9.62
1.2	0.74	1.47	2.21	2.95	4.42	5.89	7.37
1.4	0.58	1.16	1.75	2.33	3.49	4.66	5.82
1.5	0.47	0.94	1.41	1.89	2.83	3.77	4.72
1.8	0.33	0.65	0.98	1.31	1.96	2.62	3.27
2.1	0.24	0.48	0.72	0.96	1.44	1.92	2.41
2.4	0.18	0.37	0.55	0.74	1.11	1.47	1.84
2.7	0.15	0.29	0.44	0.58	0.87	1.16	1.46
3.0	0.12	0.24	0.35	0.47	0.71	0.94	1.18

DRAINAGE

Precast Concrete Inspection Chambers and Gullies to BS 5911: Part 230: 1994

Nominal sizes:	375 diameter, 750, 900 mm deep
	450 diameter, 750, 900, 1050, 1200 mm deep
Depths:	from the top for trapped or un-trapped units:
	centre of outlet 300 mm
	invert (bottom) of the outlet pipe 400 mm
Depth of water seal for trapped gullies:	
	85 mm, rodding eye int. dia. 100 mm
Cover slab:	65 mm min

Bedding Flexible Pipes: PVC-u Or Ductile Iron

Type 1 =	100 mm fill below pipe, 300 mm above pipe: single size material
Type 2 =	100 mm fill below pipe, 300 mm above pipe: single size or graded material
Type 3 =	100 mm fill below pipe, 75 mm above pipe with concrete protective slab over
Type 4 =	100 mm fill below pipe, fill laid level with top of pipe
Type 5 =	200 mm fill below pipe, fill laid level with top of pipe
Concrete =	25 mm sand blinding to bottom of trench, pipe supported on chocks, 100 mm concrete under the pipe, 150 mm concrete over the pipe

DRAINAGE

Bedding Rigid Pipes: Clay or Concrete
(for vitrified clay pipes the manufacturer should be consulted)

Class D:	Pipe laid on natural ground with cut-outs for joints, soil screened to remove stones over 40 mm and returned over pipe to 150 m min depth. Suitable for firm ground with trenches trimmed by hand.
Class N:	Pipe laid on 50 mm granular material of graded aggregate to Table 4 of BS 882, or 10 mm aggregate to Table 6 of BS 882, or as dug light soil (not clay) screened to remove stones over 10 mm. Suitable for machine dug trenches.
Class B:	As Class N, but with granular bedding extending half way up the pipe diameter.
Class F:	Pipe laid on 100 mm granular fill to BS 882 below pipe, minimum 150 mm granular fill above pipe: single size material. Suitable for machine dug trenches.
Class A:	Concrete 100 mm thick under the pipe extending half way up the pipe, backfilled with the appropriate class of fill. Used where there is only a very shallow fall to the drain. Class A bedding allows the pipes to be laid to an exact gradient.
Concrete surround:	25 mm sand blinding to bottom of trench, pipe supported on chocks, 100 mm concrete under the pipe, 150 mm concrete over the pipe. It is preferable to bed pipes under slabs or wall in granular material.

PIPED SUPPLY SYSTEMS

Identification of Service Tubes From Utility to Dwellings

Utility	Colour	Size	Depth
British Telecom	grey	54 mm od	450 mm
Electricity	black	38 mm od	450 mm
Gas	yellow	42 mm od rigid 60 mm od convoluted	450 mm
Water	may be blue	(normally untubed)	750 mm

ELECTRICAL SUPPLY/POWER/LIGHTING SYSTEMS

Electrical Insulation Class En 60.598 BS 4533

Class 1:	luminaires comply with class 1 (I) earthed electrical requirements
Class 2:	luminaires comply with class 2 (II) double insulated electrical requirements
Class 3:	luminaires comply with class 3 (III) electrical requirements

Protection to Light Fittings

BS EN 60529:1992 Classification for degrees of protection provided by enclosures.
(IP Code – International or ingress Protection)

1st characteristic: against ingress of solid foreign objects		
The figure	2	indicates that fingers cannot enter
	3	that a 2.5 mm diameter probe cannot enter
	4	that a 1.0 mm diameter probe cannot enter
	5	the fitting is dust proof (no dust around live parts)
	6	the fitting is dust tight (no dust entry)
2nd characteristic: ingress of water with harmful effects		
The figure	0	indicates unprotected
	1	vertically dripping water cannot enter
	2	water dripping 15° (tilt) cannot enter
	3	spraying water cannot enter
	4	splashing water cannot enter
	5	jetting water cannot enter
	6	powerful jetting water cannot enter
	7	proof against temporary immersion
	8	proof against continuous immersion
Optional additional codes:		A–D protects against access to hazardous parts
	H	High voltage apparatus
	M	fitting was in motion during water test
	S	fitting was static during water test
	W	protects against weather
Marking code arrangement:		(example) IPX5S = IP (International or Ingress Protection)
		X (denotes omission of first characteristic)
		5 = jetting
		S = static during water test

RAIL TRACKS

	kg/m of track	lb/ft of track
Standard gauge Bull-head rails, chairs, transverse timber (softwood) sleepers etc.	245	165
Main lines Flat-bottom rails, transverse prestressed concrete sleepers, etc.	418	280
Add for electric third rail	51	35
Add for crushed stone ballast	2600	1750
	kg/m²	**lb/sq ft**
Overall average weight – rails connections, sleepers, ballast, etc.	733	150
	kg/m of track	**lb/ft of track**
Bridge rails, longitudinal timber sleepers, etc.	112	75

RAIL TRACKS

Heavy Rails

British Standard Section No.	Rail height (mm)	Foot width (mm)	Head width (mm)	Min web thickness (mm)	Section weight (kg/m)
Flat Bottom Rails					
60A	114.30	109.54	57.15	11.11	30.62
70A	123.82	111.12	60.32	12.30	34.81
75A	128.59	114.30	61.91	12.70	37.45
80A	133.35	117.47	63.50	13.10	39.76
90A	142.88	127.00	66.67	13.89	45.10
95A	147.64	130.17	69.85	14.68	47.31
100A	152.40	133.35	69.85	15.08	50.18
110A	158.75	139.70	69.85	15.87	54.52
113A	158.75	139.70	69.85	20.00	56.22
50 'O'	100.01	100.01	52.39	10.32	24.82
80 'O'	127.00	127.00	63.50	13.89	39.74
60R	114.30	109.54	57.15	11.11	29.85
75R	128.59	122.24	61.91	13.10	37.09
80R	133.35	127.00	63.50	13.49	39.72
90R	142.88	136.53	66.67	13.89	44.58
95R	147.64	141.29	68.26	14.29	47.21
100R	152.40	146.05	69.85	14.29	49.60
95N	147.64	139.70	69.85	13.89	47.27
Bull Head Rails					
95R BH	145.26	69.85	69.85	19.05	47.07

Light Rails

British Standard Section No.	Rail height (mm)	Foot width (mm)	Head width (mm)	Min web thickness (mm)	Section weight (kg/m)
Flat Bottom Rails					
20M	65.09	55.56	30.96	6.75	9.88
30M	75.41	69.85	38.10	9.13	14.79
35M	80.96	76.20	42.86	9.13	17.39
35R	85.73	82.55	44.45	8.33	17.40
40	88.11	80.57	45.64	12.3	19.89
Bridge Rails					
13	48.00	92	36.00	18.0	13.31
16	54.00	108	44.50	16.0	16.06
20	55.50	127	50.00	20.5	19.86
28	67.00	152	50.00	31.0	28.62
35	76.00	160	58.00	34.5	35.38
50	76.00	165	58.50	–	50.18
Crane Rails					
A65	75.00	175.00	65.00	38.0	43.10
A75	85.00	200.00	75.00	45.0	56.20
A100	95.00	200.00	100.00	60.0	74.30
A120	105.00	220.00	120.00	72.0	100.00
175CR	152.40	152.40	107.95	38.1	86.92

RAIL TRACKS

Fish Plates

British Standard Section No.	Overall plate length		Hole diameter	Finished weight per pair	
	4 Hole (mm)	6 Hole (mm)	(mm)	4 Hole (kg/pair)	6 Hole (kg/pair)
For British Standard Heavy Rails: Flat Bottom Rails					
60A	406.40	609.60	20.64	9.87	14.76
70A	406.40	609.60	22.22	11.15	16.65
75A	406.40	–	23.81	11.82	17.73
80A	406.40	609.60	23.81	13.15	19.72
90A	457.20	685.80	25.40	17.49	26.23
100A	508.00	–	pear	25.02	–
110A (shallow)	507.00	–	27.00	30.11	54.64
113A (heavy)	507.00	–	27.00	30.11	54.64
50 'O' (shallow)	406.40	–	–	6.68	10.14
80 'O' (shallow)	495.30	–	23.81	14.72	22.69
60R (shallow)	406.40	609.60	20.64	8.76	13.13
60R (angled)	406.40	609.60	20.64	11.27	16.90
75R (shallow)	406.40	–	23.81	10.94	16.42
75R (angled)	406.40	–	23.81	13.67	–
80R (shallow)	406.40	609.60	23.81	11.93	17.89
80R (angled)	406.40	609.60	23.81	14.90	22.33
For British Standard Heavy Rails: Bull head rails					
95R BH (shallow)	–	457.20	27.00	14.59	14.61
For British Standard Light Rails: Flat Bottom Rails					
30M	355.6	–	–	–	2.72
35M	355.6	–	–	–	2.83
40	355.6	–	–	3.76	–

FRACTIONS, DECIMALS AND MILLIMETRE EQUIVALENTS

FRACTIONS, DECIMALS AND MILLIMETRE EQUIVALENTS

Fractions	Decimals	(mm)	Fractions	Decimals	(mm)
1/64	0.015625	0.396875	33/64	0.515625	13.096875
1/32	0.03125	0.79375	17/32	0.53125	13.49375
3/64	0.046875	1.190625	35/64	0.546875	13.890625
1/16	0.0625	1.5875	9/16	0.5625	14.2875
5/64	0.078125	1.984375	37/64	0.578125	14.684375
3/32	0.09375	2.38125	19/32	0.59375	15.08125
7/64	0.109375	2.778125	39/64	0.609375	15.478125
1/8	0.125	3.175	5/8	0.625	15.875
9/64	0.140625	3.571875	41/64	0.640625	16.271875
5/32	0.15625	3.96875	21/32	0.65625	16.66875
11/64	0.171875	4.365625	43/64	0.671875	17.065625
3/16	0.1875	4.7625	11/16	0.6875	17.4625
13/64	0.203125	5.159375	45/64	0.703125	17.859375
7/32	0.21875	5.55625	23/32	0.71875	18.25625
15/64	0.234375	5.953125	47/64	0.734375	18.653125
1/4	0.25	6.35	3/4	0.75	19.05
17/64	0.265625	6.746875	49/64	0.765625	19.446875
9/32	0.28125	7.14375	25/32	0.78125	19.84375
19/64	0.296875	7.540625	51/64	0.796875	20.240625
5/16	0.3125	7.9375	13/16	0.8125	20.6375
21/64	0.328125	8.334375	53/64	0.828125	21.034375
11/32	0.34375	8.73125	27/32	0.84375	21.43125
23/64	0.359375	9.128125	55/64	0.859375	21.828125
3/8	0.375	9.525	7/8	0.875	22.225
25/64	0.390625	9.921875	57/64	0.890625	22.621875
13/32	0.40625	10.31875	29/32	0.90625	23.01875
27/64	0.421875	10.71563	59/64	0.921875	23.415625
7/16	0.4375	11.1125	15/16	0.9375	23.8125
29/64	0.453125	11.50938	61/64	0.953125	24.209375
15/32	0.46875	11.90625	31/32	0.96875	24.60625
31/64	0.484375	12.30313	63/64	0.984375	25.003125
1/2	0.5	12.7	1.0	1	25.4

IMPERIAL STANDARD WIRE GAUGE (SWG)

IMPERIAL STANDARD WIRE GAUGE (SWG)

SWG No.	Diameter (inches)	Diameter (mm)	SWG No.	Diameter (inches)	Diameter (mm)
7/0	0.5	12.7	23	0.024	0.61
6/0	0.464	11.79	24	0.022	0.559
5/0	0.432	10.97	25	0.02	0.508
4/0	0.4	10.16	26	0.018	0.457
3/0	0.372	9.45	27	0.0164	0.417
2/0	0.348	8.84	28	0.0148	0.376
1/0	0.324	8.23	29	0.0136	0.345
1	0.3	7.62	30	0.0124	0.315
2	0.276	7.01	31	0.0116	0.295
3	0.252	6.4	32	0.0108	0.274
4	0.232	5.89	33	0.01	0.254
5	0.212	5.38	34	0.009	0.234
6	0.192	4.88	35	0.008	0.213
7	0.176	4.47	36	0.008	0.193
8	0.16	4.06	37	0.007	0.173
9	0.144	3.66	38	0.006	0.152
10	0.128	3.25	39	0.005	0.132
11	0.116	2.95	40	0.005	0.122
12	0.104	2.64	41	0.004	0.112
13	0.092	2.34	42	0.004	0.102
14	0.08	2.03	43	0.004	0.091
15	0.072	1.83	44	0.003	0.081
16	0.064	1.63	45	0.003	0.071
17	0.056	1.42	46	0.002	0.061
18	0.048	1.22	47	0.002	0.051
19	0.04	1.016	48	0.002	0.041
20	0.036	0.914	49	0.001	0.031
21	0.032	0.813	50	0.001	0.025
22	0.028	0.711			

PIPES, WATER, STORAGE, INSULATION

WATER PRESSURE DUE TO HEIGHT

Imperial

Head (Feet)	Pressure (lb/in^2)		Head (Feet)	Pressure (lb/in^2)
1	0.43		70	30.35
5	2.17		75	32.51
10	4.34		80	34.68
15	6.5		85	36.85
20	8.67		90	39.02
25	10.84		95	41.18
30	13.01		100	43.35
35	15.17		105	45.52
40	17.34		110	47.69
45	19.51		120	52.02
50	21.68		130	56.36
55	23.84		140	60.69
60	26.01		150	65.03
65	28.18			

Metric

Head (m)	Pressure (bar)		Head (m)	Pressure (bar)
0.5	0.049		18.0	1.766
1.0	0.098		19.0	1.864
1.5	0.147		20.0	1.962
2.0	0.196		21.0	2.06
3.0	0.294		22.0	2.158
4.0	0.392		23.0	2.256
5.0	0.491		24.0	2.354
6.0	0.589		25.0	2.453
7.0	0.687		26.0	2.551
8.0	0.785		27.0	2.649
9.0	0.883		28.0	2.747
10.0	0.981		29.0	2.845
11.0	1.079		30.0	2.943
12.0	1.177		32.5	3.188
13.0	1.275		35.0	3.434
14.0	1.373		37.5	3.679
15.0	1.472		40.0	3.924
16.0	1.57		42.5	4.169
17.0	1.668		45.0	4.415

1 bar	=	14.5038 lbf/in^2	
1 lbf/in^2	=	0.06895 bar	
1 metre	=	3.2808 ft or 39.3701 in	
1 foot	=	0.3048 metres	
1 in wg	=	2.5 mbar (249.1 N/m^2)	

PIPES, WATER, STORAGE, INSULATION

Dimensions and Weights of Copper Pipes to BSEN 1057, BSEN 12499, BSEN 14251

Outside Diameter (mm)	Internal Diameter (mm)	Weight per Metre (kg)	Internal Diameter (mm)	Weight per Metre (kg)	Internal Diameter (mm)	Weight per Metre (kg)
	Formerly Table X		Formerly Table Y		Formerly Table Z	
6	4.80	0.0911	4.40	0.1170	5.00	0.0774
8	6.80	0.1246	6.40	0.1617	7.00	0.1054
10	8.80	0.1580	8.40	0.2064	9.00	0.1334
12	10.80	0.1914	10.40	0.2511	11.00	0.1612
15	13.60	0.2796	13.00	0.3923	14.00	0.2031
18	16.40	0.3852	16.00	0.4760	16.80	0.2918
22	20.22	0.5308	19.62	0.6974	20.82	0.3589
28	26.22	0.6814	25.62	0.8985	26.82	0.4594
35	32.63	1.1334	32.03	1.4085	33.63	0.6701
42	39.63	1.3675	39.03	1.6996	40.43	0.9216
54	51.63	1.7691	50.03	2.9052	52.23	1.3343
76.1	73.22	3.1287	72.22	4.1437	73.82	2.5131
108	105.12	4.4666	103.12	7.3745	105.72	3.5834
133	130.38	5.5151	–	–	130.38	5.5151
159	155.38	8.7795	–	–	156.38	6.6056

Dimensions of Stainless Steel Pipes to BS 4127

Outside Diameter (mm)	Maximum Outside Diameter (mm)	Minimum Outside Diameter (mm)	Wall Thickness (mm)	Working Pressure (bar)
6	6.045	5.940	0.6	330
8	8.045	7.940	0.6	260
10	10.045	9.940	0.6	210
12	12.045	11.940	0.6	170
15	15.045	14.940	0.6	140
18	18.045	17.940	0.7	135
22	22.055	21.950	0.7	110
28	28.055	27.950	0.8	121
35	35.070	34.965	1.0	100
42	42.070	41.965	1.1	91
54	54.090	53.940	1.2	77

PIPES, WATER, STORAGE, INSULATION

Dimensions of Steel Pipes to BS 1387

Nominal Size (mm)	Approx. Outside Diameter (mm)	Outside Diameter				Thickness		
		Light		Medium & Heavy		Light (mm)	Medium (mm)	Heavy (mm)
		Max (mm)	Min (mm)	Max (mm)	Min (mm)			
6	10.20	10.10	9.70	10.40	9.80	1.80	2.00	2.65
8	13.50	13.60	13.20	13.90	13.30	1.80	2.35	2.90
10	17.20	17.10	16.70	17.40	16.80	1.80	2.35	2.90
15	21.30	21.40	21.00	21.70	21.10	2.00	2.65	3.25
20	26.90	26.90	26.40	27.20	26.60	2.35	2.65	3.25
25	33.70	33.80	33.20	34.20	33.40	2.65	3.25	4.05
32	42.40	42.50	41.90	42.90	42.10	2.65	3.25	4.05
40	48.30	48.40	47.80	48.80	48.00	2.90	3.25	4.05
50	60.30	60.20	59.60	60.80	59.80	2.90	3.65	4.50
65	76.10	76.00	75.20	76.60	75.40	3.25	3.65	4.50
80	88.90	88.70	87.90	89.50	88.10	3.25	4.05	4.85
100	114.30	113.90	113.00	114.90	113.30	3.65	4.50	5.40
125	139.70	–	–	140.60	138.70	–	4.85	5.40
150	165.1*	–	–	166.10	164.10	–	4.85	5.40

* 165.1 mm (6.5in) outside diameter is not generally recommended except where screwing to BS 21 is necessary
All dimensions are in accordance with ISO R65 except approximate outside diameters which are in accordance with ISO R64
Light quality is equivalent to ISO R65 Light Series II

Approximate Metres Per Tonne of Tubes to BS 1387

Nom. Size (mm)	BLACK						GALVANIZED					
	Plain/screwed ends			Screwed & socketed			Plain/screwed ends			Screwed & socketed		
	L (m)	M (m)	H (m)	L (m)	M (m)	H (m)	L (m)	M (m)	H (m)	L (m)	M (m)	H (m)
6	2765	2461	2030	2743	2443	2018	2604	2333	1948	2584	2317	1937
8	1936	1538	1300	1920	1527	1292	1826	1467	1254	1811	1458	1247
10	1483	1173	979	1471	1165	974	1400	1120	944	1386	1113	939
15	1050	817	688	1040	811	684	996	785	665	987	779	661
20	712	634	529	704	628	525	679	609	512	673	603	508
25	498	410	336	494	407	334	478	396	327	474	394	325
32	388	319	260	384	316	259	373	308	254	369	305	252
40	307	277	226	303	273	223	296	268	220	292	264	217
50	244	196	162	239	194	160	235	191	158	231	188	157
65	172	153	127	169	151	125	167	149	124	163	146	122
80	147	118	99	143	116	98	142	115	97	139	113	96
100	101	82	69	98	81	68	98	81	68	95	79	67
125	–	62	56	–	60	55	–	60	55	–	59	54
150	–	52	47	–	50	46	–	51	46	–	49	45

The figures for 'plain or screwed ends' apply also to tubes to BS 1775 of equivalent size and thickness
Key:
L – Light
M – Medium
H – Heavy

PIPES, WATER, STORAGE, INSULATION

Flange Dimension Chart to BS 4504 & BS 10

Normal Pressure Rating (PN 6) 6 Bar

Nom. Size	Flange Outside Dia.	Table 6/2 Forged Welding Neck	Table 6/3 Plate Slip on	Table 6/4 Forged Bossed Screwed	Table 6/5 Forged Bossed Slip on	Table 6/8 Plate Blank	Raised Face Dia.	Raised Face T'ness	Nr. Bolt Hole	Size of Bolt
15	80	12	12	12	12	12	40	2	4	M10 × 40
20	90	14	14	14	14	14	50	2	4	M10 × 45
25	100	14	14	14	14	14	60	2	4	M10 × 45
32	120	14	16	14	14	14	70	2	4	M12 × 45
40	130	14	16	14	14	14	80	3	4	M12 × 45
50	140	14	16	14	14	14	90	3	4	M12 × 45
65	160	14	16	14	14	14	110	3	4	M12 × 45
80	190	16	18	16	16	16	128	3	4	M16 × 55
100	210	16	18	16	16	16	148	3	4	M16 × 55
125	240	18	20	18	18	18	178	3	8	M16 × 60
150	265	18	20	18	18	18	202	3	8	M16 × 60
200	320	20	22	–	20	20	258	3	8	M16 × 60
250	375	22	24	–	22	22	312	3	12	M16 × 65
300	440	22	24	–	22	22	365	4	12	M20 × 70

Normal Pressure Rating (PN 16) 16 Bar

Nom. Size	Flange Outside Dia.	Table 6/2 Forged Welding Neck	Table 6/3 Plate Slip on	Table 6/4 Forged Bossed Screwed	Table 6/5 Forged Bossed Slip on	Table 6/8 Plate Blank	Raised Face Dia.	Raised Face T'ness	Nr. Bolt Hole	Size of Bolt
15	95	14	14	14	14	14	45	2	4	M12 × 45
20	105	16	16	16	16	16	58	2	4	M12 × 50
25	115	16	16	16	16	16	68	2	4	M12 × 50
32	140	16	16	16	16	16	78	2	4	M16 × 55
40	150	16	16	16	16	16	88	3	4	M16 × 55
50	165	18	18	18	18	18	102	3	4	M16 × 60
65	185	18	18	18	18	18	122	3	4	M16 × 60
80	200	20	20	20	20	20	138	3	8	M16 × 60
100	220	20	20	20	20	20	158	3	8	M16 × 65
125	250	22	22	22	22	22	188	3	8	M16 × 70
150	285	22	22	22	22	22	212	3	8	M20 × 70
200	340	24	24	–	24	24	268	3	12	M20 × 75
250	405	26	26	–	26	26	320	3	12	M24 × 90
300	460	28	28	–	28	28	378	4	12	M24 × 90

PIPES, WATER, STORAGE, INSULATION

Minimum Distances Between Supports/Fixings

Material	BS Nominal Pipe Size		Pipes – Vertical	Pipes – Horizontal on to low gradients
	(inch)	(mm)	Support distance in metres	Support distance in metres
Copper	0.50	15.00	1.90	1.30
	0.75	22.00	2.50	1.90
	1.00	28.00	2.50	1.90
	1.25	35.00	2.80	2.50
	1.50	42.00	2.80	2.50
	2.00	54.00	3.90	2.50
	2.50	67.00	3.90	2.80
	3.00	76.10	3.90	2.80
	4.00	108.00	3.90	2.80
	5.00	133.00	3.90	2.80
	6.00	159.00	3.90	2.80
muPVC	1.25	32.00	1.20	0.50
	1.50	40.00	1.20	0.50
	2.00	50.00	1.20	0.60
Polypropylene	1.25	32.00	1.20	0.50
	1.50	40.00	1.20	0.50
uPVC	–	82.40	1.20	0.50
	–	110.00	1.80	0.90
	–	160.00	1.80	1.20
Steel	0.50	15.00	2.40	1.80
	0.75	20.00	3.00	2.40
	1.00	25.00	3.00	2.40
	1.25	32.00	3.00	2.40
	1.50	40.00	3.70	2.40
	2.00	50.00	3.70	2.40
	2.50	65.00	4.60	3.00
	3.00	80.40	4.60	3.00
	4.00	100.00	4.60	3.00
	5.00	125.00	5.50	3.70
	6.00	150.00	5.50	4.50
	8.00	200.00	8.50	6.00
	10.00	250.00	9.00	6.50
	12.00	300.00	10.00	7.00
	16.00	400.00	10.00	8.25

PIPES, WATER, STORAGE, INSULATION

Litres of Water Storage Required Per Person Per Building Type

Type of Building	Storage (litres)
Houses and flats (up to 4 bedrooms	120/bedroom
Houses and flats (more than 4 bedrooms)	100/bedroom
Hostels	90/bed
Hotels	200/bed
Nurses homes and medical quarters	120/bed
Offices with canteen	45/person
Offices without canteen	40/person
Restaurants	7/meal
Boarding schools	90/person
Day schools – Primary	15/person
Day schools – Secondary	20/person

Recommended Air Conditioning Design Loads

Building Type	Design Loading
Computer rooms	500 W/m² of floor area
Restaurants	150 W/m² of floor area
Banks (main area)	100 W/m² of floor area
Supermarkets	25 W/m² of floor area
Large office block (exterior zone)	100 W/m² of floor area
Large office block (interior zone)	80 W/m² of floor area
Small office block (interior zone)	80 W/m² of floor area

PIPES, WATER, STORAGE, INSULATION

Capacity and Dimensions of Galvanized Mild Steel Cisterns – BS 417

Capacity (litres)	BS type (SCM)	Length (mm)	Dimensions Width (mm)	Depth (mm)
18	45	457	305	305
36	70	610	305	371
54	90	610	406	371
68	110	610	432	432
86	135	610	457	482
114	180	686	508	508
159	230	736	559	559
191	270	762	584	610
227	320	914	610	584
264	360	914	660	610
327	450/1	1220	610	610
336	450/2	965	686	686
423	570	965	762	787
491	680	1090	864	736
709	910	1070	889	889

Capacity of Cold Water Polypropylene Storage Cisterns – BS 4213

Capacity (litres)	BS type (PC)	Maximum height (mm)
18	4	310
36	8	380
68	15	430
91	20	510
114	25	530
182	40	610
227	50	660
273	60	660
318	70	660
455	100	760

PIPES, WATER, STORAGE, INSULATION

Minimum Insulation Thickness to Protect Against Freezing for Domestic Cold Water Systems (8 Hour Evaluation Period)

Pipe size (mm)	Insulation thickness (mm)					
	Condition 1			Condition 2		
	λ = 0.020	λ = 0.030	λ = 0.040	λ = 0.020	λ = 0.030	λ = 0.040
Copper pipes						
15	11	20	34	12	23	41
22	6	9	13	6	10	15
28	4	6	9	4	7	10
35	3	5	7	4	5	7
42	3	4	5	8	4	6
54	2	3	4	2	3	4
76	2	2	3	2	2	3
Steel pipes						
15	9	15	24	10	18	29
20	6	9	13	6	10	15
25	4	7	9	5	7	10
32	3	5	6	3	5	7
40	3	4	5	3	4	6
50	2	3	4	2	3	4
65	2	2	3	2	3	3

Condition 1: water temperature 7°C; ambient temperature –6°C; evaluation period 8 h; permitted ice formation 50%; normal installation, i.e. inside the building and inside the envelope of the structural insulation
Condition 2: water temperature 2°C; ambient temperature –6°C; evaluation period 8 h; permitted ice formation 50%; extreme installation, i.e. inside the building but outside the envelope of the structural insulation
λ = thermal conductivity [W/(mK)]

Insulation Thickness for Chilled And Cold Water Supplies to Prevent Condensation

On a Low Emissivity Outer Surface (0.05, i.e. Bright Reinforced Aluminium Foil) with an Ambient Temperature of +25°C and a Relative Humidity of 80%

Steel pipe size (mm)	t = +10			t = +5			t = 0		
	Insulation thickness (mm)			Insulation thickness (mm)			Insulation thickness (mm)		
	λ = 0.030	λ = 0.040	λ = 0.050	λ = 0.030	λ = 0.040	λ = 0.050	λ = 0.030	λ = 0.040	λ = 0.050
15	16	20	25	22	28	34	28	36	43
25	18	24	29	25	32	39	32	41	50
50	22	28	34	30	39	47	38	49	60
100	26	34	41	36	47	57	46	60	73
150	29	38	46	40	52	64	51	67	82
250	33	43	53	46	60	74	59	77	94
Flat surfaces	39	52	65	56	75	93	73	97	122

t = temperature of contents (°C)
λ = thermal conductivity at mean temperature of insulation [W/(mK)]

PIPES, WATER, STORAGE, INSULATION

Insulation Thickness for Non-domestic Heating Installations to Control Heat Loss

Steel pipe size (mm)	t = 75			t = 100			t = 150		
	Insulation thickness (mm)			Insulation thickness (mm)			Insulation thickness (mm)		
	$\lambda = 0.030$	$\lambda = 0.040$	$\lambda = 0.050$	$\lambda = 0.030$	$\lambda = 0.040$	$\lambda = 0.050$	$\lambda = 0.030$	$\lambda = 0.040$	$\lambda = 0.050$
10	18	32	55	20	36	62	23	44	77
15	19	34	56	21	38	64	26	47	80
20	21	36	57	23	40	65	28	50	83
25	23	38	58	26	43	68	31	53	85
32	24	39	59	28	45	69	33	55	87
40	25	40	60	29	47	70	35	57	88
50	27	42	61	31	49	72	37	59	90
65	29	43	62	33	51	74	40	63	92
80	30	44	62	35	52	75	42	65	94
100	31	46	63	37	54	76	45	68	96
150	33	48	64	40	57	77	50	73	100
200	35	49	65	42	59	79	53	76	103
250	36	50	66	43	61	80	55	78	105

t = hot face temperature (°C)
λ = thermal conductivity at mean temperature of insulation [W/(mK)]

Decoding Eurocode 7

A. Bond et al.

'Well presented, clear and unambiguous ... I have no hesitation in recommending Decoding Eurocode 7 to both students and practitioners as an authoritative, practical text, representing excellent value for money.' – *Tony Bracegirdle, Engineering Structures*

'A beautifully written and presented book.' – *Yul Tammo, Cornwall County Council*

Decoding Eurocode 7 provides a detailed examination of Eurocode 7 Parts 1 and 2 and an overview of the associated European and International standards. The detail of the code is set out in summary tables and diagrams, with extensive. Fully annotated worked examples demonstrate how to apply it to real designs. Flow diagrams explain how reliability is introduced into design and mind maps gather related information into a coherent framework.

Written by authors who specialise in lecturing on the subject, *Decoding Eurocode 7* explains the key principles and application rules of Eurocode 7 in a logical and simple manner. Invaluable for practitioners, as well as for high-level students and researchers working in geotechnical fields.

August 2008: 616pp
hbk: 978-0-415-40948-3: **£80.00**

To Order: Tel: +44 (0) 1235 400524 **Fax:** +44 (0) 1235 400525
or Post: Taylor and Francis Customer Services,
Bookpoint Ltd, Unit T1, 200 Milton Park, Abingdon, Oxon, OX14 4TA UK
Email: book.orders@tandf.co.uk

For a complete listing of all our titles visit:
www.tandf.co.uk

Index

Manual for Detailing Reinforced Concrete Structures to EC2

J. Calavera

Detailing 210 structural elements, each with a double page spread, this book is an indispensible guide to designing for reinforced concrete. Drawing on Jose Calavera's many years of experience, each entry provides commentary and recommendations for the element concerned, as 0077ell as summarising the appropriate Eurocode legislation and referring to further standards and literature.

Includes a CD ROM with AutoCad files of all of the models, which you can adapt to suit your own purposes.

Section 1. General Rules for Bending, Placing and Lapping of Reinforcement
Section 2. Constructive Details

1. Foundations 2. Walls 3. Columns and Joints 4. Walls Subjected to Axial Loads. Shear Walls and Cores 5. Beams and Lintels 6. Slabs 7. Flat Slabs 8. Stairs 9. Supports 10. Brackets 11. Pavements and Galleries 12. Cylindrical Chimneys, Towers and Hollow Piles 13. Rectangular Silos, Caissons and Hollow Piles 14. Reservoirs, Tanks and Swimming Pools 15. Special Details for Seismic Areas

October 2011: 512pp
Hb: 978-0-415-66348-9: **£80.00**

To Order: Tel: +44 (0) 1235 400524 **Fax:** +44 (0) 1235 400525
or Post: Taylor and Francis Customer Services,
Bookpoint Ltd, Unit T1, 200 Milton Park, Abingdon, Oxon, OX14 4TA UK
Email: book.orders@tandf.co.uk

For a complete listing of all our titles visit:
www.tandf.co.uk

Taylor & Francis
Taylor & Francis Group

Reynolds's Reinforced Concrete Designer Handbook

11th Edition

A. Threlfall et al.

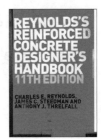

This classic and essential work has been thoroughly revised and updated in line with the requirements of new codes and standards which have been introduced in recent years, including the new Eurocode as well as up-to-date British Standards.

It provides a general introduction along with details of analysis and design of a wide range of structures and examination of design according to British and then European Codes.

Highly illustrated with numerous line diagrams, tables and worked examples, *Reynolds's Reinforced Concrete Designer's Handbook* is a unique resource providing comprehensive guidance that enables the engineer to analyze and design reinforced concrete buildings, bridges, retaining walls, and containment structures.

Written for structural engineers, contractors, consulting engineers, local and health authorities, and utilities, this is also excellent for civil and architecture departments in universities and FE colleges.

July 2007: 416pp
Hb: 978-0-419-25820-9: **£99.99**
Pb: 978-0-419-25830-8: **£41.99**

To Order: Tel: +44 (0) 1235 400524 **Fax:** +44 (0) 1235 400525
or Post: Taylor and Francis Customer Services,
Bookpoint Ltd, Unit T1, 200 Milton Park, Abingdon, Oxon, OX14 4TA UK
Email: book.orders@tandf.co.uk

For a complete listing of all our titles visit:
www.tandf.co.uk

A Guide to Field Instrumentation in Geotechnics

R. Bassett

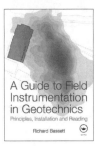

Geotechnical instrumentation is used for installation, monitoring and assessment on any sizeable project, particularly in urban areas, and is used for recording, controlled remedial work, and safety.

This unique and up-to-date book deals with the conceptual philosophy behind the use of instruments, and then systematically covers their practical use. It is divided into displacement dominated systems and stress recording systems. The limitations are discussed and the theoretical background for data assessment and presentation are covered in some detail, with some relevant background material in theoretical soil mechanics. Relevant advanced electronic techniques such as laser scanning in surveying and fibre-optics are also included and communication and data recovery systems are discussed.

It is written for senior designers, consulting engineers and major contractors who need a major introduction to the general purpose, availability and analysis of field instruments before details of their own project can be progressed, and serves as a text book to any specialist geotechnical MSc or professional seminar course in which instrumentation forms a major part.

November 2011: 376pp
Hb: 978-0-415-67537-6: **£90.00**

To Order: Tel: +44 (0) 1235 400524 **Fax:** +44 (0) 1235 400525
or Post: Taylor and Francis Customer Services,
Bookpoint Ltd, Unit T1, 200 Milton Park, Abingdon, Oxon, OX14 4TA UK
Email: book.orders@tandf.co.uk

For a complete listing of all our titles visit:
www.tandf.co.uk

Failures in Concrete Structures: Case Studies in Reinforced and Prestressed Concrete

R. Whittle

Some lessons are only learned from mistakes. But, it's much cheaper to learn from someone else's mistakes than to have to do so from your own. Drawing on over 50 years of working with concrete structures, Robin Whittle examines the problems which he has seen occur and shows how they could have been avoided.

The first and largest part of the book tells the stories of a number of cases where things have gone wrong with concrete structures. Each case is analysed to identify its cause and how it might have been prevented. It then looks at how failures in structural modelling can lead to big problems if they are not identified before construction is undertaken. Beyond this it examines how contract arrangements can encourage or prevent problems in the designing and building processes. It concludes with an examination of the role research and development in preventing failures.

By identifying the differences between shoddy economisations and genuine efficiency savings, this book offers savings in the short term which won't be at the expense of a structure's long-term performance. Invaluable reading if you're designing or building concrete structures and want to avoid problems which could be expensive or embarrassing further down the line.

December 2011: 208pp
Hb: 978-0-415-56701-5: **£60.00**

To Order: Tel: +44 (0) 1235 400524 **Fax:** +44 (0) 1235 400525
or Post: Taylor and Francis Customer Services,
Bookpoint Ltd, Unit T1, 200 Milton Park, Abingdon, Oxon, OX14 4TA UK
Email: book.orders@tandf.co.uk

For a complete listing of all our titles visit:
www.tandf.co.uk

Taylor & Francis
Taylor & Francis Group

Software and eBook Single-User Licence Agreement

We welcome you as a user of this Spon Price Book Software and eBook and hope that you find it a useful and valuable tool. Please read this document carefully. **This is a legal agreement** between you (hereinafter referred to as the "Licensee") and Taylor and Francis Books Ltd. (the "Publisher"), which defines the terms under which you may use the Product. **By breaking the seal and opening the document inside the back cover of the book containing the access code you agree to these terms and conditions outlined herein. If you do not agree to these terms you must return the Product to your supplier intact, with the seal on the document unbroken.**

1. **Definition of the Product**
   The product which is the subject of this Agreement, *Spon's Civil Engineering and Highway Works Price Book 2012* Software and eBook (the "Product") consists of:
1.1 Underlying data comprised in the product (the "Data")
1.2 A compilation of the Data (the "Database")
1.3 Software (the "Software") for accessing and using the Database
1.4 An electronic book containing the data in the price book (the "eBook")

2. **Commencement and licence**
2.1 This Agreement commences upon the breaking open of the document containing the access code by the Licensee (the "Commencement Date").
2.2 This is a licence agreement (the "Agreement") for the use of the Product by the Licensee, and not an agreement for sale.
2.3 The Publisher licenses the Licensee on a non-exclusive and non-transferable basis to use the Product on condition that the Licensee complies with this Agreement. The Licensee acknowledges that it is only permitted to use the Product in accordance with this Agreement.

3. **Multiple use**
   For more than one user or for a wide area network or consortium, use is only permissible with the purchase from the Publisher of a multiple-user licence and adherence to the terms and conditions of that licence.

4. **Installation and Use**
4.1 The Licensee may provide access to the Product for individual study in the following manner: The Licensee may install the Product on a secure local area network on a single site for use by one user.
4.2 The Licensee shall be responsible for installing the Product and for the effectiveness of such installation.
4.3 Text from the Product may be incorporated in a coursepack. Such use is only permissible with the express permission of the Publisher in writing and requires the payment of the appropriate fee as specified by the Publisher and signature of a separate licence agreement.
4.4 The Product is a free addition to the book and no technical support will be provided.

5. **Permitted Activities**
5.1 The Licensee shall be entitled:
   5.1.1 to use the Product for its own internal purposes;
   5.1.2 to download onto electronic, magnetic, optical or similar storage medium reasonable portions of the Database provided that the purpose of the Licensee is to undertake internal research or study and provided that such storage is temporary;
5.2 The Licensee acknowledges that its rights to use the Product are strictly set out in this Agreement, and all other uses (whether expressly mentioned in Clause 6 below or not) are prohibited.

6. **Prohibited Activities**
   The following are prohibited without the express permission of the Publisher:
6.1 The commercial exploitation of any part of the Product.
6.2 The rental, loan, (free or for money or money's worth) or hire purchase of this product, save with the express consent of the Publisher.
6.3 Any activity which raises the reasonable prospect of impeding the Publisher's ability or opportunities to market the Product.
6.4 Any networking, physical or electronic distribution or dissemination of the product save as expressly permitted by this Agreement.
6.5 Any reverse engineering, decompilation, disassembly or other alteration of the Product save in accordance with applicable national laws.
6.6 The right to create any derivative product or service from the Product save as expressly provided for in this Agreement.
6.7 Any alteration, amendment, modification or deletion from the Product, whether for the purposes of error correction or otherwise.

7. General Responsibilities of the License

7.1 The Licensee will take all reasonable steps to ensure that the Product is used in accordance with the terms and conditions of this Agreement.

7.2 The Licensee acknowledges that damages may not be a sufficient remedy for the Publisher in the event of breach of this Agreement by the Licensee, and that an injunction may be appropriate.

7.3 The Licensee undertakes to keep the Product safe and to use its best endeavours to ensure that the product does not fall into the hands of third parties, whether as a result of theft or otherwise.

7.4 Where information of a confidential nature relating to the product of the business affairs of the Publisher comes into the possession of the Licensee pursuant to this Agreement (or otherwise), the Licensee agrees to use such information solely for the purposes of this Agreement, and under no circumstances to disclose any element of the information to any third party save strictly as permitted under this Agreement. For the avoidance of doubt, the Licensee's obligations under this sub-clause 7.4 shall survive the termination of this Agreement.

8. Warrant and Liability

8.1 The Publisher warrants that it has the authority to enter into this agreement and that it has secured all rights and permissions necessary to enable the Licensee to use the Product in accordance with this Agreement.

8.2 The Publisher warrants that the Product as supplied on the Commencement Date shall be free of defects in materials and workmanship, and undertakes to replace any defective Product within 28 days of notice of such defect being received provided such notice is received within 30 days of such supply. As an alternative to replacement, the Publisher agrees fully to refund the Licensee in such circumstances, if the Licensee so requests, provided that the Licensee returns this copy of *Spon's Civil Engineering and Highway Works Price Book 2012* to the Publisher. The provisions of this sub-clause 8.2 do not apply where the defect results from an accident or from misuse of the product by the Licensee.

8.3 Sub-clause 8.2 sets out the sole and exclusive remedy of the Licensee in relation to defects in the Product.

8.4 The Publisher and the Licensee acknowledge that the Publisher supplies the Product on an "as is" basis. The Publisher gives no warranties:

    8.4.1 that the Product satisfies the individual requirements of the Licensee; or

    8.4.2 that the Product is otherwise fit for the Licensee's purpose; or

    8.4.3 that the Data are accurate or complete or free of errors or omissions; or

    8.4.4 that the Product is compatible with the Licensee's hardware equipment and software operating environment.

8.5 The Publisher hereby disclaims all warranties and conditions, express or implied, which are not stated above.

8.6 Nothing in this Clause 8 limits the Publisher's liability to the Licensee in the event of death or personal injury resulting from the Publisher's negligence.

8.7 The Publisher hereby excludes liability for loss of revenue, reputation, business, profits, or for indirect or consequential losses, irrespective of whether the Publisher was advised by the Licensee of the potential of such losses.

8.8 The Licensee acknowledges the merit of independently verifying Data prior to taking any decisions of material significance (commercial or otherwise) based on such data. It is agreed that the Publisher shall not be liable for any losses which result from the Licensee placing reliance on the Data or on the Database, under any circumstances.

8.9 Subject to sub-clause 8.6 above, the Publisher's liability under this Agreement shall be limited to the purchase price.

9. Intellectual Property Rights

9.1 Nothing in this Agreement affects the ownership of copyright or other intellectual property rights in the Data, the Database of the Software.

9.2 The Licensee agrees to display the Publishers' copyright notice in the manner described in the Product.

9.3 The Licensee hereby agrees to abide by copyright and similar notice requirements required by the Publisher, details of which are as follows:

"© 2012 Taylor & Francis. All rights reserved. All materials in *Spon's Civil Engineering and Highway Works Price Book 2012* are copyright protected. All rights reserved. No such materials may be used, displayed, modified, adapted, distributed, transmitted, transferred, published or otherwise reproduced in any form or by any means now or hereafter developed other than strictly in accordance with the terms of the licence agreement enclosed with *Spon's Civil Engineering and Highway Works Price Book 2012*. However, text and images may be printed and copied for research and private study within the preset program limitations. Please note the copyright notice above, and that any text or images printed or copied must credit the source."

9.4 This Product contains material proprietary to and copyedited by the Publisher and others. Except for the licence granted herein, all rights, title and interest in the Product, in all languages, formats and media throughout the world, including copyrights therein, are and remain the property of the Publisher or other copyright holders identified in the Product.

10. Non-assignment
This Agreement and the licence contained within it may not be assigned to any other person or entity without the written consent of the Publisher.

11. Termination and Consequences of Termination.
11.1 The Publisher shall have the right to terminate this Agreement if:
  11.1.1 the Licensee is in material breach of this Agreement and fails to remedy such breach (where capable of remedy) within 14 days of a written notice from the Publisher requiring it to do so; or
  11.1.2 the Licensee becomes insolvent, becomes subject to receivership, liquidation or similar external administration; or
  11.1.3 the Licensee ceases to operate in business.
11.2 The Licensee shall have the right to terminate this Agreement for any reason upon two month's written notice. The Licensee shall not be entitled to any refund for payments made under this Agreement prior to termination under this sub-clause 11.2.
11.3 Termination by either of the parties is without prejudice to any other rights or remedies under the general law to which they may be entitled, or which survive such termination (including rights of the Publisher under sub-clause 7.4 above).
11.4 Upon termination of this Agreement, or expiry of its terms, the Licensee must destroy all copies and any back up copies of the product or part thereof.

12. General
12.1 **Compliance with export provisions**
The Publisher hereby agrees to comply fully with all relevant export laws and regulations of the United Kingdom to ensure that the Product is not exported, directly or indirectly, in violation of English law.
12.2 **Force majeure**
The parties accept no responsibility for breaches of this Agreement occurring as a result of circumstances beyond their control.
12.3 **No waiver**
Any failure or delay by either party to exercise or enforce any right conferred by this Agreement shall not be deemed to be a waiver of such right.
12.4 **Entire agreement**
This Agreement represents the entire agreement between the Publisher and the Licensee concerning the Product. The terms of this Agreement supersede all prior purchase orders, written terms and conditions, written or verbal representations, advertising or statements relating in any way to the Product.
12.5 **Severability**
If any provision of this Agreement is found to be invalid or unenforceable by a court of law of competent jurisdiction, such a finding shall not affect the other provisions of this Agreement and all provisions of this Agreement unaffected by such a finding shall remain in full force and effect.
12.6 **Variations**
This agreement may only be varied in writing by means of variation signed in writing by both parties.
12.7 **Notices**
All notices to be delivered to: Spon's Price Books, Taylor & Francis Books Ltd., 2 Park Square, Milton Park, Abingdon, Oxfordshire, OX14 4RN, UK.
12.8 **Governing law**
This Agreement is governed by English law and the parties hereby agree that any dispute arising under this Agreement shall be subject to the jurisdiction of the English courts.

If you have any queries about the terms of this licence, please contact:

Spon's Price Books
Taylor & Francis Books Ltd.
2 Park Square, Milton Park, Abingdon, Oxfordshire, OX14 4RN
Tel: +44 (0) 20 7017 6672
Fax: +44 (0) 20 7017 6702
www.tandfbuiltenvironment.com

Multiple-user use of the Spon Press Software and eBook

To buy a licence to install your Spon Press Price Book Software and eBook on a secure local area network or a wide area network, and for the supply of network key files, for an agreed number of users please contact:

Spon's Price Books
Taylor & Francis Books Ltd.
2 Park Square, Milton Park, Abingdon, Oxfordshire, OX14 4RN
Tel: +44 (0) 207 017 6672
Fax: +44 (0) 207 017 6072
www.pricebooks.co.uk

Number of users	Licence cost
2–5	£450
6–10	£915
11–20	£1400
21–30	£2150
31–50	£4200
51–75	£5900
76–100	£7100
Over 100	Please contact Spon for details

Software Installation and Use Instructions

System requirements

Minimum

- Pentium processor
- 256 MB of RAM
- 20 MB available hard disk space
- Microsoft Windows 98/2000/NT/ME/XP/Vista
- SVGA screen
- Internet connection

Recommended

- Intel 466 MHz processor
- 512 MB of RAM (1,024MB for Vista)
- 100 MB available hard disk space
- Microsoft Windows XP/Vista
- XVGA screen
- Broadband Internet connection

Microsoft® is a registered trademark and Windows™ is a trademark of the Microsoft Corporation.

Installation

Spon's Civil Engineering and Highway Works Price Book 2012 Electronic Version is supplied solely by internet download. No CD-ROM is supplied.

In your internet browser type in www.pricebooks.co.uk/downloads and follow the instructions on screen. Then type in the unique access code which is sealed inside the back cover of this book.

When the access code is successfully validated, click on the download links to download and then save the files.

Please note: you will only be allowed one download of these files and onto one computer.

A folder called *PriceBook* will be added to your desktop, which will need to be unzipped. Then click on the application called *install*.

Use

- The installation process will create a folder containing the price book program links as well as a program icon on your desktop.
- Double click the icon (from the folder or desktop) installed by the Setup program.
- Follow the instructions on screen.

Technical Support

Further guidance on the installation is provided on www.pricebooks/downloads

The *Electronic Version* is a free addition to the book. For help with the running of the software please visit www.pricebooks.co.uk

The software used in *Spon's Civil Engineering and Highway Works Price Book 2012 Electronic Version* is furnished under a single user licence agreement. The software may be used only in accordance with the terms of the licence agreement, unless the additional multi-user licence agreement is purchased from Spon Press, 2 Park Square, Milton Park, Abingdon, Oxon OX14 4RN Tel: +44 (0) 20 7017 6000

Free Updates

with three easy steps…

1. Register today on www.pricebooks.co.uk/updates

2. We'll alert you by email when new updates are posted on our website

3. Then go to www.pricebooks.co.uk/updates
    and download the update.

All four Spon Price Books – *Architects' and Builders'*, *Civil Engineering and Highway Works*, *External Works and Landscape* and *Mechanical and Electrical Services* – are supported by an updating service. Three updates are loaded on our website during the year, in November, February and May. Each gives details of changes in prices of materials, wage rates and other significant items, with regional price level adjustments for Northern Ireland, Scotland and Wales and regions of England. The updates terminate with the publication of the next annual edition.

As a purchaser of a Spon Price Book you are entitled to this updating service for the 2012 edition – free of charge. Simply register via the website www.pricebooks.co.uk/updates and we will send you an email when each update becomes available.

If you haven't got internet access we can supply the updates by an alternative means. Please write to us for details: Spon Price Book Updates, Spon Press Marketing Department, 2 Park Square, Milton Park, Abingdon, Oxfordshire, OX14 4RN.

Find out more about Spon books
Visit www.sponpress.com for more details.